# PATHWAYS THROUGH AN ECLECTIC UNIVERSE

*COVER ILLUSTRATION:*

Composite mosaic image of a portion of the Orion Nebula, taken with the WFPC2 camera on the *Hubble Space Telescope* (*HST*) on 2002 January 22 through the filters F502N, F656N, F658N and F547W to obtain the emission in O III (coded blue), Hα (coded green) and N II (coded red), with suitable continuum subtraction for each line. These emission-line images were used, together with their equivalents taken 6.84 years earlier, to measure proper motions within the nebula and to search for variability in the emission from the ionized gas over this time interval (see Beckman, Gutiérrez, & Giammanco, this volume, p. 106, and also Giammanco, Gutiérrez, & Beckman, this volume, p. 111).

We are pleased to acknowledge that the component images we have used to produce this illustration were taken within an *HST* observation program with C. Robert O'Dell as P.I., who has kindly agreed that we use the material for the cover. The images were taken directly from the Multimission Archive at the Space Telescope Science Institute (MAST). STScI is operated by the Association of Universities for Research in Astronomy, Inc., under NASA contract NAS5-26555. Combining the original images to produce the final version as shown on the cover was the work of Leonel Gutiérrez and Nik Szymanek (www.ccdland.com).

# ASTRONOMICAL SOCIETY OF THE PACIFIC CONFERENCE SERIES

## A SERIES OF BOOKS ON RECENT DEVELOPMENTS IN ASTRONOMY AND ASTROPHYSICS

**Volume 390**




Recent ASPCS Volumes may be found as e-books with color images at http://www.aspbooks.org

A listing of recently published volumes may be found at the back of this volume.

For a complete listing of ASPCS and IAU Volumes published by the ASP see http://www.astrosociety.org/pubs.html

ASTRONOMICAL SOCIETY OF THE PACIFIC
CONFERENCE SERIES

**Volume 390**

# PATHWAYS THROUGH AN ECLECTIC UNIVERSE

Proceedings of a conference held at
Santiago del Teide, Tenerife, Spain
23–27 April 2007
to celebrate John Beckman's 40 years in astrophysics

Edited by

**J. H. Knapen**
*Instituto de Astrofísica de Canarias, La Laguna, Tenerife, Spain*

**T. J. Mahoney**
*Instituto de Astrofísica de Canarias, La Laguna, Tenerife, Spain*

and

**A. Vazdekis**
*Instituto de Astrofísica de Canarias, La Laguna, Tenerife, Spain*

SAN FRANCISCO

ASTRONOMICAL SOCIETY OF THE PACIFIC
390 Ashton Avenue
San Francisco, California, 94112-1722, USA

Phone: 415-337-1100
Fax: 415-337-5205
E-mail: service@astrosociety.org
Web Site: www.astrosociety.org
E-books: www.aspbooks.org


ASP Conference Series - First Edition



ISBN: 978-1-58381-650-9
e-Book ISBN: 978-1-58381-651-6

Library of Congress (LOC) Cataloging in Publication (CIP) Data:
Main entry under title
Library of Congress Control Number (LCCN): 2008926836

Printed in the United States of America by Sheridan Books, Ann Arbor, Michigan

# Bernard E. J. Pagel (1930–2007)
# A Note of Appreciation

When bidding the participants welcome at the start of the conference I made special mention of a number of those present who had been helpful to me at different points of my astronomical career. Among these, and indeed the first person to whom I expressed my gratitude in this way, was Bernard Pagel. It came as a particularly hard and unexpected blow when we learned that, so soon after the meeting, which he had attended with all his usual vigor and enthusiasm, Bernard was seriously ill, and much too soon afterward, that he had left us.

Bernard's mastery of the elements of understanding required to infer element abundances made him an obligatory point of reference for generations of specialists in the field. This was based on two pillars: an encyclopedic knowledge of the relevant topics and a consummate grasp of the theoretical fundamentals. He was always more than up to date on all the theoretical and observational advances, and knew who was making them, or likely to make them. To those around him he was in this way a human advance on astro-ph long before the electronic form was invented, and with the advantage of the additional critical faculty. His commentaries were of inestimable aid in sorting the wheat from the chaff, both theoretical and observational.

At this conference there were a number of those present who owe much to Bernard, and risking sins of serious omission I will name some of them in order to give a feel for the scope of his influence. All of the research topics I will outline

Bernard either started, or contributed to strongly, or inspired. Mike Edmunds' work over the years has had a great deal to do with the use of modeling of element production and distribution on galactic scales, notably with the effects of gas flows on this global evolution. More recently, he has been interested in the distribution of the elements in the intergalactic medium, an interest also shared with Bernard in his later period. Angeles Díaz has contributed to our understanding of galactic evolution by paying special attention to diagnostics that allow us to distinguish the contributions from dwarfs, giants, and supergiants, and to break the age–metallicity ambiguity, as well as deepening our understanding of the ionized interstellar medium around star-forming regions. Pepe Vílchez has made basic advances in the use of interstellar emission lines for abundance determinations and on that basis has worked extensively on many aspects of galaxy evolution. Rafa Rebolo worked on the primordial light element abundances, and his students and collaborators have tackled a variety of problems related to the evolution of the elements in a Galactic context. Paolo Molaro has made continuous contributions to our knowledge of the light element abundances for the past 20 years, strongly influenced by Bernard's understanding and knowledge of this material.

As this conference is oriented towards myself, I think it reasonable to dedicate a paragraph to what I have learned from Bernard. Primarily, I should say that whatever I understand about the chemical evolution of galaxies has been due to him. He gave me the agenda, starting from the pioneer articles of Beatrice Tinsley, and leading me via his deep practical understanding of their consequences. Above all he showed me what was important to observe in order to test the models, and how the observations could put good constraints on those models. In the first place came work on light element abundances, notably beryllium, where he explained how one could use beryllium to set constraints on the Galactic production of lithium, and hence separate the primordial from the Galactic contribution. Secondly, I remember my efforts, together with Emilio Casuso, to set constraints on models of galaxy evolution via photometric imaging. Bernard was totally scathing about this, saying that it was impossible in principle. Of course he was right, but what we learned laid the foundations of more sophisticated models and of the use of spectroscopic constraints, later taken up with considerable success by Alejandro Vazdekis, working with Reynier Peletier. Thirdly, there was the question of infall. Learning about the G-dwarf problem at Bernard's knee and understanding that the infall of low metallicity hydrogen is one of the few ways of resolving the problem, Emilio and I produced models for the chemical evolution of the solar neighborhood, which, when tested against the available evidence, seemed to show that not only is this infall needed, but that the mean rate of infall has remained virtually constant over the lifetime of the Galactic disk, while varying strongly on shorter timescales. Without going into any detail about this, the consequences turned out to be very much in line with what Françoise Combes was deriving from very different considerations; namely, the way hydrogen is accreted onto galaxies in cosmological models, basically along the large filaments remaining when the galaxies had condensed out of the primordial inhomogeneities. There are several corollaries of direct interest. One is the probable influence of the accreting hydrogen on the gas in the disks of galaxies. This was developed into an interesting model of warp formation by Martin López-Corredoira and Juan Betancort-Rijo. Another is the probable observability of the accreting hydrogen in the form of high velocity clouds around the Milky Way and other galaxies, which is being pursued now with success using the THINGS H I survey by Valeria Buenrostro and Elias Brinks. As we can see, the inspiring influence of Bernard has spread widely and given rise to interesting lines of research in which I have been privileged to participate.

It would not do to write this note only about Bernard's impact as a leader and inspirer of research. His personality was striking and impressed all of us. His criticisms were always laced with irony and his humor was a dry version of the finest British strain. His high intellectual standards did not imply a dry personality, though. His love of good wine and good food, and his knowledgeable conversation about almost everything, meant that to spend a working period with him was always a warm and most agreeable experience. He personified the best of a number of interlocking traditions, British, Jewish, and Continental European, though he was clearly universal and universally admired. It is painful but also a pleasure, to write this brief appreciation in this volume. Bernard honored the conference with his vigorous presence as he had invigorated our lives.

John E. Beckman

# Contents

**Session III. Star Formation Histories**

**Session VI. Eclectics**

# Preface

## eclectic

> In ancient use, the distinguishing epithet of a class of philosophers who neither attached themselves to any recognized school, nor constructed independent systems, but 'selected such doctrines as pleased them in every school' (Liddell and Scott); Diogenes Laertius speaks of an 'eclectic sect' founded by Potamon of Alexandria in the second century after Christ. In modern times this designation has been for similar reasons given to or assumed by various philosophers, notably V. Cousin; and it is also applied to those who combine elements derived from diverse systems of opinion or practice in any science or art. So also eclectic method, system, etc.
>
> *Oxford English Dictionary*

> Moreover, it is not long ago, that a new Eclectic school was set up by Potamo, of Alexandria, who picked out of the doctrines of each school what pleased him most. And as he himself says, in his Elementary Instruction, he thinks that there are certain criteria of truth: first of all the faculty which judges and this is the superior one; the other that which is the foundation of the judgement, being a most exact appearance of the objects. And the first principles of everything he calls matter, and the agent, and the quality, and the place. For they show out of what, and by what, and how, and where everything is done. The end is that to which everything is referred; namely, a life made perfect with every virtue, not without the natural and external qualities of the body.
>
> *Diogenes Laertius, The Lives and Opinions of Eminent Philosophers*

Over the past decades, rapid advances in astronomical instrumentation and facilities have continuously improved the quality and quantity of the data underlying our understanding of the Universe. The collective work of thousands of individual researchers has translated this progress into the development of detailed models of the physics of planets, stars and galaxies.

This book contains the proceedings of a conference, held on 23–27 April 2007 in Tenerife, aimed at discussing recent progress on star formation in galaxies, taking into account advances in related areas such as the structure, dynamics, and evolution of galaxies, their stellar and dark matter components, and central black holes as clues to their histories, the physics of the interstellar medium, and the feedback between the star formation and the host galaxy on all spatial scales.

The study of star formation in galaxies is of paramount importance for a range of topics in modern astrophysics. It is important to explore star formation in a wide range of environments with varying metallicity, pressure, and gas density, to study the processes statistically, and to investigate their effects on the surrounding interstellar medium. Concentrating on galaxies which can be observed in detail allows the construction of an observationally and phenomenologically sound framework for the interpretation of observations of the collective output of star formation as witnessed in galaxies at high redshift.

The conference served as a celebration of the fact that John Beckman has been an active astrophysicist for 40 years, a career in which he has worked on many different topics, ranging from solar physics to cosmology. Because of this a number of contributions of varied kinds were made by friends and associates of John, and some of these, in fields as varied as the history and philosophy of astronomy, or other branches of astronomy not related to the interaction of star formation and galaxy formation, can be found in these proceedings.

The Scientific Organizing Committee consisted of Eduardo Battaner, John Beckman, Daniela Calzetti, Françoise Combes, Bruce Elmegreen, Rosa González Delgado, Rob Kennicutt, Johan Knapen, David Lambert, Rafael Rebolo, and Isaac Shlosman, with Elmegreen, Kennicutt and Knapen acting as co-chairs. A Science Advisory Committee was also formed, co-chaired by Francisco Sánchez and Johan Knapen, and with Claude Bertout, Ramón Canal, Mike Edmunds, Franco Giovannelli, Ricardo González Peón, Javier Gorgas, Bengt Gustafsson, Gerhard Hensler, Paul Hodge, Mariano Moles, Casiana Muñoz Tuñón, Bernard Pagel, Manuel Peimbert, Reynier Peletier, Carlos Quiroga Blanco, Jürgen Renn, Ian Robson, Elena Terlevich, José Manuel Vílchez, Rosie Wyse, and Almudena Zurita as members. Finally, but importantly, we should highlight the efforts of the Local Organizing Committee which ensured the smooth organization and running of the conference. The LOC was co-chaired by Artemio Herrero, Alejandro Vazdekis, and Johan Knapen, and members were Judith de Araoz, Ruymán Azzollini, Eva Bejarano, Valeria Buenrostro, Jesús Burgos, Sébastien Comerón, Lucio Crivellari, Kambiz Fathi, Leonel Gutiérrez, Tanja Karthaus, Terry Mahoney, Inma Martínez Valpuesta, José Montesinos, Jorge Pérez, Mónica Relaño, and Almudena Zurita. The conference was financially supported by the Consejo Superior de Investigaciones Científicas, the Ministerio de Educación y Ciencia, and the Instituto de Astrofísica de Canarias, with in-kind support from the conference hotel, the Hotel Playa la Arena.

La Laguna, 6 March 2008

# Participants

R. Aladro, University of La Laguna, c/ Los Nispereros 4, puerta 2, E-38205 La Laguna (Tenerife), Spain ⟨ rebeca_aladro@yahoo.es ⟩

J. Alves, Calar Alto Observatory, c/ Jesus Durban Remón 2-2, E-04004 Almeria, Spain ⟨ jalves@caha.es ⟩

R. Amorín Barbieri, Instituto de Astrofísica de Canarias, C/ Vía Láctea s/n, E-38205 La Laguna (Tenerife), Spain ⟨ ricardo.amorin@iac.es ⟩

E. Athanassoula, Observatoire de Marseille, 2 Place Le Verrier, F-13248 Marseille cedex 04, France ⟨ lia@oamp.fr ⟩

R. Azzollini, Instituto de Astrofísica de Canarias, C/ Vía Láctea s/n, E-38205 La Laguna (Tenerife), Spain ⟨ ruyman@iac.es ⟩

M. Balcells, Instituto de Astrofísica de Canarias, C/ Vía Láctea s/n, E-38205 La Laguna (Tenerife), Spain ⟨ balcells@iac.es ⟩

N. Bastian, University College London, Department of Physics & Astronomy, Gower Street, London WC1E 6BT, UK ⟨ bastian@star.ucl.ac.uk ⟩

E. Battaner, Universidad de Granada, Departamento de Física Teórica y del Cosmos, Facultad de Ciencias, Edificio Mecenas, Campus de Fuentenueva, E-18071 Granada, Spain ⟨ battaner@ugr.es ⟩

J. Beckman, Instituto de Astrofísica de Canarias & CSIC, C/ Vía Láctea s/n, E-38205 La Laguna (Tenerife), Spain ⟨ jeb@iac.es ⟩

I. Berentzen, Zentrum für Astronomie at Heidelberg, Astronomisches Rechen-Institut, Mönchhofstr. 12-14, D-69120 Heidelberg, Germany ⟨ iberent@ari.uni-heidelberg.de ⟩

T. Böker, European Space Agency, Keplerlaan 1, NL-2200 AG Noordwijk, The Netherlands ⟨ tboeker@rssd.esa.int ⟩

I. Bonnell, University of St Andrews, School of Physics and Astronomy, North Haugh, St Andrews, KY16 9SS, Scotland, UK ⟨ iab1@st-and.ac.uk ⟩

V. Buenrostro, Instituto de Astrofísica de Canarias, C/ Vía Láctea s/n, E-38205 La Laguna (Tenerife), Spain ⟨ valeria@iac.es ⟩

D. Calzetti, University of Massachusetts, Department of Astronomy, LGRT-B 524, 710 North Pleasant Street, Amherst, MA 01003-9305, USA ⟨ calzetti@astro.umass.edu ⟩

R. Capuzzo-Dolcetta, University of Rome, La Sapienza, Department of Physics, Ed. E. Fermi, P.le A. Moro 2, I-00185, Roma, Italy ⟨ roberto.capuzzodolcetta@uniroma1.it ⟩

A. Cardwell, Isaac Newton Group of Telescopes, Apartado de Correos 321, E-38700 Santa Cruz de La Palma, Spain ⟨ a.cardwell@open.ac.uk ⟩

C. Carignan, Laboratoire d'Astrophysique Expérimentale, Département de Physique, Université de Montréal, C.P. 6128, Succ. Centre-ville, Montréal, Québec, H3C 3J7, Canada ⟨ claude.carignan@umontreal.ca ⟩

V. Casasola, Dipartimento di Astronomia, Università di Padova, Vicolo dell'Osservatorio 2, I-35122, Padova, Italy ⟨ viviana.casasola@unipd.it ⟩

J. Cenarro, Instituto de Astrofísica de Canarias, C/ Vía Láctea s/n, E-38205 La Laguna (Tenerife), Spain ⟨ cenarro@iac.es ⟩

I. Chilingarian, Observatoire de Paris, LERMA, 61 Av. de l'Observatoire, F-75014 Paris, France ⟨ chil@sai.msu.su ⟩

E. Churchwell, University of Wisconsin-Madison, Department of Astronomy, 475 North Charter St., Madison, WI 53706, USA ⟨ churchwell@astro.wisc.edu ⟩

C. Clarke, Institute of Astronomy, University of Cambridge, Madingley Road, Cambridge CB3 0HA, UK ⟨ cclarke@ast.cam.ac.uk ⟩

F. Combes, Observatoire de Paris, LERMA, 61 Av. de l'Observatoire, F-75014 Paris, France ⟨ francoise.combes@obspm.fr ⟩

S. Comerón, Instituto de Astrofísica de Canarias, C/ Vía Láctea s/n, E-38205 La Laguna (Tenerife), Spain ⟨ sebastien@iac.es ⟩

C. Conselice, University of Nottingham, School of Physics and Astronomy, University Park, Nottingham NG7 2RD, UK ⟨ conselice@nottingham.ac.uk ⟩

A. de Lorenzo-Cáceres, Instituto de Astrofísica de Canarias, C/ Vía Láctea s/n, E-38205 La Laguna (Tenerife), Spain ⟨ adlcr@iac.es ⟩

P. Di Matteo, Observatoire de Paris, LERMA, 61 Av. de l'Observatoire, F-75014 Paris, France ⟨ paola.dimatteo@obspm.fr ⟩

Á. Díaz, Universidad Autónoma de Madrid, Departamento de Física Teórica, C-XI Cantoblanco, E-28049 Madrid, Spain ⟨ angeles.diaz@uam.es ⟩

C. Dobbs, University of Exeter, School of Physics, Stocker Road, Exeter, EX4 4QL, UK ⟨ dobbs@astro.ex.ac.uk ⟩

R. Domínguez Tenreiro, Universidad Autónoma de Madrid, Departamento de Física Teórica, C-XI Cantoblanco, E-28049 Madrid, Spain ⟨ rosa.dominguez@uam.es ⟩

Y. Dubois, Centre d'Études de Saclay, L'orme des Merisiers, F-91191 Gif-sur-Yvette Cedex, France ⟨ ydubois@cea.fr ⟩

M. Edmunds, Cardiff University, School of Physics and Astronomy, Queens Buildings, 5 The Parade, Cardiff CF24 3YB, UK ⟨ mge@astro.cf.ac.uk ⟩

F. Egusa, Institute of Astronomy, University of Tokyo, 2-21-1 Osawa, Mitaka, Tokyo 181-0015, Japan ⟨ fegusa@ioa.s.u-tokyo.ac.jp ⟩

B. Ercolano, Harvard-Smithsonian Center for Astrophysics, 60 Garden Street, Cambridge, MA 02138, USA ⟨ bercolano@cfa.harvard.edu ⟩

P. ERWIN, Max Planck Institut für Extraterrestrische Physik, Giessenbachstr., D-85748 Garching, Germany ⟨ erwin@mpe.mpg.de ⟩

C. ESTEBAN, Instituto de Astrofísica de Canarias, C/ Vía Láctea s/n, E-38205 La Laguna (Tenerife), Spain ⟨ cel@iac.es ⟩

N. EVANS, The University of Texas at Austin, Department of Astronomy, 1 University Station, C1400, Austin, TX 78712-0259, USA ⟨ nje@astro.as.utexas.edu ⟩

J. FALCÓN-BARROSO, European Space Agency, Keplerlaan 1, NL-2200 AG Noordwijk, The Netherlands ⟨ jfalcon@rssd.esa.int ⟩

K. FATHI, Instituto de Astrofísica de Canarias, C/ Vía Láctea s/n, E-38205 La Laguna (Tenerife), Spain ⟨ fathi@iac.es ⟩

D. FENECH, Jodrell Bank Observatory, University of Manchester, Macclesfield, Cheshire, SK11 9DL, UK ⟨ dfenech@jb.man.ac.uk ⟩

S. FERNÁNDEZ ACOSTA, GEMINI Observatory, Colina el Pino s/n, La Serena, Chile ⟨ sfernand@gemini.edu ⟩

I. FUENTES-CARRERA, Observatoire de Paris, GEPI, 5 Place Jules Janssen, F-92190 Meudon, France ⟨ isaura.fuentes@obspm.fr ⟩

C. GALLART, Instituto de Astrofísica de Canarias, C/ Vía Láctea s/n, E-38205 La Laguna (Tenerife), Spain ⟨ carme@iac.es ⟩

D. GARCÍA-APPADOO, European Southern Observatory, Alonso de Cordova 3107, Casilla 19001, Vitacura, Santiago, Chile ⟨ dgarcia@eso.org ⟩

R. GARCÍA LÓPEZ, Instituto de Astrofísica de Canarias, C/ Vía Láctea s/n, E-38205 La Laguna (Tenerife), Spain ⟨ rgl@iac.es ⟩

C. GIAMMANCO, Universität Bern, Physikalisches Institut, Sidlerstrasse 5, CH-3012 Bern, Switzerland ⟨ giammanco@space.unibe.ch ⟩

F. GIOVANNELLI, INAF, Istituto di Astrofisica Spaziale e Fisica Cosmica - Roma, Area di Ricerca CNR di Tor Vergata, Via Fosso del Cavaliere 100, I-00133 Roma, Italy ⟨ franco.giovannelli@iasf-roma.inaf.it ⟩

R. GONZÁLEZ DELGADO, Instituto de Astrofísica de Andalucía, Camino Bajo de Huétor 50, E-18008 Granada, Spain ⟨ rosa@iaa.es ⟩

J. GRIMES, Johns Hopkins University, Dept. of Physics and Astronomy, 3400 N Charles St, Baltimore, MD 21218, USA ⟨ jgrimes@pha.jhu.edu ⟩

P. GROSBØL, ESO, Karl-Schwarzschild-Str. 2, D-85748 Garching bei München, Germany ⟨ pgrosbol@eso.org ⟩

B. GROVES, Leiden Observatory, Leiden University, Postbus 9513, NL-2300 RA Leiden, The Netherlands ⟨ brent@strw.leidenuniv.nl ⟩

S. GUPTA, Bhilai Steel Plant, 1B, Street 57, Sector 8, Bhilai I 490008, c.g., India ⟨ snp.gupta@gmail.com ⟩

L. GUTIÉRREZ, Instituto de Astrofísica de Canarias, C/ Vía Láctea s/n, E-38205 La Laguna (Tenerife), Spain ⟨ leonel@astrosen.unam.mx ⟩

S. HAAN, Max Planck Institute for Astronomy, Königstuhl 17, D-69117 Heidelberg, Germany ⟨ haan@mpia.de ⟩

F. Hammer, Observatoire de Paris, GEPI, 5 Place Jules Janssen, F-92190 Meudon, France ⟨ francois.hammer@obspm.fr ⟩

B. Henriques, Astronomy Centre, University of Sussex, Falmer, Brighton BN1 9QH, UK ⟨ b.m.henriques@sussex.ac.uk ⟩

G. Hensler, University of Vienna, Institute of Astronomy, Türkenschanzstr. 17, A-1180 Vienna, Austria ⟨ hensler@astro.univie.ac.at ⟩

J. Iglesias Páramo, Instituto de Astrofísica de Andalucía, Camino Bajo de Huétor 50, E-18008 Granada, Spain ⟨ jiglesia@iaa.es ⟩

P. James, Liverpool John Moores University, Astrophysics Research Institute, Twelve Quays House, Egerton Wharf, Birkenhead, CH41 1LD, UK ⟨ paj@astro.livjm.ac.uk ⟩

R. Kennicutt, Institute of Astronomy, University of Cambridge, Madingley Road, Cambridge CB3 0HA, UK ⟨ robk@ast.cam.ac.uk ⟩

M. Kissler-Patig, ESO, Karl-Schwarzschild-Str. 2, D-85748 Garching bei München, Germany ⟨ mkissler@eso.org ⟩

J. Knapen, Instituto de Astrofísica de Canarias, C/ Vía Láctea s/n, E-38205 La Laguna (Tenerife), Spain ⟨ jhk@iac.es ⟩

M. Koleva, Observatoire de Lyon, CRAL, 9 Avenue Charles André, F-69561 Saint Genis Laval cedex, France ⟨ mina.koleva@gmail.com ⟩

S. Komugi, Institute of Astronomy, University of Tokyo, 2-21-1 Osawa, Mitaka, Tokyo 181-0015, Japan ⟨ skomugi@ioa.s.u-tokyo.ac.jp ⟩

P. Kroupa, Argelander Institute for Astronomy, University of Bonn, Auf dem Hügel 71, D-53121 Bonn, Germany ⟨ pavel@astro.uni-bonn.de ⟩

M. Krumholz, Princeton University, Department of Astrophysical Sciences, Princeton NJ 08544-1001, USA ⟨ krumholz@astro.princeton.edu ⟩

D. Lambert, University of Texas at Austin, McDonald Observatory, 1 University Station, C1400, Austin TX 78712-0259, USA ⟨ dll@astro.as.utexas.edu ⟩

M. Lehnert, Observatoire de Paris, GEPI, 5 Place Jules Janssen, F-92190 Meudon, France ⟨ matthew.lehnert@obspm.fr ⟩

M. López Corredoira, Instituto de Astrofísica de Canarias, C/ Vía Láctea s/n, E-38205 La Laguna (Tenerife), Spain ⟨ martinlc@iac.es ⟩

Á. López-Sánchez, CSIRO / Australia Telescope National Facility, PO Box 76, Epping, NSW 1710, Sydney, Australia ⟨ Angel.Lopez-Sanchez@csiro.au ⟩

A. Lundgren, APEX, ESO, Casilla 19001, Santiago 19, Chile ⟨ alundgre@eso.org ⟩

V. Luridiana, Instituto de Astrofísica de Andalucía, Camino Bajo de Huétor 50, E-18008 Granada, Spain ⟨ vale@iaa.es ⟩

T. Mahoney, Instituto de Astrofísica de Canarias, C/ Vía Láctea s/n, E-38205 La Laguna (Tenerife), Spain ⟨ tjm@iac.es ⟩

I. Martínez Valpuesta, Instituto de Astrofísica de Canarias, C/ Vía Láctea s/n, E-38205 La Laguna (Tenerife), Spain ⟨ imv@iac.es ⟩

T. Maschberger, Argelander Institute for Astronomy, University of Bonn, Auf dem Hügel 71, D-53121 Bonn, Germany ⟨ tmasch@astro.uni-bonn.de ⟩

L. Mazzuca, NASA Goddard Space Flight Center, Code 441, Greenbelt, MD 20771, USA ⟨ lisa.m.mazzuca@nasa.gov ⟩

I. Meschin, Instituto de Astrofísica de Canarias, C/ Vía Láctea s/n, E-38205 La Laguna (Tenerife), Spain ⟨ imeschin@iac.es ⟩

M. Michałowski, Dark Cosmology Centre, Niels Bohr Institute, University of Copenhagen, Juliane Maries Vej 30, DK-2100 Copenhagen Ø, Denmark ⟨ michal@astro.ku.dk ⟩

P. Molaro, INAF-Osservatorio Astronomico di Trieste, Trieste Via G.B. Tiepolo 11, I-34143, Italy ⟨ molaro@oats.inaf.it ⟩

M. Mollá, CIEMAT, ed. 2, Avda. Complutense 22, E-28040 Madrid, Spain ⟨ mercedes.molla@ciemat.es ⟩

N. Nesvadba, Observatoire de Paris, 5, Place Jules Janssen, F-92190 Meudon, France ⟨ nicole.nesvadba@obspm.fr ⟩

J. Oñorbe, Universidad Autónoma de Madrid, Departamento de Física Teórica, C-XI Cantoblanco, E-28049 Madrid, Spain ⟨ jose.onnorbe@uam.es ⟩

H. Otí, LAEFF-INTA, Apartado 78, E-28691 Villanueva de la Cañada, Madrid, Spain ⟨ otih@laeff.inta.es ⟩

B. Pagel, Astronomy Centre, University of Sussex, Falmer, Brighton BN1 9QH, UK ⟨ bejp@sussex.ac.uk ⟩

R. Peletier, Kapteyn Astronomical Institute, Rijksuniversiteit Groningen, Postbus 800, NL-9700 AV Groningen, The Netherlands ⟨ r.f.peletier@astro.rug.nl ⟩

I. Pérez, Kapteyn Astronomical Institute, Rijksuniversiteit Groningen, Postbus 800, NL-9700 AV Groningen, The Netherlands ⟨ isa@astro.rug.nl ⟩

I. Pérez Fournon, Instituto de Astrofísica de Canarias, C/ Vía Láctea s/n, E-38205 La Laguna (Tenerife), Spain ⟨ ipf@iac.es ⟩

J. Pflamm-Altenburg, Argelander Institute for Astronomy, University of Bonn, Auf dem Hügel 71, D-53121 Bonn, Germany ⟨ jpflamm@astro.uni-bonn.de ⟩

A. Pipino, University of Oxford, Astrophysics, Denys Wilkinson Building, Keble Road, Oxford OX1 3RH, UK ⟨ axp@astro.ox.ac.uk ⟩

M. Pohlen, School of Physics and Astronomy, Cardiff University, The Parade, Cardiff CF24 3AA, UK ⟨ michael.pohlen@astro.cf.ac.uk ⟩

C. Quiroga Blanco, Torres Sofer II - Dpto. 5 C, Cochabamba, Bolivia ⟨ carlosjulio.quiroga@gmail.com ⟩

R. Rebolo, Instituto de Astrofísica de Canarias, C/ Vía Láctea s/n, E-38205 La Laguna (Tenerife), Spain ⟨ rrl@iac.es ⟩

M. Relaño, Universidad de Granada, Departamento de Física Teórica y del Cosmos, Facultad de Ciencias, Edificio Mecenas, Campus de Fuentenueva, E-18071 Granada, Spain ⟨ mrelano@ugr.es ⟩

J. Renn, Max-Planck-Institut für Wissenschaftsgeschichte, Boltzmannstr. 22, D-14195 Berlin, Germany ⟨renn@mpiwg-berlin.mpg.de⟩

D. Reverte-Payá, Instituto de Astrofísica de Andalucía, Camino Bajo de Huétor 50, E-18008 Granada, Spain ⟨drp@iaa.es⟩

M. Rodrigues, Observatoire de Paris, GEPI, 5 Place Jules Janssen, F-92190 Meudon, France ⟨myriam.rodrigues@obspm.fr⟩

P. San José, K2000 S.A., Vía Lusitana 31, Ático A, Madrid, Spain ⟨pedrosjs@wanadoo.es⟩

M. Sarzi, University of Hertfordshire, Centre for Astrophysics Research, Hatfield, AL10 9AB, UK ⟨sarzi@star.herts.ac.uk⟩

K. Sheth, Spitzer Science Center, MS 220-6, Caltech, Pasadena, CA 91125, USA ⟨kartik@astro.caltech.edu⟩

I. Shlosman, University of Kentucky, Department of Physics and Astronomy, Lexington, KY 40506 USA ⟨shlosman@pa.uky.edu⟩

J. Silk, University of Oxford, Astrophysics, Denys Wilkinson Building, Keble Road, Oxford OX1 3RH, UK ⟨silk@astro.ox.ac.uk⟩

L. Smith, Space Telescope Science Institute, 3700 San Martin Drive, Baltimore, MD21218, USA ⟨lsmith@stsci.edu⟩

A. Sołtan, Copernicus Astronomical Center, Bartycka 18, 00-716 Warsaw, Poland ⟨soltan@camk.edu.pl⟩

T. Speltincx, University of Oulu, Division of Astronomy, Department of Physical Science, P.O. Box 3000, FIN-90014, Finland ⟨Tom.Speltincx@oulu.fi⟩

M. Stringer, Theoretical Astrophysics, California Institute of Technology, Caltech MC130-33, 1200 E, California Blvd., Pasadena, CA 91125, USA ⟨stringer@caltech.edu⟩

H. Thrall, Jodrell Bank Observatory, University of Manchester, Macclesfield, Cheshire, SK11 9DL, UK ⟨Hannah.Thrall@postgrad.manchester.ac.uk⟩

G. Trancho, Gemini Observatory Southern Operations Center, c/o AURA, Casilla 603, La Serena, Chile ⟨gtrancho@gemini.edu⟩

I. Trujillo, Instituto de Astrofísica de Canarias, C/ Vía Láctea s/n, E-38205 La Laguna (Tenerife), Spain ⟨itc@iac.es⟩

Y. Tsamis, University College London, Dept. of Physics and Astronomy, Gower Street, London WC1E 6BT, UK ⟨ygt@star.ucl.ac.uk⟩

M. Vallbé i Mumbrú, Instituto de Astrofísica de Canarias, C/ Vía Láctea s/n, E-38205 La Laguna (Tenerife), Spain ⟨marc.vallbe@iac.es⟩

A. Vazdekis, Instituto de Astrofísica de Canarias, C/ Vía Láctea s/n, E-38205 La Laguna (Tenerife), Spain ⟨vazdekis@iac.es⟩

J. Vílchez, Instituto de Astrofísica de Andalucía, Camino Bajo de Huétor 50, E-18008 Granada, Spain ⟨jvm@iaa.es⟩

G. Vladilo, INAF-Osservatorio Astronomico di Trieste, Trieste Via G.B. Tiepolo 11, I-34143, Italy ⟨vladilo@oats.inaf.it⟩

C. VLAHAKIS, Leiden Observatory, Leiden University, P.O. Box 9513, NL-2300 RA Leiden, The Netherlands ⟨ vlahakis@strw.leidenuniv.nl ⟩

M. WESTMOQUETTE, University College London, Dept. of Physics & Astronomy, Gower Street, London, WC1E 6BT, UK ⟨ msw@star.ucl.ac.uk ⟩

C. WINGE, Gemini Observatory, c/o AURA, Casilla 603, La Serena, Chile ⟨ cwinge@gemini.edu ⟩

Y. YAMADA, National Astronomical Observatory of Japan, 2-21-1 Osawa, Mitaka, Tokyo 181-8588, Japan ⟨ yamadays@cc.nao.ac.jp ⟩

Y. YANG, Observatoire de Paris, GEPI, 5 Place Jules Janssen, F-92190 Meudon, France ⟨ yanbin.yang@obspm.fr ⟩

S. ZHUKOVSKA, Institute of Theoretical Astrophysics, Albert-Überle-str. 2, Heidelberg, D-69120, Germany ⟨ zhukovska@ita.uni-heidelberg.de ⟩

A. ZURITA, Universidad de Granada, Departamento de Física Teórica y del Cosmos, Facultad de Ciencias, Edificio Mecenas, Campus de Fuentenueva, E-18071 Granada, Spain ⟨ azurita@ugr.es ⟩

# Session I

# The Star Formation Process

*Pathways through an Eclectic Universe*
*ASP Conference Series, Vol. 390, © 2008*
*J. H. Knapen, T. J. Mahoney, and A. Vazdekis, eds.*

# The IMF of Simple and Composite Populations

Pavel Kroupa

*Argelander-Institut für Astronomie, Universität Bonn, Germany*

**Abstract.** The combination of a finite timescale for star formation, rapid early stellar evolution, and rapid stellar dynamical processes imply that the stellar initial mass function (IMF) cannot be inferred for *any* star cluster independently of its age (the CLUSTER IMF THEOREM). The IMF can nevertheless be constrained statistically by evolving many theoretical populations drawn from one parent distribution and testing these against observed populations. It follows that all known well-resolved stellar populations are consistent with having been drawn from the same parent mass distribution. The IMF UNIVERSALITY HYPOTHESIS cannot therefore be discarded despite the existence of the CLUSTER IMF THEOREM. This means that the currently existing star formation theory fails to describe the stellar outcome because it predicts a dependency of the IMF on the physical boundary conditions, which is not observed. The INTEGRATED GALACTIC IMF THEOREM, however, predicts a variation of galaxy-wide IMFs in dependence on the galaxy's star formation rate even if the IMF UNIVERSALITY HYPOTHESIS is valid. This variation has now been observed in SDSS galaxy data. Detailed analysis of the binary properties in the very low mass star and brown dwarf mass regime on the one hand and in the stellar regime on the other shows there to be a discontinuity in the IMF near $0.1\,M_\odot$ such that brown dwarfs follow a separate distribution function. Very recent observations of the stellar population within 1 pc of the nucleus of the Milky Way do suggest a top-heavy IMF, perhaps hinting at a variation in the star formation outcome with tidal field and temperature, thereby violating the IMF UNIVERSALITY HYPOTHESIS under these physically extreme conditions. Another violation of this hypothesis appears to emerge for extremely metal-poor stars, such that the primordial IMF appears to have been depleted in low mass stars.

## 1. Introduction

The stellar initial mass function (IMF), $\xi(m)$, where $m$ is the stellar mass, is the parent distribution function of the masses of stars formed in *one* event. Here, the number of stars in the mass interval $(m, m + dm)$ is $dN = \xi(m)\,dm$. The IMF is, strictly speaking, an abstract theoretical construct because any *observed* system of $N$ stars merely constitutes a particular *representation* of this universal distribution function (Elmegreen 1997; Maíz Apellániz & Úbeda 2005). The probable existence of a *unique* $\xi(m)$ can be inferred from observations of an ensemble of systems, each consisting of $N$ stars (e.g., Massey 2003). If, after corrections for (a) stellar evolution, (b) unknown multiple stellar systems, and (c) stellar-dynamical biases, the individual distributions of stellar masses are similar *within the expected statistical scatter*, then we (the community) deduce that the hypothesis that the stellar mass distributions are not the same can be excluded. That is, we make the case for a *universal, standard* or *canonical stellar*

*IMF* within the physical conditions probed by the relevant physical parameters (metallicity, density, mass) of the populations at hand.

A detailed related overview of the IMF can be found in Kroupa (2007a), and a review with an emphasis on the metal-rich problem is available in Kroupa (2007b). Zinnecker & Yorke (2007) provide an in-depth review of the formation and distribution of massive stars. Elmegreen (2007) discusses the possibility that star formation occurs in different modes with different IMFs.

## 2. The Canonical or Standard Form of the Stellar IMF

The *canonical stellar IMF* is a two-part power law, $\xi(m) \propto m^{-\alpha_i}$, the only structure found with confidence so far being the change of index from the Salpeter/Massey value to a smaller one near $0.5\,M_\odot$:[1]

$$\begin{array}{ll} \alpha_1 = 1.3 \pm 0.3 & 0.08 \lesssim m/M_\odot \lesssim 0.5 \\ \alpha_2 = 2.3 \pm 0.5 & 0.5 \lesssim m/M_\odot \lesssim 150. \end{array} \quad (1)$$

It has been corrected for bias through unresolved multiple stellar systems in the low mass ($m < 1\,M_\odot$) regime using a multi-dimensional optimization technique. The general outline of this technique is as follows (Kroupa, Tout, & Gilmore 1993): first the correct form of the stellar mass–luminosity relation is extracted using observed stellar binaries *and* theoretical constraints on the location, amplitude, and shape of the minimum of its derivative near $m = 0.3\,M_\odot, M_V \approx 12, M_I \approx 9$, in combination with the observed shape of the nearby and deep Galactic field luminosity function (LF). Having established the semi-empirical mass–luminosity relation of stars, which is an excellent fit to the most recent observational constraints by Delfosse et al. (2000), a model of the Galactic field is then calculated assuming a parameterized form for the mass function and different values for the scale height of the Galactic disk, and different binary fractions in it. Measurement uncertainties and age and metallicity spreads must also be considered in the theoretical stellar population. Optimization in this multi-parameter space (mass function parameters, scale height and binary population) against observational data leads to the canonical stellar mass function for $m < 1\,M_\odot$.

One important result from this work is the finding that the stellar LF has a universal sharp peak near $M_V \approx 12, M_I \approx 9$ that results from changes in the internal constitution of stars that drive a non-linearity in the stellar mass–luminosity relation.

A consistency check is then performed as follows: the above mass function is used in creating young populations of binary systems that are born in modest star clusters consisting of a few hundred stars. Their dissolution into the Galactic field is computed with an $N$-body code, and the resulting theoretical field is compared to the observed LFs (Kroupa 1995a,b). Further confirmation of the form of the canonical IMF comes from independent sources, most notably by Reid, Gizis, & Hawley (2002) and also Chabrier (2003).

---

[1]The uncertainties in $\alpha_i$ are estimated from the alpha plot (Sect. 3.2.), as shown in figure 5 in Kroupa (2002), to be about 95% confidence limits.

In the high mass regime, Massey (2003) reports the same *slope or index* $\alpha_3 = 2.3 \pm 0.1$ for $m \gtrsim 10\,M_\odot$ in many OB associations and star clusters in the Milky Way (MW), and in the Large and Small Magellanic Clouds (LMC, SMC, respectively). It is therefore suggested to refer to $\alpha_2 = \alpha_3 = 2.3$ as the *Salpeter/Massey slope* or *index*, given the pioneering work of Salpeter (1955), who derived this value for stars with masses 0.4–10 $M_\odot$. However, multiplicity corrections remain to be done once we learn more about how the components are distributed in massive stars (cf. Preibisch et al. 1999; Zinnecker 2003). Scalo (1986) found $\alpha_{\rm MW\,disk} \approx 2.7$ ($m \gtrsim 1\,M_\odot$) from a very thorough analysis of OB star counts in the MW disk. Similarly, the star count analysis of Reid et al. (2002) leads to $2.5 \lesssim \alpha_{\rm MW\,disk} \lesssim 2.8$, and Tinsley (1980), Kennicutt (1983; his "extended Miller–Scalo IMF"), Portinari et al. (2004), and Romano et al. (2005) find $2.5 \lesssim \alpha_{\rm MW\,disk} \lesssim 2.7$. That $\alpha_{\rm MW\,disk} > \alpha_2$ naturally is shown in Section 5.

The evidence for a universal upper mass cut-off near 150 $M_\odot$ (Weidner & Kroupa 2004; Figer 2005; Oey & Clarke 2005; Koen 2006; Maíz Apellániz et al. 2007) seems to be rather well established in populations with metallicities ranging from that of the LMC ($Z \approx 0.008$) to the super-solar Galactic Center ($Z \gtrsim 0.02$), such that the stellar mass function simply stops at that mass. This mass needs to be understood theoretically (see discussion in Kroupa & Weidner 2005 and Zinnecker & Yorke 2007).

Below the hydrogen-burning limit (see also Sect. 4) there is substantial evidence that the IMF flattens further to $\alpha_0 \approx 0.3 \pm 0.5$ (Martín et al. 2000; Chabrier 2003; Moraux et al. 2004). Therefore, the canonical IMF most likely has a peak at 0.08 $M_\odot$. Brown dwarfs, however, comprise only a few per cent of the mass of a population and are therefore dynamically irrelevant. The logarithmic form of the canonical IMF, $\xi_{\rm L}(m) = \ln(10)\; m\; \xi(m)$, which gives the number of stars in $\log_{10} m$ intervals, also has a peak near 0.08 $M_\odot$. However, the *system* IMF (of stellar single and multiple systems combined to system masses) has a maximum in the mass range 0.4–0.6 $M_\odot$ (Kroupa et al. 2003).

The above canonical or standard form has been derived from detailed considerations of star counts, thereby representing an *average* IMF: for low mass stars it is a mixture of stellar populations spanning a large range of ages (0–10 Gyr) and metallicities ([Fe/H] $\gtrsim -1$). For the massive stars it constitutes a mixture of different metallicities ([Fe/H] $\gtrsim -1.5$) and star-forming conditions (from OB associations to very dense starburst clusters: R136 in the LMC). Therefore, it can be taken as a canonical form, and the aim is to test the

> IMF UNIVERSALITY HYPOTHESIS: The canonical IMF constitutes the parent distribution of all stellar populations.

Negation of this hypothesis would imply a variable IMF. Note that the work of Massey (2003) has already established the IMF to be invariable for $m \gtrsim 10\,M_\odot$ and for densities $\rho \lesssim 10^5$ stars/pc$^3$ and metallicities $Z \gtrsim 0.002$.

## 3. Universality of the IMF: Resolved Populations

The strongest test of the IMF UNIVERSALITY HYPOTHESIS (Sect. 2) is obtained by studying populations that can be resolved into individual stars. Since one also seeks coeval populations at the same distance and with the same metallicity to minimize uncertainties, star clusters and stellar associations would seem to be the test objects of choice. But before contemplating such work some lessons from stellar dynamics are useful.

### 3.1. Star Clusters and Associations

To access a pristine population one would consider observing star clusters that are younger than a few Myr. However, such objects carry rather massive disadvantages: the pre-main sequence stellar evolutionary tracks are unreliable (Baraffe et al. 2002; Wuchterl & Tscharnuter 2003) such that the derived masses are uncertain by at least a factor of about two. The remaining gas and dust lead to patchy obscuration. Very young clusters evolve rapidly: the dynamical crossing time is $t_{\rm cr} = 2\, r_{\rm cl}/\sigma_{\rm cl}$, where the cluster radii are typically $r_{\rm cl} < 1\,$pc and for cluster masses $M_{\rm cl} > 10^3\, M_\odot$ the velocity dispersion $\sigma_{\rm cl} > 2\,{\rm km\,s^{-1}}$ such that $t_{\rm cr} < 1\,$Myr. The inner regions of populous clusters have $t_{\rm cr} \approx 0.1$ Myr, and thus significant mixing and relaxation (relaxation time $t_{\rm relax} \approx t_{\rm cr}\, 0.1\, N/{\rm ln}N$) occurs there by the time the residual gas is expelled by the winds and photo-ionizing radiation from massive stars, if they are present, this being the case in clusters with $N \gtrsim$ a few $\times$ 100 stars. Much of this is summarized in Kroupa (2005). For example, the 0.5–2 Myr old Orion Nebula Cluster (ONC), which is known to be supervirial with a virial mass about twice the observed mass (Hillenbrand & Hartmann 1998), has already expelled its residual gas and is expanding rapidly, thereby probably having lost its outer stars (Kroupa, Aarseth, & Hurley 2001). The supervirial state of young clusters makes measurements of their mass-to-light ratios a bad measure of the stellar mass within them (Bastian & Goodwin 2006; Goodwin & Bastian 2006), and rapid dynamical mass segregation likewise makes "naive" measurements of the $M/L$ ratio wrong (Boily et al. 2005; Fleck et al. 2006).

Massive stars ($m > 8\, M_\odot$) are either formed at the cluster center or get there through dynamical mass segregation, i.e., energy equipartition (Bonnell, Larson, & Zinnecker 2006). The latter process is *very rapid*, operating on a timescale $t_{\rm msgr} \approx (m_{\rm av}/m_{\rm massive})t_{\rm relax}$, where $m_{\rm av}, m_{\rm massive}$ are the average and massive star masses, respectively, and can occur within 1 Myr. A cluster core of massive stars is therefore either primordial or forms rapidly because of energy equipartition in the cluster, and it is dynamically highly unstable, decaying within a few $t_{\rm cr,\, core}$. The ONC, for example, should not be hosting a Trapezium as it is extremely volatile. The implication for the IMF is that the ONC and other similar clusters, and the OB associations that stem from them, must be very depleted in their massive star content (Pflamm-Altenburg & Kroupa 2006).

Important for measuring the IMF are corrections for the typically high multiplicity fraction of the very young population. However, these are very uncertain because the binary population is in a state of change (figure 1 in Kroupa 2000).

The determination of an IMF relies on the assumption that all the stars in a very young cluster formed together. Trapping of older field or OB association stars by the forming cluster has been found to be possible for ONC-type clusters (Pflamm-Altenburg & Kroupa 2007) and also for massive $\omega$ Cen-type clusters (Fellhauer, Kroupa, & Evans 2006). Additionally, the sample of cluster stars may be contaminated by enhanced fore- and background densities of field stars due to focusing of stellar orbits during cluster formation (Pflamm-Altenburg & Kroupa 2007).

Thus, whether it be at the low mass end or the high mass end, the "IMF" estimated from very young clusters cannot be the true IMF. Statistical corrections for the above effects need to be applied, and comprehensive $N$-body modeling is required.

Old open clusters in which most stars are on or near the main sequence are not better stellar samples: they are dynamically highly evolved, since they have left their previous concentrated and gas-rich state and so they contain only a fraction of the stars originally born in the cluster (Kroupa & Boily 2002; Weidner et al. 2007; Baumgardt & Kroupa 2007). The binary fraction is typically high and comparable to the Galactic field, but does depend on the initial density and the age of the cluster, also the mass ratio distribution of companions. So simple corrections cannot be applied equally for all old clusters. The massive stars have died, and secular evolution begins to affect the remaining stellar population (after gas expulsion) through energy equipartition. Baumgardt & Makino (2003) have quantified the changes in the mass function for clusters of various masses and in different Galactic orbits. Near the half-mass radius the local mass function resembles the global mass function in the cluster, but the global mass function becomes significantly depleted of its lesser stars already by about 20% of the cluster's disruption time.

Given that we are never likely to learn the exact dynamical history of a particular cluster, it follows that we can *never* ascertain the IMF for any individual cluster. This can be summarized concisely with the following theorem:

CLUSTER IMF THEOREM: The IMF cannot be extracted for any individual star cluster.

PROOF: For clusters younger than about 0.5 Myr star formation has not ceased and the IMF is therefore not assembled yet, the cluster cores consisting of massive stars having already dynamically ejected members (Pflamm-Altenburg & Kroupa 2006). For clusters with an age between 0.5 and a few Myr the expulsion of residual gas has led to loss of stars (Kroupa et al. 2001). Older clusters are either still losing stars due to residual gas expulsion or are evolving secularly through evaporation driven by energy equipartition (Baumgardt & Makino 2003). There thus exists no time when all the stars are assembled in an observationally accessible volume (i.e., a star cluster). END OF PROOF.

Note that the CLUSTER IMF THEOREM implies that individual clusters cannot be used to make deductions on the similarity or otherwise of their IMFs unless a complete dynamical history of each cluster is available.

Notwithstanding this pessimistic theorem, it is nevertheless necessary to observe and study star clusters of any age. Combined with thorough and realistic $N$-body modeling, the data do lead to essential *statistical* constraints on the IMF UNIVERSALITY HYPOTHESIS (Sect. 2). Such an approach is discussed in the next section.

### 3.2. The Alpha Plot

Scalo (1998) conveniently summarized a large part of the available observational constraints on the IMF of resolved stellar populations with the *alpha plot*, as used by Kroupa (2001, 2002) for explicit tests of the IMF UNIVERSALITY HYPOTHESIS (Sect. 2) given the CLUSTER IMF THEOREM (Sect. 3.1). One example is presented in figure 1 in Kroupa (2007b), which demonstrates that the observed scatter in $\alpha(m)$ can be readily understood as being due to Poissonian uncertainties (see also Elmegreen 1997, 1999) and dynamical effects, as well as arising from biases through unresolved multiple stars. Furthermore, there is no evident systematic change in $\alpha$ at a given $m$ with metallicity or density of the star-forming cloud. More exotic populations such as the Galactic bulge have also been found to have a low mass mass function indistinguishable from the canonical form (e.g., Zoccali et al. 2000). Thus, the IMF UNIVERSALITY HYPOTHESIS cannot be falsified for known resolved stellar populations.

### 3.3. Very Ancient Resolved Populations

Witnesses of the early formation phase of the MW are its globular clusters. Such $10^{4-6}\,M_\odot$ clusters formed with individual star formation rates of 0.1–1 $M_\odot$/yr and densities $\approx 5 \times 10^{3-5}\,M_\odot/\mathrm{pc}^3$. These are relatively high values, when compared with the current star formation activity in the MW disk. For example, a $5 \times 10^3\,M_\odot$ Galactic cluster forming in 1 Myr corresponds to a star formation rate of $0.005\,M_\odot$/yr. The alpha plot, however, does not support any significant systematic difference between the IMF of stars formed in globular clusters and present-day low mass star formation. For massive stars, it can be argued that the mass in stars more massive than $8\,M_\odot$ cannot have been larger than about half the cluster mass, because otherwise the globular clusters would not be as compact as they are today. This constrains the IMF to have been close to the canonical IMF (Kroupa 2001).

A particularly exotic star formation mode is thought to have occurred in dwarf spheroidal (dSph) satellite galaxies. The MW has about 19 such satellites at distances from 50 to 250 kpc (Metz & Kroupa 2007). These objects have stellar masses and ages comparable to those of globular clusters but are 10–100 times larger and are thought to have 10–100 times more mass in dark matter than in stars. They also show evidence for complex star formation activity and metal enrichment histories and must therefore have formed under rather exotic conditions. Nevertheless, the mass functions in two of these satellites are found to be indistinguishable from those of globular clusters in the mass range 0.5–$0.9\,M_\odot$, thus again showing consistency with the canonical IMF (Grillmair et al. 1998; Feltzing, Gilmore, & Wyse 1999).

### 3.4. The Galactic Bulge and Center

For low mass stars the Galactic bulge has been shown to have a mass function indistinguishable from the canonical form (Zoccali et al. 2000). However, abundance patterns of bulge stars suggest the IMF to have been top heavy (Ballero, Kroupa, & Matteucci 2007), which may be a result of extreme starburst conditions valid in the formation of the bulge (Zoccali et al. 2006).

Even closer to the Galactic center, Hertzsprung–Russell diagram modeling of the stellar population within 1 pc of Sgr A* suggests the IMF to have always been top-heavy there (Maness et al. 2007). Perhaps this is the long sought after evidence for a variation in the IMF under extreme conditions, in this case a strong tidal field and higher temperatures.

### 3.5. Population III: The Primordial IMF

Most theoretical workers agree that the primordial IMF ought to be top-heavy because the ambient temperatures were much higher and the lack of metals did not allow gas clouds to cool and fragment into sufficiently small cores. The existence of extremely metal-poor low mass stars with chemical peculiarities is interpreted to mean that low mass stars could form under extremely metal-poor conditions, but that their formation was suppressed in comparison to later star formation (Tumlinson 2007). Modeling of the formation of stellar populations during cosmological structure formation suggests that low mass Population III stars should be found within the Galactic halo *if* they formed. Their absence to date would imply a primordial IMF depleted in low mass stars (Brook et al. 2007).

## 4. Very Low Mass Stars and Brown Dwarfs

The origin of brown dwarfs (BDs) and some very low mass stars (VLMSs) is being debated fiercely. One camp believes these objects to form as stars do, because the star formation process does not know where the hydrogen burning mass limit is (e.g., Eislöffel & Steinacker 2007). The other camp believes that BDs cannot form exactly like stars through continued accretion because the conditions required for this to occur in molecular clouds are far too rare (e.g., Reipurth & Clarke 2001; Goodwin & Whitworth 2007).

If BDs and VLMSs form like stars then they should follow the same pairing rules. In particular, BDs and G dwarfs would pair in the same manner as M dwarfs and G dwarfs. Kroupa et al. (2003) tested this hypothesis by constructing $N$-body models of Taurus–Auriga-like groups and Orion Nebula-like clusters, finding that it leads to far too many star–BD and BD–BD binaries with the wrong semimajor axis distribution. Instead, star–BD binaries are rare while BD–BD binaries have a semimajor axis distribution significantly narrower than that of star–star binaries. The hypothesis of a star-like origin of BDs must therefore be discarded. BDs and some VLMSs form a separate population, which is, however, linked to that of the stars.

Thies & Kroupa (2007) re-address this problem with a detailed analysis of the underlying mass function of stars and BDs given observed mass functions of four populations, Taurus, Trapezium, IC 348, and the Pleiades. Correcting for unresolved binaries in all four populations, by taking into account the different pairing rules of stellar and VLMS and BD binaries, a significant discontinuity of the mass function emerges. BDs and VLMSs therefore form a truly separate population from that of the stars and can be described by a single power-law mass function, $\xi_{\rm BD}$, with index $\alpha_0 \approx 0.3$ and $\xi_{\rm BD}(0.075\,M_\odot) = (0.2\text{–}0.3)\xi(0.075\,M_\odot)$, where $\xi$ is the canonical stellar IMF. This implies that about one BD forms per five stars in all four populations.

This strong correlation between the number of stars and BDs, and the similarity of the BD mass function in the four populations implies that the formation of BDs is closely related to the formation of stars. Indeed, the truncation of the binary binding energy distribution of BDs at a high energy suggests that energetic processes must be operating in the production of BDs, as discussed in Thies & Kroupa (2007).

## 5. Composite Populations: The Integrated Galactic IMF

The vast majority of all stars form in embedded clusters and so the correct way to proceed to calculate a galaxy-wide stellar IMF is to add up all the IMFs of all star clusters born in one "star formation epoch". Such "epochs" may be identified with the Zoccali et al. (2006) starburst events creating the Galactic bulge. In disk galaxies they may be related to the timescale of transforming the interstellar matter to star clusters along spiral arms. Addition of the clusters born in one epoch gives the *integrated galactic initial mass function*, the IGIMF (Kroupa & Weidner 2003).

> IGIMF DEFINITION: The IGIMF is the IMF of a composite population which is the integral over a complete ensemble of simple stellar populations.

Note that a *simple population* has a mono-metallicity and a mono-age distribution and is therefore a star cluster. Age and metallicity distributions emerge for massive populations with $M_{\rm cl} \gtrsim 10^6\, M_\odot$ (e.g., $\omega$ Cen), indicating that such objects, which also have relaxation times comparable to or longer than a Hubble time, are not "simple". A *complete ensemble* is a statistically complete representation of the initial cluster mass function (ICMF) in the sense that the actual mass function of $N_{\rm cl}$ clusters lies within the expected statistical variation of the ICMF.

> IGIMF THEOREM: The IGIMF is steeper than the canonical IMF if the IMF UNIVERSALITY HYPOTHESIS holds.

PROOF: Weidner & Kroupa (2006) calculate that the IGIMF is steeper than the canonical IMF for $m \gtrsim 1\, M_\odot$ if the IMF UNIVERSALITY HYPOTHESIS holds. The steepening becomes negligible if the power-law ICMF is flatter than $\beta \approx 1.8$. END OF PROOF.

It may be argued that IGIMF = IMF (e.g., Elmegreen 2006) because when a star cluster is born, its stars are randomly sampled from the IMF up to the most massive star possible. On the other hand, the physically motivated ansatz by Weidner & Kroupa (2005, 2006) of taking the mass of a cluster as the constraint and of including the observed correlation between the maximal star mass and the cluster mass, yields an IGIMF equal to the canonical IMF for $m \lesssim 1.5\, M_\odot$ but which is systematically steeper above this mass. By incorporating the observed maximal cluster mass vs star formation rate of galaxies for the youngest clusters (Weidner, Kroupa, & Larsen 2004), it follows for $m \gtrsim 1.5\, M_\odot$ that low surface brightness (LSB) galaxies have very steep IGIMFs while normal or $L_*$ galaxies have Scalo-type IGIMFs, i.e., $\alpha_{\rm IGIMF} = \alpha_{\rm MW\,disk} > \alpha_2$ ( Sect. 2) follows naturally. This systematic shift of $\alpha_{\rm IGIMF}$ ($m \gtrsim 1.5\, M_\odot$) with galaxy type implies that less massive galaxies have a significantly suppressed supernova II rate per low mass star. They also show a slower chemical enrichment such that the observed metallicity–galaxy mass relation can be nicely accounted for (Koeppen, Weidner, & Kroupa 2007).

Strikingly, the IGIMF variation has now been directly measured by Hoversten & Glazebrook (2007) using galaxies in the SDSS. Lee et al. (2004) have indeed found LSBs to have bottom-heavy IMFs, while Portinari et al. (2004) and Romano et al. (2005) find the MW disk to have a steeper than Salpeter IMF for massive stars that is, in comparison with Lee et al. (2004), much flatter than the IMF of LSBs, as required by the IGIMF THEOREM.

## 6. Origin of the IMF: Theory vs Observations

General physical concepts such as coalescence of protostellar cores, mass-dependent focusing of gas accretion onto protostars, stellar feedback, and fragmentation of molecular clouds lead to predictions of systematic variations in the IMF with changes in the physical conditions of star formation (Murray & Lin 1996; Elmegreen 2004; Tilley & Pudritz 2005; but see Casuso & Beckman 2007 for a simple cloud coagulation/dispersal model leading to an invariant mass distribution). Thus, the thermal Jeans mass of a molecular cloud decreases with temperature and increasing density, implying that for higher metallicity (= stronger cooling) and density the IMF should shift on average to smaller stellar masses (e.g., Larson 1998; Bonnell, Larson, & Zinnecker 2006). The entirely different notion that stars regulate their own masses through a balance between feedback and accretion also implies smaller stellar masses for higher metallicity due in part to more dust and thus more efficient radiation pressure (Adams & Fatuzzo 1996; Adams & Laughlin 1996)

Klessen, Spaans, & Jappsen (2007) report state-of-the-art calculations of star formation under physical conditions as found in molecular clouds near the Sun and they are able to reproduce the canonical IMF. Applying the same computational technology to the conditions near the Galactic center they obtain a theoretical IMF in agreement with the previously reported apparent decline in the stellar mass function in the Arches cluster below about $6\,M_{\odot}$. Kim et al. (2006) published their observations of the Arches cluster on the astro-ph preprint archive shortly after Klessen et al. (2007) and performed the necessary state-of-the art $N$-body calculations for the dynamical evolution of this young cluster, thus revising our knowledge significantly. In contradiction to Klessen et al. (2007), they find that the mass function continues to increase down to their 50% completeness limit ($1.3\,M_{\odot}$) with a power-law exponent only slightly shallower than the canonical Massey/Salpeter value once mass segregation is corrected for. This situation is demonstrated in Figure 1.

Observations of cloud cores appear to suggest that the canonical IMF is frozen in already at the pre-stellar cloud-core level (Motte, André, & Neri 1998; Motte et al. 2001). Nutter & Ward-Thompson (2007) and Alves, Lombardi, & Lada (2007) find, however, the pre-stellar cloud cores to be distributed according to the same shape as the canonical IMF, but shifted to larger masses by a factor of about three or more. This is taken perhaps to mean a star formation efficiency per star of 30% or less, independently of stellar mass. The interpretation of such observations in view of multiple star formation in each cloud core is being studied by Goodwin et al. (2007), while Krumholz (2007, these proceedings, p. 16) outlines the current theoretical understanding of how massive stars form out of the massive pre-stellar cores.

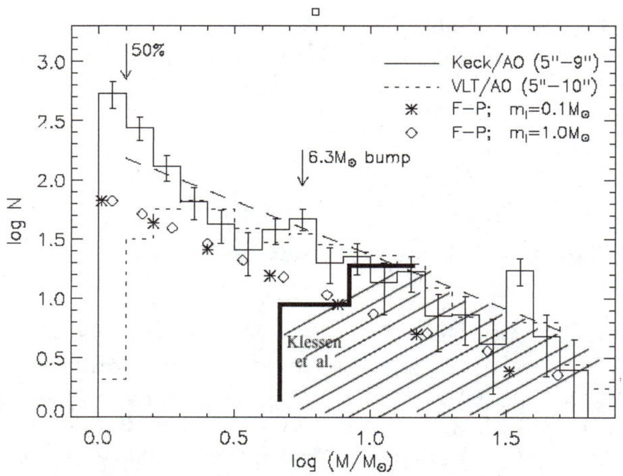


Figure 1. The observed mass function of the Arches cluster near the Galactic center by Kim et al. (2006), shown as the thin histogram, is confronted with the theoretical mass function for this object calculated with the SPH technique by Klessen et al. (2007). The latter has a down-turn incompatible with the observations, thereby ruling out a theoretical understanding of the stellar mass spectrum ('one counter-example suffices to bring down a theory'). One possible reason for the theoretical failure may be the assumed turbulence driving. For details on the figure see Kim et al. (2006).

## 7. Conclusions

The IMF UNIVERSALITY HYPOTHESIS, the CLUSTER IMF THEOREM, and the IGIMF THEOREM have been stated. Furthermore,

1. The stellar LF has a pronounced maximum at $M_V \approx 12, M_I \approx 9$ which is universal and well understood as a result of stellar physics. Thus, by counting stars on the sky we can look into their interiors.

2. Unresolved multiple systems must be accounted for when the mass functions of different stellar populations are compared.

3. BDs and some VLMSs form a separate population which correlates with the stellar content; there is a discontinuity in the mass function near the star/BD mass transition.

4. The canonical IMF (Eq. 1) fits the solar neighborhood star counts and *all* resolved stellar populations available to date. Recent data near the Galactic center suggest a top-heavy IMF, perhaps hinting at a possible variation with conditions, although the results on the Arches cluster near the Galactic center are not entirely supportive of this statement.

5. *Simple* stellar populations are found in individual star clusters with $M_{cl} \lesssim 10^6\ M_{\odot}$. These have the canonical IMF.

6. *Composite* populations describe entire galaxies. They are a result of many epochs of star cluster formation and are described by the IGIMF THEOREM.

7. The IGIMF above $\approx 1\,M_\odot$ is steep for LSB galaxies, flattening to the Scalo slope ($\alpha_{\rm IGIMF} \approx 2.7$) for $L_*$ disk galaxies. This is nicely consistent with the IMF UNIVERSALITY HYPOTHESIS in the context of the IGIMF THEOREM.

8. Therefore, the IMF UNIVERSALITY HYPOTHESIS cannot be excluded for conditions $\rho \lesssim 10^5\,{\rm stars/pc}^3$, $Z \gtrsim 0.002$ and non-extreme tidal fields, despite the CLUSTER IMF THEOREM.

9. Modern star formation computations appear to give wrong results concerning the shape of the stellar IMF.

10. The stellar IMF appears to be frozen in at the pre-stellar cloud-core stage, thereby probably being a result of the processes leading to the formation of self-gravitating molecular clouds.

**Acknowledgments.** I am very grateful to the organizers for a most enjoyable meeting. I also thank Thomas Maschberger and Jan Pflamm-Altenburg for helpful discussions and comments, and my collaborators for important input. Parts of this work were supported by the DFG.

## References


Adams, F. C., & Fatuzzo, M. 1996, ApJ, 464, 256
Adams, F. C., & Laughlin, G. 1996, ApJ, 468, 586
Alves, J., Lombardi, M., & Lada, C. J. 2007, A&A, 462, L17
Ballero, S., Kroupa, P., & Matteucci, F. 2007, MNRAS, 467, 117
Baraffe, I., Chabrier, G., Allard, F., & Hauschildt, P. H. 2002, A&A, 382, 563
Baumgardt, H., & Kroupa, P. 2007, MNRAS, 380, 1589
Baumgardt, H., & Makino, J. 2003, MNRAS, 340, 227
Bastian, N., & Goodwin, S. P. 2006, MNRAS, 369, L9
Boily, C. M., Lançon, A., Deiters, S., & Heggie, D. C. 2005, ApJ, 620, L27
Bonnell, I. A., Larson, R., & Zinnecker, H. 2006, Proto Stars and Planets V (astro-ph/0603447)
Brook, C. B., Kawata, D., Scannapieco, E., Martel, H., & Gibson, B. K. 2007, ApJ, 661, 10
Casuso, E., & Beckman, J. E. 2007, ApJ, 656, 897
Chabrier, G. 2003, PASP, 115, 763
Delfosse, X., Forveille, T., Ségransan, D., Beuzit, J.-L., Udry, S., Perrier, C., & Mayor, M. 2000, A&A, 364, 217
Eislöffel, J., & Steinacker, J. 2007, in The Formation of Low-Mass-Protostars and Proto-Brown Dwarfs, at the 14th Cambridge Workshop on Cool Stars, Stellar Systems and the Sun in Pasadena (astro-ph/0701525)
Elmegreen, B. G. 1997, ApJ, 486, 944
Elmegreen, B. G. 1999, ApJ, 515, 323
Elmegreen, B. G. 2004, MNRAS, 354, 367
Elmegreen, B. G. 2006, ApJ, 648, 572
Elmegreen, B. G. 2007, ApJ, 668, 1064
Fellhauer, M., Kroupa, P., & Evans, N. W. 2006, MNRAS, 372, 338

Feltzing, S., Gilmore, G., & Wyse, R. F. G. 1999, ApJ, 516, L17
Figer, D. F. 2005, Nature, 434, 192
Fleck, J.-J., Boily, C. M., Lançon, A., & Deiters, S. 2006, MNRAS, 369, 1392
Goodwin, S. P., & Bastian, N. 2006, MNRAS, 373, 752
Goodwin, S. P., & Whitworth, A. 2007, A&A, 466, 943
Grillmair, C. J., et al. 1998, AJ, 115, 144
Hillenbrand, L. A., & Hartmann, L. W. 1998, ApJ, 492, 540
Hoversten, E. A., & Glazebrook, K. 2007. ApJ, in press
Kennicutt, R. C., Jr. 1983, ApJ, 272, 54
Kim, S. S., Figer, D. F., Kudritzki, R. P., & Najarro, F. 2006, ApJ, 653, L113
Klessen, R. S., Spaans, M., & Jappsen, A.-K. 2007, MNRAS, 374, L29
Koen, C. 2006, MNRAS, 365, 590
Koeppen, J., Weidner, C., & Kroupa, P. 2007, MNRAS 375, 673
Kroupa, P. 1995a, ApJ, 453, 350
Kroupa, P. 1995b, ApJ, 453, 358
Kroupa, P. 2000, New Astron., 4, 615
Kroupa, P. 2001, MNRAS, 322, 231
Kroupa, P. 2002, Science, 295, 82
Kroupa, P. 2005, in ESA SP-576: The Three-Dimensional Universe with Gaia, 629 (astro-ph/0412069)
Kroupa, P. 2007a, in ASP Conf. Ser., Resolved Stellar Populations, ed. D. Valls-Gabaud & M. Chavez (San Francisco: ASP), in press (astro-ph/0703124)
Kroupa, P. 2007b, in The Metal Rich Universe, ed. G. Israelian & G. Meynet (Cambridge: Cambridge University Press), astro-ph/0703282
Kroupa, P., & Boily, C. M. 2002, MNRAS, 336, 1188
Kroupa, P., & Weidner, C. 2003, ApJ, 598, 1076
Kroupa, P., & Weidner, C. 2005, IAUS, 227, 423 (astro-ph/0507582)
Kroupa, P., Aarseth, S., & Hurley, J. 2001, MNRAS, 321, 699
Kroupa, P., Tout, C. A., & Gilmore, G. 1993, MNRAS, 262, 545
Kroupa, P., Bouvier, J., Duchêne, G., & Moraux, E. 2003, MNRAS, 346, 354
Larson, R. B. 1998, MNRAS, 301, 569
Lee, H.-C., Gibson, B. K., Flynn, C., Kawata, D., & Beasley, M. A. 2004, MNRAS, 353, 113
Maíz Apellániz, J., & Úbeda, L. 2005, ApJ, 629, 873
Maíz Apellániz, J., Walborn, N. R., Morrell, N. I., Niemela, V. S., & Nelan, E. P. 2007, ApJ, 660, 1480
Maness, H., et al. 2007, ApJ, 669, 1024
Martín, E. L., Brandner, W., Bouvier, J., Luhman, K. L., Stauffer, J., Basri, G., Zapatero Osorio, M. R., & Barrado y Navascués, D. 2000, ApJ, 543, 299
Massey, P. 2003, ARA&A, 41, 15
Metz, M., & Kroupa, P. 2007, MNRAS, 376, 387
Moraux, E., Bouvier, J., & Clarke, C. 2004, in SF2A-2004: Semaine de l'Astrophysique Francaise, ed. F. Combes, D. Barret, T. Contini, F. Meynadier and L. Pagani, p. 251
Motte, F., Andre, P., & Neri, R. 1998, A&A, 336, 150
Motte, F., André, P., Ward-Thompson, D., & Bontemps, S. 2001, A&A, 372, L41
Murray, S. D., & Lin, D. N. C. 1996, ApJ, 467, 728
Nutter, D., & Ward-Thompson, D. 2007, MNRAS, 374, 1413
Oey, M. S., & Clarke, C. J. 2005, ApJ, 620, L43
Pflamm-Altenburg, J., & Kroupa, P. 2006, MNRAS, 373, 295
Pflamm-Altenburg, J., & Kroupa, P. 2007, MNRAS, 375, 855
Portinari, L., Sommer-Larsen, J., & Tantalo, R. 2004, MNRAS, 347, 691
Preibisch, T., Balega, Y., Hofmann, K., Weigelt, G., & Zinnecker, H. 1999, New Astron., 4, 531
Reid, I. N., Gizis, J. E., & Hawley, S. L. 2002, AJ, 124, 2721

Reipurth, B., & Clarke, C. 2001, AJ, 122, 432
Romano, D., Chiappini, C., Matteucci, F., & Tosi, M. 2005, A&A, 430, 491
Salpeter, E. E. 1955, ApJ, 121, 161
Scalo J. M. 1986, Fund. Cosmic Phys., 11, 1
Scalo, J. 1998, in ASP Conf. Ser., 142: The Stellar Initial Mass Function (38th Herstmonceux Conference), ed. G. Gilmore & D. Howell (San Francisco: ASP), p. 201
Thies, I., & Kroupa, P. 2007, ApJ, 671, 767
Tilley, D. A., & Pudritz, R. E. 2005, Protostars and Planets, V, 8473
Tinsley, B. M. 1980, Fund. Cosmic Phys., 5, 287
Tumlinson, J. 2007, ApJ, 665, 1361
Weidner, C., & Kroupa, P. 2004, MNRAS, 348, 187
Weidner, C., & Kroupa, P. 2005, ApJ, 625, 754
Weidner, C., & Kroupa, P. 2006, MNRAS, 365, 1333
Weidner, C., Kroupa, P., & Larsen, S. S. 2004, MNRAS, 350, 1503
Weidner, C., Kroupa, P., Nürnberger, D., & Sterzik, M. 2007, MNRAS, 376, 1879
Wuchterl, G., & Tscharnuter, W. M. 2003, A&A, 398, 1081
Zinnecker, H. 2003, IAUS, 212, 80
Zinnecker, H., & Yorke, H. W. 2007 (astro-ph/0707.1279)
Zoccali, M., Cassisi, S., Frogel, J. A., Gould, A., Ortolani, S., Renzini, A., Rich, R. M., & Stephens, A. W. 2000, ApJ, 530, 418
Zoccali, M., et al. 2006, A&A, 457, L1

## Discussion

*Krumholz*: Your model of the evaporation of the ONC would seem to predict that there should be an evaporating halo of BDs and low mass stars around the ONC. People have looked for such a halo around Taurus and have not seen it. Can you explain the lack of a halo around Taurus, and would you predict that one should be seen in the ONC?

*Kroupa*: Yes, one should be seen around the ONC. High precision proper motion and line-of-sight velocity measurements should show a radially expanding population of stars. For Taurus this is not expected, because the residual gas is removed over time through the accumulating radiative heating and outflows from the emerging stellar population. Evolution is thus expected to be adiabatic, i.e., Taurus-like groups don't pop and so a radially expanding halo would not be expected, apart from the difficulty of detecting such halos given the small number of stars in the Taurus groups.

*Hensler*: Your IGIMF compares dwarf and spiral galaxies in states of self-regulation of SF. How would starburst dwarf galaxies that form their stars in massive star clusters fit into your relation?

*Kroupa*: There is some observational evidence reported at this meeting that these dIrrs may actually have fairly continuous SFHs. Pflamm-Altenburg is working on the implication of non-linearities in the luminosity–mass relation for star clusters, which may mimic the apparent existence of one massive cluster, with few lower-mass clusters in dwarf galaxies that are forming stars.

*Pathways through an Eclectic Universe*
*ASP Conference Series, Vol. 390, © 2008*
*J. H. Knapen, T. J. Mahoney, and A. Vazdekis, eds.*

# From Massive Cores to Massive Stars

Mark R. Krumholz
*Department of Astrophysical Sciences, Princeton University, Princeton, NJ 08544-1001, USA*

**Abstract.** The similarity between the mass and spatial distributions of pre-stellar gas cores in star-forming clouds and young stars in clusters provides strong circumstantial evidence that these gas cores are the direct progenitors of individual stars. Here, I describe a physical model for the evolution of massive cores into stars, starting with the initial phases of collapse and fragmentation, through disk formation and fragmentation, the later phases of stellar feedback, and finally interaction of the newly formed stars with their environments. This model shows that direct mapping from cores to stars is the natural physical outcome of massive core evolution and thereby allows us to explain many of the properties of young star clusters as direct imprints of their gas-phase progenitors.

## 1. Introduction

Massive stars form in regions of extremely high column density, hidden behind hundreds of magnitudes of visual extinction. As radio and submillimeter interferometers have matured over the past decade, they have revealed the properties of these regions with ever higher detail, to the point where today it is not entirely unreasonable to speak of observationally determined "initial conditions" for the problem of star formation. One of the most striking results of this exploration has been the extent to which the properties of young star clusters are directly mirrored in the conditions found in pre-stellar molecular gas.

At the most basic level, young clusters and the molecular clumps from which they form have similar bulk properties, such as column density ($\sim$1 g cm$^{-2}$), size ($\sim$1 pc), and velocity dispersion (a few km s$^{-1}$) (McKee & Tan 2003). More interestingly, the dense cores within these clumps also mirror the properties of stars. Cores are bound, centrally condensed objects with characteristic sizes $\sim$0.1 pc or smaller and masses comparable to those of individual stars (Sridharan et al. 2005; Beuther, Sridharan, & Saito 2005). Observations in many regions with a variety of techniques find that the core and star mass functions have the same shape, differing only in that cores are a factor of 2–4 more massive (e.g., Motte, André, & Neri 1998; Johnston et al. 2001; Reid & Wilson 2006; Alves, Lombardi, & Lada 2007; Alves, this volume, p. 72). Moreover, cores appear to be mass-segregated: cores with masses greater than a few $M_\odot$ are found only in the centers of their parent clumps, but the core mass function is otherwise independent of position (Elmegreen & Krakowski 2001; Stanke et al. 2006). This is remarkably similar to the pattern in young clusters, where there is no segregation for stars smaller than a few $M_\odot$, but more massive stars are almost exclusively in cluster centers (Hillenbrand & Hartmann 1998; Huff & Stahler 2006).

The coincidence in both the mass and spatial distributions of cores and stars makes a strong circumstantial case that young stars' properties might be direct imprints of core properties. (The somewhat higher core masses are to be expected, since outflows should prevent $\sim 50\%$ of a core's mass from reaching a star—Matzner & McKee 2000.) However, we cannot make such an inference without a theoretical understanding of how cores might evolve into stars. The goal of this paper is to summarize work in the last few years that has focused on the most problematic part of this process, understanding the evolution of massive cores. In the following sections I sketch a model for the evolution of these objects, beginning with observed core properties and using numerical and analytical arguments to understand how they collapse into stars.

## 2. Initial Collapse and Fragmentation

The first phase of evolution for a massive core begins when it starts to collapse but has not yet formed any stars. One might initially suspect that there is no plausible way for a massive core to collapse to a single star or a small multiple system, since the Jeans mass in cold molecular clumps is only roughly solar. If massive cores do truly fragment into Jeans mass-sized objects once their collapse begins, then there can be no direct mapping from cores to stars. Such behavior is exactly what purely hydrodynamical numerical simulations find: as objects collapse and their density rises, the Jeans mass falls, and the objects break into smaller and smaller pieces that always have masses comparable to the Jeans mass at their current density. The fragmentation process ceases only when the assumed equation of state stiffens, so massive cores generate numerous small stars (e.g., Bate & Bonnell; Dobbs, Bonnell, & Clarke 2005).

However, while a purely hydrodynamical approach to fragmentation is analytically very simple and numerically very cheap, it neglects the important effect of radiation feedback from embedded, forming protostars. Krumholz (2006) points out that, even before embedded stars begin nuclear burning, just the gravitational energy released as gas accretes onto them can significantly heat the surrounding gas, raising the Jeans mass and suppressing fragmentation. Krumholz, Klein, & McKee (2007c) follow up this point by simulating the collapse of turbulent, massive cores whose initial masses, sizes, central concentrations, and levels of turbulence are chosen to match those of observed cores (as seen, for example, by Sridharan et al. 2005 and Beuther et al. 2005). The simulations combine a protostellar evolution model with a new adaptive mesh refinement gravity–radiation–hydrodynamics algorithm (Krumholz, Klein, & McKee 2004; Krumholz et al. 2007a) to model accurately the effects of radiative heating. The simulations show that radiative heating strongly suppresses fragmentation of massive cores, allowing the great majority of the mass in a core to collapse into one or two stars. In contrast, a control run omitting radiative heating qualitatively reproduces the earlier hydrodynamical result that massive cores collapse into dozens of small fragments. Figure 1 illustrates the difference made by the inclusion of radiative heating.

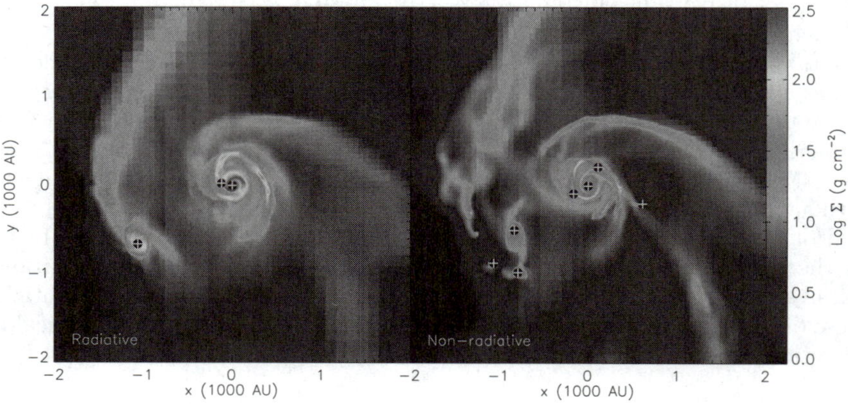


Figure 1. Column density in two simulations of the collapse and fragmentation of a 100 $M_\odot$ core. The left and right panels show two runs from Krumholz et al. (2007c, runs 100A and 100ISO) at 20 kyr of evolution; one uses radiative transfer and one does not, but otherwise the simulations have identical initial conditions and resolution. The plus signs indicate the positions of stars. Note the significantly greater number of stars in the non-radiative run, with several more condensations on the verge of collapse to stars.

The most important point to take from this is that when more detailed physics than simple hydrodynamics is included, simulations and analytical arguments both show that the observed massive cores are unlikely to fragment into many pieces. Thus, a model in which there is a direct mapping from the masses and positions of massive cores to those of massive stars passes its first test: the cores will collapse largely monolithically, rather than fragmenting into small stars. A secondary point is that we must be careful about drawing conclusions based on simulations with very simple physics. Models that do not include radiative transfer produce qualitatively different results from those that do.

## 3. Disk and Binary Formation

Since collapsing turbulent cores have non-zero angular momentum, they naturally form protostellar disks. Observationally, these are a potential signpost of the star formation process. Both the simulations described above and analytical models (Kratter & Matzner 2006) find that the disks formed by massive cores are likely to be strongly gravitationally unstable. This instability causes the disks to develop large amplitude $m = 1$ spiral modes, and potentially even fragment to form companions to the primary star (although a majority of the mass still goes into the primary, not the fragments). Krumholz, Klein, & McKee (2007b) show that the strong $m = 1$ spiral structure present in such unstable disks should be observable with next-generation telescopes such as ALMA and the EVLA, and that a systematic offset between the disk's "zero" velocity and the central star's velocity produced by the instability might also be detectable. The detection of a disk with these signatures inside a massive core would be strong evidence

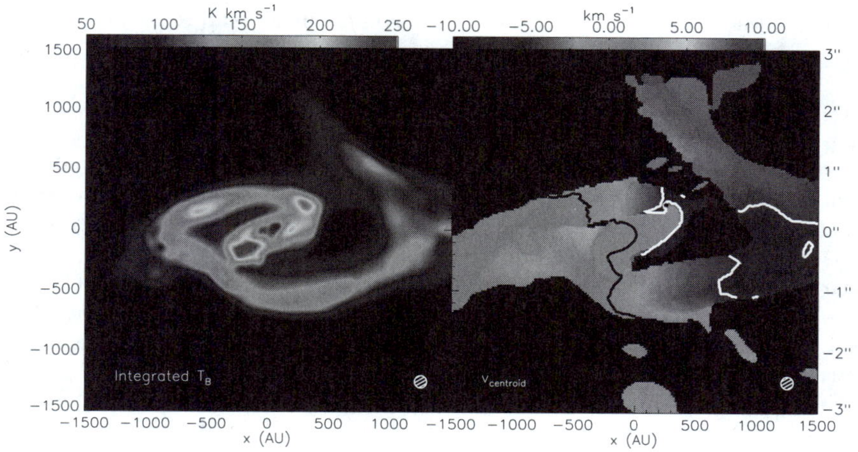


Figure 2. The velocity-integrated brightness temperature (*left panel*) and centroid velocity (*right panel*) in a simulated 0.1 arcsec resolution ALMA observation of a massive protostellar disk 500 pc away in the $CH_3CN$ 220.7472 GHz line. The disk is from the simulation illustrated in the left panel of Fig. 1 at an evolution time of 27 kyr, when the central star is $8.3\,M_\odot$ in mass. The white contour in the right panel corresponds to $-5\,\mathrm{km\,s^{-1}}$, and the black contour shows $0\,\mathrm{km\,s^{-1}}$, measured relative to the velocity of the central star. The disk is systematically offset to negative velocities relative to the star, so while there is a large region with velocity $< -5\,\mathrm{km\,s^{-1}}$, there are no pixels with velocities $> 5\,\mathrm{km\,s^{-1}}$. Black pixels correspond to locations where ALMA would not detect the line at $> 3\sigma$ confidence.

in favor of the model that massive cores collapse to form massive stars. Figure 2 shows a simulated ALMA observation of such a disk, computed using the technique of Krumholz et al. (2007b).

Disks around massive stars are also important for their role in forming companions to massive stars. Most massive stars have close companions (Lada 2006), and gravitational instability in massive protostellar disks provides a natural explanation for this because even radiatively heated massive disks suffer some fragmentation (though vastly less than if radiation is omitted). These fragments initially form with masses around solar at distances $> 100$ AU from the primary (Kratter & Matzner 2006; Krumholz et al. 2007c), but they subsequently migrate inward to separations $< 10$ AU as the disk accretes. The final separations of these disk-formed companions from the primary, and whether some of them merge with it, has not yet been determined.

An interesting fate awaits migrating stars that get close to the primary but do not merge with it. Massive protostars go through a phase of deuterium shell burning, during which their radii swell to tenths of an AU in size. Krumholz & Thompson (2007) point out that this may lead a primary to overflow its Roche lobe and transfer mass onto close companions. Such mass transfer is almost always unstable, terminating only once the mass ratio of the system reaches unity. This provides a natural mechanism for the origin of the heretofore unexplained massive "twins", binaries consisting of two massive stars with a mass ratio of almost exactly unity (e.g., Pinsonneault & Stanek 2006).

## 4. Radiation Pressure Feedback

A direct core-to-star mapping is possible only if most of a core's mass is able to accrete onto the massive star forming within it. However, spherically symmetric calculations indicate that the huge radiation output of massive stars should exert a force stronger than gravity on dust grains suspended in the gas around them for stars of $\sim 20\, M_\odot$ or larger (Larson & Starrfield 1971; Yorke & Kruegel 1977; Wolfire & Cassinelli 1987). How massive stars can form despite this radiation barrier is a classical problem in astrophysics. Fortunately, rotation and the formation of a disk can significantly mitigate this effect, because gas in a disk self-shields against the radiation due to high optical depths (Nakano 1989; Nakano, Hasegawa, & Norman 1995; Jijina & Adams 1996), while at the same time a disk collimates the radiation field and beams it away preferentially in the polar direction, thereby reducing the radiation force felt by gas in the equatorial plane (Yorke & Sonnhalter 2002). However, even with these effects it is not entirely clear that stars can grow to arbitrary masses by accretion.

Two additional effects may help. First, radiation hydrodynamic simulations show that the first effect of radiation pressure is that massive stars blow radiation bubbles above and below accretion disks (Yorke & Sonnhalter 2002; Krumholz, Klein, & McKee 2005a). However, in three dimensions these bubbles do not halt accretion, because gas that reaches the bubble wall flows along the wall onto the accretion disk. Bubbles may also collapse due to Rayleigh–Taylor instability, allowing accretion to continue through optically thick channels while radiation escapes through optically thin regions around them (Krumholz et al. 2005a). If magnetic fields are present and sufficiently strong, this effect will be enhanced by photon bubble instability (Turner, Quataert, & Yorke 2007), which arranges the gas into dense lumps separated by low density gaps through which radiation leaks, effectively reducing the radiation pressure force experienced by the bulk of the gas.

Second, protostellar outflows provide a third escape valve for radiation. Massive protostars appear to generate hydromagnetic outflows just like low mass stars, with the difference that for massive stars the outflow cavities are largely dust-free because the base of the outflow is close enough to the star for the dust within it to have been destroyed by sublimation. Such outflow cavities therefore present optically thin channels through which radiation can leak out of the optically thick cores. Radiative transfer calculations show that this can lead to order-of-magnitude reductions in the radiation force on accreting gas near the equatorial plane, again allowing accretion to continue where it might otherwise have been halted (Krumholz, McKee, & Klein 2005b).

While a definitive numerical simulation including the effects of radiation forces, magnetic fields, and protostellar outflows in three dimensions has not yet been done and is probably at best barely within the capabilities of present-day supercomputers, it seems clear that each of these effects will help massive stars form by accretion. Thus, we can tentatively say that there is no barrier to most of the gas in a protostellar core accreting onto a massive star. Conversely, however, simulations of massive star formation that do not include radiation force effects are simply ignoring this problem entirely. Preliminary simulations indicate that models of star formation that depend on Bondi–Hoyle accretion are likely to fail once radiation pressure is included (Edgar & Clarke 2004), so models of this sort must be viewed with caution.

## 5. Competitive Accretion

Thus far we have seen that massive protostellar cores will not fragment strongly, that what fragmentation they do show is consistent with the observed multiplicity properties of massive stars, and that radiation pressure will not prevent most of the mass in a core from accreting onto a star. There remains, however, one more way in which a direct mapping from cores to stars could fail: if, once stars accrete their parent cores, they were subsequently to accrete a great deal more mass, then there would be no direct relationship between core and stellar masses. This process of accretion onto stars from gas that was not originally part of a bound core is known as competitive accretion (Bonnell et al. 2001a,b).

Purely hydrodynamic simulations of star cluster formation show that this process is the dominant mechanism by which stars gain mass. In effect, all stars are born at masses of the order of the Jeans mass, but some of them fall to the center of the collapsing gas cloud, and the deep potential well then channels gas to them. They subsequently accrete this gas and grow in mass, producing a full range of initial masses (Bonnell, Vine, & Bate 2004; Bonnell & Bate 2006).

However, direct observational estimates of the rate of competitive accretion generally find that it is too small to make a significant contribution to final stellar masses (André et al. 2007). Krumholz, McKee, & Klein (2005c, 2006) point out that competitive accretion is possible in simulations only because the simulated gas clumps are in the process of global collapse, which creates deep potential wells within which the gas is dense and non-turbulent and Bondi–Hoyle accretion is rapid. These deep, dense, quiescent gas wells have not been observed, however. This is probably because clumps are not in a state of global collapse. Such a collapse necessarily converts order unity of the mass in a gas clump into stars and ends star formation in 1–2 free-fall times (e.g., Bonnell et al. 2004). However, there is strong evidence that the star formation process cannot be anywhere near that fast. The gas clump from which the ONC formed likely had a density $\sim 10^5\,\mathrm{cm}^{-3}$ (Elmegreen 2000), implying a free-fall time of 0.1–0.2 Myr, but the estimated ages of the stars in Orion point to a formation process lasting 1–3 Myr, implying a minimum formation timescale of five free-fall times even if one assumes the fastest plausible formation time, with $\sim$15 free-fall times being more likely (Tan, Krumholz, & McKee 2006). Furthermore, the total galactic mass of infrared dark clouds, which have densities $\sim 10^3\,\mathrm{cm}^{-3}$, is such that, if they were collapsing to form stars on a free-fall timescale, the total galactic star formation rate would have to be $\sim$100 times higher than its observed value. The same conclusion holds for dense gas clumps traced by HCN($1 \rightarrow 0$) emission, which have densities $\sim 10^4$–$10^5\,\mathrm{cm}^{-3}$ (Krumholz & Tan 2007). This implies that, even in regions with mean densities of $10^5\,\mathrm{cm}^{-3}$ star formation cannot be anywhere near as rapid as would be required for competitive accretion to occur.

The final piece of evidence against competitive accretion comes from simulations that include more detailed physics. Li & Nakamura (2006) and Nakamura & Li (2007) simulate the formation of a star cluster using magnetohydrodynamics rather than simple hydrodynamics, and including the effects of outflows driven by the protostars forming within the cluster. They find that the protostellar outflows drive turbulent motions, preventing global collapse and ensuring that conditions are too turbulent for competitive accretion processes to alter the masses of protostars significantly after they have consumed their parent cores. Moreover, these simulations, unlike ones where competitive accretion occurs, produce star formation rates in good agreement with observations. We can therefore tentatively conclude that the core accretion hypothesis passes this final test: once stars have accreted their parent cores, they will not gain much additional mass from the gas to which they were not bound at birth. A direct core to star mapping will survive.

## 6. Summary and Conclusion

For the past decade observations have increasingly pointed to an intimate link between young stars and the dense stellar-mass gas clouds known as pre-stellar cores. Cores and stars have very similar mass and spatial distributions, so it is tempting to explain the properties of young star clusters as imprinted at birth. However, this hypothesis requires that cores map directly onto stars. Here, I provide a physical model for such a mapping. When massive cores first collapse, they do not fragment strongly because the first stars that form within them heat the gas and suppress fragmentation. As a result, massive cores collapse to only a few stars. Most of the fragmentation that does happen occurs in unstable self-gravitating disks, which should be directly observable. Disk fragmentation ensures that massive stars will essentially always have companions, and some of these companions will turn into twins of the primary star. As the massive star grows, it will begin to generate a huge radiation force opposing accretion. However, a combination of instabilities and the leakage of radiation through protostellar outflow cavities allows the radiation to escape and accretion to continue unimpeded to high masses. Finally, once stars have accreted their parent cores, they will be unable to gain additional mass that was not originally part of the core. Thus, the hypothesis that stars come directly from cores, with a one-to-one mapping of core mass to star mass, is in agreement with a physical model for how massive cores evolve. The best explanation for many properties of young star clusters appears to be that they are set at birth, when the cluster is still a dark cloud.

**Acknowledgments.** My work on massive stars is done in collaboration with R. Klein, K. Kratter, C. Matzner, C. McKee, J. Tan, and T. Thompson. Support for this work was provided by NASA through Hubble Fellowship grant #HSF-HF-01186 awarded by the Space Telescope Science Institute, which is operated by the Association of Universities for Research in Astronomy, Inc., for NASA, under contract NAS 5-26555. The simulations described were made possible by grants of high performance computing resources from the Arctic Region Supercomputing Center; the NSF San Diego Supercomputer Center through NPACI program grant UCB267; the National Energy Research Scientific Computing Center, which is supported by the Office of Science of the US Department of Energy under Contract No. DE-AC03-76SF00098, through ERCAP grant 80325; and the US Department of Energy at the Lawrence Livermore National Laboratory under contract W-7405-Eng-48.

## References

Alves, J., Lombardi, M., & Lada, C. J. 2007, A&A, 462, L17
André, P., Belloche, A., Motte, F., & Peretto, N. 2007, A&A, 472, 519
Bate, M. R., & Bonnell, I. A. 2005, MNRAS, 356, 1201
Beuther, H., Sridharan, T. K., & Saito, M. 2005, ApJ, 634, L185
Bonnell, I. A. & Bate, M. R. 2006, MNRAS, 370, 488
Bonnell, I. A., Bate, M. R., Clarke, C. J., & Pringle, J. E. 2001a, MNRAS, 323, 785
Bonnell, I. A., Clarke, C. J., Bate, M. R., & Pringle, J. E. 2001b, MNRAS, 324, 573
Bonnell, I. A., Vine, S. G., & Bate, M. R. 2004, MNRAS, 349, 735
Dobbs, C. L., Bonnell, I. A., & Clark, P. C. 2005, MNRAS, 360, 2
Edgar, R. & Clarke, C. 2004, MNRAS, 349, 678
Elmegreen, B. G. 2000, ApJ, 530, 277
Elmegreen, B. G., & Krakowski, A. 2001, ApJ, 562, 433
Hillenbrand, L. A., & Hartmann, L. W. 1998, ApJ, 492, 540
Huff, E. M., & Stahler, S. W. 2006, ApJ, 644, 355
Jijina, J., & Adams, F. C. 1996, ApJ, 462, 874
Johnstone, D., Fich, M., Mitchell, G. F., & Moriarty-Schieven, G. 2001, ApJ, 559, 307
Kratter, K. M., & Matzner, C. D. 2006, MNRAS, 373, 1563
Krumholz, M. R. 2006, ApJ, 641, L45
Krumholz, M. R., Klein, R. I., & McKee, C. F. 2005a, in IAUS 227: Massive Star Birth: A Crossroads of Astrophysics, ed. R. Cesaroni, M. Felli, E. Churchwell, & M. Walmsley, p. 231
Krumholz, M. R., Klein, R. I., McKee, C. F., & Bolstad, J. 2007a, ApJ, 667, 626
—. 2007b, ApJ, 665, 478
—. 2007c, ApJ, 656, 959
Krumholz, M. R., McKee, C. F., & Klein, R. I. 2004, ApJ, 611, 399
—. 2005b, ApJ, 618, L33
—. 2005c, Nat, 438, 332
—. 2006, ApJ, 638, 369
Krumholz, M. R., & Tan, J. C. 2007, ApJ, 654, 304
Krumholz, M. R., & Thompson, T. A. 2007, ApJ, 661, 1034
Lada, C. J. 2006, ApJ, 640, L63
Larson, R. B., & Starrfield, S. 1971, A&A, 13, 190
Li, Z.-Y., & Nakamura, F. 2006, ApJ, 640, L187
Matzner, C. D., & McKee, C. F. 2000, ApJ, 545, 364
McKee, C. F., & Tan, J. C. 2003, ApJ, 585, 850
Motte, F., André, P., & Neri, R. 1998, A&A, 336, 150
Nakamura, F. & Li, Z.-Y. 2007, ApJ, 662, 395
Nakano, T. 1989, ApJ, 345, 464
Nakano, T., Hasegawa, T., & Norman, C. 1995, ApJ, 450, 183
Pinsonneault, M. H., & Stanek, K. Z. 2006, ApJ, 639, L67
Reid, M. A., & Wilson, C. D. 2006, ApJ, 650, 970
Sridharan, T. K., Beuther, H., Saito, M., Wyrowski, F., & Schilke, P. 2005, ApJ, 634, L57
Stanke, T., Smith, M. D., Gredel, R., & Khanzadyan, T. 2006, A&A, 447, 609
Tan, J. C., Krumholz, M. R., & McKee, C. F. 2006, ApJ, 641, L121
Turner, N. J., Quataert, E., & Yorke, H. W. 2007, ApJ, 662, 1052
Wolfire, M. G., & Cassinelli, J. P. 1987, ApJ, 319, 850

Wu, J., Evans, N. J., II, Gao, Y., Solomon, P. M., Shirley, Y. L., & Vanden Bout, P. A. 2005, ApJ, 635, L173
Yorke, H. W., & Kruegel, E. 1977, A&A, 54, 183
Yorke, H. W., & Sonnhalter, C. 2002, ApJ, 569, 846

## Discussion

*Bonnell*: Comment: I am concerned that the information on the densest regions comes from the most distant (extragalactic) sources. Question: The disk fragmentation produces wider, low mass companions. How do they evolve to within 0.1 AU?

*Krumholz*: 1) While the original Gao & Solomon HCN work was indeed on extragalactic sources, Wu et al. (2005, ApJ, 635, L173) followed this up with Galactic HCN observations. They find that individual HCN clumps in the Milky Way, which are resolved, fall on the same HCN mass star formation rate law. This strongly suggests that the results do not depend much on resolution. 2) The fragments are at ~300 AU, and have masses smaller than $0.5\,M_{\odot}$, when they first form. However, they are embedded in a disk that is rapidly accreting and much more massive than the embedded fragment. As a result, the fragment is dragged inward by the disk while simultaneously gaining mass from it. I don't have the resolution in my simulations to follow the migration in to 0.1 AU, but I can say that the disk-formed stars migrate inward to the smallest radius I can resolve. I see no physical reason why the migration would stop inside that.

*Shlosman*: Whether the disk fragments depends how deep one can drive $Q$ below unity. Gravitational viscosity will turn on new source viscosity. So the efficiency of this heating will depend on such sensitive issues as the viscosity of a numerical code. To what extent have you tested the code in this respect? Do you reproduce the energy cascade and to what spatial scales?

*Krumholz*: In our simulations numerical viscosity is negligible compared to angular momentum transport by gravitational torques, except in the inner ~10 cells where we resolve the disk very poorly. This statement is based on detailed tests of the code against analytical solutions published in Krumholz, McKee & Klein (2004). However, in the massive star simulations, the dominant heating source is illumination by the central star, not viscous dissipation. We resolve the heating and cooling reasonably well in the outer disk where the fragmentation occurs: about 30–40 cells per disk radius, ~10–15 cells per disk scale height. Fragmentation can occur because, at a given radius, the disk is essentially isothermal. This favors fragmentation.

*Hensler*: There is a fundamental debate as to what extent turbulence affects clump formation and collapse. How do you insert turbulence? Do you maintain it from external sources? How does decascading of turbulence change the evolution on small scales, i.e., core mass, accretion, accretion disk formation, etc.?

*Krumholz*: My simulations mostly focus on single cores, so I don't artificially drive turbulence at all. I insert it as an initial condition, at a level such that my initial cores are virialized, but then let it decay freely. In the simulations

I used to test competitive accretion, I did artificially drive the turbulence to set up the box, but I stopped driving once I began the test. In that run, the turbulence was critical to halting competitive accretion. In the Nakamura & Li simulation I showed, there is no artificial driving, just whatever turbulent driving is provided self-consistently by outflows from stars forming in the simulation box. In this run, the turbulence is again critical in keeping the gas clump in equilibrium, forming stars at a slow rate consistent with observations. Without the turbulence, the star formation rate is much too high; this is seen in the competitive accretion simulations.

Caricature of John Beckman, drawn by Eduardo Battaner in 1983, and kept all those years by Mike Edmunds. It was made during the course "Current topics on galaxies", in Granada, organized jointly by the Royal Greenwich Observatory and the Instituto de Astrofísica de Andalucía.

*Pathways through an Eclectic Universe*
*ASP Conference Series, Vol. 390, © 2008*
*J. H. Knapen, T. J. Mahoney, and A. Vazdekis, eds.*

# Competitive Accretion and the Formation of Massive Stars

I. A. Bonnell
*SUPA, School of Physics and Astronomy, North Haugh,*
*University of St Andrews, St Andrews, Scotland, KY16 9SS, UK*

**Abstract.** Competitive accretion in a stellar cluster is a probable mechanism to explain the formation of high mass stars and their properties in the context of low mass star formation. The fragmentation of a turbulent cloud produces the individual stars with masses of the order the Jeans mass or mean stellar mass in the cluster. Once individual stars fall together to form a small-$N$ cluster, their mutual potential funnels gas down to the center. The higher gas densities in the cluster centers, and the fact that this gas is continually replenished, result in much higher accretion rates for the few stars located there. These stars become massive stars primarily due to their location in the center of a cluster. This process requires a distributed gas reservoir with initially low velocity dispersion as expected in a turbulent medium. Competitive accretion can explain the distribution of stellar masses, the mass segregation of young stellar clusters, and the high binary frequency and properties of massive stars. Finally, competitive accretion predicts a direct relationship between the total stellar mass of a cluster and the most massive star therein.

## 1. Introduction

The formation of massive stars is currently one of the major unknowns in the field of star formation. Unlike their low mass brethren, massive stars are rare, form relatively quickly in distant, deeply embedded molecular clouds, and then drastically affect their pre-natal environment with their winds and ionizing feedback. The two current models to form massive stars are that the massive star forms directly from a massive gas clump, or that they form due to ongoing "competitive" accretion in a clustered environment. The difference between the two models lies primarily in how and when the mass is gathered to form the massive star. Does it occur in the pre-stellar stage due to the turbulence, or does it occur subsequently due to accretion in a common potential? The difficulty lies in that we are not able to fully ascertain the properties of a cloud in which massive stars form, and we cannot therefore sufficiently constrain the initial conditions for the models.

The one feature of massive star formation that we do know reasonably well is that they are thought to form almost ubiquitously in stellar clusters, and that these stellar clusters are the minimum unit needed to fully represent the stellar initial mass function (IMF). It is therefore crucial that any model includes the environment of a stellar cluster and should be able to form massive stars in the context of forming low mass stars and a full IMF.

## 2. Massive Clumps and Monolithic Massive Star Formation

The most straightforward model for massive star formation has to be one where the mass is decided in the pre-stellar regime and there is a direct mapping of a clump mass onto a star's mass (McKee & Tan 2003; Krumholz, Klein, & McKee 2005b). This model is attractive not only for its simplicity but also because observations have shown that the clump mass function in molecular clouds is very similar to the stellar IMF (Motte, Andre, & Neri 1998; Alves, Lombardi, & Lada 2007). Furthermore, such a clump mass function is a prediction of turbulent fragmentation in molecular clouds (Padoan & Nordlund 2003). Potential difficulties include the fact that more massive clumps are increasingly susceptible to fragmentation (Lada et al. 2007), that clump lifetimes are likely to depend on mass thus changing the real slope of the cluster mass function (CMF; Clark, Klessen, & Bonnell 2007), and that generally turbulent theory does not predict massive clumps in the crowded regions that will develop into the centers of clusters. Models envisioning such a massive, turbulently supported clump in the center of clusters are one possible solution (McKee & Tan 2003; Krumholz et al. 2005b; Krumholz, this volume, p. 16), but as yet we have no idea how such physical conditions could arise without the clump fragmenting into lower mass stars during the long accumulation stage (Dobbs, Bonnell, & Clark 2005).

## 3. Competitive Accretion

The alternative model for massive star formation, *competitive accretion*, relies primarily on the physics of gravity and the fact that massive stars form in clustered environments. In this scenario, fragmentation produces lower mass stars while the few higher mass stars form subsequently due to continued accretion in stellar clusters where the overall system potential funnels gas down to the center of the potential, to be accreted by the proto-massive stars located there (e.g., Bonnell et al. 2001a; Bonnell & Bate 2006; Fig. 1). This model produces not only massive stars, but also a full IMF and mass-segregated stellar clusters (Bonnell, Larson, & Zinnecker 2007).

The basis of this model is that gravitational fragmentation of turbulent (or non-turbulent) clouds always produces objects that have initial masses typical of the thermal Jeans mass of the dense cloud (Klessen, Burkert, & Bate 1998; Klessen & Burkert 2000; Bate, Bonnell, & Bromm 2003; Bonnell, Vine, & Bate 2004; Bate & Bonnell 2005). Thus, turbulence drives structures into the molecular cloud while the thermal physics determines the fragmentation scale and thus the characteristic mass for star formation (Larson 2005; Jappsen et al. 2005; Bonnell, Clarke, & Bate 2006).

Competitive accretion relies on the inefficiency of fragmentation such that there is a large common reservoir of gas from which the protostars can accrete (e.g., Bonnell et al. 2001a; Bonnell et al. 2004). Observations of pre-stellar structures and of young stellar clusters both support this view with the large majority of the total mass being in a distributed gaseous form (Motte et al. 1998; Johnstone et al. 2000; Johnstone, Di Francesco, & Kirk 2004; Lada & Lada 2003; Bastian & Goodwin 2006). The second requirement is that the gas be free to move under the same gravitational acceleration as the stars. If the gas is fixed

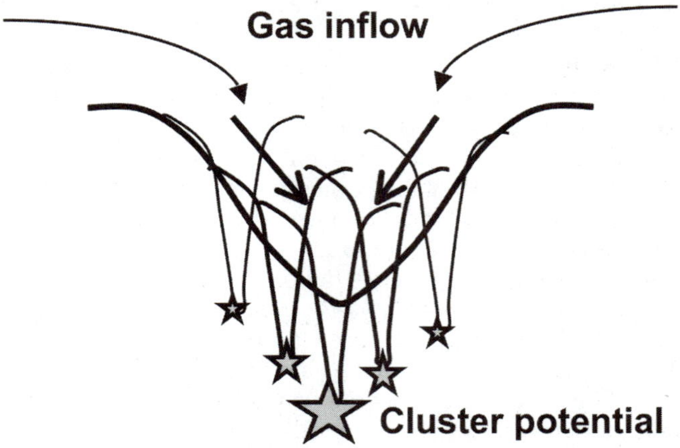


Figure 1. A schematic diagram of the physics of accretion in a stellar cluster: the gravitational potentials of the individual stars form a larger scale potential that funnels gas down to the cluster core. The stars located there are therefore able to accrete more gas and become higher mass stars. The gas reservoir can be replenished by infall into the large scale cluster potential. (*From Bonnell et al. 2007.*)

in place due to magnetic fields then accretion will be limited. When these two requirements are fulfilled, the dynamical timescales for accretion and evolution are similar, such that a significant amount of gas can be accreted.

Competitive accretion can occur in two different regimes. Firstly, at the point of fragmentation the relative star–gas velocities are low (as are the star–star velocities). The accretion rates are then determined by the tidal limits of each star's potential relative to the overall system potential. Secondly, once stars dominate the potential and virialize in the cluster, the accretion rates are primarily determined by the Bondi–Hoyle type accretion rates producing a Salpeter-type IMF (Zinnecker 1984; Bonnell et al. 2001b). It should be noted that the most massive star can be an exception to this as it sits in the center of the cluster potential and accretes quasi-spherically.

## 4. Accretion in a Turbulent Medium

Recently, Krumholz et al. (2005a) have cast some doubt on this process by claiming that accretion in such environments cannot significantly increase a star's mass and therefore does not play a role in establishing the stellar IMF. In part, this is correct, as with competitive accretion most stars do not continue to accrete. It is the few that do that are important in terms of forming higher mass stars and the IMF. What the Krumholz analysis overlooks is that Bondi–Hoyle accretion,

$$\dot{M}_* \approx 4\pi\rho\frac{(GM_*)^2}{v^3}, \qquad (1)$$

is highly dependent on the exact values of the gas density $\rho$, stellar mass $M_*$, and velocity dispersion $v$, such that with slight variations in these properties, the accretion can vary from $10^{-9}\,M_\odot\,\mathrm{yr}^{-1}$ to $10^{-4}\,M_\odot\,\mathrm{yr}^{-1}$. The higher values take more typical gas densities found in the cores of stellar clusters and velocity dispersion of $0.5\,\mathrm{km\,s}^{-1}$, in keeping with the expectation of turbulence and the internal stellar velocity dispersion expected on these scales (Bonnell & Bate 2006).

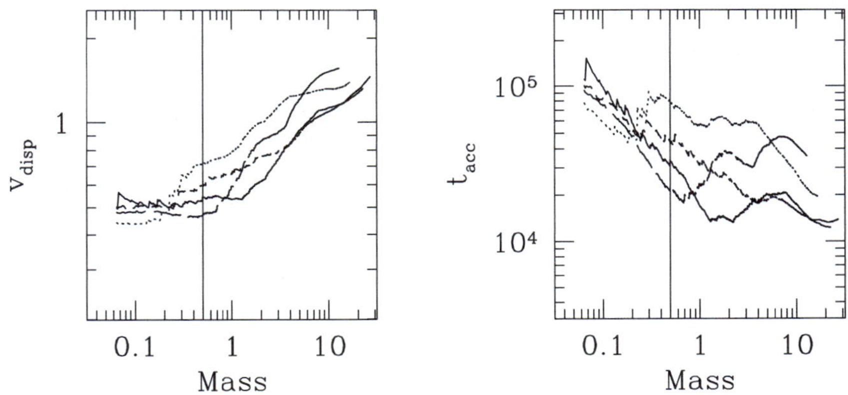


Figure 2. The 3D velocity dispersion of the accreted gas is plotted (*left*) as a function of the cumulative accreted gas mass (in $M_\odot$) for four massive stars. The velocity dispersion is low for low masses due to the nature of turbulence and the low velocity dispersions in small-$N$ clusters, where much of the accretion occurs. The velocity dispersion increases with accreted mass due to the velocity–sizescale relation for turbulent clouds. The accretion timescale, $t_{\mathrm{acc}} \approx M_*/\dot{M}_*$, is plotted as a function of accreted mass (*right*) showing that accretion occurs on relatively short timescales and thus determines the final masses of high mass stars. (*From Bonnell & Bate 2006.*)

In Fig. 2 we plot the 3D velocity dispersion of the gas that is subsequently accreted to form the four most massive stars in one of our simulations (Bonnell & Bate 2006). The initial fragment masses are $0.5\,M_\odot$, and the gas has relatively low velocity dispersion as expected at the point of fragmentation. The gas, plotted as function of the mass in order of its accretion history, increases in velocity dispersion, dispersion increases with mass, as one expects for a turbulent medium, until nearly reaching the system velocity dispersion of $2\,\mathrm{km\,s}^{-1}$. The most important point is that the velocity dispersion is initially low, in keeping with turbulence on small scales. Thus, the accretion rates can be large. By the time the velocity dispersion is large, the star has accreted sufficient mass in order to maintain a large accretion rate.

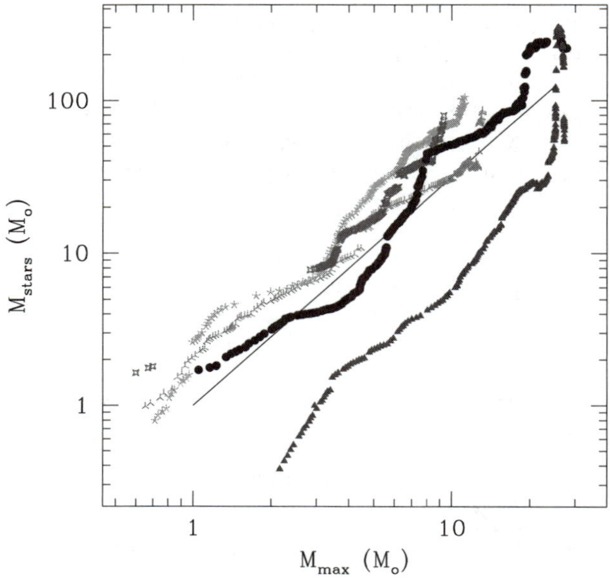


Figure 3. The total mass in stars is plotted against the mass of the most massive star in the system. Competitive accretion predicts that as the total mass grows, the mass of the most massive star also increases as $M_{\rm max} \propto M_{\rm stars}^{2/3}$. *(From Bonnell et al. 2004.)*

## 5. Predictions

There are several observational predictions that can be gathered from the numerical simulations of competitive accretion. The most striking of these is the relationship between the cluster properties and the mass of the most massive star. Stellar clusters form when newly formed stars fall together to produce small-$N$ groups. These groups grow to form larger clusters by accreting gas and stars from the surrounding environment. The gas accompanies the stars into the cluster potential with a significant fraction being accreted by the more massive stars. This naturally produces a correlation between the number of stars in the cluster (or the total mass in stars) and the mass of the most massive star therein (Bonnell et al. 2004). This prediction (Fig. 3), where

$$M_{\rm max} \propto M_{\rm cluster}^{2/3}, \tag{2}$$

agrees remarkably well with the analysis of observed young stellar clusters by Weidner & Kroupa (2006).

Competitive accretion also provides an explanation for the frequency and properties of binary systems among massive stars (Bonnell & Bate 2005). Binary systems form readily through fragmentation and three-body capture in the small-$N$ systems in the center of the forming clusters. The continued accretion hardens binary systems while increasing the masses of the component. Thus, wide (100 AU) low mass binary systems that undergo continued accretion evolve into high mass close binary systems (Fig. 4). The frequency of such high mass

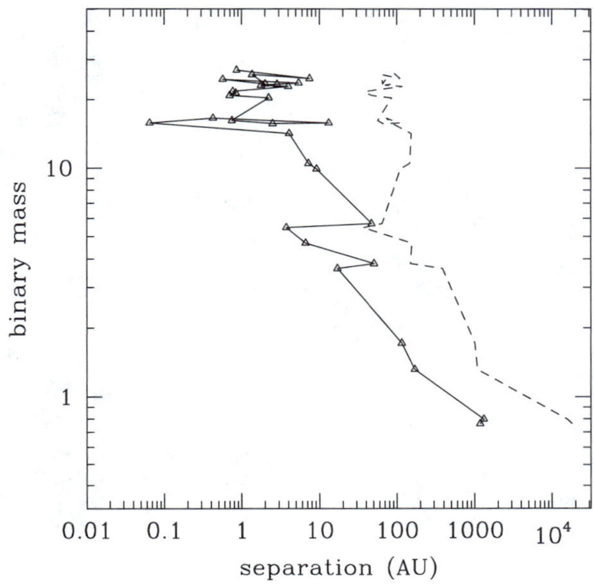


Figure 4. The formation of a massive binary system due to continued accretion. The binary forms as a wide low mass system and, as it accretes, it evolves to higher masses and smaller separations until it becomes a close, massive binary. The dashed line represents the separation including the gravitational softening of 200 AU while the solid line is an extrapolation of the periastron separation given the energy and angular momentum contained in the binary. (*From Bonnell & Bate 2005.*)

binary systems is 100 per cent in these systems with the majority of these being close with massive companions. They generally are found to have fairly high eccentricities which, along with their small separations, implies that some of these should merge to form higher mass stars. Stellar densities required to perturb such a binary and force it to merge are estimated to be of order $10^6$ stars $\mathrm{pc}^{-3}$.

Recent simulations with higher spatial and mass resolution have shown that circumstellar disks should also be a common outcome of competitive accretion, especially around more massive stars. This is at first surprising as earlier simulations had shown that stellar interactions are common which to first order should disrupt the disks. The explanation lies in that many of the disks are actually circumbinary disks, and that even when they are perturbed by a stellar encounter, the continued accretion reforms the disk on relatively short timescales. Any accretion is likely to contain some angular momentum such that disks must be a ubiquitous part of any massive star formation process. The expectation is that the mass in a circumstellar disk should be relatively large, limited by gravitational instabilities and later by feedback processes.

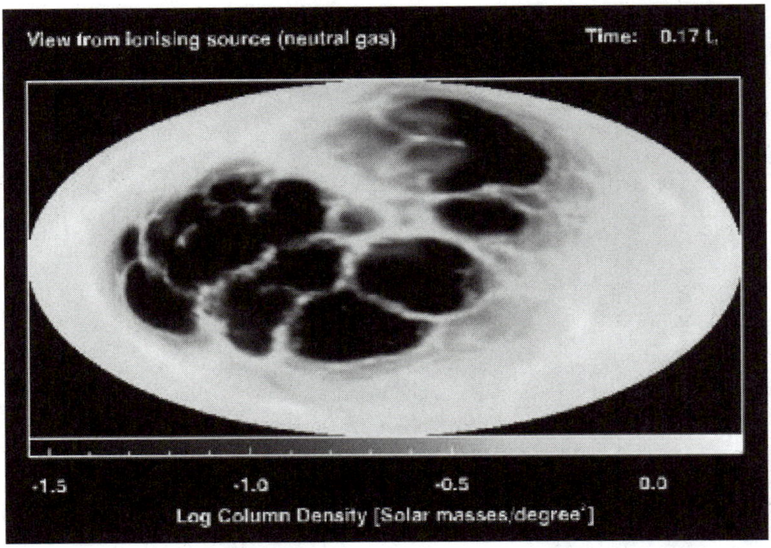


Figure 5. A $\theta, \phi$, column density map of the neutral gas distribution as seen from the ionizing source in the cluster center. Note that the evacuated regions channel the radiation away from the accreting gas. (*From Dale et al. 2005.*)

## 6. Feedback and Triggering

We know that feedback from young stars is likely to play an important role in the star formation process, potentially either stopping or even triggering subsequent star-forming events. This is even more important in the case of massive stars where their ionizing radiation, winds, and radiation pressure are likely to play a crucial role in their formation process. Work to date has centered on the effects of the ionizing radiation and radiation pressure from young massive stars. Dale et al. (2005) have included ionization from O stars into an earlier simulation of massive star formation. One of the conclusions from this study is that although the total accretion rates are somewhat reduced, the feedback tends to escape along any weak points in the environment (Fig. 5). This allows accretion to occur relatively unimpeded unless the feedback is sufficiently strong to completely evacuate the region (see also Krumholz et al. 2005b). Stellar winds from massive stars are also found to be externally collimated by the surrounding medium such that their effect on the overall evolution is limited (J. E. Dale & I. A. Bonnell, in preparation).

## 7. Conclusions

Competitive accretion is a viable mechanism to explain the formation of massive stars. Its significant advantage over other models is that it forms massive stars in the context of a full IMF, and the formation of mass segregated stellar clusters. It is the action of gravity, and the gravitational potential of the full cluster, which is responsible for the formation of the massive star. Competitive accretion starts with the accretion of low relative velocity gas due to the nature of turbulence and the low velocity dispersion in small-$N$ clusters. Accretion proceeds to higher relative velocity gas once the star has attained sufficient mass to maintain a high accretion rate. This model predicts that there is a strong causal link between the formation of stellar clusters and that of the most massive star therein. It can also reproduce the binary properties of massive stars and predicts disks to be a common occurrence around massive stars. Finally, although feedback can affect the star-forming environment, the channeling away of the feedback from the accretion flows suggest that it will not be a significant impediment to massive star formation.

The computations reported here were performed using the UK Astrophysical Fluids Facility (UKAFF).

**References**


Alves, J., Lombardi, M., & Lada, C. 2007, A&A, 462, L17

Bastian, N., & Goodwin, S. P., 2006, MNRAS, 369, 9

Bate, M. R., & Bonnell, I. A. 2005, MNRAS, 356, 1201

Bate, M. R., Bonnell, I. A., & Bromm, V. 2003, MNRAS, 339, 577

Bonnell, I. A., & Bate, M. R. 2005, MNRAS, 362, 915

Bonnell, I. A., & Bate, M. R. 2006, MNRAS, 370, 488

Bonnell, I. A., Bate, M. R., Clarke, C. J., & Pringle, J. E. 2001a, MNRAS, 323, 785

Bonnell, I. A., Clarke, C. J., Bate, M. R., & Pringle, J. E. 2001b, MNRAS, 324, 573

Bonnell, I. A., Clarke, C. J., & Bate, M. R. 2006, MNRAS, 368, 1296

Bonnell, I. A., Larson, R. B., & Zinnecker, H. 2007, in Protostars and Planets V, ed. B. Reipurth, D. Jewitt, & K. Keil (Tucson: University of Arizona Press), p. 149

Bonnell, I. A., Vine, S. G., & Bate, M. R. 2004, MNRAS, 349, 735

Clark, P. C. Klessen, R. S., & Bonnell, I.A. 2007, MNRAS, 379, 57

Dale, J. E., Bonnell, I. A., Clarke, C. J., & Bate, M. R. 2005, MNRAS 358, 291

Dobbs, C. L., Bonnell, I. A., & Clark, P. C. 2005, MNRAS, 360, 2

Jappsen, A.-K., Klessen, R. S., Larson, R. B., Li, Y., & Mac Low, M.-M. 2005, A&A, 435, 611

Johnstone, D., Wilson, C. D., Moriarty-Schieven, G., Joncas, G., Smith, G., Gregersen, E., & Fich, M. 2000, ApJ, 545, 327

Johnstone, D., Di Francesco, J., & Kirk, H. 2004, ApJ, 611, L45

Klessen, R. S., & Burkert, A. 2000, ApJS, 128, 287

Klessen, R. S., Burkert, A., & Bate, M. R. 1998, ApJ, 501, L205

Krumholz, M., Klein, R. I., & McKee, C. F. 2005b, ApJ, 618, L33

Krumholz, M., Klein, R. I., & McKee, C. F. 2005a, in IAUS 227, Massive Star Birth: A Crossroads of Astrophysics, ed. R. Cesaroni et al. (Cambridge: Cambridge University Press), p. 231

Lada, C. J., & Lada, E. 2003, ARA&A, 41, 57

Lada, C. J., Muench, A., Rathborne, J., Alves, J., & Lombardi, M. 2007, ApJ, in press

Larson, R. B., 2005, MNRAS, 359, 211

McKee, C. F., & Tan, J. C. 2003, ApJ, 585, 850

Motte, F., Andre, P., & Neri, R. 1998, A&A, 336, 150

Padoan, P., & Nordlund, Å. 2002, ApJ, 576, 870

Weidner, C., & Kroupa, P. 2006, MNRAS, 365, 1333

Zinnecker, H. 1984, MNRAS, 210, 43

*Pathways through an Eclectic Universe*
*ASP Conference Series, Vol. 390,* © *2008*
*J. H. Knapen, T. J. Mahoney, and A. Vazdekis, eds.*

# Molecular Cloud Formation and Magnetic Fields in Spiral Galaxies

C. L. Dobbs and D. J. Price

*School of Physics, University of Exeter, Stocker Road, Exeter, UK*

I. A. Bonnell

*School of Physics & Astronomy, North Haugh, University of St Andrews, St Andrews, Fife, UK*

**Abstract.** We present ongoing hydrodynamic and MHD simulations of molecular cloud formation in spiral galaxies. The hydrodynamic results show the formation of molecular gas clouds where spiral shocks compress atomic gas to high densities. The spiral shocks also produce structure in the spiral arms, provided the gas is cold ($< 1000$ K). When both hot and cold components of the ISM are modeled, this structure is enhanced. Properties such as the clump mass spectra and spatial distribution will be compared from clouds identified in these simulations. In particular, the multiphase simulations predict the presence of much more interarm molecular gas than when a single phase is assumed. We also discuss very recent results from galactic-scale MHD calculations. From observational comparisons of the magnetic and thermal pressure, magnetic fields are expected to be a major factor in explaining the dynamics of the ISM, from kpc scales to those of star formation. We describe the difference in structure of the spiral arms and the evolution of the global magnetic field for a range of field strengths.

## 1. Introduction

Numerical calculations have now become a major component of research into molecular clouds, galactic dynamics, and the interstellar medium (ISM). Here, we present a summary of recent results which examine the influence of spiral shocks on the ISM.

The presence of young stars in spiral arms is often associated with spiral shock triggering of star formation, following theoretical analysis of the gaseous response to spiral density waves (Roberts 1969; Shu, Milione, & Roberts 1973). Several observations show an increased star formation rate or efficiency in the spiral arms (e.g., Cepa & Beckman 1990; Seigar & James 2002), indicative of spiral triggering. The primary objective of our numerical analysis, however, is to determine how the gas accumulates into giant molecular clouds (GMCs) in which star formation occurs, e.g., through the agglomeration of smaller clouds or instabilities in the ISM. Furthermore, we discuss the degree of interarm structure, the topic of several recent numerical calculations (Kim & Ostriker 2002; Chakrabarti, Laughlin, & Shu 2003; Dobbs & Bonnell 2006). Such substructure, in the form of spurs and feathering, is commonly observed in spiral galaxies (La Vigne, Vogel, & Ostriker 2006), and is often associated with star formation.

## 2. Calculations

We perform global hydrodynamical and magnetohydrodynamical (MHD) calculations of a 3D galactic disk. All the calculations are non-self gravitating and use the Smoothed Particle Hydrodynamics (SPH) code. The modifications adopted to incorporate magnetic fields are described in Price & Bate (2007). The gas is subject to a galactic potential which includes a four-armed spiral component (Cox & Gómez 2002). The calculations described here are all isothermal, although we investigate different gas temperatures and apply a two-phase medium.

## 3. Hydrodynamical Results

We show the structure of a section of the disk from three purely hydrodynamic simulations in Figure 1. The panels show cases when the ISM is cold (50 K), warm ($10^4$ K), and a two-phase medium consisting of half 100 K and half $10^4$ K gas. Where the gas is warm ($10^4$ K), there is no substructure in the disk, and the spiral arms are smooth and continuous. On the other hand, where the gas is cold, the spiral arms are clumpy, with substructure along the trailing side of the arms. At later times in the simulation, these clumps produce more prominent, widely spaced spurs (Dobbs & Bonnell 2006). The formation of the clumps is due to the dynamics of the spiral shock. Gas from initially dissimilar radii agglomerates into clumps where orbit crowding occurs in the spiral arms (Dobbs, Bonnell, & Pringle 2006). Furthermore, the passage of the ISM through the shock induces a velocity dispersion in the gas, providing the gas is clumpy, which is the case when there is cold gas (Dobbs & Bonnell 2006).

The molecular hydrogen column density is also overplotted on the total column density on Figure 1. The density of $H_2$ was determined post-process using an algorithm from Bergin et al. (2004). When the gas is ~100 K, the amount of molecular gas in the disk is 10%. However, for the simulation with warm gas, no molecular hydrogen forms, and the gas remains atomic. This is because the density of gas in the shock is too low for $H_2$ to form, and any $H_2$ that does form is immediately photodissociated. For the case where the gas is 50 K, the molecular gas is located in highly structured clouds, predominantly along the spiral arms. For the two-phase result, the molecular clouds tend to be smaller, since the extra pressure from the warm gas confines the cold gas to smaller clumps. There is more $H_2$ in the interarm regions, since the clumps are more dense, and less readily photodissociated. Consequently, twice as much molecular gas is present compared to the same time frame of the single phase simulation of cold gas. The difference between the simulations is also reflected in the cloud mass spectrum. For the two-phase results, there are more lower mass clouds and the mass spectrum is $dN/dM \propto M^{-2.35}$ compared to $dN/dM \propto M^{-1.7}$ for a single phase medium (Dobbs & Bonnell 2007).

## 4. MHD Results

In Fig. 2, we show a section of the disk for different magnetic field strengths. For all the panels, the temperature of gas in the disk is 100 K. The shock becomes weaker as the magnetic field strength increases, and hence the structure in the

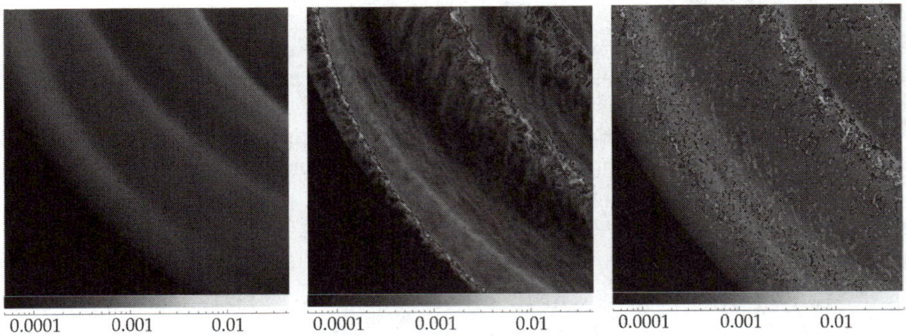


Figure 1. The spiral arm structure is shown for a 5 kpc × 5 kpc subsection of the galactic disk. The gas is 50 K (*left*), $10^4$ K (*middle*), and a two-phase medium of half 100 K and half $10^4$ K (*right*). The scale indicates the column density of molecular gas (in g cm$^{-2}$), which is overplotted in red and yellow (see online version). For the $10^4$ K case, no molecular hydrogen is present. These panels primarily show that when cold gas is present, this component of the ISM is organized into clumpy structures by the spiral shock, which contain a substantial fraction of $H_2$.

disk decreases. The spiral arms also appear wider. Thus, increasing the magnetic field has a similar effect to increasing the temperature in reducing the strength of the shock and smoothing out structure. Similar to the hydrodynamic results, we found that when the gas is warm ($10^4$ K), there is no substructure or interarm spurs. However, when the gas is cold, substructure still occurs in the disk for a realistic value of $\beta$ (where $\beta$ is the ratio of the gas to magnetic pressure).

The nature of the magnetic field also changes for different temperatures and field strengths. When the temperature is high, the field is very regular. However, when the gas is cold, the magnetic field contains a substantial random component. This random component is associated with the velocity dispersion induced by the spiral shock. Observations of the diffuse (warm) ISM indicate that the magnetic field contains both regular and disordered components, which are comparable in magnitude (Beck 2007). In two phase simulations we find that the magnetic field of the warm gas has similar random and regular components (Dobbs & Price 2007). The magnetic field in the cold gas is more disordered, although currently there are no corresponding observations of the degree of order for cold H I.

## 5. Conclusions

For non self-gravitating galactic disks, the thermal and magnetic pressure determine the density and degree of structure along the spiral arms, and thus whether molecular clouds can form. If the ISM is assumed to be a two-phase medium, the structure in the cold gas is enhanced from the pressure of the warm component, and consequently more molecular clouds are present in the interarm regions. The spiral shock not only affects the spiral structure, but also induces a velocity dispersion in the gas and a random component of the magnetic field.

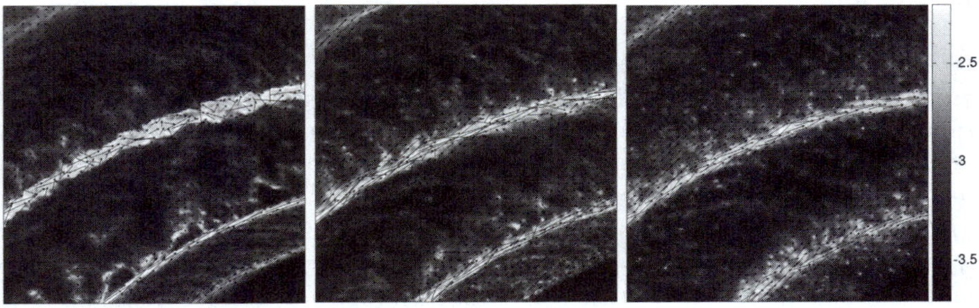


Figure 2. The spiral arm structure of a 4 kpc × 4 kpc subsection of the disk is shown when the gas is cold (100 K) for varying magnetic field strengths. The ratio of the thermal to magnetic pressure ($\beta$) is 10 (*left*), 1 (*middle*) and 0.1 (*right*). The right-hand scale indicates the logarithmic column density (in $\mathrm{g\,cm^{-2}}$) and the the vectors indicate the magnetic field integrated through $z$. There is less structure along the spiral arms for increasing field strength, and the magnetic field is more ordered.

**Acknowledgments.** Some of this work was conducted as part of the award 'The formation of stars and planets: Radiation hydrodynamical and magnetohydrodynamical simulations' made under the European Heads of Research Councils and European Science Foundation EURYI (European Young Investigator) Awards scheme, supported by funds from the Participating Organisations of EURYI and the EC Sixth Framework Programme. The computations reported here were performed using the UK Astrophysical Fluids Facility (UKAFF). DJP is supported by a PPARC postdoctoral research fellowship.

## References

Beck, R. 2007, in EAS Publications Series, 23, Magnetic Field Structure from Synchrotron Polarization (Cedex A: EDP), 19

Bergin, E. A., Hartmann, L. W., Raymond, J. C., & Ballesteros-Paredes, J. 2004, ApJ, 612, 921

Cepa, J., & Beckman, J. E. 1990, ApJ, 349, 497

Chakrabarti, S., Laughlin, G., & Shu, F. H. 2003, ApJ, 596, 220

Cox, D. P., & Gómez, G. C. 2002, ApJS, 142, 261

Dobbs, C. L., & Bonnell, I. A. 2006, MNRAS, 367, 873

Dobbs, C. L., & Bonnell, I. A. 2007, MNRAS, 376, 1747

Dobbs, C. L., Bonnell, I. A., & Pringle J. E. 2006, MNRAS, 371, 1663

Dobbs, C. L., & Price, D. J. 2007, MNRAS, submitted

Kim, W., & Ostriker, E. C. 2002, ApJ, 570, 132

La Vigne, M. A., Vogel, S. N., & Ostriker, E. C. 2006, ApJ, 650, 818

Price, D. J., & Bate, M. R. 2007, MNRAS, 377, 77

Roberts, W. W. 1969, ApJ, 158, 123

Seigar, M. S., & James, P. A. 2002, MNRAS, 337, 1113

Shu, F. H., Milione, V., & Roberts, W. W. 1973, ApJ, 183, 819

## Discussion

*Sheth*: In the observations we find $M_{\rm spurs} \sim 10^6$–$10^7\,M_\odot$, $\Sigma_{\rm H_2} \sim 100\,M_\odot\,{\rm pc}^{-2}$, and a separation of spurs of $\sim 100$–$300\,{\rm pc}$. Given these how do your simulations compare? It seems that the model is showing many more spurs than are seen. And the spurs seem too transient in your simulations, also apparently inconsistent with the observations.

*Dobbs*: The spurs increase in size and spacing over time. At their peak amplitude, the separation is $\sim 600\,{\rm pc}$ and $\Sigma_{\rm H_1+H_2} \sim 10\,M_\odot\,{\rm pc}^{-2}$. However, they may still appear to be frequent compared to M51, probably the large separation reflects the larger radii of our simulations compared to the inner regions of M51. The spurs are less transient and of higher surface density when the gas is situated in two phases, and (although not shown here) when self gravity is present.

*Krumholz*: An interesting observation regarding molecular clouds is that they all appear to have column densities of $\sim 10^{22}\,{\rm H}$ atoms ${\rm cm}^{-2}$, even though $\sim 10^{21}\,{\rm H\,atoms\,cm}^{-2}$ ($A_V = 1.4$) is all that is required for CO to form. This is seen in a number of Local Group galaxies, including the Milky Way. Have you computed the column density distribution for the molecular clouds formed in your simulations and compared to this result?

*Dobbs*: The column densities of our molecular clouds are distributed from $10^{21}$ to $10^{22}\,{\rm cm}^{-2}$. This, however, is sensitive to the cut-off density used to identify clouds (see Dobbs & Bonnell 2007). Possibly observations have similar selection criteria which leads to an apparently constant $\Sigma$.

Sébastien Comerón (*left*) and John Beckman.

*Pathways through an Eclectic Universe*
*ASP Conference Series, Vol. 390, © 2008*
*J. H. Knapen, T. J. Mahoney, and A. Vazdekis, eds.*

# Star Cluster Formation

Linda J. Smith[1,2]

[1]*Space Telescope Science Institute and European Space Agency, 3700 San Martin Drive, Baltimore, MD 21218, USA*
[2]*Department of Physics and Astronomy, University College London, Gower St., London WC1E 6BT, UK*

**Abstract.** Recent observations of two young star clusters in the Small Magellanic Cloud are presented. These clusters formed in very different environments 3–4 Myr ago. NGC 346 in the main body of the Small Magellanic Cloud contains ~15 coeval subclusters and appears to be a good example of cluster formation via the collapse and hierarchical fragmentation of a giant molecular cloud. On the other hand, NGC 602 is in a relatively isolated environment in the wing of the Small Magellanic Cloud and its formation was probably triggered by colliding H I shells. The dynamics of the ionized gas in the H II regions associated with both clusters indicate that they are quiescent. This suggests that at the low metallicity of the Small Magellanic Cloud, the stellar winds from the cluster O stars are weak, and the nebulae are shaped by their ionizing radiation rather than their winds.

## 1. Introduction

It is well known that star-forming regions in galaxies cover a range of scales, from compact clusters to diffuse OB associations, to large complexes. Studies of these sites, however, often show that they contain subclusters or sub-substructures. In extreme environments such as merging or starburst galaxies, young massive clusters are seen to congregate together to form clumps, containing dozens of clusters (e.g., the Antennae galaxies; Zhang, Fall, & Whitmore 2001; Bastian et al. 2006). This structure within structure suggests a hierarchical picture of star formation where young clusters represent the dense inner regions of the hierarchy (e.g., Elmegreen et al. 2006; Elmegreen 2007). In this contribution, I will present some recent observational results on two young clusters in the Small Magellanic Cloud (SMC), and discuss the formation mechanisms of these clusters.

When a cluster is formed, stellar winds remove the residual gas on a timescale of ~1 Myr and, depending on the efficiency of star formation, this can unbind the cluster (Bastian & Goodwin 2006). At low metallicity, the early evolution of clusters may be different because of the reduced stellar wind power. Studies of this effect are important for understanding the formation and duration of clusters in young, low metallicity galaxies. Recently formed clusters in the SMC are excellent objects to study in this context because they are resolved and their massive star contents can be determined.

## 2. The SMC Cluster NGC 346

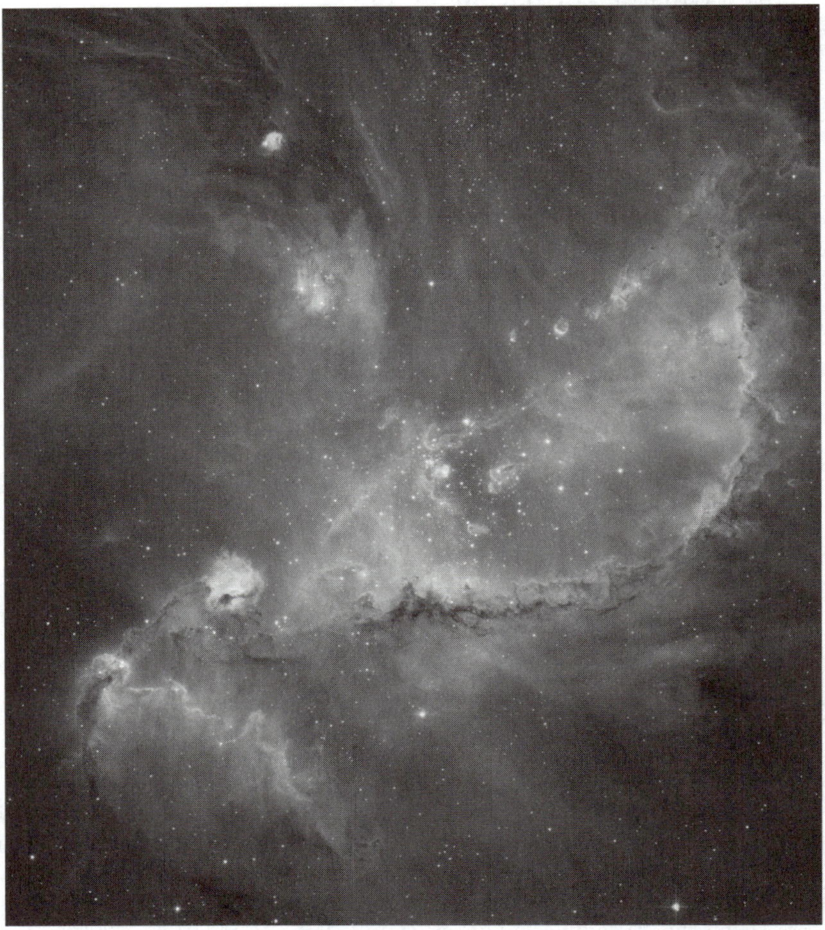

Figure 1. *HST*/ACS Hα image of the SMC cluster NGC 346, covering an area of 200″ by 200″ (or 60 pc ×60 pc). North is up and east is to the left. (*From Nota et al. 2006.*)

The young cluster NGC 346 is the most active star-forming region in the SMC and is located towards the northern end of the SMC bar. It contains at least 30 O stars (Massey, Parker, & Garmany 1989) which ionize N66, the largest H II region in the SMC. Relaño, Peimbert, & Beckman (2002) find a total stellar ionizing flux of $4 \times 10^{39}$ erg s$^{-1}$, and that 45 per cent of these ionizing photons escape from the nebula. In Fig. 1, an image of NGC 346 in the light of Hα is shown. This image, taken from Nota et al. (2006), was obtained with the Advanced Camera for Surveys (ACS) on board the *Hubble Space Telescope* (*HST*). It shows that the cluster still contains some of its residual gas despite its age of ~3 Myr

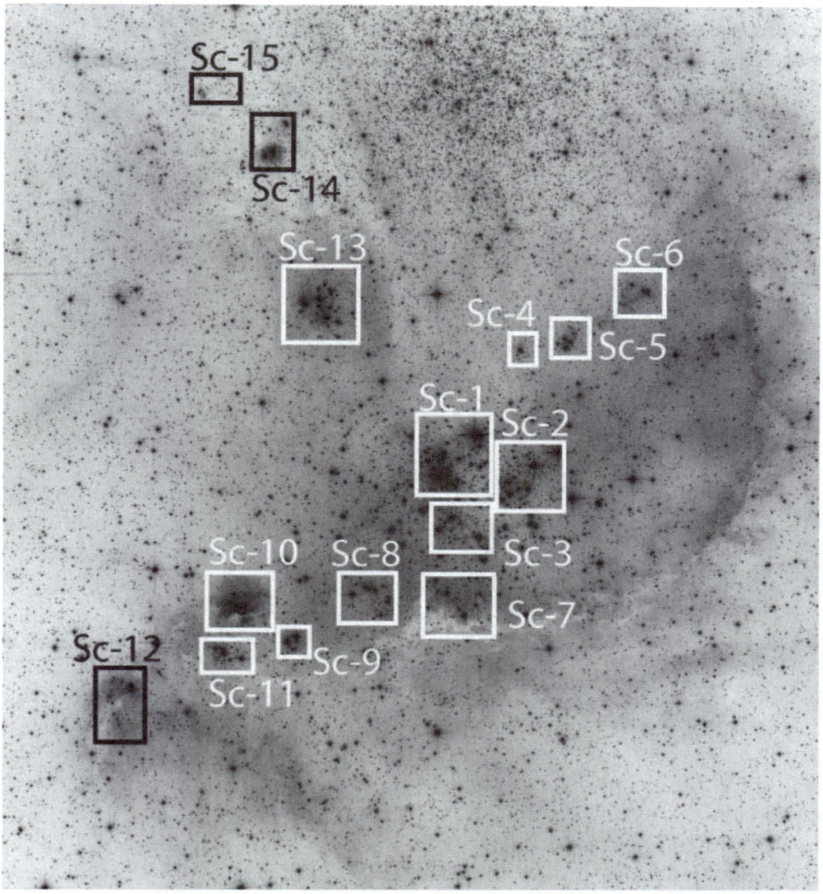


Figure 2. *HST*/ACS F555W image of the SMC cluster NGC 346 showing the 15 identified subclusters.

(Bouret et al. 2003). Although at least two supernova remnants are present close to NGC 346 (Ye, Turtle, & Kennicutt 1991; Reid et al. 2006), it appears that NGC 346 is probably too young to have encountered a supernova event. This is in accord with the gas motions (see later) and the low diffuse X-ray flux (Nazé et al. 2002).

Nota et al. (2006) and Sabbi et al. (2007a,b) present photometric analyses of NGC 346 from *HST* F555W and F814W images. We find that the young stellar population of NGC 346 has an age of $3 \pm 1$ Myr and that there is a rich population of low mass (0.6–3 $M_\odot$) pre-main sequence stars that likely formed at the same time as the central cluster.

The high spatial resolution of the observations reveals that the young stellar population is not uniformly distributed within the ionized nebula. We identify at least 15 subclusters that differ in size and stellar content; the locations of these subclusters are shown in Figure 2. The central cluster itself is made up of at least three subclusters (Sc 1, 2, and 3), and has a radius of 2.3 pc and a total stellar mass of a few $\times 10^4\,M_\odot$. Within the uncertainties of the isochrone fitting process, we find that the subclusters all have an age of $3 \pm 1$ Myr and thus are coeval. The subclusters are embedded in nebulosity and coincide with clumps of neutral gas and CO emission (Rubio et al. 2000; Contursi et al. 2000).

The pre-main sequence stars are concentrated in the subclusters at the center of NGC 346 and are also present, in smaller numbers, in most of the small compact subclusters (e.g., Sc 10, 12, 13, and 14). In a recent study, Simon et al. (2007) present *Spitzer* observations of NGC 346. They find 111 embedded young stellar objects (YSOs), demonstrating that star formation is still continuing in this region today. The YSOs in their sample have masses between 2–17 $M_\odot$ (if they are single objects) and the most massive objects are located in the central subclusters, presenting strong evidence for primordial mass segregation. They find that all but one of the 15 subclusters identified by Sabbi et al. (2007a) contain YSOs.

Recently, Sabbi et al. (2007b) have determined the present-day mass function (PDMF) for NGC 346 from their *HST* optical photometry. We find that the PDMF is Salpeter over the mass range of 0.8–60 $M_\odot$, and that it varies as a function of the radial distance from the center of the cluster, indicating that mass segregation, probably of primordial origin, is present.

The crossing time from the central subcluster to the outer subclusters (Sc 6, 12, and 15) is $\sim$2 Myr (for a sound speed of 10 km s$^{-1}$) and thus NGC 346 is not much older than one crossing time. The subcluster structure we observe and the coevality of the subclusters strongly suggest that NGC 346 was formed as the result of the collapse and subsequent hierarchical fragmentation of a giant molecular cloud into multiple seeds of star formation (Elmegreen 2000; Klessen & Burkert 2000; Bonnell & Bate 2002; Bate, Bonnell, & Bromm 2003; Bonnell, Bate, & Vine 2003). According to this model, the fragmentation of the cloud is due to supersonic turbulent motions in the gas. The turbulence induces the formation of shocks in the gas, and produces filamentary structures. The chaotic nature of the turbulence produces local density enhancements in the filamentary structures. High density regions that become self-gravitating then collapse to form stars, and this happens simultaneously at different locations within the cloud (Bonnell et al. 2003).

To test these theories further and to study stellar wind feedback at the low metallicity of the SMC, we have obtained high resolution observations of the ionized gas in NGC 346/N66. These data were obtained with the University College London Echelle Spectrograph (UCLES) on the Anglo–Australian Telescope (AAT), and cover the H$\alpha$ and [O III]$\lambda$5007 emission lines at a spectral resolution of 7.5 km s$^{-1}$. We observed a total of six slit positions across N66 covering subclusters Sc 1, 2 (center), 6, 7, 8, 10, 13, and 14. Surprisingly, we find that the ionized gas is quiescent with no evidence for large scale gas motions. Overall, we find that the H$\alpha$ profiles are single with a velocity dispersion of $\sim$14 km s$^{-1}$ and a constant velocity along the length of each slit position (1 arcmin). As an example, the H$\alpha$ profiles at the extremes of the slit crossing Sc 6 are shown in Figure 3. Even at the center where the most massive O stars are located, we detect no significant motions. These results strongly suggest that at the low metallicity of the SMC, stellar winds are much reduced, and the

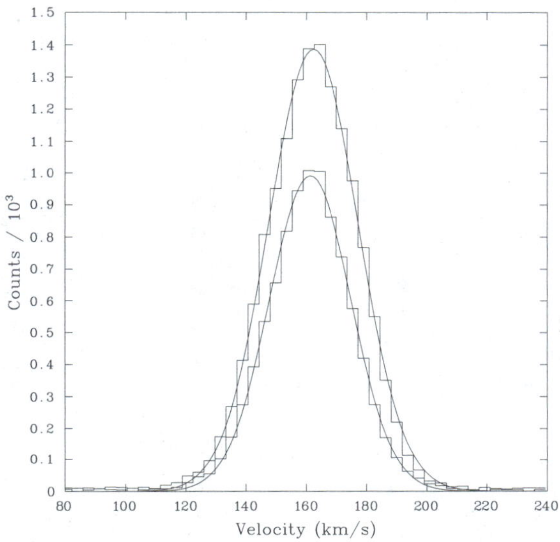


Figure 3. AAT+UCLES Hα profiles for the slit crossing the subcluster Sc 6 in NGC 346. The two profiles, with Gaussian fits superimposed, represent data from the extremes of the 1 arcmin long slit and show that the nebular gas is quiescent.

dominant form of interaction is via stellar radiation rather than by winds. This is in accord with Bouret et al. (2003) who measured the mass loss rates of six O stars within the central cluster and found that their winds are considerably weaker than their Galactic counterparts.

Finally, considering relative velocities within the nebula, we find that there are very small velocity differences ($\leq 3\,\mathrm{km\,s^{-1}}$) between the subclusters. This suggests that if the subclusters merge with the main cluster in the center, it will take at least ~7 Myr.

## 3. The SMC Cluster NGC 602

The young stellar cluster NGC 602 is located in the south-eastern "wing" of the SMC. This region is close to the Magellanic bridge and has a low gas and stellar content. The environment of NGC 602 is thus quite different to that of NGC 346. NGC 602 contains about a dozen OB stars which ionize the H II region N90. Again, we have obtained deep imaging with *HST*/ACS in the F555W, F814W, and F658N (Hα) filters of the cluster and nebula. A composite image, presented by Carlson et al. (2007), is shown in Figure 4. The morphology of N90 is that of a partial shell and NGC 602 is composed of two stellar groups—the dominant group near the nebular center, and a small grouping to the north-west.

*Spitzer* images presented in Carlson et al. (2007) show that the "elephant trunks" and dusty pillars visible in Fig. 4 contain ~25 YSOs. From isochrone fitting to the ACS color–magnitude diagram, we find that the age of NGC 602

Figure 4. *HST*/ACS H$\alpha$ image of the SMC cluster NGC 602, covering an area of 200″ by 200″ (or 60 pc ×60 pc). North is up and east is to the left. (*From Carlson et al. 2007.*)

is ~4 Myr, and that it contains a substantial population of low mass pre-main sequence stars. Thus we find that NGC 602 has a similar age to NGC 346 and that star formation is still continuing. Like NGC 346, we find from AAT high resolution spectroscopy that the ionized gas associated with NGC 602 is quiescent. This again suggests that the main interaction of hot stars with their surroundings is via radiation rather than winds at the low metallicity of the SMC.

NGC 602 seems to have formed in relative isolation and under conditions that are marginal for star formation in the wing of the SMC. The *HST* image in Fig. 4 shows that many background galaxies are visible both inside and outside the nebula. This transparency suggests a lack of large scale molecular concentrations in the immediate vicinity of the nebula. To address the formation mechanism of NGC 602, we have compared the kinematics of the nebular H$\alpha$ emission with that of H I gas from the 21 cm survey of Stanimirović et al. (1999). The results will be presented in L. Nigra et al. (in preparation). The H I data of Stanimirović et al. (1999) show that there is a dense H I cloud to the south that has the same velocity as the nebular H$\alpha$ emission, suggesting that this is the progenitor cloud. Analysis of the dynamics of H I shells in the vicinity of NGC 602 suggests that the formation of the cluster was triggered by the interaction of two expanding H I shells about 7 Myr ago.

## 4. Conclusions

The two young star clusters NGC 346 and NGC 602 in the SMC have similar ages but probably different formation histories. NGC 346 contains many coeval subclusters, suggesting that it was formed as the result of the collapse and subsequent hierarchical fragmentation of a giant molecular cloud. On the other hand, the low density of the environment of NGC 602 in the wing of the SMC suggests a different formation mechanism to gravitational collapse. The formation of the cluster was probably triggered by the collision of two H I shells.

We find that the ionized gas associated with both clusters is quiescent. The nebulae thus appear to be shaped by the ionizing radiation from the cluster O stars rather than their stellar winds. This is in accord with the lower mass loss rates of O stars in NGC 346 (Bouret et al. 2003). The central subclusters in NGC 346 contain residual gas and a large number of pre-main sequence stars. This suggests that the early evolution of stellar clusters at low metallicity may be different because the retention of residual gas will allow stars to continue forming, and thus the star formation efficiency may increase.

**Acknowledgments.** I would like to thank my collaborators on the SMC clusters project: A. Nota, L. Carlson, J. Gallagher, M. Meixner, E. Sabbi, M. Sirianni, M. Tosi, M. S. Oey, A. Pasquali, and R. Walterbos.

### References

Bate, M. R., Bonnell, I. A., & Bromm, V. 2003, MNRAS, 339, 577
Bastian, N., & Goodwin, S. P. 2006, MNRAS, 369, L6
Bastian, N., Emsellem, E., Kissler-Patig, M., & Maraston, C. 2006, A&A, 445, 471
Bonnell, I. A., & Bate, M. R. 2002, MNRAS, 336, 659
Bonnell, I. A., Bate, M. R., & Vine, S. 2003, MNRAS, 343, 413
Bouret, J.-C., Lanz, T., Hillier, D. J., Heap, S. R., Hubeny, I., Lennon, D. J., Smith, L. J., & Evans, C. J. 2003, ApJ, 595, 1182
Carlson, L. R., et al. 2007, ApJ, 665, L109
Contursi, A., et al. 2000, A&A, 362, 310
Elmegreen, B. G. 2000, AJ, 530, 227
Elmegreen, B. G., Elmegreen, D. M., Chandar, R., Whitmore, B. C., & Regan, M. 2006, ApJ, 644, 879
Elmegreen, B. G. 2007 in ASP Conf. Ser., Mass Loss from Stars and the Evolution of Stellar Clusters, ed. A. de Koter, L. J. Smith, & R. Waters (San Francisco: ASP), in press
Klessen, R. S., & Burkert, A. 2000, ApJS, 128, 287
Massey, P., Parker, J. W., & Garmany, C.D. 1989, AJ, 98, 1305
Nazé, Y., Hartwell, J. M., Stevens, I. R., Corcoran, M. F., Chu, Y.-H., Koenigsberger, G., Moffat, A. F. J., & Niemela, V. S. 2002, ApJ, 580, 225
Nota, A., et al. 2006, ApJ, 640, L29
Reid, W. A., Payne, J. L., Filipović, M. D., Danforth, C. W., Jones, P. A., White, G. L., & Staveley-Smith, L. 2006, MNRAS, 367, 1379

Relaño, M., Peimbert, M., & Beckman, J. E. 2002, ApJ, 564, 704
Rubio, M., Contursi, A., Lequeux, J., Probst, R., Barbá, R., Boulanger, F., Cesarsky, D., & Maoli, R. 2000, A&A, 359, 1139
Sabbi, E. et al. 2007a, AJ, 133, 44
Sabbi, E. et al. 2007b, AJ, in press
Simon, J. D. et al. 2007, ApJ, in press
Stanimirović, S., Staveley-Smith, L., Dickey, J. M., Sault, R. J., & Snowden, S. L. 1999, MNRAS, 302, 417
Ye, T., Turtle, A. J., & Kennicutt, R. C. 1991, MNRAS, 249, 722
Zhang, Q., Fall, S. M., & Whitmore, B. C. 2001, ApJ, 561, 727

## Discussion

*Kroupa*: 1) How much gas is still within 2.4 pc of the center of NGC 346? 2) The subclusters have relative velocities of $\sim 2\,\mathrm{km\,s^{-1}}$, so they could merge in 10 Myr (roughly).

*Smith*: 1) Ionized and molecular gas is still present in the central subclusters of NGC 346. We have not estimated the mass of residual gas yet but we will do so in the future and compare it to the total mass of stars. 2) Yes, I agree that on a long timescale, the subclusters could merge with the main cluster in the center.

*James*: Is it surprising that no supernovae have been seen yet in NGC 346, given the large number of O stars and an age of ~3 Myr.

*Smith*: If a supernova explosion had occurred, we would expect to observe large scale gas motions but we do not. Also, very little if any diffuse X-ray emission is observed to be associated with NGC 346 (Nazé et al. 2002).

*Hensler*: What is the average separation of the subclusters in NGC 346? Does the sound travel time allow them to communicate, i.e., to trigger their star formation within the age spread between the subclusters?

*Smith*: The crossing time from the most distant subclusters is ~2 Myr. We can determine ages for the subclusters to an accuracy of 1–2 Myr. Given those considerations, it appears very unlikely that the star formation was triggered in this region.

Mike Edmunds giving his conference dinner speech.

*Pathways through an Eclectic Universe*
*ASP Conference Series, Vol. 390, © 2008*
*J. H. Knapen, T. J. Mahoney, and A. Vazdekis, eds.*

# Dynamical Masses of Young Star Clusters: Constraints on the Stellar IMF and Star Formation Efficiency

Nate Bastian

*Department of Physics and Astronomy, University College London, Gower Street, London WC1E 6BT, UK*

**Abstract.** Through the use of detailed light profiles and dynamical measurements of young clusters we investigate claims that the stellar initial mass function (IMF) within clusters varies greatly. We find a strong age dependence in the clusters which have been claimed to have non-standard stellar IMFs, and suggest that the lack of equilibrium of these clusters is responsible for their "strange" light-to-mass ratios and *not* IMF variations. The most likely culprit is the rapid removal of residual gas left over from the star formation process, which leaves the clusters severely out of dynamical equilibrium. By comparing the observations to $N$-body simulations we quantify to what degree a cluster is out of equilibrium and consequently its survival chances. We find that $> 60\%$ of young clusters will be disrupted, due to gas removal, within the first 20–50 Myr of their lives.

## 1. The Stellar Initial Mass Function in Clusters

Many recent works have attempted to constrain the stellar initial mass function (IMF) inside massive clusters by comparing their dynamical mass estimates (found through measuring the velocity dispersion and effective radius) to the measured light. These studies have come to different conclusions, with some claiming standard Kroupa-type (Kroupa 2002) IMFs (e.g., Maraston et al. 2004) while others have claimed extreme non-standard IMFs (e.g., the top or bottom of the IMF is over-populated with respect to a Kroupa IMF, Smith & Gallagher 2001). However, the results appear to be correlated with the age of the clusters, as older clusters ($>$ 80 Myr) all appear to be well fitted by a Kroupa-type IMF, whereas younger clusters display significant scatter in their best-fitting IMF (Bastian et al. 2006). This has led to the suggestion that the younger clusters are out of virial equilibrium, thus undercutting the fundamental assumption that is necessary to derive dynamical masses. We will return to this point in Sections 2 and 3. Focusing on the older clusters, we see that they all have standard IMFs (see Fig. 2), arguing that—at least in massive clusters—the IMF does not vary significantly.

## 2. Dynamical Equilibrium of Young Clusters

One explanation of why the youngest clusters are not in dynamical equilibrium is that young clusters are expected to expel their remaining gas (left over from the star formation process) on extremely short timescales, which will leave the cluster severely out of equilibrium (e.g., Goodwin 1997). In order to search

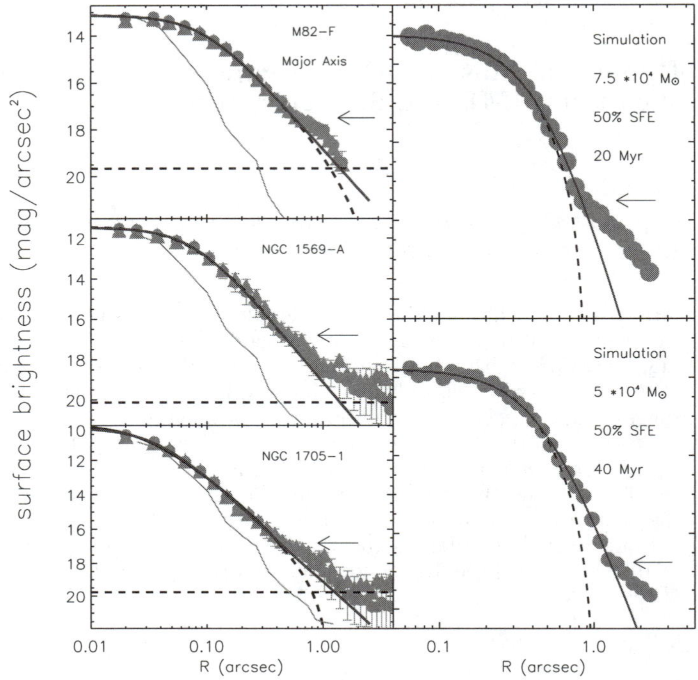


Figure 1. Surface brightness profiles for three young clusters (*left:* M82-F, NGC 1569-A, and NGC 1705-1) and two $N$-body simulations which include the rapid removal of gas left over from a non-100% star formation efficiency (*right*). The solid (red) and dashed (blue) lines are the best-fitting EFF (Elson, Fall, & Freeman 1987) and King (1962) profiles respectively. Note the excess of light at large radii with respect to the best-fitting EFF profile in both the observations and models. This excess light is due to an unbound expanding halo of stars caused by the rapid ejection of the remaining gas after the cluster forms. *Hence, excess light at large radii strongly implies that these clusters are not in dynamical equilibrium.* For details of the modeling and observations see Bastian & Goodwin (2006) and Goodwin & Bastian (2006). *(Reproduced with permission from Bastian & Goodwin 2006).*

for such an effect we compared the luminosity profiles of three young clusters with that of $N$-body simulations of clusters undergoing violent relaxation due to rapid gas loss (Bastian & Goodwin 2006). The simulations (Fig. 1, *right panel*) make the generic prediction of excess light at large radii (with respect to the best-fitting EFF profile, Elson, Fall, & Freeman 1987), due to an unbound expanding halo of stars that remains associated with the cluster for ~20–50 Myr. These stars are unbound due to the rapid decrease in potential energy as the gas is removed on timescales shorter than a crossing time (e.g., Goodwin 1997). Observations of the three young clusters also show excess light at large radii (Fig. 1, *left panel*), strongly suggesting that they are experiencing violent relaxation (Bastian & Goodwin 2006). Hence these clusters are not in dynamical equilibrium.

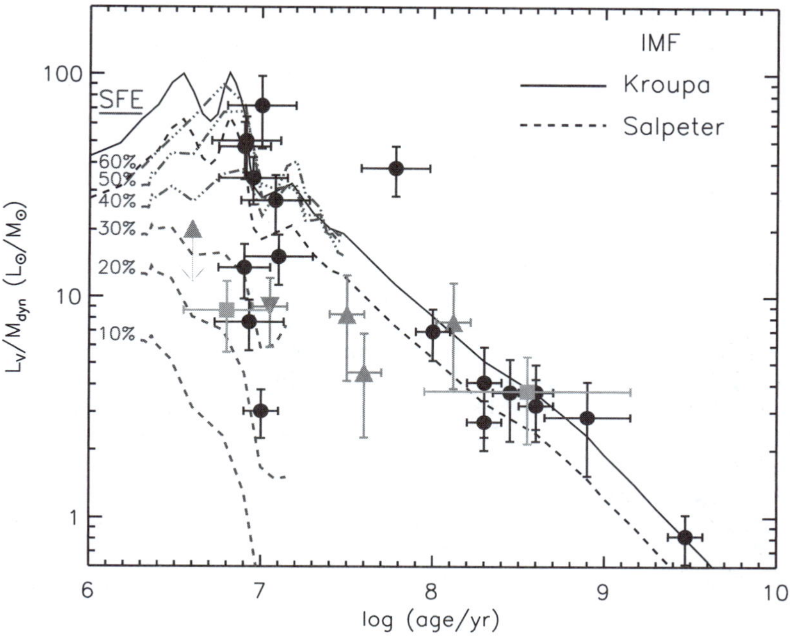


Figure 2. The light-to-mass ratio of young clusters. The data are from Goodwin & Bastian (2006) and references therein. The solid (black) line is the prediction of SSP models with a Kroupa (2002) stellar IMF. The dashed lines are the SSP model tracks folded with the effects of rapid gas removal following non-100% SFEs. Dashed lines represent the SFEs where the clusters will become completely unbound. The SFE in the simulations measures the degree to which the cluster is out of virial equilibrium after gas loss, and so is an *effective* SFE (see Goodwin & Bastian 2006). *(Based on figure 5 from Bastian & Goodwin 2006).*

## 3. The Star Formation Efficiency and Infant Mortality

Assuming that young clusters are out of equilibrium due to rapid gas loss (the extent of which is determined by the star formation efficiency, SFE) one can fold these effects (see figure 3 in Bastian & Goodwin 2006) into simple stellar population (SSP) models (Goodwin & Bastian 2006). The results are shown as solid (black) and dashed (red) lines in Fig. 2 for various SFEs, where we have assumed that all gas is lost instantaneously at 2 Myr. The dashed lines show the results for SFEs below 30%, for which the cluster will become completely unbound. Solid lines represent SFEs above 30% where a bound core may remain. Note that the observed SFEs of the clusters are in the range 10–60% (Goodwin & Bastian 2006).

We also note that seven out of the 12 clusters with ages below 20 Myr appear unbound (i.e., SFE $<$ 30%), suggesting that ~60% of clusters will become unbound in the first 20–50 Myr of their lives (Goodwin & Bastian 2006), i.e., what has been termed “infant mortality”. This is in close agreement with cluster population studies of M51 that found an infant mortality rate of 68% (Bastian et al. 2005) and comparable to the open cluster dispersal rate of ~87% (Lada & Lada 1991; see also Whitmore 2003 and Pellerin et al. 2007).

## 4. Conclusions

Through detailed comparisons of the luminosity profiles of young clusters with $N$-body simulations of clusters including the effects of rapid gas loss, we argue that young clusters are not in virial equilibrium. This undercuts the fundamental assumption needed to determine dynamical masses. This suggests that the claimed IMF variations are probably due to the internal dynamics of the clusters and are not related to the IMF. By limiting the sample to the oldest clusters (which appear to be in equilibrium) we see that they are all well fitted by a Kroupa-type IMF arguing that, at least in massive star clusters, the IMF does not vary significantly.

By combining the above $N$-body simulations with SSP models we can derive the (effective) SFE of clusters. We find that ~60% of young clusters appear to be unbound, in good agreement with other estimates of the infant mortality rate. Note, however, that even if a cluster survives this phase it may not survive indefinitely due to internal and external effects (e.g., Gieles et al. 2005).

**Acknowledgments.** NB gratefully acknowledges Simon Goodwin for the large part he has played in the studies presented here.

### References

Bastian, N., Gieles, M., Lamers, H. J. G. L. M., Scheepmaker, R. A., & de Grijs, R. 2005, A&A, 431, 905
Bastian, N., & Goodwin, S.P. 2006, MNRAS, 369, L9
Bastian, N., Saglia, R. P., Goudfrooij, P., Kissler-Patig, M., Maraston, C., Schweizer, F., & Zoccali, M. 2006, A&A, 448, 881
Elson, R. A. W., Fall, M. S., & Freeman, K. C. 1987, ApJ, 323, 54 (EFF)
Gieles, M., Bastian, N., Lamers, H. J. G. L. M., & Mout, J. N. 2005, A&A, 441, 949
Goodwin, S. P. 1997, MNRAS, 284, 785
Goodwin, S. P., & Bastian, N. 2006, MNRAS, 373, 752
King, I. 1962, AJ, 67, 471
Kroupa, P. 2002, Science, 295, 82
Lada, C. J., & Lada, E. A. 1991, in ASP Conf. Ser., 13, The Formation and Evolution of Star Clusters, ed. K. Joes (San Francisco: ASP), 3
Maraston, C., Bastian, N., Saglia, R. P., Kissler-Patig, M., Schweizer, F., & Goudfrooij, P. 2004, A&A, 416, 467
Pellerin, A., Meyer, M., Harris, J., & Calzetti, D. 2007, ApJ, 658, L87
Smith, L. J., & Gallagher, J. S. 2001, MNRAS, 326, 1027
Whitmore, B. C. 2003, in A Decade of HST Science, ed. M. Livio, K. Noll, & M. Stiavelli (Cambridge: Cambridge University Press), 153

## Discussion

*Böker*: Are you aware of any simulations of rapid gas removal for clusters that are located close to the galaxy nucleus, i.e., at the bottom of the potential well? Can the gas fall back onto the cluster?

*Bastian*: I am not aware of any such simulations. Observationally, it is clear that there are multiple-aged populations within nuclear clusters, so gas must be falling onto the cluster, stimulating further star formation. Seth et al. (2006, AJ, 132, 2539) have recently shown that some nuclear clusters have a blue compact disk of recent star formation around them. Does this gas originate from the cluster formation (i.e., the residual gas) or outer gas driven into the cluster? At the moment this is unknown.

Isabel Pérez (*left*) and Mónica Relaño.

*Pathways through an Eclectic Universe*
*ASP Conference Series, Vol. 390, © 2008*
*J. H. Knapen, T. J. Mahoney, and A. Vazdekis, eds.*

# Star Formation in Molecular Clouds?

Neal J. Evans II
*The University of Texas at Austin, Department of Astronomy, 1 University Station, C1400, Austin, TX 78712-0259, USA*

**Abstract.** Using studies of nearby star formation with *Spitzer*, I will argue that star formation is restricted to dense cores within molecular clouds. The nature of these dense cores and their connection to star formation will be discussed. Their distribution over masses and over the cloud is similar to that of stars, and their efficiency of forming stars is much higher than that of the whole cloud. Moving to regions forming more massive stars, we find that the mass distribution of the dense clumps is similar to that of OB associations. The infrared luminosity per unit mass of dense gas is high and comparable to that seen in starburst galaxies. The relation between star formation and dense gas appears to be linear. Understanding the Kennicutt–Schmidt law requires an understanding of what controls the conversion of gas into the dense entities where stars actually form.

## 1. Introduction

The image that comes to mind when we talk about star formation across disciplinary lines is that of the "sight-challenged" villagers and the elephant. Each feels a different part of the elephant and comes to different conclusions about what it is. Discussions of star formation within the Milky Way use language unfamiliar to those who work on star formation in other galaxies and vice versa. If we are to progress, we need to find a common language.

Studies of low mass star formation in nearby (typically within 500 pc) clouds can speak in detail about the distribution of star formation, the timescales for various phases, and the current efficiency in stellar mass per cloud mass. Detailed, predictive theoretical models exist for the formation of individual stars (e.g., Shu et al. 1987). There is a strong focus on detailed studies of the flow of material from envelope to disk to star and on the connection to planet formation (e.g., many articles in Reipurth, Jewitt, & Keil 2007). We can also provide observational constraints on the origin of the initial mass function (IMF). However, these studies are not directly relevant to star formation on galactic scales because low mass star formation is not observable at large distances.

Regions forming more massive stars begin to show up at about 440 pc (Hirota et al. 2007) in the Orion complex but most lie at distances of kpc, with the most impressive being in the Galactic ring, located about 5 kpc from the center of the Milky Way (Clemens, Sanders, & Scoville 1988). For the most part, we are really talking about formation of clusters of stars, including both high and low mass stars, though most forming clusters do not survive (Lada & Lada 2003). Our knowledge of these regions is much less detailed, owing to the lack of spatial resolution, and theoretical predictions are less well developed. Considerable theoretical progress is being made, but there are fundamental disagreements on some issues (cf. Krumholz, McKee, & Klein 2005; Bonnell & Bate 2006; Martel, Evans, & Shapiro 2005). For most of the distant regions, we do not separate individual stars, but rather we characterize the star formation in terms of surrogates, such as the far-infrared luminosity ($L_{\rm IR}$) and the mass of molecular material. While less detailed, these crude measures are more compatible with what can be done in other galaxies.

Discussion of star formation in other galaxies speaks nearly exclusively in terms of "Schmidt laws". The original version compared the scale heights of young stars and the H I gas and concluded that the star formation rate is proportional to the square of the *volume* density (Schmidt 1959). Were this to be done today, recognizing that star formation is restricted to molecular gas, one would probably infer a linear relationship. However, *surface* density is more amenable to study in other galaxies and modern versions find relations between the surface densities of star formation and gas (Kennicutt 1998). A power law above a threshold provides a remarkably good fit:

$$\Sigma({\rm SF})(M_\odot\,{\rm yr}^{-1}\,{\rm kpc}^{-2}) = (2.5 \pm 0.7) \times 10^{-4}(\Sigma({\rm gas})/1\,M_\odot\ {\rm pc}^{-2})^{1.4\pm0.15}$$

over a wide range of galaxy types and star formation rates. Several explanations have been offered for the value of the exponent (Elmegreen & Scalo 2004; Krumholz & McKee 2005; Shu et al. 2007). Models of galaxy formation and evolution generally use a Kennicutt–Schmidt relation to simulate star formation, which is treated as "sub-grid physics".

To return to the elephant analogy, each villager feels a different part of the elephant, but also speaks a different language, which is only partly understood, and often misinterpreted, by the others. But the need for communication is great as rapid progress is occurring in studies of massive star formation in our Galaxy. At the same time, studies of galaxy formation are progressing rapidly, including both archaeological studies of the oldest stars and direct look-back studies of high-$z$ luminous starbursts. Can we find a common language?

I will summarize recent progress in studies of both local, low mass and clustered, high mass star formation in our Galaxy, and then I will discuss some connections that can be made.

## 2. Low mass Star Formation in Nearby Clouds

For convenience, I will focus here on recent results from the *Spitzer* legacy project, 'From Molecular Cores to Planet-Forming Disks', known as 'Cores to Disks' or c2d (Evans et al. 2003). In particular, I will concentrate on studies of five large nearby clouds. For three of these clouds (Perseus, Serpens, and Ophiuchus), we have complementary information on the dust emission at 1 mm using Bolocam (Enoch et al. 2006, 2007; Young et al. 2006) and molecular line emission (Ridge et al. 2006). These studies provide the first complete coverage of molecular clouds in tracers of gas, dense clores, and young stellar (or substellar) objects (YSOs) down to luminosities of about 0.01 $L_\odot$.

Comparison of the CO and $^{13}$CO maps with the Bolocam maps shows that the dense ($n > 2 \times 10^4$ cm$^{-3}$) cores traced by dust continuum emission at millimeter wavelengths are distributed very non-uniformly over the cloud. Dense cores are highly clustered, and many lie along filaments traced by $^{13}$CO. The dense cores strongly prefer regions of high overall extinction, with 75% of dense cores found within extinction contours that contain only 0.1 to 0.3 of the total cloud mass (Enoch et al. 2007). Comparison of the dense core distribution to that of YSOs shows a strong correlation, especially for the early stages of star formation. The dense cores are clearly the sites of star formation; most of the cloud is completely inactive.

Until spectral types are available for more of the YSOs, we cannot construct IMFs for the clouds, but all data so far are consistent with a typical IMF. The core mass distribution for *starless* cores, averaging over all three clouds, is remarkably consistent with the usual IMF, but shifted to larger masses by a factor $< 4$, suggesting that the IMF is set by the core mass distribution with an efficiency $> 25\%$ (M. L. Enoch et al., in preparation). Similar conclusions have been reached by Alves, Lombardi, & Lada (2007) using extinction mapping to trace cores; their method produces less uncertain masses but probes lower densities on average. Of course, the similarity of the IMF and the core mass distribution could be fortuitous if cores of different masses have different lifetimes (Clark, Klessen, & Bonnell 2007).

We define the actual efficiency of star formation as the fraction of the mass that ends up in stars (or substellar objects):

$$\epsilon_\star = M_\star/(M_\star + M_{\rm gas}) \approx M_\star/M_{\rm gas}.$$

The *final* efficiency, after the gas is consumed or dissipated, is impossible to determine directly since the gas lifetime vastly exceeds human lifetimes, but we can constrain it. In practice we measure the *current* efficiency using the current mass in stars and in gas. To get a star formation rate ($\dot{M}_\star$), we assume a mean stellar mass of $0.5\,M_\odot$ (consistent with actual determinations in one cloud; J. Alcalà et al., in preparation), and a timescale of 2 Myr (essentially the half-life of infrared excesses that the c2d observations are sensitive to). It is possible for these nearby regions to separate the efficiency, as defined above, from the question of the *speed* of star formation. We define the timescale for gas depletion, as in extragalactic studies:

$$t_{\rm dep} = M({\rm gas})/\dot{M}_\star$$

and the speed is $1/t_{\rm dep}$. Krumholz & Tan (2007) define the speed as

$${\rm SFR}_{\rm ff} = t_{\rm ff}/t_{\rm dep} = t_{\dot{M}_\star {\rm ff}}/M({\rm gas}).$$

This is the speed ($1/t_{\rm dep}$) normalized to the maximum likely speed ($1/t_{\rm ff}$), or the star formation rate normalized to the maximum possible in a free fall time.

The preliminary results for these quantities for the five large clouds studied by c2d are summarized here. The total mass in YSOs is about 2% to 4% of the cloud mass (e.g., Harvey et al. 2007; N. J. Evans et al., in preparation) consistent with earlier conclusions that star formation is inefficient. The total mass in dense cores is only 1% to 4% of the total cloud mass included in the $A_V = 2$ contour. Low star formation efficiency begins with inefficient core formation. The values of $\dot{M}_\star$ range from 6.0 to 71 $M_\odot$/Myr, a significant range. Star formation is slow: the depletion timescales for the whole molecular clouds are $t_{\rm dep} = 50$–90 Myr, much longer than most estimates of cloud lifetimes of 5–10 Myr. While further star formation is likely, the final efficiencies are likely to remain low. In contrast, the depletion times for dense cores, the actual sites of star formation, are 0.6–3.5 Myr, and masses in dense cores are similar to current masses in YSOs. Star formation is fast and efficient *in dense gas.*

What about the speed relative to that obtained from assuming free fall? In calculating $SFR_{ff}$, we use the mean density of the relevant unit to calculate the free fall time. The $SFR_{ff}$ is 0.02 to 0.03 for the cloud as a whole, consistent with long-standing arguments against efficient star formation at the free fall rate (e.g., Zuckerman & Evans 1974). Considering the clump that is forming a cluster, the $SFR_{ff}$ increases to 0.1 to 0.3, but if we use the mean density of an individual dense core, the $SFR_{ff}$ is only 0.03 to 0.12. Star formation remains "slow" compared to a free fall time. This slowness cannot be blamed entirely on a slow prestellar phase; comparison of the numbers of dense, starless (prestellar) cores to the numbers of protostellar cores yields lifetimes of a few free fall times once the density exceeds about $2 \times 10^4$ cm$^{-3}$ (M. L. Enoch et al., in preparation).

Does the surface density relation (Kennicutt 1998) have any relevance to local star formation? This one is hard to test locally because of incompleteness, but we can make a rough estimate. With a local gas surface density of 6.5 $M_\odot$ pc$^{-2}$ (L. Blitz, personal communication), of which about 85% is H I (Dame 1993; Levine, Blitz, & Heiles 2006), the Kennicutt relation would predict $\Sigma(\mathrm{SF}) = 3.4 \times 10^{-3}$ $M_\odot$ yr$^{-1}$ kpc$^{-2}$. Lada & Lada (2003) estimate that embedded clusters contribute, to within a factor of three, $\Sigma(\mathrm{SF}) = 3 \times 10^{-3}$ $M_\odot$ yr$^{-1}$ kpc$^{-2}$ within a radius of 0.5 kpc. This is surely an underestimate until we have a more complete census of all clouds within the local area, but it is already remarkably close to the prediction. A more extensive study of this topic can be found in Blitz & Rosolowsky (2006). It is interesting to note that the low mass star formation in the clouds studied by the c2d project would not be apparent to the usual tracers used in extragalactic studies, with the possible exception of $L_{IR}$.

The lessons we should take from these very detailed studies of nearby clouds are as follows. The fundamental units of star formation are dense cores, not molecular clouds, per se (hence the question mark in the title). The cores are not located randomly over cloud faces, but are concentrated in clumps, which are forming clusters, and filaments, which tend to form smaller aggregates. The core mass distribution *may* determine the IMF. Star formation in molecular clouds is very inefficient (2–4%), but quite efficient in dense cores ($> 25\%$). Star formation is slow in terms of a free fall time, especially for the cloud as a whole, but somewhat faster in terms of a free fall time (and much faster in absolute terms) in dense cores.

## 3. Massive, Clustered Star Formation

With the exception of a few, relatively nearby regions of massive, clustered star formation, such as the Orion region (e.g., Hillenbrand 1997; Lada & Lada 2003; Allen et al. 2007), we do not have such detailed information on stellar IMF, ages, efficiencies, etc. Instead we accept cruder measures, but these provide a bridge to star formation in other galaxies. Theories are not well developed yet, but progress is being made, as discussed at the meeting by Bonnell, Dobbs, and Krumholz.

There have been a number of surveys for dense gas, but they have all been directed toward signposts of star formation, such as water masers (Plume, Jaffe, & Evans 1992) or *IRAS* sources (Sridharan et al. 2002; Beuther et al. 2002). An unbiased survey of 1 mm continuum emission now under way using Bolocam will provide a more complete census (Williams et al. 2007) and deeper surveys with SCUBA-2 and far-infrared surveys with *Herschel* will follow. We should soon have a much more complete picture of star formation sites in the Milky Way.

For now, I will focus on results from the survey toward water masers (Plume et al. 1992), which has been followed up by many other studies obtaining gas densities from multitransition CS observations (Plume et al. 1997), images of dust emission at 350 $\mu$m (Mueller et al. 2002), maps of CS $J = 5 \rightarrow 4$ emission (Shirley et al. 2003), and maps of HCN emission (Wu et al. 2005). While these references call the objects "dense cores", I will adopt the convention of McKee and co-workers (e.g., Krumholz & Tan 2007) and refer to them as dense *clumps*. The dense clumps are the sites of cluster formation and we reserve the term core to refer to the structure that forms one or a few stars.

The dense clumps can be fitted to power law density profiles with exponents very similar to those of dense cores forming low mass stars, but the density, measured at the same radius, is typically 100 times higher and the line widths are typically 16 times wider. Some show evidence in the line profiles for inward motions (Wu & Evans 2003), similar to those seen in some low mass cores. In general, they are scaled up versions of low mass dense cores. This point is important, as some authors use "clump" to mean something less dense than a core. Massive, dense clumps are *denser* than the low mass cores. The clumps within them are presumably even denser, but they have been hard to separate cleanly.

The mass function of the dense clumps is steeper that that of molecular clouds or of less dense clumps traced by CO isotopes (Shirley et al. 2003), but it is incomplete below about 1000 $M_{\odot}$. Above that level, it is similar to the distribution of the total masses of OB associations (Massey, Johnson, & Degioia-Eastwood 1995; McKee & Williams 1997). These dense clumps are very likely to be the sites where clusters and OB associations form. However, it has been difficult to study the forming stars themselves because of the heavy extinction and large distances. The GLIMPSE legacy project (Benjamin et al. 2003) using the IRAC instrument on *Spitzer* is revealing clusters in some of these, but the strong diffuse emission in the IRAC bands makes it difficult to extract detailed information on the stellar content (M. Nordhaus et al., in preparation).

Our observational measures are the luminosity in a molecular line or in dust continuum at long wavelengths, which are tracers of the mass of dense gas, and the bolometric luminosity measured by integrating over the full spectrum of dust emission. The line luminosities of tracers of dense gas, in particular CS and HCN, trace very well the virial mass of dense gas (e.g., J. Wu et al., in preparation). The integrated infrared luminosity ($L_{\rm IR}$) traces the star formation rate ($\dot{M}_{\star}$) given enough time for the IMF to be reasonably sampled (Krumholz & Tan 2007). Both these measures are subject to fluctuations about mean values of at least a factor of three. Despite the variations, $L_{\rm IR}$ correlates well with $L_{\rm mol}$ or $M_{\rm vir}$.

Without a detailed census of the stellar content, we cannot compute the star formation efficiency in the same sense ($\epsilon_\star$) as we could for low mass star formation, but the similarity of the mass function to that of OB associations suggests reasonably high efficiency in the dense gas.

We will use an efficiency measure common in the extragalactic context as the star formation rate per unit mass of gas ("$\epsilon$"$= \dot{M}_\star/M(\mathrm{gas})$). Note that this "efficiency" is really the speed ($1/t_{\rm dep}$) unnormalized to the free fall time. In observational terms, this is measured by $L_{\rm IR}/L_{\rm mol}$. As for low mass star formation, the "$\epsilon$" is very low for molecular clouds as a whole, but much higher for dense gas (e.g., a factor of 30 higher in one study, Mueller et al. 2002). Krumholz & Tan (2007) have argued that star formation is also slow relative to a free fall time ($\mathrm{SFR_{ff}}$), though again there is some evidence that $\mathrm{SFR_{ff}}$ is higher in the densest gas, probed for example by the CS $J = 5 \to 4$ transition.

These massive dense cores provide the connection point between detailed studies of star formation in the Milky Way and extragalactic star formation.

## 4. Star Formation in Galaxies

Since we seek a connection between studies of star formation in our Galaxy and extragalactic star formation, we will focus on studies using common tracers. In dusty galaxies, the star formation rate is traced by $L_{\rm IR}$. This is basically a calorimetric method: the dust absorbs all the energy from young stars and re-radiates in the infrared. The advantage is that there is no need for uncertain extinction corrections, which are needed for ultraviolet or H$\alpha$ measures of star formation. However, one is assuming that *all* the light is indeed absorbed, so some star formation can be missed (this problem can apparently be alleviated by combining infrared and H$\alpha$ observations, as described by Calzetti, this volume, p. 121). Also, a calorimeter measures any heat input, so heating by older stars or an AGN can confuse matters. In practice, this method is applied only to systems where star formation overwhelms the heating by older stars. Assessing the contribution of AGN can be trickier, especially for high-$z$ systems. A common calibration, based on a continuous burst model, is that

$$\dot{M}_\star(M_\odot\,\mathrm{yr}^{-1}) = 1.7 \times 10^{-10} L_{\rm IR}(L_\odot)$$

(Kennicutt 1998), with an averaging time of 10 to 100 Myr. This number may vary by a factor of at least two depending on the star formation rate and age of a starburst (see Krumholz & Tan 2007).

How does $L_{\rm IR}$ correlate with various tracers of gas? If CO is used to trace the molecular gas, the relation is strongly non-linear. For CO, the "$\epsilon$" increases by a factor of 100 as $L_{\rm IR}$ rises from $10^{10}$ to $10^{14}$ $L_\odot$ (Solomon & Vanden Bout 2005). Even higher "$\epsilon$" may be observed in high-$z$ submillimeter galaxies (Greve et al. 2005).

If the $L_{\rm IR}$ versus $L_{\rm mol}$ plot is made using HCN $J = 1 \to 0$, instead of CO, the correlation is better, the scatter is less, and the relation is *linear* (Gao & Solomon 2004a,b). In this case, the "$\epsilon$" is constant over more than three orders of magnitude in $L_{\rm mol}$. The HCN line is sensitive to relatively dense gas, which

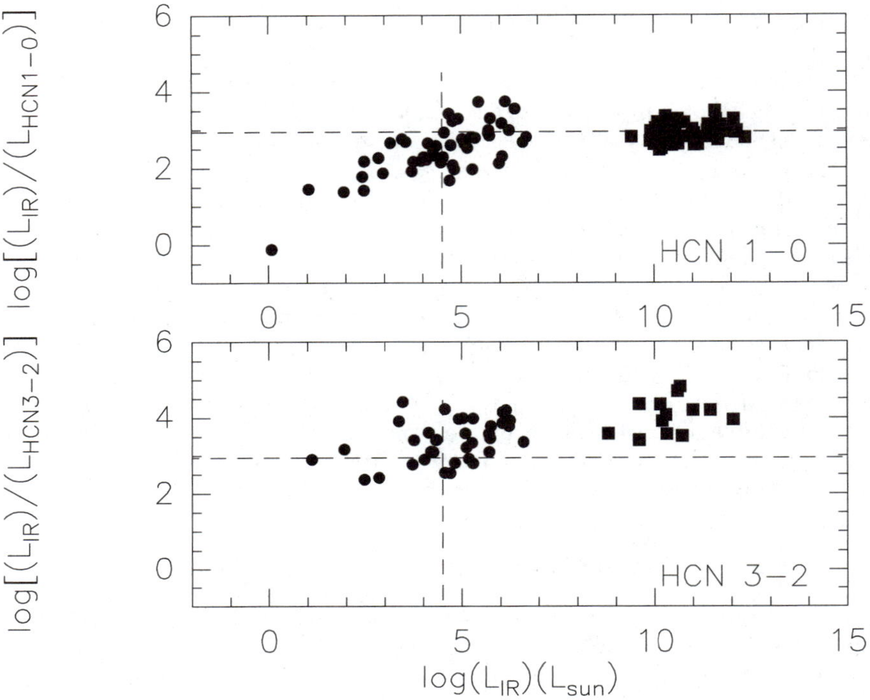


Figure 1. Correlations between the distance-independent ratio $L_{\rm IR}/L_{\rm mol}$ vs $L_{\rm IR}$ for HCN $J = 1 \rightarrow 0$ and HCN $J = 3 \rightarrow 2$. The squares are HCN $J = 1 \rightarrow 0$ observations of galaxies (Gao & Solomon 2004a) in the first panel, and HCN 3→2 observations of galaxies (Paglione, Jackson, & Ishizuki 1997) in the second panel. The circles are the equivalent data for massive dense clumps in our Galaxy. The horizontal dashed line in the top plot indicates the average $L_{\rm IR}/L_{\rm HCN}$ ratio for galaxies; the vertical dashed line in the top plot shows the cutoff at $L_{\rm IR} = 10^{4.5}\,L_\odot$. These two lines are also shown in the bottom plot to indicate the relative shifts of $L_{\rm IR}/L_{\rm mol}$ between HCN $J = 1 \rightarrow 0$ and HCN $J = 3 \rightarrow 2$.

reminds us that it was only the dense molecular gas that is involved in star formation in well-studied local molecular clouds.

In order to compare directly the situation in local clouds to that in starburst galaxies, Wu et al. (2005) surveyed the HCN $J = 1 \rightarrow 0$ line in Galactic clumps. The results showed that, above a threshold in $L_{\rm mol}$ or $L_{\rm IR}$, the Galactic clumps fit a linear relation between $L_{\rm IR}$ and $L_{\rm mol}$, essentially identical to that of the starburst galaxies. Above the threshold in $L_{\rm IR}$ of about $10^{4.5}\,L_\odot$, the "efficiency", $L_{\rm IR}/L_{\rm mol}$, is very similar in the clumps to that in starbursts (Fig. 1). Below the threshold, $L_{\rm IR}/L_{\rm mol}$ drops rapidly with decreasing $L_{\rm IR}$. While data on the HCN $J = 3 \rightarrow 2$ line are scarce for galaxies, the currently available data tell a similar story, but with a higher ratio of $L_{\rm IR}/L_{\rm mol}$ (Fig. 1). A higher ratio is not surprising because the HCN $J = 3 \rightarrow 2$ line traces still denser gas.

HCN is not the only dense gas tracer. Many other tracers, including CS and $HCO^+$ have been used in studies of dense gas in our Galaxy. Graciá-Carpio et al. (2006) recently suggested that $HCO^+$ could be a better tracer in galaxies, but Papadopoulos (2007) came to the opposite conclusion and noted the need for more observations of higher $J$ transitions to constrain the conditions in the dense gas. A recent multitransition study of CS, HCN, and $HCO^+$ in two local ULIRGs (Greve et al. 2006) found many similarities to the properties of dense clumps in the Galaxy. They concluded that, for the transitions they studied, $HCO^+$, HCN, and CS probe progressively denser gas.

## 5. An Emerging Picture

The comparisons in the last section suggest that local studies of massive star formation can indeed provide insights into starbursts. Qualitatively, a modest starburst can happen if a galaxy has a lot of molecular gas, with some fraction of it being dense. For an extreme starburst, a large fraction of the molecular gas must be dense. At the extreme, the entire ISM might look like the dense, cluster-forming clumps in our Galaxy.

Quantitatively, we offer some "Schmidt laws" (Wu et al. 2005):

$$\dot{M}_\star(M_\odot\,\mathrm{yr}^{-1}) = 1.4 \times 10^{-7} L_{\mathrm{HCN}}\ (\mathrm{K\,km\,s^{-1}\,pc^{-2}})$$

$$\dot{M}_\star(M_\odot\,\mathrm{yr}^{-1}) = 1.2 \times 10^{-8} M(\mathrm{dense})\ (M_\odot)$$

where $L_{\mathrm{HCN}}$ is the monochromatic line luminosity of the HCN $J = 1 \rightarrow 0$ line in observer's units, and we have assumed that $\dot{M}_\star = 2.0 \times 10^{-10} L_{\mathrm{IR}}$ and $M(\mathrm{dense})(M_\odot) = 7\,L_{\mathrm{HCN}}\ (\mathrm{K\,km\,s^{-1}\,pc^{-2}})$ (Wu et al. 2005). While these are not yet of practical use, CARMA and ALMA will be supplying many more observational data on the lines in the coming years. In the meantime, we should try to understand the relationship between these *linear* relations for dense gas and the non-linear relations for less dense gas.

First, we should ask why $L_{\mathrm{IR}}/L_{\mathrm{HCN}}$ is constant from $L_{\mathrm{IR}} = 10^{4.5}$ to $10^{13}\,L_\odot$, but drops sharply for lower $L_{\mathrm{IR}}$. At first glance, the constancy is quite puzzling for Galactic clumps even if the star formation rate is linear with the amount of dense gas because one would expect the mass of the most massive star to increase with the number of stars formed and the stellar luminosity is a non-linear function of the stellar mass. Thus, one might expect $L_{\mathrm{IR}}$ to increase non-linearly with $M(\mathrm{dense})$. In fact, it does exactly that *below* the threshold of $L_{\mathrm{IR}} \sim 10^{4.5}\,L_\odot$. This is a clue. Wu et al. (2005) proposed that there is a "basic unit" of massive clustered star formation. For clumps below the mass of the basic unit, the IMF is not fully sampled and $L_{\mathrm{IR}}$ increases non-linearly with $M(\mathrm{dense})$. Once the mass exceeds the threshold, the IMF is fairly well sampled. Further increases in mass produce more stars but no further change in $L_{\mathrm{IR}}/M(\mathrm{dense})$. In this picture, the difference between star formation in Galactic molecular clouds, normal galaxies, starbursts, and ULIRGs is simply how many such basic units (or how much dense gas) they contain. To be concrete, a $L_{\mathrm{IR}}$ of $10^5 L_\odot$ corresponds to a stellar mass of about 30 to $40\,L_\odot$. The basic unit contains 300 to $1000\,M_\odot$ of dense ($n > 10^5\ \mathrm{cm}^{-3}$) gas. This picture also explains why the scatter in $L_{IR}/L_{HCN}$ is greater in the Galactic clumps than in galaxies: the most massive star formed will be subject to stochastic fluctuations, which can produce order of magnitude fluctuations in the ratio. These tend to average out for a whole galaxy.

Second, we can ask why $L_{IR}/L_{CO}$ increases with $L_{IR}$. While CO traces the overall mass in Galactic clouds, it fails completely to trace the mass in dense clumps and cores. This failure is not surprising as CO is optically thick and thermalized easily. CO does not trace the gas that is relevant to star formation. Roughly speaking $L_{HCN}/L_{CO}$ provides an estimate of the fraction of dense gas, as long as that fraction is not too large. Indeed, this ratio increases with $L_{IR}$, as expected in this picture (Gao & Solomon 2004a).

## 6. Future Directions

Further studies are needed in both the Galactic and extragalactic context. Systematic, unbiased surveys of the Galactic plane for dense clumps will remove the biases in the samples used so far and provide a much larger data set. Already the Bolocam Galactic Plane Survey has found thousands of clumps (Williams et al. 2007) and deeper surveys are on the way. With large samples in different environments, we can begin to understand the dependence of the relations on chemistry, metallicity, environment, etc. Extragalactic studies should test the relations at lower $L_{IR}$, aided by the tremendous sensitivity of ALMA. Further studies of the very luminous end will also be important, and higher $J$ lines of dense gas tracers can be studied in many more sources.

We also need to understand the non-linear Kennicutt relations and their relationship to the linear relations for dense gas. Theories need to explain *both* relations. It is important to realize that the gas surface density, which appears in the Kennicutt relations, is two steps removed from the actual star-forming entities: dense clumps and cores. We need to understand how the large-scale surface density controls the formation of molecular clouds and what controls the formation of dense clumps and cores within those clouds. The first step may be best studied with high resolution observations of other galaxies and detailed comparison to simulations with sufficient resolution to separate individual molecular clouds. The second step, from molecular clouds to dense clumps and cores, will be hard to study in other galaxies because of resolution issues, though insights may be available from studies of other galaxies to see how $L_{HCN}/L_{CO}$, for example, depends on conditions. For the most part, though, we need to study the formation of dense cores from molecular clouds with more unbiased and complete studies of molecular clouds in our Galaxy.

I will close by thanking the organizers for providing a forum that allowed some of us touching different parts of the elephant to struggle at least toward a common language.

**Acknowledgments.** I am grateful to Jingwen Wu, whose work underlies the latter part of this paper. Many of my c2d colleagues have provided information in advance of publication, most notably Melissa Enoch. Leo Blitz provided guidance on local surface densities of gas. This work has been supported by NSF Grants AST-0307250 and AST-0607793. Additional support came from NASA Origins grants NNG04GG24G and NNX07AJ72G. The c2d project was part of the *Spitzer* Legacy Science Program, with support provided by NASA through contracts 1224608, 1230779, and 1230782 issued by the Jet Propulsion Laboratory, California Institute of Technology, under NASA contract 1407.

**References**

Allen, L., et al. 2007, in Protostars and Planets V, ed. B. Reipurth, D. Jewitt, & K. Keil (Tucson: Univ. Arizona Press), 361
Alves, J., Lombardi, M., & Lada, C. J. 2007, A&A, 462, L17
Benjamin, R. A., et al. 2003, PASP, 115, 953
Beuther, H., Schilke, P., Menten, K. M., Motte, F., Sridharan, T. K., & Wyrowski, F. 2002, ApJ, 566, 945
Blitz, L., & Rosolowsky, E. 2006, ApJ, 650, 933
Bonnell, I. A., & Bate, M. R. 2006, MNRAS, 370, 488
Clark, P. C., Klessen, R. S., & Bonnell, I. A. 2007, MNRAS, 379, 57
Clemens, D. P., Sanders, D. B., & Scoville, N. Z. 1988, ApJ, 327, 139
Dame, T. M. 1993, in AIP Conf. Proc., vol. 278, Back to the Galaxy (New York: AIP), 267
Elmegreen, B. G., & Scalo, J. 2004, ARA&A, 42, 211
Enoch, M. L., et al. 2006, ApJ, 638, 293
Enoch, M. L., Glenn, J., Evans, N. J., II, Sargent, A. I., Young, K. E., & Huard, T. L. 2007, ApJ, 666, 982
Evans, N. J., II, et al. 2003, PASP, 115, 965
Gao, Y., & Solomon, P. M. 2004a, ApJ, 606, 271
Gao, Y., & Solomon, P. M. 2004b, ApJS, 152, 63
Graciá-Carpio, J., García-Burillo, S., Planesas, P., & Colina, L. 2006, ApJ, 640, L135
Greve, T. R., et al. 2005, MNRAS, 359, 1165.
Greve, T. R., Papadopoulos, P. P., Gao, Y., & Radford, S. J. E. 2006, ArXiv Astrophysics e-prints, arXiv:astro-ph/0610378
Harvey, P. M., et al. 2007, ApJ, 663, 1139
Hillenbrand, L. A. 1997, AJ, 113, 1733
Hirota, T., et al. 2007, PASJ, in press (ArXiv e-prints, 705, arXiv:0705.3792)
Kennicutt, R. C., Jr. 1998, ARA&A, 36, 189
Krumholz, M. R., & McKee, C. F. 2005, ApJ, 630, 250
Krumholz, M. R., McKee, C. F., & Klein, R. I. 2005, Nat, 438, 332
Krumholz, M. R., & Tan, J. C. 2007, ApJ, 654, 304
Lada, C. J. & Lada, E. A. 2003, ARA&A, 41, 57
Levine, E. S., Blitz, L., & Heiles, C. 2006, ApJ, 643, 881
Martel, H., Evans, N. J., II, & Shapiro, P. R. 2006, ApJS, 163, 122
Massey, P., Johnson, K. E., & Degioia-Eastwood, K. 1995, ApJ, 454, 151
McKee, C. F., & Williams, J. P. 1997, ApJ, 476, 144
Mueller, K. E., Shirley, Y. L., Evans, N. J., II, & Jacobson, H. R. 2002, ApJS, 143, 469
Paglione, T. A. D., Jackson, J. M., & Ishizuki, S. 1997, ApJ, 484, 656
Papadopoulos, P. P. 2007, ApJ, 656, 792
Plume, R., Jaffe, D. T., & Evans, N. J., II, 1992, ApJS, 78, 505
Plume, R., Jaffe, D. T., Evans, N. J., II, Martin-Pintado, J., & Gomez-Gonzalez, J. 1997, ApJ, 476, 730
Reipurth, B., Jewitt, D., & Keil, K. (eds) 2007, Protostars and Planets V (Tucson: Univ. Arizona Press)
Ridge, N. A., et al. 2006, AJ, 131, 2921
Schmidt, M. 1959, ApJ, 129, 243
Shirley, Y. L., Evans, N. J., II, Young, K. E., Knez, C., & Jaffe, D. T. 2003, ApJS, 149, 375
Shu, F. H., Adams, F. C., & Lizano, S. 1987, ARA&A, 25, 23

Shu, F. H., Allen, R. J., Lizano, S., & Galli, D. 2007, ApJ, 662, L75
Solomon., P. M. & Vanden Bout, P. A. 2005, ARA&A, 43, 677
Sridharan, T. K., Beuther, H., Schilke, P., Menten, K. M., & Wyrowski, F. 2002, ApJ, 566, 931
Williams, J. P., et al. 2007, BAAS, 210, #12.11
Wu, J., Evans, N. J., II, Gao, Y., Solomon, P. M., Shirley, Y. L., & Vanden Bout, P. A. 2005, ApJ, 635, L173
Wu, J., & Evans, N. J., II, 2003, ApJ, 592, 79
Young, K. E., et al. 2006, ApJ, 644, 326
Zuckerman, B., & Evans, N. J., II 1974, ApJ, 192, L149

## Discussion

*Kennicutt*: You showed that HCN and CO measurements imply that the fraction of molecular mass contained in dense cores increased from a few percent in Galactic clouds to nearly 100% in extreme starbursts. Yet we apply the same $CO/H_2$ conversion factor to estimate molecular masses across this entire mass spectrum. Do you think this is valid, and if so why does this work?

*Evans*: The use of CO to measure mass works at some level, but it should not be taken too literally. We know from studies in our Galaxy that it fails to trace mass in dense cold regions. If the molecular medium becomes denser in a galaxy, it would be natural to expect that CO would underestimate the total molecular mass. There is some recent theoretical work that shows a fairly large range of conversion factors (Bell, Viti, & Williams 2007, MNRAS, 378, 983).

*Sheth*: Comment: I wanted to mention that CARMA is now operational and will be excellent at HCN or 1 mm observations of nearby galaxies with superb image fidelity and mapping speed. Question: is there anything more you can say about the distribution of the dense cores—they are not uniformly distributed in a cloud but if we can say anything about their distances perhaps we can get an idea of the conversion from CO to dense cores.

*Evans*: Thanks for reminding us that CARMA is operating. It will provide crucial data in the northern hemisphere. We have quantified the non-uniform distribution using the correlation function (Enoch et al. 2006, ApJ, 638, 293). Morphologically, the dense gas is strongly clustered and lies along filaments visible in extinction or in tracers of lower density gas. But we lack a full understanding of how the dense cores separate out from the turbulent flows.

*Tsamis*: How do you characterize infall in massive SF regions? Do we see a similar picture in nearby low mass clouds? What form do these cores have: Bonner–Ebert spheres, or some other form? Are the diagnostics the same—i.e., "blue asymmetric" CS, $HCO^+$, etc.?

*Evans*: We do see "blue asymmetric" profiles in some cases (Wu et al. 2005, ApJ, 635, L173), and these can be modeled with a scaled up version (higher density, faster infall) of the Shu model. This modeling is by no means simple, so work is ongoing. As to the pre-star formation state, low mass cores are reasonably represented by Bonner–Ebert spheres, but we have less information about the massive regions. However we know they are dominated by turbulence, so they could not be *thermal* Bonner–Ebert spheres.

*Pathways through an Eclectic Universe*
*ASP Conference Series, Vol. 390, © 2008*
*J. H. Knapen, T. J. Mahoney, and A. Vazdekis, eds.*

# Infrared Bubbles: The Environments of Massive Stars

Ed Churchwell

*Department of Astronomy, University of Wisconsin at Madison, 475 N. Charter Street, Madison, WI 53706, USA*

**Abstract.** High spatial resolution radio continuum and mid-infrared images of a prototypical stellar wind-blown bubble have been used to infer general properties of this class of object. In particular, I explore the spatial relationships between warm thermally emitting dust, the polycyclic aromatic hydrocarbons-dominated photo-dissociation region, and ionized hydrogen gas, and use these to compare with bubble evolution models. This bubble appears to be $<10^6$ years old. Candidate YSOs and the ionizing star are identified using a grid of spectral energy distribution models for YSOs and main sequence and giant stars. The location of the candidate YSOs on the rim of the H II region and the association with an infrared dark cloud suggest that several of the YSO candidates have been triggered by the expansion of the H II region.

## 1. Introduction

Many studies have taught us that H II regions are essentially fully ionized regions filled with gas of about $10^4$ K surrounding O and B stars. The size of an H II region is determined by the UV photon flux from the ionizing star, the age of the star, and the density of the ambient interstellar medium (ISM). This "broad brush" picture appears to be correct for ionizing stars that have weak winds, such as B dwarfs. However, O stars, early B supergiants, and WR stars have strong winds (Vink, de Koter & Lamers 2001 and references therein) and H II regions ionized by these stars are expected to have very much more complicated internal structures than those ionized by stars with insignificant winds. Strong stellar winds fundamentally alter the velocity, density, temperature, and ionization structures of H II regions; all four parameters are expected to change by orders of magnitude (at a given age) over small radial distances at several theoretically predicted boundaries within wind-blown bubbles. Analytic evolution models by Castor, McCray, & Weaver (1975), Weaver et al. (1977), Whitworth et al. (1994), and Capriotti & Kozminsky (2001) predict that wind-blown bubbles around O stars and B supergiants should have the following distinct regions: most of the a hypersonic stellar wind-evacuated central cavity surrounding the central star where densities are very low; a hot ($\sim10^7$ K), low-density, shocked stellar wind region surrounding the central evacuated cavity that occupies most of the volume of the H II region; a thin conduction zone where temperature sharply decreases and density rapidly increases (with radius); a thin, dense shell of shocked $H^+$ gas surrounded by a very thin shell of non-shocked $H^+$ gas at about $10^4$ K. The outer thin shell of $10^4$ K $H^+$ gas is the region classically thought to fill the whole volume of the H II region. Recent numerical models by Freyer, Hensler, & Yorke (2003, 2006), Kröger, Hensler, & Freyer (2006), and Kröger et al. (2007) of H II regions produced by a star or stars with strong winds basically confirm the predictions of the analytic models but show important differences that could not be included in analytic models. The numerical models show that a thick turbulent mixing layer forms relatively quickly and persists for most of the lifetime of the nebula at the boundary between the $10^7$ K gas and the $10^4$ K gas. The implications of this is that one should find knots of dense and cool gas mixed with the lower density, hot gas throughout the mixing layer. The thickness of the $10^4$ K shell is a strong function of the stellar luminosity; it is thicker for bubbles blown by stars with lower wind luminosities. The predictions of both the analytic and numerical models have been only partially tested observationally toward a couple of bubbles driven by WR stars (see Freyer et al. 2006 and references therein).

Mid-infrared (MIR) surveys of the Galaxy beginning with *MSX* (*Midcourse Space Experiment*; Price et al. 2001), and *ISO* (*Infrared Space Observatory*; Kessler et al. 1996) detected some of the largest and brightest bubbles. The Galactic Legacy Infrared Mid-Plane Survey Extraordinaire (GLIMPSE; Benjamin et al. 2003) using the *Spitzer Space Telescope* (Werner et al. 2004) Infrared Array Camera (IRAC; Fazio et al. 2004), which has about ten times the spatial resolution and about one hundred times the sensitivity of *MSX*, has detected many more bubbles than any other survey to the present time. Churchwell et al. (2006, 2007) have catalogued almost 600 bubbles in an incomplete visual examination of the GLIMPSE images (an area of 274 square degrees in the longitude range $|l| \leq 65°$). Churchwell et al. (2006, 2007) argued that almost all the MIR bubbles in their sample are produced by the impact of winds and radiation from O and B stars on ambient interstellar gas and dust based on the similarity in distribution of O and B stars and bubbles and the association with radio H II regions and stellar clusters. About 2/3 of the bubbles appear to be produced by late B stars that do not produce a detectable radio H II region and do not have winds strong enough to produce the hot, shocked gas filled H II regions expected around O stars.

Here, I will focus on high spatial resolution observations from NIR to radio wavelengths of a single Galactic MIR bubble that illustrates general features seen in MIR bubbles that enclose radio H II regions ionized by stars with strong winds.

## 2. The Data

In the following, I will present GLIMPSE images from *Spitzer*/IRAC bands (3.6, 4.5, 5.8, and 8.0 $\mu$m), a *Spitzer*/MIPS (24 $\mu$m) image from the MIPSGAL survey (Carey et al. 2005), and a 20 cm radio continuum image from the MAGPIS survey (Helfand et al. 2006) of a prototypical wind-blown MIR bubble. I will focus on the N49 bubble from the Churchwell et al. (2006) catalog, which is an almost spherically symmetric, isolated bubble that has several young stellar objects (YSOs) located in projection on the periphery of this bubble. The relative distributions of infrared and radio continuum emission from N49 are typical of most high-excitation bubbles in the Churchwell et al. (2006, 2007) catalogs, and N49 is therefore a good object to explore general emission properties of bubbles excited by luminous stars with strong winds. N49 has a kinematic distance of $5.7 \pm 0.6$ kpc. At 1.4 GHz, it has an integrated flux density of 2.8 Jy and an angular diameter (FWHM) of $3'$ ($\sim$5 pc) (Helfand et al. 2006). The ionizing UV photon flux required to maintain ionization of the H II region is $\sim 7.8 \times 10^{48}$ photons $s^{-1}$, equivalent to a single O6.5 V star (Martins, Schaerer, & Hillier 2005). Results reported here are based on work by C. Watson et al. (in preparation).

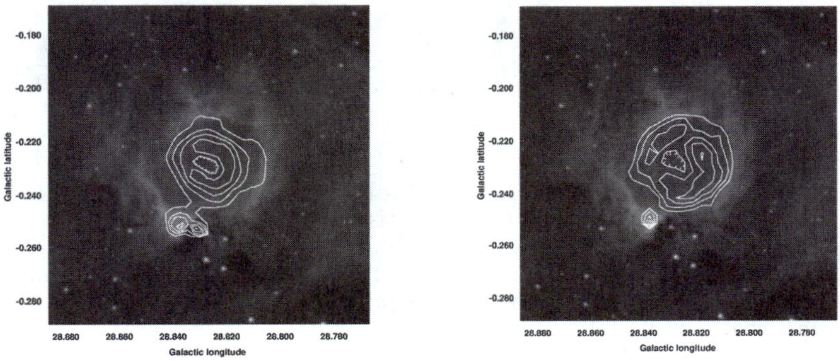


Figure 1. *Left:* 8 $\mu$m grayscale image of N49 with 24 $\mu$m contours superimposed. *Right:* 20 cm continuum contours superimposed on the 8 $\mu$m image. Note the bright shell structure at both 20 cm and 24 $\mu$m and the bright ultra-compact H II region seen at both 20 cm and 24 $\mu$m at $(l, b) = (28.84, -0.25)$. No radio continuum counterpart is seen toward the two YSO candidates, seen as lower longitude extensions to the ultra-compact H II region at 24 $\mu$m.

## 3. Emission Properties of Bubble N49

The left panel of Fig. 1 shows an image of N49 at 8.0 $\mu$m (grayscale) that primarily shows polycyclic aromatic hydrocarbon (PAH) emission and a few stars. Superimposed on the grayscale image are contours showing 24 $\mu$m emission that delineate warm thermal dust emission (the brightness is related to both the amount, for an optically thin source, and temperature of the dust along the line of sight). The right panel of Fig. 1 shows the same grayscale image but with 20 cm continuum contours (showing the distribution of ionized hydrogen) superimposed.

Figure 1 illustrates several properties that seem to be common among MIR bubbles. These are: 1) the 24 $\mu$m and radio continuum have about the same distribution; 2) both the radio continuum and 24 $\mu$m emission are generally confined within a shell of 8.0 $\mu$m emission that has a bright and sharp inner face and tapers off with increasing radius from the center of the bubble; and 3) in order of declining brightness, the bubble is brightest by far at 24 $\mu$m, distantly followed by 8 $\mu$m, which is distantly followed by radio continuum.

These emission properties reveal much about what is happening in N49 and, by extension, in other bubbles ionized by a hot O star with a strong stellar wind. First, it is clear that dust is mixed with the ionized hydrogen since the distributions of both 20 cm and 24 $\mu$m emission are essentially the same.

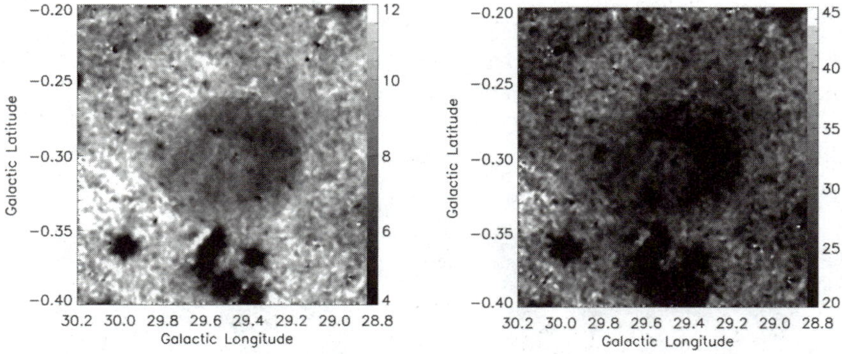


Figure 2. *Left:* Ratio of the 5.8 to 4.5 $\mu$m images of N49. The darkened region (low ratios) at the center indicates the area where PAHs have been destroyed, and the lighter region surrounding the dark central region is the PAH-dominated PDR of the bubble. *Right:* The ratio of the 8.0 to 4.5 $\mu$m images of N49 shows the same features as in the left panel. The black spots are stars that have no PAH emission features.

Second, PAHs are seen in a shell outside the H II region, implying that PAHs are destroyed by direct stellar UV radiation (see the next section). Third, the 8 $\mu$m shell dominated by PAH emission appears to define the photo-dissociation region (PDR) of the bubble. It declines in brightness with increasing radius as expected for geometrical dilution plus extinction of the radiation field of soft UV radiation that can pass through the hydrogen ionization front. Fourth, there is a small evacuated cavity immediately surrounding the central star (apparent in Fig. 1). This seems to be the wind-evacuated cavity predicted by both analytic and numerical models.

## 4. PAH Destruction Radius and Thickness of the PDR Shell

Let us now estimate the PAH destruction radius and the extent of the PDR shell around the H II region. To do this we will use the fact that the *Spitzer*/IRAC 4.5 $\mu$m band does not include any PAH feature, but both the 5.8 and 8.0 $\mu$m bands do. As a consequence, ratios of [5.8]/[4.5] and [8.0]/[4.5] should show the onset of the PAH emission with radius as a jump in these ratios. This technique was used by Povich et al. (2007) in M17 and C. Watson et al. (in preparation) for three bubbles, including N49. In Fig. 2, these ratios are shown in grayscale, and it is quite obvious where the PAH emission sets in with radius. However, in Fig. 1, some 8.0 $\mu$m emission is apparent inside the supposed PAH destruction radius. Does this mean that PAHs are not completely destroyed inside the "destruction" radius? Or possibly we are seeing either thermal emission at 8 $\mu$m from the hot dust that produces the 24 $\mu$m emission or PAH emission from the front and back faces of the 8 $\mu$m shell. To investigate which of these might be true, we assumed that the 8 $\mu$m emission is confined to a spherical shell of a defined thickness and that the 8 $\mu$m emission is proportional to the line-of-sight thickness of the shell (front and back face) as a function of radius. The calculated

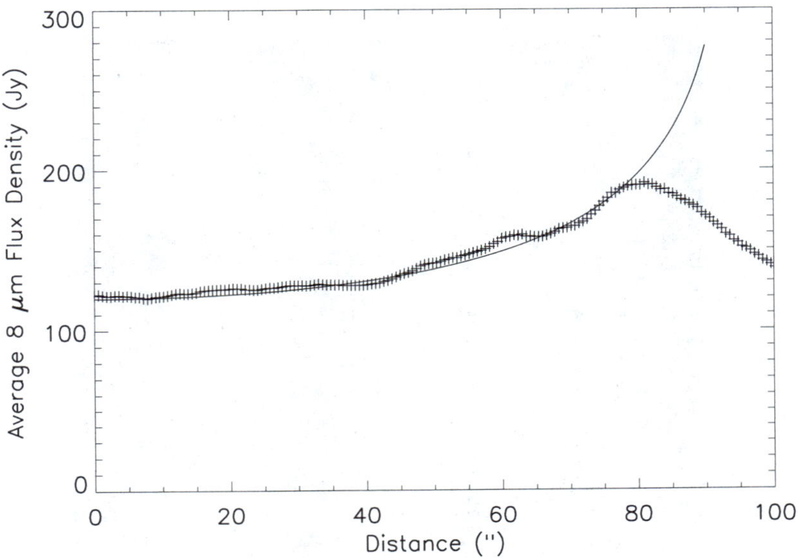


Figure 3. The expected 8 $\mu$m brightness distribution with radius normalized to the observed brightness at the center of a spherical shell having a thickness of ~30% that of the bubble radius is shown as a solid curve. It is assumed that the brightness is simply proportional to the line of sight through the front and back face of the shell. The data are indicated by + symbols.

brightness at 8 $\mu$m with radius, under these simple assumptions, normalized to the brightness at the center of the bubble is shown in Figure 3. The observed brightness is over-plotted in Fig. 3 as well. Figure 3 shows that the 8 $\mu$m emission is accounted for by the front and back faces of the shell. A very small fraction of the 8 $\mu$m emission may be due to thermal dust, but it is so small that we cannot accurately measure it. We can only say that the dust temperature must be less than ~150 K. The turn-down of the observations at a radius of ~80$''$ in Fig. 3 shows the decline in brightness due to attenuation and geometrical dilution of the soft UV radiation field with distance from the center of the bubble.

## 5. YSOs and Ionizing Stars

To identify YSO candidates, GLIMPSE Point Source Archive sources combined with 24 $\mu$m photometry of MIPSGAL images were fitted with spectral energy distributions (SEDs) from a large, pre-computed grid of YSO models (Robitaille et al. 2006). The grid consists of about 200 000 2D Monte Carlo radiation transfer models (Whitney et al. 2003a, b; Whitney et al. 2004) spanning the range of stellar mass and evolutionary stage as seen at ten viewing angles. The fitting algorithm includes a grid of Kurucz (1993) stellar model atmospheres, which permits a robust distinction between YSOs and extincted photospheres of main

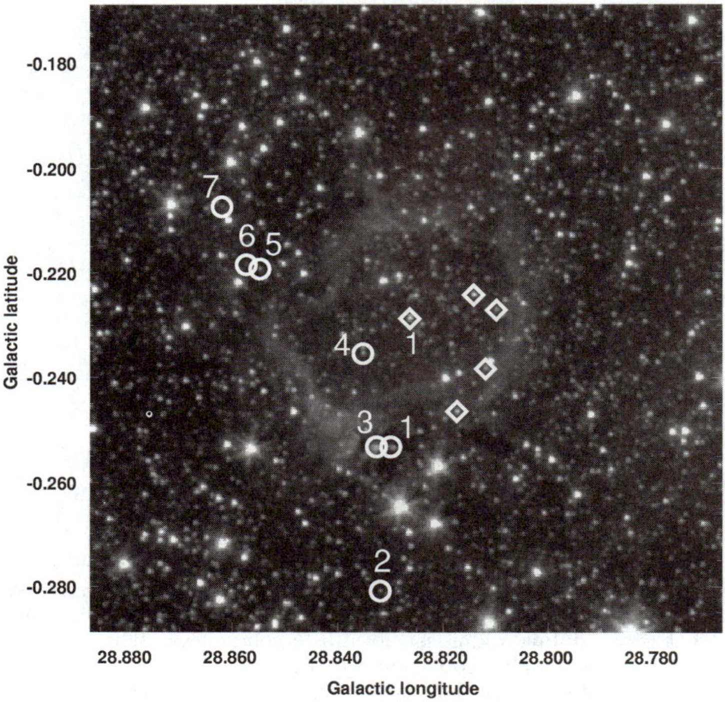


Figure 4. IRAC 3.6 $\mu$m image of N49 with candidate YSOs indicated by circles and candidate ionizing stars indicated by diamonds. The star labeled by diamond 1 is likely the sole ionizing star for this bubble. The best fits to its SED suggest an O8 III star.

sequence and giant stars because YSOs have mid-IR excesses from circumstellar dust.

722 GLIMPSE archive sources were analyzed within a radius of 3.6′ (6 pc) from the center of N49. Seven sources were fitted with high confidence as YSOs; they range in mass from $\sim$29 $M_\odot$ to $\sim$0.5 $M_\odot$ and are indicated by labeled circles in Figure 4. The two most prominent YSOs are numbers 1 and 3. Both lie on the rim of N49 toward an infrared dark cloud (IRDC). The extinction toward both YSOs is small ($A_V \lesssim 4.5$ mag), indicating that they are not deeply embedded in the IRDC. YSO #3 has a spectacular extended 4.5 $\mu$m excess that has been attributed to a shocked outflow. Both sources are bright at 24 $\mu$m. Due to bright MIR diffuse background emission and confusion, other potential YSOs may have been hidden from GLIMPSE. Both the location on the rim of N49 and the association with the IRDC suggest that YSOs 1 and 3 may be the most massive members of a cluster that has been triggered by the expansion of N49. From Fig. 4, we see that all candidate YSOs except 2 and 7 are located close to the rim of N49. Five candidate ionizing stars have been identified

(diamonds in Fig. 5). The star at the center (#1) has an SED consistent with an O8 III star and is likely the one primarily responsible for ionizing the H II region. The others are offset from the center and could be late O or early B dwarfs or foreground cooler stars. The O8 III star lies behind ~10 mag of visual extinction and supplies all of the UV photon flux necessary to ionize the nebula. The 24 $\mu$m emission inside the bubble is probably the result of heating primarily by trapped Lyman $\alpha$ photons, although some stellar ionizing photons must also be absorbed by dust within the H II region. Spectral classification is probably only accurate to plus or minus one subclass, so at this point the 24 $\mu$m emission does not necessarily require an additional ionizing star.

Since several well-developed YSOs lie on the rim of N49 that appear to have been triggered by the expansion of the bubble, I infer that the dynamical age of the bubble has to be at least a few times $10^5$ yr. The analytic relation between radius $R(t)$, wind luminosity $L_{\rm W}$, age $t$, and ambient density $n_0$ is $R(t) \propto n_0^{-0.2} L_{\rm W}^{0.2} t^{0.6}$ (Weaver et al. 1977). Given $R$ and $L_{\rm W}$, one can estimate the dynamical age in terms of the ambient density. I used stellar properties from Vink et al. (2001) to estimate $L_{\rm W}$ for an O8 III star. I find that for a bubble of radius 2.5 pc, the ambient ISM density has to be at least $10^4\,{\rm cm}^{-3}$ for an age of $3.8 \times 10^5$ yr; a density of $10^5\,{\rm cm}^{-3}$ would require an age of $8.3 \times 10^5$ yr. The density would have to be even higher for an older age. This quickly places an upper limit on the age of the bubble because the ambient density becomes unrealistically high as age increases. From this, we conclude that the bubble is probably $<10^6$ years old. It turns out that the numerical models of Freyer et al. (2006) for a 35 $M_\odot$ star, when scaled to an ambient density of $10^4\,{\rm cm}^{-3}$, are quite consistent with the measured radius at an age of $\gtrsim 4 \times 10^5$ yr. It appears that the analytic and numerical models of bubble evolution are in reasonable agreement.

## 6. Summary

I have used the N49 bubble as a prototype to explore typical properties of stellar wind-blown bubbles. MIR and radio continuum images have been used to place constraints on the spatial relationships between warm thermally emitting dust, the PAH dominated PDR region, and ionized hydrogen gas. Stars in the GLIMPSE Archive within a radius of 3.6′ (well beyond the PDR shell) have also been examined using the model grid and fitting tool of Robitaille et al. (2006, 2007) to identify candidate YSOs and possible ionizing star or stars. The main conclusions from this study are as follows:

- The distributions of 24 $\mu$m and 20 cm emission are almost identical, indicating that warm dust exists within the H II region. An unsolved problem is why the dust is not swept out of the H II region by the stellar wind very soon after the onset of the wind.
- Both 20 cm and 24 $\mu$m emission show a central cavity centered on the O8 III star that is believed to be its wind-evacuated cavity.
- Band-ratios of [8.0]/[4.5] and [5.8]/[4.5] delineate the PDR region that encloses the H II region. PAHs are destroyed by direct stellar UV radiation interior to the PDR shell.

- The PDR shell is defined by PAHs that dominate its MIR emission. PAHs are excited by soft ($\lambda > 912\,\text{Å}$) UV photons that can pass through the hydrogen ionization front.

- Based upon morphology and environment, the identified YSO candidates appear to be triggered by the expanding bubble.

- N49 appears to be very young ($<10^6$ yr) and surrounded by dense ambient ISM ($\gtrsim 10^4\,\text{cm}^{-3}$). Models need to be calculated for bubble evolution with higher ambient ISM densities than are currently available.

**Acknowledgments.** I thank Matthew Povich for helping me with image preparation, converting the text to LaTeX format, and for informative discussions. This paper would not be possible without the GLIMPSE team who reduced the data, especially Marilyn Meade and Brian Babler. Stephan Jansen has faithfully maintained our computers. Most of the new results referred to here are from a forthcoming paper by Christer Watson et al. (in preparation).

**References**

Benjamin, R. A., et al. 2003, PASP, 115, 953
Capriotti, E. R., & Kozminsky, J. F. 2001, PASP, 113, 677
Carey, S. J., et al. 2005, BAAS, 207, 63.33
Castor, J., McCray, R., & Weaver, R. 1975, ApJ, 200, L107
Churchwell, E., et al. 2006, ApJ, 649, 759
Churchwell, E., et al. 2007, ApJ, submitted
Fazio, G. G., et al. 2004, ApJS, 154, 10
Freyer, T., Hensler, G., & Yorke, H. W. 2003, ApJ, 594, 888
Freyer, T., Hensler, G., & Yorke, H. W. 2006, ApJ, 638, 262
Helfand, D. J. et al. 2006, AJ, 131, 2525
Kessler, M. F. et al. 1996, A&A, 315, 27
Kröger, D., Hensler, G., & Freyer, T. 2006, A&A, 450, L5
Kröger, D., Freyer, T., Hensler, G., & Yorke, H. W. 2007, submitted
Kurucz, R. L. 1993, ATLAS9 Stellar Atmosphere Programs and 2 km/s grid. Kurucz CD-ROM No. 3 (Cambridge: Smithsonian Astrophysical Observatory), 13
Martins, F., Schaerer, D., & Hillier, D. J. 2005, A&A, 436, 1049
Povich, M. S., et al. 2007, ApJ, 660, 346
Price, S. D., et al. 2001, AJ, 121, 2819
Robitaille, T. P., et al. 2006, ApJS, 167, 256
Robitaille, T. P., et al. 2007, ApJS, 169, 328
Vink, J. S., de Koter, A., & Lamers, H. J. G. L. M. 2001, A&A, 369, 574
Weaver, R., McCray, R., Castor, J., Shapiro, P., & Moore, R. 1977, ApJ, 218, 377
Werner, M. W., et al. 2004, ApJS, 154, 1
Whitney, B . A., Wood, K., Bjorkman, J. E., & Wolff, M. J. 2003a, ApJ, 591, 1049
Whitney, B. A., Wood, K., Bjorkman, J. E., & Cohen, M. 2003b, ApJ, 598, 1079
Whitney, B. A., Indebetouw, R., Bjorkman, J. E., & Wood, K. 2004, ApJ, 617, 1177
Whitworth, A. P., et al. 1994, MNRAS, 268, 291

**Discussion**

*Ercolano*: Comment: the absence of PAHs in H II regions has implications for recent photoionization model grids aimed at deriving metallicity calibrations from strong lines, where PAHs are included as a source of photoelectric heating.

*Churchwell*: Absence of PAHs does not preclude photoelectric heating since 24 $\mu$m emission shows that dust exists in H II regions. Destruction of PAHs will release metals within H II regions, although these are likely to be cleared out by stellar winds.

*Beckman*: 1) What is the physical size scale of the bubbles? 2) Do you have any kinematical information about the bubbles?

*Churchwell*: 1) They range from a few tenths of a pc to radii up to $\sim$15–20 pc. Large ones are difficult to detect against the MIR background so there could well be larger ones that we are not sensitive to. 2) A few have radio recombination line observations but these generally do not have high enough spatial resolution to map the internal kinematics of the bubbles.

*Kroupa*: Can you estimate the mass of the gas removed from the central cluster, and on which timescale this occurs?

*Churchwell*: Unfortunately we do not have spectroscopic data and we do not detect emission from this volume directly. Neither do we have density information (yet) so we have not been able to obtain accurate mass determinations. I plan to obtain spectra that will be sensitive to a wide range of ionization states in the next *Spitzer* observing cycle, which I hope will permit reasonable estimates.

*Evans*: Do you have any constraint on dust temperatures in the shocked wind bubbles?

*Churchwell*: We have tried, but one has to separate PAH emission due to the front and back surfaces of the bubbles to obtain the thermal emission at 8 $\mu$m. When we do this, the residual thermal dust emission at 8 $\mu$m is so low that the derived temperature is only a limit, which may not be very interesting. We can only say that $T_d$ <150 K.

John Beckman (*left*) and Ed Churchwell.

*Pathways through an Eclectic Universe*
*ASP Conference Series, Vol. 390, © 2008*
*J. H. Knapen, T. J. Mahoney, and A. Vazdekis, eds.*

# Molecular Cloud Structure and the Origin of Stellar Masses

João Alves

*Calar Alto Observatory, Spain*

**Abstract.** Near-infrared dust extinction mapping is opening a new window on molecular cloud research. Applying a straightforward technique to near-infrared large-scale data of nearby molecular complexes, one can easily construct density maps with dynamic ranges in column density covering $3\sigma \sim 0.5\,\mathrm{mag} < A_V < 50\,\mathrm{mag}$ or $10^{21}\,\mathrm{cm}^{-2} < N < 10^{23}\,\mathrm{cm}^{-2}$. These maps are unique in capturing the low column density distribution of gas in molecular cloud complexes, where most of the mass resides, and at the same time allow the identification of dense cores ($n \sim 10^4\,\mathrm{cm}^{-3}$) which are the precursors of stars. For example, the application of this technique to the nearby Pipe Nebula complex revealed the presence of 159 dense cores (the largest sample of such objects in a single complex) whose mass spectrum presents the first robust evidence for a departure from a single power law. The form of this mass function is surprisingly similar in shape to the stellar initial mass function but scaled to a higher mass by a factor of about three. This suggests that the distribution of stellar birth masses (initial mass function) is the direct product of the dense core mass function and a uniform star formation efficiency of 30% ± 10%, and that the stellar initial mass function may already be fixed during or before the earliest stages of core evolution.

## 1. Introduction

Little is understood about the internal structure of molecular clouds and consequently the initial conditions that give rise to star and planet formation. This is largely because molecular clouds are primarily composed of molecular hydrogen, which is virtually inaccessible to direct observation. Therefore, the traditional methods used to derive the basic physical properties of such molecular clouds make use of observations of trace $H_2$ surrogates; namely, those rare molecules with sufficient dipole moments to be easily detected by radio spectroscopic techniques, and interstellar dust, whose thermal emission can be detected by far-infrared and radio continuum techniques (e.g., Motte, Andre, & Neri 1998; Lehtinen et al. 2001). However, the interpretation of results derived from these methods is not always straightforward (e.g., Alves, Lada, & Lada 1999; Chandler & Richer 2000). Several poorly constrained effects inherent in these techniques (e.g., deviations from local thermodynamic equilibrium, opacity variations, chemical evolution, small-scale structure, depletion of molecules, unknown emissivity properties of the dust, unknown dust temperature) make the construction of an unambiguous picture of the physical structure of these objects a very difficult task.

There is then a need for a less complicated and more robust tracer of $H_2$ not only to access the physical structure of these objects but also to accurately

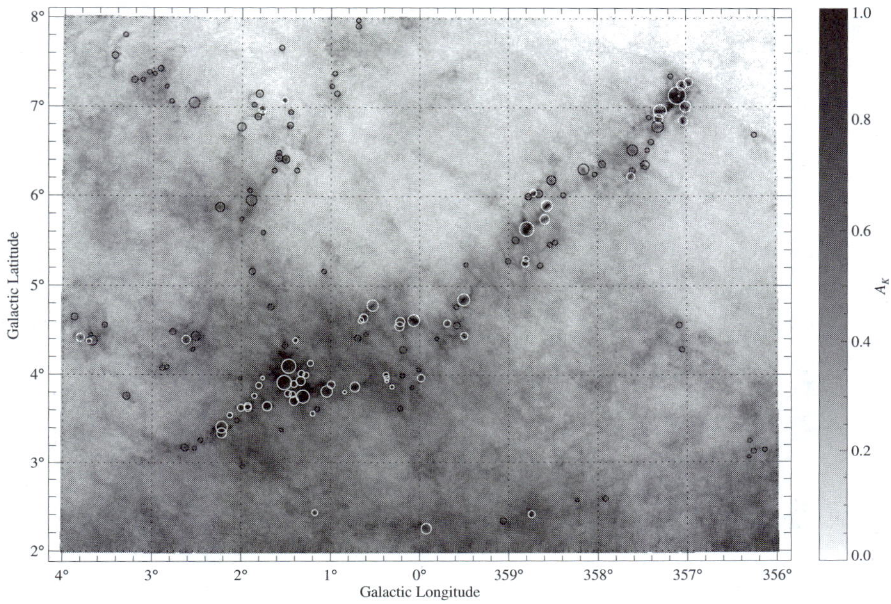


Figure 1. Dust extinction map of the Pipe Nebula molecular complex from Lombardi et al. (2006). This map was constructed from near-infrared observations of about four million stars in the background of the complex. Approximately 160 individual cores are identified within the cloud and are marked by an open circle proportional in size to the core radius. Most of these cores appear as distinct, well separated entities. (*Figure from Lombardi et al. 2006.*)

calibrate molecular abundances and dust emissivity inside these clouds. The deployment of sensitive infrared array cameras on large telescopes, however, has fulfilled this need by enabling direct and precise measurements of dust extinction toward thousands of individual background stars observed through a molecular cloud (Lada et al. 1994; Alves, Lada, & Lada 2001; Lombardi & Alves 2001). Such straightforward measurements are free from the complications that plague molecular line or dust emission data and enable detailed maps of cloud structure to be constructed.

## 2. Results from the Pipe Nebula

In Fig. 1 we present the $8^\circ \times 6^\circ$ extinction map of the Pipe Nebula complex constructed with four million stars from the 2MASS database (from Lombardi, Alves, & Lada 2006). This molecular complex is one of the closest to Earth ($\sim 130$ pc) of this size and mass and is particularly well positioned along a relatively clean line of sight to the rich star field of the Galactic bulge, which, given the close distance of the Pipe Nebula, allowed us to achieve spatial resolutions of $\sim$0.03 pc, or about three times smaller than the typical dense core size. The complex also exhibits very low levels of star formation, suggesting that its dense cores likely represent a fair sample of the initial conditions of star formation. We

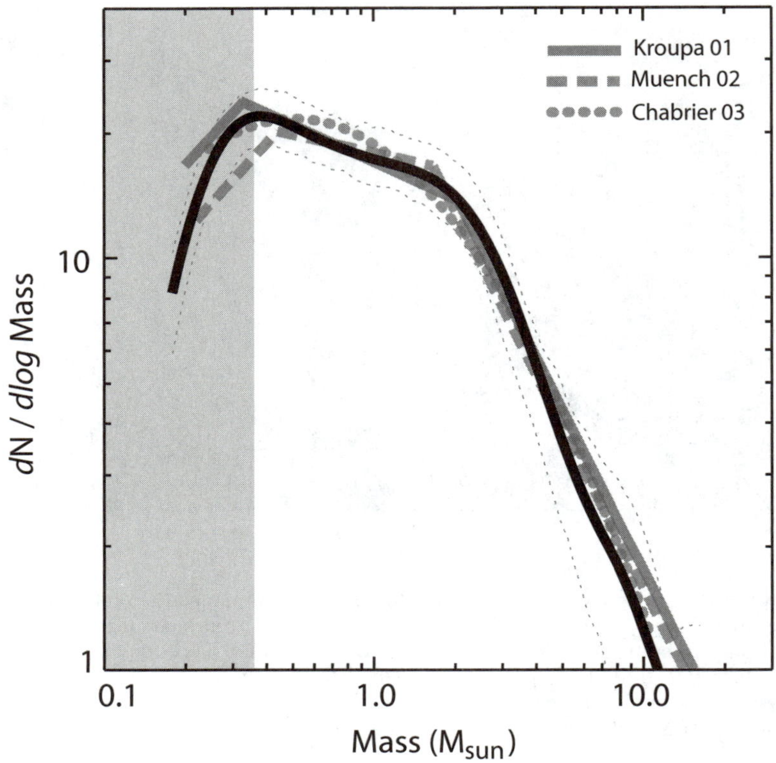


Figure 2. Probability density function of Pipe Nebula core masses (black). The gray region indicates sample incompleteness while the two thin dashed line indicates the 90% confidence limits. The core mass distribution seems to be characterized by two power laws and a well defined break point at around $2\,M_{\odot}$. Also plotted are the field star IMFs of Kroupa (2001; solid gray) and Chabrier (2003; dotted gray), and the Muench et al. (2002; dashed gray) IMF, all scaled up by a factor of ~3 in mass. (*Figure from Alves, Lombardi, & Lada 2007.*)

present in Fig. 2 the probability density function for the 159 core masses identified in the Pipe Nebula complex (marked in Fig. 1). The core mass distribution seems to be characterized by two power laws and a well defined break point at $\sim 2\,M_{\odot}$. For comparison, we also present the field star initial mass function (IMF) determined by Kroupa (2001; solid gray) and the Chabrier (2003) IMF (dotted gray), as well as the IMF for the young Trapezium cluster (Muench et al. 2002; dashed gray). All these stellar IMFs were shifted to higher masses by an average scale factor of about three. The point of this comparison is to illustrate simply the overall similarity between different stellar IMFs from different environments and constructed in different ways, and the dense core mass function of the Pipe Nebula. Considering the overall uncertainties, apparent differences between the different distributions are likely not significant.

## 3. Conclusions and Future Work

The close similarity in shape between the dense core mass function and the IMF supports a general concept of a one-to-one correspondence between the individual dense cores and soon-to-be-formed stars. Our study provides the first robust evidence for a departure from a single power law in the core mass function. We find that the location of the break indicates that there is a factor of about three difference in mass scale between the two distributions, which implies that a uniform star formation efficiency, which we estimate to be 30%±10%, will likely characterize the star formation in these dense cores across the entire span of (stellar) mass.

Recently, and on a larger scale, Casuso & Beckman (2007) were able to reproduce the mass spectrum of clouds and clumps simulating the dynamical processes that lead to the coagulation and the disruption of interstellar molecular clouds. Of critical importance, they found that a range of input conditions will yield the same observed cloud mass spectrum. An extension of this work to the cores (sizes $\sim 0.1\,\mathrm{pc}$, masses $\sim 1\,M_\odot$) is needed. Breaking the code behind the origin of the IMF might not be too far off in the future.

### References

Alves, J., Lada, C., & Lada, E. 1999, ApJ 515, 265
Alves, J., Lada, C., & Lada, E. 2001, Nature 409, 159
Alves, J., Lombardi, M., & Lada, C. 2007, A&A, 462, L17
Casuso, E. & Beckman, J. E. 2007, ApJ 656, 897
Chabrier, G. 2003, ApJ 586, 133
Chandler, C. & Richer, J. 2000, ApJ 530, 851
Kroupa, P. 2001, MNRAS, 322, 231
Lada, C. J., Lada, E. A., Clemens, D. P., & Bally, J. 1994, ApJ 429, 694
Lehtinen, K., Haikala, L. K., Mattila, K., Lemke, D. 2001, A&A 367, 311
Lombardi, M. & Alves, J. 2001, A&A 377, 1023
Lombardi, M., Alves, J., Lada, C. 2006, A&A 454, 781
Motte, F., Andre, P., & Neri, R. 1998, A&A 336, 150
Muench, A. A., Lada, E. A., Lada, C. J., & Alves, J. 2002, ApJ 573, 366

*Pathways through an Eclectic Universe*
*ASP Conference Series, Vol. 390, © 2008*
*J. H. Knapen, T. J. Mahoney, and A. Vazdekis, eds.*

# The Formation of Extreme Mass Ratio Binary Stars: A Tribute to Eduardo Delgado Donate

C. J. Clarke

*Institute of Astronomy, Cambridge, CB3 OHA, UK*

**Abstract.** Eduardo Delgado was due to have presented a poster at this meeting on his latest results on the formation of extreme mass ratio binaries. Tragically, Eduardo was among those killed in a hiking accident in Tenerife earlier this year. The organizers of this meeting kindly invited me, as his PhD supervisor, and as a longstanding collaborator, to incorporate a report on his most recent work into a more general tribute to his life and work. I will reflect on Eduardo's scientific career, the problems that motivated him and his achievements, focusing particularly on a problem which had intrigued us both for several years and on which Eduardo was making important progress at the time of his death. Finally, I will mention the personal qualities that Eduardo brought to his work and the acute sense of loss that is shared by all those—friends and collaborators—who were privileged to know him.

## 1. Cambridge Days: Predictions and Puzzles in Multiple Star Formation

Eduardo came to Cambridge from his native Tenerife in 2000 in order to work on hydrodynamic (SPH) star formation simulations under my supervision. He was funded by an EU PhD studentship as part of the 'Young Stellar Cluster' network. Anyone who has been involved in these networks will know that it is not always easy to find students of suitable caliber to fill these posts, given their restrictive nationality and residence requirements, and can therefore imagine my delight at receiving such a strong application from Eduardo. He went on to fulfill my high expectations, quickly developing into a fine numericist and astrophysicist. (I should add that I was often struck by the fact that Eduardo's written English was far more eloquent than anything that his British contemporaries could manage—although I'm afraid that this says something not only about Eduardo but also about the British educational system!)

Eduardo arrived in Cambridge when Matthew Bate was still a postdoc there and Matthew was a great help to Eduardo in helping him to master his SPH code. This was an exciting time for Eduardo to be working in numerical star formation, as this was the time that Matthew was producing the first turbulent fragmentation calculations which have generated so much interest (and controversy!) ever since. It was clearly going to be a challenge for Eduardo to do anything in this field which was going to have a distinct impact. I think the collection of well cited papers that Eduardo produced during his thesis shows that he met this challenge most successfully.

Eduardo became known in the field for his series of turbulent fragmentation calculations which explored the production of multiple star systems and brown dwarfs in small-$N$ ensembles. My own vote for the part of his thesis that will

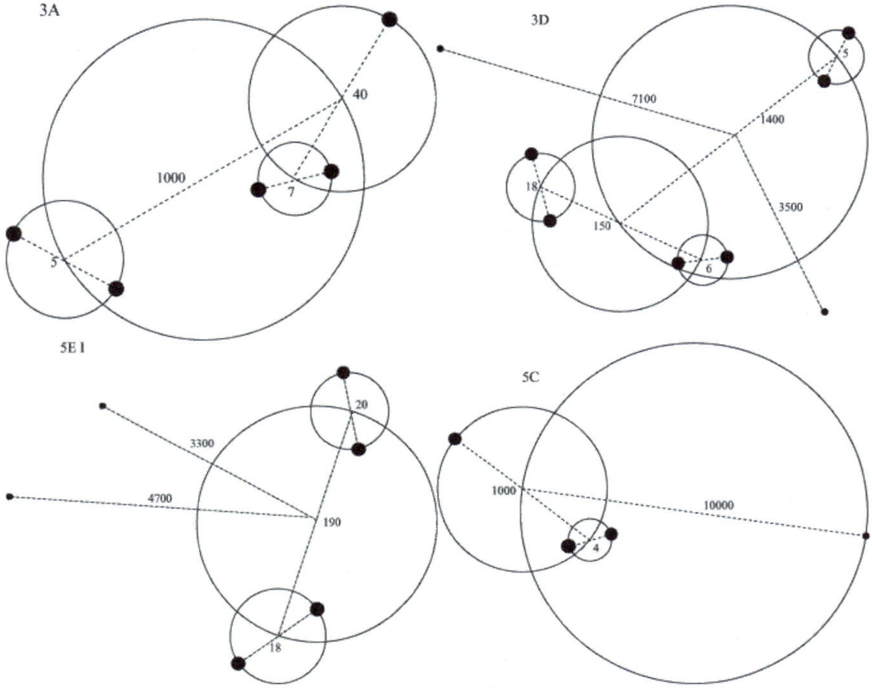


Figure 1. Examples of some of the system architectures of complex multiple star systems formed in the hydrodynamical simulations of Delgado et al. (2004). The numbers refer to separations in AU.

be his biggest contribution to the field would go to his calculations which were the first to make predictions about the properties of higher order multiple systems (see some of the complex system architectures generated in Fig. 1). He was able to do this because rather than focusing his computational effort on system scale (as Matthew Bate did with his 50 solar mass calculation), Eduardo followed small (5 solar mass) clouds and—in compensation—was able to follow the calculation for much longer. Thus Eduardo's simulations could follow even wide systems to the point of dynamical stability. Evidently this is important, not only because wide binaries are numerous in nature (half of all binaries with solar-type primaries are wider than 30 AU) but because in this way one can start to make pronouncements about higher order multiples (which necessarily involve wide, as well as close, components).

Thus Eduardo entered uncharted waters in terms of simulations (and indeed his paper on this, Delgado et al. 2004, is still the only one on this topic in existence). Moreover, the characterization of multiple star systems is far from complete observationally, as it is a laborious task to collate complete statistics on components with a wide dynamic range of separations (e.g., Tokovinin et al. 2006). Indeed, the motivation for such observational studies has always been undermined by the fact that—until Eduardo's work—there were no theoretical predictions with which to compare the data.

This brings me to one of Eduardo's most important characteristics as a scientist, that in addition to being technically accomplished as a numericist, he had a strong interest in—and respect for—observations. One would never hear Eduardo implying that observations must be wrong if they contradicted his simulations—or that observers should simply go and 'look harder'. He understood, better than most, that it's actually the areas of disagreement that illuminate your calculations. And, moreover, he understood that the range of diagnostic information that is contained in multiple star statistics would ultimately provide a very stringent—arguably the most stringent—test of star formation theories.

Thus Eduardo became intrigued by an obvious area of discrepancy between observations and all star formation simulations. In the simulations, all the binaries were within a factor of two of being of unit mass ratio (with the exception of a few weakly bound very low mass "outliers" at large separations). Moreover, this result extended to every level of the hierarchy (i.e., in a triple, a nearly equal mass binary would be bound to a third star with mass nearly equal to the binary). Evidently, nature does not work like that! It is true that there are categories of binaries that favor more nearly equal masses, arguably OB stars (Garcia & Mermilliod 2001), very low mass binaries (Bouy et al. 2003), and short period spectroscopic binaries among solar type stars (Mazeh et al. 1992). However, for the bulk of solar type binaries, the median mass ratio is $\sim$0.4 (Duquennoy & Mayor 1991), a result which—given that it is based partly on pairs from the visual binary literature which are almost certainly selectively incomplete at low mass ratios—must, if anything, *underestimate* the incidence of pairs with very disparate masses.

Early in 2003, Eduardo started writing a paper with me, alas never to be published, entitled 'On the problem of forming extreme mass ratio binaries'. I told him to start by writing a few pages setting up the problem (which he did) and then, by the time he'd written that, his simulations would have run to the point where he could unveil the solution (or at any rate *a* solution) in the concluding sections. But the latter proved to be impossible. None of the fixes to initial conditions which he attempted (such as adjusting the power spectrum of the initial velocity field or changing the geometry) were ultimately successful. He would excitedly tell me that, yes, he'd managed to make a binary with $q$ of 0.1, only to return a week later to tell me that, no, it had now evolved to $q$ of 0.7... Thus it turned out that the problem was not the initial creation of such systems, but the effect of continued accretion onto the protobinary, which drove up the mass ratio.

It has been known since at least the work of Artymowicz (1981)—and note that Eduardo went on to spend two years in Stockholm working with Pawel Artymowicz as a postdoc, although never on this problem—that accretion of gas whose specific angular momentum exceeds that of the binary leads to preferential accretion onto the secondary. This commonsense result, which was apparent in these early ballistic calculations and was backed up by the subsequent SPH simulations of Bate & Bonnell (1997) and Bate (2000), occurs simply because the secondary's Roche lobe is further from the system's center of mass, and is thus more likely to intercept high angular momentum material. Since in the case of any plausible binary formation scenario, higher angular momentum material will fall in later, the binary mass ratio, $q$, should therefore rise with time.

## 2. Numerical Controversy: Latest Results

While Eduardo and I abandoned this question, discouraged (and Eduardo took up a position in Stockholm where he worked mainly on codes for planet formation), there meanwhile appeared the paper of Ochi, Sugimoto, & Hanawa (2005). This claimed, using a grid-based code, that actually material of high specific angular momentum is preferentially accreted onto the binary *primary* and thus that the mass ratio should *fall.* This study did not contradict the widely acknowledged result that the flow preferentially enters the secondary's Roche lobe, but argued instead that, after half an orbit around the secondary, it crosses via the L1 point into the primary's Roche lobe and is then accreted. Ochi et al. conjectured that the failure of SPH calculations to demonstrate this behavior stemmed from the excessively viscous nature of SPH, which might cause the SPH particles to spiral in sufficiently, in half an orbit, so that they could avoid the L1 region. In the same spirit, proponents of SPH countered that Ochi et al.'s flows were too warm (with a sound speed of 0.25× the orbital velocity of the binary), and that this caused artificial acceleration of the flow through the L1 point. Clearly these issues had to be investigated further, using both Lagrangian and Eulerian codes. At the time of his death, Eduardo was close to completing a suite of high resolution SPH calculations which were beginning to illuminate the problem and whose results I summarize below.

His repetition, at higher resolution, of the SPH calculation of Bate & Bonnell (1997), which modeled accretion onto a protobinary of a flow with fixed specific angular momentum, demonstrated that the rate of increase of $q$ is indeed somewhat resolution dependent and is over-estimated at low resolution (about $5 \times 10^4$ particles in the disks) compared with the highest resolution (a hundred times more particles, which has achieved numerical convergence) by about a factor two. This can be traced to the fact that at low resolution, particles in the outer part of the primary's Roche lobe are not behaving in a fully fluid-like manner and, through precession of their elliptical orbits, are able to tip back through the L1 point into the secondary's Roche lobe. Nevertheless, $q$ *increases* (for this value of specific angular momentum of the flow) *at all resolutions*, in contrast to the results of Ochi et al.

Eduardo found that at high resolution, the shocks are well delineated (see left hand panel of Fig. 2) and, indeed, that the Jacobi constant (i.e., the Bernoulli function in the co-rotating frame) is well conserved along streamlines except in shocks. This result goes against the hypothesis of Ochi et al. that the SPH results are driven by artificial viscosity in the secondary's disk flow, since under these circumstances the action of artificial viscosity should cause changes in the Jacobi constant, even away from shocks. Crucially, however, the shock geometry is quite different in Eduardo's simulations compared with those of Ochi et al. In the SPH calculation, loosely bound material in the secondary's Roche lobe encounters a shock with material flowing in from the direction of the primary (see dense structure to the east of the secondary in left hand panel of Fig. 2) and it is dissipation in this shock which prevents material flowing through L1 and into the primary's Roche lobe. This shock is simply not there in Ochi et al.'s calculation, so that material entering via L2 retains enough energy to smoothly transit through L1.

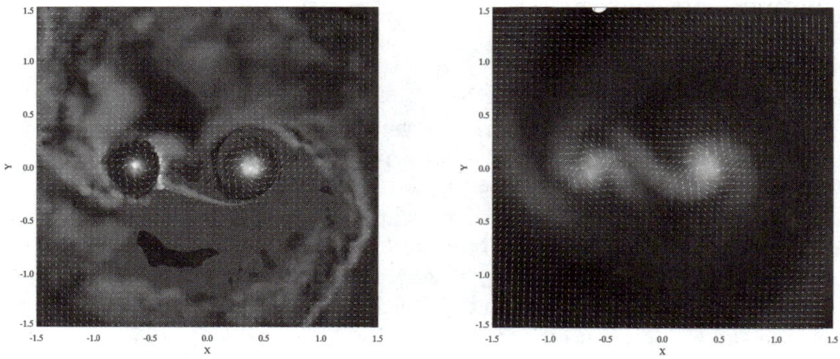


Figure 2. Density and flow structure in the co-rotating frame around an accreting protobinary, contrasting a cold three-dimensional simulation (*left*) and a warm two-dimensional simulation (*right*). The secondary is to the left. Note the pronounced shock to the east of the secondary in the left hand panel. Unpublished simulations by E. Delgado.

Thus while both simulations are apparently self-consistent, *given the location of the shocks*, one has to enquire why the shock morphology is so different in the two cases. One possibility is that—although the SPH and Eulerian simulations are matched in terms of specific angular momentum of the accretion flow and mass ratio of the binary—there are two potentially important differences. Firstly, the Eulerian calculation is much warmer (about a factor 25 in temperature). Secondly, it is strictly two-dimensional whereas the SPH calculation—although highly flattened near the binary—is fully three-dimensional and introduces particles on the surface of a sphere at large radius. An indication that one, or both, of these effects may be important is provided by one of Eduardo's simulations which mimicked as closely as possible the precise conditions of Ochi et al., being both 2D and warm. The right hand panel of Fig. 2 demonstrates that the shock has in this case moved away from the secondary's Roche lobe, in qualitative agreement with the flow morphology of Ochi et al. Although *quantitative* agreement between the SPH and Eulerian codes is not achieved even here, the *sign* of the change is encouraging. This shift in flow morphology is reflected in a change in the relative accretion rates on to the primary and secondary. Although $q$ still increases in these warm 2D SPH calculations, $\dot{q}$ is reduced by an order of magnitude compared with the cold 3D calculation.

This was the situation at the time of Eduardo's death and there is evidently still much for his collaborators to do. We now suspect that gas temperature and/or flow dimensionality are important determinants of the shock morphology, and we must now discover which of these two effects is the critical one. One then has to judge which set of conditions is more realistic. Ultimately, this will tell us whether the old "simple" idea is correct—i.e., that the flow simply enters the secondary's Roche lobe and remains therein—or whether the more complex outcome of Ochi et al.—involving flow between Roche lobes—is closer to the truth.

All this has implications for the extreme mass ratio binary problem which Eduardo had been trying to solve for years. If the result is closer to that of Ochi et al., then this could solve the problem at a stroke. If the results instead support the old SPH results, then the puzzle remains (although if low resolution simulations have over-estimated the *magnitude* of the growth of $q$, then binaries may remain at low $q$, even if the sign of $\dot{q}$ is positive). Otherwise, one has to think harder about how to prevent the infall of gas with high specific angular momentum onto protobinaries. Although it is tempting to invoke feedback for this, it is not obvious why this should be so important in the low mass binaries which we are largely trying to explain.

## 3. Eduardo the Person

Having talked about the problems that motivated Eduardo as a scientist, it leads me to how much can one discern of Eduardo the person in all this? An unexpected death leads one to analyze what one saw of a person in the daily contacts of studentship and collaboration. Eduardo was always calm, always gentle, always attentive. He was never arrogant, never distorted the problem at hand, or his relations with those around him, by the obtrusion of his own ego. He commanded the respect of all those who came into contact with him through this obvious intellectual and personal integrity. He was a joy to work with.

I see great continuity between those characteristics as a collaborator and the personal testimonies from his friends, many of which are movingly collected on his blogspot (http://eduardo-delgado.blogspot.com/). I was aware that Eduardo was very popular with his peers in Cambridge and that he belonged to a big crowd of friends of all nationalities whose joyful social life never got in the way of Eduardo's scientific activities. But these messages—and the evidence of the conversations I have had with so many—speak of a very special warmth of affection and regard. Unfortunately I can only dimly decipher the majority in Spanish but they clearly tell the same story as those I can read (in Italian and English). So many speak of the gentleness of his eyes, the warmth of his friendship, his calmness (one Italian friend exclaimed 'You even played football with calm'—incredible!). He was, to put it simply, much loved.

All this is very particular to Eduardo, but I should end on a note that perhaps has relevance to all those gathered for this meeting and to astronomers everywhere. In the last months of his life, Eduardo was particularly happy. A former Cambridge colleague, who stayed with him on an observing run at Christmas time, told me 'You know, Edu was just *so* happy when I last saw him'. He had everything he dreamed of, living on the island he loved, starting new projects in his new postdoc with Casiana Muñoz, also continuing his independent research (he felt he was so close to cracking the extreme mass ratio binary problem). We're perhaps not all in one of those special moments in life when everything seems to be working, but I reckon that if any one of us suffered the same awful and random misfortune as Eduardo, then they would be able to say of us that we'd spent our lives doing what we loved. This is a rare privilege. So, in keeping with the celebratory mood of this conference, let us recall that, while mourning Eduardo no less, we can also celebrate a life spent—like ours—in doing something we believe to be intrinsically valuable. To do this in a spirit of calm and integrity was Eduardo's gift. We miss him immensely.

Figure 3. Edu emanating his characteristic calm.

## References


Artymowicz, P. 1983, Acta Astron. 33, 223

Bate, M. R. 2000, MNRAS, 314, 33

Bate, M. R., & Bonnell, I. A. 1997, MNRAS, 285, 33

Bouy, H., Brandner, W., Martin, E., Delfosse, X., Allard, F. & Basri, G. 2003, AJ, 126, 1526

Delgado-Donate, E. J., Clarke, C. J., Bate, M. R., & Hodgkin, S. T. 2004, MNRAS, 351, 617

Duquennoy, A., & Mayor, M. 1991, A&A, 248, 485

Mazeh, T., Goldberg, D., Duquennoy, A., & Mayor, M. 1992, ApJ, 401, 265

Garcia, B., & Mermilliod, J. 2001, A&A, 368, 122

Ochi, Y., Sugimoto, K., & Hanawa, T. 2005, ApJ, 623, 922

Tokovinin, A., Thomas, S., Sterzik, M., & Udry, S. 2006, A&A, 450, 681

*Pathways through an Eclectic Universe*
*ASP Conference Series, Vol. 390, © 2008*
*J. H. Knapen, T. J. Mahoney, and A. Vazdekis, eds.*

# Internal Chemo–dynamical Modeling of Gas Exchange within Galaxies and with their Environment

Gerhard Hensler,[1] Simone Recchi,[1,2] Danica Kroeger,[3] and Tim Freyer[3]

[1] *Institute of Astronomy, University of Vienna, Türkenschanzstr. 17, 1180 Vienna, Austria,*
[2] *INAF—Osservatorio Astronomico di Trieste, 34131 Trieste, Italy*
[3] *Inst. of Theor. Physics & Astroph., U of Kiel, 24098 Kiel, Germany*

**Abstract.** With a few, but important, examples in the form of models of combined radiative and wind-driven H II regions and galactic winds we demonstrate the importance of refined small to medium scale studies of chemo-dynamical effects. These processes determine the internal dynamics and energetics of the interstellar medium and affect its observational signatures, e.g., by abundance contributions, but are not yet reliably and satisfactorily explored.

## 1. Introduction

Galaxies experience energy deposit and element release into their interstellar medium (ISM) from massive stars during their lives and from their explosions. These are main drivers of the energetics and dynamics in galaxies, and in both stages they release characteristic elements that are preferably contained in the hot gas phase. Nevertheless, because of the close coupling of the hot, warm, and cool ISM, energetic and dynamical processes lead to chemo-dynamical effects that alter the energy budget and the element content in the different gas phases (Hensler 2003). Understanding these processes from stellar to galactic scales is therefore vital for understanding and modeling the evolution of galaxies, and also their interaction with the environment.

## 2. Massive Stars

### 2.1. Effects of Massive Stars during their Lives

Massive stars play a crucial role in the evolution of galaxies, as they are the primary source of metals, and they dominate the turbulent energy input into the ISM by their massive and fast stellar winds, the ultraviolet radiation, and supernova explosions. The radiation field of these stars, at first, photo-dissociates the ambient molecular gas and forms a so-called photo-dissociation region (PDR) of neutral hydrogen. Subsequently, the Lyman continuum photons of the star ionize the H I gas and produce an H II region that expands into the neutral ambient medium.

As these stars have short lifetimes of only a few million years, H II regions indicate the sites of star formation (SF) and are targets to measure the current SF rate in a galaxy. Furthermore, the emission line spectrum produced by the ionized gas allows the accurate determination of the current chemical composition of the gas in a galaxy. Although the physical processes of the line excitation are quite well understood and accurate atomic data are available, so that the spectral analysis of H II regions (e.g., Stasińska 1979; Evans & Dopita 1987) serves as an essential tool to study the evolution of galaxies, their reliability as diagnostic tool must also be studied with particular emphasis, for example, on temperature fluctuations (Peimbert 1967; Stasińska 2002) and line excitations.

While the simple concept of a uniform medium in ionization equilibrium with the radiation from stars (the Strömgren sphere) is successful in describing several global features of the object and allows us to predict the emission-line spectrum quite reliably, it has long been realized that H II regions are complex and dynamical objects. Dynamical modeling of H II regions based purely on the energy deposit of the stellar radiation field had therefore been started already long ago (e.g., Yorke 1986, and references therein) providing both a first insight into the formation of dynamical structures and allowing us to derive more realistic scaling laws for observational comparisons.

Massive stars thus not only release energy by their radiation but also act by means of strong stellar winds that drive a turbulent ISM. Both processes, however, are metal-dependent, which has three major consequences for galaxies with present-day solar metallicities: first, the classical H II region has lost its simplicity, because, second, it engulfs a vehemently expanding stellar wind bubble (SWB) and, third, it peels off the stellar outermost shells, thus incorporating nuclear burning regions into the wind mass loss.

## 2.2. Star Formation Feedback

Quiescent SF as a self-regulated process is a widely accepted concept. The stellar feedback can adapt both signs, positive as a triggering mechanism in a self-propagating manner as in superbubble shells (Ehlerova et al. 1997; Fukuda & Hanawa 2000) and negative as self-regulation. Primarily, the correlation between the surface density of disk galaxies' H I gas and the vertically integrated SF rate derived from the H$\alpha$ flux (Kennicutt 1998) serves as the best proof of SF self-regulation. Even more refined observations that trace more complex molecular gas (Gao & Solomon 2004) find the same surface density relation. Surprisingly, this relation can be reproduced by different numerical models and with different prescriptions from purely collapsing cloud cores (Li, Mac Low, & Klessen 2005) via thermodynamic cloud models (Krumholz & McKee 2005) to sophisticated global chemo-dynamical treatments of galaxy models (Burkert, Truran, & Hensler 1992; Harfst, Theis, & Hensler 2006) and even seems to hold when neglecting self-regulation processes in cosmological simulations (Kravtsov 2003). This fact emphasizes that the gas dependence of processes within the SF matter cycle, although expected to behave non-linearly, cancels out and results in a universal law. Such a dependence is plausible because Köppen, Theis, & Hensler (1995), for example, have demonstrated that, because of the square-dependence of collisionally excited radiative cooling on the gas volume density, the SF rate also adapts to that over a large density range and is almost independent of the heating rate and other fiducial parameters. This also holds for a multiphase ISM, accounting for the production of hot gas by type II supernovae (SNe II) and for gas transitions by evaporation and condensation (Köppen, Theis, & Hensler 1998). From this result, it seems reasonable that the vertical accumulation of gas mass should also lead to an SF rate dependence on the gas column density because the self-gravitational potential determines the disk stratification. The Kennicutt law could therefore reflect a sum over the volumes, while the 1.4 power is also close to the pure free-fall criterion. The timescales must then definitely tell us about both the process and the carrier of self-regulation.

### 2.3. Evolution and Structure of Bubbles around Massive Stars

However, this question needs further exploitation. Already the energy deposit by massive stars described by a parameter $\epsilon$, called the energy transfer efficiency, is well derived neither during their normal lives nor for their SNII explosions and the expanding SN remnants. The reason is that for both stages the environmental state, and thus the preceding processes, and also dynamical effects, play a significant role. Analytical estimates for purely radiative H II regions yield $\epsilon$ of the order of a few percent (Lasker 1967). Although the additional stellar wind power $L_{\rm w}$ can easily be evaluated from model and observational values (Kudritzki & Puls 2000), the fraction of this power that is transferred, for example, into thermal energy, i.e., the thermal $\epsilon$, or into turbulent energy is not obvious from first principles. The picture is that the fast stellar wind creates shocks that form the SWB filled with very hot plasma, which expands into the H II region so that also this H II shell has to expand into the surrounding ISM due to its overpressure.

The structure and evolution of SWBs and the transition to their surrounding H II regions can be described by a set of equations (Weaver et al. 1977) from which the wind-energy deposit can be derived theoretically under the simplifying assumptions of a point source of a constant and spherically symmetric strong wind that interacts with a homogeneous ambient ISM. The $\epsilon$'s amount to significant fractions of $L_{\rm w}$ (see Freyer, Hensler, & Yorke 2003, and references therein). It has become evident that stellar parameters such as mass-loss rate, terminal velocity, effective temperature, and luminosity vary strongly during stellar lifetimes. While most previous studies dealt with the evolution of either H II regions or SWBs separately and with constant values, little is known about the interaction of these two structures.

Although the analytical and semi-analytical solutions for the evolution of SWBs have improved over the years and the numerical simulations have been done with increasing complexity, e.g., 2D calculations of SWBs (see, for example, Różyczka 1985, and later papers by this author) and/or combined 1D radiation–hydrodynamical models of an H II region coupled with the dynamical SWB (e.g., by García-Segura & Mac Low 1995, for further references see Freyer et al. 2003), a variety of physical effects remain to be included in order to achieve better agreement between models and observations with regard to the evolution of the hot phase in bubbles (Mac Low 2000; Chu 2000).

To gain further insight into the evolution of radiation-driven + wind-blown bubbles around massive stars, we have performed a series of radiation-hydrodynamic simulations with a 2D cylindrical-symmetric nested-grid scheme for stars of masses of $15\,M_\odot$ (D. Kroeger et al., in preparation), $35\,M_\odot$ (Freyer, Hensler, & Yorke 2006), $60\,M_\odot$ (Freyer et al. 2003), and $85\,M_\odot$ (Kroeger et al. 2007). The main issues are:

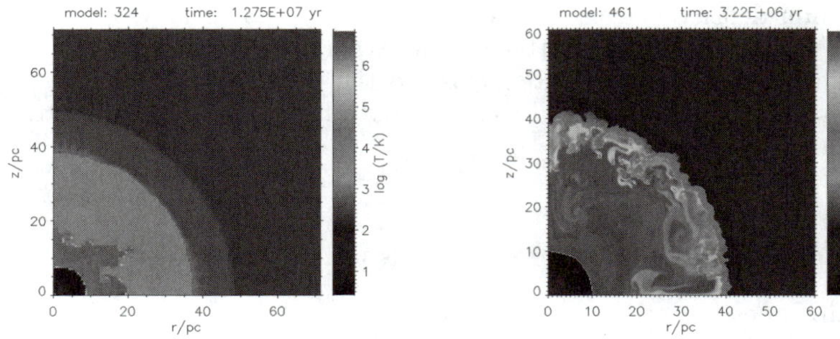


Figure 1. Temperature distributions of a 15 $M_\odot$ (*left panel*) and an 85 $M_\odot$ (*right panel*) model at late stages of their stellar evolution. The temperatures are color-coded to the right of the plots but differently (from Kroeger 2006). For discussion see text.

1. The H II region formed around the SWB has a complex structure mainly affected by dynamical processes such as shell instabilities, vortices, mixing effects, etc. It is more compressed by stronger winds (see Fig. 1).

2. Finger-like and spiky structures of different densities and temperatures are formed in the photoionized region (Freyer et al. 2003).

3. The regions contributing to the H II emission line are not solely limited to the photoionized shell around the SWB but also form from photo-evaporated gas at the trailing surface of the SWB shock front (see the 85 $M_\odot$ model in Fig. 1).

4. Because dispersion of this cooler photo-evaporated gas into the hot SWB leads to mixing, the stellar material expelled by the wind also partly emerges in the spectra.

5. As a consequence, the metal-enrichment of the wind during the Wolf–Rayet stage, which is generally assumed to remain only in the hot SWB, affects the observationally discernible abundance of the H II gas for a long time. With these models (Kroeger, Hensler, & Freyer 2006) it could be proven for the first time that the metal released by Wolf–Rayet stars can be mixed within short timescales from the hot SWB into the warm ionized gas and should become observationally accessible. As the extreme case we found a 22% enhancement of carbon for the 85 $M_\odot$ star, but negligible amounts for N and O.

6. As expected from the distribution of H II gas, the radially projected H$\alpha$ brightness shows a decrease towards the center and a slight brightening towards the limb but not as strong as expected according to the increase of the line of sight with impact parameter (Freyer et al. 2003). This effect depends on the bubble age and starts from the central brightening. It also demonstrates that neglecting the heat conduction and the homogeneous initial density does not allow sufficient brightening of heat conductive interfaces so that, secondly, only the photo-evaporated backflow can contribute to the H$\alpha$ luminosity in present models. In reality, condensations that become embedded in the hot SWB are exposed to heat conduction.

7. The sweeping-up of the slow red supergiant wind by the fast Wolf–Rayet wind produces remarkable morphological structures and emission signatures that agree well with the observed X-ray luminosity and temperature, as well as with the limb brightening of the radially projected X-ray intensity profile (for details see Freyer et al. 2006).

8. $\epsilon$'s for both radiative and kinetic energies remain much lower than analytically derived (by more than an order of magnitude) and amount to only a few per mil (Hensler 2007). There is almost no dependence on the stellar mass, which is in principle expected because the energy impact by Lyman continuum photons and by wind luminosity increases with stellar mass. In contrast, since the gas compression is made stronger by a more energetic wind, the energy loss by radiation is also more efficient.

Nonetheless, a word of caution and unfortunately of discouragement has to be expressed here because stellar evolutionary models are not unique but depend on the authors. In order to get a quantifiable comparison of the models by García-Segura, Mac Low, & Langer (1996a) and García-Segura, Langer, & Mac Low (1996b) with our 2D radiation-hydrodynamic simulations (Freyer et al. 2003, 2006) for the 35 and 60 $M_\odot$ studies we used the same stellar parameters. Since no stellar parameters were available from the same group for the 15 and 85 $M_\odot$ models, we had to make use of the Geneva models (Schaller et al. 1992). A comparison of the age-dependent 60 $M_\odot$ parameters of both groups has revealed enormous differences in the energetics of almost an order of magnitude. Also, the Wolf–Rayet and luminous blue variable stages occur in contrary sequence (Kroeger 2006).

Furthermore, H II regions show a great diversity of shapes giving direct evidence that the ISM is not a smooth, homogeneous, and uniform medium. In addition, the complexity of the internal structure and of the flow pattern of the gas indicates that the evolution of the ionized region is strongly affected by existing irregularities and hydrodynamic instabilities. What is more, stellar wind bubbles and supernova remnants (SNRs) of neighboring stars collide and may overlap to form superbubbles with diameters on a scale of several 100 pc, which are prominent by their X-ray emission in galaxies with active SF. This picture of SWB/H II aggregates has also to be extended by PDRs, which are mainly produced in the early evolutionary phases of ultra-compact H II regions. Since they are heavily dust-enshrouded they emerge from GLIMPSE observations (Benjamin et al. 2003), which allow detailed diagnostics of these common stellar bubbles (Churchwell et al. 2006).

## 3. Galactic Mass Exchange

### 3.1. Galactic Outflows

Finally, massive stars explode as SNe II, creating an SNR that sweeps up the ambient medium. These phases have been understood and modeled in detail energetically and dynamically. Cumulative SNe II form superbubbles that expand vehemently and can drive a galactic outflow. Since the hot superbubble gas carries SN II elements and has to interact dynamically with the surrounding

ISM, it has huge energetic and chemical effects on galaxies. Therefore, the questions that arise for their chemo-dynamical contribution to galaxy evolution are the same as for the H II/SWB regions. Although investigations have been performed for $\epsilon$ of SNe (Thornton et al. 1998) and superbubbles (e.g., Strickland et al. 2004) they are yet too simplistic (e.g., only 1D) for quantitative results.

Because of their low gravitational potential dwarf galaxies are sensitive to energetic processes such as superbubbles so that these can easily develop to a galactic outflow as perceptible, for example, in NGC 1705 (Hensler et al. 1998), NGC 1569 (Martin et al. 2002), I Zw 18 (Martin 1996), and many other blue compact dwarf galaxies (BCDs). Dynamical models (Mac Low & Ferrara 1999) can still provide only a qualitative understanding. Nevertheless, the element abundances stemming from different stellar progenitors should enable one to trace back the history of element enhancement vs depletion by different processes (Hensler, Rieschick, & Köppen 1999). Since in general the SF varies in BCDs and, furthermore, produces massive star clusters, one can imagine that a first superbubble has to exert most of its energy in its expansion while the successive SN II explosions can be funneled through the chimney and through this more easily expel its metal-enriched gas. This can only happen if the next SF episode follows shortly enough after the former one. For this scenario it is thus important to explore the timescale to refill superbubbles vs the SF timescale. For the refill of superbubbles in the gas layer of typical BCDs we (Recchi & Hensler 2006) found a range of a few 100 Myr. This means that the quantity of galactic winds depends not only on the SF rate but also on its modes, bursting vs gasping, but affects the chemo-dynamical evolution (Hensler & Rieschick 2002).

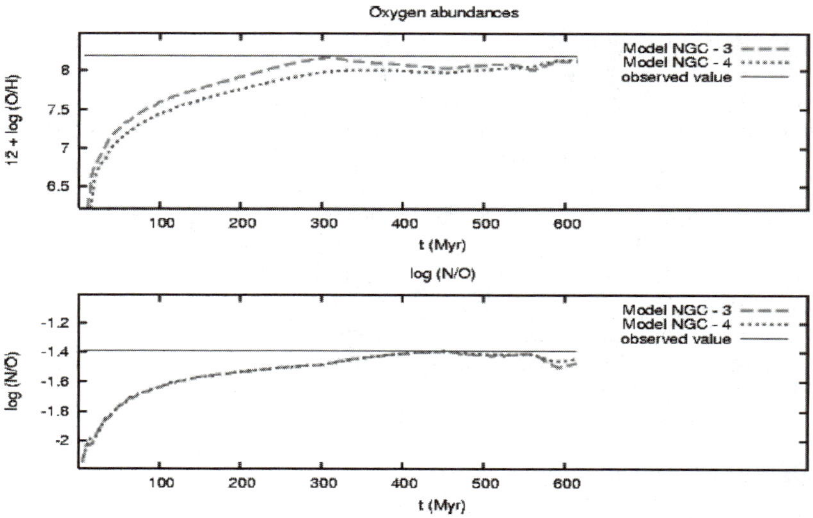


Figure 2. Comparison of abundance evolution in two models representing NGC 1569: NGC-3 with a single strong and short SF burst, NGC-4 with two former SF epochs, and a mild recent burst. (For details see Recchi & Hensler 2006).

For the BCDs that have been best studied—I Zw 18 (Recchi et al. 2004) and NGC 1569 (Recchi et al. 2006)—a 2D dynamical model with characteristic tracer elements has been performed aimed at achieving a best fit to the observations. The main results can be briefly summarized by saying that galactic winds always occur. Models with bursting SF are generally unable to account for the chemical and morphological properties of NGC 1569, because they either underproduce O or inject too much energy into the system, enough to lose too large a gas fraction. The best agreement with the chemical composition is found for long-lasting continuous episodes of SF of some hundred Myr of age and a recent more intensive short SF burst (see Fig. 2). In most models with gasping SF, the final chemical composition of the galaxy mostly reflects the chemical enrichment from old stellar populations. In fact, if the first episodes of SF are powerful enough to create a galactic wind or to heat up a large fraction of the gas surrounding the SF region, the metals produced by the last burst of SF are released into too hot a medium or are directly ejected from the galaxy through the wind. They do not have the chance to pollute the surrounding medium and contribute to the chemical enrichment of the galaxy. This result confirms the conclusion that most of the O from the last SF epoch is stored in the hot X-ray gas (Martin et al. 2002).

### 3.2. Gas Infall

Although the element abundance properties of most BCDs favor only a young stellar population of at most 1–2 Gyr, they mostly consist of an underlying old population, as is the case for NGC 1569. Furthermore, most of the objects are embedded in H I envelopes from which at least NGC 1569 definitely suffers gas infall (Stil & Israel 2002; Mühle et al. 2005). This has led Köppen & Hensler (2005) to exploit the influence of gas infall with metal-poor gas into an old galaxy with continuous SF in particular abundance patterns. Their models could match not only the observational regime of BCDs in [12 + log(O/H) – log(N/O)] space but also explain the shark-fin shape of their data distribution by means of the ratio of mass infall to existing cloud mass, in the sense that for larger infall fractions the loop is more extended.

Analytical investigations and overly simplified numerical models, however, are unable to distinguish between the different gas phases and their metal content in order to derive abundances in the observed objects.

Also, mixing effects at interfaces between gas phases and turbulence (such as those in the combined SWB/H II complexes) contribute to the observed values by the enhancement or, respectively, dilution of metals. Since gas infall affects not only the chemistry but also the SF (Hensler et al. 2004) and dynamics of outflows, we have extended the former models (Recchi et al. 2004, 2006) to include infalling clouds (Recchi & Hensler 2007). In these models, two kinds of cloud contributions are considered: only initially existing and continuously formed, respectively. The following issues are raised. Owing to dynamical processes and thermal evaporation, the clouds survive only a few tens of Myr. The internal energy of cloudy models is typically reduced by 20–40% compared to models with a smooth density distribution. The clouds delay the development of large scale outflows, thus helping in retaining a larger amount of gas inside the galaxy. However, especially in models with continuous creation of infalling clouds, their ram pressure pierces the expanding supershells so that through these holes freshly produced metals can more easily escape and vent into the galactic wind. Moreover, assuming a pristine chemical composition for the clouds, their interaction with the superbubble dilutes the hot gas, reducing its metal content. The resulting final metallicity is therefore generally smaller (by ~0.2–0.4 dex) than that attained by smooth models.

**Acknowledgments.** The authors gratefully acknowledge contributions and discussions by Jay S. Gallagher, Stefan Hirche, Joachim Koeppen, Andreas Rieschick, Christian Theis, Wolfgang Vieser, and Harald W. Yorke. Part of the work was funded by the DFG under grants HE 1487/17, /25, /28. GH cordially thanks Johan Knapen and the organizers for the invitation to this conference. Participation was made possible by grants from the conference and from the University of Vienna under BE518001.

## References

Benjamin, R. A., et al. 2003, PASP, 115, 953
Burkert, A., Truran, J., & Hensler, G. 1992, ApJ, 391, 651
Chu, Y.-H. 2000, Rev. Mex. Astron. Astrofis., 9, 262
Churchwell, E., et al. 2006, ApJ, 649, 759
Ehlerova, S., Palous, J., Theis, C., & Hensler, G. 1997, A&A, 328, 111
Evans, I. N., & Dopita, M. A. 1987, ApJ, 319, 662
Freyer, T., Hensler, G., & Yorke, H. W. 2003, ApJ, 594, 888
Freyer, T., Hensler, G., & Yorke, H. W. 2006, ApJ, 638, 262
Fukuda, N., & Hanawa, T. 2000, ApJ, 533, 911
Gao, Y., & Solomon, P. M. 2004, ApJ, 606, 271
García-Segura, G., & Mac Low, M.-M. 1995, ApJ, 455, 145
García-Segura, G., Mac Low, M.-M., & Langer, N. 1996a, A&A, 305, 229
García-Segura, G., Langer, N., & Mac Low, M.-M. 1996b, A&A, 316, 133
Harfst, S., Theis, C., & Hensler, G. 2006, A&A, 499, 509
Hensler, G. 2003, ASP Conf. Ser., 304, CNO in the Universe, ed. C. Charbonnel, D. Schaerer, & G. Meynet (San Francisco: ASP), p. 371
Hensler, G. 2007, EAS Publ. Ser., 7, ed. E. Emsellem et al., p. 113
Hensler, G., Rieschick, A., & Köppen, J. 1999, ASP Conf. Ser., 187, The Evolution of Galaxies on Cosmological Timescales, ed. J. E. Beckman & T. J. Mahoney (San Francisco: ASP), p. 214
Hensler, G. & Rieschick, A. 2002, ASP Conf. Ser., 285, Modes of Star Formation and the Origin of Field Populations, ed. E. K. Grebel & W. Brandner (San Francisco: ASP), p. 341
Hensler, G., Dickow, R., Junkes, N., & Gallagher, J. S. III 1998, ApJ, 502, L17
Hensler, G., Köppen, J., Pflamm, J., & Rieschick, A. 2004, IAUS 217, Recycling Intergalactic and Interstellar Matter, ed. P.-A. Duc, J. Braine, & E. Brinks (San Francisco: ASP), p. 178
Kennicutt, R. J. 1998, ApJ, 498,
Köppen, J., & Hensler, G. 2005, A&A, 434, 531
Köppen, J., Theis, C., & Hensler, G. 1995, A&A, 296, 99
Köppen, J., Theis, C., & Hensler, G. 1998, A&A, 328, 121
Kravtsov, A. V. 2003, ApJ, 590, L1
Kroeger, D. 2006, PhD thesis, University of Kiel, Germany
Kroeger, D., Hensler, G., & Freyer, T. 2006, A&A, 450, L5
Kroeger, D., Freyer, T., Hensler, G., & Yorke, H. W. 2007, A&A, submitted
Krumholz, M. R. & McKee, C. F. 2005 ApJ, 630, 250
Kudritzki, R. P., & Puls, J. 2000, ARA&A, 38, 613
Lasker, B. M. 1967, ApJ, 149, 23
Li, Y., Mac Low, M.-M., & Klessen, R. S. 2005, ApJ, 620, L19
Mac Low, M.-M. 2000, Rev. Mex. Astron. Astrofis., 9, 273

Mac Low, M.-M., & Ferrara, A. 1999, ApJ, 513, 142
Martin, C. L. 1996, ApJ, 465, 680
Martin, C. L., Kobulnicky, H. A., & Heckman, T. M. 2002, ApJ, 574, 663
Mühle, S., Klein, U., Wilcots, E. M., & Hüttemeister, S. 2005, AJ, 130, 524
Peimbert, M. 1967, ApJ, 150, 825
Recchi, S., & Hensler, G. 2006, A&A, 445, L39
Recchi, S., & Hensler, G. 2007, A&A, submitted
Recchi, S., Matteucci, F., D'Ercole, A., & Tosi, M. 2004, A&A, 426, 37
Recchi, S., Hensler, G., Angeretti, L., & Matteucci, F. 2006, A&A, 445, 875
Różyczka, M. 1985, A&A, 143, 59
Schaller, G., Schaerer, D., Maeder, A., & Meynet, G. 1992, A&AS, 96, 269
Stasińska, G. 1979, A&AS, 32, 429
Stasińska, G. 2002, Rev. Mex. Astron. Astrofis., 12, 62
Stil, J. M. & Israel, F. P. 2002, A&A, 392, 473
Strickland, D. K., et al. 2004, ApJ, 606, 829
Thornton, K., Gaudlitz, M., Janka, H.-Th., & Steinmetz, M. 1998, ApJ, 500, 95
Weaver, R., et al. 1977, ApJ, 218, 377; erratum, 220, 742
Yorke, H. W. 1986, ARA&A, 24, 49

Gerhard Hensler.

*Pathways through an Eclectic Universe*
*ASP Conference Series, Vol. 390, © 2008*
*J. H. Knapen, T. J. Mahoney, and A. Vazdekis, eds.*

# Metals in H II Regions: The Determination of Chemical Abundances

José M. Vílchez
*Instituto de Astrofísica de Andalucía, Apdo. 3004, E-18080 Granada, Spain*

**Abstract.** H II regions are sites of active star formation in galaxies harboring massive hot stars within an extended ionized gaseous component. Their relative structural simplicity and their characteristic emission line spectrum make them ideal laboratories for studying the physical and chemical properties of the gas and stars in our Galaxy and beyond. Chemical abundances can be derived for nearby H II regions and also for those located in more distant galaxies, thus helping us to study their chemical enrichment and some fundamental relations governing their evolution. In this paper I present an overview of abundances in H II regions in galaxies as a multiple spatial scale problem, ranging from spatially resolved spectra of Galactic H II regions to integrated spectra of more distant galaxies.

## 1. Introduction

### 1.1. Astrophysical Motivation

Metallicity is one of the fundamental parameters governing the evolution of galaxies. The metallicity of the interstellar medium (ISM) of star-forming galaxies can be derived from the flux of metallic emission lines present in their spectra. H II regions populate these galaxies which can be studied at different spatial scales; thanks to their characteristic emission line spectrum, H II regions are observed out to very large distances. In the case of spiral galaxies, it is well known that they show radial chemical abundance gradients (e.g., Pagel, Edmunds, & Smith 1980; McCall, Rybski, & Shields 1985; Vílchez et al. 1988; Garnett 1989; Pilyugin, Vílchez, & Contini 2004, and references therein). For these galaxies, precious information can be obtained from the determination of their abundance gradients and their maximum abundance (of elements like oxygen, nitrogen, carbon, etc.) that can be reached in the centers. This information should be very relevant, for example, to our understanding of the formation mechanisms of galaxies or to the derivation of their metal yields.

The oxygen abundance of the ISM in the Milky Way (MW) has been derived making use, for example, of measurements of the interstellar O I$\lambda$1356 Å absorption line (e.g., Sofia & Meyer 2001). Massive stars have been used to derive the abundance gradient in the MW and in nearby galaxies (e.g., Smartt & Rolleston 1997; Smartt et al. 2001; Urbaneja et al. 2005, and references therein) as well as the abundances of massive star clusters near the center of our Galaxy, such as the Arches cluster (Najarro 2007). Planetary nebulae are also useful tools for investigating abundance gradients (e.g., Stasinska et al. 2006). Giant H II regions are a powerful tool for deriving the chemical properties of the ISM in nearby and distant galaxies: these objects are luminous enough to be observed in very distant galaxies, thus allowing their chemical abundances to be derived.

The study of abundances in Galactic H II regions has been performed on the smallest spatial scales as observed with *HST*. However, for giant extragalactic H II regions, such as those observed in the disks of spiral galaxies, the spectroscopic information available gives us only average properties for their ionized gas and star clusters. The situation appears more complicated when observing the so-called H II galaxies, for which the spectra obtained can be integrated over spatial scales of kiloparsecs. This spatial scale limitation becomes very serious in the case of distant star-forming galaxies, the spectroscopic observations of these objects corresponding typically to the integrated flux of the entire galaxy. Therefore, physical properties such as the electron temperature or the electron density of the ionized gas, which can be derived from corresponding line ratios, must also be understood as average values and they should not be used independently of the corresponding spatial scales for which they were derived.

The number of results on abundance gradients of relatively nearby spiral galaxies has increased in recent years; however, this is not the case for the study of more distant galaxies, for which the slope of their abundance gradients and maximum central metallicity are still not well known. This multi-spatial scale nature of the derivation of chemical abundances in H II regions is an essential ingredient for the interpretation of the spectroscopic observations, particularly in the case of distant galaxies. It seems clear that, given the huge range of spatial scales involved, before deriving chemical abundances from integrated spectra we should understand the effect that the loss of spatial information has on the final results. This way, we could minimize the uncertainties in the physical interpretation of the results, e.g., when using line ratio proxies from collisionally excited lines (CEL) measured in the optical (Bresolin 2007b).

The derivation of chemical abundances from the spectra of H II regions can be carried out using the permitted, faint recombination lines (ORL) or either the brighter CEL, easily seen in their spectra, as can be seen in Fig. 1 for the Orion Nebula (adapted from Sánchez et al. 2007). The use of the CEL observed in the optical range requires an accurate derivation of the electron temperature, given the strong dependence of these lines on this variable. On the other hand, the emissivities of the infrared CEL or of the optical ORL do not depend very much on the electron temperature. Unfortunately, measurements of the faint ORL in external galaxies are still very scarce and require the use of very large aperture telescopes; conversely, observation of the infrared CEL must be performed from space observatories. Relevant work on the use of the ORL (Esteban 2002, and references therein) and of the infrared CEL (Bernard-Salas et al. 2005, and references therein) for the derivation of chemical abundances can be found in the recent literature.

### 1.2. Sources of Uncertainty: Some Words of Caution

It this section I comment on several aspects that should be taken into account in the process of the derivation of chemical abundances in H II regions. The standard, direct abundance derivation for H II regions based on the optical CEL relies on the accurate measurement of the electron temperature. However, the measurement of the electron temperature is very difficult for (cooler) metal-rich

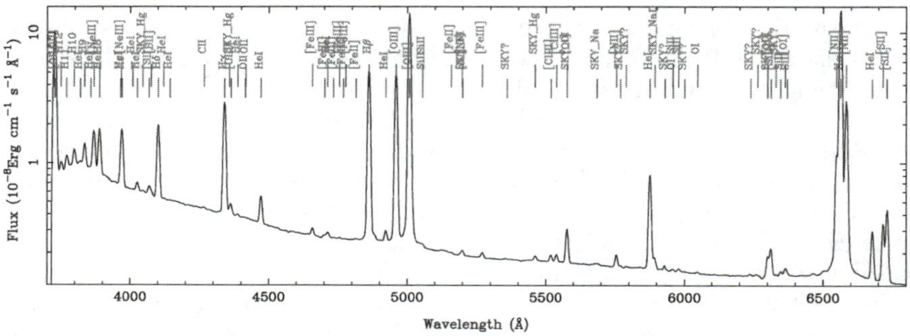


Figure 1. The integrated spectrum of the Orion Nebula as observed with the IFU of the PPAK spectrograph. See Section 2 for details.

H II regions, such as those populating the inner zones of spirals (e.g., Bresolin 2007a). In addition, the determination of accurate oxygen abundances, O/H, of H II regions can be subject to several corrections; some of them are related to the following physical scenarios:

1. The existence of electron temperature fluctuations within the H II region (Peimbert 1967),
2. The presence of chemical inhomogeneities within the H II region (e.g., Tsamis & Pequignot 2005),
3. The possible depletion of oxygen on dust grains (e.g., García-Rojas et al. 2007; Simon-Diaz 2007).

All in all, it seems that the corrections of the oxygen abundance due to the effects mentioned above can reach, on average, values of order 0.2 dex. These corrections, though important and non-negligible, are comparable to the typical errors of the O/H abundance derived for distant galaxies.

On the other hand, perhaps one of the more relevant aspects—frequently overlooked—in physical models of H II regions is the role played by the exact geometry of the region (see Ercolano, Bastian, & Stasinska 2007; Ercolano, this volume, p. 110, for the MOCASSIN 3D code). The inclusion of this factor could provide new light, for example, on our understanding of some of the discrepancies observed between the abundances derived from CEL and ORL in the optical, versus the corresponding abundances derived from CEL in the infrared.

Another source of uncertainty that appears relevant for the subject of this paper is the fact that recent model predictions imply that direct (i.e., from measurements of the electron temperature) derivation of the O/H abundance for high metallicity H II regions could produce wrong values (Stasinska 2005). These predictions seem to be based on the presence of strong internal temperature gradients in H II region models for high metallicity. Observations of high metallicity H II regions in M101 (e.g., Bresolin 2007a; Haegele et al. 2007), however, do not appear to show this effect. In any case, high metallicity is not the only cause that could be invoked to produce strong electron temperature gradients inside H II regions; these temperature gradients can be produced even in the case of H II regions of nearly primordial metallicities, as illustrated by H II region models ionized by a Population III star (see Abel, Wyse, & Bryan 2007). Given the high sensitivity of the [O III]λ4363 Å line emissivity to the electron temperature it is very difficult to observe this line for regions with metallicity around twice solar or higher: this seems true even for the largest aperture telescopes presently available. However, other temperature-sensitive lines in the optical range, such as [N II]λ5755 Å and/or [S III]λ6312 Å can be efficiently measured in this high metallicity domain. In addition, within the standard two-zone ionization structure scheme (frequently adopted for the derivation of abundances in H II regions), each of the two zones has a different characteristic electron temperature; therefore, an accurate derivation of both electron temperatures is needed. In the case that only one of the two temperatures is known, a physically based procedure has to be applied in order to estimate the other electron temperature (e.g., Garnett 1992; Pilyugin et al. 2006; Pilyugin 2007): Fig. 2 illustrates the effect of this correction.

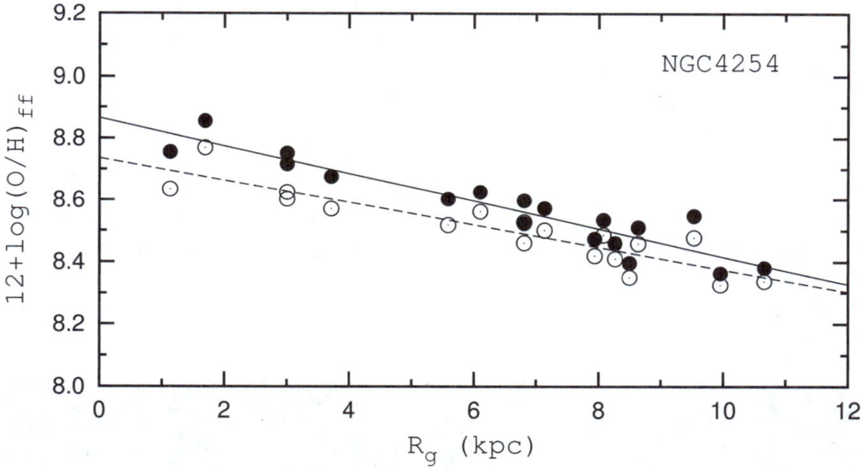


Figure 2. The O/H abundance gradient of NGC 4254. The abundance points derived taking into account the correction for the existence of two electron temperatures within the H II regions (black dots) are shown with respect to the uncorrected points (empty circles). Assuming 0.08 dex for the fraction of oxygen locked into dust grains, the maximum ISM oxygen abundance can reach up to 12 + log(O/H)$_{max}$ ~ 8.95. The effect in the gradient slope is non-negligible. See the text for details.

Finally, I want to comment here another aspect which has been (frequently) neglected: we observers many times (naively) expect to derive well-behaved, smooth radial abundance gradients; however, these well-behaved gradients may not exist at all, especially in those galaxies subject to substantial gas flows (e.g., see the potential effects on these gradients of cyclonic/anticyclonic gas flows at corotation, according to the models of Vorobyov 2007)! It is important to bear this point in mind since it should affect the derivation of the slope of the gradient, and possibly also the derivation of the maximum O/H value. Therefore, hydro models could provide new constraints on the formation mechanisms of disk galaxies and on their chemical evolution. Thus, the existence of 2D—or even 3D—galactic structures associated with possible mass flows in the disks appears to be of extraordinary relevance for chemical evolution studies. A corollary is that the next step forward in the study of chemical abundances in galaxies would require using the powerful 3D spectroscopic capabilities now provided by present (and future) integral field unit (IFU) instruments: new observations should produce realistic abundance maps of the galaxies, allowing for the detection of possible asymmetries in the abundance gradients and/or constraining the chemical homogeneity of the ISM (e.g., Vílchez & Iglesias-Páramo 1998; Tsamis & Pequignot 2005, and references therein) across entire galaxies.

## 2. Physical and Chemical Properties from 2D Spectroscopy

An original study in this respect has been carried out using the IFU of the PPAK spectrograph at Calar Alto (CAHA, Almeria, Spain). A low resolution spectroscopic survey of the Orion Nebula has been performed constructing an IFU mosaic of an area of ~5 arcmin × 6 arcmin centered on the Trapezium, including the ionization front to the south-east. The survey includes 8182 individual spectra, each sampling a circular area of ~2.7 arcsec diameter.

The integrated spectrum (see Fig. 1) was used to obtain the main integrated properties of the nebula, including the electron density and temperature, the extinction, the H$\alpha$ integrated flux, etc. Individual spectra were used to build up line intensity maps. In Fig. 3 we show a sample of these results. The ionization structure, the dust extinction, the electron density and temperature, and the helium and oxygen abundance maps were derived (see Sanchez et al. 2007 for details). All of them show a considerable degree of structure, as already shown in previous studies.

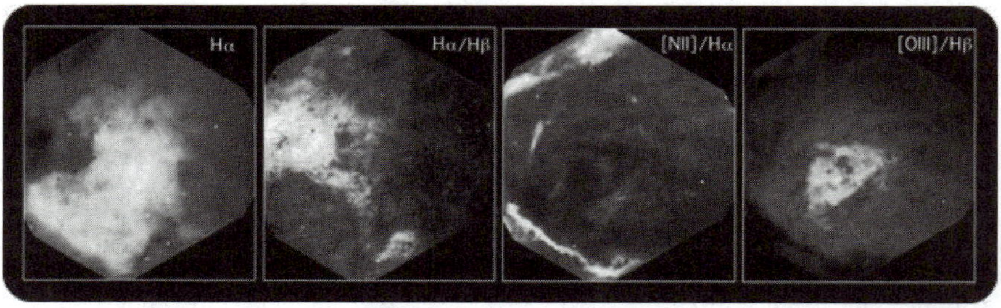


Figure 3. The Orion Nebula as seen by PPAK at the 3.5 m (Calar Alto). *From left to right:* The spatial distribution of: the emission in H$\alpha$; the Balmer decrement as measured by H$\alpha$/H$\beta$; the excitation as indicated by the [N II]/H$\alpha$ and [O III]/H$\beta$ ratios. The size of the field is ~5 arcmin × 6 arcmin centered on the Trapezium. North is up and east is left (see Sánchez et al. 2007 for details).

Another example of the application of 2D spectroscopy to H II region abundance studies is the survey across the disk of the spiral galaxy M33 for which spectra have recently been obtained with the multi-object, wide field spectrograph AF2/WYFFOS at the 4.2 m William Herschel Telescope (WHT) by Magrini et al. (2007). Line intensities, extinction, and electron density were determined for the whole sample of 72 objects. A direct derivation of the electron temperature was obtained for a subset of H II regions. For these regions chemical and physical properties were derived, and from them the metallicity gradient of M33. The presence of abundance gradients for several elements was inferred from the radial behavior of several emission-line ratios, and measured from the chemical abundances directly derived. The oxygen abundances of our H II regions, located in the radial region from ~2 to ~7.2 kpc, gave an oxygen gradient of $-0.054 \pm 0.011\,\mathrm{dex\,kpc^{-1}}$. The overall oxygen gradient for M33 has been obtained using our results, as well as previous oxygen abundance values (obtained from direct electron temperature determination), together with the abundances of young massive stars in M33; the gradient appear to present two slopes: $-0.19\,\mathrm{dex\,kpc^{-1}}$ for the central regions ($R < 3$ kpc), and $-0.04\,\mathrm{dex\,kpc^{-1}}$ for the outer regions ($R \geq 3$ kpc). Nonetheless, the N/O abundance ratio appears to show no substantial variation; this ratio presents a rather flat behavior as illustrated in Fig. 4, where the N/O versus O/H relation is shown for M33 H II regions.

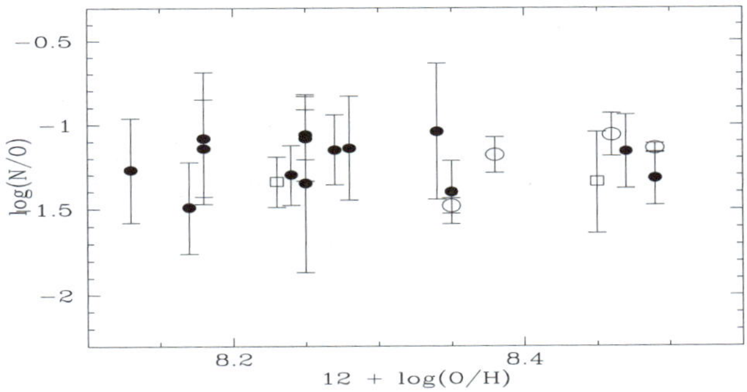


Figure 4. The N/O versus O/H relationship obtained from a sample of H II regions in M33. Observations were performed with AF2 at the 4.2 m WHT (La Palma).

## 3. The Maximum O/H and the Metallicity–Luminosity Relation

In our recent work we have followed a direct methodology to determine the maximum oxygen abundance in a spiral galaxy. Galaxy disks typically show negative O/H radial abundance gradients—a result that has been well known since the seventies (e.g., Shields & Searle 1978; Pagel et al. 1979). The method we have applied required first to obtain reliable O/H gradients across the disks of the galaxies, and second extrapolating them to their respective galaxy centers. In doing that, the oxygen abundance will reach its maximum value $(\mathrm{O/H})_{\mathrm{max}}$ when the galactocentric radius equals zero. This derivation can have strong implications, e.g., for the determination of the oxygen yield in these galaxies, or for the correct understanding of the metallicity–luminosity (or the mass–metallicity) relation, thus helping us to constrain models of galaxy evolution. In

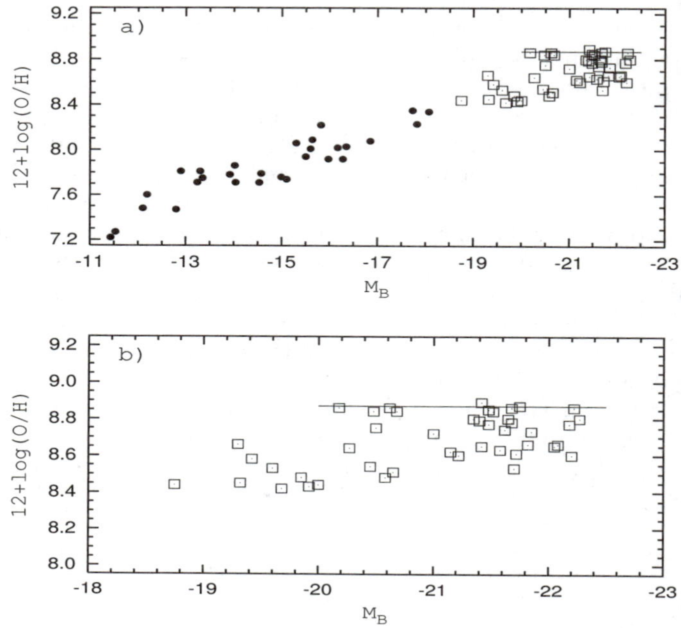


Figure 5. The metallicity–luminosity relation obtained for our extended sample of galaxies (*a*). Note the apparent flattening of the relation for absolute blue magnitude $M_B \leq -20$ (*b*). Filled dots: dwarf galaxies; open symbols: spiral galaxies.

addition, it could be a useful tool for the interpretation of the observations of high redshift emission-line galaxies.

An open question is how does the maximum metallicity depend on galaxy mass or galaxy luminosity? In order to answer this question, we first have to derive a reliable value for the maximum ISM oxygen abundance of a disk galaxy. The abundance of oxygen has been derived for an extended sample of giant H II regions using our compilation of high quality spectroscopic data for spiral galaxies. The O/H abundance gradient derived for the Milky Way gives an average oxygen abundance at the radius of the solar vicinity which agrees very well with the abundance of oxygen derived from measurements of the interstellar O I$\lambda$1356 Å absorption line; an abundance of 284 to 390 oxygen atoms per million hydrogen atoms (Meyer et al. 1998; Sofia & Meyer 2001; Cartledge et al. 2004) is derived, equivalent to 12 + log(O/H) between 8.45 to 8.59.

For the sample of galaxies, the extrapolation to their centers of the O/H gradients we have obtained provides maximum gas-phase oxygen abundance values which do not appear to exceed $12 + \log(\mathrm{O/H})_{\mathrm{max}} \sim 8.87$ for the largest galaxies (see Pilyugin et al. 2006). Assuming 0.08 dex as the maximum fraction of oxygen locked into dust grains, then the maximum ISM oxygen abundance can attain up to $12 + \log(\mathrm{O/H})_{\mathrm{max}} \sim 8.95$ or slightly larger, if temperature fluctuations are present (e.g., García-Rojas 2006). This abundance value would represent an upper limit for metallicity in the metallicity–luminosity relation. This relation (Pilyugin et al. 2004, 2006) appears to show a flattening of the O/H abundance in the luminosity range corresponding to the most massive galaxies, as can be seen in Figure 5. Recent studies (Tremonti et al. 2004; Lamareille et al. 2006) have explored this range of the metallicity–luminosity relation in extended samples of galaxies, although their O/H values appear to be a bit larger than ours. Similarly, the flattening of the metallicity–luminosity relation appears to be consistent with the findings of Erb et al. (2006) for high redshift star-forming galaxies.

**Acknowledgments.** I want to thank the LOC/SOC for an exciting and well organized conference. It is a pleasure to acknowledge here my collaborators on chemical abundances for their help and continuous support. This research has been partially funded by project "Estallidos" AYA2004-08260-C03 of the Spanish PNAYA.

## References

Abel, T., Wise, J. H., Bryan, G. L. 2007, ApJ, 659, L87
Bernard-Salas, J. et al. 2005, in AIP Conf., Planetary Nebulae as Astronomical Tools, 804, p. 56
Bresolin, F. 2007a, ApJ, 656, 186
Bresolin, F. 2007b, in The Metal Rich Universe (Cambridge: CUP), in press
Cartledge, S. I. B., et al. 2004, ApJ, 613, 1037
Erb, D., et al. 2006, ApJ, 644, 813
Ercolano, B., Bastian, N., & Stasinska, G. 2007, MNRAS, in press
Esteban, C. 2002, RMexAASC, 12, 56
García-Rojas, J. 2006, PhD Thesis, University of La Laguna
García-Rojas, J. et al. 2007, in The Metal Rich Universe (Cambridge: CUP), in press
Garnett, D. 1989, ApJ, 345, 282
Garnett, D. 1992, AJ, 103, 1330
Haegele, G., Diaz, A. I., Cardaci, M., Terlevich, E., & Terlevich, R. 2007, MNRAS, 378, 163
Lamareille, F. et al. 2006, A&A, 448, 907
Magrini, L., Vílchez, J. M., Mampaso, A., Corradi, R., & Leisy, P. 2007, A&A, in press
McCall, M. L., Rybski, P. M., & Shields, G. 1985, ApJS, 57, 1
Meyer, D. M., Jura, M., & Cardelli, J. A. 1998, ApJ, 493, 222
Najarro, F. 2007, in The Metal Rich Universe (Cambridge: CUP), in press
Pagel, B. E. J., et al. 1979, MNRAS, 189, 95
Pagel, B. E. J., Edmunds, M. G., & Smith, G. 1980, MNRAS, 193, 219
Peimbert, M. 1967, ApJ, 150, 825
Pilyugin, L. S. 2007, MNRAS, 375, 685
Pilyugin, L. S., Vílchez, J. M., & Thuan, T. X. 2006, MNRAS, 370, 1928
Pilyugin, L. S., Vílchez, J. M., & Contini, T. 2004, A&A, 425, 849
Sanchez, S., Cardiel, N., Verheijen, M., Martin-Gordon, D., Vílchez, J. M., & Alves, J. 2007, A&A, 465, 207
Simon-Diaz, S. 2007, in The Metal Rich Universe (Cambridge: CUP), in press
Shields, G. A., & Searle, L. 1978, ApJ, 222, 821
Smartt, S., et al. 2001, A&A, 367, 86
Smartt, S., & Rolleston, W. R. J. 1997, ApJ, 481, L47
Smith, H. E. 1975, ApJ, 199, 591
Sofia, U., & Meyer, D. M. 2001, ApJ, 554, L221
Stasinska, G., et al. 2006, in ESO Astrophysics Symposia PNe beyond the Milky Way, (Berlin: Springer), p. 234

Stasinska, G. 2005, A&A, 434, 507
Tremonti, C., et al. 2004, ApJ, 613, 898
Tsamis, Y. G., & Pequignot, D. 2005, MNRAS, 364, 687
Urbaneja, M. A., et al. 2005, ApJ, 635, 311
Vílchez, J. M., & Iglesias-Páramo, J. 1998, ApJ, 508, 248
Vílchez, J. M., Pagel, B. E. J., Díaz, A. I., Terlevich, E., & Edmunds, M. G. 1988, MNRAS, 235, 633
Vorobyov, E. I. 2006, MNRAS, 370, 1046

## Discussion

*Edmunds*: To push a bit further on the IR/optical H II region N/O abundance discrepancy. What is happening? Is it the optical results or the IR results that are changing, or are they meeting in the middle?

*Vílchez*: What I can say is that the situation has much improved in recent years, and that in the majority of the cases both collisional line determinations in the IR and optical seem to agree within the errors.

*Ercolano*: Temperature gradients are a problem at high metallicity and also at very low metallicities (due to Ly$\alpha$ collisional excitation cooling, the same way as the decrease of photoionization through geometrical dilation). However, at intermediate ($0.2 \leq Z/Z_{\odot} \leq 1$) metallicities temperature gradients are much shallower and present a lesser problem to the determination of abundances.

*Vílchez*: I fully agree. Let me comment here only that the presence of temperature gradients has been known for many years now.

*Churchwell*: Is the current best abundance gradient determined for the Milky Way consistent with the temperature gradient measured independently from radio recombination lines?

*Vílchez*: This is a very good point. Unfortunately, apart from the reference you have mentioned, as well as few classical papers (e.g., Shaver et al. 1983, MNRAS, 204, 53), not enough data exists on radio recombination line temperatures. Indeed, we are very interested in this aspect and are trying to study it in more depth, observationally and theoretically.

*Pagel*: The flat N/O in M33 is a result you and I got 20 years ago. However, in the Milky Way, there have been discrepancies between optically and IR derived N/O ratios, possibly related to the $N^{+}$–$N^{++}$ ionization equilibrium. Would you like to comment?

*Vílchez*: You are right. The data are not so "clean" for the Milky Way, but new measurements in the IR and improved N/O determinations appear to reach a measure of agreement for collisionally excited lines.

*Pathways through an Eclectic Universe*
*ASP Conference Series, Vol. 390, © 2008*
*J. H. Knapen, T. J. Mahoney, and A. Vazdekis, eds.*

# Collisional Enhancement of H I Lines in Ionized Nebulae

Valentina Luridiana

*Instituto de Astrofísica de Andalucía (CSIC), Apdo. 3004, E-18008 Granada, Spain*

**Abstract.** Collisional excitation of neutral hydrogen atoms can significantly increase the intensity of Balmer lines with respect to pure recombination. If this effect is not taken into account, the abundance analysis of these objects returns biased values, and the bias may be significant if accuracies better than a few percent are required. The most affected objects are young, metal-poor H II regions, due to their comparatively large temperatures. Until now, estimates of collisional enhancement have been based on tailored modeling of individual H II regions. In this contribution, I describe an ongoing effort to develop a general calibration, suitable for application to large samples of objects. Emphasis is placed on the uncertainties affecting the resulting predictions.

## 1. Introduction

Gaseous nebulae are characterized by bright emission spectra, produced by different ions through diverse mechanisms. Two of the main mechanisms are recombination and collisional excitation. Recombination is the main formation mechanism for the hydrogen and helium spectra, but the lines of both elements can also be excited through collisions with thermal electrons. In the case of hydrogen, collisional excitation enhances the lines by small, but not always negligible, amounts—typically a few percent. This mechanism has received relatively little attention, both because its accurate estimation is difficult, and because the large electron energies it requires make it small anyway.

Early in 1985, Davidson & Kinman called attention to the fact that H I collisional excitation could introduce a bias in abundance determinations, but only a few quantitative estimations have been produced since (Stasińska & Izotov 2003; Luridiana et al. 2003; Izotov, Thuan, & Stasińska 2007; Peimbert, Luridiana, & Peimbert 2007). A common drawback of these works is that they are restricted to specific objects, while general prescriptions would be desirable for applications to large samples, or to bypass the need for a tailored modeling of individual regions. A further drawback of the existing literature on the subject is the lack of estimates of the uncertainty. Uncertainty is an important piece of information as it puts a lower limit on the accuracy attainable in abundance determinations; since hydrogen is the reference element to which the abundances of all the other elements are referred, it is desirable that its spectrum be fully understood and correctly interpreted, lest biased abundances be derived.

## 2. Features of the Balmer Collisional Spectrum

The collisional excitation rate of Balmer line H$n$ depends on three factors: the effective collision strength $\Upsilon(1,n';T_e)$ (with $n' \geq n$), the Boltzmann factor $\exp(-\Delta E(1,n')/kT_e)$, and the probability $P(n',\mathrm{H}n)$ for an electron on level $n'$ to cascade through H$n$. These factors concur to favor lower series over higher series photons (e.g., H$\alpha$ over H$\beta$). The resulting purely collisional H$\alpha$/H$\beta$ ratio depends on temperature but is consistently larger than the case B ratio (Fig. 1).

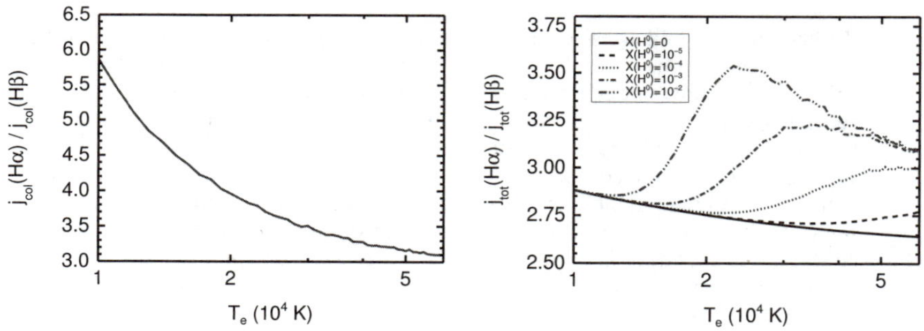


Figure 1. Purely collisional (*left*) and total (*right*) H$\alpha$/H$\beta$ ratios as a function of $T_e$. The total H$\alpha$/H$\beta$ ratio is given for different H ionization fractions.

## 3. Collisions in Model Nebulae

The collisional rate at any given point in a nebula is determined by the electron temperature $T_e$ and the ionization fraction of neutral hydrogen $X(\mathrm{H}^0)$. Typical $T_e$ values in nebulae are in the range $T_e \approx 10\,000$–$20\,000$ K; typical $X(\mathrm{H}^0)$ values are in the range $X(\mathrm{H}^0) \approx 10^{-5}$–$10^{-3}$ in the inner parts, approaching $X(\mathrm{H}^0) \lesssim 1$ in the outer parts. In this section I will describe the collisional intensity predicted from the interplay between these two variables in the models of a multidimensional grid of H II regions. The models of the grid are appropriate to represent giant H II regions and have been computed with the photoionization code CLOUDY (v. 07.02.00; Ferland et al. 1998), with ionizing sources computed with the evolutionary code by Cerviño, Mas-Hesse, & Kunth (2002), while the collisional contribution has been computed with the collisional data by Anderson et al. (2000, 2002) and the models' output structure.

Figure 2 shows how the predicted collisional intensities vary with selected parameters. The collisional enhancement in the reference model, around which the grid is constructed and whose position is marked by a gray vertical line, is $\mathrm{H}\alpha_{\mathrm{col}}/\mathrm{H}\alpha_{\mathrm{tot}} = 0.025$ (0.011 dex) and $\mathrm{H}\beta_{\mathrm{col}}/\mathrm{H}\beta_{\mathrm{tot}} = 0.014$ (0.006 dex).

*Integration Region* Panel a) of Fig. 2 shows the effect of variations in the integration region. The points labeled "1D", "2D" and "3D" represent the collisional intensities of the reference model seen through a point-like slit, a long slit, and a slit covering the complete nebula, respectively. The distinction is relevant because each line has a different distribution within the nebula. The sequence of results indicates that the dominant contribution to the collisional emissivity comes from a more external region of the nebula than the dominant contribution to the recombination emissivity. The reference integration adopted in the following model sequences is the 3D one. Note that the distinction between the various types of integration is relevant only in spherical or thick-shell models, but not in the case of plane–parallel nebular geometries.

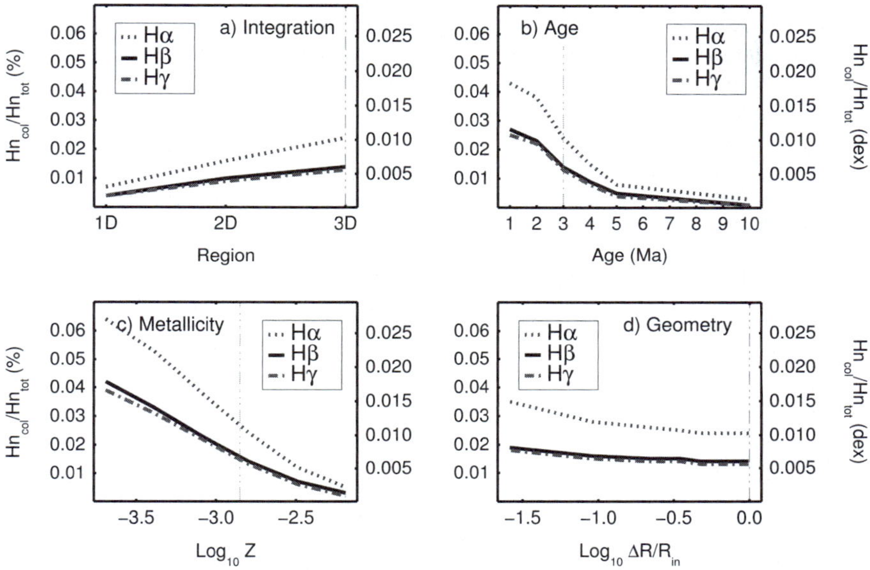


Figure 2. Dependence of the fractional collisional contribution to Hα, Hβ, and Hγ on several parameters, as labeled and as discussed in the text.

*Age* Panel b) shows the effect of the ionizing cluster's age. The monotonic decrease in the fractional collisional intensity is a consequence of the decrease in the models' temperature, which in turn is a consequence of the progressive softening of the spectrum as the cluster ages. At larger metallicities, larger numbers of Wolf–Rayet stars form, hardening the ionizing flux at around 3 Ma. This evolution should, in principle, bring about a simultaneous increase of collisional intensities, but at larger metallicities collisions become more and more negligible, due to the decrease in $T_e$ (see below). From a practical point of view, then, it can be assumed that the collisional contribution decreases monotonically with the H II region's age.

*Metallicity* Panel c) shows the effect of the gas metallicity, which is varied from $Z = 2 \times 10^{-4}$ to $Z = 6 \times 10^{-3}$. The collisional contributions in this sequence of models vary roughly linearly with log $Z$ over one order of magnitude, eventually leveling off at the largest metallicities. The main trigger of this behavior is the gas temperature, which decreases monotonically with increasing $Z$. Note that the relevant temperature here is not the model's average temperature $T_0$, but rather the temperature of the outermost zone, due to the emissivity profile of collisional Balmer lines. Since this profile very nearly coincides with the emissivity profile of [O II] lines (Peimbert et al. 2007; Luridiana, in preparation), the sequence of the [O II] temperatures is a very good predictor of the sequence of collisional contribution.

*Geometry* Panel d) shows the effect of the assumed geometry. The sequence depicts a progression from an effectively plane–parallel symmetry ($\Delta R/R_{\rm in} \ll 1$) to a spherical symmetry ($\Delta R/R_{\rm in} \simeq 1$) at constant radius and filling factor. The differences among the models of this sequence reside mostly in the internal portion; while the innermost layers in the reference (spherical) model reach temperatures larger than 20 000 K due to the inefficient cooling (oxygen is mostly O IV and O V), the innermost layers of plane–parallel models have a smaller ionization degree (due to the combined effect of the increased density and the geometrically diluted radiation field), and they can be efficiently cooled by oxygen. On the other hand, the outer layers are cooled by the same mechanism and display the same $X(\mathrm{H}^0)$–$T_{\rm e}$ interplay. Since collisional Balmer lines form mostly in the outer zone, collisional intensities are constant along this sequence.

## 4. Uncertainties

Estimations of collisional enhancement are badly needed for high precision applications, such as the determination of primordial helium. However, predictions are affected by several sources of uncertainty, whose amplitude must also be known as they directly limit the accuracy of abundance determinations. I mention here a few of these sources: a) the collision strengths $\Upsilon(1, n; T_{\rm e})$ are affected by large uncertainties, especially at large $n$; b) only excitations to the lowest $n$ are taken into account in this work, due to the lack of a complete, homogeneous set of $\Upsilon(1, n; T_{\rm e})$, and the progressively larger uncertainties at large $n$; c) simple photoionization models are not always good approximations of real H II regions; in particular, the temperature structure of real objects is much more complex.

**Acknowledgments.** I wish to thank G. Ferland, M. Cerviño, K. Butler, and M. and A. Peimbert for helpful discussions on the topic, and J. Maíz for help with the plots. The work was supported by PNAYA grants 2004-02703 and 2004-07466. A special note of thanks to John and the organizers for a most enjoyable meeting.

### References

Anderson, H., et al. 2000, J. Phys. B, Atomic and Molecular Physics, 33, 1255
Anderson, H., et al. 2002, J. Phys. B, Atomic and Molecular Physics, 35, 1613
Cerviño, M., Mas-Hesse, J. M., & Kunth, D. 2002, A&A 392, 19
Davidson, K., & Kinman, T. D. 1985, ApJS, 58, 321
Ferland, G., et al. 1998, PASP, 110, 761
Izotov, Y. I., Thuan, T. X., & Stasińska, G. 2007, ApJ 662, 15
Luridiana, V., Peimbert, A., Peimbert, M., & Cerviño, M. 2003, ApJ, 592, 846
Peimbert, M., Luridiana, V., & Peimbert, A. 2007, ApJ, 666, 636
Stasińska, G., & Izotov, Y. 2003, A&A, 397, 71

## Discussion

*Ercolano*: How does the uncertainty of 3% introduced by collisional excitation of the Balmer line fit in with potentially larger uncertainties in temperature determinations?

*Luridiana*: Roughly speaking, both the bias and the additional error bar associated with $H^0$ collisions are comparable to those associated with temperature fluctuations.

*Churchwell*: Are the $H^0$ densities in H II regions consistent with the collisional excitation required to give the suggested 2–3% contribution to H$\beta$ emission over that of radiative excitation?

*Luridiana*: Yes, they are collisions with free electrons so they go only linearly with $N(H^0)$, which is sufficient to produce the quoted rates. The electron densities considered range from ~100 to 10 000 but as a matter of fact the result does not depend on density (the larger the density, the more neutral, but also the colder the medium so that the collisional rate remains essentially constant).

From left to right: Valentina Luridiana, John Beckman, and Adriana de Lorenzo-Cáceres.

*Pathways through an Eclectic Universe*
*ASP Conference Series, Vol. 390, © 2008*
*J. H. Knapen, T. J. Mahoney, and A. Vazdekis, eds.*

# Medium-term Variability in Emission from the Orion Nebula

J. E. Beckman,[1,2] L. Gutiérrez,[1,3] and C. Giammanco[4]

[1]*Instituto de Astrofísica de Canarias, La Laguna, Tenerife, Spain*
[2]*Consejo Superior de Investigaciones Científicas, Spain*
[3]*Universidad Nacional Autónoma de México, Campus Ensenada, Mexico*
[4]*Physikalisches Institut der Universität Bern, Switzerland*

**Abstract.** Measuring the surface brightness changes in *HST* images in H$\alpha$ and [O III]5007 Å taken with a time interval of 6.84 yr, we show how to detect separately the changes due to density variations and those due to temperature variations. Following O'Dell et al., we used the brightness changes to show and measure proper motions in streams and fronts at up to 140 km s$^{-1}$. The strongest density variations coincide with the proper motions, while in quiescent zones we can detect mean temperature variations of tens of degrees along the line of sight in the time interval between the observations.

## 1. The Data and Selection of Zones for Variability Analysis

The images used were through three narrow-band filters, one for the [O III] line, one for H$\alpha$ and one close by to calibrate out the [N II] emission (O'Dell & Doi 1999). Ten zones chosen for variability analysis were free from (a) saturated pixels, (b) edges of CCDs, and (c) scattered light contamination from $\theta$'Ori C. The change in surface brightness $\Delta I/I$ between the 1995 and 2002 observations for any point on an image and for any filter is given by $\Delta I/I = 1 - I_{1995}/I_{2002}$. On single pixel scales the maps of $\Delta I/I$ are noise-dominated, but at lower frequencies, on scales of a few arcsec, significant structure (though not of large amplitude) is clearly seen (Fig. 1).

## 2. Proper Motions and Filtering the Difference Images

We divided the image into small zones and for each area found, by maximizing the cross correlation, the shift $\delta$ in pixels for the time interval of 6.84 yr. A single scale value of 16 km s$^{-1}$ pixel$^{-1}$ converts this into a linear transverse velocity. Combining the vectors $\delta$ over all the zones, we produced a proper motion map of a large fraction of the nebula. We designed a low-pass filter with unit amplitude at low frequencies and a cosine roll-off to zero amplitude at high frequencies to filter out much of the noise. In fields with clear proper motion, rises (or falls) in H$\alpha$ brightness coincide with rises (or falls) in [O III] brightness, but in quiescent fields the changes tend to be in opposite senses for the two emission lines.

## 3. Deriving the Temperature Variations

Giammanco, Gutiérrez, & Beckman (2007) give formulae for the change in surface brightness for each emission line as a combined function of temperature and density changes. These show that, where density changes are negligible, a rise in temperature will cause a (quantified) rise in the [O III] brightness but a (quantified) fall in the H$\alpha$ brightness (and vice versa for a fall in temperature). Density changes, on the other hand, yield identical brightness changes in sign and amplitude for both lines. From these equations we could derive the fractional temperature and density changes across the face of the nebula between the two epochs.

Using the *HST* emission line images described we inferred the presence of proper motions in the gas with speeds of tens of km s$^{-1}$, in good agreement with work by Doi, O'Dell, & Hartigan (2002) and O'Dell & Doi (2003), though our proper motion map covers more of the nebula. In quiescent zones H$\alpha$ and [O III] frequently show variations in opposite senses implying dominant effects of temperature variations (Fig. 1, *left*), as predicted by Giammanco et al. (2007). In zones with notable proper motion the surface brightness changes between the epochs are of comparable amplitude and in the same sense (Fig. 1, *right*). Comparing the respective amplitudes we can separate density and temperature effects, deriving values for the latter of a few tens of K during the 6.84 yr interval.

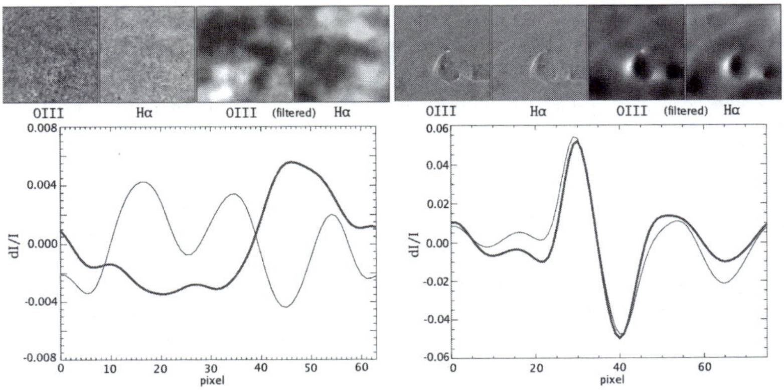


Figure 1. This figure shows the changes in [O III] and H$\alpha$ in two fields: quiescent (*left*) and with high proper motion (*right*), before and after filtering, as well as a scan across the field.

**Acknowledgments.** This work was supported by project P3/86 of the IAC and AYA2004-08251-CO2-01 of the Spanish Ministry of Education and Science.

### References

Giammanco, C., Gutiérrez, L., & Beckman, J. E. 2007, submitted
Doi, T., O'Dell, C. R, & Hartigan, P. 2002, ApJ, 124, 445
O'Dell, C. R., & Doi, T. 1999, PASP, 111, 1316
O'Dell, C. R., & Doi, T. 2003, AJ, 125, 277

*Pathways through an Eclectic Universe*
*ASP Conference Series, Vol. 390, © 2008*
*J. H. Knapen, T. J. Mahoney, and A. Vazdekis, eds.*

# Galactic Interstellar Gas Cloud Mass Functions: A Simple Quantitative Approach

Emilio Casuso[1] and John E. Beckman[1,2]

[1]*Instituto de Astrofisica de Canarias,E-38200 La Laguna, Tenerife, Spain*
[2]*Consejo Superior de Investigaciones Científicas, Madrid, Spain*

**Abstract.** We present here a simple approach to understanding the gas cloud mass distribution function by simulating formation and destruction of gas clouds and gas clumps in the interstellar medium. We include as relevant processes coagulation to form bigger clouds, as well as disruption by collisions and the removal of gas by collapse to form stars. We evolve initial sets of pre-existing gas clumps with a range of initial distribution functions (flat, Gaussian, fractal) for their physical parameters, and with different geometrical forms (spherical or elongated) for the individual clouds, and constrain them within an imaginary box representing gravitational bounding, applying the kinematic laws of non-elastic collisions. The results agree well with observations of the mass distribution function of Galactic giant gas clouds, notably if we choose a Gaussian for the initial distribution function, and initial gas clouds which are quasi-spherical.

The mass functions (MFs) of astrophysical objects are among their most important collective parameters. There are obvious reasons to look for the links between the MFs of gas clouds and clumps within the interstellar medium (ISM) in galaxies and in the intergalactic medium (IGM) and those of the galaxies, star clusters and stars which eventually form from the gas. In recent studies there has been a tendency to believe that there are characteristic mass scales associated with stellar formation and by implication with ISM clouds (see Larson 1998; Casuso & Beckman 2002).

The dynamical processes which lead to the coagulation and the disruption of interstellar molecular clouds, and to their eventual condensation in stars, are the processes which also determine the quasi-equilibrium distribution of their present-day masses. The treatment presented here, in spite of its dynamical simplicity, yields certain advances compared with previous models. These are: (a) the reproduction not only of the global form of the mass function but also of the specific structures within it, namely peaks localized in mass; (b) the reproduction of the fall of the MF towards lower masses; and (c) the results require quite realistic inputs and normal physical assumptions. Of the models examined, the best fit to the observations was found with the following model: input Gaussian distributions of temperature, with parameters (40 K, 40 K, 80 K), of density, with parameters ($0.025\times10^5$, $0.25\times10^6$, $0.8\times10^5$, units $M_\odot\,\mathrm{pc}^{-3}$), of mass, with parameters ($1\times10^8\,M_\odot$, $1\times10^9\,M_\odot$, Jeans Mass), and of relative velocity, with parameters ($0.1\,\mathrm{pc\,Myr}^{-1}$, $1\,\mathrm{pc\,Myr}^{-1}$, $0.05\,\mathrm{pc\,Myr}^{-1}$), and with spherical clouds. The best fit in this model occurs after a time interval of 3 Myr, but the fit deteriorates very little over the next 100 Myr and this result is significant because the initial conditions need to be set up again after any

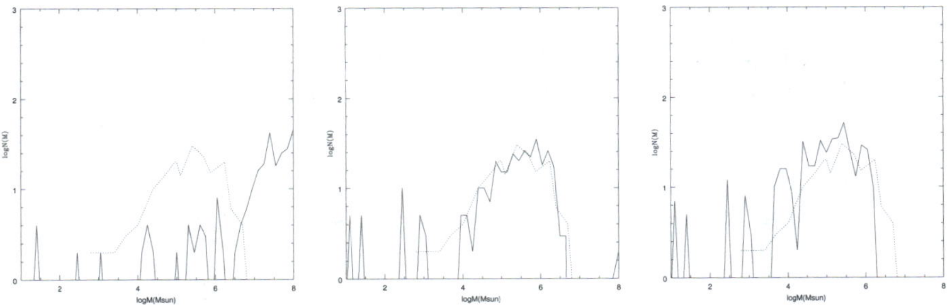


Figure 1. Present-day mass distribution function of giant molecular clouds in our Galaxy. Data are taken from Solomon et al. (1987, *dotted line*). For our numerical model (*full line*) we assume here spherical gas clouds. *Left*: Width of Gaussian, center of Gaussian, and critical value, for: temperatures (40 K, 40 K, 80 K), densities ($0.025 \times 10^5\, M_\odot\, \mathrm{pc}^{-3}$, $0.25 \times 10^6\, M_\odot\, \mathrm{pc}^{-3}$, $0.8 \times 10^5\, M_\odot\, \mathrm{pc}^{-3}$), masses ($1 \times 10^8\, M_\odot$, $1 \times 10^9\, M_\odot$, Jeans mass), velocities ($0.1\,\mathrm{pc\,Myr}^{-1}$, $1\,\mathrm{pc\,Myr}^{-1}$, $0.05\,\mathrm{pc\,Myr}^{-1}$), assuming Gaussian initial distribution functions for all parameters and using an age of 2 Myr. *Center*: The same but for an age of 3 Myr. *Right*: The same but for an age of 100 Myr.

major Galaxy-wide disturbance. From the local distribution of stellar ages we know that major incidents of star formation capable of disrupting the cloud distribution in the disk have occurred at intervals of less than 1 Gyr, and that the most recent burst occurred less than 1 Gyr ago (Casuso & Beckman 2007). We also deduce from this model that the outputs obtained when the input distribution is varied, either using a flat or a fractal initial distribution, do not differ in major ways from the output of the Gaussian model, so that we can conclude that the form of input distribution is not the major contributor to the form of the output distribution. The initial range of cloud masses, however, does make a significant difference to the output. In any case, if the basic conditions of density and velocity distribution are maintained for periods longer than a few Myr, the output mass distribution remains robustly stable so that a range of input conditions will yield the observed molecular cloud mass distribution, both in its overall form and in the structure of its individual peaks.

## References

Casuso, E., & Beckman, J. E. 2002, PASJ, 54, 405

Casuso, E., & Beckman, J. E. 2007, ApJ, 656, 897

Larson, R. B. 1998, MNRAS, 301, 569

Solomon, P. M., Rivolo, A. R., Barret, J., & Yahil, A. 1987, ApJ, 319, 730

*Pathways through an Eclectic Universe*
*ASP Conference Series, Vol. 390, © 2008*
*J. H. Knapen, T. J. Mahoney, and A. Vazdekis, eds.*

# The Effects of Spatially Distributed Ionization Sources on the Temperature Structure of H II Regions

B. Ercolano

*Harvard–Smithsonian Center for Astrophysics, 60 Garden Street, Cambridge, MA 02138, USA*

**Abstract.** Spatially resolved studies of star-forming regions show that the assumption of spherical geometry is not realistic in most cases, with a major complication posed by the gas being ionized by multiple non-centrally located stars or star clusters. Geometrical effects including the spatial configuration of ionizing sources affect the temperature and ionization structure of these regions. We try to isolate the effects of multiple non-centrally located stars via the construction of 3D photoionization models using the 3D Monte Carlo photoionization code MOCASSIN with very simple gas density distributions, but various spatial configurations for the ionization sources. Our first aim is to study the resulting temperature structure of the gas and investigate the behavior of temperature fluctuations within the ionized region. We show that geometry affects the temperature structures in our models differently according to metallicity. For the geometries and stellar populations considered in our study, at intermediate and high metallicities, models with ionizing sources distributed in the full volume, whose Strömgren spheres rarely overlap, show *smaller* temperature fluctuation than their central ionization counterparts, with fully overlapping concentric Strömgren spheres. The reverse is true at low metallicities. Finally the true temperature *fluctuations* due to the stellar distribution (as opposed to the large-scale temperature gradients due to other gas properties) are small in all cases and not a significant cause of error in metallicity studies.

Emission line spectra from H II regions are often used to study the metallicity of star-forming regions, as well as providing a constraint for temperatures and luminosities of the ionizing sources. Empirical metallicity diagnostics must often be calibrated with the aid of photoionization models. However, most studies so far have been carried out by assuming spherical or plane–parallel geometries, with major limitations on allowed gas and dust density distributions and with the spatial distribution of multiple, non-centrally located ionizing sources not being accounted for. We compare integrated emission line spectra from our models and quantify any systematic errors caused by the simplifying assumption of a single, central location for all ionizing sources. We find that the dependence of the metallicity indicators on the ionization parameter causes a clear bias, due to the fact that models with a fully distributed configuration of stars always display lower ionization parameters than their fully concentrated counterparts. The errors found imply that the geometrical distribution of ionization sources may partly account for the large scatter in metallicities derived using model-calibrated empirical methods.

For details please see B. Ercolano, N. Bastian, & G. Stasińska (2007, MNRAS, 379, 945)

*Pathways through an Eclectic Universe*
*ASP Conference Series, Vol. 390, © 2008*
*J. H. Knapen, T. J. Mahoney, and A. Vazdekis, eds.*

# Determination of Time Dependence in the Temperature Structure of Ionized Interstellar Gas

C. Giammanco,[1] L. Gutiérrez,[2,3] and J. E. Beckman[3,4]

[1] *Universtität Bern, Switzerland*
[2] *UNAM, Mexico*
[3] *Instituto de Astrofísica de Canarias, E-38200 La Laguna, Spain*
[4] *CSIC, Spain*

**Abstract.** It is well known that the temperature structure of ionized interstellar gas departs considerably from homogeneity. This is detected in the form of "temperature fluctuations" characterized by a parameter represented as $t^2$. In this paper we present the formalism to check whether these small deviations from the equilibrium temperature are also variable with time.

## 1. Equations

From Peimbert (1967) we have the classical equation:

$$I_{\alpha,\beta} = D_{\alpha,\beta} \int n^2 T_{\rm o}^{\alpha} [1 + t^2 \cdot (\alpha - 1)/2] d\omega dl, \tag{1}$$

where $n = n_{\rm e} = n_{\rm H}$, $T_{\rm o}$ is the mean temperature in the line of sight and $t^2$ is the temperature fluctuation parameter. If now we suppose that after a time interval $\Delta t$, $T_{\rm o}$ changes by a quantity $\Delta T$ and the integral $\int n^2 d\omega dl$ by a quantity $\Delta N^2$, and the variations in $t^2$ are negligible, we can write:

$$\begin{aligned} I(\Delta t) \quad &\sim \quad I(0) + D T_{\rm o}^{\alpha} \left(1 + \frac{\alpha - 1}{2} t^2\right)^{\alpha} \Delta N^2 \\ &+ D \left[\int n^2(0) d\omega dl\right] \alpha T_{\rm o}^{\alpha-1} \Delta T \left(1 + \frac{\alpha - 1}{2} t^2\right)^{\alpha}, \end{aligned} \tag{2}$$

from which finally we derive

$$\frac{\Delta I}{I_o} \sim \frac{\Delta N^2}{\int n^2 d\omega dl} + \alpha \frac{\Delta T}{T_{\rm o}}. \tag{3}$$

Following a similar derivation for the forbidden lines, we obtain an analogous expression where $\alpha$ depends on the energy gap of the considered transition, so that for [O III]$\lambda$5007 the numerical value of $\alpha$ is 2.4, which is positive. The value for H$\alpha$ or H$\beta$, however, is negative so that by measuring surface luminosity variations in hydrogen and forbidden oxygen lines it is possible to determine whether they are generated by temperature or density variation.

## 2. Spatial Scale

Beckman, Gutiérrez, & Giammanco (this volume, p. 106) found some evidence of temporal temperature variation in Orion. The typical variation they found is about 0.4%. Let us consider the spatial temperature fluctuations and the temporal variation to come from the same phenomena. Let Orion be composed of cold elements at temperature $T$ and by warm elements at temperature $T + dt$ (distributed in a Poissonian way around a mean number $N_{\rm w}$). Let $N$ be the total number of thermally independent elements $N_{\rm cold} + N_{\rm warm}$, so that the spatial $t^2$ fluctuations across the line of sight take the form:

$$t^2 \sim \frac{N_{\rm w}}{N}\left(\frac{dt}{T}\right)^2 . \tag{4}$$

We also define $N_1$ as the number of warm elements at the first epoch of Beckman et al., and $N_2$ the number of warm element at the second epoch. So we can write:

$$\frac{\Delta T}{T_{\rm o}} \sim \frac{N_1 - N_2}{N}\frac{dt}{T} . \tag{5}$$

Since $N_1$ and $N_2$ are variables of the same distribution, the difference between them gives an estimate of the variance. Finally, we find:

$$(N_1 - N_2) \sim 2\sqrt{N_{\rm w}} \tag{6}$$

and from Eqs (4), (5), and (6) we can estimate the number of thermally independent elements existing in the line of sight. From this, and using the estimate given by Wen & O'Dell (1995) for the thickness of Orion in the center of the nebula (0.13 pc, or 58 arcsec), we can find a typical dimension for the thermally independent element: (0.018±0.002) arcsec. We must compare this number with the Fourier analysis made by O'Dell, Peimbert, & Peimbert (2003). They found that the size scale for the maximum power of the Fourier spectrum of a temperature image of a chosen Orion field is almost 0.28 arcsec, but they also see that a large amount of power is found on larger scales. This is typical behavior for Fourier power-spectra when the image is composed of frequencies smaller than the Nyquist frequency, as is the case of our prediction. In conclusion, using our analysis plus the O'Dell et al. (2003) results, we have the indication that in Orion there are probably phenomena that involve scales smaller than one arcsecond. In order to get a clear picture, we need either better spatial resolution, or, more plausibly, higher time resolution.

**Acknowledgments.** We acknowledge the IAC, the organizers of the congress, and the Swiss Kommission für Astronomie for financial support.

### References

O'Dell, C. R., Peimbert, M., & Peimbert, A. 2003, ApJ, 125, 2590
Peimbert, M. 1967, ApJ, 150, 825
Wen, Z., & O'Dell, C. R. 1995, ApJ, 438, 784

*Pathways through an Eclectic Universe*
*ASP Conference Series, Vol. 390, © 2008*
*J. H. Knapen, T. J. Mahoney, and A. Vazdekis, eds.*

# Young Stellar Clusters in Nearby Spiral Galaxies

P. Grosbøl

*European Southern Observatory, Karl-Schwarzschild-Str. 2, 85748 Garching, Germany*

H. Dottori

*Instituto de Física, Univ. Federal do Rio Grande do Sul, Av. Bento Gonçalves 9500, 91501-970 Porto Alegre, RS, Brazil*

**Abstract.** The detection and study of very young stellar clusters in disk galaxies are often disturbed by dust in their spiral arms. The $K$-band, with its low dust attenuation, offers a much better and unbiased view of such stellar clusters. We present the statistics of young clusters in a sample of 46 nearby spiral galaxies, with linear resolutions of better than 100 pc, for which deep $K$-band photometry is available.

Many grand-design spirals (25% of the sample) show strong concentration of young stellar clusters in their arm regions. The clusters have absolute magnitudes $M_K$ up to $-15.5^{\rm m}$. Both the maximum absolute magnitude and the total number of clusters correlate with the total luminosity of the host galaxy and the relative amplitude of its two-armed spiral pattern.

## 1. Introduction

From the time when stellar populations in external galaxies were first studied (e.g., Baade 1944; van den Bergh 1960), it was realized that blue, young objects (such as H II regions and OB associations) were concentrated in spiral arms of disk galaxies. The strong and greatly varying attenuation by dust in the arm regions makes it difficult to make complete statistical studies of such very young stellar clusters in visual bands. It was noticed by Grosbøl & Patsis (1998) that many spiral galaxies had bright knots along their spiral arms in $K$-band images. Grosbøl, Dottori, & Gredel (2006) identified such knots in NGC 2997 with very young stellar clusters (ages $< 10$ Myr) using $K$-band spectra obtained at ISAAC/VLT. The low attenuation by dust in the $K$-band makes it possible to study an unbiased sample of very young clusters in spiral galaxies even though many of them are fully embedded in dust lanes in visual maps.

## 2. Data and Reduction

A set of 46 spiral galaxies was selected from an initial sample of 95 disk galaxies for which deep $K$-band maps had been obtained with SOFI/NTT. They had seeing $< 1''$, linear resolutions $< 100$ pc, inclination angles $< 65°$, and blue total absolute magnitudes $M_B < -19^{\rm m}$. The reduction procedure is described in Grosbøl, Patsis, & Pompei (2004). Sources in the $K$-band maps were identified by SExtractor (Bertin & Arnouts 1996). The classifier class_star, provided by SExtractor, was used to separate foreground stars from more diffuse sources

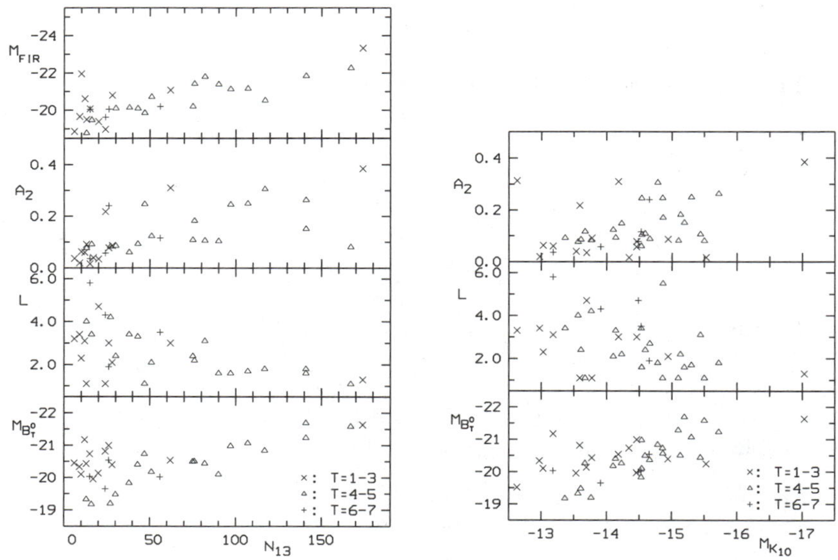


Figure 1. Distributions of $N_{13}$ and $M_{K_{10}}$ with respect to $M_{B_T^0}$, $L$, $A_2$, and $F_{\rm FIR}$.

such as stellar clusters associated with the galaxy and background galaxies. The contamination from background galaxies was reduced by including only those sources that were within the disk of the main galaxy.

## 3. Conclusions

The $K$-band gives an unbiased view of very young stellar clusters in spiral galaxies. Around 25% of the sample (mainly grand-design spirals) show a strong concentration of young clusters in their arm regions. General properties of the clusters are (see also Fig. 1): 1) their absolute $K$ magnitudes, $M_K$, extend up to $-15.5^{\rm m}$, 2) both the absolute magnitude $M_{K_{10}}$ of the tenth brightest diffuse object, and total number of young clusters $N_{13}$ with $M_K < -13^{\rm m}$ correlate with total blue absolute magnitude $M_{B_T^0}$ and amplitude, $A_2$, of the two-armed spiral pattern of the host galaxy, 3) $N_{13}$ is also correlated with the far-infrared flux $F_{\rm FIR}$ (i.e., dust), and 4) only galaxies with luminosity classes ($L$) I–II have larger populations; i.e., $N_{13}$.

### References


Baade, W. 1944, ApJ, 100, 1378
Bertin, E., & Arnouts, S. 1996, A&AS, 117, 393
Grosbøl, P., Dottori, H., & Gredel, R. 2006, A&A, 453, L25
Grosbøl, P., & Patsis, P. A. 1998, A&A, 336, 840
Grosbøl, P., Patsis, P. A., & Pompei, E. 2004, A&A, 423, 849
van den Bergh, S. 1960, ApJ, 131, 215

*Pathways through an Eclectic Universe*
*ASP Conference Series, Vol. 390, © 2008*
*J. H. Knapen, T. J. Mahoney, and A. Vazdekis, eds.*

# Star Clusters as Diaries of Galaxies

Th. Maschberger and P. Kroupa

*Argelander-Institut für Astronomie, Auf dem Hügel 71, D-53121 Bonn, Germany; Rhine Stellar Dynamics Network RSDN*

**Abstract.** Most if not all stars form in star clusters. Thus the distribution of star clusters preserves the information on the star formation history of a galaxy. Massive clusters form only during episodes of high star formation activity whereas periods of low star formation activity cannot produce them. We present here the method of Maschberger & Kroupa (2007) to derive the star formation history of a galaxy from its star cluster content.

Weidner, Kroupa, & Larsen (2004) showed that the mass of the brightest young (= most massive) cluster in a galaxy follows from the present star formation rate (SFR). The proposed scenario is that a complete (in a statistical sense) population of star clusters emerges during a "formation epoch" which lasts for $\delta t \approx 10\,\mathrm{Myr}$. The total mass of the population is given by the SFR, $M_{\mathrm{tot}} = \mathrm{SFR} \times \delta t$.

A sample of most massive clusters is characterized by the average mass of the most massive cluster, $\bar{M}_{\mathrm{max}}$ and is also parameterized by $M_{\mathrm{tot}}$, written symbolically $\bar{M}_{\mathrm{max}} = f(M_{\mathrm{tot}}) = f(\mathrm{SFR} \times \delta t)$.

The length of the formation epoch is determined from the observational data: the brightest young cluster in a galaxy is normally the most massive one of the current formation epoch. Independently, the present SFR can be determined, e.g., from FIR or H$\alpha$ data. With this, a region which contains a certain fraction of the most massive clusters can be defined. A fit of this region to the data then allows us to determine the duration of the formation epoch, since the height of the region depends on $\delta t$. This gives $\delta t = 10\,\mathrm{Myr}$.

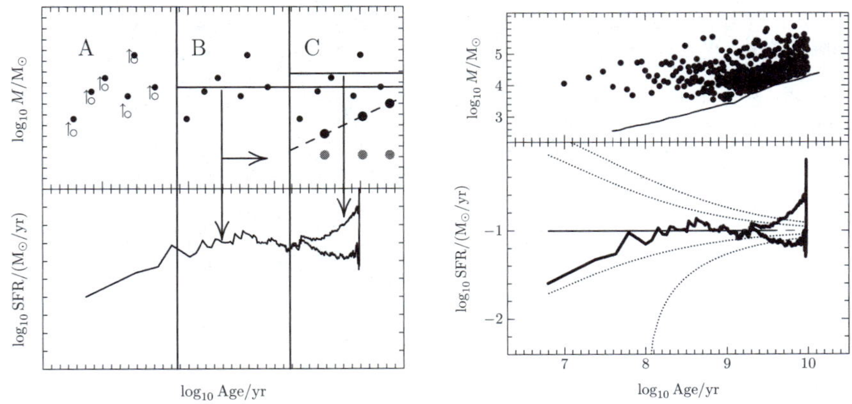


Figure 1. *Left:* The method schematically. *Right:* Model for the LMC.

In this notion the star formation history (SFH) of a galaxy can be seen as a sequence of formation epochs, each containing a most massive cluster. This leads to the general concept to infer an SFH using the available relation between $\bar{M}_{\rm max}$ and the SFR (Fig. 1, *left panel*):

[A] Correct the observed cluster masses for dynamical evolution using, e.g., the formulae of Boutloukos & Lamers (2003).

[B] Calculate the average mass, $\bar{M}_{\rm max}$, of the most massive clusters in the averaging window. Then use SFR $= f^{-1}(\bar{M}_{\rm max})$ to calculate the SFR. Move the averaging window in time to obtain a continuous SFH.

[C] Due to the evolution of star clusters and the observational limiting magnitude not every formation epoch contains a cluster. During such an epoch the SFR cannot be determined, but an upper and lower limit for it can be derived by assuming either no star formation ($M_{\rm max} = 0$), giving the lower limit, or star formation is not detectable ($M_{\rm max}$ corresponding to the observational limiting magnitude), giving the upper limit.

To study the capabilities of our method we generated a model with parameters similar to those for the Large Magellanic Cloud (LMC): constant SFR of $0.1\,M_\odot\,{\rm yr}^{-1}$ (the total mass of the LMC is $2 \times 10^9\,M_\odot$, Kim et al. 1998) and cluster disruption parameter $t_4 = 7.9 \times 10^9\,{\rm yr}$, Boutloukos & Lamers (2003).

In the upper part of Fig. 1, right panel, the most massive clusters are shown, together with the observational limiting mass (solid line, limiting magnitude $-3.5\,{\rm mag}$, Hunter et al. 2003). The lower part shows the reconstructed SFH (thick solid lines), which branches into two parts at early times due to the observational limit. The horizontal line shows a constant SFR, resulting from averaging up to the time where the line stops. If the reconstructed SFH were to cross the (red) dotted lines, then we would conclude that the underlying SFH has variations, but since this is not the case the reconstructed shape of the SFH agrees with the assumed constant SFR.

## References


Boutloukos, S. G., & Lamers, H. J. G. L. M. 2003, MNRAS, 338, 717

de Grijs, R., Anders, P., Lamers, H. J. G. L. M., Bastian, N., Fritze-v. Alvensleben, U., Parmentier, G., Sharina, M. E., & Yi, S. 2005, MNRAS, 359, 874

Hunter, D. A., Elmegreen, B. G., Dupuy, T. J., & Mortonson, M. 2003, AJ, 126, 1836

Kim, S., Staveley-Smith, L., Dopita, M. A., Freeman, K. C., Sault, R. J., Kesteven, M. J., & Mc Connell, D. 1998, ApJ, 503, 674

Maschberger, Th., & Kroupa, P. 2007, MNRAS, 379, 34

Weidner, C., Kroupa, P., & Larsen, S. S. 2004, MNRAS, 350, 1503

*Pathways through an Eclectic Universe*
*ASP Conference Series, Vol. 390, © 2008*
*J. H. Knapen, T. J. Mahoney, and A. Vazdekis, eds.*

# Modeling Dust Evolution in the Interstellar Medium

S. Zhukovska and H.-P. Gail

*ZAH, Institut für Theoretische Astrophysik, 69120 Heidelberg, Germany*

**Abstract.** The results of modeling dust evolution in the interstellar medium of the Galaxy are presented. We combine a chemical evolution model of the Galaxy with a model for dust evolution in the interstellar medium. To study dust growth in molecular clouds we develop a new model in which dust production is determined by characteristic growth timescales of dust species in molecular clouds. We study carbon, silicates, iron, and SiC dust species, discriminating between dust injected by asymptotic giant branch stars and supernovae, and dust grown in the interstellar medium. The observed abundance ratios of presolar dust grains formed in supernova ejecta and in asymptotic giant branch outflows require that for the ejecta from supernovae the fraction of refractory elements condensed into dust be quite small.

## 1. Introduction

The composition and origin of the dust population in the interstellar medium (ISM) is a long term issue for discussion. It is known from observations and presolar grain studies that dust grains are mainly condensed in stellar winds of asymptotic giant branch (AGB) stars and supernova (SN) explosions. The timescales of dust destruction by SN shocks are much shorter than those of dust injection into the ISM by stars; therefore, an additional efficient source of dust production is required to reproduce the observed dust-to-gas ratio and depletion patterns in the ISM, which is most likely dust growth in molecular clouds. To study the evolution and origin of dust in the ISM we construct a detailed dust model that discriminates between dust condensed in molecular clouds and stellar dust from AGB stars and SNe.

## 2. Model

We combine a standard model for the chemical evolution of the Galactic disk with a detailed model for dust in a way similar to that of Dwek (1998). We construct a two-infall model in which the chemical evolution of the ISM is determined by the feedback from SNe II, SNe Ia, and AGB stars and infall of gas. Our model reproduces sufficiently well a standard set of observational constraints for the chemical evolution models, as well as observations of dust in the solar vicinity (for details see Zhukovska, Gail, & Trieloff 2008).

We study the evolution of four main dust species (carbon, silicate, SiC, and iron dust) and for the first time discriminate between dust from AGB stars, SNe type II and Ia and dust produced in molecular clouds. Our dust model is based on three essential elements: 1) metallicity-dependent dust yields for dust

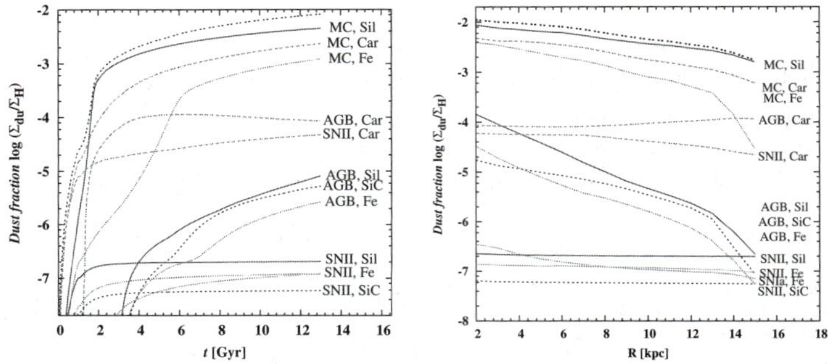


Figure 1. *Left panel:* Predicted time evolution of dust abundances relative to hydrogen in the solar vicinity. *Right panel:* Radial profiles of dust abundances at the present time. Black lines show the change in the dust-to-gas ratio.

production by AGB stars, which result from the first consistent calculations of dust condensation in stellar winds with M-, S-, and C-star chemistry combined with the calculation of stellar evolution on the AGB (Ferrarotti & Gail 2006); 2) a new approach to determine the SN type II dust production efficiency, fitting the SN type II dust fraction predicted by the model at the instant of the formation of the Solar System to that measured from presolar dust grains in meteorites, which is available for silicate, SiC, and carbon grains; 3) a new theory of dust growth in molecular clouds due to the accretion of key elements from the gas phase that considers the dust production rate by molecular clouds in terms of characteristic growth timescales and the degree of condensation of key elements into dust (Zhukovska et al. 2008).

## 3. Results

Isotopically normal dust grown in molecular clouds dominates the dust mass in our model, except for the very earliest time (see Fig. 1). Stellar grains with isotopic anomalies are always a minor dust component, which is dominated by graphite grains from AGB stars in our model. The composition of interstellar dust strongly varies with Galactic radius, as seen in the left panel of Figure 1.

Our estimations of the efficiency of SN type II dust production indicate that SNe mainly produce carbon dust with a condensation degree of 0.04–0.2, while for silicate and SiC dust it is very low, at about $10^{-4}$. This result is in good agreement with low condensation degrees obtained from fitting the SN spectra.

**Acknowledgments.** This work was supported by the DFG, Sonderforschungbereich 439 "Galaxies in Young Universe". SZ is a Fellow of the IMPRS Heidelberg.

### References

Dwek, E. 1998, ApJ, 501, 643
Ferrarotti, A. S., & Gail, H.-P. 2006, A&A, 447, 553
Zhukovska, S., Gail, H.-P., & Trieloff, M. 2008, A&A, 479, 453

# Session II

# Star Formation Rates

*Pathways through an Eclectic Universe*
*ASP Conference Series, Vol. 390, © 2008*
*J. H. Knapen, T. J. Mahoney, and A. Vazdekis, eds.*

# Star Formation Rate Determinations

Daniela Calzetti
*Department of Astronomy, University of Massachusetts, Amherst, Massachusetts, USA*

**Abstract.** I review determinations of star formation rate indicators from the ultraviolet to the infrared in the context of their use for galaxies and galaxy surveys. Mid-infrared star formation rate indicators have generated interest in recent years, thanks to *Spitzer*'s capabilities and the opportunities offered by the upcoming *Herschel Space Telescope*. I discuss what we have learned in the mid-infrared from studies of local galaxies, combining *Spitzer* with *HST* and other data.

## 1. Introduction

Star formation is one of the fundamental processes driving the evolution of galaxies. It depletes galaxies of gas (and dust) by converting it to stars, controls the metal enrichment of the interstellar (ISM) and intergalactic medium (IGM), regulates the radiative and mechanical feedback into the ISM and IGM, and shapes the mix of stellar populations. Star formation is the true link between the invisible Universe (driven by gravity and traced by models of galaxy formation and evolution; e.g., Katz 1992; Springel & Hernquist 2003; Springel, di Matteo, & Hernquist 2005; Stinson et al. 2006) and the visible Universe that is typically measured, one example being the redshift evolution of the cosmic star formation rate density in comoving space (e.g., Madau et al. 1996; Lilly et al. 1996; Smail, Ivison, & Blain 1997; Madau, Pozzetti, & Dickinson 1998; Hughes et al. 1998; Steidel et al. 1999; Eales et al. 1999; Yan et al. 1999; Barger, Cowie, & Richards 2000; Wilson et al. 2002; Giavalisco et al. 2004; Norman et al. 2004; Chapman et al. 2005; Hopkins & Beacom 2006). Thus, it becomes paramount to characterize and physically quantify the laws of star formation (Kennicutt 1998a; Kennicutt et al. 2007) and to obtain accurate measures of the star formation rates (SFRs) of galaxies.

SFR indicators have been defined at basically all wavelengths across the spectrum, from the X-ray to the radio (e.g., Kennicutt 1998b; Kewley et al. 2002; Ranalli, Comastri, & Setti 2003; Bell 2003; Hopkins 2004; Kewley et al. 2004; Calzetti et al. 2005; Schmitt et al. 2006; Calzetti et al. 2007, and references therein). The common characteristic of all indicators is that they all probe the *massive SFR*. To convert this measure to an SFR, an assumption concerning the stellar initial mass function (IMF) needs to be made; this problem is covered in other presentations at this conference, together with the stochastic effects present at low star formation levels (see, for example, Kroupa, this volume, p. 3).

This review mostly concentrates on the robustness of converting the luminosity, $L(\lambda)$, of a whole galaxy into an SFR indicator, and the potential impact of contributions to $L(\lambda)$ that are not strictly linked to the current star formation. Examples of such "extraneous" contributions are evolved stellar populations, which may contribute to the light emitted at a given wavelength but are not part of the star-forming population, or the presence of dust.

For the sake of brevity, I will only touch upon SFR indicators at ultraviolet (UV), optical, and far-infrared (FIR) wavelengths. UV and optical/near-infrared indicators only probe the stellar light that emerges from galaxies *unabsorbed by dust.* Infrared SFR indicators are complementary to UV–optical indicators, because they measure star formation via the stellar light that has been reprocessed by dust and emerges beyond a few $\mu$m.

## 2. The Ultraviolet

The ultraviolet ($\lambda \sim$ 912–3000 Å) emission from galaxies is extensively used as an SFR indicator, especially in the medium/high redshift regime, where the restframe UV moves into the optical observer's frame. Recently, *GALEX* (Martin et al. 2005), with its UV survey capability, has reawakened interest in this wavelength regime for studies of the local Universe (e.g., Salim et al. 2007).

The advantage of using the UV is that it directly probes the bulk of the emission from young, massive stars. Its main disadvantage, however, is its high sensitivity to dust reddening and attenuation. For reference, a star-forming galaxy with $A_V$=1 mag has $A_{1500\,\text{Å}} \approx 3$ mag, but the specific value will depend on both the dust amount *and* the dust geometry in the galaxy. To make matters worse, there is a well-established correlation between attenuation and SFR in galaxies and star-forming regions (Fig. 1; Wang & Heckman 1996; Heckman et al. 1998; Calzetti 2001; Hopkins et al. 2001; Sullivan et al. 2001; Buat et al. 2002; Calzetti et al. 2007); thus, the galaxies that are most actively forming stars and are therefore likely to be important in the cosmic SFR census are also those whose UV luminosity requires the largest corrections for the effects of dust attenuation. An increase of a factor 100 in SFR implies on average an increase in the attenuation by $A_V \approx$1–3 mag for star-forming galaxies. The quantitative details may depend on the exact nature of the star-forming system, but the qualitative trend is common to most (Fig. 1). At fixed metallicity (and dust-to-gas ratio, Draine et al. 2007), this trend is a consequence of the correlation between SFR density and gas density (Kennicutt 1998a; Kennicutt et al. 2007). A number of authors (e.g., Calzetti, Kinney, & Storchi-Bergmann 1994; Meurer, Heckman, & Calzetti 1999; Calzetti et al. 2000; Hopkins et al. 2001; Sullivan et al. 2001; Buat et al. 2002, 2005; Bell 2003; Hopkins 2004; Iglesias-Páramo et al. 2006; Salim et al. 2007) have derived techniques to correct the observed UV emission from galaxies for the effects of dust attenuation. The effectiveness of these techniques, however, depends in general on the nature of the galaxy population they are being applied to.

The additional complication with using the UV as an SFR indicator is the long timescale (of order 100 Myr) in which stars are relatively luminous in the non-ionizing UV wavelength range. The cross-calibration of SFR(UV) with indicators at other wavelengths may be complicated by the different timescales

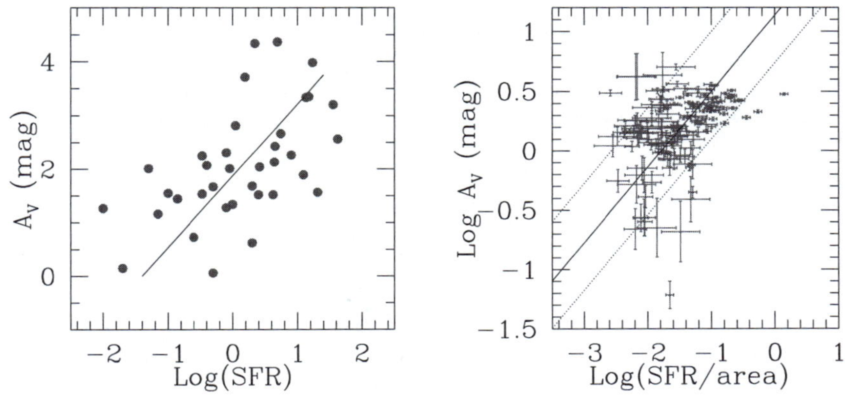


Figure 1. *Left:* The attenuation $A_V$ (in magnitudes), derived from hydrogen nebular line emission (H$\alpha$/H$\beta$ or H$\beta$/Br$\gamma$, Calzetti, Kinney, & Storchi-Bergmann 1996) as a function of the log(SFR) (in units of $M_\odot\,\mathrm{yr}^{-1}$) for a sample of local starbursts; the continuous line shows the mean trend of the data points. *Right:* A similar plot to the one on the left (but on a log–log scale), this time for 164 star-forming regions in 21 nearby galaxies. All the galaxies have metallicity close to the solar value. $A_V$ (from the H$\alpha$/P$\alpha$ ratio) is in units of magnitudes, and the SFR density on the horizontal axis is in units of $M_\odot\,\mathrm{yr}^{-1}\,\mathrm{kpc}^{-2}$. The continuous line is the expected trend after combining the Schmidt–Kennicutt law with the dust-to-gas ratio of the galaxies (Calzetti et al. 2007). The dotted lines are the 90% boundary to the data points.

involved. This may be the case when comparing SFR(UV) with, for example, SFR(H$\alpha$) (see next section), where the latter has a timescale about 10 times shorter than the former. In this case, the star formation history of the system plays a significant role. If an unresolved object is only measured, say, at 1500 Å, it will be impossible to know whether the emission is coming from a system that has been forming stars at the constant level of $1\,M_\odot\,\mathrm{yr}^{-1}$ for the past 100 Myr, or a system that is no longer forming stars but has been passively evolving for the past 50 Myr after a $4\times10^8\,M_\odot$ burst of star formation (Leitherer et al. 1999).

The ~10 times longer stellar timescale probed by the UV relative to tracers of ionizing photons accounts for the difference in the properties of the dust attenuation between starburst and quiescent star-forming systems. Starburst galaxies show a well-defined correlation in the FIR/UV versus UV color plane (Fig. 2 and Meurer et al. 1999). The FIR/UV ratio is a measure of the UV attenuation suffered by the system: the higher the dust attenuation, the larger the amount of UV stellar energy reprocessed into the FIR by dust. Variations in the UV color probe the amount of dust reddening present in the system. Thus, in starbursts, the UV reddening is a good tracer of the total UV dust attenuation (Calzetti 2001). Quiescent star-forming galaxies and regions show a ~10 times larger spread in the FIR/UV ratio than starburst galaxies, at fixed UV color; the spread is in the sense that the starburst galaxies form the upper envelope to the quiescently star-forming systems (Fig. 2 and Buat et al. 2002, 2005; Bell

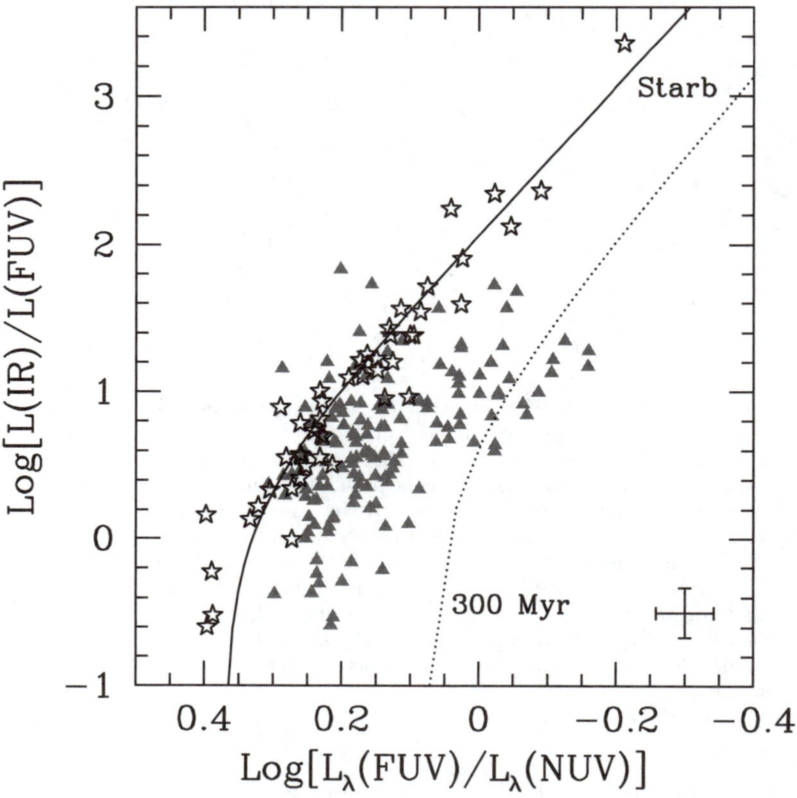


Figure 2. The FIR/UV ratio (the ratio of the FIR to the far-UV luminosity) versus the UV color (given here as the ratio between the *GALEX* far-UV and near-UV fluxes) for starburst galaxies and quiescently star-forming regions. The starburst galaxies are shown with star symbols. The filled triangles are star-forming regions within the galaxy NGC 5194 (Calzetti et al. 2005). Redder UV colors correspond on average to larger FIR/UV ratios. The continuous line shows the best fit to the starburst galaxies, which is well represented by a model of a progressively more attenuated (from left to right) constant star-forming population. The dotted line shows the same dust attenuation trend for a 300 Myr old stellar population; this model represents a reasonable lower envelope to star-forming regions in the NGC 5194.

et al. 2002; Gordon et al. 2004; Kong et al. 2004; Seibert et al. 2005; Calzetti et al. 2005). Thus, a far larger range of UV dust attenuations seems to be present in star-forming objects than in starbursts, for the same UV color. This is possible if in quiescent star-forming systems the reddening of the UV colors is not only a probe of dust, but also of ageing of the individual star-forming regions contributing to the UV emission in the system (regions up to an age of $\sim$ 100–300 Myr, Calzetti et al. 2005). Specifically, older clusters can contribute significantly to the measured UV emission in these galaxies, as they are not as attenuated by dust as younger systems. This will complicate the definition of SFR(UV) in quiescent star-forming galaxies, and change its calibration relative to the starburst galaxies (e.g., Seibert et al. 2005; Cortese et al. 2006; Salim et al. 2007).

## 3. The Optical and Near Infrared

SFR indicators in the optical and near-infrared wavelength range ($\lambda \sim 3000$–$25\,000$ Å) are derived from the many hydrogen recombination lines (mainly H$\alpha$, H$\beta$, P$\beta$, P$\alpha$, Br$\gamma$) and from forbidden line emission (chiefly [O II] and [O III]). These indicators thus trace the ionizing photons in the system. The short lifespan of massive, ionizing stars confines the timescale probed by the optical indicators to about 10 Myr, implying that they are tracers of the *current* SFR.

Calibrations for these indicators, in particular for H$\alpha$ ($\lambda$6563 Å) and for [O II] ($\lambda$3727 Å), have been presented by many authors (to name a few, Kennicutt 1998b; Gallagher, Hunter, & Bushouse 1989; Jansen, Franx, & Fabricant 2001; Rosa-González et al. 2002; Kewley et al. 2002; Charlot & Longhetti 2001; Kewley et al. 2004; Moustakas, Kennicutt, & Tremonti 2006, and references therein).

There are a number of effects that impact the SFRs derived from optical and near-infrared line emission. First and foremost, dust extinction, which affects the blue lines more than the red ones. For example, if extinction corrections are neglected, the median SFR derived for a generic sample of nearby galaxies will be underestimated by roughly a factor of three if using H$\alpha$, and by about twice as much if using [O II] (Rosa-González et al. 2002). These mean values do not convey the fact that brighter galaxies are also more extinguished, as discussed in the previous section. In starburst galaxies, dust attenuation is about a factor of two larger for the ionized gas than for the stellar continuum at the same wavelength (Calzetti et al. 1994): the dust attenuation in the H$\beta$($\lambda$4861 Å) line is the same as that in the stellar continuum at ~2300 Å.

Line emission SFR indicators are also sensitive to the upper end of the stellar IMF, i.e., the number of ionizing stars formed, much more so than the UV stellar continuum. Changing the upper stellar mass of the IMF from 100 $M_\odot$ to 30 $M_\odot$ decreases the number of ionizing photons by a factor of about five, and the UV stellar continuum at 1500 Å by a factor 2.3 (from the models of Leitherer et al. 1999). Other factors to take into account for the line emission SFR indicators are the underlying stellar absorption in the case of the hydrogen recombination lines (Rosa-González et al. 2002), and the metallicity and ionization conditions for the forbidden lines (e.g., Moustakas et al. 2006).

For reference, for a solar metallicity stellar population with constant SFR over the past 100 Myr, forming stars with an IMF consisting of two power laws, with slope $-1.3$ in the range 0.1–0.5 $M_\odot$ and slope $-2.3$ in the range 0.5–120 $M_\odot$, the SFR(H$\alpha$) is given by:

$$\mathrm{SFR}\,(M_\odot\,\mathrm{yr}^{-1}) = 5.3 \times 10^{-42} L(\mathrm{H}\alpha)\ (\mathrm{erg\,s}^{-1}). \tag{1}$$

Variations of ±20% in the calibration in this relation are present for younger/older ages and metallicities down to ~1/5 solar. The ~50% difference between the calibration in Equation 1 and that of Kennicutt (1998b) is mainly due to differences in assumptions concerning the stellar IMF and the age of the stellar populations (100 Myr in our case versus infinite age in Kennicutt 1998b).

## 4. The Infrared

The earliest calibrations of the infrared ($\lambda \sim$ 5–1000 $\mu$m) dust emission as an SFR indicator date back to *IRAS* (for a review and an independent calibration see Kennicutt 1998b). The connection between star formation and infrared emission is based on the fact that, at least in the nearby Universe, star-forming regions tend to be dusty and the dust absorption cross-section peaks in the UV (also the peak of emission of young, massive stars).

Since the very beginning, however, it was clear that the calibration of the FIR emission as a SFR tracer carried a number of issues (Hunter et al. 1986; Lonsdale Persson & Helou 1987; Rowan–Robinson & Crawford 1989; Devereux & Young 1990; Sauvage & Thuan 1992). First, not all of the luminous energy produced by recently formed stars is reprocessed by dust in the infrared; in this case, the FIR only recovers part of the SFR, and the fraction recovered depends, at least partially, on the amount of dust in the system. Second, evolved, i.e., non-star-forming, stellar populations also heat the dust that emits in the FIR wavelength region, thus affecting the calibration of SFR(FIR) in a stellar population-dependent manner. Third, SFR(FIR) is a "calorimetric" measure (the full energy census in the range ~5–1000 $\mu$m needs to be included), and the FIR spectral energy distribution (SED) depends on the stellar field intensity (Helou 1986; Draine & Li 2007); thus, extrapolations of SFR(FIR) from a too sparsely sampled SED are subject to large uncertainties.

The use of monochromatic (i.e., one band or wavelength) infrared emission for measuring SFRs offers one definite advantage over the bolometric infrared luminosity: it removes the need for uncertain extrapolations of the dust SED across the full wavelength range. The interest in calibrating monochromatic mid-infrared SFR diagnostics stems from their potential application to both the local Universe (e.g., in the investigation of the scaling laws of star formation; Kennicutt et al. 2007) and intermediate and high redshift galaxies observed with *Spitzer*, *Herschel*, and future millimeter/radio facilities (e.g., Daddi et al. 2005; Wu et al. 2005).

Studies using *ISO* data provided the first window on the use of the monochromatic mid-IR (MIR, $\lambda \sim$ 5–40 $\mu$m) emission as an SFR indicator; investigations with the higher angular resolution and more sensitive *Spitzer* 8 $\mu$m and 24 $\mu$m data have expanded on the *ISO* results. The dust emission in the MIR wavelength range is characterized by both continuum and bands. The continuum is due to dust heated by a combination of single-photon and thermal equilibrium processes, with the latter becoming more and more prevalent over the former at longer wavelengths (Draine & Li 2007). The interest in calibrating the MIR dust continuum as an SFR indicator stems from the consideration that the dust heated by hot, massive stars can have high temperatures and will then emit at short infrared wavelengths. The MIR bands are generally attributed to polycyclic aromatic hydrocarbons (PAHs, Leger & Puget 1984; Sellgren 1984), large molecules transiently heated by single UV and optical photons in the general radiation field of galaxies or near B stars (Peeters, Spoon, & Tielens 2004; Mattioda, Allamandola, & Hudgins 2005; Li & Draine 2007), and which can be destroyed, fragmented, or ionized by harsh UV photon fields (Boulanger et al. 1988; Pety et al. 2005).

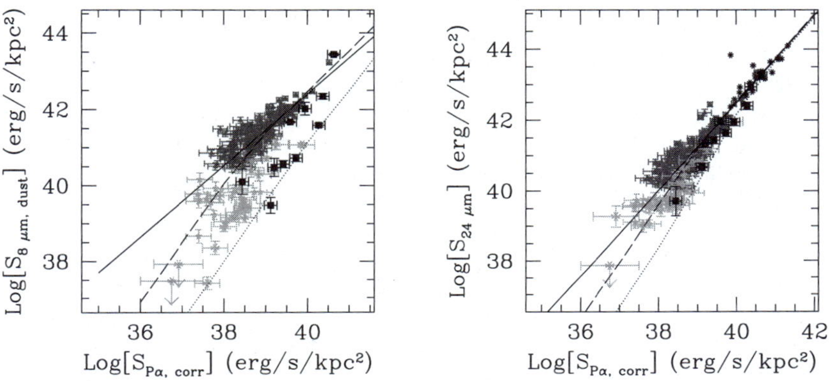


Figure 3. *Left:* The luminosity surface density (LSD = luminosity/area) at 8 $\mu$m versus the P$\alpha$ LSD of ten starburst galaxies and 220 star-forming regions in 33 nearby galaxies (Calzetti et al. 2007). The 8 $\mu$m emission is from *Spitzer* images, and has the stellar continuum subtracted. The P$\alpha$ line emission ($\lambda = 1.876\,\mu$m) is from *HST*/NICMOS images, has been corrected for dust extinction using the H$\alpha$/P$\alpha$ ratio and is used here as an unbiased tracer of the massive SFR. Of the 220 regions, the ~180 regions in high metallicity galaxies, with 12+log(O/H)>8.3, are marked in dark gray, and the ~40 regions in low metallicity galaxies are in light gray. The starburst galaxies (Engelbracht et al. 2005) are the black squares. The continuous line is the best fit to the high metallicity star-forming regions (dark gray), and has slope 0.94. Models for a young stellar population with increasing amounts of star formation and dust are shown as a dashed line ($Z = Z_\odot$) and a dotted line ($Z = 1/10\,Z_\odot$), using the stellar population models of Leitherer et al. (1999) and the dust models of Draine & Li (2007). The spread in the data points around the best fit line is well accounted for by a spread in stellar population ages in the range 2–8 Myr. *Right:* The same as the left panel, with the vertical axis now reporting the 24 $\mu$m LSD, also from *Spitzer*. The black asterisks are the luminous infrared galaxies (LIRGs) from Alonso-Herrero et al. (2006). The best fit line (continuous line) has slope 1.2.

Roussel et al. (2001) and Förster Schreiber et al. (2004) showed that the emission in the 6.75 $\mu$m ISO band correlates with the number of ionizing photons in galaxy disks and in the nuclear regions of galaxies. Conversely, Boselli, Lequeux, & Gavazzi (2004) found that the MIR emission in a more diverse sample of galaxies (types Sa through Im–BCDs) correlates more closely with tracers of evolved stellar populations not linked to the current star formation. Additionally, Haas, Klaas, & Bianchi (2002) found that the *ISO* 7.7 $\mu$m emission is correlated with the 850 $\mu$m emission from galaxies, suggesting a close relation between the *ISO* band emission and the cold dust heated by the general (non-star-forming) stellar population. *Spitzer* data of the nearby galaxies NGC 300 and NGC 4631 show that the 8 $\mu$m band emission highlights the rims of H II regions and is depressed inside the regions, indicating that the PAH dust is heated in the photo-dissociation regions surrounding H II regions and is destroyed within the regions (Helou et al. 2004; Bendo et al. 2006). Analysis of the MIR emission from the First Look Survey galaxies shows that the correlation between the 8 $\mu$m band emission and tracers of the ionizing photons is shallower than unity (Wu et al. 2005), in agreement with the correlations observed for H II regions in the nearby star-forming galaxy NGC 5194 (Calzetti et al. 2005).

A recent analysis of the 8 $\mu$m emission from 220 star-forming regions in 33 nearby galaxies from the project SINGS (the *Spitzer* Infrared Nearby Galaxies Survey; Kennicutt et al. 2003) has shown that this MIR band is sensitive to both metallicity and star formation history (Fig. 4 *left*, and Calzetti et al. 2007). The dependence on metallicity (e.g., Boselli et al. 2004; Engelbracht et al. 2005; Draine et al. 2007) implies that regions with values about $1/10\, Z_{\odot}$ are about 30 times underluminous at 8 $\mu$m relative to regions with solar metallicity and the same SFR. The dependence on star formation history is due to the contribution to the 8 $\mu$m emission from dust heated by non-ionizing stellar populations that are unrelated to the current star formation (Peeters et al. 2004; Calzetti et al. 2007). However, for a restricted choice of parameters, i.e., actively star-forming regions with similar (e.g., solar) metallicity, the 8 $\mu$m emission correlates almost linearly with the SFR (Fig. 4, *left*).

The same analysis shows that the 24 $\mu$m emission is, conversely, a good SFR tracer for galaxies, in the absence of strong AGN (Fig. 4, *right*, and Calzetti et al. 2007). Similar conclusions had been reached by Calzetti et al. (2005) and by Pérez-González et al. (2006) for star-forming regions in NGC 5194 and NGC 3031, respectively, and by Wu et al. (2005) and Alonso-Herrero et al. (2006) for bright galaxies. In particular, Calzetti et al. (2005) have concluded that the 24 $\mu$m emission is more closely related to the H$\alpha$ (ionizing photons) emission than to the UV (non-ionizing stellar continuum) emission. A single correlation can be defined over 3.5 orders of magnitude in luminosity surface density (luminosity/area), yielding the following SFR calibration:

$$\mathrm{SFR}(M_{\odot}\,\mathrm{yr}^{-1}) = 1.27 \times 10^{-38}\,[L(24\ \mu\mathrm{m})\ (\mathrm{erg\,s}^{-1})]^{0.8850}. \tag{2}$$

This calibration is very similar to that of Alonso-Herrero et al. (2006), who have used a sample of (ULIRGs) and NGC 5194. The non-linear trend of $L(24\,\mu\mathrm{m})$ with SFR is a direct consequence of the increasing dust temperature for more actively star-forming objects (Draine & Li 2007); higher dust temperatures correspond to higher fractions of the dust emission emerging in the MIR.

The 24 $\mu$m emission has a much lower sensitivity to metallicity than the 8 $\mu$m emission, decreasing by just a factor of two to four for a tenfold decrease in metallicity. This small decrease is fully accounted by the increased transparency (lower dust-to-gas ratio) of the medium for lower metal abundances (Draine et al. 2007). In addition, there seems to be very little contribution to the 24 $\mu$m emission from non-ionizing stellar populations, at least within the limits of the Calzetti et al. (2007) analysis.

The sensitivity of the SFR indicator to variations in the dust content of the system can be removed by combining two tracers: one that probes the dust-obscured star formation and one that probes the unobscured one. Combining

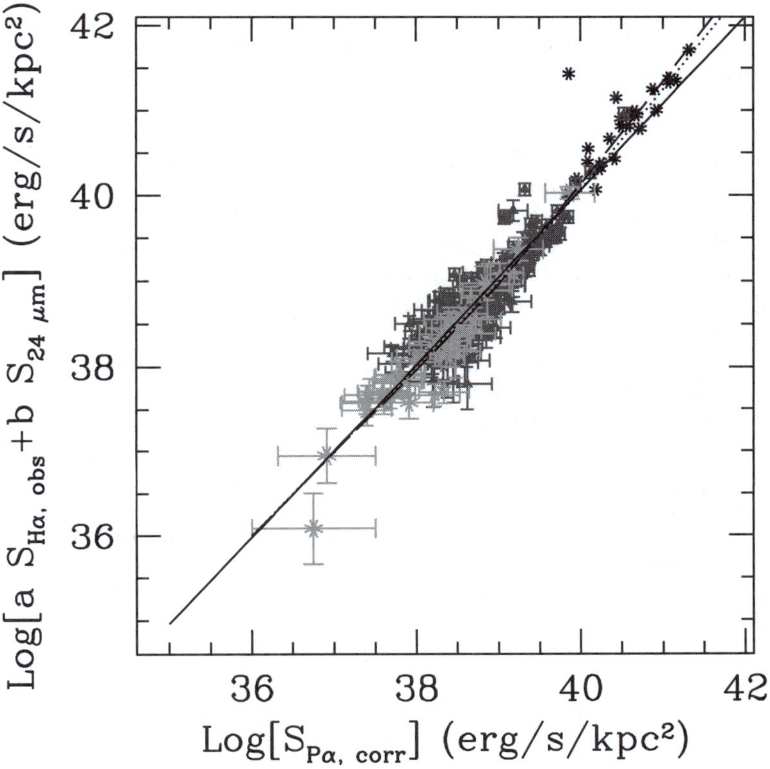


Figure 4. As in Fig. 4, for the combined H$\alpha$ and 24 $\mu$m LSD. For this SFR indicator, data and models are degenerate in metallicity. The slope of the best fit line is unity, and the data follow the one-to-one line up to the LIRGs. For the brightest galaxies, however, there are deviations from this trend, reflecting the increasing temperature of the dust in a regime where the 24 $\mu$m emission (i.e., the dust-obscured SFR) dominates over the H$\alpha$ emission (the unobscured SFR). The dashed and dotted lines are for the same models as those described in Figure 3.

the 24 $\mu$m luminosity with the *observed* H$\alpha$ luminosity (Figure 4) yields a new SFR indicator, with calibration (Kennicutt et al. 2007; Calzetti et al. 2007):

$$\mathrm{SFR}(M_\odot\,\mathrm{yr}^{-1}) = 5.3 \times 10^{-42}[L(\mathrm{H}\alpha)_{\mathrm{obs}} + (0.031 \pm 0.006)L(24\,\mu\mathrm{m})], \quad (3)$$

from Equation 1. The calibration in Equation 3 is unaffected not only by metallicity variations (Fig. 4), but also by stellar population mix. However, it shows some deviation from a simple, linear correlation at high luminosity surface densities (Fig. 4, and Calzetti et al. 2007), which simply reflects the non-linear trend of $L(24\,\mu\mathrm{m})$ with SFR once luminosities are sufficiently high that the infrared emission dominates over the line emission.

## 5. Summary

The SFRs based on the UV stellar continuum and the optical/near-infrared line emission probe the stellar light that emerges from a galaxy unabsorbed by dust. They both require corrections for the effects of dust, which can change the mean values of local samples by factors of a few. In addition, comparisons of calibrations between SFR(UV) and SFR(line) require care, as the former probes stellar timescales that are at least an order of magnitude longer than the latter.

In the infrared, the calorimetric SFR(FIR) is based on measuring the stellar energy reprocessed by dust and suffers from difficult to quantify contributions from non-star-forming stellar populations. Often SFR(FIR) needs to be extrapolated from sparse wavelength sampling, due to observational limitations, which leads to uncertainties.

Monochromatic mid-infrared SFR indicators offer a valid complement to the stellar-based SFR tracers and to SFR(FIR). SFR(MIR) tends to be more closely associated with regions of ionizing photon emission than to regions of UV emission. In the absence of AGN, $L(24\,\mu\mathrm{m})$ and a linear combination of $L(\mathrm{H}\alpha)$ and $L(24\,\mu\mathrm{m})$ provide relatively robust SFR tracers, with uncertainties of order a factor two. Conversely, the $8\,\mu$m emission is strongly dependent on both variations in metallicity (factor of 30 variation in luminosity for a factor $\sim$10 variation in metallicity) and in stellar population mix (a factor of a few variation in luminosity). Equations 2 and 3 should be generally applicable to regions and galaxies dominated by current star formation.

**Acknowledgments.** This work would not have been possible without the combined efforts of the SINGS team, whom the author thanks heartfully. Special thanks are due to the SINGS project's Principal Investigator, Robert Kennicutt, and collaborators Chad Engelbracht, Bruce Draine, and Fabian Walter.

The author is grateful to Johan Knapen for all his excellent efforts in organizing this stimulating conference, and to John Beckman for offering the perfect excuse for it.

### References

Alonso-Herrero, A., Rieke, G. H., Rieke, M. J., Colina, L., Perez-Gonzalez, P. G., & Ryder, S. D. 2006, ApJ, 650, 835
Barger, A. J., Cowie, L. L., & Richards, E. A. 2000, AJ, 119, 2092
Bell, E. F. 2003, ApJ, 586, 794
Bell, E. F., Gordon, K. D., Kennicutt, R. C., & Zaritsky, D. 2002, ApJ, 565, 994
Bendo, G. J., et al. 2006, ApJ, 652, 283
Boselli, A., Lequeux, J., & Gavazzi, G. 2004, A&A, 428, 409
Boulanger, F., Beichmann, C., Desert, F.-X., Helou, G., Perault, M., & Ryter, C. 1988, ApJ, 332, 328
Buat, V., Boselli, A., Gavazzi, G., & Bonfanti, C. 2002, A&A, 383, 801
Buat, V., et al. 2005, ApJ, 619, L51
Calzetti, D. 2001, PASP, 113, 1449
Calzetti, D., Armus, L., Bohlin, R. C., Kinney, A. L., Koornneef, J., & Storchi-Bergmann, T. 2000, ApJ, 533, 682
Calzetti, D., et al. 2005, ApJ, 633, 871
Calzetti, D., et al. 2007, ApJ, 666, 870
Calzetti, D., Kinney, A. L., & Storchi-Bergmann, T. 1994, ApJ, 429, 582
Calzetti, D., Kinney, A. L., & Storchi-Bergmann, T. 1996, ApJ, 458, 132

Chapman, S. C., Blain, A. W., Smail, I., & Ivison, R. J. 2005, ApJ, 622, 772
Charlot, S., & Longhetti, M. 2001, MNRAS, 323, 887
Cortese, L., Boselli, A., Buat, V., Gavazzi, G., Boissier, S., Gil de Paz, A., Seibert, M., Madore, B. F., & Martin, D. C. 2006, ApJ, 637, 242
Daddi, E., et al. 2005, ApJ, 631, L13
Devereux, N. A., & Young, J. S. 1990, ApJ, 350, L25
Draine, B. T., & Li A. 2007, ApJ, 657, 810
Draine, B. T., et al. 2007, ApJ, 663, 866
Eales, S., Lilly, S., Gear, W., Dunne, L., Bond, J.R., Hammer, F., Le Fevre, O., & Crampton, D. 1999, ApJ, 515, 518
Engelbracht, C. W., Gordon, K. D., Rieke, G. H., Werner, M. W., Dale, D. A., & Latter, W. B. 2005, ApJ, 628, 29
Förster Schreiber, N. M., Roussel, H., Sauvage, M., & Charmandaris, V. 2004, A&A, 419, 501
Gallagher, J. S., Hunter, D. A., & Bushouse, H. 1989, AJ, 97, 700
Giavalisco, M., et al. 2004, ApJ, 600, L103
Gordon, K.D., et al. 2004, ApJS, 154, 215
Haas, M., Klaas, U., & Bianchi, S. 2002, A&A, 385, L23
Heckman, T. M., Robert, C., Leitherer, C., Garnett, D. R., & van der Rydt, F. 1998, ApJ, 503, 646
Helou, G. 1986, ApJ, 311, L33
Helou, G., et al. 2004, ApJS, 154, 253
Hopkins, A. M. 2004, ApJ, 615, 209
Hopkins, A. M., & Beacom, J. F. 2006, ApJ, 651, 142
Hopkins, A. M., Connolly, A. J., Haarsma, D. B., & Cram, L. E. 2001, AJ, 122, 288
Hughes, D. H., et al. 1998, Nature, 394, 241
Hunter, D. A., Gillett, F. C., Gallagher, J. S., Rice, W. L., & Low, F. J. 1986, ApJ, 303, 171
Iglesias-Paramo, J., et al. 2006, ApJS, 164, 38
Jansen, R. A., Franx, M., & Fabricant, D. 2001, ApJ, 551, 825
Katz, N. 1992, ApJ, 391, 502
Kennicutt, R. C. 1998a, ApJ, 498, 541
Kennicutt, R. C. 1998b, ARA&A, 36, 189
Kennicutt, R. C., et al. 2003, PASP, 115, 928
Kennicutt, R. C., Calzetti, D., et al. 2007, ApJ, submitted
Kewley, L. J., Geller, M. J., Jansen, R. A., & Dopita, M. A. 2002, AJ, 124, 3135
Kewley, L. J., Geller, M. J., & Jansen, R. A. 2004, AJ, 127, 2002
Kong, X., Charlot, S., Brinchmann, J., & Fall, S. M. 2004, MNRAS, 349, 769
Leger, A., & Puget, J. L. 1984, A&A, 137, L5
Leitherer, C., Schaerer, D., Goldader, J. D., González Delgado, R. M., Robert, C., Kune, D. F., de Mello, D. F., Devost, D., & Heckman, T. M. 1999, ApJS, 123, 3
Li, A., & Draine, B. T. 2002, ApJ, 572, 762
Lilly, S. J., Le Fèvre, O., Hammer, F., & Crampton, D. 1996, ApJ, 460, L1
Lonsdale Persson, C. J., & Helou, G. X. 1987, ApJ, 314, 513
Madau, P., Ferguson, H. C., Dickinson, M. E., Giavalisco, M., Steidel, C. C., & Fruchter, A. 1996, MNRAS, 283, 1388
Madau, P., Pozzetti, L., & Dickinson, M. E. 1998, ApJ, 498, 106
Martin, D. C., et al. 2005, ApJ, 619, L1
Mattioda, A. L., Allamandola, L. J., & Hudgins, D. M. 2005, ApJ, 629, 1183
Meurer, G. R., Heckman, T. M., & Calzetti, D. 1999, ApJ, 521, 64
Moustakas, J., Kennicutt, R. C., & Tremonti, C. A. 2006, ApJ, 642, 775
Norman, C., et al. 2004, ApJ, 607, 721
Peeters, E., Spoon, H. W. W., & Tielens, A. G. G. M. 2004, ApJ, 613, 986
Pérez-González, P. G. et al. 2006, ApJ, 648, 987

Pety, J., Teyssier, D., Fossé, D., Gerin, M., Roueff, E., Abergel, A., Habart, E., & Cernicharo, J. 2005, A&A, 435, 885
Ranalli, P., Comastri, A., & Setti, G. 2003, A&A, 399, 39
Rosa-González, D., Terlevich, E., & Terlevich, R. 2002, MNRAS, 332, 283
Roussel, H., Sauvage, M., Vigroux, L., & Bosma, A. 2001, A&A, 372, 427
Rowan-Robinson, M., & Crawford, J. 1989, MNRAS, 238, 523
Salim, S., et al. 2007, ApJS, in press (astroph/0704.3611)
Sauvage, M., & Thuan, T. X. 1992, ApJ, 396, L69
Schmitt, H. R., Calzetti, D., Armus, L., Giavalisco, M., Heckman, T. M., Kennicutt, R. C., Leitherer, C., & Meurer, G. R. 2006, ApJ, 643, 173
Seibert, M. et al. 2005, ApJ, 619, L55
Sellgren, K. 1984, ApJ, 277, 623
Smail, I., Ivison, R. J., & Blain, A. W. 1997, ApJ, 490, L5
Springel, V., & Hernquist, L. 2003, MNRAS, 339, 289
Springel, V., di Matteo, T. & Hernquist, L. 2005, MNRAS, 361, 776
Steidel, C.C., Adelberger, K. L., Giavalisco, M., Dickinson, M., & Pettini, M. 1999, ApJ, 519, 1
Stinson, G., Seth, A., Katz, N., Wadsley, J., Governato, F., & Quinn, T. 2006, MNRAS, 373, 1074
Sullivan, M., Mobasher, B., Chan, B., Cram, L., Ellis, R., Treyer, M., & Hopkins, A. 2001, ApJ, 558, 72
Wang, B., & Heckman, T. M. 1996, ApJ, 457, 645
Wilson, G., Cowie, L. L., Barger, A. J., & Burke, D. J. 2002, AJ, 124, 1258
Wu, H., Cao, C., Hao, C.-N., Liu, F.-S., Wang, J.-L., Xia, X.-Y., Deng, Z.-G., & Young, C. K.-S. 2005, ApJ, 632, L79
Yan, L., McCarthy, P. J., Freudling, W., Teplitz, H. I., Malumuth, E. M., Weymann, R. J., & Malkan, M. A. 1999, ApJ, 519, L47

## Discussion

*Krumholz*: Can you say a bit more about why 8 $\mu$m flux traces star formation rates so poorly? Is it because the 8 $\mu$m emission is dominated by single photon heating of small PAHs?

*Calzetti*: That is correct. It is widely believed that the 8 $\mu$m band emission detected by *Spitzer*/IRAC is due to single-photon heating of small PAHs. Thus, even a non-star-forming stellar population can heat this dust component. Hence, it is relatively poorly correlated with SFR.

*Beckman*: I much appreciated your very clear and lucid exposition. In such a short time you cannot expect to go into detail about the data, so I will ask a question about that: to what extent are the results you present dependent on the beamwidths of the different instrument used?

*Calzetti*: All measurements were corrected for the effects of the different angular resolution of the different images. Thus, we have tried to minimize the impact of beamwidths. One component we cannot include, however, is the diffuse emission from normal star-forming galaxies (as opposed to starburst galaxies or star-forming regions), which may somewhat change our calibrations. This is the topic of a future investigation.

*Hammer*: How did you get such a correlation between [O II] and H$\alpha$? It seems that [O II] is not a tracer of the SFR, and especially at high $z$, it might be dangerous to consider it so. For example, we know that many distant galaxies are LIRGs and their [O II] emission is comparable to that of non-LIRGs. Please comment, especially on the physical reason to use [O II] as a tracer of the SFR.

*Calzetti*: I agree that the use of [O II] produces a highly uncertain correlation with SFR if that emission line is the only information available on a distant galaxy. I would not be surprised if uncertainties in the SFR from [O II] for an individual object were of the order of one magnitude, depending on the extinction, metallicity, and ionization conditions of the galaxy (see detailed discussions in Kewley et al. 2004 and Moustakas et al. 2006).

Torsten Böker and Daniela Calzetti.

*Pathways through an Eclectic Universe*
*ASP Conference Series, Vol. 390, © 2008*
*J. H. Knapen, T. J. Mahoney, and A. Vazdekis, eds.*

# Circumnuclear Regions of Star Formation in Early-type Galaxies

Ángeles I. Díaz,[1] Elena Terlevich,[2] Guillermo F. Hägele,[1] and Marcelo Castellanos[1]

[1]*Departamento de Física Teórica, Universidad Autónoma de Madrid, Spain*
[2]*Instituto Nacional de Astrofísica, Óptica y Electrónica, Puebla, Mexico*

**Abstract.** Circumnuclear star-forming regions, also called hotspots, are often found in the inner regions of some spiral galaxies where intense processes of star formation are taking place. In the UV, massive stars dominate the observed circumnuclear emission, even in the presence of an active nucleus, contributing between 30 and 50% to the total H$\beta$ emission of the nuclear zone. Spectrophotometric data of moderate resolution ($3000 < R < 11\,000$) are presented from which the physical properties of the ionized gas (electron density, oxygen abundances, ionization structure, etc.) have been derived.

## 1. Introduction

The inner (~1 kpc) zones of some spiral galaxies show high star formation rates frequently arranged in a ring or pseudo-ring pattern around their nuclei. In general, circumnuclear star-forming regions (CNSFRs)—also referred to as "hotspots"—and luminous and large disk H II regions are very much alike, but the former look more compact and show higher peak surface brightnesses (Kennicutt, Keel, & Blaha 1989). In many cases they contribute substantially to the emission of the entire nuclear region. Their large H$\alpha$ luminosities, typically higher than $10^{39}$ erg s$^{-1}$, point to relatively massive star clusters as their ionization source. These regions, then, constitute excellent places to study how star formation proceeds in high metallicity, high density circumnuclear environments.

The importance of an accurate determination of the abundances of high metallicity H II regions cannot be overestimated since these object constitute most of the H II regions in early spiral galaxies (Sa to Sbc) and the inner regions of most late-type ones (Sc to Sd), without which our description of the metallicity distribution in galaxies cannot be complete. The question of how high the highest oxygen abundance is in the gaseous phase of galaxies is still outstanding, and the extrapolation of known radial abundance gradients would point to CNSFRs as the most probable sites for these high metallicities. Accurate measurements of elemental abundances of high metallicity regions are therefore crucial to obtaining reliable calibrations of empirical abundance estimators, widely used but poorly constrained, whose choice can severely bias results obtained for quantities of the highest relevance for the study of galactic evolution such as the luminosity–metallicity ($L$–$Z$) relation for galaxies. CNSFRs are also ideal cases for studying the behavior of abundance estimators in the high metallicity regime.

## 2. Observations and Measurements

We have obtained moderate-resolution observations of 12 CNSFRs in three "hot spot" galaxies (NGC 2903, NGC 3351, and NGC 3504). These three galaxies are early barred spirals and show high star formation rates in their nuclear regions. They are quoted in the literature as among the spirals with the highest overall oxygen abundances (Pérez-Olea 1996). Figure 1 shows an *HST* archive image of the circumnuclear region of one of the observed galaxies, NGC 3351, taken with the WFPC2 camera through the F606W filter.

Our observations were obtained with the ISIS double spectrograph mounted on the 4.2 m William Herschel Telescope (WHT) at the Roque de los Muchachos Observatory and provided spectrophotometry of the spectral ranges $\lambda\lambda$3650–7000 Å in the blue and $\lambda\lambda$8850–9650 Å in the near-IR, with spectral resolutions of ~2.0 Å and 1.5 Å FWHM, respectively. Different slit positions were used to observe several distinct H II regions, as exemplified in Figure 1. The data were reduced using the IRAF (Image Reduction and Analysis Facility) package following standard procedures. The blue and red spectra corresponding to one of the observed regions are also shown in Figure 1. The presence of a conspicuous underlying stellar population in most observed regions, more evident in the blue spectra, complicates the measurements of emission lines (see the blue spectrum in Fig. 1). A two-component (emission and absorption) Gaussian fit was performed in order to correct the Balmer lines for underlying absorption.

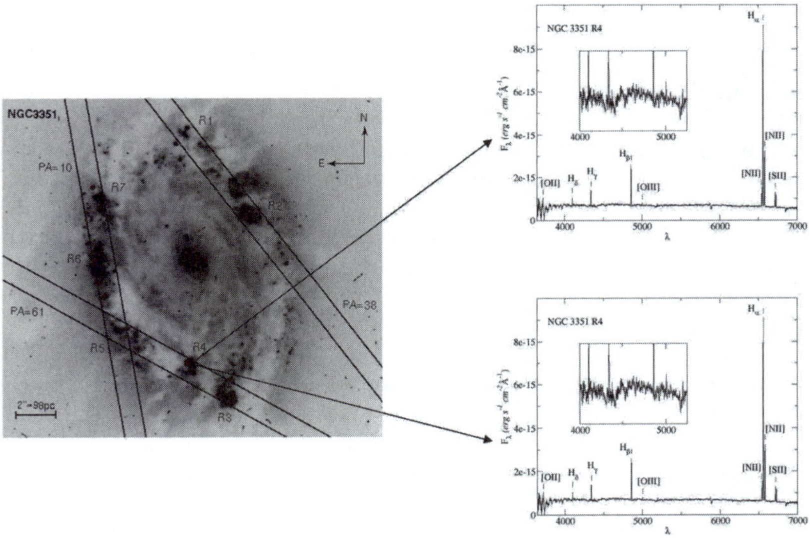


Figure 1. Observed CNSFRs in NGC 3351. The different slit positions are superimposed on an image taken from the *HST* archive and obtained with the WFPC2 camera with the F606W filter. The position angles of every slit position are indicated. The blue and red spectra corresponding to one of the observed regions are also shown.

## 3. Results and Discussion

Electron densities for each observed region were derived from the [S II] λλ6717, 6731 Å line ratio, following standard methods. They were found to be, in all cases, $\leq 600\,\mathrm{cm}^{-3}$, higher than those usually derived in disk H II regions, but still below the critical value for collisional de-excitation. The low excitation of the regions, as evidenced by the weakness of the [O III] λ5007 Å line (see the blue spectrum in Fig. 1) precludes the detection and measurement of the auroral [O III] λ4363 Å necessary for the derivation of the electron temperature. We have therefore used a semi-empirical procedure for the derivation of abundances. Firstly, we have produced a calibration of the [S III] temperature as a function of the abundance parameter $SO_{23}$ defined as:

$$SO_{23} = \frac{I([\mathrm{O\,II}]\lambda 3727,29) + I([\mathrm{O\,III}]\lambda 4959,5007)}{I([\mathrm{S\,II}]\lambda 6716,31) + I([\mathrm{S\,III}]\lambda 9069,9532)}.$$

This parameter is similar to the $S_3O_3$ proposed by Stasińska (2006) but is, in the first order, independent of geometrical (ionization parameter) effects. To perform this calibration we have compiled all the data so far available with sulfur emission line detections of both the auroral and nebular lines at λ6312 Å and λλ9069, 9532 Å, respectively (see Díaz et al. 2007). The calibration is shown in Fig. 2, together with the quadratic fit to the high metallicity H II region data:

$$T_{\mathrm{e}}([\mathrm{S\,III}]) = 0.596 - 0.283 \log SO_{23} + 0.199 (\log SO_{23})^2.$$

For one of the observed regions, R1+R2, the [S III] λ6312 Å line has been detected and measured yielding a value $T_{\mathrm{e}}([\mathrm{S\,III}]) = 8400^{+4650}_{-1250}$ K, represented as a solid black circle in Fig. 2 (*left*). We have used this calibration to derive $T_{\mathrm{e}}([\mathrm{S\,III}])$ for our observed CNSFRs. In all cases the values of $\log O_{23}$ are within the range used in performing the calibration, thus requiring no extrapolation of the fit. These temperatures, in turn, have been used to derive the $S^+/H^+$ and $S^{++}/H^+$ ionic ratios. Once the sulfur ionic abundances have been derived, we have estimated the corresponding oxygen abundances, assuming that sulfur and oxygen temperatures follow the relation given by Garnett (1992) and $T_{\mathrm{e}}$ ([O II]) $\simeq T_{\mathrm{e}}([\mathrm{S\,III}])$. Finally, we have derived the $N^+/O^+$ ratio assuming that $T_{\mathrm{e}}([\mathrm{O\,II}]) \simeq T_{\mathrm{e}}([\mathrm{N\,II}]) \simeq T_{\mathrm{e}}([\mathrm{S\,III}])$.

The observed CNSFRs turn out to be of high metallicity, but they show marked differences with respect to high metallicity disk H II regions. Even though their derived oxygen and sulfur abundances are similar, they show values of the $O_{23}$ and the N2 parameters whose distributions are shifted to lower and higher values, respectively, with respect to the high metallicity disk sample. Hence, if purely empirical methods were used to estimate the oxygen abundances for these regions, higher values would in principle be obtained. This would seem to be in agreement with the fact that CNSFRs, when compared to the disk high metallicity regions, show the highest [N II]/[O II] ratios.

CNSFRs also show lower ionization parameters than their disk counterparts, as derived from the [S II]/[S III] ratio. Their ionization structure also seems to be different, with CNSFRs showing radiation field properties more similar to those of H II galaxies than to those of disk high metallicity H II regions. This can be

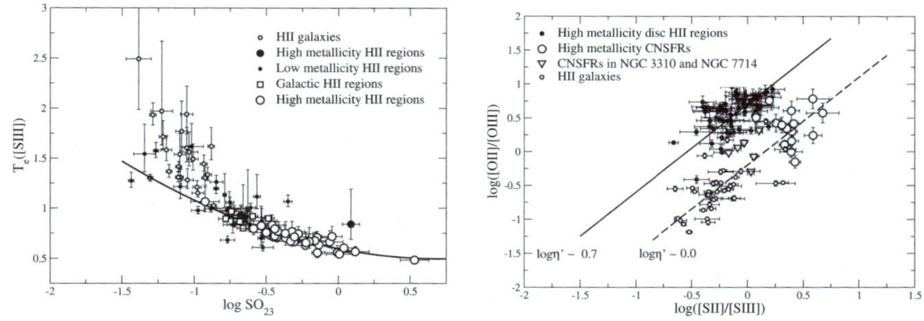


Figure 2. *Left:* Empirical calibration of the [S III] electron temperature as a function of the abundance parameter $SO_{23}$. The solid line represents a quadratic fit to the high metallicity H II region data. *Right:* The [O II]/[O III] vs [S II]/[S III] logarithmic ratios for different ionized regions. References for the data in the two figures can be found in Díaz et al. (2007).

seen in Fig. 2 (*right*), a diagram of the [O II]/[O III] vs [S II]/[S III] emission line ratios, that works as a diagnostic for the nature and temperature of the radiation field. In this plot, diagonal lines of slope unity show the locus of ionized regions with constant ionization temperature. CNSFRs are seen to segregate from disk H II regions. The former cluster around a value of $T_{\rm ion} \sim 40\,000$ K, while the latter cluster around $T_{\rm ion} \sim 35\,000$ K. Also shown are the data corresponding to H II galaxies. Indeed, CNSFRs seem to share more the locus of H II galaxies than that of disk H II regions.

One possible concern about these CNSFRs is that, given their proximity to the galactic nuclei, they could be affected by hard radiation coming from a low luminosity AGN. Alternatively, the spectra of these regions harboring massive clusters of young stars might be affected by the presence of shocked gas. Diagnostic diagrams of the kind presented by Baldwin, Phillips, & Terlevich (1981) can be used to investigate the possible contribution by either a hidden AGN or the presence of shocks to the emission line spectra of the observed CNSFR. When plotted on one of these diagnostic diagrams, log ([N II]/H$\alpha$) vs log ([O III])/H$\beta$, some of our CNSFRs are found close to the transition zone between H II region and LINER spectra but only one region in NGC 3504 may show a hint of a slight contamination by shocks.

**Acknowledgments.** We acknowledge financial support from DGICYT grant AYA-2004-02860-C03. AID thanks the hospitality of the Institute of Astronomy, Cambridge, during her sabbatical financed by the Spanish MEC through grant PR2006-0049.

## References

Baldwin, J. A., Phillips, M. M., & Terlevich, R. 1981, PASP, 93, 5
Díaz, A. I., Terlevich, E., Castellanos, M., & Hägele, G. F. 2007, MNRAS, submitted
Garnett, D. R. 1992, AJ, 103, 1330
Kennicutt, R. C., Keel, W. C., & Blaha, C. A. 1989, AJ, 97, 1022
Pérez-Olea, D. 1996, PhD Thesis, Univ. Autónoma de Madrid

Stasińska, G. 2006, A&A, 454, L127

## Discussion

*Pagel*: The circumnuclear H II regions presumably occupy a peculiar position in, for example, [O III] vs [N II] diagrams.

*Díaz*: These regions actually show two kinematic components in the ionized gas, both in H I and [O III], the wider of which shows a velocity dispersion which agrees with that shown by stellar lines. If that is taken into account, the gas with the wider component shows [O III]/H$\beta$ higher than that shown by the gas with the narrow component. But I can say nothing about the [N II]/H$\alpha$ ratio since we do not have comparable high resolution data for these lines. At any rate, everything points to these CNSFRs being in the low excitation region in the diagnostic diagram and not in the zone occupied by LINER-like objects.

*Ercolano*: The fact that abundance indicators for high metallicity overestimate the abundances may be due to geometrical effects of the spatial distribution of stars (see B. Ercolano, N. Bastian, & G. Stasińska [2007, MNRAS, 379, 945] and the poster, this volume, p. 110).

*Díaz*: If this is the case, then since CNSFRs show average values of abundance calibrators different from disk high metallicity regions, this would indicate different spatial distributions for these two types of star-forming regions. Semi-empirical abundance derivation should then be considered as more reliable. This would lead us to conclude that it is very difficult to find oxygen abundance higher than solar. Also, one has to explain the high N/O ratios in CNSFRs, by up to five times solar.

Ángeles Díaz.

*Pathways through an Eclectic Universe*
*ASP Conference Series, Vol. 390, © 2008*
*J. H. Knapen, T. J. Mahoney, and A. Vazdekis, eds.*

# A Consistent Picture of Gas Kinematics within Nuclear Rings

L. M. Mazzuca

*NASA Goddard Space Flight Center, Greenbelt, MD 20771, USA*

R. A. Swaters and S. Veilleux

*University of Maryland, Department of Astronomy, College Park, MD 20742, USA*

J. H. Knapen

*Instituto de Astrofísica de Canarias, E-38200 La Laguna, Spain*

**Abstract.** We present two-dimensional H$\alpha$ velocity fields for ten well-defined nuclear rings observed with the DensePak fiber-optic array on the KPNO WIYN telescope. In this paper, we present detailed results for two of those rings, in NGC 1530 and NGC 4314. We conclude that in general regions of enhanced H$\alpha$, thus massive star formation, in the nuclear rings correspond to regions with (1) high luminosities, of order $10^{40}$–$10^{42}$ erg s$^{-1}$, (2) low residual velocities, after subtracting disk rotation, of order 10 km s$^{-1}$, and (3) low velocity dispersions in the range 20–50 km s$^{-1}$. In the rings, star formation takes place in regions where the shear is small and the gas is cool enough to allow molecular condensation. The lack of strong non-circular motions, coupled with a direct relationship between the position angles and ellipticities of the rings and those of their host galaxies, indicates that the rings are in the same plane as the disk and are circular.

## 1. Introduction

Nuclear rings are composed of individual H II regions and associated with starbursting events. The rings exist due to gas inflow from the outer parts of the galaxy to the central regions and are often related to the presence of a bar (e.g., Schwarz 1981; Combes & Gerin 1985; Shlosman, Begelman, & Frank 1990). Kinematics have revealed that bars can perturb the potential and cause non-circular motions in the circumnuclear regions, as first noted by Rubin, Peterson, & Ford (1975) and Peterson et al. (1976). Although small in scale (~1 kpc in radius), nuclear rings may disturb the axisymmetry of the gas as the rings intermingle with the dust lanes of the parent bar (e.g., Knapen et al. 1995; Heller & Shlosman 1996).

The non-circular motions within nuclear rings must be limited in amplitude, as strong velocity changes will prohibit star formation, which is prevalent in the rings. The existence of very large velocity residuals can indicate shocks and shears, which create an environment too turbulent to allow gas to form concentrated zones that may condense to form stars. In this paper we present part of a larger H$\alpha$ kinematic nuclear ring study (L. M. Mazzuca et al., in preparation) aimed at characterizing the level of non-circular motions in the nuclear rings via velocity residuals, dispersions, and morphological characteristics.

## 2. Observations and Methodology

The two rings presented in this paper were obtained using the DensePak fiber-optic array and bench spectrograph on the 3.5 m WIYN telescope From the multiple pointings accrued for each ring, we constructed sparse as well as contiguous smooth (interpolated) velocity field maps, which were registered with the H$\alpha$ images of Mazzuca et al. (2007). Figure 1 shows the smoothed velocity contours superimposed on the H$\alpha$ continuum-subtracted images of two well-known nuclear rings: those in NGC 1530 and NGC 4314.

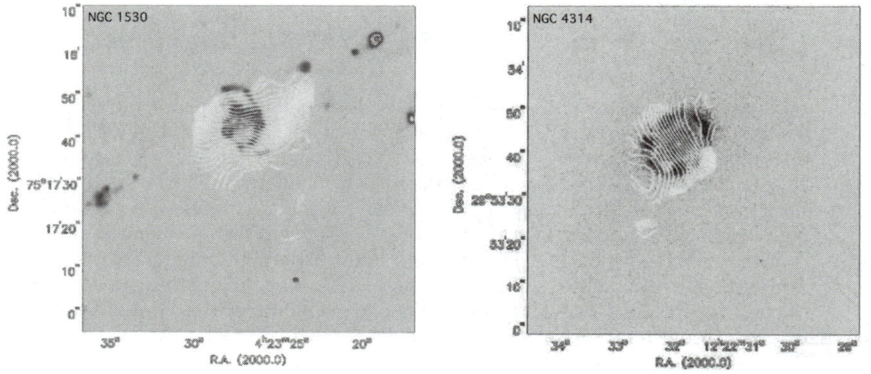


Figure 1. Velocity contours superimposed on H$\alpha$ continuum-subtracted images for NGC 1530 and NGC 4314.

## 3. Results

The smoothed velocity fields in Fig. 1 indicate that non-circular velocities may exist in the circumnuclear regions. To quantify the deviations better, we plot the velocity data as a function of azimuthal position, where the radial coverage is divided to show velocities along the inner and outer halves of the ring (Fig. 2). Uncertainties associated with the measured velocity, wavelength calibration, and position of the H$\alpha$ peak within a given fiber beam were added in quadrature. The measured uncertainties range from 0.5 km s$^{-1}$ to 5 km s$^{-1}$. Within the nuclear rings, we find no cases that show strong (i.e., greater than the systematic scatter around the model curve) departures from circular motion.

We made residual velocity maps to identify any non-circular motions along each of the rings. We find low residual velocities (both approaching and receding) near the central ridge of the rings (averaging $\sim$10 km s$^{-1}$ and $\sim$20 km s$^{-1}$ for NGC 4314 and NGC 1530, respectively). A harmonic analysis (L. M. Mazzuca et al., in preparation) independently confirms low non-circular motions for the two galaxies shown here and for another half dozen in our kinematic sample.

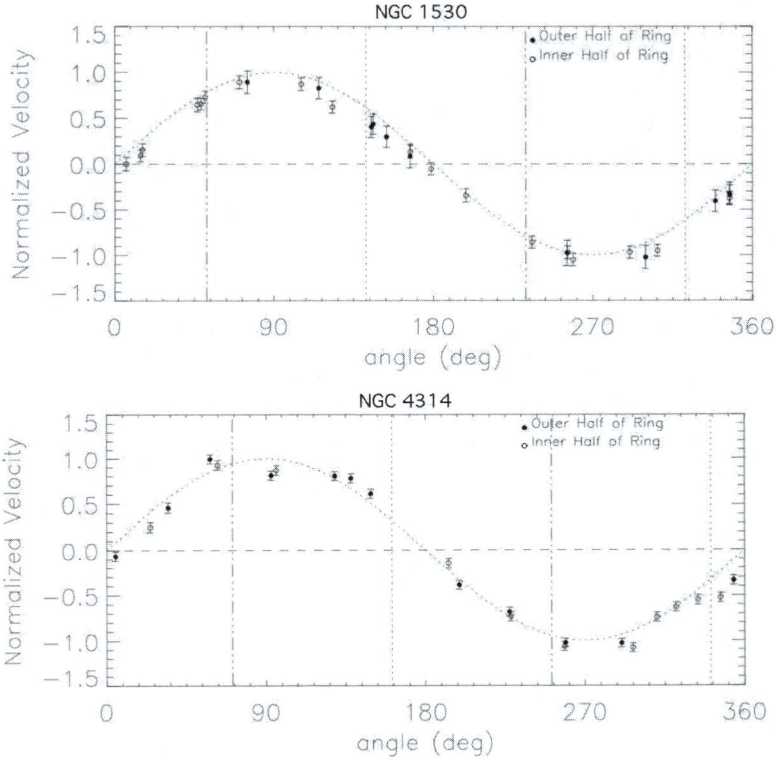


Figure 2. Radial velocities for the nuclear rings of NGC 1530 and NGC 4314. All velocities are plotted with respect to the systemic velocity, normalized by the rotation velocity, and adjusted by removing the inclination in order to allow comparison with the curve representing circular motion. The bar major axis (*dash-dot lines*) and minor axis (*dotted lines*) are drawn for context. Note that there is a phase shift of 90°, plus the difference between the kinematic PA of the velocity data points and the photometric PA used to create the model curve, to allow the cross-over point to occur at 180°.

Dispersion due to random motions is another indicator of possible axisymmetric behavior. We produced such maps, accounting for instrumental broadening, as shown in Figure 3. Dispersion values within the rings range from ~20 km s$^{-1}$ for NGC 4314 to 40–50 km s$^{-1}$ for NGC 1530.

## 4. Discussion and Conclusions

In our imaging survey (Mazzuca et al. 2007) we showed that nuclear rings comprise several distinct young massive star-forming H II regions, each with luminosities of order $10^{40}$ erg s$^{-1}$. We compare the luminous H II regions, as seen in H$\alpha$, to the corrected dispersion maps in which values range from $\sim$ 20 km s$^{-1}$ to $\sim$ 50 km s$^{-1}$ and find that the regions of low dispersion seen throughout the rings

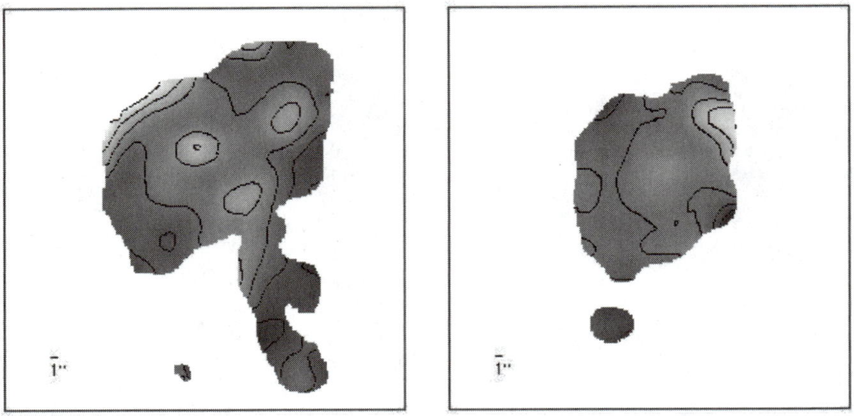


Figure 3. Dispersion maps for NGC 1530 (*left*) and NGC 4314 (*right*), corrected for instrumental broadening.

coincide with the presence of strong H$\alpha$ emission and young luminous stars. We compared the total luminosity in the nuclear rings with the average dispersion and find the low dispersion seen in our sample to be indicative of H II regions with luminosities from $10^{40}$–$10^{42}$ erg s$^{-1}$. The low dispersions are also correlated spatially with the strong H$\alpha$ emission seen within the nuclear rings and active nuclei. Such low dispersions are indicative of regions with cool gas (low random motions) that allow the triggering of star formation.

In L. M. Mazzuca et al. (in preparation) we compare the disk and nuclear ring position angles (PAs), as well as the kinematic and photometric PAs of the rings, as variations between the two can trace non-circular motions. We find a general linear relationship between the various PAs. The strong agreement between the ring photometric and galaxy disk PAs, as well as between their ellipticities, was also found in the H$\alpha$ imaging study of Mazzuca et al. (2007) for 22 nuclear rings, which includes the two highlighted here. Overall, our results confirm that the nuclear rings are in the same plane as the disk, and are circular.

## References

Combes, F., & Gerin, M. 1985, A&A, 150, 327

Falcón-Barroso, J., et al. 2006, MNRAS, 369, 529

Heller, C. H., & Shlosman, I. 1996, ApJ, 471, 143

Knapen, J. H., Beckman, J. E., Shlosman, I., Peletier, R. F., Heller, C. H., & de Jong, R. S. 1995, ApJ, 443, L73

Mazzuca, L. M., Sarzi, M., Knapen, J. H., Veilleux, S., & Swaters, R. 2006, ApJ, 649, L79

Mazzuca, L. M., Veilleux, S., Knapen, J. H., Regan, M. W. 2007, ApJ, in press

Peterson, C. J., Thonnard, N., Rubin, V. C., & Ford, W. K., Jr. 1976, ApJ, 208, 662

Rubin, V. C., Peterson, C. J., & Ford, W. K., Jr. 1975, ApJ, 199, 39

Schwarz, M. P. 1981, ApJ, 247, 77

Shlosman, I., Begelman, M. C., & Frank, J. 1990, Nature, 345, 679

**Discussion**

*Peletier*: I am happy to see that your DensePak data for NGC 4314 and NGC 7742 agree well with SAURON results from Falcón-Barroso et al. (2006). Since you also have stellar continuum data, do you have any results on the stellar kinematics in your galaxy?

*Mazzuca*: We combined DensePak and SAURON data to perform a stellar kinematic study of the circumnuclear region of NGC 7742. Results were published in Mazzuca et al. (2006).

*Díaz*: You said that regions with the lowest velocity dispersions are those showing the highest H$\alpha$ emission. Do you have any reason for that?

*Mazzuca*: Low velocity dispersions within nuclear rings, which are known to have high H$\alpha$ luminosities ($10^{40}$–$10^{42}$ erg/s) emerging from the young stellar clusters, imply that the gas is cool in the rings (of order $10^3$ K) and relatively non-turbulent. Literature on the nature of bars reveals that this cool gas emanates from the bar dust lanes and is not prone to star formation until it reaches the ring.

*Combes*: Have you tried to compare your detailed kinematics with CO interferometric maps? There are detailed interferometric velocity fields for NGC 1530.

*Mazzuca*: We do compare the CO data from Jogee et al. (NGC 5248), Allard et al. (M100), Benedict et al. (NGC 4314), Knapen et al., Zurita et al., and Reynard & Downes (NGC 1530) with our sample and find good correlation in general. For example, Reynard & Downes find that the residuals in NGC 1530 are comparable to our values (50 km s$^{-1}$ maximum), and claim that strong velocity gradients ($> 100$ km s$^{-1}$) will prevent enhanced star formation.

From left to right: Lisa Mazzuca, Martin Stringer, Marc Sarzi, and Kartik Sheth.

*Pathways through an Eclectic Universe*
*ASP Conference Series, Vol. 390, © 2008*
*J. H. Knapen, T. J. Mahoney, and A. Vazdekis, eds.*

# Star Formation and Molecular Gas in M83

A. A. Lundgren,[1] H. Olofsson,[2] T. Wiklind,[3] and R. Beck[4]

[1]*European Southern Observatory, Casilla 19001, Santiago 19, Chile*

[2]*Stockholm Observatory, AlbaNova, SE-106 91 Stockholm, Sweden*

[3]*ESA Space Telescope Division, STScI, 3700 San Martin Drive Baltimore, MD 21 218, USA*

[4]*Max-Planck-Institut für Radioastronomie, Auf dem Hügel 69, D-53121 Bonn, Germany*

**Abstract.** We compare star formation rates and efficiencies, obtained from different star formation tracers, before and after compensation for internal extinction due to gas and dust, in the barred spiral galaxy M83 (NGC 5236). We use far-UV continuum, $B$-band continuum, and H$\alpha$ line images to estimate the star formation rate in the disk of this galaxy at 25″ resolution. We use the total gas map, derived from CO($J = 1 \rightarrow 0$) and H I data, to compensate for internal extinction. We find that the star formation rate estimates based on the different data agree well when corrections for internal extinction have been made. The star formation rate is particularly high in the nuclear region, at the bar ends, and in the spiral arms. Combined with the total gas mass map, the star formation rates provide estimates of the star formation efficiencies in the disk at 25″ resolution. The star formation efficiency map is more patchy than the star formation rate map. There is evidence of an increased star formation efficiency along the spiral arms, i.e., not only do the arms present regions of higher gas mass surface density, the efficiency for making stars is also increased in them.

## 1. Introduction

The formation of stars out of cosmic gas is certainly one of the most important phenomena within contemporary astrophysics. It stretches from the formation of isolated low mass stars, with an associated protoplanetary disk, over the large scale distribution of stars of various types and ages in galaxies of different characteristics, to the formation of the first stars in the Universe at least 13 Gyr ago ($z \approx 6$–7).

We have initiated a detailed study of the barred grand design spiral galaxy M83 (distance about 4.5 Mpc) aiming at its star formation properties. M83 is a place of vigorous stellar births and deaths. About 25% of the massive star formation rate in the local Universe ($D < 10h_{50}^{-1}$ Mpc, about 200 galaxies) is confined to four galaxies: M82, M83, NGC 253, and NGC 4945 (Heckman 1997). Also, six supernovae (SNe) have been observed in M83 during the last century (1923A, 1945B, 1950B, 1957D, 1968L, and 1983N), more than in any other Messier galaxy, and second only to NGC 6946, in which eight SNe have been observed.

## 2. Star Formation Rate Measures

Star formation rate measures are almost exclusively based on the contribution from main sequence stars (often the most massive ones) to broad band fluxes, ionization radiation, or dust heating. The main theoretical problem is to convert reliably from the minor fraction of contributing stars, and the timescale on which they contribute, to the large fraction of stars that are born per unit time. This requires stellar evolutionary models combined with stellar spectra and an initial mass function (IMF). We use SFRs that are based on a Salpeter IMF, and the mass limits 0.1 and 100 $M_\odot$. The main observational problem is to isolate the fraction of the radiation that is due to the main sequence stars, and to make proper corrections for the internal extinction inside the galaxy. We will here briefly describe the SFR measures used in this paper. Kennicutt (1998) discusses the problems associated with estimating SFRs.

The UV continuum is exclusively generated by young, massive stars on the main sequence and is thus an excellent measure of the current SFR. The $B$-band luminosity SFR measure is similar to the UV continuum measure. The main difference is that the $B$-flux is dominated by stars in the mass interval 1.0–2.5 $M_\odot$ and thus probes the SFR over the last $4 \times 10^8$–$6 \times 10^9$ yr. The H$\alpha$ line intensity is assumed to be the direct effect of ionizing radiation from young massive stars, i.e., it becomes a measure of the Lyman continuum flux. The SFR can also be estimated from many other tracers, such as the far-infrared luminosity, and thermal/non-thermal radio continuum flux. These tracers are discussed in Lundgren et al. (2007). To calculate the SFR we use the expressions in Kennicutt (1998) for UV continuum and H$\alpha$ line intensity and Gallagher, Hunter, & Tutukov (1984) for the $B$-band.

## 3. Data

We have used FUV data taken from the Multimission Archive (MAST) at STScI, and the H$\alpha$ and $B$-band data were obtained with the Danish 1.54 m telescope, equipped with DFOSC (Danish Faint Object Spectrograph and Camera), at the European Southern Observatory (ESO), La Silla, Chile.

The CO line emission data of M83 were taken using the Swedish–ESO Submm Telescope (SEST). Earlier results based on this data set have been presented in Lundgren et al. (2004a,b), where we determined the distribution and kinematics of the molecular gas, identified important dynamical features of the disk, and combined the molecular data with 21 cm H I (from VLA, Tilanus & Allen 1993) column densities in order to derive the total gas surface density.

The $H_2$ column density is derived in the commonly accepted way from the velocity-integrated CO($J = 1 \rightarrow 0$) intensity ($I_{\rm CO} = \int T_{\rm mb} {\rm d}v$). We adopted the conversion factor $X_{\rm CO} = 2.3 \times 10^{20}$ (K km s$^{-1}$)$^{-1}$ cm$^{-2}$.

We have chosen to use our total gas mass surface density map to estimate the extinction. It should be kept in mind here that our gas mass surface densities have an angular resolution of 25″, while, for example, the H$\alpha$ fluxes are line-of-sight measurements towards H II regions, a non-trivial complication. However, if the radiation sources are randomly distributed within the material responsible for the extinction it is possible to get an estimate of the extinction that is rea-

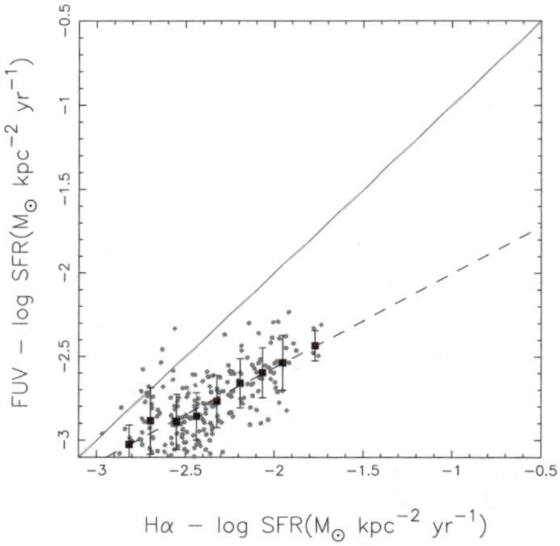


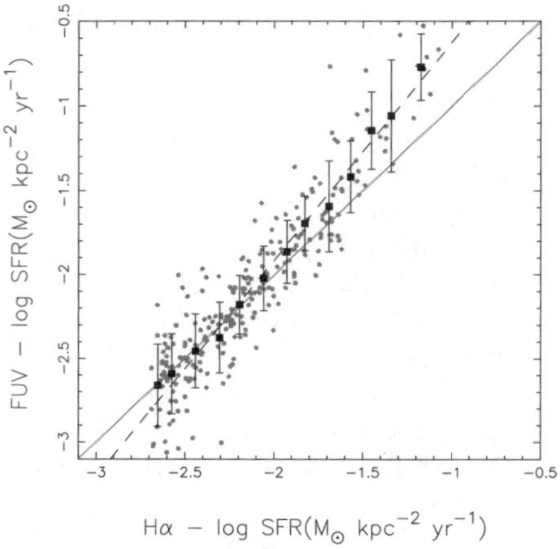


Figure 1. $\Sigma_{\rm SFR}$s for FUV as a function of $\Sigma_{\rm SFR}$s for H$\alpha$. *Left:* Not corrected for extinction. *Right:* Corrected for extinction. The error bars represent the $\pm 1\sigma$ scatter of the binned data. The data are not corrected for extinction. The dashed black line shows a fit to the binned data and the gray line shows the one-to-one relation.

sonable, at least in a statistical sense. With this assumption the extinction can be obtained approximately as $A_V = 2.5 \log\left[\tau_{\rm ext}/(1-e^{-\tau_{\rm ext}})\right]$, where, according to Melnick (1979), $\tau_{\rm ext} = N_{\rm H}/\left(2.05\times10^{21}\,{\rm cm}^{-2}\right) = \Sigma_{\rm g}/\left(22.4\,M_\odot\,{\rm pc}^{-2}\right)$. The standard Galactic relation between extinction and hydrogen column density is used (Cox 2000, we have corrected $\Sigma_{\rm g}$ for the He abundance). This results in $A_V = \tau_{\rm ext}/2$ for low optical depths and $A_V = 2.5\log\tau_{\rm ext}$ for high optical depths. Thus, the correction is only weakly dependent on $\tau_{\rm ext}$ for higher optical depths and hence probably gives a rather conservative estimate of the extinction. The assumption of well-mixed gas and stars is probably quite good in the case of M83, since the gas in the disk is clearly dominated by the $H_2$ gas, which also dominates the total gas (Lundgren et al. 2004a).

## 4. Comparison of Star Formation Rates

We compare the results of the various SFR measures in terms of SFR surface densities, $\Sigma_{\rm SFR}$ (in units of $M_\odot\,{\rm yr}^{-1}\,{\rm kpc}^{-2}$). The data are also convolved to the same angular resolution as the total gas map (25″) in order to be corrected for extinction internal to M83. Figure 1 shows, as an example, the improvement of the correlation for the $\Sigma_{\rm SFR}$s estimated from the FUV data as a function of those from the H$\alpha$ data.

We finally adopt the *median* of the FUV, $B$, and H$\alpha$ results as the best estimate of the current $\Sigma_{\rm SFR}$ with a resolution of 25″. This has the advantage of not overestimating the SFR in regions of high extinction (where the FUV data tend to give very high values), while at the same time giving a reasonable "average" in other regions.

The star formation efficiency (SFE) is obtained as $\Sigma_{\rm SFR}/\Sigma_{\rm g}$ converted to the $\tau_8$ time scale, where $\tau_8$ corresponds to $10^8$ yr (Fig. 2). Clearly, the SFE is enhanced

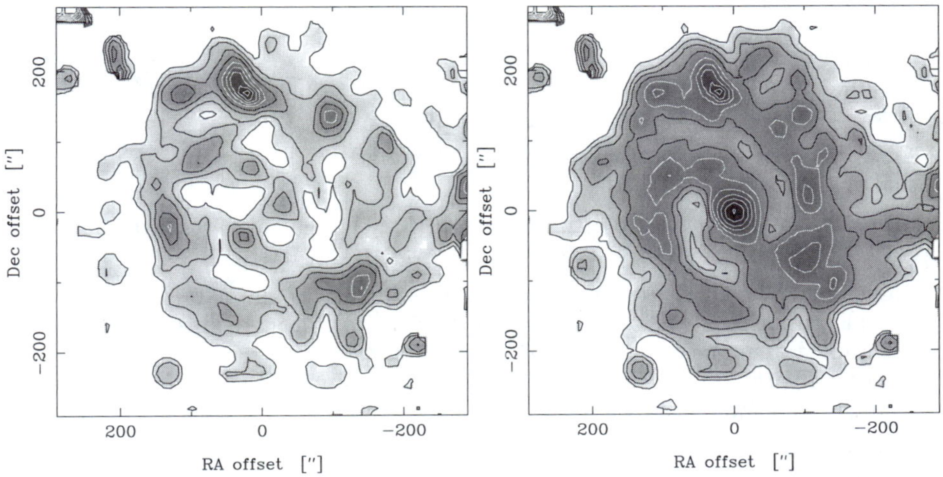


Figure 2. The SFE($\tau_8$) maps of M83 (*left:* uncorrected for extinction and *right:* corrected for extinction). The scale is logarithmic, and the contours range from −2.5 to 0 with an increment of 0.25 (first white at −1.5) in units of $(10^8\,\mathrm{yr})^{-1}$. Features outside the main disk are most likely exaggerated, due to a lack of CO observations.

in the spiral arms, an effect present also in the data that are not corrected for extinction. This effect is pronounced in the extinction-corrected map, where also the nucleus (even though it is very likely overestimated here) and the bar ends become prominent. In comparison, the SFE maps appear more patchy than the SFR maps. This could be an effect of regions where efficient star formation has consumed, and dispersed, a significant fraction of the molecular gas.

The SFR estimates are excellently fitted with a Schmidt law of the form $\Sigma_{\mathrm{SFR}} \propto \Sigma_{\mathrm{g}}^N$ with an exponent $N \approx 1.0$ (no extinction correction applied) and 1.9 (extinction correction applied). However, equally good fits are obtained for alternative Schmidt laws like $\Sigma_{\mathrm{SFR}} \propto (\Sigma_{\mathrm{g}}\Omega)^N$ and $\Sigma_{\mathrm{SFR}} \propto \Sigma_{\mathrm{H_2}}^N$.

## References

Cox, A. N. 2000, Allen's Astrophysical Quantities, 4th edn (New York: AIP Press, Springer)
Heckman, T. M. 1997, AIP Conf. Ser., 393, 271
Gallagher, J. S, Hunter, D. A., & Tutukov, A. V. 1984, ApJ, 284, 544
Kennicutt, R. C. 1998, ARA&A, 36, 189
Lundgren, A. A., Wiklind, T., Olofsson, H., & Rydbeck, G. 2004a, A&A, 413, 505
Lundgren, A. A., Olofsson, H., Wiklind, T., & Rydbeck, G. 2004b, A&A, 422, 865
Lundgren, A. A., Olofsson, H., Wiklind, T., & Beck, R. 2007, A&A, (submitted)
Melnick, J. 1979, ApJ, 228, 112
Thilker, D. A., et al. 2005, ApJ, 619, L79
Tilanus, R. P. J., & Allen, R. J. 1993, A&A, 274, 707

## Discussion

*Edmunds*: In the $CO/H_2$ conversion factor and extinction correction did you take account of the abundance gradient??

*Lundgren*: M83 has a relatively weak abundance gradient, so we did not take that into account.

*Combes*: Have you tried to estimate the SFE in the central parts, where Thilker et al. (2005) found star formation with *GALEX*?

*Lundgren*: The SFE in the center is very difficult to estimate because of all the uncertainties, such as change in the conversion factor, the breakdown of the well-mixed assumption, and the different timescale of star formation. We have compared with Thilker et al., and in the nucleus (and most of the disk) the two studies give similar results.

*James*: I am interested in the central peak in SFE you find, which you say is due to the method breaking down. Does this effect occur independently of which SF indicator (FUV, $H\alpha$, $B$) is used?

*Lundgren*: All star formation tracers show an increased SFE towards the center. The FUV clearly gives too high an SFR (leading to an overestimated SFE), most probably because the assumption of a well-mixed medium does not work any more. The $H\alpha$ and $B$ SFEs agree very well in the nucleus (but not in the disk since $B$ is contaminated by a contribution from evolved stars).

*Krumholz*: Did you try correlating your SFRs with CO alone as well as with CO + H I? If so, was the fit better or worse?

*Lundgren*: In the inner part of the disk, $H_2$ dominates over H I by several factors, so there it hardly matters if we include H I or not. In the outskirts of the galaxy and beyond the OLR, the H I mass surface density starts to dominate. We do see that if H I is not included, both the SFR, and in particular the SFE, are overestimated in the outer part of the disk. However, this may also be an effect of, for example, the conversion factor ($X_{CO}$) that we use underestimating the $H_2$ content at large galactic radii.

Andreas Lundgren.

*Pathways through an Eclectic Universe*
*ASP Conference Series, Vol. 390, © 2008*
*J. H. Knapen, T. J. Mahoney, and A. Vazdekis, eds.*

# The Schmidt Law: Is it Universal and What Are its Implications?

Robert C. Kennicutt, Jr.

*Institute of Astronomy, University of Cambridge, Madingley Road, Cambridge CB3 0HA, UK*

**Abstract.** Galaxies exhibit an immense diversity in their star formation properties, yet these properties appear to be linked by a few underlying scaling laws. One of the most fundamental of these laws is the star formation law relating the surface densities of star formation and cold gas in galaxies. Virtually the entire range of global star formation rates in galaxies can be reproduced by the simple combination of a Schmidt power law relation at high density with a sharp turnover in star formation rate below a threshold surface density. Thanks to a new generation of multi-wavelength observations of nearby galaxies, it is now possible to test the validity of this description as a function of physical scale and interstellar environment. These new observations also offer new clues to the physical processes that underlie and drive the observed star formation law. This paper reviews recent progress in our observational and theoretical understanding of the star formation law in the Schmidt power law and threshold regimes.

## 1. Introduction

Despite the many recent advances in our observations and theoretical understanding of star formation on the local scale, our understanding of star formation on galactic scales is at a relatively embryonic stage. A deeper understanding is crucial for a complete theory of star formation itself, because many of the agents that trigger cloud formation and regulate the star formation rate (SFR) appear to operate on the scales of hundreds of parsecs or larger. Such an understanding is also a necessary prerequisite for a complete theory of galaxy formation and evolution; indeed our poor understanding of star formation and the associated feedback processes between young stars and the interstellar medium (ISM) stand as the primary impediments to galaxy formation and evolution models today. This is hardly surprising. The physical process involved are complex and challenge the best of our theorists and numerical simulators.

Despite this imposing challenge, nature has kindly provided us with tantalizing empirical hints to these processes, in the form of a suite of remarkably tight scaling laws relating the large-scale SFR to the large-scale properties of the interstellar gas. The most well-known of these is the Schmidt law relating the SFR surface densities to the total gas surface density (e.g., Schmidt 1963; Kennicutt 1998b). More recently a similar, linear scaling relation has been discovered between the total SFR and the mass of dense molecular gas (e.g., Gao & Solomon 2004). These relations hold over large ranges of densities, host galaxy properties, and star formation environments, and when combined with low density threshold effects provide a nearly complete empirical prescription for the large-scale SFR.

In this paper I begin by reviewing the observations of the star formation law in nearby galaxies and some of the outstanding questions that remain to be addressed by observation and theory. I then present some recent results from the *Spitzer* Infrared Nearby Galaxies Survey (SINGS) on the spatially resolved Schmidt law in galaxies.

## 2. Basic Observations and Unanswered Questions

Schmidt (1959) originally formulated the star formation law as a power-law relation between the volume densities of star formation and gas, but since volume densities are not readily measurable most current applications relate the projected surface densities of these quantities:

$$\Sigma_{\rm SFR} = A\ \Sigma_{\rm gas}^N \tag{1}$$

For the sake of convenience whenever I refer to "density" in this paper I always mean the surface density. For most applications the gas density $\Sigma_{\rm gas}$ refers to the total surface density of cold hydrogen (atomic and molecular), but it is important of course to test the form of the relation for different gas components, as will be discussed later.

Since the publication of Schmidt's original papers there have been dozens of empirical studies correlating the surface densities of various star formation tracers with the surface densities of H I, molecular gas, or both (see Kennicutt 1997, 1998b for a review of papers up to that time). A breakthrough occurred about 10 years ago, when sufficient data on extinction-corrected SFRs, CO maps, and H I maps became available to measure the correlation over a large dynamic range in surface densities. The results are summarized in Fig. 1, which is adapted from Kennicutt (1998b). The left panel plots the disk-averaged SFRs and gas surface densities for normal spirals (solid circles), infrared-luminous starbursts (squares), and the central regions of normal galaxies (open circles). The best fitting relation has a slope $N = 1.4 \pm 0.1$. We return to the right-hand panel later.

Although the Schmidt power law seems to provide an excellent description of the SFR at high gas surface densities, at low densities one often observes a rapid transition or threshold below which the SFR declines more rapidly. These thresholds are most readily observed as radial truncations to disks observed in H$\alpha$; star formation often persists beyond this threshold radius but at a much lower rate (e.g., Kennicutt 1989; Martin & Kennicutt 2001). In some gas-poor galaxies this threshold regime extends across the entire radius of the disk (e.g., Kennicutt 1989; van der Hulst et al. 1993). It is the combination of a relatively mild-slope power-law behavior at high density with a steeper, nonlinear behavior at low density that can account for the enormous range in SFRs and SFR densities among galaxies.

This nonlinear dependence on total gas surface density is not the only scaling law of its kind. In a series of papers Gao & Solomon (2004) have correlated the infrared luminosities of actively star-forming galaxies with the total masses of dense molecular gas as traced primarily in the HCN line. This reveals a

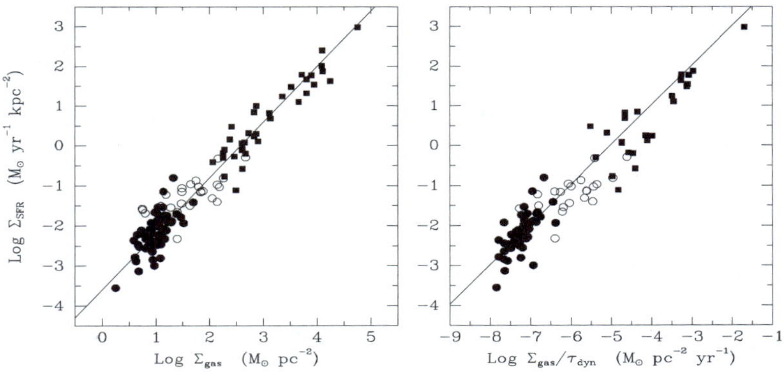


Figure 1. *Left:* The global Schmidt law in galaxies parameterized in terms of the SFR surface density and the total (atomic plus molecular) gas surface density. *Right:* The same data fitted to a "Silk law", where the SFR surface density is plotted against the ratio of gas surface density to mean disk orbit time. (*Both figures adapted from Kennicutt 1998b, copyright AAS.*)

comparably tight but linear ($N = 1$) relation, again observed over a wide range of SFRs. The HCN is thought to trace the mass of dense stellar cores in star-forming regions, and hence this result suggests that the fraction of core mass that is ultimately converted to stars is relatively insensitive to the scale and environment in which the star formation takes place.

Beyond this rudimentary picture a plethora of questions remain unanswered. Are these strong correlations universal, or are they partly the result of selection effects in the datasets assembled to date? Is part of the scatter observed in the relations physical, and if so are the residuals systematically correlated with other properties of the galaxies? Does the disk-integrated Schmidt law also describe the local relationship between SFR and gas densities *within* galaxies, and if so on what linear scale does this law break down. Is the Schmidt law a causal physical relationship, or is it merely the empirical consequence of a deeper underlying correlation? For example, the right-hand panel of Fig. 1 shows the same SFR densities plotted not as a function of gas surface density but rather the ratio of the gas density to the local dynamical (disk orbit) time, following a relation proposed by Silk (1997) and others. The latter correlation is nearly as tight as the Schmidt law, so which (if either) is the fundamental correlation, and which is merely the empirical consequence of the other? Likewise, is the observed correlation of the SFR with the total surface density of gas fundamental, or is it the result of a stronger correlation with the molecular or atomic gas surface densities by themselves? And of course an overarching unanswered question is the underlying physical origin of the scaling law.

A similar set of questions surrounds the apparent star formation thresholds that are observed in many galaxies. Are the strong radial transitions observed in the H II region disks also seen in other star formation tracers? What is the relationship between the SFR density and gas density in the subcritical regime? Is it determined by a different Schmidt-type law, or does the relationship between

SFR and gas surface densities become more stochastic in this regime? And again, what is the physical origin of the observed thresholds? Is it largely dictated by gravitational stability processes, ISM phase transitions, magneto-gravitational instabilities, or more complicated processes? In the remainder of this paper I review some recent observations directed at answering some of these questions, with emphasis on some new results from my collaborators.

## 3. The Disk-Averaged Star Formation Law

The global star formation laws shown in Fig. 1 show a considerable dispersion (approximately $\pm 0.4$ dex rms), and it is not clear how much of this dispersion is physical, rather than reflecting observational uncertainties in the measured SFRs and gas densities. For the Kennicutt (1998b) study the largest observational errors were in the extinction corrections applied to the H$\alpha$-based SFRs (usually a constant correction factor of 1.1 mag), and uncertainties arising from the application of a constant CO/$H_2$ conversion factor ($X$-factor) to compute molecular surface densities. Recently I initiated a project to enlarge and improve on the Kennicutt (1998b) dataset. Spatially resolved SFR, CO, and H I data are now available for more than twice as many galaxies as a decade ago, and for most of these we can apply individual H$\alpha$ extinction corrections, either from optical spectroscopic measurements or from Pa$\alpha$ line imaging at 1.87 $\mu$m, which is nearly free of dust extinction. In addition we have applied individual corrections to the H$\alpha$ fluxes for [N II] forbidden line contamination, and have used H$\alpha$ or infrared maps of the galaxies to precisely measure the size of the star-forming region over which the SFR and gas surface densities are averaged (in the 1998 study we used the RC3 diameters to approximate these radii).

This work is still in progress, but a preliminary plot of the intermediate results is shown in Figure 2. The new data (solid points) confirm the relation found by Kennicutt (1998b), with an $N = 1.4$ Schmidt law providing a good fit to the new data. Interestingly, although there are fewer discrepant points than found in the Kennicutt (1998b) analysis (mainly because of the improved extinction and [N II] corrections and better disk diameters), the overall dispersion of the Schmidt law is similar to that seen earlier. This suggests either that the observed dispersion is real, or that other systematic effects (e.g., a variable CO/$H_2$ $X$-factor) are responsible for most of the dispersion.

For the first time we have enough data to isolate a possible second parameter influencing the relation—metallicity. The solid points in Fig. 2 correspond to luminous spiral and irregular galaxies with $M_B < -17$, while fainter dwarf irregular galaxies are shown as open circles. S–Irr galaxies follow a metallicity–luminosity relation, and this transition luminosity corresponds to a metallicity of $Z \sim 0.2$–$0.3\,Z_\odot$ (Skillman, Kennicutt, & Hodge 1989). The galaxies with low SFR intensities generally follow the main Schmidt law, but the most active star-forming low metallicity galaxies tend to lie well above (or to the left of) the main law defined by the more luminous galaxies.

One interpretation of this difference is that the gas surface densities in the starbursting dwarf galaxies are systematically underestimated, due to a breakdown in the CO/$H_2$ $X$-factor. Indeed in many of these galaxies CO is either undetected or very weak, despite the intense star formation bursts, and the au-

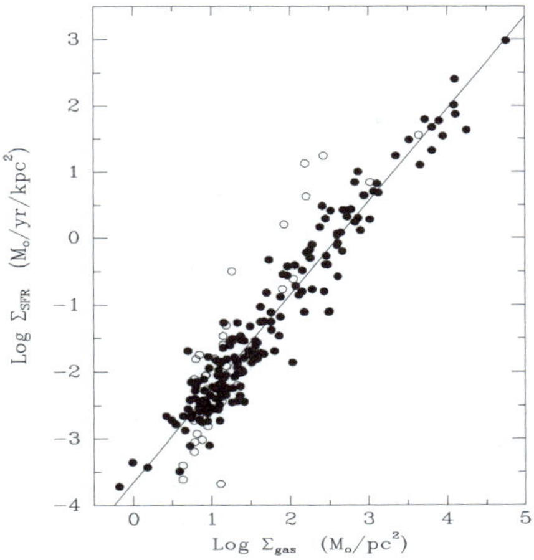


Figure 2. Disk-averaged SFR surface-density vs gas-density relation for normal and starburst galaxies, from a new study in progress. Solid points denote luminous spiral and irregular galaxies with $M_B < -17$, while fainter dwarf irregular galaxies are shown as open circles. This corresponds to a transition in gas-phase metallicity at about 0.2–0.3 $Z_{\odot}$. The solid line shows a slope $N = 1.4$ for reference (not a fit to the data).

thors of these papers often suggest that CO underestimates the true molecular gas content. However, other explanations are possible. The most intense starbursts could be observed at an evolved stage where UV radiation has dissociated and dispersed much of the gas, or the Schmidt law in these systems could be much steeper than in normal galaxies. I am hoping that we will be able to distinguish between these alternatives when all of the data are analyzed, but in any case we can identify at least one systematic source of scatter in the Schmidt law, at least when a constant $X$-factor is assumed.

## 4. The Spatially Resolved Schmidt Law in Galaxies

One of the limitations of disk-averaged studies of the star formation law is that many parameters apart from gas density and SFR vary systematically between galaxies, so isolating underlying physical mechanisms for the observed star formation laws can be problematic. Moreover the disk-integrated measurements average out enormous local variations in the gas and SFR surface densities within disks and may well mask important behaviors that are only manifested on a local scale. Fortunately the rapid advances in multi-wavelength observations of nearby galaxies now make it possible to study the behavior of the star formation law on a spatially resolved basis within galaxies, either as a function of radius (azimuthally averaged) in disks or on a point by point basis.

Until recently the most comprehensive spatially resolved studies of galaxies were studies of radially averaged SFR and gas density profiles by Kennicutt (1989) and Martin & Kennicutt (2001). These were based on H$\alpha$ measurements, usually with a constant factor applied to correct for extinction, a precarious practice at best. These data did provide quantitative measurements of the radial star formation thresholds and revealed a Schmidt law at high surface density that was consistent with the disk-averaged law ($N \sim 1.3$–$1.4$), but with large systematic uncertainties because of extinction. The problem has been revisited by Wong & Blitz (2002), Boissier et al. (2003), Komugi et al. (2005), and Schuster et al. (2006) using radial profiles of galaxies, and by Kuno et al. (1995), Zhang, Fall, & Whitmore (2001), and Heyer et al. (2004) using point-by-point measurements. Most of these studies yield reasonable fits to a Schmidt power law with slope $N \sim 1.5 \pm 0.3$. A few studies derive much shallower or steeper laws (e.g., Kuno et al. 1995; Boissier et al. 2003), but in all of those cases the authors applied values of the $CO/H_2$ ($X$) conversion factor that were either very different from the normally applied value or they used a variable $X$-factor. This underscores the sensitivity of the derived star formation law on the methods used to convert measured CO line intensities to molecular hydrogen surface densities.

The other major challenge in carrying out such studies is the need to accurately correct for extinction in determining the local SFR surface densities. Most of the studies above used constant extinction corrections, which clearly is a poor approximation at best. The problem is especially vexing for applications to the Schmidt law, because one would naively expect the column density of dust—or the amount of extinction expressed in magnitudes—to correlate roughly linearly with the column density or surface density of gas. The effect of this bias depends on the surface density regime of interest. If we assume approximately the same ratio of extinction to column density as observed in the galaxy the absence of an extinction correction would cause the Schmidt index $N$ to be underestimated by as much as 1 for high column densities ($N_{\rm H} > 10^{22}\,{\rm cm}^{-2}$ or $\Sigma_{\rm H} > 100\,M_{\odot}\,{\rm pc}^{-2}$), but to have a negligible effect at low densities. Wong & Blitz (2002) investigated the importance of this correction for the inner disks of several nearby galaxies and found that introducing a density-dependent extinction correction increased the slope of the derived Schmidt law from $N \sim 0.8$ to $N \sim 1.4$. In a similar study Komugi et al. (2005) showed that extinction corrections increased the slope of their derived Schmidt law from $N = 1.14$ to $N = 1.33$. However. we would like to avoid if possible the use of the gas surface density itself to correct the SFRs, to avoid an obvious circularity in the calibration of the relation.

The advent of spatially resolved infrared mapping of galaxies with *Spitzer* now allows us to attack the extinction problem head on. As part of the SINGS project we have been exploring the use of hybrid optical + infrared extinction-free SFR indices (see Daniela Calzetti's review, this volume, p. 121). In particular the use of a combination of H$\alpha$ and 24$\,\mu$m fluxes of H II regions provides a surprisingly robust means of deriving extinction-corrected H$\alpha$ and ionizing fluxes. Variations in geometry limit the precision of these corrections to a few tenths of a magnitude, but this is well suited to a statistical problem like the Schmidt law.

As a pilot application of this approach we have combined SINGS H$\alpha$, Pa$\alpha$ and 24$\,\mu$m imaging of M51 with high resolution CO maps from the Berkeley Illinois Maryland Array Survey of Nearby Galaxies (BIMA SONG; Helfer et al. 2003) and H I maps from the VLA H I Nearby Galaxy Survey (THINGS; Walter et al. 2005), to study the form of the star formation law on a point-by-point basis in the galaxy (Kennicutt et al. 2007). The left panel of Fig. 3 shows an H$\alpha$ image of M51 with the measuring apertures overlaid.

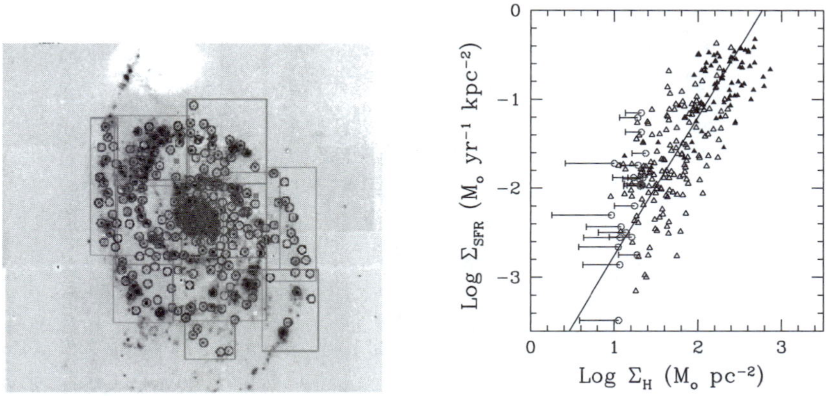


Figure 3. *Left:* H$\alpha$ image of M51 with 520 pc measuring apertures overlaid. *Right:* Relationship between SFR surface density and total hydrogen surface density for the 257 star-forming regions in M51. (*Figures taken from Kennicutt et al. 2007, copyright AAS.*)

The right-hand panel of Fig. 3 shows the relationship between the extinction-corrected SFR per unit area and the total gas surface density for 257 star-forming regions in M51. Somewhat to our surprise there is a strong correlation, corresponding to a Schmidt law with slope $N = 1.57 \pm 0.04$. Apparently, the disk-averaged law shown in Figs 1 and 2 is rooted in a local relationship between the SFR and gas surface densities on a region-by-region basis. We were restricted in this analysis to star-forming regions, rather than mapping the SFR at every pixel, in part because the signal/noise of our CO map is not sufficient to detect molecular gas between H II regions (only two CO peaks were observed outside H II regions), and because our method for deriving extinction-corrected SFRs breaks down outside of H II regions. However, when we make the same comparison using 45″ (1850 pc) diameter apertures from the CO survey of Lord & Young (1990), which fully samples the disk, we also observe a tight Schmidt law, in this case with slope $N \sim 1.4$.

Figure 4 shows the SFR surface densities correlated separately with H I (blue asterisks) and CO-inferred $H_2$ surface densities (triangles). In this molecular-rich galaxy the Schmidt law clearly is driven by a correlation between the SFR and the $H_2$ surface density, though we cannot distinguish whether this is because the molecular component is always more tightly correlated with the SFR, or whether it is simply because molecular gas happens to be the dominant mass phase in this galaxy. We hope to test this by applying the same analysis to galaxies where H I is a more significant constituent of the gas disk.

The large scatter in the Schmidt law shown in Figs 3 and 4 immediately raises the question of whether a second parameter other than gas density may be influencing the local SFRs. We tested for systematic trends with radius and location in vs between spiral arms and found no evidence for systematic differences. We suspect instead that the dispersion in the relation is partly or entirely due to random variations in extinction geometry, cluster age, and the

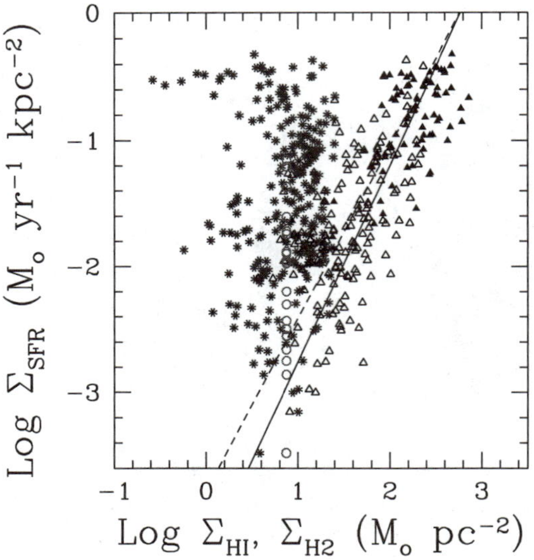


Figure 4. Relationship between the SFR surface density and the H I (*asterisks*) and $H_2$ (*triangles*) surface densities for the same regions in M51. (*Figure taken from Kennicutt et al. 2007, copyright AAS.*)

population of the uppermost initial mass functions (IMFs) in the clusters, all of which will influence the amount of ionizing radiation per unit stellar mass being formed in the associations. We certainly cannot rule out an intrinsic dispersion in the Schmidt law itself, in addition to these effects.

## 5. Star Formation Thresholds

Spatially resolved measurements of the SFR and gas surface densities also show evidence for a turnover or threshold in the star formation law at low gas densities. Qualitative evidence for such thresholds came with the first spatially resolved H I maps of spiral and irregular galaxies, which often show emission extending far beyond the active star-forming disks of the galaxies. Such behavior could only be explained if the slope $N$ of the Schmidt law were much steeper (larger) than the modest values of order $N \sim 1.5$ that are observed at higher gas densities. Kennicutt (1989) quantified this result by using H$\alpha$ images of 15 galaxies to measure the radii at which the Schmidt law showed the threshold breaks; beyond the threshold radii star formation often persists, but with a much lower rate per unit surface density of gas. The results showed that the thresholds did not occur at a constant gas surface density, but instead varied by more than a factor of 10 in $\Sigma_{crit}$. He tentatively attributed the thresholds instead to a critical gravitational stability level (Toomre $Q$ parameter) in the gas disk (also see Zasov & Simakov 1988). This analysis was subsequently extended to a larger sample by Martin & Kennicutt (2001), who also found evidence for $Q$ thresholds, but with some evidence for systematic variations in $Q_{\rm crit}$ with galaxy type and a breakdown of the $Q$ picture completely in low mass galaxies with slowly rising rotation curves (also see below).

Over the past ten years this problem has received considerable attention, both observational and theoretical. A plethora of studies have confirmed the basic observation of thresholds, at least in the star formation traced by H II regions, in a variety of galaxy samples. However, these new observations also suggest that the association of the thresholds with a critical surface density in the cold gas is considerably more complicated than suggested by the original observations. Probably the most important of the new results is a growing body of evidence for a breakdown of the constant $Q$ interpretation of the thresholds in low mass galaxies. Several authors have shown that in spiral galaxies with rising rotation curves, star formation can proceed efficiently even when the standard $Q$-criterion would predict that the disks should be in the sub-threshold regime (e.g., Thornley & Wilson 1995; Martin & Kennicutt 2001). Likewise, in irregular galaxies, which tend to have rising rotation curves over the entire optical disk, the typical values of $\Sigma_{\rm gas}/\Sigma_{\rm crit}$ are lower by approximately a factor of two (e.g., Hunter, Elmegreen, & Baker 1998), consistent with the observations of low mass spirals.

There is some disagreement over the interpretation of these observations. One interpretation is that the onset of star formation is still determined by local gravitational instabilities, but the threshold criterion is more complex than a simple isothermal disk $Q$ criterion. Alternatives proposed include two-fluid instabilities (e.g., Jog & Solomon 1984; Elmegreen 1995; Martin & Kennicutt 2001) or shear (e.g., Kenney, Carlstrom, & Young 1993; Hunter et al. 1998). Others have suggested that non-gravitational mechanisms may be more fundamental in regulating the star formation. A promising alternative is a phase instability in the ISM, as proposed for example by Elmegreen & Parravano (1994) and Schaye (2004). Distinguishing between these different threshold mechanisms can be challenging, because one can readily be triggered by the other, so one needs observations spanning a wide range of physical conditions to be able to separate them.

Most recently the very existence of the star formation thresholds has been called into question by some authors, following the observation with the *Galaxy Evolution Explorer* (*GALEX*) of extended ultraviolet disks in many nearby spiral galaxies (e.g., Thilker et al. 2005; Gil de Paz et al. 2005; Boissier et al. 2007). Since the UV emission extends well beyond the H$\alpha$ threshold radii, some have suggested that the ionized gas thresholds do not represent transitions in the SFR at all. While admitting to some biases, I have not found these arguments to be convincing. H II regions extending well beyond the threshold radii have always been observed, and careful cross-comparisons with the ultraviolet sources shows no significant differences in the source properties (Zaritsky & Christlein 2007), apart from those produced by differential extinction between the UV and visible, small number sampling of O-stars in the faint clusters, and the much larger age range in the UV clusters. On the other hand these observations show clearly the superiority of the ultraviolet imaging for studying the extent and distributions of massive star formation in low surface brightness regimes. This may make it possible for the first time to study the form of the star formation law in the sub-threshold regime, though this will require careful analysis of the ultraviolet data to isolate the truly young clusters that are still associated with their parent gas clouds.

## 6. Concluding Remarks

In the latter part of this review I focused not on what we understand about the star formation law but rather on the many areas where observations and theory have yet to converge. While this might seem discouraging at first, it actually is a very encouraging sign, the result of new data that are allowing us to confront the nature of the star formation law in a fundamental way and begin to probe its underlying physics. As will be underscored in the panel discussion later in this conference (see p. 330), the combination of new multi-wavelength datasets with new physical insights makes this an exciting time for studying star formation in galaxies, and we can look forward to enormous progress during the next half of John Beckman's career.

**Acknowledgments.** Foremost it is a pleasure to acknowledge the support and contributions of my collaborators in this work, most notably Daniela Calzetti and the other members of the SINGS team. The work reported here was partly supported by NSF grant AST-03073856. The *Spitzer Space Telescope* Legacy Science Program "The *Spitzer* Infrared Nearby Galaxies Survey" was made possible by NASA through contract 1224769 issued by JPL/Caltech under NASA contract 1407.

### References

Boissier, S., Prantzos, N., Boselli, A., & Gavazzi, G. 2003, MNRAS, 346, 1215
Boissier, S., et al. 2007, ApJS, 173, 524
Calzetti, D., et al. 2005, ApJ, 633, 871
Elmegreen, B. G. 1995, MNRAS, 275, 944
Elmegreen, B. G., & Parravano, A. 1994, ApJ, 435, L121
Gao, Y., & Solomon, P. M. 2004, ApJ, 606, 271
Gil de Paz, A., et al. 2005, ApJ, 627, L29
Gordon, K. D., Clayton, G. C., Witt, A. N., & Misselt, K. A. 2000, ApJ, 533, 236
Helfer, T. T., Thornley, M. D., Regan, M. W., Wong, T., Sheth, K., Vogel, S. N., Blitz, L., & Bock, D. C.-J. 2003, ApJS, 145, 259
Heyer, M. H., Corbelli, E., Schneider, S. E., & Young, J. S. 2004, ApJ, 602, 723
Hunter, D. H., Elmegreen, B. G., & Baker, A. L. 1998, ApJ, 493, 595
Jog, C. J., & Solomon, P. M. 1984, ApJ, 276, 127
Kenney, J. D. P., Carlstrom, J. E., & Young, J. S. 1993, ApJ, 418, 687
Kennicutt, R. C. 1989, ApJ, 344, 685
Kennicutt, R. C. 1997, in The Interstellar Medium in Galaxies, ed. J. M. van der Hulst (Dordrecht: Kluwer) Ap & Sp Sci Lib, Vol. 219, p. 171
Kennicutt, R. C. 1998a, ARA&A, 36, 189
Kennicutt, R. C. 1998b, ApJ, 498, 541
Kennicutt, R. C., et al. 2003, PASP, 115, 98
Kennicutt, R. C., et al. 2007, ApJ, submitted
Komugi, S., Sofue, Y., Nakanishi, H., Onodera, S., & Egusa, F. 2005, PASJ, 57, 733
Kuno, N., Nakai, N., Handa, T., & Sofue, Y. 1995, PASJ, 47, 745
Lord, S. D., & Young, J. S. 1990, ApJ, 356, 135
Martin, C. L., & Kennicutt, R. C. 2001, ApJ, 555, 301

Misiriotis, A., Xilouris, E. M., Papamastorakis, J., Boumis, P., & Goudis, C. D. 2007, A&A, 459, 113
Schaye, J. 2004, ApJ, 609, 667
Schmidt, M. 1959, ApJ, 129, 243
Schmidt, M. 1963, ApJ, 137, 758
Schuster, K. F., Kramer, C., Hitschfeld, M., García-Burillo, S., & Mookerjea, B. 2007, A&A, 461, 143
Scoville, N. Z., Polletta, M., Ewald, S., Stolovy, S. R., Thompson, R., & Rieke, M. 2001, AJ, 122, 3017
Silk, J. 1997, ApJ, 481, 703
Skillman, E. D., Kennicutt, R. C., & Hodge, P. W. 1989, ApJ, 347, 875
van der Hulst, J. M., Skillman, E. D., Smith, T. R., Bothun, G. D., McGaugh, S. S., & de Blok, W. J. G. 1993, AJ, 106, 548
Thilker, D. A., et al. 2005, ApJ, 619, L79
Thornley, M. D., & Wilson, C. D. 1995, ApJ, 447, 616
Walter, F., et al. 2005, in ASP Conf. Ser., 331, Extra-Planar Gas, ed. R. Braun (San Francisco: ASP), p. 269
Wong, T., & Blitz, L. 2002, ApJ, 569, 167
Zaritsky, D., & Christlein, D. 2007, AJ, 134, 135
Zasov, A. V., & Simakov, S. G. 1988, Astrophysics, 29, 518
Zhang, Q., Fall, S. M., & Whitmore, B. C. 2001, ApJ, 561, 727

## Discussion

*Beckman*: I'd just like to mention the broad-band data on the outer parts of disk profiles, which in many cases do not show any break in brightness slope out to ten scale lengths. How is this compatible with breaks in the H$\alpha$ and/or *GALEX* UV profiles?

*Kennicutt*: Outer breaks in the H$\alpha$ profiles of disks, which are indicative of some type of star formation threshold, have been observed in galaxies for some time. Star formation persists outside of these breaks but at a much lower rate per unit area. More recently the *GALEX* images reveal extended star formation in these outer disks. This star formation is much easier to trace in the UV than H$\alpha$, because lifetimes of the UV-emitting clusters are much longer, and the light is emitted by a more heavily populated part of the IMF. There have been some claims that the breaks seen in H$\alpha$ are absent in the UV, while others have argued that there is little difference in the distributions of H II regions and UV light, when one takes into account biases introduced by extinction and stellar aging. Preliminary work by Quinton Goddard and our group in Cambridge tends to support the latter interpretation, but more work is needed before we can draw firm conclusions. In any case it is clear that the UV continuum is a much better tracer of the star formation in low surface brightness environments, for the reasons given above.

*Díaz*: Regarding H II galaxies a relation has been found between $L(\mathrm{H}\beta)$, H$\beta$ luminosity, and gas velocity dispersion that could suggest a relation between SFR and total mass for this kind of low metallicity dwarf galaxies. Could this sort of relation be used in combination with your data to investigate the reason for the deviation of these objects from the Schmidt law?

*Kennicutt*: I am unsure whether the emission-line velocity dispersions of these objects reflect virialized motions. I suspect they reflect motions of gas introduced by stellar winds and supernovae. Nevertheless, I agree it would be interesting to perform such a test, as it might indicate whether some of the H II galaxies deviate from the relation because they are evolved starbursts, where most of the cold gas has been ionized or dispersed.

*Krumholz*: Comment: in low surface brightness, low metallicity dwarfs, there is evidence that $X_{CO}$ is different. Question: have you looked at the Silk law computed using the local shear instead of the local dynamical time? This will be different in arm and interarm regions.

*Kennicutt*: Indeed, there are many studies of CO in dwarfs that indicate a low $X_{CO}$ factor. At this point it is not clear whether adopting these lower conversion values removes all of the discrepancy in the Schmidt relation. As to your question, we did not yet test a shear-based law of the kind you suggest, though most of our regions are located in spiral arms, and we see no significantly different behavior for the interarm regions.

*Sheth*: Regarding the Silk law, while it breaks down on a local scale, can it explain the thresholds, because $\tau_{dyn}$ is much higher at higher radii? Can a single Silk law index describe the $\Sigma_{SFR}$?

*Kennicutt*: The Silk law does reproduce the global SFR densities of galaxies surprisingly well, and I do not believe it really was intended to serve as a local prescription. The simple form of the law will not produce a threshold, if you adopt an alternate prescription replacing $(\Omega_{gas} - \Omega_P)$ where $\Omega_P$ is the pattern speed of of a spiral wave, that would produce a transition near corotation.

Rob Kennicutt (*left*) and Matthew Lehnert.

*Pathways through an Eclectic Universe*
*ASP Conference Series, Vol. 390, © 2008*
*J. H. Knapen, T. J. Mahoney, and A. Vazdekis, eds.*

# The Star Formation Law as a Function of Galactic Properties

S. Komugi, Y. Sofue, S. Onodera, F. Egusa, and K. Muraoka

*Institute of Astronomy, University of Tokyo, Japan*

H. Nakanishi

*Australia Telescope National Facility, Australia*

J. S. Young

*Department of Astronomy, University of Massachusetts, USA*

**Abstract.** The relation between molecular gas quantity and star formation rate, known as the Schmidt law, is tested for its dependency on galactic morphology, with an emphasis on the presence of bars. The galaxy sample is based on a survey of the $^{12}$CO($J = 1 \rightarrow 0$) emission in nearby galaxies completed at the Nobeyama Radio Observatory 45 m telescope, combined with previous surveys with similar resolution. These data were compared to star formation rates derived using internal extinction corrected H$\alpha$. The slope of the Schmidt law is found to vary considerably with sample distance, from 0.7 to 1.4. The Schmidt law is categorized according to the presence of bars. Consequently, we find that barred galaxies occupy the denser regime of the Schmidt law, but with no apparent difference in the star formation efficiency, indicating that bars stimulate central inflow of gas, but that star formation still occurs according to a common star formation law as with non-barred galaxies.

## 1. Introduction

Past studies have shown that the Schmidt law, or the power law relation between gas surface density $\Sigma$ and star formation rate (SFR), is to some extent a universal characteristic of spiral galaxies in general (e.g., Kennicutt 1998; Wong & Blitz 2002; Komugi et al. 2005, 2006, 2007). Dynamical properties of galaxies, however, are also known to have effects on the nature of star formation. It is therefore natural to expect that these dynamical characteristics (such as Hubble type and the presence of bars) also affect the transition law from gas to stars. This preliminary study aims to address this issue by categorizing the Schmidt law according to the presence of galactic scale bars.

## 2. Data

We observed a total of 69 galaxies in the $^{12}$CO($J = 1 \rightarrow 0$) line with the Nobeyama Radio Observatory (NRO) 45 m telescope (beamsize 16″) during three observing runs from May 2005 to January 2006. The galaxies were observed with a single pointing at the center.

Combined with previous measurements at the NRO 45 m (Nishiyama & Nakai 2001) and similar angular resolution (22″) single dish data from the IRAM 30 m telescope (Combes et al. 1988; Planesas, Gómez-González, & Martín-Pintado 1989; Braine et al. 1993; Böker, Lisenfeld, & Schinnerer 2003), our data set comprises 189 nearby (mainly spiral) galaxies, of which 162 are detected at better than 5$\sigma$ level. The morphology of the samples spans a wide range, from early to late types. Most of the galaxies lie within 30 Mpc, clustered around 20 Mpc representing the Virgo cluster (see S. Komugi et al., in preparation, for details). The SFR in units of $M_\odot\,\mathrm{pc}^{-2}\,\mathrm{yr}^{-1}$, was derived from narrow-band H$\alpha$ imaging by Young et al. (1996) and the calibration by Kennicutt (1998). Internal extinction was corrected using the H$\alpha$/H$\beta$ ratio from spectroscopic data from Ho, Filippenko, & Sargent (1997). SFRs for 66 galaxies were corrected for extinction using this method, and the present paper focuses on this subsample.

## 3. The Schmidt Law and the Effect of Distance

Figure 1 is the relation between SFR and surface gas density $\Sigma$ ($M_\odot\,\mathrm{pc}^{-2}$) assuming a uniform CO–$H_2$ conversion factor of $X_{\mathrm{CO}} = 2.8 \times 10^{20}\,(\mathrm{cm}^{-2})$ for samples with non-irregular morphology. A linear fit is given for subsamples with distances larger or smaller than 15 Mpc.

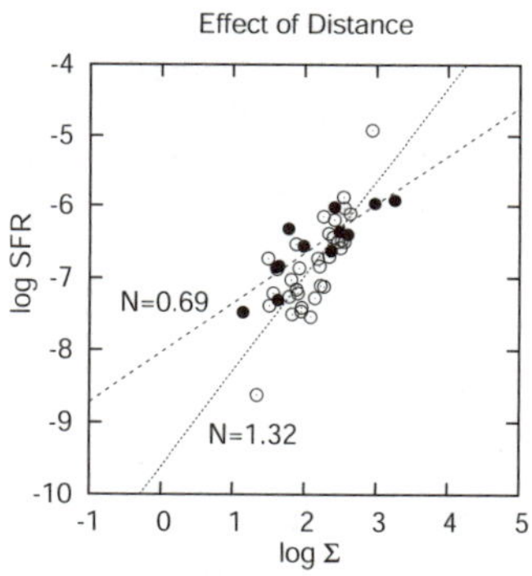


Figure 1. The obtained Schmidt law. Galaxies with a distance of 15 Mpc or less are filled circles, and with distances greater than 15 Mpc open circles. The dashed and the straight lines are the best fits for the closer and farther galaxies, respectively.

The observing beams of both the NRO 45 m and the IRAM 30 m telescopes subtend different scales depending on the distance to the galaxy. It is important to know what effect this can have on the $\Sigma$–SFR relation. The Schmidt law index, $N$, is defined by SFR $\propto \Sigma^N$. A best fit to the closer galaxies gives a slope of $N = 0.69 \pm 0.13$, and $N = 1.32 \pm 0.19$ for the farther galaxies. The two are significantly different.

The arbitrary value of 15 Mpc is chosen so that the 16″ beam corresponds to a scale of 1.2 kpc, or 600 pc from the nucleus. Sakamoto et al. (1999), Regan et al. (2001), Sofue et al. (2003) and others have shown that many spirals have a central exponential molecular disk, with a characteristic radius of roughly this size. If this is a universal characteristic, observing galaxies closer than 15 Mpc will truncate this exponential structure. Assuming that the star-forming nuclear disk is also exponential (as is often observed), the Schmidt law index of $N = 0.69$ can be explained by this effect of observationally truncating an exponential structure with an original Schmidt law index of $N = 1.32$ (S. Komugi et al., in preparation). The categorization of the Schmidt law for the presence of bars must therefore be done separately for these sample distances.

## 4. The Effect of Bars

The primary goal of this study is to see the effect of bars on the Schmidt law sequence. Bars are understood in general to feed gas into the central kpc of galaxies (e.g., Kuno et al. 2007). Under this assumption, one of two cases may be observed: 1) bars concentrate gas into the central kpc, but a common Schmidt law explains the SFR in the same way as non-barred galaxies, or 2) due to the central condensation of gas, bars induce additional star formation to that which can be predicted by the Schmidt law, in which case the star formation efficiency (SFE) will rise and a different sequence in the Schmidt law can be seen. The Schmidt law categorized by the existence of bars can be seen in Figure 2.

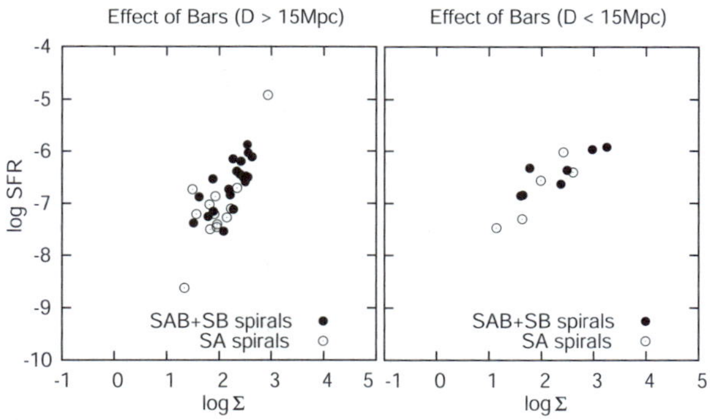


Figure 2. The Schmidt law categorized by barred/non-barred galaxies. *Left:* Galaxies with distance of more than 15 Mpc. *Right:* Galaxies with distance of less than 15 Mpc.

Apparently, barred galaxies occupy the denser regime, confirming that bars condense gas into the central region. A sequential offset between barred and non-barred galaxies is not seen, however, and the two morphological types seem to follow a common Schmidt law. Case 1 is supported, therefore. This can be further visualized in terms of the SFE, in Figure 3. The SFE as a function of $\Sigma$ seems to be independent of the existence of bars, although with considerable dispersion.

These observations imply that in a statistical sense, bars do concentrate gas into the central kpc but do not raise the SFE, and that star formation occurs according to a common Schmidt law as in non-barred galaxies.

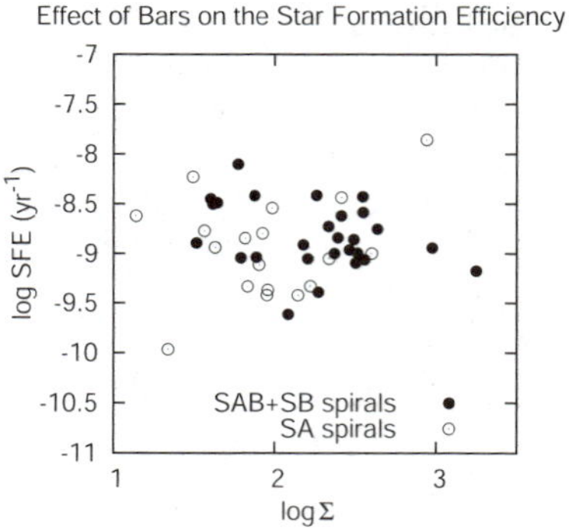


Figure 3. The SFE for all the galaxies in Figure 1. The SFE is derived as SFR/$\Sigma$.

## References


Böker, T., Lisenfeld, U., & Schinnerer, E. 2003, A&A, 406, 87

Braine, J., Combes, F., Casoli, F., Dupraz, C., Gerin, M., Klein, U., Wielebinski, R., & Brouillet, N. 1993, A&AS, 97, 887

Combes, F., Dupraz, C., Casoli, F., & Pagani, L. 1988, A&A, 203, L9

Ho, L. C., Filippenko, A. V., & Sargent, W. L. W. 1997, ApJS, 112, 315

Kennicutt, R. C., Jr. 1998, ApJ, 498, 541

Komugi, S., Sofue, Y., Nakanishi, H., Onodera, S., & Egusa, F. 2005, PASJ, 57, 733

Komugi, S., Sofue, Y., & Egusa, F. 2006, PASJ, 58, 793

Komugi, S., Kohno, K., Tosaki, T., Nakanishi, H., Onodera, S., Egusa, F., & Sofue, Y. 2007, PASJ, 59, 55

Kuno, N., et al. 2007, PASJ, 59, 117

Nishiyama, K., & Nakai, N. 2001, PASJ, 53, 713

Planesas, P., Gómez-González, J., & Martín-Pintado, J. 1989, A&A, 216, 1

Regan, M. W., Thornley, M. D., Helfer, T. T., Sheth, K., Wong, T., Vogel, S. N., Blitz, L., & Bock, D. C.-J. 2001, ApJ, 561, 218

Sakamoto, K., Okumura, S. K., Ishizuki, S., & Scoville, N. Z. 1999, ApJS, 124, 403

Sofue, Y., Koda, J., Nakanishi, H., & Onodera, S. 2003, PASJ, 55, 59

Wong, T., & Blitz, L. 2002, ApJ, 569, 157

Young, J. S., Allen, L., Kenney, J. D. P., Lesser, A., & Rownd, B. 1996, AJ, 112, 1903


## Discussion

*Shlosman*: Wouldn't it be natural to look for star formation along the bars and not in the central regions, if you want to find how the bars affect the efficiency of SF? Inside the bar the nature of the gas flow changes and the shear is known to affect the star formation. For example, in strong bars the star formation is heavily damped because of a strong shear.

*Komugi*: Of course. A reliable estimate of extinction is difficult, however (see Lundgrens talk on M83), altering the Schmidt law. Considering that my study is sensitive to SFR calibrators we believe that extinction corrected (via H$\alpha$/H$\beta$) H$\alpha$ should be used. Once we have narrow-band H$\beta$ images, a point-to-point Schmidt law in bars would absolutely be the thing to do. We only infer the central region in this case because the spectroscopic slits on the centers were only $2'' \times 4''$ in size and therefore not applicable to the bars.

*Kennicutt*: How were the SFRs in your study measured, and how did you correct for extinction?

*Komugi*: 1) SFRs for normal galaxies in Komugi's (2005) central kpc: $H\alpha/H\beta$ from Ho et al. (1997). Disks: assume $A_V \propto N(H_2)$ (see Lundgren's talk). 2) SFRs for the CO survey: $H\alpha/H\beta$ leading to $E(B-V)$ from Ho et al. (1997) with $2'' \times 4''$ aperture spectrometry.

*Erwin*: In dividing your sample into barred and non-barred galaxies, how did you treat the RC3 "S" classification, which is distinct from the SA (unbarred), SAB (weakly barred) and SB (strongly barred) classifications?

*Komugi*: I re-checked the samples, and found only 3 S's, for one of which $H\beta$ was available, so they are not plotted on the Schmidt law.

From left to right: Shinya Komugi, Fumi Egusa and Yoshihiko Yamada. Claire Dobbs is in the background to the right.

*Pathways through an Eclectic Universe*
*ASP Conference Series, Vol. 390, © 2008*
*J. H. Knapen, T. J. Mahoney, and A. Vazdekis, eds.*

# Star Formation in Nuclear Rings

Torsten Böker
*European Space Agency, Dept. RSSD, Keplerlaan 1,*
*2200 AG Noordwijk, Netherlands*

**Abstract.** We discuss a new method for estimating the relative ages of young star clusters or "hot spots" in circumnuclear star formation rings, using near-infrared emission line fluxes, and present a first application of the method to the spiral galaxy NGC 613.

## 1. Introduction

Star formation rings are a natural consequence of gas orbiting within a bar-shaped potential: dynamical resonances with the rotating bar potential cause orbiting gas clouds to lose angular momentum and thus "spiral in" towards the nucleus. In the process, the gas is compressed and forms two elongated and curved dust lanes on either side of, and slightly offset from, the bar major axis. These dust lanes "connect" the end of the bar to the nuclear ring. However, the fate of the gas after it enters the ring orbits is still unclear.

One possibility is that the gas accumulates in the ring until a critical density is reached. Presumably driven by turbulence, the ring becomes unstable to gravitational collapse, and star formation is induced at random times and locations within the ring. In a competing model, a quasi-instantaneous "delta burst" of star formation is induced in (and only in) two well-defined regions along the ring where the gas density becomes sufficiently high to ignite star formation. The precise location of these "overdensity regions" (ODRs) depends on the details of the potential, but they are often found close to, and somewhat downstream from where the gas enters the ring. Any cluster formed in this way will age passively as it continues on the ring orbit, leading to systematic age gradients along each half ring, and away from the ODRs.

In an attempt to decide between these competing scenarios, we have observed five nuclear rings with the SINFONI integral-field spectrograph on the Very Large Telescope (VLT). The morphologies, star formation rates, and kinematic properties of the rings are discussed in detail in T. Böker et al. (2008, AJ, 135, 479).

## 2. Measuring Cluster Ages

Measuring the stellar cluster ages along a circumnuclear ring is difficult, even in a relative sense. Most previous studies have relied either on stellar absorption indices or equivalent width measurements of hydrogen recombination lines. The advantage of both diagnostics is that they are insensitive to extinction (but

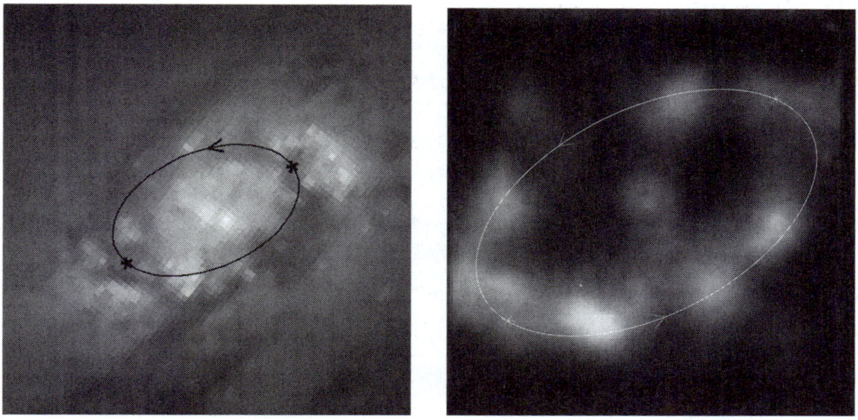

Figure 1. *Left:* *HST*/WFPC2 F606W map of the inner 15″ × 15″ of NGC 613. *Right:* False color map of the NIR emission line morphology in the inner 8″ × 8″, with blue representing He I, green Brγ, and red [Fe II].

not to continuum emission from hot dust). On the other hand, both require accurate knowledge of the continuum emission *from the hot spot itself*, i.e., after subtraction of the underlying bulge and/or disk, which often is rather uncertain.

Here, we propose to use instead the fluxes of three near-infrared (NIR) emission lines that are often prominent in the spectra of star forming regions: He I (2.06 $\mu$m), Brγ (2.16 $\mu$m), and [Fe II] (1.64 $\mu$m). Because line *fluxes* are independent of the underlying continuum level, they can be measured accurately without any knowledge of the disk/bulge structure. The (rather safe) assumption here is that the emission lines are produced only in the ring itself, i.e., that the underlying bulge/disk can be considered quiescent. For an instantaneous burst, population synthesis models such as STARBURST99 predict a rather different time evolution for the three line fluxes: He I is produced only in the most massive stars, and hence disappears after only a few Myr, more rapidly than Brγ, which becomes undetectable after about 15 Myr. The [Fe II] emission, in turn, traces the supernova rate which remains more or less constant until about 35 Myr after the burst. The relative strength of these lines thus provides a robust qualitative estimate for the relative cluster ages.

Figure 1 demonstrates the use of the method in the case of NGC 613, a large, relatively nearby ($d$ = 17.5 Mpc), strongly barred Sbc spiral. The left panel contains the *HST*/WFPC2 *V*-band (F606W) image of the central 15″ (1.3 kpc). Two prominent dust lanes are easily identified, and the inferred approximate locations of the ODRs are marked by star symbols. The right panel shows a false-color image constructed from the SINFONI emission line maps of He I, Brγ, and [Fe II], assigned to the blue, green, and red color channel, respectively. At least for the southern half of the ring, the color map appears to provide evidence for systematically older clusters as one moves away from the ODR. For a full discussion of the NGC 613 results, as well as of those for the other sample galaxies, see T. Böker et al. (2008, AJ, 135, 479).

*Pathways through an Eclectic Universe*
*ASP Conference Series, Vol. 390, © 2008*
*J. H. Knapen, T. J. Mahoney, and A. Vazdekis, eds.*

# GHaFaS: Galaxy H alpha Fabry–Perot Spectrometer for the WHT

C. Carignan and O. Hernandez

*Laboratoire d'Astrophysique Expérimentale, Département de Physique, Université de Montréal, Montréal, Québec, Canada*

J. E. Beckman and K. Fathi

*Instituto de Astrofísica de Canarias, E-38200 La Laguna, Tenerife, Spain*

**Abstract.** A new scanning Fabry–Perot system will soon be available at the Nasmyth focus of the 4.2 m William Herschel Telescope. It has been designed by the Laboratoire d'Astrophysique Expérimentale in Montréal and is being built in collaboration with astronomers at the Instituto de Astrofísica de Canarias. The instrument will see first light at the beginning of July 2007.

## Description of the Project

GHaFaS is an improved version of the scanning Fabry–Perot (FP) system FaNTOmM (Hernandez et al. 2003), which is a resident instrument on the Observatoire du Mont Mégantic (OMM) 1.6 m telescope, and which has also been used on the Canada–France–Hawaii (CFH) and ESO La Silla 3.6 m telescopes. The complete system is composed of a focal reducer, a calibration unit, a filter wheel for the order sorter filters, an FP etalon and an IPCS camera. The IPCS is composed of a Hamamatsu intensifier MCP tube, which intensifies every generated electron coming from the photocathode by a factor $10^7$. Each photon event, recorded on a DALSA CCD, is then analyzed by a centering algorithm. With this amplification, the camera has essentially no readout noise. Because of this, a zero noise IPCS is to be preferred over CCDs at very low flux level (Gach et al. 2002), even if the GaAs IPCS has a DQE of only 25%. Moreover, because of its fast scanning capability, it can average out the variation of atmospheric transmission which is not possible with the long integration times needed per channel for CCDs in order to beat the readout noise.

In the last three years, around 150 galaxies have been observed with the FaNTOmM system on the OMM, CFH and ESO La Silla telescopes in the context of three large surveys: the SINGS sample (Daigle et al. 2006), a survey of barred galaxies, the BHabar sample (Hernandez et al. 2005), and a sample of Virgo spirals (Chemin et al. 2006). While the first scientific justification of the Montréal group was to derive high spatial resolution optical rotation curves for mass modeling purposes, the data were also used by astronomers from the Instituto de Astrofísica de Canarias to constrain the role of gravitational perturbations as well as feedback from individual H II regions on the evolution of structures in galaxies and by a Berkeley–Munich group and the GÉPI group in Paris to compare those local samples to high-$z$ galaxies.

GHaFaS will come with its own custom-designed focal reducer developed to be optically and mechanically compatible with the Nasmyth focus of the William Herschel Telescope (WHT). The system has its own control and data acquisition systems. It will have a $4 \times 4$ arcmin$^2$ field with 0.45 arcsec pixels and $\sim$5 km/s velocity resolution. Full acquisition and reduction software (mainly based on IDL routines) will be provided by the Montréal group. The project will be done in three phases. For Phase I (July 2007), the optical system (focal reducer, filter wheel and calibration unit) will be delivered to the WHT and used with the FaNTOmM camera for this first run. For Phase II (end of 2007), an improved GaAs IPCS will be added to the system. Phase III (early 2008) will provide an FP controller and possibly a monochromator to calibrate the data at the observing wavelength.

## Science to be Done with GHaFaS

Two-dimensional kinematics is a very powerful technique for studying the structure and evolution of galaxies. The distribution of dark matter, circumnuclear star formation and fueling of active galactic nuclei, the detection of counter-rotating and kinematically decoupled components, and the effects of interaction between massive stars and the interstellar medium are among the physical phenomena that can be studied with this technique: see, for example, Fathi et al. (2007) and Relaño et al. (2007).

The first large program for which we plan to use GHaFaS consists in observing a sample of 46 carefully selected nearby galaxies that are all included in the SINGS, THINGS, *GALEX*, and other CO and optical archives. Due to the angular size of some of the objects, two to four fields may be necessary to reach the 25th magnitude radius. This totals 72 fields that will require $\sim$18 clear nights of observing with GHaFaS on the WHT. Priority will be given to enlarge our Virgo sample of galaxies. The full sample ranges from elliptical (with emission) to irregular galaxies, 2/3 of which are intermediate-type objects, since this is where highly star-forming regions will be observed. It will be possible to use this sample for many scientific projects ranging from large scale mass modeling using the velocity fields in order to derive the dark matter density profiles to the study of the internal kinematics of the individual H II regions.

## References

Chemin, L., et al. 2006, MNRAS, 366, 812
Daigle, O., Carignan, C., Amram, P., Hernandez, O., Chemin, L., Balkowski, C., & Kennicutt, R. 2006, MNRAS, 367, 469
Fathi, K., Beckman, J. E., Zurita, A., Relaño, M., Knapen, J. H., Daigle, O., Hernandez, O., & Carignan, C. 2007, A&A, 466, 905
Gach, J.-L., et al. 2002, PASP, 114, 1043
Hernandez, O., Carignan, C., Amram, P., Chemin, L., & Daigle, O. 2005, MNRAS, 360, 1201
Hernandez, O., Gach, J.-L., Carignan, C., & Boulesteix, J. 2003, SPIE, 4841, 1472
Relaño, M., Beckman, J. E., Daigle, O., & Carignan, C. 2007, A&A, 467, 1117

*Pathways through an Eclectic Universe*
*ASP Conference Series, Vol. 390, © 2008*
*J. H. Knapen, T. J. Mahoney, and A. Vazdekis, eds.*

# Molecular Gas and Star Formation in M81

V. Casasola,[1,2] F. Combes,[2] G. Galletta,[1] and D. Bettoni[3]

[1] *Università di Padova, Italy*
[2] *Observatoire de Paris-LERMA, France*
[3] *INAF-Osservatorio di Padova, Italy*

**Abstract.** We present IRAM 30 m observations of the central ∼1.6 kpc of the spiral galaxy M81. The molecular gas appears weak and with an unusual excitation physics. We discuss a possible link between low CO emission and weak FUV surface brightness.

## 1. Introduction

The galaxy M81, studied at all wavelengths, is the prototype of CO-poor galaxies. The CO emission appears confined only to the H II regions in the spiral arms, and the central region seems devoid of molecular gas. This lack of CO emission is even more surprising if we consider that, in general, interacting galaxies, like M81, have higher CO luminosity than non-interacting galaxies. This atypical physics of the molecular gas makes M81 an optimum candidate to study the $X$ conversion factor between CO and $H_2$ molecules.

## 2. Detection and Distribution of the Molecular Gas

Observing the inner ∼1.6 kpc of M81 with the IRAM 30 m telescope, we found that there are both regions completely devoid of molecular gas and regions in which the CO is clearly detected and the signal is strong. In agreement with Brouillet, Baudry, & Combes (1988) and Sakamoto et al. (2001), the central ∼300 pc are devoid of CO(1→0) emission, and the more important molecular gas emission is a clump structure at the distance of around 460 pc from the nucleus in the northeastern direction (Fig. 1, *left*).

## 3. Line Ratio and Clumping Properties

Studying the line intensities showed very low temperatures both in the CO(1→0) and CO(2→1) transitions: for the detections $\geq 2\sigma$ the average brightness temperature in CO(1→0) is 33.44 mK, while in CO(2→1) it is 21.47 mK. The average $R_{21}$ ratio is 0.68, a quite low and atypical value for similar galactic nuclei. Usually, for the inner kpc of galaxies $R_{21} \sim 0.9$ on average, and an even higher value is to be expected for interacting/perturbed galaxies (Braine & Combes 1992). In addition, the line ratio typically decreases from galactic nuclei to spiral arms (García-Burillo, Guelin, & Cernicharo 1993). For M81 the $R_{21}$ line ratio spans a wide range of values, with no clear radial trend.

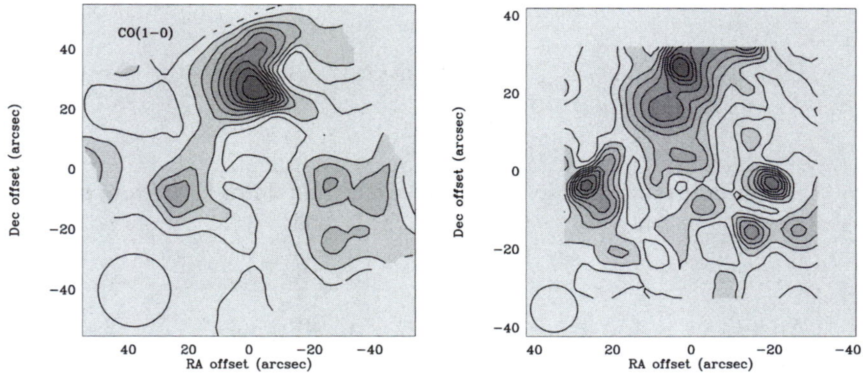


Figure 1. *Left:* Integrated $^{12}CO(1\rightarrow 0)$ emission. The contour levels are between 0.701 to 7.01 K km s$^{-1}$ in steps of 0.701 K km s$^{-1}$. The beam of 22″ is indicated at the bottom left. *Right:* Integrated $^{12}CO(2\rightarrow 1)$ emission. The contour levels are between 0.410 to 4.10 K km s$^{-1}$ in steps of 0.410 K km s$^{-1}$. The beam of 12″ is indicated at the bottom left.

The detailed study of the CO(2→1) emission and distribution, shown in Fig. 1 (*right*), allowed us to resolve individual molecular structures of diameters ~250 pc and masses $\sim 10^5\,M_\odot$. Molecular structures with these dimensions define giant molecular associations (GMAs). The $X = N(H_2)/I_{CO}$ conversion factor computed for all the individual resolved GMAs in M81, assumed virialized, takes a value from ~10 to ~18 times larger than the *standard* Galactic value ($X = 2.3 \times 10^{20}$ mol cm$^{-2}$ (K km s$^{-1}$)$^{-1}$; Strong et al. 1988).

## 4. Heating of the Gas

In the center of M81, the molecular gas is absent or subthermally excited. This could be due to a flux of cosmic rays and a far-ultraviolet surface brightness too low to heat the molecular gas component of the interstellar medium.

## 5. Conclusions

M81 seems to be a galaxy with unusual physics in the molecular gas in the central region when compared to galaxies with similar distances and morphological type: the CO is absent or very weak, the $R_{21}$ line ratio is very low, and the $X$ conversion factor is extraordinarily large. We interpret these results as due to a lack of excitation of the gas, rather than to the absence of molecular gas.

### References

Braine, J., & Combes, F. 1992, A&A, 264, 433
Brouillet, N., Baudry, A., & Combes, F. 1988, A&A, 196, L17
García-Burillo, S., Guelin, M., & Cernicharo, J. 1993, A&A, 274, 123
Sakamoto, K., Fukuda, H., Wada, K., & Habe, A. 2001, AJ, 122, 1319
Strong, A. W., et al. 1988, A&A, 207, 1

*Pathways through an Eclectic Universe*
*ASP Conference Series, Vol. 390, © 2008*
*J. H. Knapen, T. J. Mahoney, and A. Vazdekis, eds.*

# Discovery of Four New Ultra-compact Nuclear Rings in Three Spiral Galaxies

Sébastien Comerón,[1] Johan H. Knapen,[1] and John E. Beckman[1,2]

[1] *Instituto de Astrofísica de Canarias, E-38200, La Laguna, Spain*
[2] *Consejo Superior de Investigaciones Científicas, Spain*

**Abstract.** Ultra-compact nuclear rings (UCNRs) are ring-shaped star-forming regions around galactic nuclei that range in radius from a few to 200 pc. Only a handful are known in the literature, but thanks to the high resolution of *HST*, we are now able to study their frequency and properties in detail. We have discovered UCNRs in three nearby galaxies. In NGC 2985 we have found an intense ring of about 50 pc radius. In NGC 4800 we have found two rings, a patchy and intense ring of 130 pc radius and a faint one very close to the nucleus with a radius of 30 pc. These rings have probably been formed in connection with inner Lindblad resonances. In NGC 4579 we have found one sharply defined ring whose asymmetric morphology indicates that it is a resonance ring induced by a superposition of $m = 1$ and $m = 2$ perturbations. The rings we have found are important because they are tracers of the dynamics very close to the nuclei of their host galaxies. Only very few were known prior to this study and we have now virtually doubled the number of known UCNRs.

## 1. The Galaxies and Images

NGC 2985 is an SAab LINER galaxy at a distance $D = 22.4$ Mpc. NGC 4579 is an SABb AGN host galaxy at a distance $D = 15.2$ Mpc. NGC 4800 is a galaxy classified as SAb where we have discovered a faint near-IR bar. The galaxy is at a distance $D = 16.8$ Mpc. We retrieved images from the *HST* archive. We applied the structure map technique to a red image to enhance the dust structure and we subtracted the continuum from H$\alpha$ narrow-band images. We also obtained UV images. The resulting images can be seen in Fig. 1, and full details are given in Comerón et al. (2008).

## 2. Results

*NGC 2985*: The ring has a 50 pc radius. The structure map traces the inflow of dust from outer regions to the UCNR where the flow seems to stop. There is no bar in this galaxy, so the ILR that causes the ring must originate elsewhere, e.g., in an interaction with a nearby galaxy (possibly NGC 3027 or KGD059).
*NGC 4579*: The ring diameter is 270 pc. The eccentricities of the ring suggest that it may be caused by a superposition of $m = 1$ and $m = 2$ perturbations. The galaxy is weakly barred.
*NGC 4800*: The inner ring has a radius of 30 pc and the outer one 130 pc. The

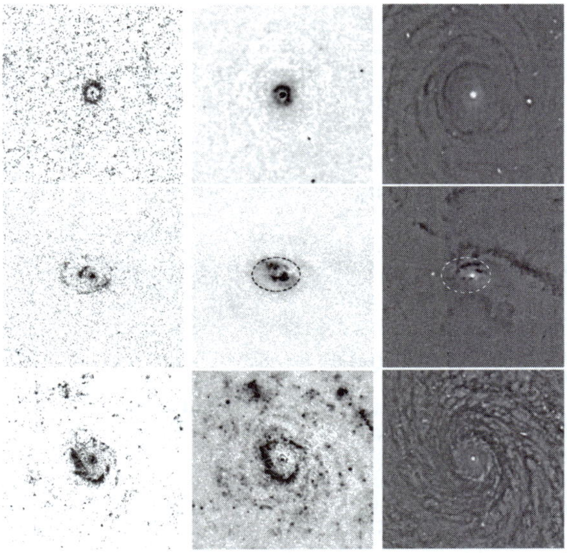

Figure 1. Images of the centers of the three galaxies in our study: NGC 2985 (*top*), NGC 4579 (*middle*), and NGC 4800 (*lower panels*). The left hand images are through UV filters (F218W, F250W, and F300W, respectively), the center column shows H$\alpha$ images and the right one shows the structure maps. Each square has 1 kpc on a side, corresponding to 9 arcsec for NGC 2985, 12 arcsec for NGC 4579, and 13.5 arcsec for NGC 4800. North is at the top and east at the left.

structure map traces very well the dust spiraling inwards. In this galaxy a faint bar is visible in the near IR.

## 3. Conclusions

We have discovered a total of four small nuclear rings with diameters between 60 and 270 pc in three spiral galaxies, thus virtually doubling the number of known UCNRs. We also confirm the known UCNRs in NGC 5248. This was done from archival *HST* images in a small pilot study, suggesting that UCNRs may be much more common than currently thought. In fact, we found four UCNR host galaxies in a sample of 38 face-on spiral galaxies, which yields a first estimate for the UCNR fraction of some 10% (Comerón et al. 2008). In future work, we plan to identify and study tens of UCNRs and use them as a powerful tool to investigate the properties of the innermost regions of galaxies, including the fuelling of any AGN.

### References

Comerón, S., Knapen, J. H., Beckman, J. E., & Shlosman, I. 2008, A&A, 478, 403

*Pathways through an Eclectic Universe*
*ASP Conference Series, Vol. 390,* © *2008*
*J. H. Knapen, T. J. Mahoney, and A. Vazdekis, eds.*

# A Statistical Study of Circumnuclear Morphology of Sigma-Drop Galaxies

Sébastien Comerón,[1] Johan H. Knapen,[1] and John E. Beckman[1,2]

[1] *Instituto de Astrofísica de Canarias, E-38200, La Laguna, Spain*
[2] *Consejo Superior de Investigaciones Científicas, Spain*

**Abstract.** Sigma-drop galaxies are characterized by a drop in the stellar velocity dispersion in their centers. One of the most likely theories to explain this phenomenon is that there is a young stellar population that has recently formed, and that has inherited the kinematics of the gas. We compare the morphology of the dust and the H$\alpha$ emission in 20 $\sigma$-drop galaxies and in a matched control sample. We also study the total emission in CO and in H I in order to investigate any correlation between the occurrence of $\sigma$-drops and the amount of gas. We have found more circumnuclear spiral arms as outlined in dust, and more star formation in nuclear rings in the sigma-drop galaxies. That indicates that sigma-drops are probably related to starbursts, driven by gas inflows.

## 1. The Samples

We have matched a control sample to a sample of $\sigma$-drop galaxies selected from the literature using four different parameters: $B$-magnitude, distance, axis ratio and morphological type. We started by classifying $\sigma$-drop host galaxies into several bins for each parameter. To create the control sample, we then took random galaxies and populated the bins until we got the best possible match between the two samples. The limits of the bins we selected are for distance: $R <$ 10 Mpc, 10–20 Mpc, 20–30 Mpc, and 30–50 Mpc, for luminosity: $B < -21.00$, $-21.00 < B < -20.00$, $-20.00 < B < -19.00$, and $B > -19.00$, for the relationship between axes: $d/D$ = 0.2–0.4, 0.4–0.6, 0.6–0.8, and 0.8–1.0. For morphological type, we have made a bin for lenticular galaxies and bins for galaxy types 0, 1, 2, 3, 4, and 5–6.

## 2. Classification of H$\alpha$ Emission

We classified the galaxies from the H$\alpha$ emission in the center (within a radius $r_c$, which is 2% of the radius of the $B$ 25 mag isophote, and in the inner kpc) studying archival *Hubble Space Telescope* (*HST*) images. We divided the inner emission into the categories of *peak* (peaked emission region centered on the nucleus), *amorphous* (emission region without any significant peak), *peak+UCNR* (peaked emission surrounded by an ultra-compact nuclear ring, as discussed in Comerón et al. 2008), and *others* (peaked emission plus UCNR+bipolar structure). For the inner kpc, the classification is *patchy* (individual regions of H$\alpha$ emission, not centrally peaked), *ring*, *diffuse* (the H$\alpha$ emission cannot be ascribed to individual

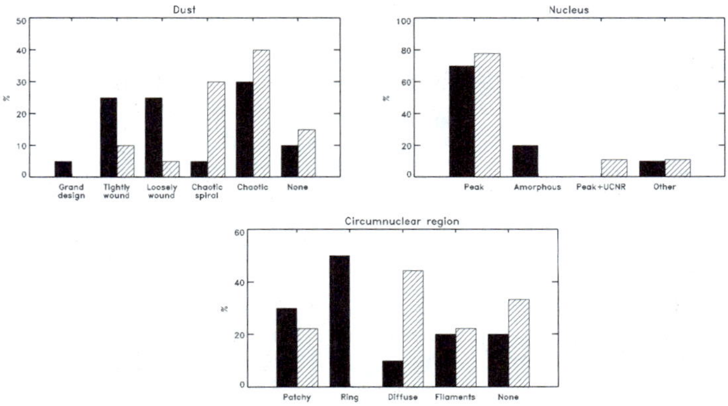


Figure 1. Classification of the galaxies from their dust morphology, and from their nuclear and circumnuclear H$\alpha$ emission. Filled columns correspond to $\sigma$-drop galaxies and dashed columns to control galaxies.

and well-defined regions), *filaments*, and *none* (less than 1% of the area of the inner kpc region has H$\alpha$ emission). The results are shown in Fig. 1 and can be summarized by stating that nuclear star-forming rings are found only in $\sigma$-drop galaxies.

## 3. Classification of Dust Structure

Using structure maps which we produced from archival *HST* images, we classified the inner dust structure of the galaxies within $r_c$. The classification criteria we used are those of Peeples & Martini (2006). The results are shown in Fig. 1 and can be summarized as an excess of dust spiral arms in sigma-drop hosts.

## 4. CO and H I Emission

We studied the CO and H I emission from the literature and find no significant difference either between the gas content of the samples.

## 5. Results and Conclusions

We find that the presence of a $\sigma$-drop correlates with that of dust spiral arms, which indicate gas inflow. We also find that this inflow is probably trapped by inner Lindblad resonances, where it accumulates until the density is high enough to start a starburst. This then creates the kinematically cold stellar population which is detected as a $\sigma$-drop.

### References

Peeples, M., & Martini, P. 2006, ApJ, 652, 1097
Comerón, S., Knapen, J. H., Beckman, J. E., & Shlosman, I. 2008, A&A, 478, 403

*Pathways through an Eclectic Universe*
*ASP Conference Series, Vol. 390, © 2008*
*J. H. Knapen, T. J. Mahoney, and A. Vazdekis, eds.*

# Kinematic Analysis of the Double-Barred Galaxy NGC 357

Adriana de Lorenzo-Cáceres, Alexandre Vazdekis & J. Alfonso L. Aguerri

*Instituto de Astrofísica de Canarias, E-38200 La Laguna, Spain*

**Abstract.** We present a kinematic analysis of the double-barred early-type galaxy NGC 357. The radial velocity and velocity dispersion profiles along both the nuclear bar axis and the semi-major axis clearly reveal the presence of the two bars. The results indicate that the bulge is pressure supported whereas the $|v_r|/\sigma$ ratio shows a plateau along the nuclear bar.

## 1. Introduction

Galaxies with two bars are very interesting from a kinematic point of view, since they show very different features that might provide unique clues on the role of these structural components in the formation of these galaxies. The two bars can be kinematically decoupled or not, so the angle between them might not be a fixed one (Friedli & Martinet 1993). Moreover, these bars are thought to be the mechanisms to transport gas efficiently to the central regions, triggering the star formation there and feeding active galactic nuclei (Shlosman et al. 1990). As a first step in our analysis we present here a detailed kinematic study of the double-barred early-type galaxy NGC 357 (Erwin 2004; Aguerri et al. 2005). We will present in a subsequent paper a detailed stellar population analysis that together with the results shown here will provide us with a more complete picture of this galaxy.

## 2. The Data

High quality long-slit spectra were obtained with EMMI at the New Technology Telescope at La Silla, Chile. The slit was oriented along the semi-major axis and the nuclear and outer bars. The spectral range covers a wide region, from 3990 Å to 5460 Å, with a resolution of ~1.3 Å FWHM.

## 3. Kinematics

The radial velocity and the velocity dispersion were obtained by cross-correlating the data with the spectrum of a star with similar spectral type. We get a recession velocity of $(2400 \pm 180)\,\mathrm{km\,s^{-1}}$ for NGC 357. Since we have spectra along the different structural components (nuclear bar, outer bar, and semi-major axis) we can measure their kinematic properties. Figure 1 shows the local $|v_r|/\sigma$ ratio along the inner bar. As we expected, the bulge (the inner ~ 4 arcsec) has $|v_r|/\sigma \sim 0$, indicating that it is pressure supported. For the nuclear bar, with a length of ~8 arcsec, this ratio shows a nearly constant and high value with respect to the surrounding areas. This profile is basically due to the velocity dispersion, which in the nuclear bar region is nearly flat with a slight downward trend in value. Looking at the velocity dispersion profiles (not shown here due to space limitation) along the nuclear bar, the outer bar, and the semi-major axis, we do see these drops in the case of the nuclear bar. The mean velocity dispersion values are shown in Table 1.

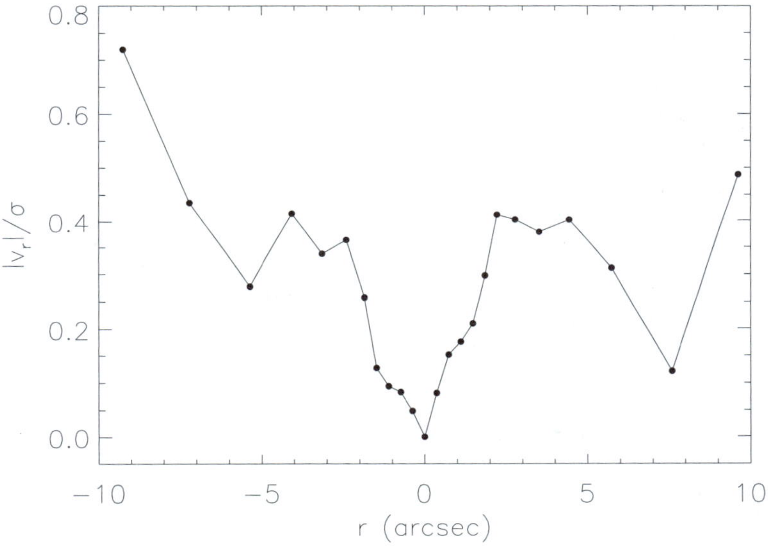


Figure 1. Local radial velocity/velocity dispersion ($|v_{\rm r}|/\sigma$) ratio for NGC 357 along the nuclear bar axis. The bulge corresponds to the inner ~4 arcsec and the nuclear bar to ~8 arcsec. The values decrease in the inner regions, as they should for a pressure-supported bulge, whereas higher and nearly constant values are obtained for the nuclear bar, increasing outwards.

Table 1. Mean velocity dispersion for each structural component of NGC 357

| Bulge at $0 \leq r$ ($''$) $\leq 1.8$ | Nuclear bar at $2 \leq r$ ($''$) $\leq 3$ | Outer bar at $10 \leq r$ ($''$)$\leq 25$ | Semi-major axis at $15 \leq r$ ($''$) $\leq 50$ |
|---|---|---|---|
| $180\,{\rm km\,s^{-1}}$ | $165\,{\rm km\,s^{-1}}$ | $99\,{\rm km\,s^{-1}}$ | $105\,{\rm km\,s^{-1}}$ |

## References

Aguerri, J. A. L., Elias-Rosa, N., Corsini, E. M., & Muñoz-Tuñón, C. 2005, A&A, 434, 109
Erwin, P. 2004, A&A, 415, 941
Friedli, D., & Martinet, L. 1993, A&A, 277, 27
Shlosman, I., Begelman, M. C., & Frank, J. 1990, Nature, 345, 679

*Pathways through an Eclectic Universe*
*ASP Conference Series, Vol. 390, © 2008*
*J. H. Knapen, T. J. Mahoney, and A. Vazdekis, eds.*

# Star Formation Efficiency in Galaxy Interactions and Mergers: A Statistical Study

P. Di Matteo, F. Combes, A. L. Melchior, and B. Semelin

*Observatoire de Paris, 61, Avenue de l'Observatoire, F-75014, Paris, France*

**Abstract.** We investigate the enhancement of star formation efficiency in galaxy interactions and mergers by numerical simulations of several hundred galaxy collisions. Only major mergers are considered in this study. Our principal results are: (1) the amount of gas available in the galaxy is not the main parameter governing the star formation efficiency in the burst phase, (2) there is a negative correlation between the amplitude of the star-forming burst and the tidal forces exerted per unit of time, which is due to the large amount of gas dragged out of the galaxy by tidal tails in strong interactions, and (3) mergers are not always starburst triggers, and they do not always convert high gas mass quantities into new stars.

## 1. Initial Conditions

The simulations performed and the numerical method adopted are fully described in Di Matteo et al. (2007). All morphological types along the Hubble sequence are considered in the initial conditions of the two colliding galaxies, with varying bulge-to-disk ratios and gas mass fractions. The present work is limited to coplanar encounters, with galaxies having null relative disk inclinations and spins parallel to the orbital angular momentum.

## 2. Results

The total amount of gas available just before the final coalescence phase does not seem to be the main parameter that influences the amplitude of the burst during the merger (see Fig. 1, *left*). In turn, the amplitude of the star formation rate in the coalescence phase is strongly anti-correlated with the effects of tidal forces at pericenter. Indeed, strong tidal effects can remove a large fraction of the gas mass from the disk, reducing the fuel for the subsequent burst phase (see Fig. 1, *right*).

Our results show that interactions can lead to intense starbursts, but, in general, they are *not a sufficient condition to convert large quantities of gas mass into new stars* (see Fig. 2).

### References

Di Matteo, P., Combes, F., Melchior, A. L., & Semelin, B. 2007, A&A, 468, 61

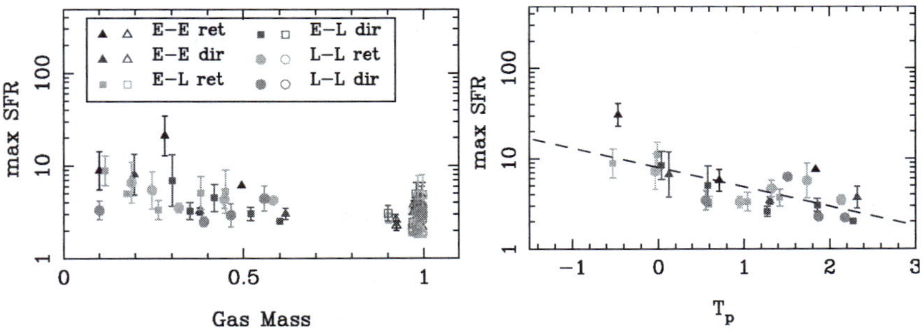


Figure 1. *Left panel:* Maximum star formation rate, as a function of the gas content (rescaled to its initial value), for flybys (empty symbols) and mergers (solid symbols). Different symbols correspond to different morphologies of interacting galaxies, as explained in the legend. *Right panel:* Maximum star formation rate, in the merging phase, relative to the isolated case, as a function of tidal parameter $T_{\rm p}$, which measures the tidal effects acting on a galaxy. Different symbols correspond to the different morphologies of the interacting galaxies, as explained in the legend in the left panel.

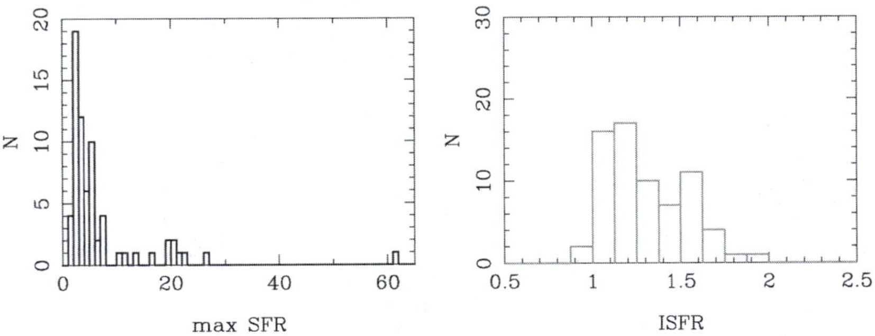


Figure 2. *Left panel:* Histogram of the maximum star formation rate (relative to the isolated case) for mergers. *Right panel:* Histogram of the integrated star formation rate (relative to the isolated case) for mergers.

*Pathways through an Eclectic Universe*
*ASP Conference Series, Vol. 390, © 2008*
*J. H. Knapen, T. J. Mahoney, and A. Vazdekis, eds.*

# Determination of Star Formation Timescale and Pattern Speed of Spiral Galaxies

F. Egusa, Y. Sofue, and S. Komugi

*Institute of Astronomy, the University of Tokyo, Japan*

H. Nakanishi

*Australia Telescope National Facility, Australia*

**Abstract.** We propose a method to determine the star formation timescale and pattern speed simultaneously in spiral galaxies. Though they are important and fundamental parameters in galactic dynamics and star formation scenarios, it has hitherto been difficult to derive their values directly from observations. Our method utilizes azimuthal offsets between arms of H II regions and molecular clouds, and has been successfully applied to three nearby galaxies so far through the use of CO and H$\alpha$ images. As derived star formation timescales fall into the narrow range of 4–13 Myr, which is consistent with typical timescales for molecular clouds to collapse gravitationally, a dominant mechanism of global star formation in spiral arms might be the gravitational collapse of molecular clouds.

## 1. Method

We define a star formation timescale ($t_{\rm SF}$) to be the average time needed for molecular clouds to form massive stars. In spite of its importance in any star formation scenario, its value has not been determined theoretically or observationally. An easy way to estimate its value is to calculate the Jeans timescale ($\propto [G\rho]^{-1/2}$) for molecular clouds, and its typical value is about 6 Myr for $\rho = 100\,{\rm cm}^{-3}$. Meanwhile, a pattern speed ($\Omega_{\rm P}$) is defined as an angular rotational velocity of a spiral pattern, and is especially important for spiral density wave theory (Lin & Shu 1964). As it cannot be derived from observations, several indirect methods, such as the Tremaine–Weinberg method (Tremaine & Weinberg 1984), have been proposed.

An advantage of our method is that we can derive $t_{\rm SF}$ and $\Omega_{\rm P}$ simultaneously. Assuming circular rotation, the angular offset ($\theta$) and the angular rotational velocity ($\Omega$) should have a linear correlation as $\theta = (\Omega - \Omega_{\rm P}) \cdot t_{\rm SF}$. As these two values can be derived directly from observations, it is easy to make a 2D plot of them and fit it with a line using the least $\chi^2$ method. The $t_{\rm SF}$ and $\Omega_{\rm P}$ are determined by the gradient and abscissa intercept of the fitted line, respectively (see Egusa, Sofue, & Nakanishi 2004 for details).

## 2. Application and Results

We applied our method to four nearby spiral galaxies (NGC 4254, NGC 4303, NGC 4321, and NGC 5194). All these galaxies are classified as "grand-design";

that is, their arm classes are 9 or 12 according to Elmegreen & Elmegreen (1987). CO data are from our mosaic observation with the Nobeyama Millimeter Array (NMA) for NGC 4254 and from BIMA SONG (Helfer et al. 2003) for the rest. H$\alpha$ images are from Koopmann, Kenney, & Young (2001) for NGC 4254, Knapen et al. (2004) for NGC 4303, and Martin & Kennicutt (2001) for the other two. Rotational velocities are derived from a CO velocity field for NGC 4254, and from Sofue et al. (1999) for the BIMA SONG galaxies.

Three of the four galaxies have a clear offset between the CO and H$\alpha$, so that we can fit each $\theta$–$\Omega$ plot (Fig. 1) to derive $t_{\rm SF}$ and $\Omega_{\rm P}$. The corotation radius ($R_{\rm CR}$) occurs where the disk matter and spiral pattern rotate at the same speed, and can be calculated from the derived $\Omega_{\rm P}$ and rotation curves. The results listed in Table 1 indicate that CO lines are detected only within the corotation radii, and that the star formation process in spiral arms might be dominated by the gravitational collapse of molecular clouds.

The remaining galaxy (NGC 4321) has no offset between the CO and H$\alpha$, which might be due to large non-circular motions from the bar potential or material (not density) spiral arms being dragged by the bar.

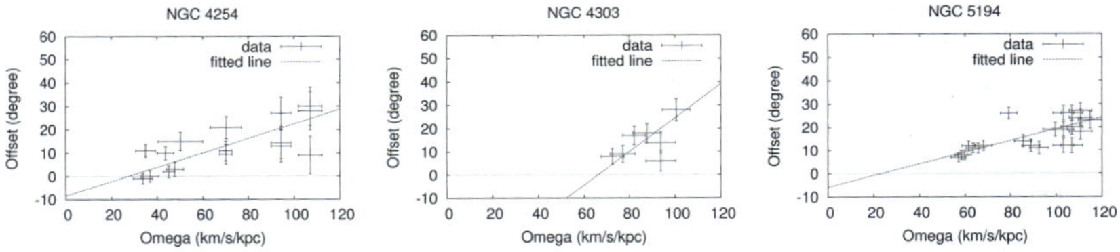


Figure 1. The $\theta$–$\Omega$ plots and fitted lines.

Table 1. Results from the fitting

| | NGC 4254 | NGC 4303 | NGC 5194 |
|---|---|---|---|
| $t_{\rm SF}$ (Myr) | $5.3 \pm 0.9$ | $12.5 \pm 5.5$ | $4.3 \pm 0.5$ |
| $\Omega_{\rm P}$ (km/s/kpc) | $27 \pm 10$ | $66 \pm 50$ | $23 \pm 10$ |
| $R_{\rm CR}$ (kpc) | $\sim 5$ | – | $\sim 9$ |

## References

Egusa, F., Sofue, Y., & Nakanishi, H. 2004, PASJ, 56, L45
Elmegreen, D. M., & Elmegreen, B. G. 1987, ApJ, 314, 3
Helfer, T. T., Thornley, M. D., Regan, M. W., Wong, T., Sheth, K., Vogel, S. N., Blitz, L., & Bock, D. C.-J. 2003, ApJS, 145, 259
Knapen, J. H., Stedman, S., Bramich, D. M., Folkes, S. L., & Bradley, T. R. 2004, A&A, 426, 1135
Koopmann, R. A., Kenney, J. D. P., & Young, J. 2001, ApJS, 135, 125
Lin, C. C., & Shu, F. H. 1964, ApJ, 140, 646
Martin, C. L., & Kennicutt, R. C. 2001, ApJ, 555, 301
Sofue, Y., Tutui, Y., Honma, M., Tomita, A., Takamiya, T., Koda, J., & Takeda, Y. 1999, ApJ, 523, 136
Tremaine, S., & Weinberg, M. D. 1984, ApJ, 282, L5

*Pathways through an Eclectic Universe*
*ASP Conference Series, Vol. 390, © 2008*
*J. H. Knapen, T. J. Mahoney, and A. Vazdekis, eds.*

# Fabry–Perot Interferometry and Spiral Galaxy Dynamics

K. Fathi,[1] J. E. Beckman,[1] C. Carignan,[2] and O. Hernandez[2]

[1]*Instituto de Astrofísica de Canarias, E-38200 La Laguna, Spain*
[2]*Laboratoire d'Astrophysique Expérimentale, Université de Montréal, Montréal, Québec, Canada*

**Abstract.** We present two-dimensional Fabry–Perot observations of the emission-line distribution and kinematics in nearby spiral galaxies. We have developed and demonstrated the utility of a number of analysis tools which have general applicability, but which we have, so far, applied to only one galaxy (M74, Fathi et al. 2007). In this galaxy, we have found kinematic signatures of radial motions caused by an $m = 2$ perturbation. Such a perturbation may well be responsible for the inflow of material forming the nuclear ring and the inner rapidly rotating disk-like structure. The latter, in turn, could help build a pseudo-bulge. In the second paper in this series, we will apply the kinematic analysis tools to a sample of nine late-type spiral galaxies observed with the FaNTOmM Fabry–Perot spectrometer at the Canada–France–Hawaii Telescope.

## 1. Introduction

In studying galactic dynamics, kinematic measurements over two spatial dimensions provide the necessary wealth of information. Currently used methods for achieving such observations have either limited angular resolution (radio or submillimeter astronomy) or cover a relatively small field of view (integral field units). We use Fabry–Perot interferometry, which combines high spatial sampling with a large field of view, an ideal combination for studying extended objects such as galaxies. We aim to study a sample of nine late-type spiral galaxies (NGC 3049, NGC 4294, NGC 4519, NGC 4654, NGC 5371, NGC 5921, NGC 5964, NGC 7479, and NGC 7741), all observed with the FaNTOmM Fabry–Perot interferometer (Hernandez et al. 2003) at the 3.6 m Canada–France–Hawaii Telescope. The observations yield the ideal combination of the spatial coverage and the resolution by covering a field of $\sim 4' \times 4'$ with angular resolution of 0.48 arcsec/pixel. FaNTOmM is optimized for targeting the 6562.8 Å H$\alpha$ emission line, which we use to study the dynamics of the star-forming gas.

## 2. Methods

The data are reduced homogeneously, both following the standard data reduction packages and the improved IDL-based reduction package developed by Daigle et al. (2006). In Fig. 1, we illustrate the derived maps for one example in our sample of nine galaxies (NGC 4519). The velocity fields are used to analyze the rotational components by means of tilted-ring and harmonic decomposition techniques (e.g., Fathi et al. 2005). The rotation curve is in turn used to derive the angular frequencies and the effects of gravitational perturbations on the evolution of structures in the disk. In conjunction with H II region catalogs, we are then able to extract and quantify the kinematic effects of individual H II regions and to separate these from the role of the gravitational perturbations on the observed velocity fields. Moreover, by averaging the velocity dispersion maps along elliptic annuli, we are able to derive the H$\alpha$-emitting gas velocity dispersion profiles to study the effect of feedback and disk heating mechanisms as a function of galactocentric radius. The data analysis methods have been applied to the nearby late-type spiral galaxy M74, and the utility of the codes is presented in detail in Fathi et al. (2007).

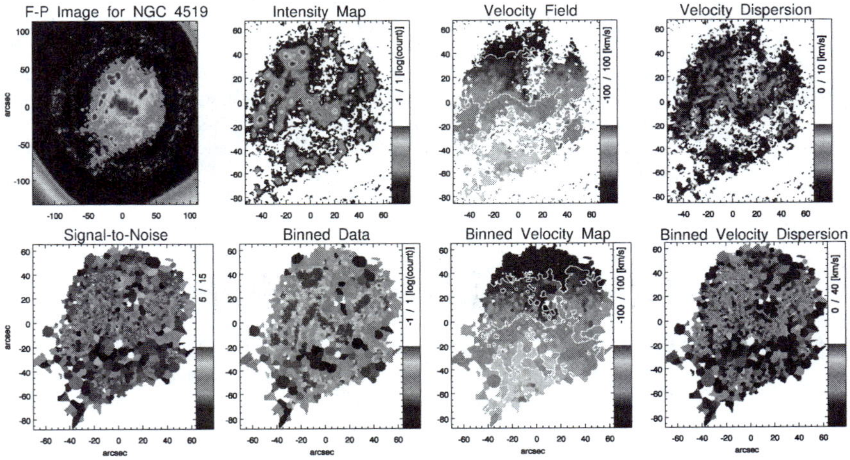


Figure 1. Fabry–Perot H$\alpha$ observations of one of the nine sample galaxies, NGC 4519. Illustrated are the results from the two different sets of reduction procedures. In the top row are the results from the reduction and derivation of the kinematics for all individual pixels, but only trusting the fits for pixels that have a signal-to-noise ratio higher than 3. The bottom row shows the results of the two-dimensional binning scheme, which increases the signal-to-noise ratio, but reduces spatial resolution. The maps in the top row are ideal for studying individual H II regions, whereas the maps in the bottom row are optimized for analyzing the large scale galaxy dynamics.

**Acknowledgments.** We thank the secretarial staff of the IAC, the LOC, and the SOC for an eclectic and stimulating conference.

## References

Daigle, O., Carignan, C., Hernandez, O., Chemin, L., & Amram, P. 2006, MNRAS, 368, 1016

Fathi, K., van de Ven, G., Peletier, R. F., Emsellem, E., Falcón-Barroso, J., Cappellari, M., & de Zeeuw, P. T. 2005, MNRAS, 364, 773

Fathi, K., Beckman, J. E., Zurita, A., Relaño, M., Knapen, J. H., Daigle, O., Hernandez, O., & Carignan, C. 2007, A&A, 466, 905

Hernandez, O., Gach, J.-L., Carignan, C., & Boulesteix, J. 2003, SPIE, 4841, 1472

*Pathways through an Eclectic Universe*
*ASP Conference Series, Vol. 390, © 2008*
*J. H. Knapen, T. J. Mahoney, and A. Vazdekis, eds.*

# The Evolution of Supernova Remnants in the M82 Central Starburst

D. Fenech, T. W. B. Muxlow, A. Pedlar, R. Beswick, and M. K. Argo

*Jodrell Bank Observatory, University of Manchester, Macclesfield, UK*

**Abstract.** An 8 day deep MERLIN observation of the starburst galaxy M82 at 5 GHz has been used to study over 50 of the compact sources in the central region, including supernova remnants and H II regions, producing the most sensitive images of this galaxy to date. In addition, global very long baseline interferometry (VLBI) observations at 18 cm have enabled detailed observations of the most compact sources in this starburst and have been used to follow their expansion. Simultaneous MERLIN 18 cm observations have been combined with the latest epoch of global VLBI data to produce high fidelity, high resolution images of the majority of the SNRs in M82 for the first time at this frequency. We present here an overview of these observations and results.

## 1. Introduction

M82 is considered a prototypical starburst galaxy and, as one of the closest (distance $\sim 3.2$ Mpc), provides an ideal opportunity to study a star-forming environment in detail. The first detailed radio studies of M82 revealed a number of compact sources in the central region first identified as either radio supernovae or supernova remnants (SNRs) (Kronberg & Sramek 1992). Their subsequent lack of variability confirmed their identity as SNRs. Further MERLIN (Multi Element Linked Interferometer Network) and VLA (Very Large Array) observations have shown the central kiloparsec of M82 to contain at least 50 compact sources including SNRs and H II regions.

## 2. Deep MERLIN 5 GHz Observations

The MERLIN array was used to make observations of M82 between 2002 April 1 and 28 at 4.994 GHz. This deep integration provided over 175 hours on source, resulting in an image rms sensitivity of $\sim 17\,\mu$Jy/beam. This has been used to produce detailed images of 55 sources in the central region of M82, showing shell and partial shell structures. The sizes of the sources have also been measured showing diameters ranging from 0.3 to 8.1 pc with a mean size of 3.1 pc (D. Fenech et al., in preparation).

The supernova rate determined from these SNR observations is $\sim$0.07/yr, which can be used to infer a star formation rate of $\sim 1.8\,M_{\odot}$/yr. Previous MERLIN observations made in 1992 (see Muxlow et al. 1994) over 36 hours have been used to produce astrometrically aligned images to investigate the expansion and variability of these SNRs.

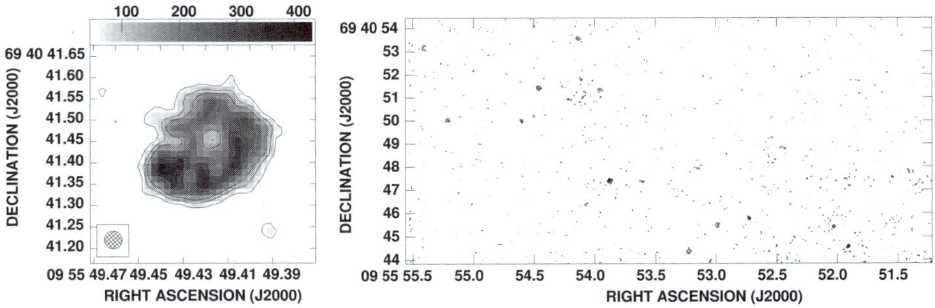


Figure 1. *Left:* Image of the SNR 40.68+55.1 at 5 GHz restored with a 41 mas beam. Contours plotted at $-1, 1, 2, \ldots, 8 \times 51\,\mu$Jy/beam. Scale ranges from 55-425 $\mu$Jy/beam. *Right:* Field image of the combined 18 cm data at 35 mas resolution. Contours plotted at $-1, 1, 2, \ldots, 20 \times 130\,\mu$Jy/beam.

## 3. Global 18 cm VLBI and MERLIN Observations

In March 2005, 16 radio stations across America and Europe were used to make global VLBI observations of M82 at 18 cm. These observations enable detailed study of the four most compact sources in M82 with an angular resolution of 4 mas ($\sim$0.05 pc at the distance of M82). These data are the fifth epoch of 18 cm observations of M82, allowing an investigation of these sources over 19 years. Comparison of the previous four epochs has yielded expansion velocities for the compact sources ranging from $1500 \pm 400\,\mathrm{km\,s^{-1}}$ for 41.95+57.5 to $\sim$10,000 $\mathrm{km\,s^{-1}}$ for 43.31+59.2 (Beswick et al. 2006).

The MERLIN array was used to observe M82 at 18 cm simultaneously with the VLBI observations. This has enabled the combination of the two data sets to produce high fidelity images of the central kiloparsec of M82. The fully imaged data have an angular resolution of $\sim$50 mas, close to that of the 5 GHz MERLIN observations. The combined 18 cm and MERLIN 5 GHz data will be used along with other radio wavelength observations to study the spectral energy distributions for the majority of the SNRs seen in M82.

**Acknowledgments.** MERLIN is a UK national facility run by the University of Manchester on behalf of PPARC. DF and MKA acknowledge financial support via a PPARC-funded studentship.

### References

Beswick, R., Riley, J., Marti-Vidal, I., Pedlar, A., Muxlow, T. W. B., McDonald, A., Wills, K., Fenech, D., & Argo, M. 2006, MNRAS, 369, 1221

Kronberg, P., & Sramek, R. 1992, in MPE Conf. 235, X-Ray Emission from Active Galactic Nuclei and the Cosmic X-Ray Background, ed. W. Brinkman & J. Truemper, p. 247

Muxlow, T. W. B., Pedlar, A., Wilkinson, P., Axon, D., Sanders, E., & de Bruyn, A. 1994, MNRAS, 266, 455

*Pathways through an Eclectic Universe*
*ASP Conference Series, Vol. 390, © 2008*
*J. H. Knapen, T. J. Mahoney, and A. Vazdekis, eds.*

# Disentangling Interacting Galaxy Pairs and Entangling Them Back

I. Fuentes-Carrera,[1] P. Amram,[2] C. Balkowski,[1] H. Flores,[1] M. Puech,[1] Y. Yang,[1] M. Rosado,[3] and H. Hernández-Toledo[3]

[1] *GEPI, Observatoire de Paris, CNRS, Université Paris Diderot, Place Jules Janssen, 92190 Meudon, France*
[2] *Laboratoire d'Astrophysique de Marseille, 2 Place Le Verrier, Marseille, Cedex 4, France*
[3] *Instituto de Astronomía, Universidad Nacional Autónoma de México (UNAM), Apdo. Postal 70-264, 04510, México, D.F., Mexico*

**Abstract.** Two-dimensional kinematics of disk galaxies give valuable information on the structure and evolution of these systems, complementing—sometimes contradicting—information derived from morphological studies. We present two cases of the use of two-dimensional kinematic information to study the interacting stage of galaxies. The first case portrays an apparently isolated local galaxy. The other points out the effects that higher redshifts—and thus the lack of spatial resolution—might have on the interpretation of this type of data.

## 1. An "Isolated" LIRG with Perturbed Kinematics

Figure 1 shows the luminous infrared galaxy IC 4734 ($L_{\rm IR} \sim 10^{11.27} L_{\odot}$). Although the morphology shows no sign of recent interaction, the distribution of radial velocities in the velocity field seems to indicate that the spiral arms are departing from the classical velocity field of a rotating disk. This is also seen in the rotation curve (RC), where the outermost rotational velocities decrease compared to the central plateau. This behavior could be due to tidal perturbations, though there are no visible companions or satellite galaxies in the vicinity.

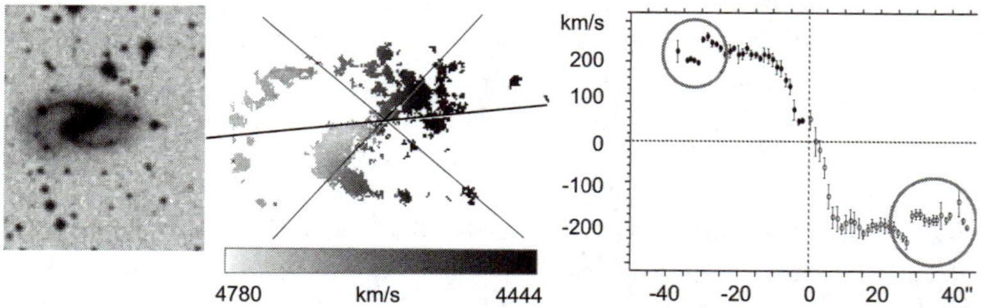


Figure 1. *Left:* DSS image of IC 4734. *Center:* Hα line-of-sight velocity map from scanning Fabry–Perot (FP) interferometric observations with CIGALE at the ESO 3.6 m telescope. *Right:* Rotation curve of the galaxy derived from the velocity map considering the points within the angular sector within the thin black lines. Circles show the velocity decrease in the outer parts.

## 2. Simulated Data of Two Local Interacting Pairs at $z = 1.5$

Figure 2 shows the velocity fields of two local interacting galaxy pairs and those that would be observed if the galaxies were located at $z = 1.5$. Two cases are presented: one with seeing-limited sampling and the other making use of adaptive optics (AO) facilities such as that of the NIR integral field spectrograph SINFONI. For both pairs the interacting galaxies blend together into one apparent object. For Arp 271, in seeing-limited sampling, a regular velocity field similar to that of a rotating disk is seen. With AO sampling, however, the velocity field can be decomposed into two velocity gradients that may be associated with two interacting disks. For KPG 468, the kinematics derived with both samplings suggest something more complex than a simple disk.

For Arp 271, the RC of the "galaxy" observed at $z = 1.5$ was derived from the seeing-limited map. The lack of spatial resolution results in the determination of an RC similar to that of a mildly perturbed rotating disk, even though the object is actually an interacting pair, leading to a serious misinterpretation of the data.

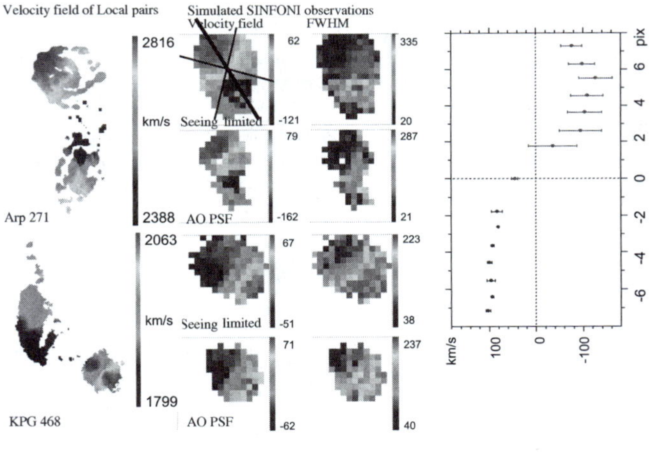


Figure 2. *Left panels:* H$\alpha$ velocity maps from FP observations of the galaxy pairs Arp 271 (Fuentes-Carrera et al. 2004) and KPG 468 (Hernández-Toledo et al. 2003). *Center and right panels:* Simulated data obtained if the pairs were at $z = 1.5$ (M. Puech et al., in preparation). The two galaxies of each pair have blended into a single "object". "Seeing-limited" corresponds to 100 mas/pixel (FWHM = 500 mas). For "AO-PSF", FWHM = 100 mas and a PSF given by simulations of the AO system. *Right panels:* RC derived from the seeing-limited velocity field of Arp 271 at $z = 1.5$ considering the points within the angular sector shown in the top central panel.

### References

Fuentes-Carrera, I., et al. 2004, A&A, 415, 451

Hernández-Toledo, H., Fuentes-Carrera, I., Rosado, M., Cruz-González, I., Franco-Balderas, F., & Dultzin-Hacyan, D. 2003, A&A, 412, 669

*Pathways through an Eclectic Universe*
*ASP Conference Series, Vol. 390, © 2008*
*J. H. Knapen, T. J. Mahoney, and A. Vazdekis, eds.*

# Interactions and Star Formation in Galaxies

Johan H. Knapen

*Instituto de Astrofísica de Canarias, E-38200 La Laguna, Spain*

Phil A. James

*Liverpool John Moores University, UK*

**Abstract.** We present results derived from H$\alpha$ images of 327 local galaxies from the HaGS (H$\alpha$ Galaxy Survey), where we have correlated the star formation rates with the interaction characteristics of the galaxies. We see a significant increase in the star formation rate in galaxies with close neighbors, an increase which has probably been triggered by galaxy–galaxy interactions. But when we analyze the equivalent width of the H$\alpha$ emission, which is the H$\alpha$ flux normalized by the continuum emission, or a measure of the current compared to the average past star formation rate, we find that galaxies with close companions show no significant enhancement at all. Our conclusion is that although close neighbors and interactions increase the star formation rate, they do this continuously. When we define a starburst as a star formation event with a star formation rate that is currently higher than it has been on average in the past, we must thus conclude that the presence of a close companion does not lead to starburst activity.

## 1. Introduction

Of the mechanisms proposed to bring gaseous fuel into the central starburst regions of a galaxy, non-axisymmetries in the gravitational potential induced by interactions or by bars are among the most promising. In fact, there is no doubt that the most extreme starbursts, such as UltraLuminous InfraRed Galaxies (ULIRGs), all occur in interacting or merging galaxies. For starbursts of lower luminosities, or indeed "normal" star formation in galaxies, a connection with interactions is tempting but not immediately obvious, and results from the literature show a confused picture (see Knapen 2004 for a review). It is also not clear what fraction of merging galaxies lead to starburst activity, and over what timescales.

## 2. Observations and Methodology

We have analyzed the complete H$\alpha$ Galaxy Survey (HaGS; James et al. 2004) sample of 327 local galaxies by studying the star formation rate (SFR, in $M_{\odot}\,\mathrm{yr}^{-1}$) and the equivalent width (EW, in nm, and derived by dividing the H$\alpha$ by the continuum emission as derived from $R$-band imaging) for subsamples of galaxies with and without a close companion. For this purpose, we have defined a galaxy to have such a companion if there is another galaxy within a radius of five times the diameter of the sample (HaGS) galaxy, which is also within $\pm 200\,\mathrm{km\,s^{-1}}$ in radial velocity and not fainter than 3 mag below the HaGS galaxy (cf. Knapen et al. 2006). This combination of criteria ensures that any companions are close enough and massive enough that gravitational effects may be expected to have some influence on the gas and stars in the sample galaxy. The companions were selected from NED.

## 3. Results

Of the 327 sample galaxies, 48 have a close companion as defined above. The median SFR for the galaxies with and without companion is 0.69 and 0.28 $M_\odot\,\mathrm{yr}^{-1}$, respectively, with the result of a Kolmogorov–Smirnov (KS) test confirming that the two populations are statistically different. The median SFR for the whole sample is 0.30 $M_\odot\,\mathrm{yr}^{-1}$.

Considering the EWs, the story is very different, however. Across the whole HaGS sample the median EW is 3.00 nm, for those HaGS galaxies with companions it is 3.05 nm, and for those without, 2.90 nm. The KS test indicates that this difference is not at all statistically significant.

## 4. Discussion and Conclusions

We can thus conclude that the presence of a close companion statistically increases the SFR of a HaGS sample galaxy, but also that it does *not* increase its EW. The EW of a galaxy is the ratio between the emission in H$\alpha$, from the massive SFR, and the $R$-band continuum, from stars that formed at least 100 Myr ago. The EW thus measures the "birthrate" parameter, a direct indication of whether a galaxy is undergoing a burst of star formation (Kennicutt et al. 2005). From the 11HUGS sample, similar to our HaGS sample, Lee (2006) defines starbursts as those galaxies with EW$>$ 10 nm.

We thus conclude that the presence of a close companion increases the SFR of a galaxy, but apparently does so over long enough timescales for the continuum light to increase by roughly the same amount as the H$\alpha$ line emission which traces the massive young stars. But the presence of companions does *not* lead to the occurrence of a star*burst*, in the sense of a higher current massive SFR than, on average, in its past. A more detailed account of this work will be published by J. H. Knapen & P. A. James (in preparation).

### References

James, P. A., et al. 2004, A&A, 414, 23

Kennicutt, R. C., Lee, J. C., Akiyama, S., Funes, J. G., & Sakai, S. 2005, AIP Conf. Proc., 783, The Demographics of Starburst Galaxies (New York: AIP), p. 3

Knapen, J. H. 2004, ASP Conf. Ser., 320, The Neutral ISM of Starburst Galaxies, ed. S. Aalto, S. Hüttemeister, & A. Pedlar (San Francisco: ASP), p. 205

Knapen, J. H., Mazzuca, L. M., Böker, T., Shlosman, I., Colina, L., Combes, F., & Axon, D. J. 2006, A&A, 448, 489

Lee, J. C. 2006, Ph.D. Thesis, Univ. Arizona

*Pathways through an Eclectic Universe*
*ASP Conference Series, Vol. 390, © 2008*
*J. H. Knapen, T. J. Mahoney, and A. Vazdekis, eds.*

# Interactions and Star Formation Activity in Wolf–Rayet Galaxies

Ángel R. López-Sánchez[1] and César Esteban

*Instituto de Astrofísica de Canarias, E-38200 La Laguna, Spain*

**Abstract.** We present some global results of the PhD thesis research carried out by López-Sánchez (2006), in which a detailed morphological, photometric and spectroscopic analysis of a sample of 20 Wolf–Rayet (WR) galaxies was carried out. The main aims are the study of the star formation and O and WR stellar populations in these galaxies, and the role that interactions between low surface companion objects have in the triggering of the bursts. We analyze the morphology, stellar populations, physical conditions, chemical abundances, and kinematics of the ionized gas, as well as the star formation activity of each system.

## 1. Introduction

Wolf–Rayet galaxies are a subtype of H II galaxies whose integrated spectra show broad emission lines attributed to WR stars, indicating the presence of an important population of massive stars and the youth of the starburst. Studying a sample of WR galaxies, Méndez (1999) and Méndez & Esteban (2000) suggested that interactions with or between dwarf objects could be the main star formation triggering mechanism in dwarf galaxies and noted that the interacting and/or merging nature of WR galaxies can be detected only when deep and high resolution images and spectra are available. Subsequent works also found a relation between massive star formation and the presence of interaction signatures. Thus, we have performed a detailed analysis of a sample of 20 WR galaxies combining deep optical and near-infrared (NIR) broad band and H$\alpha$ imaging together with optical spectroscopy (long slit and echelle) data. Additional X-ray, far-infrared and radio data were compiled from the literature.

## 2. Global Properties

The global analysis of our sample of 20 Wolf–Rayet galaxies follows from the more complete and exhaustive data set of this kind of galaxies, involving multiwavelength results and analyzing every system following the same procedures. Although we have obtained particular results of each galaxy (see López-Sánchez 2006 and references within), we only present here the main global results:

[1]Now at Australia Telescope National Facility (ATNF/CSIRO), PO Box 76, Epping NSW 1710, Australia.

1. The analysis of WR features in our sample suggests that aperture effects and the localization of the bursts with WR stars seem to play a fundamental role in the detection of this sort of massive stars in starburst galaxies.

2. Photometric data have been corrected for both extinction and nebular emission using our spectroscopic values. A good agreement between our optical and NIR colors and theoretical models is found; small discrepancies are explained by the existence of several stellar populations.

3. Physical and chemical properties are in agreement with both previous observations and models of chemical evolution of galaxies. We have compared abundances obtained by the direct method with those obtained for several empirical calibrations: Pilyugin (2001a,b) seems to give similar results whereas calibrations based on photoionization models such as those by McGaugh (1994) and Kewley & Dopita (2002) give abundances higher (~0.20 dex) than expected.

4. The comparison of the star formation rate (SFR) derived from our H$\alpha$ data (corrected for *both* extinction and [N II] emission using our spectroscopic data) is in good agreement with the SFR obtained using multiwavelength relations. We have also derived an X-ray based SFR for this kind of starburst galaxies.

5. We have determined the ionized gas mass ($M_{\rm H\,II}$), neutral gas mass ($M_{\rm H\,I}$), mass of the ionizing stellar cluster ($M_\star$), warm dust mass ($M_{\rm dust}$), Keplerian mass ($M_{\rm kep}$) and the dynamical mass ($M_{\rm dyn}$). As expected, all mass values increase with the luminosity of the galaxy. Furthermore, we find a good correlation between $M_{\rm dyn}$ and the luminosity of the galaxy, as well as a clear relation between the reddening coefficient derived from the Balmer decrement, $C({\rm H}\beta)$, and $M_{\rm dust}$. This fact suggests that extinction is mainly produced within the starburst.

6. However, our main conclusion is that *the majority of studied galaxies* (16 out of 20, ~80% of the objects) *show clear interaction features* such as plumes, tails, tidal dwarf galaxies, regions with very different chemical abundances inside galaxies, perturbed kinematics of the ionized gas or lack of neutral hydrogen gas, *confirming the hypothesis that interaction with or between dwarf objects triggers the star formation activity in Wolf–Rayet galaxies.*

**References**


Kewley, L. J., & Dopita, M. A. 2002, ApJS, 142, 35

McGaugh, S. S. 1994, ApJ, 426, 135

Méndez, D. I. 1999, PhD Thesis, University of La Laguna

Méndez, D. I., & Esteban, C. 2000, A&A, 359, 493

López-Sánchez, Á. R. 2006, PhD Thesis, University of La Laguna (see astro-ph/0704.2846 for a 16-page English resume with figures)

Pilyugin, L. S. 2001a, A&A, 369, 594

Pilyugin, L. S. 2001b, A&A, 374, 412

*Pathways through an Eclectic Universe*
*ASP Conference Series, Vol. 390, © 2008*
*J. H. Knapen, T. J. Mahoney, and A. Vazdekis, eds.*

# Integral Field Spectroscopy of Nearby Starbursts

A. Monreal-Ibero,[1,2] C. Muñoz-Tuñón,[2] N. Caon,[2] L. M. Cairós,[1] V. Melo,[2] B. García-Lorenzo,[2] P. Weilbacher,[1] G. Tenorio-Tagle,[3] and M. M. Roth[1]

[1]*AIP, An der Sternwarte 16, D-14482 Potsdam, Germany*
[2]*Instituto de Astrofísica de Canarias, E-38200 La Laguna, Spain*
[3]*INAOE, Aptdo Postal 51 y 216, 72000 Puebla, Mexico*

**Abstract.** Starbursts are events in which hundreds of solar masses per year of gas are transformed into stars on timescales much shorter than the Hubble time. They are essential in galaxy evolution as they are responsible for the largest part of the chemically enriched material in the interstellar and intergalactic medium where the metals have been deposited through stellar and supergalactic winds. The superposition of two stellar populations, young and old, and their complex structure make these systems difficult to study with traditional techniques. Instead, integral field spectroscopy, which offers simultaneous spatial and spectral resolution, is an ideal tool with which to analyze these systems. Here we outline the bases of an undergoing project carried out in collaboration between the IAC and AIP to study a sample of nearby starbursts.

## 1. Integral Field Spectroscopy

We aim to understand the nature and evolution of the star formation processes in starburst galaxies. High resolution images have revealed that these are morphologically complex objects showing several star-forming knots. A thorough analysis of these objects requires a complete spectrophotometric mapping. Integral field spectroscopy (IFS), which solves the problem of recording three variables ($\alpha$, $\delta$, and $\lambda$) on a two-dimensional detector, suits our needs very well. Its main advantage over other techniques, such as Fabry–Pérot interferometry or long-slit scanning, is the great homogeneity of the data, as they are taken simultaneously under the same instrumental and atmospheric conditions. We are using for this project two IFS-based instruments: INTEGRAL (Arribas et al. 1998) and PMAS (Roth et al. 2005).

## 2. Sample Selection

From our catalog of starburst galaxies we have selected: (1) galaxies close enough to resolve the physical units (greater than a super star cluster) able to produce the filaments observed in super galactic winds (SGWs), and (2) galaxies in the northern hemisphere from our atlas of blue compact dwarfs (BCDs, Cairós et al. 2001).

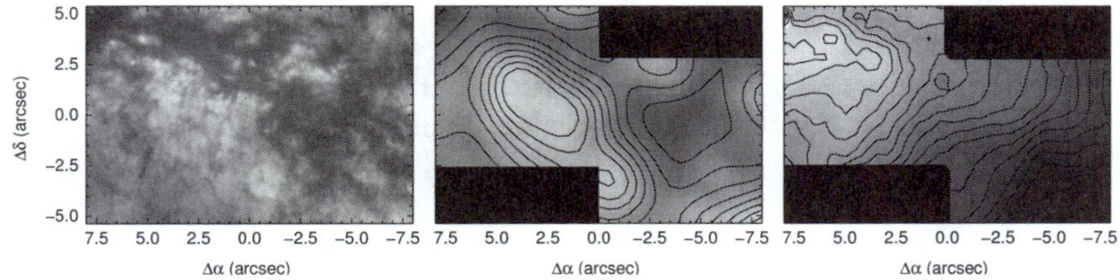


Figure 1. *Left: HST* image of the covered region of M82. *Middle:* Hα map from PMAS. *Right:* Velocity field (from 90, right, to 320 km $s^{-1}$, left).

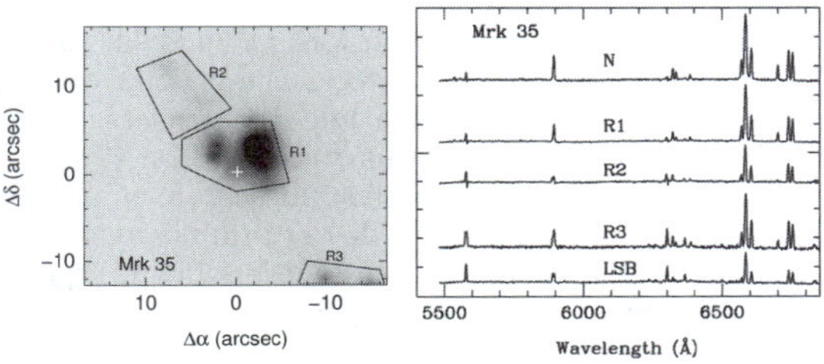


Figure 2. Spectra (*right*) of the selected regions (*left*), plus the spectra of the nucleus and of the low surface brightness region.

## 2.1. Example A: PMAS Results for M82

We observed the archetype of a galaxy with an SGW in 2006. The two pointings include the "SE zone" in Melo et al. (2005) as well as a highly extinguished heart-shaped region towards the west. Figure 1 shows the Hα flux and the derived velocity field. This map agrees well with the one presented in figure 8a of Rodríguez-Rico et al. (2004), although the spatial resolution is much better in our data.

## 2.2. Example B: INTEGRAL Results for Mrk 35

The BCD Mrk 35 was observed in February 2003 with the INTEGRAL bundle SB3. Some results are shown in Figure 2. The [O I], [He I], and [S III] emission lines are visible in the nucleus and in R1. R2, which includes several smaller SF regions towards the NE, and R3, which corresponds to two knots separated towards the SW, both display similar flat spectra. No absorption lines have been detected.

## References

Arribas, S., et al. 1998, Proc SPIE, 3355, 821

Cairós, L. M., Vílchez, J. M., González Pérez, J. N., Iglesias-Páramo, J., & Caon, N. 2001, ApJS, 133, 321

Melo, V. P., Muñoz-Tuñón, C., Maíz, J., & Tenorio-Tagle, G. 2005, ApJ, 619, 270

Rodríguez-Rico, C. A., Viallefond, F., Zhao, J.-H., Goss, W. M., & Anantharamaiah, K. R. 2004, ApJ, 616, 783

Roth, M. M. et al. 2005, PASP, 117, 620

*Pathways through an Eclectic Universe*
*ASP Conference Series, Vol. 390, © 2008*
*J. H. Knapen, T. J. Mahoney, and A. Vazdekis, eds.*

# Calibration of Star Formation Rate Tracers with Population Synthesis Models: The $L_X/L_{FIR}$ Ratio

Héctor Otí-Floranes[1] and Jose Miguel Mas-Hesse[2]

[1]*LAEFF–INTA, POB 78, E-28691 Villanueva de la Cañada, Spain*
[2]*Centro de Astrobiología (CSIC–INTA), E-28850 Torrejón de Ardoz, Spain*

**Abstract.** The study of massive star formation needs some basic tools, among which reliable star formation rate tracers are crucial ones. We are presently revising the calibration of star formation rate tracers at different wavelength ranges using last generation evolutionary population synthesis codes. The FIR luminosity produced by the heated dust in star forming regions is commonly used to characterize a starburst. The X-ray luminosity has been found to show a narrow correlation with it, and therefore it has been proposed as an additional tracer of potential interest for high $z$ galaxies, whose FIR emission is redshifted to the sub-mm range. In this communication we analyze the evolution of the X-ray to FIR luminosity ratio as predicted by the CMHK population synthesis models, both for constant stellar formation and instantaneous bursts. The results are compared to some sample data taken from the literature. The conclusion drawn from the comparison is that the empirical calibration of the soft X-ray luminosity seems a valid star formation rate tracer only for starbursts around a rather short period of time.

## 1. The $L_X/L_{FIR}$ Ratio

We have computed the X-ray and far-infrared (FIR) luminosities expected in a massive, young starburst using the CMHK (Cerviño, Mas-Hesse & Kunth) synthesis population models (Cerviño & Mas-Hesse 1994; Cerviño, Mas-Hesse, & Kunth 2002). A Salpeter initial mass function ($\phi(m) \sim m^{-2.35}$) has been assumed, with masses within the range 2–120 $M_\odot$, two different star formation regimes—instantaneous burst (IB) or constant star formation rate (CSFR)—and solar metallicities. We have analyzed the first 30 Myr after the onset of a massive starburst episode.

The X-ray luminosity is calculated in the range 0.4–2.4 keV. The models compute only the emission by the diffuse gas heated by the mechanical energy released by stellar winds and supernova remnants. The X-ray emission is modeled by Raymond–Smith thermal plasmas with a range of temperatures typical of starburst galaxies (between 0.5 and 1.0 keV). The fraction of mechanical energy $\epsilon_{\rm eff}$ effectively heating the gas up to X-ray temperatures is left as a free parameter. The contribution by X-ray binaries and stellar atmospheres has not been considered since it should be negligible (Cerviño et al. 2002).

Thermal equilibrium of dust has been assumed, which implies all energy absorbed by dust being re-emitted in the far-infrared range. The models assume that a fraction $(1 - f)$ of Lyman continuum photons are directly absorbed by the dust. The absorption of UV–optical continuum photons is parameterized through $E(B-V)$. The values considered were $f = 0.3$ and $E(B-V) = 1$.

Figure 1 shows the evolution of the soft X-ray luminosity and the $L_{\rm X}/L_{\rm FIR}$ ratio, as predicted by the models assuming different star formation regimes and efficiencies. We want to stress that this ratio is strongly dependent on the evolutionary state of the starburst, spanning around two orders of magnitude within a few million years.

The data samples from Ranalli, Comastri, & Setti (2003) and Tüllmann et al. (2006) (hereinafter, *Ran* and *Tul* respectively) were chosen in order to compare with the predictions from the models. As shown in the bottom panels of Fig. 1, the observational ratios can be well reproduced by the synthesis models considering efficiency values in the range $\epsilon_{\rm eff} = 0.01$–$0.1$, typical of starburst galaxies (Cerviño et al. 2002). Moreover, our results show that the dispersion observed in the ratios for the different samples can be explained as being mostly an evolutionary effect.

Therefore, we conclude that the empirical calibration of the soft X-ray luminosity as a star formation rate tracer has to be taken with care, since it would be valid only for starbursts around a relatively short period of time.

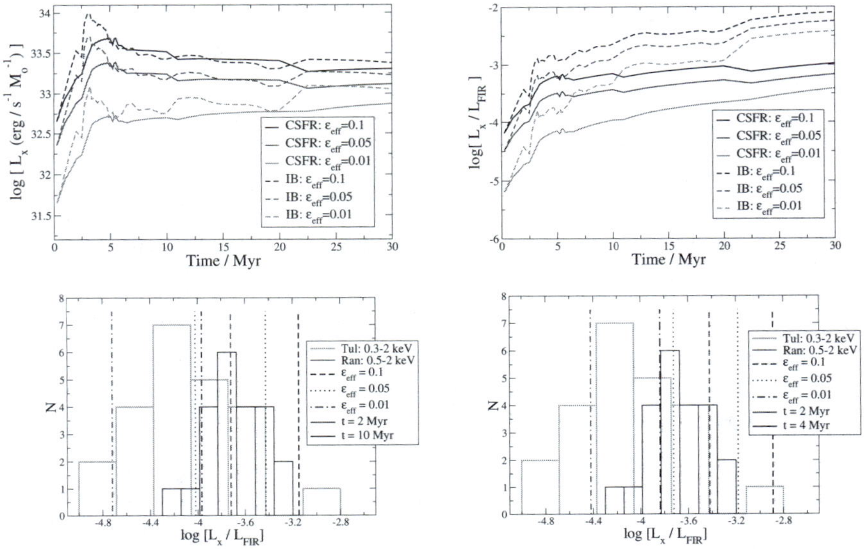


Figure 1. Evolution of $L_{\rm X}$ (*top left*) and $L_{\rm X}$ to $L_{\rm FIR}$ ratio (*top right*) predicted by the synthesis models. Histograms of the $L_{\rm X}$ to $L_{\rm FIR}$ ratio for the samples *Ran* and *Tul* considered, together with model predictions: CSFR (*bottom left*) and IB (*bottom right*) cases.

## References

Cerviño, M., & Mas-Hesse, J. M. 1994, A&A, 284, 749
Cerviño, M., Mas-Hesse, J. M., & Kunth, D. 2002, A&A, 392, 19
Ranalli, P., Comastri, A., & Setti, G. 2003, A&A, 399, 39
Tüllmann, R., et al. 2006, A&A, 457, 779

*Pathways through an Eclectic Universe*
*ASP Conference Series, Vol. 390, © 2008*
*J. H. Knapen, T. J. Mahoney, and A. Vazdekis, eds.*

# Converting H$\alpha$ Luminosities into Star Formation Rates

Jan Pflamm-Altenburg,[1,2] Carsten Weidner,[3] and Pavel Kroupa[1,2]

[1]*Argelander-Institut für Astronomie (AIfA), Auf dem Hügel 71, D-53121 Bonn, Germany*
[2]*Rhine Stellar Dynamics Network (RSDN)*
[3]*Departamento de Astronomía y Astrofísica, Pontificia Universidad Católica de Chile, Campus San Joaquin, Vicuña Mackenna 4860, 782-0436 Macul, Santiago, Chile*

**Abstract.** The recent finding that the integrated galactic initial stellar mass function, comprising all newly formed stars in all young star clusters, has, through its dependence on the star formation rate, a steeper slope in the high mass regime than the underlying canonical initial mass function of each star cluster individually offers new insights into the galactic star formation process. The classical linear relation between the star formation rate and the resultant H$\alpha$ luminosity is broken and star formation rates are always underestimated. Our new relation is likely to lead to a revision of cosmological star formation history.

## 1. The Star Formation Rate–H$\alpha$ Relation

To obtain a relation between the total galaxy-wide star formation rate (SFR) and the produced H$\alpha$ luminosity, the integrated galactic initial mass function (IGIMF) as a function of the SFR is combined with stellar evolution models. The resulting relations for four different IGIMF scenarios, introduced by Weidner & Kroupa (2005), are plotted in Fig. 1 (*right*). For a given H$\alpha$ luminosity the SFRs are always higher than in the classical linear relations by Kennicutt, Tamblyn, & Congdon (1994) (gray-shaded area, solid line). In addition, two classical models are included where the IGIMF is identical to a Salpeter IMF and a canonical IMF but using the same stellar evolution models as used in the IGIMF models.

## 2. dIrr Galaxies

Applying our SFR-$L_{\mathrm{H}\alpha}$ relation to the observed H$\alpha$–luminosities of the Sculptor dwarf irregular galaxies (Skillman et al. 2003) the SFRs (Fig. 2, *left*) and related parameters such as the gas depletion timescale (Fig. 2, *right*) change dramatically.

## References


Kennicutt, R. C., Tamblyn, P., & Congdon, C. E. 1994, ApJ, 435, 22
Skillman, E. D., Côté, S., & Miller, B. W. 2003, AJ, 125, 593
Weidner, C., & Kroupa, P. 2005, ApJ, 625, 754


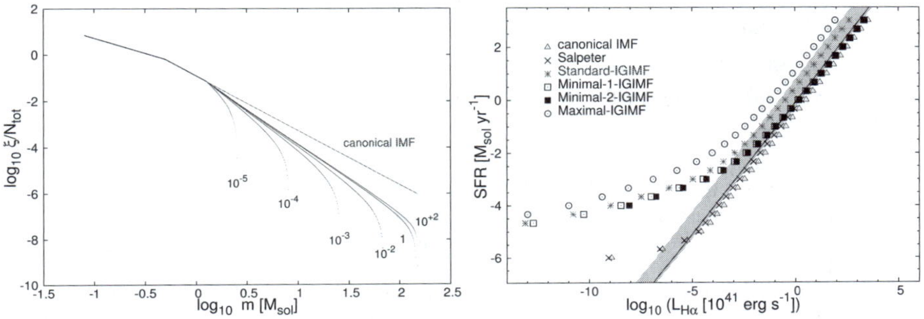


Figure 1. *Left:* The IGIMF with its dependence on the total SFR in the standard scenario. *Right:* The SFR–H$\alpha$ luminosity relation for four different IGIMF scenarios and relations by Kennicutt et al. (1994, gray-shaded area) and the widely used relation SFR / $M_\odot$ yr$^{-1}$ = $L_{\mathrm{H}\alpha}/1.26 \cdot 10^{41}$ erg s$^{-1}$ (solid line).

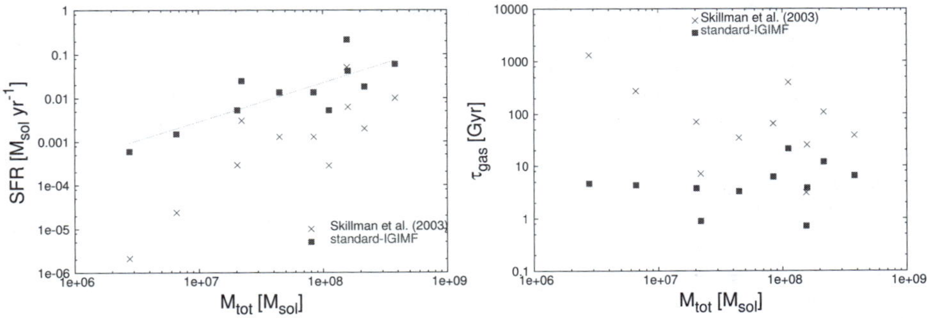


Figure 2. Derived SFRs (*left*) and gas depletion times (*right*) of the Sculptor dIrrs based on the standard IGIMF scenario and on the linear Kennicutt relation (Fig. 1, solid line) (Skillman et al. 2003). Note that the standard scenario IGIMF implies significantly higher SFRs and a constant gas depletion timescale for all dwarf galaxies.

*Pathways through an Eclectic Universe*
*ASP Conference Series, Vol. 390,* © *2008*
*J. H. Knapen, T. J. Mahoney, and A. Vazdekis, eds.*

# Two Extremely Compact Starbursts Inhabiting Clusters of Galaxies: Abell 539 and Abell 634

D. Reverte, J. M. Vílchez, J. D. Hernández-Fernández, and J. Iglesias-Páramo

*Instituto de Astrofísica de Andalucía (IAA-CSIC), Apdo 3004, E-18008 Granada, Spain*

**Abstract.** Two extremely compact Hα-emitting sources have been discovered in our deep images of the Abell 539 and Abell 634 clusters of galaxies. Follow-up spectroscopy of these two unresolved sources have revealed that they are active starburst members of their respective clusters. The Hα luminosities and extreme Hα equivalent widths (of $\simeq 430$ Å and 1010 Å, respectively) of these sources, together with their very compact appearance, have raised the question of their origin and the triggering of such active star formation in the cluster environments. Possible triggering mechanisms are pointed out, together with possible links with the recently discovered early-type ultra-compact dwarf galaxies.

## 1. Data

From our deep Hα images taken with the Wide Field Camera on the Isaac Newton Telescope, two compact emitters have shown up, associated with the clusters Abell 539 and Abell 634.

In order to know more about chemical properties and also to discard line contamination at a higher redshift (such as [O III] and Lyα), spectroscopy was carried out with the ALFOSC spectrograph on the Nordic Optical Telescope (NOT), resulting in two low metallicity starburst spectra (see Reverte et al. 2007) confirming their corresponding cluster membership.

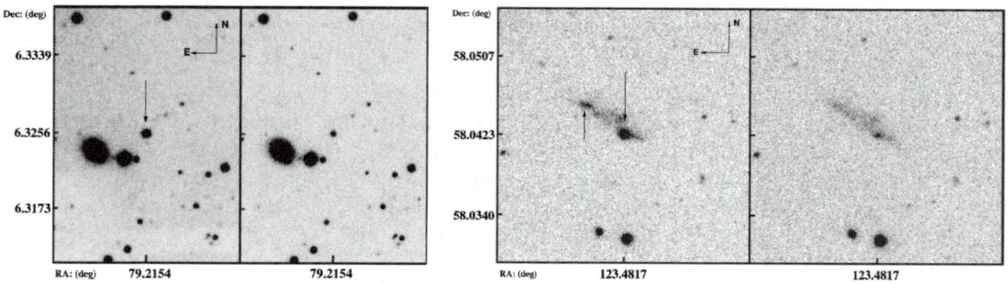


Figure 1. The first pair of images are of the DRP A539a object. The left image is the Hα+continuum emission and the right image is the continuum. The second pair of images is the same for DRP A634a. Both objects are indicated by a downward arrow.

## 2. Results and Discussion

The light profiles of these sources, named DRP A539a and DRP A634a, have shown that both objects are almost indistinguishable from a stellar profile in our images (FWHM $\sim$ 0.8 arcsec). This gives us a limit to their maximum effective radii[1] of 84 and 32 pc, respectively. The absolute magnitudes derived for DRP A539a ($M_{r'} = -16.07$) and DRP A634a ($M_{r'} = -14.50$) are brighter than those typical of early-type ultracompact dwarf galaxies (UCDGs) in clusters (Mieske et al. 2006). Nevertheless, the Starburst99 model (Leitherer et al. 1999) predicts that the optical magnitudes of an instantaneous burst of star formation can fade by more than 3 mag in about $10^8$ yr. This result, together with the limits to the sizes of DRP A539a and DRP A634a, opens up the possibility of an evolutionary link between these objects and the UCDG progenitors.

Companions and/or progenitors of a tidal interaction were not found for DRP A539a when the isophotes were analyzed on narrow- and broad-band images, and after searching for them on available database velocities and positions. On the other hand, appropriate studies of the luminosity–metallicity and Tully–Fisher predictions lead us to consider the low surface brightness extension as forming part of the galaxy DRP A634a.

The origin of the two starbursts reported here is probably associated with different physical mechanisms: DRP A539a is directly associated with the dense and hot intracluster medium which compresses intergalactic clouds and induces star formation episodes on a short timescale (Bekki & Couch 2003) before ram pressure is able to sweep up the external gas (Mori & Burkert 2000). In contrast, DRP A634a can be related to the so-called "preprocessing" of galaxies before they enter the cluster environment. In this case, the aggregates of galaxies whose final fate is to fall into the cluster inner regions are the sites where this kind of secondary evolution is taking place (Sakai et al. 2002).

**Acknowledgments.** This work has made use of ALFOSC, owned by the IAA-CSIC and operated at the NOT, and of the WFC attached to INT, operated by the Isaac Newton Group at the ORM (IAC), in La Palma. This research has made use of the NASA/IPAC Extragalactic Database. Partial funding by projects AYA2004-08260-C03-02, "Estallidos", from the Spanish PNAYA, and by TIC114 of the Junta de Andalucía are gratefully acknowledged.

### References

Bekki, K., & Couch, W. J. 2003, ApJ, 596, L13
Leitherer, C., et al. 1999, ApJS, 123, 3
Mieske, S., Hilker, M., Infante, L., & Jordán, A. 2006, AJ, 131, 2442
Mori, M., & Burkert, A. 2000, ApJ, 538, 559
Reverte, D., Vílchez, J. M., Hernández-Fernández, J. D., & Iglesias-Páramo, J. 2007, AJ, 133, 705
Sakai, S., Kennicutt, R. C., Jr., van der Hulst, J. M., & Moss, C. 2002, ApJ, 578, 842

[1]Under the assumption of a Gaussian seeing profile.

*Pathways through an Eclectic Universe*
*ASP Conference Series, Vol. 390, © 2008*
*J. H. Knapen, T. J. Mahoney, and A. Vazdekis, eds.*

# Supernovae and Star Formation in the ULIRG Arp 220

H. Thrall

*Jodrell Bank Observatory, The University of Manchester, Macclesfield, Cheshire, SK11 9DL, UK*

**Abstract.** For over a decade, the ULIRG Arp 220 has been the object of a high sensitivity VLBI campaign at a wavelength of 18 cm to observe the evolution of a cluster of unresolved radio sources, originally identified as Type IIn radio supernovae by Smith et al. (1998). The results of the most recent VLBI observations of Arp 220 are summarized here.

## 1. Introduction

Arp 220 is one of the closest and best-studied ULIRGs (ultra-luminous infrared galaxies). It is a merger of two gas-rich galaxies and has a pair of radio nuclei separated by ~370 pc. Several epochs of 18 cm VLBI observations, in combination with recent multiwavelength observations, can be used to constrain supernova rates and star formation rates (SFRs) and allow us an insight into the complex interstellar environment of this galaxy.

## 2. Recent Work

In November 2003, Arp 220 was observed at 18 cm with a global array of the world's largest telescopes. The rms noise in the resulting image is 5.5 $\mu$Jy/bm, making it the most sensitive VLBI image yet produced, by some margin. Careful analysis of this image resulted in the confirmed identification of 49 point sources across the two nuclei of Arp 220. A contour plot of the western nucleus is shown in Figure 1. For a full analysis, see Lonsdale et al. (2006). Comparisons of this image with an as yet unpublished image observed 12 months earlier revealed that four new sources had appeared in the intervening period.

Recent multiple wavelength VLBI observations (13, 6, and 3.6 cm) made by Parra et al. (2007) suggest that the older sources originally detected by Smith et al. (1998) have spectra more consistent with their being young supernova remnants interacting with the surrounding interstellar medium, while the sources which have appeared during the monitoring campaign are modeled more successfully with radio supernova models.

## 3. Supernova/Star Formation Rates

The four new sources discovered by Lonsdale et al. (2006), along with their favored identification as radio supernovae, implies a luminous supernova rate of $4 \pm 2$ per year.

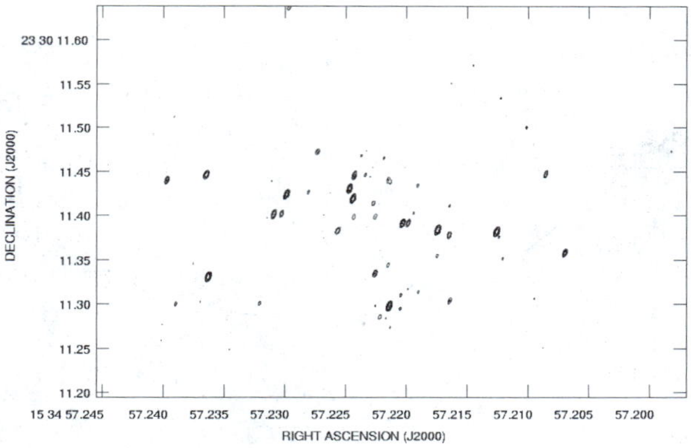


Figure 1. Contour plot of the western nucleus of Arp 220. Contours are at 1, 2, 3, 4, 8, and 20 times the source detection threshold used by Lonsdale et al. (2006).

Rovilos et al. (2005) estimated a supernova rate of $0.7\,\mathrm{yr}^{-1}$, and from this calculated an SFR of $\dot{m} = 43.5\,M_{\odot}\,\mathrm{yr}^{-1}$. This translates to a far-infrared luminosity of $L_{\mathrm{FIR}} = 5.2 \times 10^{11}\,L_{\odot}$, which is about 50% of Arp 220's total infrared luminosity ($1.4 \times 10^{12}\,L_{\odot}$: Soifer et al. 1987).

Using the updated supernova rate obtained by Lonsdale et al. (2006), an SFR of $\dot{m} = 249\,M_{\odot}\,\mathrm{yr}^{-1}$ can be estimated.

This SFR is in fact too high to be explained by the observed far-infrared luminosity. Another issue is that these supernovae are thought to be Type IIn, which are relatively rare, comprising only about 2% of core-collapse supernovae in normal galaxies, which suggests that the observed rate of appearance of compact sources could be up to 50 times higher than that predicted from the SFR. One resolution of this apparent inconsistency is that the star formation in Arp 220 is highly variable on the timescales of massive star lifetimes, consisting of a number of short ($\ll 10^6$ yr) very intense bursts, each with an SFR in the order of $1000\,M_{\odot}\,\mathrm{yr}^{-1}$. This scenario has been discussed by Anantharamaiah et al. (2000) and calculations show that the short burst scenario is a feasible explanation for the large rate of very powerful radio supernovae.

**Acknowledgments.** HT acknowledges a PPARC studentship.

### References


Anantharamaiah, K. R., Viallefond, F., Mohan, N. R., Goss, W. M., & Zhao, J. H. 2000, ApJ, 537, 613

Lonsdale, C. J., Diamond, P. J., Thrall, H., Smith, H. E., & Lonsdale, C. J. 2006, ApJ, 647, 185

Parra, R., Conway, J. E., Diamond, P. J., Thrall, H., Lonsdale, C. J., Lonsdale, C. J., & Smith, H. E. 2007, ApJ, 659, 314

Rovilos, E., Diamond, P. J., Lonsdale, C. J., Smith, H. E., & Lonsdale, C. J. 2005, MNRAS, 359, 827

Smith, H. E., Lonsdale, C. J., Lonsdale, C. J., & Diamond, P. J. 1998, ApJ, 493, L17

Soifer, B. T., Sanders, D. B., Madore, B. F., Neugebauer, G., Danielson, G. E., Elias, J. H., Lonsdale, C. J., & Rice, W. L. 1987, ApJ, 320, 238

Bernard Pagel and Pepe Vílchez (*above*), and Eduardo Battaner and Mike Edmunds (*below*).

# Session III

# Star Formation Histories

*Pathways through an Eclectic Universe*
*ASP Conference Series, Vol. 390, © 2008*
*J. H. Knapen, T. J. Mahoney, and A. Vazdekis, eds.*

# The Role of Starbursts in the Star Formation History of the Universe

Rosa M. González Delgado

*Instituto de Astrofísica de Andalucía (CSIC), E-18008 Granada, Spain*

**Abstract.** This review presents a personal view of the role that starbursts play in the star formation history of the Universe. It focuses mainly on the properties of nearby starburst galaxies selected for their strong ultraviolet and/or far-infrared emission. The similarities between local starbursts and star-forming galaxies at high redshift are also presented. I also discuss the role that luminous and ultraluminous infrared galaxies, and merging systems play in the formation and evolution of galaxies.

## 1. Introduction

### 1.1. Starburst Galaxies at High Redshift

Starburst galaxies are systems with a high star formation rate and are the preferred location for the formation of massive stars. Hence they are a relevant energy source that drives the cosmic evolution of galaxies. They are responsible for the thermal and mechanical heating of the interstellar medium and its chemical enrichment. Because they have a significantly large population of massive stars, they produce a large number of Lyman continuum photons that ionize the interstellar medium. But a fraction of these photons are absorbed by dust grains and re-emitted at far-infrared (FIR) wavelengths.

During the past decade, the role that starbursts play in the star formation history of the Universe has been highlighted mainly due to two important observational discoveries: a) a substantial population of star-forming galaxies at cosmological distances ($z \geq 2$), and b) the origin of the cosmic infrared background (CIRB). These discoveries were made thanks to the advance of 10 m class telescopes, the launch of satellites sensitive at FIR wavelengths, and the high spatial resolution and sensitive instruments on board *HST*.

Many of these high redshift star-forming galaxies have been selected with the Lyman break technique, which uses broad-band colors to target the Lyman continuum break of these galaxies (Steidel et al. 1996; Williams et al. 1996). This break has its origin in: a) the intrinsic drop in the spectral energy distribution of massive stars that dominate the ultraviolet (UV) light; b) the absorption by the neutral interstellar medium within the star-forming galaxy, and c) the opacity of the intervening intergalactic medium. Subsequent spectroscopic observations with 10 m class telescopes of these so-called Lyman break galaxies (LBGs) have shown that they are in a very active phase of star formation, and they contribute significantly to the global star formation rate (SFR) density of the Universe (Madau et al. 1996). They are powered by massive stars; thus, hot O- and B-type stars dominate the integrated spectrum of these galaxies at UV wavelengths. Their rest-UV spectra show a strong flat continuum plagued with low ionization interstellar absorption lines, some high ionization P Cygni wind lines, and a relatively weak Ly$\alpha$ emission that, in fact, is detected in absorption in many of these galaxies (Pettini et al. 2000; Shapley et al. 2003; Steidel et al. 2004; Noll et al. 2004). These spectral properties are very similar to those of local UV-bright starburst galaxies (Meurer et al. 1997; González Delgado et al. 1998; Heckman et al. 2005).

Almost 25 years ago, *IRAS* brought about a revolution in our understanding of the FIR emission of galaxies. A new class of galaxies was discovered that radiate a vast amount of their energy in the FIR. These galaxies, named luminous and ultraluminous far infrared galaxies (LIRGs and ULIRGs), have bolometric luminosities greater than $10^{11}$ and $10^{12}\,L_\odot$, respectively. They are very rare objects in the local Universe, and they contribute only a few percent of the bolometric density in the local Universe. However, more recent observations obtained with *ISO* show that bright infrared starbursts (LIRGs and ULIRGs) were more numerous in the past, and are the dominant contributor to the CIRB emission at $z \sim 1$ (Elbaz & Cesarsky 2003). Deep FIR surveys with *Spitzer* have demonstrated that the contribution of the LIRGs and ULIRGs to the total SFR density increases steadily with redshift (Le Floc'h et al. 2005; Pérez-González et al. 2005), with LIRGs and ULIRGs being the most significant contributors to the infrared galaxy population at $z \geq 1$. These galaxies appear to be high-$z$ analogs to the LIRGs and ULIRGs in the local Universe.

These similarities between high redshift and local starbursts highlight the cosmological relevance of starburst galaxies. However, the importance of studying starburst galaxies goes beyond this. Nearby starbursts are laboratories in which to study violent star formation processes and their interaction with the interstellar and intergalactic medium.

### 1.2. Starburst Galaxies: Definition and General Properties

Local starbursts are mainly selected on the basis of their nebular optical emission lines, strong UV continuum, and/or far-infrared radiation. These criteria imply that the starburst class is formed by a mixed type of star-forming systems that includes galaxies with a starburst event and real starburst galaxies (e.g., H II regions such as 30 Dor, blue compact dwarf galaxies, nuclear starbursts, and LIRGs and ULIRGs).

Recently, Heckman (2005) has proposed that the most useful parameter for defining a starburst is its intensity (SFR/area). This definition allows us to compare local starbursts with physically similar galaxies at high $z$. So, starburst galaxies must have an SFR per unit area that is larger than that in the disks of normal galaxies. Nearby starbursts have star formation intensities ranging from 1 to $100\,M_\odot\,\mathrm{yr}^{-1}\,\mathrm{kpc}^{-2}$, and similar values are found for LBGs (Meurer et al. 1997). Recently, observations with *GALEX* have provided evidence of the existence of true local starbursts ($0.1 \leq z \leq 0.3$), which are the analogs to LBGs in terms of their size and UV luminosity. They are compact ultraviolet luminous ($I_{\mathrm{FUV}} \geq 10^8\,L_\odot\,\mathrm{kpc}^{-2}$), low mass galaxies ($M_{\mathrm{star}} \sim 10^{10}\,M_\odot$) with half-light radii less than a few kpc. They have large enough SFRs to build a present-day galaxy in ∼1–2 Gyr. They also have metallicity, velocity dispersion, and extinction similar to the LBG population (Heckman et al. 2005).

## 2. UV-Selected Starbursts

Observations of starbursts at UV wavelengths are very relevant because they allow us a direct detection of massive stars and thus a direct measurement of the star formation rate. This spectral range is also quite sensitive to the star formation history of galaxies and the evolution of stellar populations up to intermediate ages (~1 Gyr). Dust, however, has a profound effect on the UV emission in starbursts. *GALEX* observations of a sample of UV-selected local galaxies have shown that only 33% of the UV emission escapes from the galaxies, the remaining 66% being absorbed by dust and re-radiated in the FIR (Buat et al. 2005). But, due to the "picket-fence" interstellar dust distribution, starbursts can be bright at UV wavelengths, as they are in the far infrared. Two good examples of nearby starbursts that are strong UV and FIR galaxies (LIRGs) are IRAS 0833+6517 and NGC 6090 (see Fig. 1).

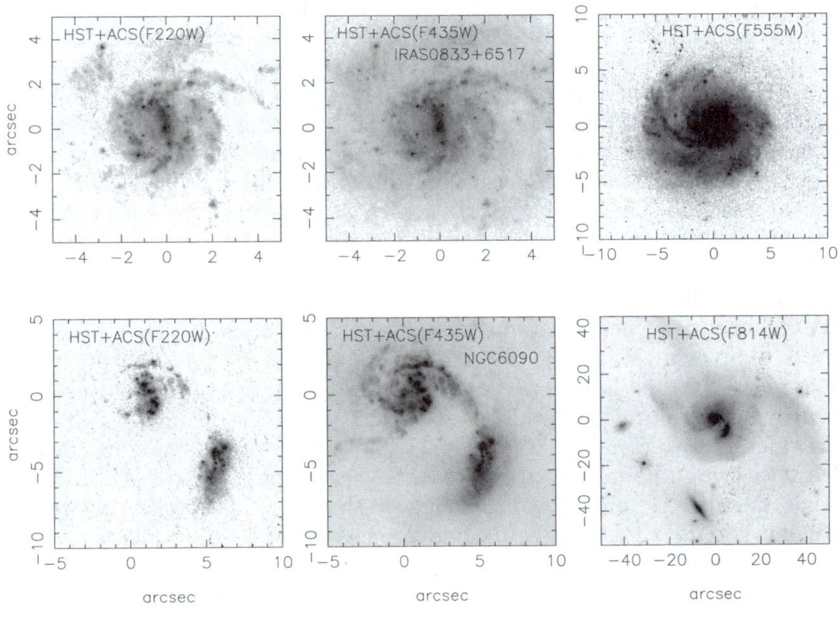


Figure 1. *HST*+ACS images taken through F220W (*left*), F435W (*center*), and F555W or F814W (*right*) filters for two bright UV-selected starbursts. Stellar cluster knots and diffuse emission are clearly detected in the UV and *B* bands.

In this section the UV continuum and Ly$\alpha$ properties of nearby starbursts are discussed. A more extended review of starbursts at UV wavelengths is in González Delgado (2006).

### 2.1. UV Continuum Properties

At UV wavelengths, many starbursts show a continuum filled with absorption lines including resonant transitions of many ionic species. These features have three different origins: stellar winds, stellar photospheres, and the interstellar medium. In contrast, at optical wavelengths the spectrum can be dominated by nebular emission lines; although photospheric lines, such us high order Balmer lines, mainly produced by B- and A-stars, can be detected also in the optical.

Massive stars develop strong stellar wind lines due to the radiation pressure in UV resonance lines. As a result, all the strong UV lines show a blueshifted absorption (about 2000–3000 $\mathrm{km\,s^{-1}}$) or a P Cygni profile. The shape of the profile reflects the stellar mass-loss rate, which is related to the stellar luminosity, and thus to the stellar mass content of the starburst. The most relevant wind lines in starbursts are: N V$\lambda$1240, Si IV$\lambda$1400, C IV$\lambda$1550, and He II$\lambda$1640. Photospheric lines form in layers that are in hydrostatic equilibrium. They originate from excited levels, mainly from C, N, O, S, Si, and Fe ions of low and high excitation potential. They are weaker than the wind lines and depend on the stellar population age and metallicity. Some of the most relevant of these lines in starbursts are: S V$\lambda$1502, C III$\lambda$1426–1428, and C III$\lambda$1176. Wind and photospheric lines are very useful diagnostics to constrain the stellar population properties (initial mass function [IMF] and age) in starbursts. *HST* observations have revealed that a Salpeter IMF with high mass stars constrains well the UV properties of starbursts, at low and high metallicities.

Most of the low ionization resonant lines are of interstellar origin. These lines are very useful to study the interaction of the massive stars with the interstellar medium, and they often reflect the hydrodynamic impact of the starbursts on the interstellar medium. Large scale outflows produced by the starburst have left their imprints on the UV interstellar lines. Outflows of a few hundred $\mathrm{km\,s^{-1}}$ are ubiquitous in nearby starbursts and in LBGs (e.g., González Delgado et al. 1998; Shapley et al. 2003).

*HST* is the first telescope to provide UV high spatial resolution images of nearby starbursts. These images show that the UV continuum in starbursts is provided by two components, compact knots and diffuse emission. The compact knots provide about 25% of the total UV emission (Meurer et al. 1995). They have magnitudes and sizes that suggest that they are stellar clusters. These clusters are distributed irregularly over the diffuse background that extends for a few 100 pc (Fig. 1). Several origins have been proposed for the UV diffuse emission: a) continuous star formation lasting for a few 100 Myr (originally, the stars form in clusters over the last few 100 Myr, but clusters dissolve with age and disperse across the field); b) the UV radiation originates in dusty compact stellar clusters, but it is scattered by dust into the field; and c) individual massive stars, unresolved even at the *HST* spatial resolution. Long slit UV spectra of starbursts taken with STIS point to the star formation origin, so that the UV diffuse radiation is created via dissipation of aging star clusters (Tremonti et al. 2001; Chandar et al. 2005). These results are very relevant because they indicate that: a) stellar clusters are the most important mode of star formation in starbursts; b) intermediate-age stars (B and A type), that result from older dissipated clusters, contribute at the optical continuum (as indicated by the detection of high order Balmer lines in absorption, see Fig. 4) in many of these nearby starbursts; and c) the UV continuum is a good SFR indicator in these objects, because the UV emission is not radiation-scattered by the clusters.

### 2.2. The Ly$\alpha$ Line: Outflows of Neutral Hydrogen Gas

The massive stars in the starbursts provide enough ionizing photons to produce the Ly$\alpha$ line in emission by recombination of photoionized hydrogen. But Ly$\alpha$ photons are attenuated by dust as a result of multiple resonant scatterings by hydrogen atoms that increase the path length of the Ly$\alpha$ photons and thus the probability that they will be absorbed by dust. Extinction also affects the line strength. However, in most of the nearby starbursts, Ly$\alpha$ is detected with a broad damped absorption or a P Cygni profile. These results suggest that dust and/or metallicity are not the key parameters for determining the strength and shape of Ly$\alpha$. UV *HST* spectra have shown that the line is asymmetric, with the peak of the emission redshifted with respect to the systemic velocity, and a deep absorption which is blueshifted by several $100\,\mathrm{km\,s^{-1}}$ with respect to the emission (González Delgado et al. 1998; Mas-Hesse et al. 2003). This absorption shift is in agreement with the blueshift observed in the interstellar lines in these galaxies. The natural explanation is that the neutral, H I, gas producing the absorption is outflowing from the starburst. Hydrodynamic calculations have proved that the properties of the Ly$\alpha$ line are linked to the evolutionary state of the starburst (Tenorio-Tagle et al. 1998).

The Ly$\alpha$ properties of LBGs are very similar to those of nearby starbursts. Shapley et al. (2003) divided a sample of $\sim$1000 LBGs into four classes, according to the equivalent width of the line, EW(Ly$\alpha$). They find that LBGs with EW(Ly$\alpha$) $\geq 20$ Å have bluer UV continuum, weaker low ionization interstellar absorption lines, smaller kinematic offset between Ly$\alpha$ and the interstellar lines, and lower SFRs than LBGs as a whole. They suggest that the appearance of LBG spectra is determined by a combination of the covering fraction of outflowing neutral gas, which contains dust, and the range of velocities over which this gas is absorbing.

In summary, like the interstellar lines, the Ly$\alpha$ line in nearby starbursts and LBGs is driven by the dynamical effects of the violent star formation processes at work in the starburst, rather than by the gravitational potential of the galaxy.

## 3. FIR-Selected Starbursts

### 3.1. Nuclear Activity, Starbursts, and Mergers

One key point of research in LIRGs and ULIRGs since they were discovered is the origin of the IR emission. Optical spectra of these galaxies show emission line ratios that indicate that they host a starburst and/or an AGN. Moreover, it has been found that the fraction of galaxies with signs of Seyfert activity increases significantly with the FIR luminosity. So, LIRGs are mainly starbursts, but about half of the ULIRGs have a Seyfert nucleus (e.g., Veilleux, Kim, & Sanders 1999). However, mid-infrared spectroscopic studies indicate that most of the cool ULIRGs ($f_{25}/f_{60} \leq 0.2$) are powered by a starburst, even if they host an AGN (e.g., Farrah et al. 2007). The origin of the power source in warm ULIRGs (most of them are QSOs) is still controversial.

Another important research issue in LIRGs and ULIRGs is related with the role that these objects play in the formation and evolution of galaxies. In the context of hierarchical galaxy formation and evolution models, mergers can trigger the star formation and the nuclear activity in galaxies, and lead to the formation of elliptical galaxies. So, as galaxies merge to form larger units, gas is driven into the nuclear region triggering a starburst and growing a supermassive black hole (BH). In this scenario, the creation and evolution of a BH (presumably associated with a QSO) is intimately connected to that of the galaxy bulge. Pieces of evidence in favor of this scenario are the tight correlation between the BH mass and the bulge stellar velocity dispersion (Ferrarese & Merritt 2000) and the similarities between the evolution of the QSO luminosity density and the SFR (Boyle & Terlevich 1998).

In the local Universe, there is significant evidence of the link between ULIRGs and galaxy interactions and mergers. Veilleux , Kim, & Sanders (2002) have found that ULIRGs are mergers of gas-rich disk galaxies in the later stages. Moreover, Borne et al. (2000) found that ULIRGs can be at different merger stages since many of them are in dense galaxy groups. LIRG morphology, however, is more diverse (Arribas et al. 2004). Recent *HST* observations with the ACS show that many LIRGs, like ULIRGs, are mergers in the later stage (e.g., NGC 6240), but some of them show very little signs of interactions (e.g., NGC 6436). Additionally, LIRG morphologies encompass all the expected merging stages (Fig. 2). At intermediate redshift, half of the LIRGs show a spiral disk morphology and only one third of these galaxies are mergers (Elbaz et al. 2005; Hammer et al. 2005).

### 3.2. Evolutionary Link between ULIRGs, Radio Galaxies, QSOs, and Elliptical Galaxies

Theoretical models predict that major disk–disk mergers can form ellipticals (e.g., Toomre & Toomre 1972). Evidence in favor of this scenario comes from the H I and optical morphology and kinematics, and stellar population properties observed in some proto-ellipticals galaxies (e.g., Schweizer 1996). The optical spectra show stellar lines typical of post-starburst galaxies. These features indicate that a major starburst occurred about 1 Gyr ago.

Kinematic observations of ULIRG mergers have proved that they are ellipticals in formation. Measurements of their stellar velocity dispersion, surface brightness, and effective radius indicate that ULIRGs closely follow the fundamental plane, and that they can form intermediate-mass ($10^{11}\,M_\odot$) elliptical galaxies (Genzel et al. 2001). In this scenario, mergers can drive an evolutionary sequence from cool ULIRGs to warm ULIRGs on their way to becoming quasars. Thus, mergers go first through a starburst phase, followed by a dust enshrouded Seyfert phase, then the dust is cleaned by the actions of AGN and the starburst wind, and finally ULIRGs are shown as QSOs (Sanders et al. 1988).

There is also evidence that radio galaxies can be part of the evolutionary sequence between ULIRGs and elliptical galaxies. Even though most of the radio galaxies are of elliptical or S0 type, optical imaging of radio galaxies shows double nuclei, tidal tails, arcs, and distorted isophotes in about half of the powerful FRII radio galaxies (Heckman et al. 1986). Kinematic evidence also supports this hypothesis (e.g., Tadhunter, Fosbury, & Quinn 1989).

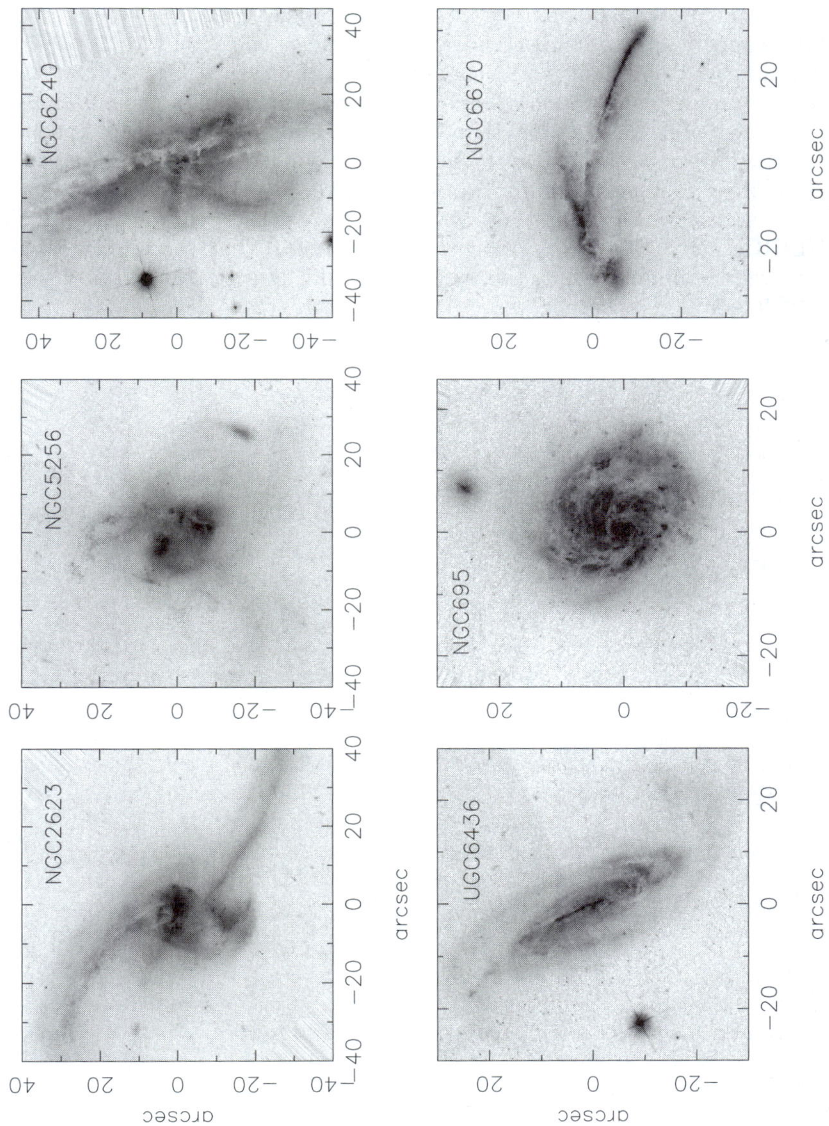


Figure 2. F435W *HST*+ACS images of LIRGs. Examples of LIRGs that are disk, interacting, or merger systems are shown.

Brotherton el al. (1999) have identified a class of QSOs, named post-starburst QSOs, the most spectacular of which is UN J1025-0040 ($z = 0.634$). Its representative spectrum is a chimera because it displays the broad Mg II$\lambda$2800 emission line and a strong continuum characteristic of classical QSOs. But the optical continuum displays a large Balmer break, prominent high order Balmer absorption lines, and Ca II K and H lines. These stellar features and the bolometric luminosity are consistent with the object having been a ULIRG a few 100 Myr ago. More cases with similar optical spectra have been identified in the SDSS data releases. These are key objects to prove the evolutionary link between ULIRGs and QSOs. *HST* images (Fig. 3) reveal that the QSO host resembles a merger remnant. Thus, these objects could be in a transition phase between ULIRGs and classical QSOs.

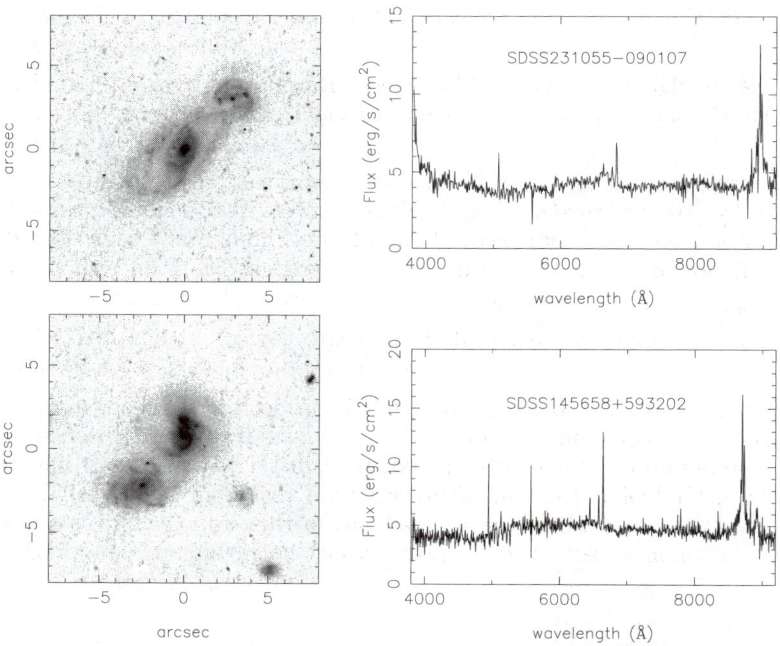


Figure 3. *HST*+ACS images and SDSS spectra of two post-starburst QSOs.

### 3.3. Stellar Populations in LIRGs, ULIRGs, and Radio Galaxies

To find an evolutionary link between these objects we are performing an optical spectroscopic program to investigate the evolution of the stellar populations in several samples of LIRGs, ULIRGs, radio galaxies, and post-starburst QSOs. We use the age of the stellar populations as a clock to test a possible evolutionary sequence among them. This technique has been used previously to find the possible evolution from Seyferts to low luminosity AGN (e.g., González Delgado et al. 2001, 2004; Cid Fernandes et al. 2005a) and from ULIRG–QSOs to classical QSOs (Canalizo & Stockton 2001).

The optical spectral range is rich in photospheric stellar lines that are very useful in constraining the young and intermediate age population properties. In particular, high order Balmer lines and the Balmer break are very useful diagnostics to date starbursts and post-starbursts (González Delgado, Leitherer, & Heckman 1999). However, the fit of the whole optical range (including continuum shape and stellar lines) instead of only the equivalent widths of a few stellar lines is the best method to constrain the stellar population properties of galaxies. This has been clearly demonstrated recently using SDSS data for several hundred thousand local early-type and star-forming galaxies by Cid Fernandes et al. (2005b). This method applies well to the objects discussed here because the extinction can drastically modify their continuum shape; and, moreover, the strength of some of the stellar lines is affected by nebular contribution.

Evolutionary synthesis models have proved to be a very useful tool to constrain the stellar population properties in galaxies. Models that incorporate intermediate (e.g., Bruzual & Charlot 2003) or high spectral resolution stellar libraries (e.g., González Delgado et al. 2005) are really required to be able to isolate the young, intermediate and old population contributions, and to distinguish between age and extinction effects.

Another key issue to determine the stellar population properties is the technique used to find the best model that fits the whole observed spectrum. There is a staggering variety of techniques and codes developed for this purpose. Here we use STARLIGHT (Cid Fernandes et al. 2005b), which has proved to give very reasonable results (Cid Fernandes et al. 2005a). The STARLIGHT technique consists in fitting the observed spectrum with a combination of several simple stellar populations (SSPs) from evolutionary synthesis models. Extinction is modeled by foreground dust, and parameterized by the $V$-band extinction, $A_V$. The output is the population vector that represents the fractional contribution of each SSP of a given age and metallicity to the model flux at a given wavelength. The fit is carried out for the minimum chi-square between the observations and the combined models.

Here, in this work, we have adopted for each metallicity a base of 14 SSPs with ages 4, 5, 10, 25, 40, 100, 280, 500, and 900 Myr, and 1.26 and 14 Gyr from the models of González Delgado et al. (2005). The Calzetti et al. (2000) dust law was adopted. We also include several additional stellar components that correspond to SSPs of ages $\leq$ 10 Myr and which are reddened by one, two, or three magnitudes. The purpose of this is to allow the code to fit young SSPs with larger extinction than the intermediate and old SSPs.

The results obtained for LIRGs and ULIRGs are summarized as follows: a) the old stellar populations (age $\geq$ few Gyr) contribute very little to the total optical continuum; b) very young ($\leq$ 10 Myr) plus intermediate (100–900 Myr) age stellar population can account for almost all the optical light; and c) extinction is significant for these objects, and the young stellar population is redder than the intermediate age one. As expected, the extinction is high. This is noted because: a) the continuum is very red even though the stellar features, such as high order Balmer lines are typical of young and/or intermediate-age populations; b) the Na I line is very deep, indicating that its origin is mainly interstellar. It means that the source is partially covered by large column densities of interstellar gas, and in consequence, the extinction is high. Results obtained for several LIRGs and the ULIRG Arp 220 are presented in Figure 4.

We have found, as in LIRGs and ULIRGs, strong evidence for intermediate (0.3–2.5 Gyr) and young ($\leq$100 Myr) stellar populations in the nuclear regions of radio galaxies (Tadhunter et al. 2005; Holt et al. 2007). These stellar populations are reddened but significantly less so than those in LIRGs and ULIRGs. They are very massive and could represent starbursts that were able to heat the dust and produce a fraction of the FIR luminosity detected with *Spitzer* for these objects (Tadhunter et al. 2007).

We have now taken 3D optical data with VIMOS at the VLT and PMAS at the CAHA observatory to determine the host stellar population properties of post-starburst QSOs. Age is the key parameter for determining whether an evolutionary link exists between ULIRGs and post-starburst QSOs. This work is in progress and the results will be reported soon.

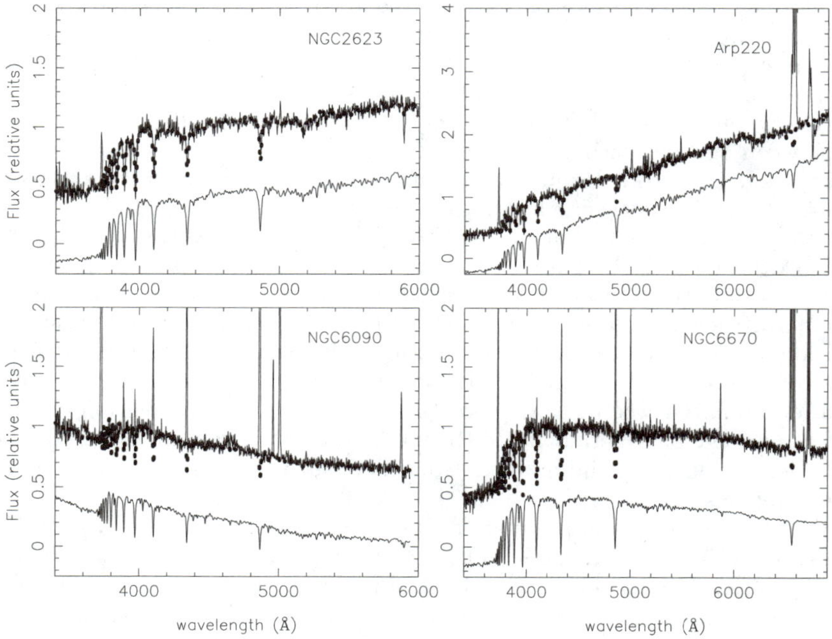


Figure 4. Optical spectra of the LIRGs NGC 6090, NGC 6670, and NGC 2623, and the archetypal ULIRG Arp 220. The best models found with STARLIGHT and the high spectral resolution evolutionary synthesis models from González Delgado et al. (2005) are plotted as dotted lines.

## 4. Summary and Future Prospects

This review provides evidence of the role that starbursts play in the star formation history of the Universe and the origin and evolution of galaxies. During this decade, many high redshift galaxies that are in a starbursting phase have been discovered. This highlights the cosmological relevance of starbursts. But it is still necessary to observe nearby starbursts because they are the preferred laboratories in which to study violent star formation processes and their interaction with the interstellar medium. Near and far starbursts can be triggered by mergers and interactions between galaxies; LIRGs and ULIRGs are good examples of these processes. The stellar population properties of LIRGs and ULIRGs indicate that they could be elliptical galaxies in formation.

The future prospects are bright. The Spanish community can contribute significantly to this field thanks to the imminent coming into operation of the GTC (first light was in July 2007) and the extensive cosmological survey ALHAMBRA (Moles et al., this volume, p. 495) and observations that we are carrying out at the 3.5 m CAHA telescope. ALHAMBRA will provide accurate redshifts for $6.6 \times 10^5$ galaxies with $I \leq 25$ and mean redshift of 0.74. Through the use of 20 contiguous medium band optical filters, many emission-line galaxies will be detected whose spectral energy distribution will be used to trace the star formation history of the Universe from local to intermediate redshift.

OSIRIS (Cepa et al. 2005) at the GTC will be a crucial instrument to probe the feedback processes taking place in starbursts and AGN. Deep imaging taken with OSIRIS tuned at the Ly$\alpha$ and H$\alpha$ emission lines, and optical lines (such as [O III], [N II], and [S II]) of starbursts (ULIRGs–LIRGs) and QSOs will be crucial to determining whether there is an evolutionary link between star formation and nuclear activity.

*Acknowledgments* I am very grateful to John and Johan for inviting me to this lovely conference. I would like to thank my collaborators Roberto Cid Fernandes, Enrique Pérez, Javier Rodríguez, Clive Tadhunter, and Joanna Holt. This work is supported by Spanish grant AYA2004-02703.

## References

Arribas, S., Bushouse, H., Lucas, R. A., Colina, L., & Borne, K. D. 2004, AJ, 127, 2522
Borne, K. D., Bushouse, H., Lucas, R. A., & Colina, L. 2000, ApJ, 529, 77
Boyle, B. J., & Terlevich, R. J. 1998, MNRAS, 293, L49
Brotherton, M. S., van Breugel, W., Stanford, S. A., Smith, R. J., Boyle, B. J., Miller, L., Shanks, T., Croom, S. M., & Filippenko, A. V. 1999, ApJ, 520, L87
Bruzual, G., & Charlot, S. 2003, MNRAS, 344, 1000
Buat, V., et al. 2005, ApJ, 619, L63
Calzetti, D., Armus, L., Bohlin, R. C., Kinney, A. L., Koornneef, J., & Storchi-Bergmann, T. 2000, ApJ, 533, 682
Canalizo, G., & Stockton, A. 2001, ApJ, 555, 719
Cepa, J., et al. 2005, RMxAC, 24, 1
Chandar, R., Leitherer, C., Tremonti, C. A., Calzetti, D., Aloisi, A., Meurer, G. R., & de Mello, D. 2005, ApJ, 628, 210
Cid Fernandes, R., González Delgado, R. M., Storchi-Bergmann, T., Martins, L. Pires, & Schmitt, H. 2005a, MNRAS, 356, 270
Cid Fernandes, R., Mateus, A., Sodré, L., Stasińska, G., & Gomes, J. M. 2005b, MNRAS, 358, 363
Elbaz, D., & Cesarsky, C. J. 2003, Science, 300, 270
Elbaz, D., Le Floc'h, E., Dole, H., & Marcillac, D. 2005, A&A, 434, L1
Farrah, D., et al. 2007, ApJ, 667, 149
Ferrarese, L., & Merritt, D. 2000, ApJ, 539, L9
Genzel, R., Tacconi, L. J., Rigopoulou, D., Lutz, D., & Tecza, M. 2001, ApJ, 563, 527
González Delgado, R. M. 2006, Ap&SS, 303, 58

González Delgado, R. M., Leitherer, C., Heckman, T., Lowenthal, J. D., Ferguson, H. C., & Robert, C. 1998, ApJ, 459, 698
González Delgado, R. M., Leitherer, C., & Heckman, T. 1999, ApJS, 125, 489
González Delgado, R. M., Heckman, T., & Leitherer, C. 2001, ApJ, 546, 845
González Delgado, R. M., Cid Fernandes, R., Pérez, E., Martins, L. P., Storchi-Bergmann, T., Schmitt, H., Heckman, T., & Leitherer, C. 2004, ApJ, 605, 127
González Delgado, R. M., Cerviño, M., Martins, L. P., Leitherer, C., & Hauschildt, P. H. 2005, MNRAS, 357, 945
Hammer, F., Flores, H., Elbaz, D., Zheng, X. Z., Liang, Y. C., & Cesarsky, C. 2005, A&A, 430, 115
Heckman, T. M., Smith, E. P., Baum, S. A., van Breugel, W. J. M., Miley, G. K., Illingworth, G. D., Bothun, G. D., & Balick, B. 1986, ApJ, 311, 526
Heckman, T. M. 2005, in ASSL 329, Starbursts: From 30 Doradus to Lyman Break Galaxies, ed. R. de Grijs & R. M. González Delgado (New York: Springer), p. 3
Heckman, T., et al. 2005, ApJ, 619, L35
Holt, J., Tadhunter, C. N., González Delgado, R. M., Inskip, K. J., Rodriguez, J., Emonts, B. H. C., Morganti, R., & Wills, K. A. 2007, MNRAS, 381, 611
Le Floc'h, E., et al. 2005, ApJ, 632, 169
Madau, P., Ferguson, H. C., Dickinson, M. E., Giavalisco, M., Steidel, C. C., & Fruchter, A. 1996, MNRAS, 283, 1388
Mas-Hesse, J. M., Kunth, D., Tenorio-Tagle, G., Leitherer, C., Terlevich, R. J., & Terlevich, E. 2003, AJ, 598, 858
Meurer, G. R., Heckman, T. M., Leitherer, C., Kinney, A., Robert, C., & Garnett, D. R. 1995, AJ, 110, 2665
Meurer, G. R., Heckman, T. M., Lehnert, M. D., Leitherer, C., & Lowenthal, J. 1997, AJ, 114, 54
Noll, S., et al. 2004, A&A, 418, 885
Pérez-González, P. G., et al. 2005, ApJ, 630, 82
Pettini, M., Steidel, C. C., Adelberger, K. L., Dickinson, M., & Giavalisco, M. 2000, AJ, 528, 96
Sanders, D. B., Soifer, B. T., Elias, J. H., Madore, B. F., Matthews, K., Neugebauer, G., & Scoville, N. Z. 1988, ApJ, 325, 74
Schweizer, F. 1996, AJ, 111, 109
Shapley, A. E., Steidel, C. C., Pettini, M., & Adelberger, K. L. 2003, ApJ, 588, 65
Steidel, C. C., Giavalisco, M, Pettini, M., Dickinson, M., & Adelberger, K. L. 1996, ApJ, 462, L17
Steidel, C. C., Shapley, A. E., Pettini, M., Adelberger, K. L., Erb, D. K., Reddy, N. A., & Hunt, M. P. 2004, ApJ, 604, 534
Tadhunter, C. N., Fosbury, R. A. E., & Quinn, P. J. 1989, MNRAS, 240, 225
Tadhunter, C., Robinson, T. G., González Delgado, R. M., Wills, K., & Morganti, R. 2005, MNRAS, 356, 480
Tadhunter, C. N., et al. 2007, ApJ, 661, L13
Tenorio-Tagle, G., Silich, S. A., Kunth, D., Terlevich, E, & Terlevich, R. J. 1998, MNRAS, 309, 332
Toomre, A., & Toomre, J. 1972, ApJ, 178, 623
Tremonti, C. .A., Calzetti, D., Leitherer, C., & Heckman, T. M. 2001, AJ, 555, 322
Veilleux, S., Kim, D.-C., & Sanders, D. B. 1999, ApJ, 522, 113
Veilleux, S., Kim, D.-C., & Sanders, D. B. 2002, ApJS, 143, 315
Williams, R. E., et al. 1996, ApJ, 112, 1335

## Discussion

*Hammer*: You mentioned two important results about LIRGs and their evolution. First, they represent a significant fraction of galaxies at $z \sim 1$ (I may quote Flores et al. 1999 or Hammer et al. 2005) and their morphologies are not always related to major mergers (see Zheng et al. 2005) as in the local Universe. How could one explain these trends, by gas?

*González-Delgado*: Another possibility is that minor mergers also play an important role at high $z$. In this case, the galaxies can show unperturbed spiral morphology, but the kinematics can yield information about this scenario.

*Silk*: You showed evidence that a young SED component occurred in radio galaxies that could be identified as the sequence LIRG–ULIRG–radio galaxy–elliptical. Another prediction would be that these radio galaxies should have compact morphologies due to interaction with the dense ISM.

*González-Delgado*: Some of these radio galaxies show compact morphology in the radio. Young stellar populations have been detected in the UV. This emission cannot be light scattered by the AGN, but the young stars can be triggered by the interaction of the radio jet with the ISM.

Juan Vladilo (*left*) and Paolo Molaro.

*Pathways through an Eclectic Universe*
*ASP Conference Series, Vol. 390, © 2008*
*J. H. Knapen, T. J. Mahoney, and A. Vazdekis, eds.*

# Recent Star Formation in Nearby Early-type Galaxies

Marc Sarzi,[1] Roland Bacon,[2] Michele Cappellari,[3] Roger L. Davies,[3] P. Tim de Zeeuw,[4] Eric Emsellem,[2] Jesús Falcón-Barroso,[5] Davor Krajnović,[3] Harald Kuntschner,[6] Richard M. McDermid,[4] Reynier F. Peletier,[7] and Glenn van de Ven[8]

[1]*Centre for Astrophysics Research, University of Hertfordshire, Hatfield, AL10 9AB, UK*

[2]*Université de Lyon 1, CRAL, Observatoire de Lyon, 9 av. Charles André, F-69230 Saint-Genis Laval, France*

[3]*Sub-Department of Astrophysics, University of Oxford, Denys Wilkinson Building, Keble Road, Oxford OX1 3RH, UK*

[4]*Sterrewacht Leiden, Universiteit Leiden, Postbus 9513, NL-2300 RA, Leiden, The Netherlands*

[5]*European Space and Technology Centre, Keplerlaan 1, NL-2200 AG Noordwijk, The Netherlands*

[6]*Space Telescope European Coordinating Facility, European Southern Observatory, Karl-Schwarzschild-Str. 2, D-85748 Garching, Germany*

[7]*Kapteyn Astronomical Institute, University of Groningen, NL-9700 AV Groningen, The Netherlands*

[8]*Institute for Advanced Study, Einstein Drive, Princeton, NJ 08540, USA*

**Abstract.** Motivated by recent progress in the study of early-type galaxies owing to technological advances, the launch of new space telescopes, and large ground-based surveys, we attempt a short review of our current understanding of the recent star formation activity in such intriguing galactic systems.

## 1. Introduction

Early-type galaxies represent one of the biggest challenges to our understanding of galaxy formation and evolution, in particular as far as their star formation history is concerned. Studies of the nearby galaxy population can provide useful clues to solving this problem, by setting tight constraints on the level of their most recent star formation activity. Traditionally, early-type galaxies used to be regarded as simple and old stellar systems mostly devoid of gas or dust. By the turn of the century, however, evidence was already mounting that a substantial fraction of elliptical and lenticular galaxies do show conspicuous gas and dust reservoirs (e.g., Goudfrooij et al. 1994; Macchetto et al. 1996, and references therein), and the presence of younger stellar components (e.g., Trager et al. 2000, and references therein). In recent years, tremendous progress has been made in uncovering the recent star formation history of early-type galaxies, owing in particular to the advent of new technologies such as integral field spectroscopic units or more sensitive receivers at radio and mm frequencies, to the launch of space observatories such as *GALEX* and *Spitzer* allowing us to probe wavelength domains otherwise inaccessible from the ground, and to large spectroscopic campaigns such as the Sloan Digital Sky Survey (SDSS). In this paper we attempt to review such recent efforts, without any pretence of completeness. We will first focus on detailed studies of the stellar populations of early-type galaxies at different wavelengths, in particular with the SAURON integral-field unit in the optical, and with *GALEX* and *Spitzer* in the UV and mid-IR (MIR) domain (Sect. 2). We will then turn to the reservoirs of neutral, molecular, and ionized gas in nearby elliptical and lenticular galaxies, and consider evidence for ongoing star formation in them (Sect. 3). Finally, we will consider some results based on the analysis of a large sample of early-type galaxies with SDSS data (Sect. 4) and conclude by discussing how all these findings could connect with each other (Sect. 5).

## 2. Stellar Population Studies and Evidence of Recent Star Formation

### 2.1. The Optical—SAURON Results

Many studies have shown that early-type galaxies are not exclusively formed by old stars (e.g., Thomas et al. 2005, and references therein). The SAURON survey of early-type galaxies (de Zeeuw et al. 2002) has further allowed us to find the location of different stellar sub-populations, in particular of young stars, and to link such structures to the photometry and kinematics of early-type galaxies. Using maps for the strength of the H$\beta$, Mg$b$ and Fe 5015 absorption lines in a representative sample of 48 early-type galaxies in H. Kuntschner et al. (in preparation), we reinforce previous indications that the integrated stellar populations of most elliptical galaxies display old stellar ages whereas lenticulars exhibit a wider spread of ages, and likewise for galaxies in clusters compared to objects in a lower density environment. Very young stars, less than $\sim 2$ Gyr old, are found to dominate the entire SAURON field in four objects in the SAURON sample, whereas in the remaining galaxies the young populations appear to be either distributed in extended disks or to cluster towards the center. In fact, such young nuclear populations are also likely to have formed in small stellar disks, since the presence of young stars in the center correlates very well with that of small (less than $\sim$500 pc) stellar components that are kinematically distinct from the rest of the galaxy (McDermid et al. 2006).

The connection between stellar ages and kinematics extends also beyond the central regions. In Emsellem et al. (2007) we assess the overall level of rotational support in a galaxy adopting a quantity, $\lambda_{\rm R}$, that is closely related to the specific angular momentum of a galaxy. In the SAURON sample, galaxies with $\lambda_{\rm R} \leq 0.1$ form a distinct class characterized by little or no global rotation and the presence of kpc-scale kinematically decoupled cores. Opposed to such slowly rotating objects are galaxies that either display faster global rotation or that are consistent with being systems supported by rotation that are viewed at small inclinations (see also the contribution by Falcón-Barroso, this volume, p. 227, for an illustration of these two kinds of objects). All SAURON galaxies that formed stars recently (which amount to 25% of the sample and include either objects with luminosity-weighted mean ages below 6.5 Gyr or with young disks and nuclei), are fast rotators, whereas slow rotators do not show evidence of secondary star formation events. In fact, beside the presence of young disks and nuclei, fast rotators can also display extended, Mg*b*-enhanced structures that appear to be flatter than the stellar isophotes (Kuntschner et al. 2006). Although not obviously younger than the rest of the galaxy, such stellar sub-components most likely originated in a secondary star formation event, consistent with the finding that fast rotators host dynamically distinct stellar components characterized by large angular momentum and radial anisotropy, as observed in stellar disks (Cappellari et al. 2007).

## 2.2. The UV—*GALEX* Results

Although the strength of optical absorption lines such as those probed by SAURON can provide sensible constraints on the presence of younger stars when compared to the predictions of *single* stellar population models, deriving the mass and age of young stellar components from line strength indices alone is an exercise that is fraught with degeneracies. In this respect probing the UV light allows us to weigh and date even a relatively small fraction of very young stars, since the UV output of a stellar population varies dramatically over the first few hundred Myr of its history. In particular, whereas the far-UV (between 1350 and 1750 Å) light is also sensitive to the radiation from old hot helium-burning horizontal branch stars, the near-UV radiation (NUV, between 1750 and 2750 Å) traces particularly well the presence of young stars.

The potential of such UV information to constrain the age and fraction of young stars of early-type galaxies has recently been shown by Jeong et al. (2007), who combined UV observations from *GALEX* with ground-based optical imaging for the E4 galaxy NGC 2974. They find a blue NUV–optical color indicative of the presence of very young stars, less than 500 Myr old, both near the center and in an outer ring. The star formation rate and mass fraction of young stars peaks in the outer ring, reaching $2 \times 10^{-3}\, M_{\odot}\, \mathrm{yr}^{-1}\, \mathrm{kpc}^{-2}$ and 0.1% respectively, whereas, overall, 1% of the stars in NGC 2974 formed in the last 500 Myr. Using the mass model of Krajnović et al. (2005), Jeong et al. also show that the location of the outer UV ring, as well as those of the central rings traced by the [O III]$\lambda\lambda$4959, 5007 emission in the SAURON data for this galaxy, are consistent with the positions of orbital resonances that would be induced by the presence of a rotating large scale bar, which is driving the observed star formation in these regions. As most of the early-type galaxies in the SAURON sample are being observed with *GALEX*, it will be interesting to see how many more objects will exhibit evidence of recent star formation in UV light, as in the case of NGC 2974.

## 2.3. The Mid-IR—*Spitzer* Results

At the opposite end of the optical spectrum, MIR observations provide an alternative way to investigate the stellar populations of early-type galaxies. The MIR spectra of passively evolving stellar populations are affected by the presence of mass-losing giant stars, in particular AGB stars. The dusty envelopes of these stars re-radiate the heat received from their parent stars in a characteristic broad region between 9 and 12 $\mu$m, mostly due to silicate grains. As a stellar population ages, the photospheric MIR continuum fades less quickly than the circumstellar 10 $\mu$m emission, so that the relative strength of the silicate features can be used to trace the age of the population.

The advent of *Spitzer* has recently allowed to detect routinely the 10 $\mu$m circumstellar emission in early-type galaxies, leading for instance Bressan et al. (2006) to confirm that 82% of their Virgo cluster galaxies have been passively evolving since their formation. Bregman, Temi, & Bregman (2006) find that early-type galaxies generally show old MIR-inferred ages even when an optical analysis based on absorption line strengths indicates relatively young luminosity-weighted mean ages. Although these results seem at odds with the larger fraction of SAURON galaxies with evidence of recent star formation, we need to keep in mind that both these MIR studies concentrated only on the brightest end of the local early-type galaxy population, and that the MIR age estimates are still subject to a severe age–metallicity degeneracy (Bressan et al. 2006).

## 3. Gaseous Reservoirs and Evidence of Ongoing Star Formation

If stellar population studies provide evidence of recent star formation in nearby early-type galaxies, we also ought to find gas reservoirs capable of fueling such activity before and while star formation occurs. Furthermore, the detection of molecular gas and of emission from H II regions should correspond to the presence of the youngest stars.

### 3.1. Neutral Gas

A number of shallow H I surveys have shown already that early-type galaxies can have massive (up to $10^{10}\,M_\odot$) and very extended (up to 200 kpc in size) disks of neutral hydrogen (e.g., Oosterloo et al. 2002, and references therein). Such investigations could only detect the most H I-rich galaxies, however, which prompted Morganti et al. (2006) to set out to explore the complete H I mass distribution of early-type galaxies, starting from the representative SAURON sample. So far, neutral gas has been found in 70% of the 12 *field* galaxies initially observed by Morganti et al. (2006), with masses between a few times $10^6\,M_\odot$ and just over $10^9\,M_\odot$. The neutral material in these objects also seems to be connected to the ionized gas observed by SAURON in the optical regions, since all galaxies where H I is detected also contain ionized gas, whereas no H I is found in galaxies without ionized gas. Additionally, in the most gas-rich systems the neutral and ionized material display similar kinematics. Considering that the presence of ionized gas may not always lead to star formation (see Sect. 3.3), these early results suggest that sufficient material to explain the observed amount of recent star formation could be generally present around early-type galaxies, and that gas accretion does not happen exclusively in peculiar early-type galaxies.

### 3.2. Molecular Gas

In order to form stars, gas ultimately has to cool and condense, leading also to the formation around dust grains of molecules such as carbon monoxide. Similarly to the case of H I, a number of CO emission surveys (e.g., the compilation by Bettoni, Galletta, & García-Burrillo 2003) have revealed that early-type galaxies can show substantial amounts of CO emission, which generally traces regularly rotating molecular disks (e.g., Young 2002, 2005). More recently, Combes, Young & Bureau (2007) have followed up with single-dish observations 43 out of the 48 early-type galaxies in the SAURON sample, detecting CO emission in 28% of them. This fraction is somewhat smaller than previously reported, but this is because Combes et al., contrary to earlier studies, also surveyed bright early-type galaxies that have a relatively lower molecular gas content. All galaxies with CO emission show evidence of recent star formation in their central regions from SAURON or OASIS data, and have radio continuum to far-IR flux ratios that further support the case for ongoing star formation. In fact, the CO-rich SAURON early-type galaxies appear to extend the well-known relation between the star formation rate surface density and the surface density of $H_2$ by two orders of magnitude down in star formation rate.

In addition to mm observations, the MIR domain provides additional ways to trace even tiny amounts of star formation activity, in particular through the detection of emission from polycyclic aromatic hydrocarbon (PAH) features (e.g., Kaneda, Onaka, & Sakon 2005). Although most investigations with *Spitzer* have focussed on the brightest end of the early-type population, the first results are nonetheless promising. For instance, Bressan et al. (2006) find evidence of recent star formation in 12% of their Virgo cluster sample, whereas Panuzzo et al. (2007) succeed in quantifying the epoch (~200 Myr ago) and intensity (adding in average $\sim 1\,M_\odot\,\mathrm{yr}^{-1}$) of the recent star formation event in the central regions of the SB0 galaxy NGC 4435, adding just about 1.5% of the stellar mass in these regions.

### 3.3. Ionized Gas

Optical nebular emission has traditionally been used to trace the warm ($\sim 10^4$ K) component of the interstellar medium and identify star formation in H II regions. Ionized gas emission in early-type galaxies is generally quite weak, however, so that its detection requires a careful subtraction of the stellar continuum.

By introducing a novel technique to address this issue, in Sarzi et al. (2006) we could reach down to a detection limit of just 0.1 Å for the equivalent width (EW) of the emission lines and detect extended ionized gas emission in 75% of the early-type galaxies in the SAURON representative sample, confirming that these systems contain gas more often than not. Although the limited SAURON wavelength range does not allow an extensive line diagnostic analysis, very low [O III]/H$\beta$ line ratios nonetheless suggest ongoing star formation in at least four objects, or 8% of the sample. In fact, that stars are forming in these galaxies is proven also by the finding of disk-like gas kinematics, regular dust morphology, young stellar populations, and molecular gas in all of them. Interestingly, very young stars, CO emission, and even PAH features (K. Shapiro, private communication) can be found in objects where the warm gas displays very high [O III]/H$\beta$ ratios (see, e.g., NGC 3489 and NGC 3156 in Sarzi et al. 2006). This finding suggests that as ionized gas in early-type galaxies is also subject to different sources of excitation than O stars, it may not always be a good tracer of star formation, in particular if this occurs at a low level. On the other hand, all the SAURON early-type galaxies with compelling evidence of ongoing star formation share relaxed gas kinematics, displaying in particular small gas velocity dispersions.

This common kinematic characteristic is found exclusively in fast rotators in the SAURON sample, however, suggesting that not all early-type galaxies are equally likely to form stars. In particular, the spatial correlation with the hot ($\sim 10^7$ K) X-ray emitting phase of the interstellar medium (Sarzi et al. 2007, and references therein) suggests that the fate of ionized gas in the most massive and slowly rotating early-type galaxies is to evaporate into the hot medium rather than to condense and form stars (see also Nipoti & Binney 2007, for a quantitative analysis). Preventing star formation in the most massive objects consistently over time through the interaction with the X-ray gas could explain the low frequency of CO and PAH detection in these systems (Sect. 3.2), and is in agreement with the notion that the most massive galaxies also host the oldest stellar populations.

## 4. Results from Larger Volumes

If the previous efforts show with tremendous detail that a substantial fraction of nearby early-type galaxies is, or has recently been, forming stars, such local studies can only offer a sketchy picture of the recent star formation history of early-type galaxies in the local Universe. Large surveys such as the SDSS, on the other hand, allow us to investigate much larger numbers of early-type galaxies, although generally offering only the integrated photometric and spectroscopic properties of these objects.

Some of the most dramatic results in this subject have been obtained by combining the SDSS optical photometry with NUV images obtained with *GALEX*. Out of 39 bright early-type galaxies from the sample of Bernardi et al. (2003), Yi et al. (2005) found six objects with NUV—$r$-band colors consistent with the formation in the last Gyr of stars amounting to approximately 1–2% of the stellar mass. Kaviraj et al. (2007) considerably extended Yi et al.'s first exercise to analyze a sample of ∼2100 morphologically selected early-type galaxies from the SDSS with $z < 0.11$, finding that at least ∼30% of their objects had formed between 1 and 3% of their mass in the last Gyr, mostly between 300 and 500 Myr ago. In an even larger effort, Schawinski et al. (2007) have compiled a catalog of ∼16 000 morphologically selected early-type galaxies with redshift between $0.05 < z < 0.10$ and SDSS, *GALEX* and 2MASS photometric and spectroscopic data, finding not only similar fractions of galaxies with recent star formation, but also that ∼4% of their early-type galaxies are currently forming stars based on their emission line ratios.

The recent work of Graves et al. (2007) further complements these studies. Yi et al. (2005) and Kaviraj et al. (2007) concentrated on red-sequence objects without strong emission to avoid AGN contamination in the UV light, whereas Schawinski et al. (2007) include galaxies with a broader range of optical colors and with gas emission but conservatively deemed as quiescent objects without reliable H$\alpha$, H$\beta$, [N II]$\lambda\lambda$6548, 6583, and [O III] emission. Graves et al. (2007) investigate more closely the red-sequence early-type galaxy population with gas emission, by including objects with relatively fainter gas emission as long as they display $H\alpha$ and [O II]$\lambda\lambda$3726, 3729 lines that indicate LINER-like emission,[1] or alternatively, which firmly exclude ongoing star formation. With such a sensitive distinction between galaxies with or without ionized gas Graves et al. (2007) find that red-sequence galaxies with LINER-like emission show integrated stellar populations that are invariably younger than their quiescent counterparts, except at the brighter and more massive end of the red-sequence where galaxies show no sign of recent star formation.

## 5. Conclusions

How do all these findings combine together? Is it possible to draw a consistent picture of the recent star formation history of local early-type galaxies based both on the detailed study of the closest galaxies and on the analysis of integrated properties for much larger samples? Unfortunately, the answer is not quite yet.

Starting with the fraction of early-type galaxies that are presently forming stars, at first glance the fact that 8% of the SAURON sample displays [O III]/$H\beta$ line ratios consistent with emission from H II regions would seem in reasonable agreement with the finding that 4% of the objects in the SDSS sample of Schawinski et al. (2007) display the same behavior. We have to keep in mind, however, that the SAURON sample is a representative but *incomplete* sample of the nearby early-type population, in which fainter galaxies are particularly underrepresented. As star formation could be more common in these systems than in more massive galaxies (Sect. 3.3), the true fraction of early-type galaxies with emission associated with H II regions could be much higher than presently found by SAURON. Similarly, the fraction of star-forming early-type galaxies estimated from more complete SDSS samples should also be regarded as a lower limit, since the SDSS spectra can only detect the most intense of starbursts. For instance, considering an EW of 0.8 Å for the faintest lines detected in the SDSS data, only one of the 48 SAURON galaxies has sufficiently strong emission within 1 $R_e$ (corresponding to the typical physical area subtended by the SDSS fibers) to be detected as one of the star-forming early-type galaxies of Schawinski et al. (2007).

As regards the presence of younger stellar components in early-type galaxies, it is reassuring that Schawinski et al. (2007) and Kaviraj et al. (2007) estimate fractions of galaxies that are presently forming stars (~4%), and that contain populations younger than a few 100 Myr (~30%), respectively, which are consistent with each other given the different timescales traced by the nebular and NUV emission. On the other hand, it is less clear how well the mass fractions and ages of the young stellar components derived by *GALEX* and SDSS data agree with the SAURON observations. Consider the fact that just 1% of 500 Myr old stars embedded in a 10 Gyr old population induces an increase of ~1 Å in the strength of the $H\beta$ absorption line. Only 10% of objects in the SAURON sample have a global (within 1 $R_e$) $H\beta$ line strength that exceeds by 1 Å that observed in the oldest objects. Although the discrepancy with the 30% fraction reported by Kaviraj et al. (2007) could be reconciled considering that more

[1]That is, characterized by low ionization emission, as in the case of the low ionization nuclear emission regions first identified by Heckman (1980).

SAURON objects show evidence of recent star formation (up to 25%, Sect. 2.1), until the mass and age fraction of these young subcomponents is determined for a sufficiently complete sample of nearby galaxies it will remain hard to ascertain whether such SAURON objects correspond to the galaxies that experienced recent star formation according to analyses based on *GALEX* and SDSS data.

Despite these uncertainties, it is clear that with the exception of the most massive objects most early-type galaxies have undergone some degree of recent star formation. The gas reservoirs fueling such activity can now be detected with relative ease, and new techniques have been developed to analyze data from different passbands to better constrain the star formation history of galaxies. The biggest challenge ahead is still to quantify more precisely the extent of such star formation activity and its impact in shaping the observed structural and dynamical properties of these systems. In fact, it is precisely to address these and other questions that we have recently started a complete integral field survey of the nearby early-type galaxy population, nicknamed ATLAS$^{3D}$, which will be complemented by single-dish mm-band observations and by deep multiwavelength optical imaging.[2]

**Acknowledgments.** Marc Sarzi is truly indebted to Johan Knapen for the opportunity to present this small review at such an interesting conference and wishes to thank Lisa Young, Martin Bureau, Alessandro Bressan, Kristen Shapiro, Sugata Kaviraj, Kevin Schawinski, and Genevieve Graves for many useful discussions.

## References

Bernardi, M., et al. 2003, AJ, 125, 1817
Bettoni, D., Galletta, G., & García-Burillo, S. 2003, A&A, 405, 5
Bregman, J. N., Temi, P., & Bregman, J. D. 2006, ApJ, 647, 265
Bressan, A., et al. 2006, ApJ, 639, L55
Cappellari, M., et al. 2007, MNRAS, 379, 418
Combes, F., Young, L. M., & Bureau, M. 2007, MNRAS, 377, 1795
de Zeeuw, P. T., et al. 2002, MNRAS, 329, 513
Emsellem, E., et al. 2007, MNRAS, 379, 401
Goudfrooij, P., Hansen, L., Jorgensen, H. E., & Norgaard-Nielsen, H. U. 1994, A&AS, 105, 341
Graves, G. J., Faber, S. M., Schiavon, R. P., & Yan, R. 2007, ApJ, 671, 243
Heckman, T. M. 1980, A&A, 87, 152
Jeong, H., Bureau, M., Yi, S. K., Krajnović, D., & Davies, R. L. 2007, MNRAS, 376, 1021
Kaneda, H., Onaka, T., & Sakon, I. 2005, ApJ, 632, L83
Kaviraj, S., et al. 2007, ApJS, 173, 619
Krajnović, D., Cappellari, M., Emsellem, E., McDermid, R. M., & de Zeeuw, P. T. 2005, MNRAS, 357, 1113
Kuntschner, H., et al. 2006, MNRAS, 369, 497
Macchetto, F., Pastoriza, M., Caon, N., Sparks, W. B., Giavalisco, M., Bender, R., & Capaccioli, M. 1996, A&AS, 120, 463
McDermid, R. M., et al. 2006, MNRAS, 373, 906
Morganti, R., et al. 2006, MNRAS, 371, 157

[2]Details can be found at http://www-astro.physics.ox.ac.uk/atlas3d/.

Nipoti, C., & Binney, J. 2007, MNRAS, 382, 1481
Oosterloo, T. A., Morganti, R., Sadler, E. M., Vergani, D., & Caldwell, N. 2002, AJ, 123, 937
Panuzzo, P., et al. 2007, ApJ, 656, 206
Sarzi, M., et al. 2006, MNRAS, 366, 1151
Sarzi, M., et al. 2007, New Astron. Rev., 51, 18
Schawinski, K., Thomas, D., Sarzi, M., Maraston, C., Kaviraj, S., Joo, S.-J., Yi, S. K, Silk, J. 2007, MNRAS, submitted
Trager, S. C., Faber, S. M., Worthey, G., & González, J. J. 2000, AJ, 120, 165
Thomas, D., Maraston, C., Bender, R., & Mendes de Oliveira, C. 2005, ApJ, 621, 673
Yi, S. K., et al. 2005, ApJ, 619, L111
Young, L. M. 2002, AJ, 124, 788
Young, L. M. 2005, ApJ, 634, 258

## Discussion

*Beckman*: 1) Where are the various components distributed (CO, H I, S I, etc.)? 2) You should be able to check the origin of the gas and stars, i.e., to distinguish extrinsic from intrinsic origin, via the metallicities of both stars and gas.

*Sarzi*: 1) The H I reservoir generally extends to very large radii but it is not necessarily confined to the galaxy outskirts, as it often extends to the central regions as well. Molecular gas and SF regions are generally more concentrated but high resolution studies for the CO are still limited and *GALEX* UV observations have revealed tidally triggered SF also in the outer regions of early-type galaxies.

*Kennicutt*: With star formation being such a common property of elliptical galaxies one would expect them to host Type II supernovae. Do you know whether they have been observed in ellipticals?

*Sarzi*: Good point! I will double check.

Marc Sarzi.

*Pathways through an Eclectic Universe*
*ASP Conference Series, Vol. 390, © 2008*
*J. H. Knapen, T. J. Mahoney, and A. Vazdekis, eds.*

# Towards a New Classification of Early-type Galaxies: An Integral-field View

Jesús Falcón-Barroso,[1] Roland Bacon,[2] Michele Cappellari,[3] Roger L. Davies,[3] P. Tim de Zeeuw,[4] Eric Emsellem,[2] Davor Krajnović,[3] Harald Kuntschner,[5] Richard M. McDermid,[4] Reynier F. Peletier,[6] Marc Sarzi,[7] and Glenn van de Ven[8]

[1]*European Space and Technology Centre, Keplerlaan 1, NL-2200 AG Noordwijk, The Netherlands*
[2]*Université de Lyon 1, CRAL, Observatoire de Lyon, 9 av. Charles André, F-69230 Saint-Genis Laval, France*
[3]*Sub-Department of Astrophysics, University of Oxford, Denys Wilkinson Building, Keble Road, Oxford OX1 3RH, UK*
[4]*Sterrewacht Leiden, Universiteit Leiden, Postbus 9513, NL-2300 RA Leiden, The Netherlands*
[5]*Space Telescope European Coordinating Facility, European Southern Observatory, Karl-Schwarzschild-Str. 2, D-85748 Garching, Germany*
[6]*Kapteyn Astronomical Institute, University of Groningen, NL-9700 AV Groningen, The Netherlands*
[7]*Centre for Astrophysics Research, University of Hertfordshire, Hatfield, Herts AL10 9AB, UK*
[8]*Institute for Advanced Study, Einstein Drive, Princeton, NJ 08540, USA*

**Abstract.** In this proceedings paper we make use of the two-dimensional stellar kinematics of a representative sample of E and S0 galaxies obtained with the SAURON integral-field spectrograph to reveal that early-type galaxies appear in two broad flavors, depending on whether they exhibit clear large-scale rotation or not. We measure the level of rotation via a new parameter ($\lambda_R$) and use it as a basis for a new kinematic classification that separates early-type galaxies into slow and fast rotators. With the aid of broad-band imaging we will reinforce this finding by comparing our kinematic results to the photometric properties of these two classes.

## 1. Introduction

The origins of the morphological classification of galaxies date back from early work by Jeans in 1929. The latter addition of the S0 galaxies as a new class by Hubble (1936) resulted in the "tuning-fork" diagram that we know and use today. The Hubble classification is a continuous sequence between elliptical (E), lenticular (S0), and spiral (S) galaxies, with the S0s occupying the transition region. Elliptical and lenticular galaxies are usually gathered into the so-called early-type category, given the large number of global photometric properties they share (de Vaucouleurs et al. 1991).

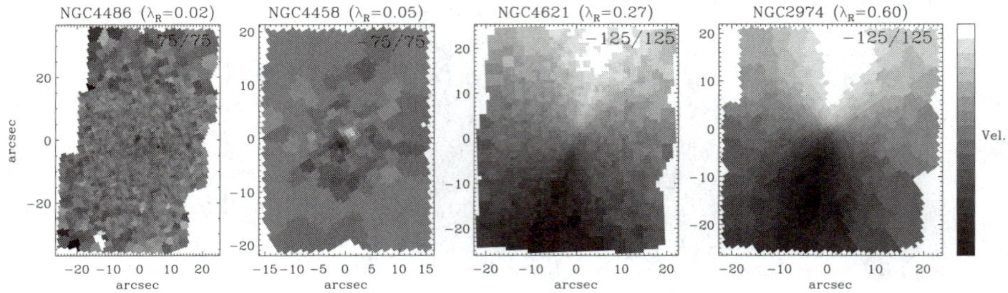


Figure 1. Stellar velocity fields for four elliptical galaxies in the SAURON sample of early-type galaxies. The maps are sorted from left to right by increasing value of the $\lambda_R$ parameter. The first two galaxies (on the left) belong to the slow rotator class, whereas the last two are fast rotators. Note the different kinematic substructures despite belonging to the same elliptical class. Numbers on the top right corner of each map indicate the velocity cuts applied to the data, which have been adjusted so as to properly emphasize the observed velocity structures.

More recently, with the advent of CCD imaging, there have been several attempts to revise the current scheme to introduce a more physical description of the objects and therefore go beyond a purely descriptive tool. With this goal in mind Kormendy & Bender (1996, hereafter KB96) updated the Hubble sequence by sorting ellipticals in terms of photometric quantities, used as a proxy for the importance of rotation. They used the diskiness or boxiness of the isophotes to define refined types: E(d) galaxies (for disky ellipticals) making the link between S0s and E(b) galaxies (for boxy ellipticals). This extension of the Hubble types has the merit of upgrading our view of Es and S0s via some easily accessible observable parameter that, at the same time, introduces some physics into the sorting criteria. It did, however, use a photometric indicator as an attempt to quantify the dynamical state of the galaxy, which may be deemed unreliable.

In this contribution, we revisit the early-type galaxy classifications above using the available full two-dimensional kinematic information coming from the unique data set obtained with the SAURON integral-field spectrograph. A more compelling study of the kinematics for this sample with a discussion of the kinematic classification presented here can be found in Emsellem et al. (2007).

## 2. A Kinematic Classification of Galaxies

The velocity fields for our sample of 48 E and S0 galaxies, the basis for this study, were presented in Emsellem et al. (2004). They revealed a large variety of kinematic structures including decoupled cores, velocity twists, misalignments, cylindrical or disk-like rotation. A close examination of the maps suggests that early-type galaxies come in two broad flavors: one which exhibits a clear large-scale and rather regular rotation pattern, and another which shows perturbed velocity structures (e.g., strong velocity twists) or central kinematically decoupled components with little rotation in the outer regions. Here, we illustrate these features in Fig. 1 using four elliptical galaxies (NGC 4486, NGC 4458, NGC 4621, NGC 2974). From the figure, it is easy to recognize the large differences in kinematic substructure between the galaxies, despite their belonging to the same E morphological class.

In order to build a robust classification based on the observed kinematics we need a simple measurable parameter that quantifies the *global* dynamical state of a galaxy, and that can be applied to all the galaxies in our sample. The ideal tool would be a physical parameter that captures the spatial information included in the kinematic maps. Since we wish to assess the level of rotation in galaxies, this parameter should follow the nature of the classic $V/\sigma$: ordered versus random motion (see Emsellem et al. 2007 for a detailed explanation of the pitfalls of the $V/\sigma$ parameter as a reliable classification parameter). Following this idea, we have defined a new quantity $\lambda_R$, such that

$$\lambda_R \equiv \frac{\langle R\,|V|\rangle}{\langle R\,\sqrt{V^2+\sigma^2}\rangle}\,,$$

measures the amount of specific (projected) angular momentum from the velocity maps. The parameter has been defined such that it is insensitive to small features in the maps and therefore provides a robust measurement of the global rotation. As we go from galaxies with low to high $\lambda_R$ values (from left to right in Fig. 1), the overall velocity amplitude naturally tends to increase. More importantly, there seems to be a *qualitative* change in the observed stellar velocity structures. Rotators with $\lambda_R < 0.1$ exhibit low stellar mean velocities at large radii with very perturbed stellar kinematics, and *all* have large-scale kinematically decoupled components. In Fig. 2 (*left*) we show the distribution of our galaxies as a function of $\lambda_R$. There are 36 fast rotators and 12 slow rotators (75% and 25% of the total sample), their median $\lambda_R$ values being $\sim$0.44 and 0.05 respectively. Within the class of slow rotators, three galaxies have $\lambda_R$ significantly below 0.03 (their mean stellar velocity maps being consistent with zero rotation everywhere). These are among the brightest galaxies in our sample (NGC 4486, NGC 4374, and NGC 5846).

## 3. Comparing Photometry and Kinematics: Misalignments and Twists

In order to assess whether our kinematic classification based on the $\lambda_R$ parameter is solid, it is desirable that the structural differences seen in the kinematics are also reflected in the photometry. In Fig. 2 (*right*) we show that slow and fast rotators display clear differences in the global alignment between their photometric and kinematic major axes ($\Psi \equiv |\mathrm{PA_{phot}} - \mathrm{PA_{kin}}|$). The figure illustrates that all fast rotators, except one, have misalignments $\Psi$ below 10°. The only exception is NGC 474, which is an interacting galaxy with well-known irregular shells (Turnbull, Bridges, & Carter 1999). In fact, the few galaxies that have $5° < \Psi < 10°$ (NGC 3377, NGC 3384, NGC 4382, NGC 4477, and NGC 7332) are most probably barred. In contrast, more than half of all slow rotators have $\Psi > 10°$, and none of these exhibits any hint of a bar. This difference in the misalignment values of slow and fast rotators cannot be entirely due to the effect of inclination, mainly because even the roundest fast rotators do not exhibit large misalignment values.

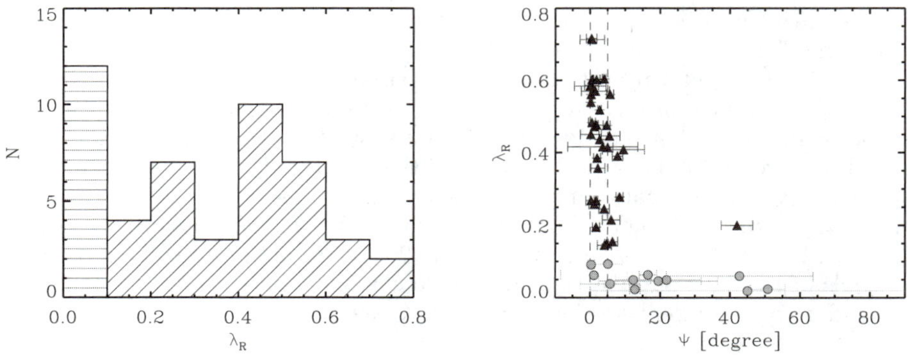


Figure 2. *Left:* Distribution of galaxies as a function of $\lambda_R$. The horizontally striped bar indicates the bin for slow rotators, whereas diagonally striped histogram corresponds to fast rotators. *Right:* $\lambda_R$ versus the kinematic misalignment $\Psi$ between the global photometric major-axis and the kinematic axis within the SAURON field. Slow rotators are represented by filled circles, and fast rotators by solid triangles. The vertical dashed lines correspond to $\Psi = 5°$. Nearly all fast rotators have small $\Psi$ values ($< 10°$), the only exception being NGC 474, the photometry of which is perturbed by the presence of irregular shells. This contrasts with slow rotators, which show significantly non-zero $\Psi$ values.

Independent evidence in support for our kinematic classification comes from the observed velocity twists in the SAURON kinematic maps: only six out of 48 galaxies exhibit strong velocity twists larger than 30° outside the inner 3″, with three out of these having large-scale counter-rotating stellar components (NGC 3414, NGC 3608, and NGC 4550). All these galaxies are in fact slow rotators. This implies that only fast rotators have a relatively well defined (apparent) kinematic major-axis, which in addition is roughly aligned with the photometric major axis. As emphasized above, slow and fast rotators exhibit qualitatively but also quantitatively different kinematic properties. This suggests that slow rotators are not just scaled down versions of fast rotators.

**Acknowledgments.** JFB would like to thank the organizers for a very fruitful meeting and wishes to congratulate J. Beckman for his outstanding contribution to astronomy in particular, but science in general, over the last 40 years.

## References

de Vaucouleurs, G., de Vaucouleurs, A., Corwin Jr, H. G., Buta, R. J., Paturel, G., & Fouque, P. 1991, Third Reference Catalogue of Bright Galaxies (Springer-Verlag: Berlin)
Emsellem, E., et al. 2004, MNRAS, 352, 721
Emsellem, E., et al. 2007, MNRAS, 379, 401
Hubble, E. 1936, The Realm of the Nebulae (New Haven: Yale University Press)
Kormendy, J., & Bender, R. 1996, ApJ, 464, L119
Turnbull, A. J., Bridges, T. J., & Carter, D. 1999, MNRAS, 307, 967

## Discussion

*Knapen*: Can you indicate to what extent SAURON results for ellipticals are representative for the general population?

*Falcón-Barroso*: This is difficult to answer properly. As you know, the SAURON survey selection criteria intended to obtain a representative sample for the E/S0 population. Then, later on, we found the result I have summarized today on the slow/fast rotator classes. Since we did not expect to find this result, it is difficult to predict what the impact is of our selection on the intrinsic distribution of slow/fast rotators. It is precisely for that reason that we have started a new project, i.e., ATLAS3D, where we will observe a complete sample of galaxies. This new sample should be free of any particular bias and will give us the true fraction of slow/fast rotators in the nearby Universe.

*Erwin*: 1) What happens if an S0 galaxy, which ought to be a fast rotator, is viewed close to face on? Does it turn into a slow rotator? 2) You have made a good case for a distinction between slow and fast rotators. But is not there still a confusion within the fast rotator class, in that it's a mixed bag of what morphologists would call ellipticals and S0s?

*Falcón-Barroso*: 1) This is indeed an obvious question to ask. Since $\Lambda_R$ is a projected quantity one may wonder what the dependence on inclination is. To understand this dependence we have built two-integral models and see how $\Lambda_R$ varies with inclination. What we saw is that a galaxy with a measured $\Lambda_R = 0.05$, typical for a slow rotator, would need to be at a nearly face-on inclination (~20 degrees) to reach an intrinsic $\Lambda_R$ value ~0.15 and therefore become a fast rotator. The probability that slow rotators are truly fast rotators is small and cannot explain the current population we see. 2) That is correct, however, remember that the slow/fast rotator classification is able to separate intrinsic properties of these galaxies unambiguously (something that cannot be done at present with the E/S0 classification). This is in fact the reason for our new scheme.

From left to right: Sergio Fernandez Acosta, Gelys Trancho, and Jesús Falcón-Barroso.

*Pathways through an Eclectic Universe*
*ASP Conference Series, Vol. 390, © 2008*
*J. H. Knapen, T. J. Mahoney, and A. Vazdekis, eds.*

# On Classical and Pseudo-bulges: The Nature of Bulges of Early-type Spirals

Reynier F. Peletier

*Kapteyn Astronomical Institute, University of Groningen, Postbus 800, NL-9700 AV Groningen, The Netherlands*

**Abstract.** Historically, galactic bulges are thought to be elliptical galaxy-like objects sitting in the middle of a generally larger disk. There are, however, more and more claims that some bulges are much more similar to disks. John Kormendy has named these pseudo-bulges. In this paper I discuss some recent integral field spectroscopy of the SAURON collaboration on a sample of 24 Sa and Sab galaxies. Using their two-dimensional maps of the stellar velocity, velocity dispersion, and absorption line strength, it is now much easier to understand the nature of nearby galactic bulges. I review several aspects of bulges, namely the surface photometry, stellar kinematics, stellar populations, the presence of interstellar matter, and their behavior in the fundamental plane of early-type galaxies.

## 1. Introduction and Definition of a Bulge

One of the remarkable features seen in an image of a spiral galaxy is its central light concentration. Recent *Spitzer Space Telescope* imaging has once again shown some beautiful examples of this, in particular the Andromeda galaxy (SSC 2006-14), the Sombrero (SSC 2005-11), M51 (SSC 2004-19) and NGC 7331 (SSC 2004-12). In the images one sees that every galaxy has a bright central concentration due to stellar light with relatively few features due to dust and star formation. This light concentration is known as a galactic bulge. In the old days bulges were known as old, metal rich, galaxy components (Whitford 1978). Nowadays, there are strong indications that there is at least a large population of galactic bulges which also contain younger stars. Details about this can be found in the long review by Kormendy & Kennicutt (2004, hereafter KK). Since research on bulges is making fast progress, however, I would like to revisit the subject in the light of new observations, especially from the SAURON instrument (Bacon et al. 2001), and discuss what we know about bulges, and what we can deduce about their formation and evolution.

Observations from the SDSS Survey (Baldry et al. 2004) show that there are two main groups of galaxies: those on the red sequence, and a group of bluer galaxies, the so-called blue cloud. The red sequence contains the elliptical galaxies in the nearby Universe. Mainly because of their similarities with elliptical galaxies, galactic bulges are traditionally thought to be elliptical galaxies in the middle of spiral disks (KK), and therefore thought to be the product of early dissipative collapse (Eggen, Lynden-Bell & Sandage 1962) and mergers (Toomre 1977), around which later a disk formed. In more recent years, however, alternative theories have been developed, in which bulges formed from disk material

(Pfenniger & Norman 1990, and others). Evidence that this process of so-called secular evolution is really happening has been summarized by KK.

To have a clean discussion, it is important to first define what we mean when we talk about a galactic bulge. Is this the same as what we measure? Traditionally, people define the bulge as the light excess above an exponential disk (Freeman 1970). This is a purely observational definition, applicable to all galaxies. We call this the *photometric* decomposition. It suffers, however, from the fact that some disks cannot be fitted well by exponential surface brightness distributions, e.g. in the presence of bars, and that we do not know what this bulge means physically. Kent (1986) therefore introduced the *morphological* decomposition, assuming that bulge and disk have different flattening. His method separates non-parametrically two components with different projected ellipticities. This method also has its problems: galaxies, especially their inner regions, have axial ratios that vary with radius. Also, there are spirals for which the axial ratio in the central regions is the same as in the outer disk (KK). For face-on galaxies this method is in any case unusable. If a spiral galaxy were to consist of a slowly rotating elliptical galaxy surrounded by a fast-rotating exponential disk one could also think of doing the decomposition using the kinematics. Due to the lack of high quality kinematic profiles, this *kinematic* decomposition has not been tried often in the literature, but it has been mentioned (e.g., Kormendy 1993). I can conclude that nowadays photometric decomposition is by far the most common. It might actually be preferable to use the kinematic decomposition, since using this method it would be much easier to understand the origin of bulges.

So, are bulges indeed similar to ellipticals, as I myself claimed a number of years ago (Peletier et al. 1999)? At first sight, one might think that it is hard to include galaxies such as NGC 4736 (Kormendy 1993), which have spiral structure in the bulge region, and which rotate faster than one would expect for an elliptical, in this category. Also, bulges of late-type spirals do not follow the de Vaucouleurs $r^{1/4}$ law (Andredakis, Peletier, & Balcells 1995) and have bright nuclear star clusters (e.g., Böker et al. 2002). What fraction of the bulges of early-type spirals (type S0a–Sb) look like ellipticals? In the rest of this paper I will discuss a few aspects of the observations of bulges that can perhaps bring us somewhat closer to answering this question and understanding the nature of bulges.

## 2. Surface Brightness Profiles

Surface brightness profiles of bulges are well fitted by $r^{1/n}$ Sérsic distributions (for a historic review see KK). The Sérsic index $n$ correlates well with morphological type (Andredakis et al. 1995) and bulge-to-disk ratio, giving values of around 4 and higher for ellipticals and S0s, decreasing for values around 1–2 for Sc's. Balcells et al. (2003) and Balcells, Graham, & Peletier (2007) noted that surface brightness profiles of early-type spirals could be fitted better if an additional central point source is included. For a sample of inclined spirals, with inclination larger than 50° of type S0–Sbc (the Balcells & Peletier sample) the Sérsic index of the bulge rarely exceeds 3, even for S0s. Central sources are also very often present in late-type spirals. Böker et al. (2002) find that 77% of their sample of 77 late-type spirals contains a nuclear cluster, which is often resolved. In intermediate-luminosity Virgo cluster ellipticals, Coté et al. (2004) report a detection rate of 82%. Motivated by theory, Carollo et al. (2002) classify their sample of bulges into so-called $r^{1/4}$—or classical—bulges and exponential bulges. In agreement with the scatter found by Andredakis et al. (1995), both classical and exponential bulges can be found at most morphological types, although for the latest types only exponential bulges are found. However, is it reasonable to make this separation into two groups, since the underlying distribution of Sérsic indices is continuous? To justify this, other, independent, arguments should also be used.

It is interesting to note about the surface photometry that the scales of bulges and disks are correlated (Courteau, de Jong, & Broeils 1996). This implies that the bulge is strongly influenced by its surrounding disk, or vice versa.

## 3. Boxy "Bulges"

A significant fraction of edge-on galaxies contains a so-called *boxy* or peanut-shaped bulge, i.e., the outer isophotes of the bulge appear strongly boxy (e.g., Jarvis 1986). It is well established that these are bars seen edge-on. $N$-body simulations show that they are due to bar orbits buckling out of the plane (e.g., Athanassoula 2005). Also, observationally, Kuijken & Merrifield (1995) have shown that along the peanut they see velocity splitting, characteristic for the presence of a bar (see also Chung & Bureau 2004). Moreover, in the azimuthally averaged surface brightness profiles, the boxy bulge is generally not identified. Since these arguments show that boxy bulges are related to bars, which are an integral part of the flat disk, I prefer to conclude that their name is very misleading. Although boxy, these components are part of the disk and should maybe be called the boxy region. They have little to do with the component in excess of the exponential disk, the bulge.

## 4. Stellar Kinematics

The $(v/\sigma)$–$\epsilon$ diagram has been a very powerful tool in characterizing the nature of galactic bulges. Kormendy & Illingworth (1982) showed that bulges fall close to the oblate rotator line in this diagram, the line of an isotropic object flattened by its rotation. They, and also Davies et al. (1983), showed that, kinematically, these bulges behave the same as intermediate mass ellipticals. The bulges for which kinematics were available at that time were all large bulges of early-type spirals. Later, Kormendy could add other bulges to the diagram, and found that some had larger $(v/\sigma)$ than expected for an isotropic rotator, showing more disk-like kinematics. In recent years, the whole picture is becoming much more clear. Many galaxies have been discovered in which a thin, rotating disk is dominating the light in the very inner parts, accompanied by a local minimum in the velocity dispersion. The first observed cases of central velocity dispersion minima date back to the late '80s and early '90s (e.g., Bottema 1989, 1993). Several others were found from long-slit data (Emsellem et al. 2001; Márquez et al. 2003). SAURON observations showed that 13 out of 24 Sa and Sab galaxies showed a central local minimum in the velocity dispersion (Fálcon-Barroso et al. 2006).

We see a weak trend that galaxies with sigma-drops are younger than those without (Peletier et al. 2007). The sigma-drops are probably due to central disks that formed from gas falling into the central regions through a secular evolution process. Simulations show that the disks will remain in place for a long time after they have been formed (Wozniak & Champavert 2006), although they will slowly heat up with time. Such central disks are also known among elliptical and S0 galaxies (e.g., NGC 4526, Emsellem et al. 2004, NGC 7332, Falcón-Barroso et al. 2004), but rare.

## 5. Stellar Populations

Up to 1990, very little was known about the stellar populations of galactic bulges, apart from that in our own Galaxy, which was thought to consist of old, super metal rich stars (e.g., Rich 1988). Balcells & Peletier (1994) showed, by looking at the bulge on the side of inclined galaxies not obscured by the dust lane, that the colors were the same as or bluer than those of elliptical galaxies of the same luminosity. Later, they showed, from color–color diagrams, that these galaxies were old and almost all coeval, $9 \pm 2$ Gyr (Peletier et al. 1999). Bulges are not visible in color maps, implying that bulges and inner disks have very similar stellar populations. When put on the fundamental plane of early-type galaxies the same sample falls on the line defined by early-type galaxies in the Coma cluster (Jørgensen, Franx & Kjaergaard 1996), confirming that they are old.

If, however, one looks at the $Mg_2$–$\sigma$ relation, a different result is obtained (Peletier et al. 2007). Although several samples of bulges fall on the $Mg_2$–$\sigma$ relation of early-type galaxies in Coma, a large fraction of the sample of early-type spirals of Prugniel, Maubon, & Simien (2001) has a lower $Mg_2$ value for the same $\sigma$. Also, several bulges of the SAURON sample of Peletier et al. (2007) fall below the relation for Coma. The difference between the samples that fall on the line and those that fall partly below it is that the samples that fall on the line consist either of inclined galaxies, or mainly of S0 galaxies. Apparently, bulges of samples for which the inclination is unbiased can have a large range in ages. This can be checked in the SAURON line strength maps (especially those of H$\beta$ absorption, Peletier et al. 2007), which confirm that the galaxies that lie far from the $Mg_2$–$\sigma$ for Coma galaxies are indeed young. This also explains the large range in ages found by others (e.g., Proctor & Sansom 2002; Moorthy & Holtzman 2006). Our late-type spirals of Ganda et al. (2006), observed with SAURON, all fall below the Coma relation.

More insight into the stellar populations can be obtained when one looks at line strength gradients. Recently, Jablonka, Gorgas, & Goudfrooij (2007; see also Gorgas, Jablonka, & Goudfrooij 2007) obtained line strength gradients of a sample of 32 edge-on galaxies, and showed that outside the central dust lane $Mg_2$ decreases as a function of distance above the plane in almost all galaxies. The same is the case for all their other indices, except for the Balmer indices. This can be explained if metallicity is decreasing vertically outwards. For the SAURON Sa–Sab galaxies, however, we find that gradients can be both positive and negative. Looking at the line strength maps this is directly understood to be due to the fact that these galaxies contain regions of recent star formation associated with dynamical resonances, i.e., rings and spiral arms. If a galaxy has star formation in the center, the $Mg_2$ gradient is positive, while in the case of no star formation, or a star-forming ring at the edge of the bulge, the gradient is negative. Similar behavior is seen in Moorthy & Holtzman (2006). Again, a difference is seen between inclined and unbiased samples. When one looks above the plane, bulges appear old. When, on the contrary, one looks in the plane itself, there are many regions with young stars.

## 6. Gas, Dust and Spiral Structure in Bulges

Spiral galaxies are dusty objects in their centers. Sb and Sc galaxies have on the average about one magnitude of extinction in $B$ in the central regions, and are basically transparent at three effective disk scale lengths (Giovanelli et al. 1995). Sa galaxies contain less extinction. Martini et al. (2003) show a large collection of central *HST* color maps of nearby spirals. The maps show that the large majority of galaxies have extinction in the central regions. There are galaxies with nuclear rings very close to the center, and well within the bulge. There are highly inclined galaxies with smooth color maps on the side of the galaxy away from the dust lane. Many galaxies show spiral structure in their inner regions, a sign that the dust is distributed in a plane. Regan et al. (2006) and Bendo et al. (2007) show that many spirals, also of type Sa, show the presence of $8\,\mu$m and $24\,\mu$m emission in their central regions, as well as molecular gas, as traced by CO. Color maps of edge-on galaxies (e.g., Howk & Savage 2000) show that dust extinction can be found high above the plane, but also that the dust seems to be associated to the disk only. The fact that color maps of bulges of galaxies that are not obscured by the disk, i.e., at inclinations between 50° and 85° are smooth also strongly suggests that the dust is found in the plane only. Since dust and gas are often seen together (e.g., Kent, Dame, & Fazio 1991; Sarzi et al. 2006), I expect that the ionized, neutral and molecular gas will also be found near the plane only, where it can be in equilibrium.

I would like to insert a caveat here about determining stellar population parameters from colors in the presence of extinction. When one is reasonably sure that the colors are not or only slightly affected by dust (e.g., Peletier et al. 1999) stellar populations can in principle be derived from colors. The same paper also shows, however, that the *central* colors of the Balcells & Peletier sample are so red that they cannot be fitted by any realistic stellar population model. Correcting for extinction here is very difficult, since the error in the correction is often so large that one has no distinguishing power left in corrected color. For dusty galaxies it is also very tricky to mask out regions with a lot of extinction and determine the colors in places where one thinks there is no extinction (e.g., Carollo et al. 2007), and most certainly the result will not be reproducible by others. SAURON line strength maps (e.g., Ganda et al. 2007) should offer a reasonable alternative.

## 7. Our New Picture of Galactic Bulges

From the stellar population distributions, by comparing samples uniformly distributed in inclination with samples biased towards high inclination we infer that galactic bulges have more than one physical component: generally they contain

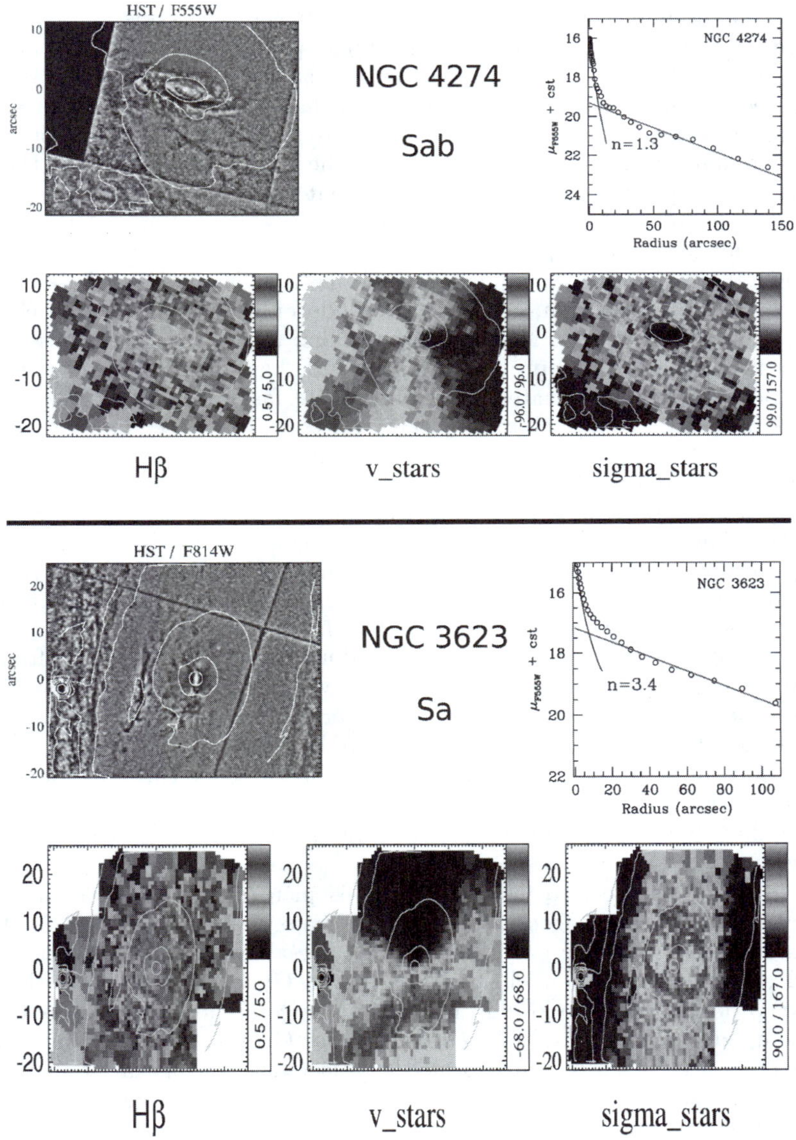


Figure 1. Diagnostic observations of NGC 4274 and NGC 3623, two early-type spirals. For each galaxy is shown in the *top left*: unsharp-masked F555W or F814W *HST* image, showing the places of non-negligible extinction (from Falcón-Barroso et al. 2006). *Top right:* major axis surface brightness profile, from the same *HST* data. A bulge–disk decomposition is also shown, with an exponential disk and a Sérsic bulge. *Bottom row (from left to right):* H$\beta$ absorption line (Peletier et al. 2007), stellar velocity, and velocity dispersion map (Falcón-Barroso et al. 2006). Overlaid on all maps is the reconstructed SAURON intensity.

a slowly rotating, elliptical-like component, and one or more fast-rotating components in the plane of the galaxy, all co-existing in the same galaxies. This picture also nicely explains the fact that bulge populations in general are very similar to those in the disk (e.g., Peletier & Balcells 1996). It is also strongly supported by the kinematics. In more than half of the SAURON early-type spirals sigma-drops occur. These correspond to central disks, which sometimes contribute most of the light corresponding to the bulge as resulting from the photometric decomposition. *HST* images show that the central disks correspond to dusty regions, often showing spiral structure. They also often, but not always, contain younger stellar populations than the regions outside the central disks.

In Fig. 1 we give two example galaxies: one that is classified, based on the photometric decomposition, as having a pseudo-bulge, and one with a classical bulge. The first example, NGC 4274, for which the surface brightness profile is best fitted by a Sérsic profile with $n = 1.3$, has a strong sigma-drop. In the region where it dominates, the stars rotate fast and have a low velocity dispersion (SAURON). Such an object would be called a pseudo-bulge by KK. The bulge is dusty, has spiral structure (*HST* image) and shows younger stellar populations (SAURON). Note also that as one goes out on the minor axis, the stellar distribution becomes smooth, and the line strengths show values corresponding to old stellar populations. This region corresponds to an elliptical-like bulge, to which KK refer as a classical bulge. NGC 4274 therefore contains a pseudo-bulge *AND* a classical bulge. Making the classical bulge a bit larger, and the pseudo-bulge smaller, one gets an object such as NGC 3623 (bottom panels in Fig. 1). Here, the classical bulge dominates the light, apart from the very inner regions. A sigma-drop is seen, but weaker than the one in NGC 4274 (Falcón-Barroso et al. 2006). The Sérsic fit for the bulge gives $n = 3.4$. Dust is associated with the central disks, but the SAURON maps show that the stellar populations in it are not different from those in the classical bulge outside it. The comparison between NGC 3623 and NGC 4274 shows that both objects are very similar, but that the inner disk to elliptical bulge ratio in both galaxies is different.

The Galactic bulge also fits well into this picture, of having a hot and a cold component living together in the same galaxy. Zoccali et al. (2003) find that stars in the Galactic bulge, measured 6 deg above the Galactic plane, are as old as Galactic globular clusters, at least 10 Gyr. This indicates that our Galaxy probably has an old elliptical-like bulge component. We do not know of a component with disk-like kinematics in our Galaxy, but near the Galactic center stars are currently being formed at a fast rate. A nuclear stellar disk is found with a scale of about 100 pc (Launhardt, Zylka, & Mezger 2002), as well as a nuclear cluster.

Fisher (2006) reports one more piece of evidence. In galaxies which he classifies as having a pseudo-bulge, the central star formation rate, as obtained from the *Spitzer* 3.6–8.0 $\mu$m color, is higher than in galaxies with classical bulges. This shows that the *Spitzer* color measures similar things as optical absorption line indices. Drory & Fisher (2007) have recently classified the bulges in a sample of low inclination S0–Sc galaxies into classical and pseudo-bulges. If they see signs of disk features, such as nuclear bars, spirals or rings, the bulge is called a pseudo-bulge, and otherwise as a classical bulge. Apart from the caveat that many galaxies probably contain both types of bulges, their result that their bulge classification correlates with color is not surprising, since these nuclear features are indicators of recent star formation. It has been shown in the past that if the bulge looks featureless the stellar populations are old.

It is clear that our knowledge about bulges is increasing rapidly. It is now up to the observers to firmly establish these new ideas, and to the theorists to understand not only what we observe for bulges but also what the implications are for galaxy formation in the Universe.

**Acknowledgments.** I would like to thank the organizers of this pleasant meeting, and wish John Beckman many more interesting years in astronomy. I am grateful for the help of my collaborators, Marc Balcells, Jesús Falcón-Barroso, and Katia Ganda in particular.

### References

Andredakis, Y. C., Peletier, R. F., & Balcells, M. 1995, MNRAS, 275, 874
Athanassoula, E. 2005, MNRAS, 358, 1477
Bacon, R., et al. 2001, MNRAS, 326, 23
Balcells, M., & Peletier, R. F. 1994, AJ, 107, 135
Balcells M., Graham A. W., Domínguez-Palmero, L., & Peletier, R. F. 2003, ApJ, 582, L79
Balcells, M., Graham, A. W., & Peletier, R. F. 2007, ApJ, 665, 1084
Baldry, I. K., et al. 2004, ApJ, 600, 681
Bendo, G. J., et al. 2007, MNRAS, 380, 1313
Böker, T., Laine, S., van der Marel, R. P., Sarzi, M., Rix, H.-W., Ho, L. C., & Shields, J. C. 2002, AJ, 123, 1389
Bottema, R. 1989, A&A, 221, 236
Bottema, R. 1993, A&A, 275, 16
Carollo, C. M., Stiavelli, M., Seigar, M., de Zeeuw, P. T., & Dejonghe, H. 2002, AJ, 123, 159
Carollo, C. M., Scarlata, C., Stiavelli, M., Wyse, R. F. G., & Mayer, L. 2007, ApJ, 658, 960
Chung, A., & Bureau, M. 2004, AJ, 127, 3192
Côté, P., et al. 2004, ApJS, 153, 223
Courteau, S., de Jong, R. S., & Broeils, A. H. 1996, ApJ, 457, L73
Davies, R. L., Efstathiou, G., Fall, S. M., Illingworth, G., & Schechter, P. L. 1983, ApJ, 266, 41
Drory, N., & Fisher, D. B. 2007, ApJ, 664, 640
Eggen, O. J., Lynden-Bell, D., & Sandage, A. R. 1962, ApJ, 136, 748
Emsellem, E., Greusard, D., Combes, F., Friedli, D., Leon, S., Pécontal, E., & Wozniak, H. 2001, A&A, 368, 52
Emsellem, E., et al. 2004, MNRAS, 352, 721
Falcón-Barroso, J., et al. 2004, MNRAS, 350, 35
Falcón-Barroso, J., et al. 2006, MNRAS, 369, 529
Fisher, D. B. 2006, ApJ, 642, L17
Freeman, K. C. 1970, ApJ, 160, 811
Ganda, K., Falcón-Barroso, J., Peletier, R. F., Cappellari, M., Emsellem, E., McDermid, R. M., de Zeeuw, P. T., & Carollo, C. M. 2006, MNRAS, 367, 46
Ganda, K., et al. 2007, MNRAS, 380, 506
Giovanelli, R., Haynes, M. P., Salzer, J. J., Wegner, G., da Costa, L. N., & Freudling, W. 1995, AJ, 110, 1059
Gorgas, J., Jablonka, P., & Goudfrooij, P. 2007, A&A, 474, 1081
Howk, J. C., & Savage, B. D. 2000, AJ, 119, 644

Jablonka, P., Gorgas, J., & Goudfrooij, P. 2007, A&A, 474, 763
Jarvis, B. J. 1986, AJ, 91, 65
Jørgensen, I., Franx, M., & Kjaergaard, P., 1996, MNRAS, 280, 167
Kent, S. M. 1986, AJ, 91, 1301
Kent, S. M., Dame, T. M., & Fazio, G. 1991, ApJ, 378, 131
Kormendy, J., & Illingworth, G. 1982, ApJ, 256, 460
Kormendy, J. 1993, in IAU Symp. 153: Galactic Bulges, ed. H. DeJonghe & H. J. Habing (Dordrecht: Kluwer), p. 209
Kormendy, J., & Kennicutt, R. C. 2004, ARA&A, 42, 603
Kuijken, K., & Merrifield, M. R. 1995, ApJ, 443, L13
Launhardt, R., Zylka, R., & Mezger, P. G. 2002, A&A, 384, 112
Márquez, I., Masegosa, J., Durret, F., González Delgado, R. M., Moles, M., Maza, J., Pérez, E., & Roth, M. 2003, A&A, 409, 459
Martini, P., Regan, M. W., Mulchaey, J. S., & Pogge, R. W. 2003, ApJS, 146, 353
Moorthy, B. K., & Holtzman, J. A. 2006, MNRAS, 371, 583
Peletier, R. F., & Balcells, M. 1996, AJ, 111, 2238
Peletier, R. F., Balcells, M., Davies, R. L., Andredakis, Y., Vazdekis, A., Burkert, A., & Prada, F. 1999, MNRAS, 310, 703
Peletier, R. F., et al. 2007, MNRAS, 379, 445
Pfenniger D., & Norman C. 1990, ApJ, 363, 391
Proctor, R. N., & Sansom, A. E. 2002, MNRAS, 333, 517
Prugniel, P., Maubon, G., & Simien, F. 2001, A&A, 366, 68
Regan, M. W., et al. 2006, ApJ, 652, 1112
Rich, R. M. 1988, AJ, 95, 828
Sarzi, M., et al. 2006, MNRAS, 366, 1151
Toomre, A. 1977, ARA&A, 15, 437
Whitford, A. E. 1978, ApJ, 226, 777
Wozniak, H., & Champavert, N. 2006, MNRAS, 369, 853
Zoccali, M., et al. 2003, A&A, 399, 931

## Discussion

*Comerón*: You are talking about $\sigma$-drops. When there is not a bar, which mechanisms do you think transport to the nucleus the material necessary to raise the star formation needed for a $\sigma$-drop?

*Peletier*: Here I would like to refer you to the talk of Athanassoula later in this conference.

*Beckman*: It is not too surprising that bulges and disks overlap in the middle of galaxies. We know that in some galaxies the spheroidal component can gradually dominate the disk component at the outside of the galaxy too. It is a matter of the proportion of each component as a function of radius.

*Peletier*: I agree. But in my opinion the number of galaxies for which the bulge contributes a significant fraction of the light, such as the Sombrero , is smaller than one thinks, since Sérsic $n$ fits generally give values smaller than 3 (Balcells et al. 2003).

*Chilingarian*: Do you see any features in the maps of [Mg/Fe] abundance ratio? Are the features seen in your maps significant?

*Peletier*: There are features in these maps. You can see them in the journal paper (Peletier et al. 2007). It is, however, not easily seen how these features should be interpreted. One should note that in order to measure [Mg/Fe] we have assumed that the galaxy light can be fitted by an SSP. If this is not the case, e.g., in star-forming rings, a more complicated analysis is needed to derive [Mg/Fe]. See Peletier et al. (2007) for more details.

*Cenarro*: Given that bulges follow the mass–metallicity relationship, low $\sigma$ bulges must have higher values of H$\beta$ just because of their lower metallicities. Did you correct for such a metallicity dependence of H$\beta$ when using the H$\beta$–$\sigma$ diagram to disentangle age differences among bulges ?

*Peletier*: No, since we want to keep this index–$\sigma$ diagram completely observational . The only reason we would prefer the H$\beta$–$\sigma$ diagram above Mgb–$\sigma$ is that gradients in the old stellar populations do not influence the interpretation. Observationally, gradients in H$\beta$ in elliptical galaxies are much smaller than those in Mgb (e.g., Kuntschner et al. 2006, MNRAS, 369, 497). To interpret the position of galaxies in the H$\beta$–$\sigma$ diagram, one still needs to compare the position of a galaxy with the locus of galaxies in an old, relaxed cluster, such as Coma.

*Gupta*: You began with the alternative definitions for bulges. After this wealth of information, are you inclined to prefer any one over the others?

*Peletier*: Since it is nice to have a definition which has a sound theoretical basis, I would prefer a kinematics-based definition. Here the hot bulge would be the component with $v/\sigma \sim 1$, and the cold bulge would be a much faster rotating component.

*Carignan*: You mentioned that a large part of the SAURON Sa sample had many properties of later morphological types. Are they real misclassifications or do only some of their properties mimic those of later types?

*Peletier*: There are no misclassifications. In fact, they only appear to be later types by some of their observed properties in the center, but, when looking at their global properties, they belong without doubt to the Sa category.

Part of the Local Organizing Committee. From left to right: Eva Bejarano, Johan Knapen, Ruyman Azzollini, Judith de Araoz, Tanja Karthaus, and Sébastien Comerón.

*Pathways through an Eclectic Universe*
*ASP Conference Series, Vol. 390,* © *2008*
*J. H. Knapen, T. J. Mahoney, and A. Vazdekis, eds.*

# The Star Formation History of Late-type Galaxies

Philip A. James and Matthew Prescott

*Astrophysics Research Institute, Liverpool John Moores University, Twelve Quays House, Egerton Wharf, Birkenhead, CH41 1LD, UK*

**Abstract.** We investigate the dominant mode of star formation in late-type galaxies using an analysis of the distribution of star formation timescales (stellar masses/star formation rates) for 117 Im and Sm galaxies. We find that these galaxies could have produced their stellar mass by extrapolation of their current rate of growth over a Hubble time. No quiescent galaxies are found, supporting continuous rather than "bursty" modes of star formation. The spatial distributions of new and old populations are mutually consistent in these late-type galaxies and the optical-to-near-infrared colors are as expected from continuous star formation activity.

## 1. Introduction

Late-type galaxies of type Im and Sm provide the opportunity to study pure "blue sequence" star-forming systems, with little or no old bulge component. The present study looks at the star formation (SF) timescale of 42 Sm and 75 Im galaxies from the H$\alpha$ Galaxy Survey (H$\alpha$GS; James et al. 2004), where this timescale is defined as the ratio of the total stellar mass to the current SF rate. The distribution of these timescales provides a test of whether SF is continuous, as found observationally by, for example, Gallagher et al. (1984) and van Zee (2001), or in temporally separated bursts. The latter have been predicted by theoretical studies such as the stochastic self-propagating star formation models of Gerola, Seiden, & Schulman (1980) and have received observational support from, for example, Almoznino & Brosch (1998) in a study of late-type Virgo cluster dwarfs.

The H$\alpha$GS dataset provides H$\alpha$ plus [N II] narrow-band and $R$-band imaging for 334 galaxies with recession velocities out to 3000 km s$^{-1}$. It includes all spiral and irregular types (S0/a to Im), enabling us to compare the SF properties of the late-type Sm and Im irregulars with those of the earlier-type spirals.

## 2. Methods

The initial analysis involved the calculation of the time required to form the total stellar mass of each galaxy studied, under the assumption of continuous SF at the current rate. This enables the identification of quiescent galaxies, for which this timescale will be much greater than a Hubble time, and galaxies currently in a starburst phase, for which it will be much less than a Hubble time.

Stellar masses were initially derived from the $R$-band total magnitudes of each of the H$\alpha$GS galaxies, with the mass-to-light ratio being derived from the models

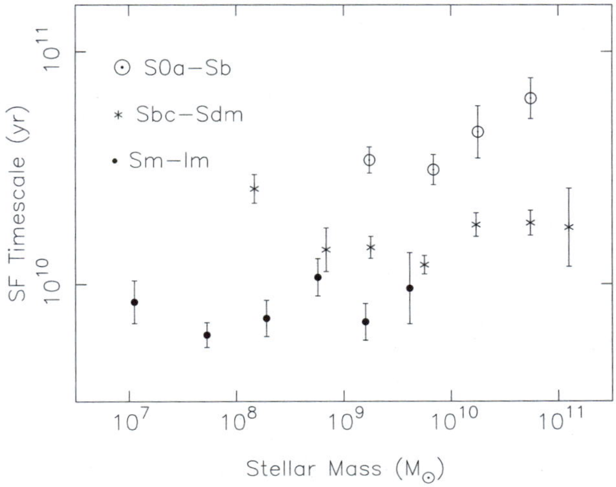


Figure 1. Mean SF timescale as a function of galaxy stellar mass.

of Bell & de Jong (2001) and a type-dependent $(B-R)$ color. A major source of uncertainty is the extinction internal to the galaxies observed, which affects both the derived SF rates and the stellar masses. Several extinction corrections were investigated, including that of Kennicutt (1983) and a type-dependent correction from James et al. (2005), with no change to the main conclusions. This analysis uses H$\alpha$ extinction corrections dependent on $R$-band total magnitudes, following the prescription of Helmboldt et al. (2004). The $R$-band extinction is then taken as one-half of that at H$\alpha$. SF timescales were calculated for each individual galaxy by dividing the stellar mass from the $R$-band absolute magnitude by the SF rate from the H$\alpha$ line fluxes.

## 3. Results and Discussion

Figure 1 shows the mean SF timescales for galaxies of different types and stellar mass ranges. This clearly shows a dependence of the mean timescale on galaxy type. The bulge-dominated early-type galaxies have high timescales, implying that their current SF rates are significantly lower than would be required to form the stellar mass within a Hubble time. The late-type Sm and Im galaxies, on the other hand, have uniformly short SF timescales, with binned means in the range 6–11 Gyr. Thus, these bulge-free galaxies could have formed their stellar mass through continuous SF activity at or slightly below the current rate. It is interesting to note that the mass ranges of the different types overlap significantly, and the shorter SF timescales can be seen to depend only on the lack of a bulge, and not on galaxy mass. All 117 Sm/Im galaxies have ongoing SF; they represent $> 50\%$ of all galaxies of these types satisfying the H$\alpha$GS selection criteria, so quiescent Sm/Im galaxies must be very rare or non-existent.

Continued SF with no indication of a fall-off at the current epoch implies a substantial reservoir of gas in the galaxies concerned. This is confirmed in Fig. 2, which shows the gas depletion timescales for the 94 members of the

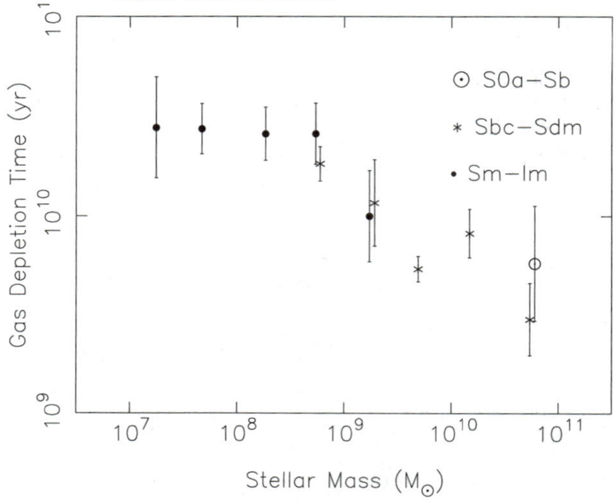


Figure 2. Mean gas-depletion timescale plotted against galaxy stellar mass.

current galaxy sample with H I masses provided by the Westerbork observations of neutral Hydrogen in Irregular and SPiral galaxies (WHISP) survey (Swaters et al. 2002). The H I masses are multiplied by 2.3 to account for helium and molecular gas; the resulting mass divided by the SF rate gives the depletion time. Figure 2 shows that spiral galaxies can have short gas depletion times, but that the Sm/Im galaxies have ample gas reservoirs to continue their current SF activity for at least a further Hubble time.

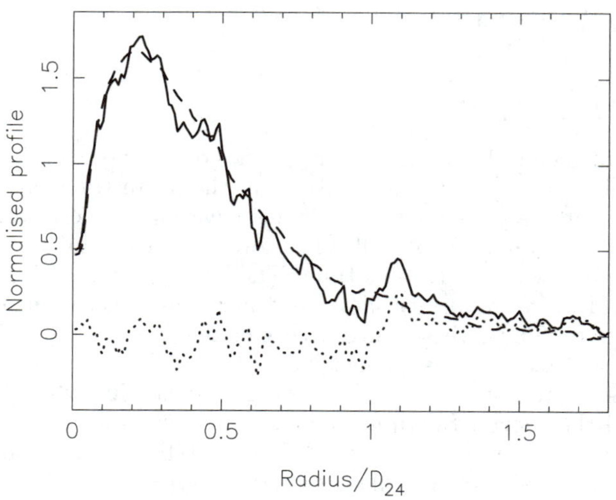


Figure 3. Mean, normalized radial distributions of H$\alpha$ (*solid line*) and $R$-band (*dashed line*) light in circular apertures; the dotted line shows the residuals between these two profiles.

We can also test whether the spatial distribution of newly formed and old stellar populations is consistent in these galaxies. Figure 3 shows the normalized distributions of the mean $R$-band (dashed) and H$\alpha$ (solid line) fluxes of all Im galaxies in the current sample. The vertical scale is flux within circular annuli (*not* surface brightness), normalized such that the total area under the curve, which represents the total flux, is unity for each galaxy. The horizontal scale is annular radius in units of the diameter of the galaxy at an $R$-band surface brightness of 24 mag arcsec$^{-2}$. Thus this is a dimensionless shape function describing the averaged light distributions of all the Im galaxies, with all galaxies, large and small, receiving equal weight. Given the small number of H II regions in many of these galaxies, the individual galaxy H$\alpha$ profiles are very spiky. In the mean, however, they look similar to the mean $R$-band light distribution, as is reflected by the small size of the residuals between the two profiles, in the sense (H$\alpha$ $-R$), shown by the dotted line in Figure 3. If we take the $R$-band light to be a fair tracer of the distribution of the old stellar population, this is consistent with the overall stellar population having been produced by SF activity distributed like that observed at present.

A final test was to model a stellar population formed by a constant SF rate over a Hubble time, using the BaSTI population synthesis code (Pietrinferni et al. 2004), and compare predicted colors and mass-to-light ratios with those observed for galaxies in the present sample. $R$ photometry was taken from the H$\alpha$GS database, and $J$ and $K$ total magnitudes were derived from the 2-Micron All-Sky Survey (Jarrett et al. 2003). There is excellent agreement in these colors: $(R-K) = 1.84$ from the model, cf. an observed mean of 1.87; $(J-K) = 0.76$ from the model, cf. 0.78 observed, for a model [Fe/H] of $-0.96$, i.e., slightly higher than that generally assumed for the Small Magellanic Cloud. The mass-to-light ratio of this same population is almost identical to that inferred earlier using the model of Bell & de Jong (2001), hence there is no inconsistency in using the mass from the latter model in calculating the SF timescale in Section 2.

## References

Almoznino, E., & Brosch, N. 1998, MNRAS, 298, 931
Bell, E. F., & de Jong, R. S. 2001, ApJ, 550, 212
Gallagher, J. S., Hunter, D. A., & Tutukov, A. V. 1984, ApJ, 284, 544
Gerola, H., Seiden, P. E., & Schulman, L. S. 1980, ApJ, 242, 517
Helmboldt, J. F., Walterbos, R. A. M., Bothun, G. D., O'Neil, K., & de Blok, W. J. G. 2004, ApJ, 613, 914
James, P. A., et al. 2004, A&A, 414, 23
James, P. A., Shane, N. S., Knapen, J. H., Etherton, J., & Percival, S. M. 2005, A&A, 429, 851
Jarrett, T. H., Chester, T., Cutri, R., Schneider, S. E., & Huchra, J. P. 2003, AJ, 125, 525
Kennicutt, R. C. 1983, ApJ, 272, 54
Pietrinferni, A., Cassisi, S., Salaris, M., & Castelli, F. 2004, ApJ, 612, 168
Swaters, R. A., van Albada, T. S., van der Hulst, J., & Sancisi, R. 2002, A&A, 390, 829
van Zee, L. 2001, AJ, 121, 2003

**Discussion**

*Sheth*: A comment: you mentioned that 2MASS is not deep enough but WISE will be an excellent dataset for this work, it will be 10 times deeper.

Phil James.

*Pathways through an Eclectic Universe*
*ASP Conference Series, Vol. 390, © 2008*
*J. H. Knapen, T. J. Mahoney, and A. Vazdekis, eds.*

# Three Types of Galaxy Disks

M. Pohlen,[1] P. Erwin,[2] I. Trujillo,[3] and J. E. Beckman[3]

[1]*Kapteyn Astronomical Institute, University of Groningen, NL-9700 AV Groningen, The Netherlands*
[2]*Max-Planck-Institut für extraterrestrische Physik, D-85748 Garching, Germany*
[3]*Instituto de Astrofísica de Canarias, E-38200 La Laguna, Spain*

**Abstract.** We present our new scheme for the classification of radial stellar surface brightness profiles for disk galaxies. We summarize the current theoretical attempts to understand their origin and give an example of an application by comparing local galaxies with their counterparts at high redshift ($z \approx 1$).

## 1. Eclectic Introduction

Galaxy formation and evolution is one of the most fascinating topics in astrophysical research today. Along several main pathways we are trying to answer the question of how galaxies are formed and how they evolve over time within a cosmological framework for structure formation. Theoretically this is often done via numerical $N$-body/SPH simulations (e.g., Governato et al. 2007) or semi-analytical modeling (e.g., Somerville & Primack 1999). According to Freeman & Bland-Hawthorn (2002) the associated observational part could be divided into doing *near-field cosmology*, i.e., looking for the fossil records of the galaxy formation and evolution process by observing local galaxies in detail, or doing *far-field cosmology*, i.e., looking at distant objects for the progenitors of modern-day galaxies.

Although it is now observationally possible to obtain spatially resolved kinematics and emission line measurements as a function of radius out to a redshift of $z \approx 1$ (Weiner et al. 2006), surface photometry is still a valuable source of information. By modeling the surface brightness distribution of a galaxy and thereby parameterizing the individual components, we can obtain a common ground to measure, compare, and sort large samples of galaxies. The structures we observe, our fossil records, should be linked to the galaxy assembly.

## 2. Sample, Data, and Profile Classification

To create a local reference dataset for comparison with high redshift galaxies we have collected imaging data for two large, complementary samples of face-on to intermediate inclined disk galaxies. The first is a diameter- and distance-limited sample of early-type galaxies (S0–Sb) split into two subsamples: 66 strong or weakly barred galaxies (Erwin, Beckman, & Pohlen 2005; Erwin, Pohlen, & Beckman 2008) and 45 unbarred galaxies (Aladro et al., this volume, p. 288) both drawn from the UGC catalog. The images are obtained with a variety

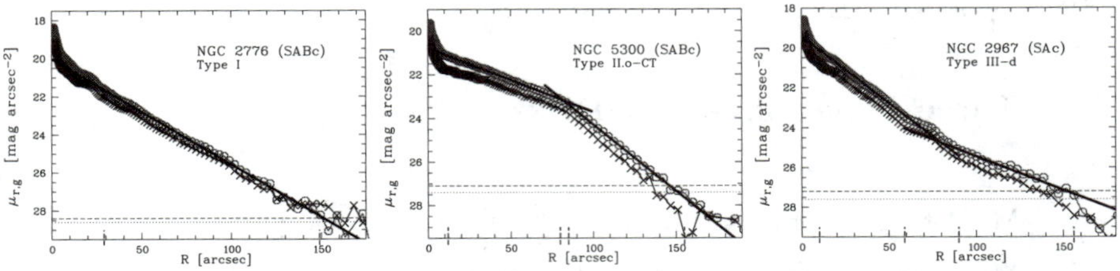


Figure 1. The three main disk types: Type I, Type II, and Type III (from left to right). Azimuthally averaged, radial SDSS surface brightness profiles in the $g'$- (*triangles*) and $r'$-band (*circles*) overlaid by $r'$-band exponential fits to the individual regions: single disk; inner and outer disk.

of different telescopes, including the Sloan Digital Sky Survey (SDSS, DR5). The second sample is a volume-limited sample ($D \lesssim 46$ Mpc) down to a limiting magnitude ($M_{\rm abs} < -18.4\,B$ mag) of late-type (Sb–Sdm) galaxies by Pohlen & Trujillo (2006). It comprises all selected galaxies from the LEDA online catalog having useful imaging data available in the SDSS (DR2), the sole data source for this study. Details about the extraction of the analyzed surface brightness profiles can be found in the cited papers.

In our analysis we concentrated not only on providing statistics on the exponential disk as given by de Jong (1996) or MacArthur et al. (2004), but rather study the shape of the profiles, as not all—indeed only the minority—are well described with a single exponential fitting function. To do so, we revised and extended the classification of surface brightness profiles introduced in the pioneering paper by Freeman's (1970) including the so-called truncation of the stellar population at the edge of the disk discovered by van der Kruit (1979). We identified three basic classes of surface brightness profiles depending on the presence of an apparent break feature or lack of one (cf. Fig. 1): 1) the well known Freeman Type I , which has an exponential profile, with no break; 2) Type II with a "down-bending break", now including the truncations; and 3) a completely new class, called Type III, also described by a broken exponential but with an up-bending profile, shallower beyond the break. The latter, discovered by Erwin et al. (2005), is also termed *antitruncated* since it shows the opposite behavior to the well known truncated profiles.

We have taken great care in ensuring that the critical sky subtraction is done properly, therefore always providing lower levels down to where we trust our profiles (cf. Fig. 1). It turns out that the uncertainty in our sky subtraction, in almost all cases, does not interfere with the basic classification of a galaxy. Nevertheless, it is worth noting that there are now completely independent measurements (from resolved star counts) of three local counterparts for each of our main types. There is NGC 300 (Bland-Hawthorn et al. 2005), where an unchanging exponential has been detected out to as far as ten scalelengths, M33 (Ferguson et al. 2007), where the profile is well fitted by a broken exponential with a down-bending break, and M31 (Ibata et al. 2005), which could be regarded as a Type III (antitruncated) disk.

In order to address the origin of the different types we have introduced an interpretative subclassification scheme for our profiles. For example, we distinguish between those Type II profiles possibly associated with a bar resonance,

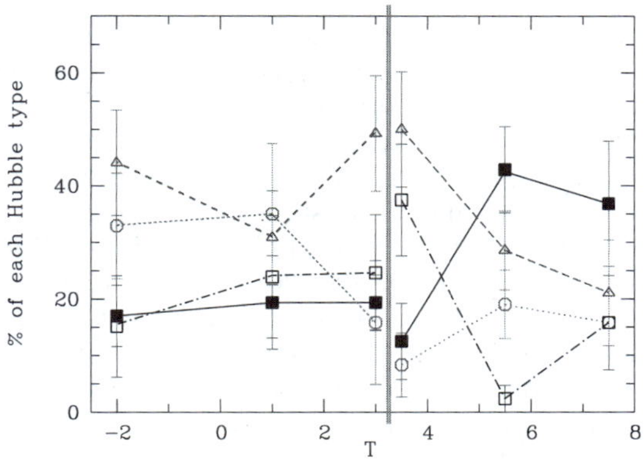


Figure 2. Frequency of profile types: Type I (*open circles*), Type II-CT (*filled squares*), Type III (*open triangles*), and Type II.o-OLR (*open squares*) in relation to the Hubble type. The galaxies are merged in six morphological bins ($T \in [-3.4\text{--}-0.5], [-0.4\text{--}1.4], [1.5\text{--}3.4], [2.5\text{--}4.4], [4.5\text{--}6.4], [6.5\text{--}8.4]$). The vertical lines separate the results of the early- and late-type samples.

called Type II.o-OLR, and those probably related to a star formation threshold, called *classical truncation* or Type II-CT (see next section, as well as Erwin et al., this volume, p. 251, and Aladro et al., this volume, p. 288). Evidence about the origin may come from the trends we observed with Hubble type (see Fig. 2). Any model of galaxy formation and evolution should be able to explain why we find, for example, more Type II-CT cases in late-type than early-type galaxies, whereas the trend is the other way round for the Type I galaxies.

## 3. Origin and Implications

Our basic classification is simply morphological in nature. We do pure taxonomy similar to grouping galaxies along the Hubble sequence. But what is the physical background for the different types? What does theory tell us about surface brightness profiles? It turns out that our current physical picture lags behind the observational evidence. The origin of purely exponential disks (Type I) in itself is not fully understood, although well known since the 1940s and '50s. One of the most successful attempts to show that an exponential disk can arise naturally out of disk formation conditions was by Yoshii & Sommer-Larson (1989), who showed that an exponential disk will form if the viscosity timescale and the star formation timescale are comparable. The currently favored model ascribes the classical truncations (Type II-CT) to the effect of a star formation threshold in disk column density as the edge is approached (Kennicutt 1989; Schaye 2004; Elmegreen & Hunter 2006; Li, Mac Low, & Klessen 2006). On the other hand, using purely collisionless $N$-body simulations Debattista et al. (2006) find down-bending breaks appearing in the disk profile as a resonance phenomenon, which may be associated with our Type II.o-OLR breaks. Observationally, there is circumstantial evidence that some of the galaxies with Type III profiles are associated with recent mergers (e.g., M31), supported now by the new $N$-body/SPH simulations of Younger et al. (2007), showing that minor mergers could produce up-bending stellar profiles in the remnant galaxy. On the other hand, by a small change of the input gas density, Elmegreen & Hunter (2006) are also able to produce Type III profiles within their evolutionary model of an isolated galaxy with star formation.

Once the nature of the galactic stellar disk is determined it will be fully justified to use the shape of the profiles and the position of the breaks for comparison of galaxies at various redshifts, as done by Pérez (2004), Trujillo & Pohlen (2005), or Tamm & Tenjes (2006). For example, Trujillo & Pohlen (2005) presented a first direct application of the new classification scheme by exploring the *Hubble* Ultra Deep Field. Using the position of the truncation as a direct estimator of the size of the stellar disk it becomes possible to observe inside-out growth of galactic disks—inherent in the current galaxy formation and evolution models—comparing the local late-type Type II galaxies with their counterparts at high $z$ (i.e., $z \approx 1$) in a straightforward manner. Although still uncertain because of the small number of galaxies analyzed, their results suggest that the radial position of the truncation has increased with cosmic time by ~1–3 kpc in the last ~8 Gyr indicating a small to moderate (~25) inside-out growth of the disk galaxies since $z \sim 1$.

**Acknowledgments.** MP would like to thank the organizers and the Kapteyn Instituut for their generous financial support. PE was supported by DFG Priority Program 1177. This work was supported by grant AYA2004.08251-CO2-01 of the Spanish Ministry of Education and Science, and project P3/86 of the IAC.

**References**

Bland-Hawthorn, J., Vlajić, M., Freeman, K. C., & Draine, B. T. 2005, ApJ, 629, 239
Debattista, V. P., Mayer, L., Carollo, C. M., Moore, B., Wadsley, J., & Quinn, T. 2006, ApJ, 645, 209
de Jong, R. S. 1996, A&A, 313, 45
Elmegreen, B. G., & Hunter, D. A. 2006, ApJ, 636, 712
Erwin, P., Beckman, J. E., & Pohlen, M. 2005, ApJ, 626, L81
Erwin, P., Pohlen, M., & Beckman, J. E. 2008, AJ, 135, 20
Ferguson, A., Irwin, M., Chapman, S., Ibata, R., Lewis, G., & Tanvir, N. 2006, in Island Universes, ed. R. S. de Jong (New York: Springer), p. 219
Freeman, K. C. 1970, ApJ, 160, 811
Freeman, K. C., & Bland-Hawthorn, J. 2002, ARA&A, 40, 487
Governato, F., et al. 2007, MNRAS, 374, 1479
Ibata, R., Chapman, S., Ferguson, A. M. N., Lewis, G., Irwin, M., & Tanvir, N. 2005, ApJ, 634, 287
Kennicutt, R. C. 1989, ApJ, 344, 685
Li, Y., Mac Low, M.-M., & Klessen, R. S. 2006, ApJ, 639, 879
MacArthur, L. A., Courteau, S., Bell, E., & Holtzman, J. A. 2004, ApJS, 152, 175
Pérez, I. 2004, A&A, 427, L17
Pohlen, M., & Trujillo, I. 2006, A&A, 454, 759
Schaye, J. 2004, ApJ, 609, 669
Somerville, R. S., & Primack, J. R. 1999, MNRAS, 310, 1087
Tamm, A., & Tenjes, P. 2006, A&A, 449, 67
Trujillo, I., & Pohlen, M. 2005, ApJ, 630, L17
van der Kruit, P. C. 1979, A&AS, 38, 15
Weiner, B. J., et al. 2006, ApJ, 653, 1027
Yoshii, Y., & Sommer-Larson, J. 1989, MNRAS, 236, 779
Younger, J. D., Cox, T. J., Seth, A. C., & Hernquist, L. 2007, ApJ, 670, 269

*Pathways through an Eclectic Universe*
*ASP Conference Series, Vol. 390, © 2008*
*J. H. Knapen, T. J. Mahoney, and A. Vazdekis, eds.*

# The Outer Structure of Galactic Disks: Connections Between Bars, Disks, and Environments

Peter Erwin

*MPI für extraterr. Physik, Giessenbachstr., D-85748 Garching, Germany*

Michael Pohlen

*Astronomical Institute, University of Groningen, PO Box 800, NL-9700 AV Groningen, The Netherlands*

John E. Beckman, Leonel Gutiérrez, and Rebeca Aladro

*Instituto de Astrofísica de Canarias, E-38200 La Laguna, Spain*

**Abstract.** Surface brightness profiles for early type (S0–Sb) disks exhibit three main classes (Type I, II, and III). Type II profiles are more common in barred galaxies, and most of the time appear to be related to the bar's outer Lindblad resonance. Roughly half of barred galaxies in the field have Type II profiles, but almost none in the Virgo cluster do. A strong *anti*correlation is found between Type III profiles ("antitruncations") and bars: Type III profiles are most common when there is no bar, and least common when there is a strong bar.

## 1. Introduction

Recent imaging studies of the disks of S0 and spiral galaxies have demonstrated that not all stellar disks have simple exponential surface brightness profiles (e.g., Erwin, Beckman, & Pohlen 2005; Pohlen & Trujillo 2006; Hunter & Elmegreen 2006; Erwin, Pohlen, & Beckman 2007). Instead, disk profiles appear to fall into three general categories: single-exponential Freeman (1970) Type I; Freeman Type II, with a shallow inner slope and a steeper outer slope (including so-called "truncations"); and Type III ("antitruncations"), with a steep inner slope and a shallow outer slope (Erwin et al. 2005). See Pohlen et al. (this volume, p. 247) for more background and illustrations of the three types.

We report here on analysis of a deep imaging study of lenticulars and early type spirals (Hubble types S0–Sb), focused on tracing the outer disk structure using azimuthally averaged surface brightness profiles. The analysis of the barred galaxy subsample (66 galaxies) is complete (Erwin, Pohlen, & Beckman 2007); analysis of the 45 *unbarred* galaxies is nearly complete (see Aladro et al., this volume, p. 288).

Pohlen et al. (this volume, p. 247) discuss the general trend of these disk types versus Hubble type, including the complementary late type sample of Pohlen & Trujillo (2006); here, we highlight what we can learn about profiles in the earlier types and their connections to bars and galaxy environment.

## 2. The Nature of Type II Profiles

For barred galaxies, we can use the size of the bar as a measuring rod. Type II profiles (downward-bending broken exponentials) fall into two fairly distinct categories: those with the break inside the bar (Type II.i) and those with the break outside (Type II.o). The first category is rare (6% of the barred galaxies), but does match profiles of some $N$-body simulations (e.g., Athanassoula & Misiriotis 2002; Valenzuela & Klypin 2003), so we suspect that this is a side-effect of the bar formation process.

Type II.o profiles are much more common (42% of barred galaxies), but more difficult to explain. One possibility is "truncations" due to a threshold in star formation. Elmegreen & Hunter (2006) recently showed that such thresholds could produce broken-exponential Type II profiles. However, the threshold (gas) surface densities predicted by such models (e.g., 3–10 $M_{\odot}\,\mathrm{pc}^{-2}$ in Schaye 2004) are quite low. While this is plausibly consistent with some breaks in late type galaxies (e.g., Pohlen et al. 2002), the surface brightnesses at the break radius which we observe correspond to stellar mass densities which are an order of magnitude or more higher: 20–2000 $M_{\odot}$ $\mathrm{pc}^{-2}$ (median $\sim 100\,M_{\odot}$ $\mathrm{pc}^{-2}$).

One potential clue is that we often find outer rings in Type II.o galaxies with the ring located at or very close to the break (Fig. 1). Since outer rings are well understood as an effect of a bar's outer Lindblad resonance (OLR; e.g. Buta & Combes 1996), we suspect that these breaks are a related phenomenon. Supporting evidence comes from recent $N$-body simulations by Debattista et al. (2006), who found that Type II.o profiles could form at the OLR of a bar-driven spiral. This leads us to suggest that the breaks in most Type II.o profiles are OLR-related, and we refer to these as "Type II.o-OLR" profiles.

Figure 2 demonstrates that the break radii in our Type II.o profiles follow the same size distribution (break radius in terms of bar radius) as the outer rings in our sample, as well as the outer rings in the sample of Buta & Crocker (1993). A handful of galaxies have breaks at larger radii ($> 3 \times R_{\mathrm{bar}}$)—corresponding to low stellar mass densities—which suggests that the star formation threshold mechanism might be at work in these galaxies, instead of an OLR-related mechanism. Figure 1 shows one galaxy where this may be the case: the break is located well *outside* the bar's outer ring, and is thus *not* connected with the OLR. We label such profiles "classical truncations" (Type II.o-CT).

Note that the CT profiles are rare in the early type galaxies (5% of the barred galaxies, 27% of the unbarred galaxies), however, Pohlen & Trujillo (2006) found that they are more common in late type spirals (see Pohlen et al., this volume, p. 247).

## 3. Frequencies of Disk Types versus Bars and Environments

A striking contrast emerges when barred and unbarred galaxies are compared: unbarred galaxies are twice as likely to have Type III profiles (60% of unbarred galaxies, 32% of barred galaxies). This carries over into the barred galaxies themselves: weakly barred (SAB) galaxies are more likely to have Type III profiles than strongly barred (SB) galaxies. Similarly, the distribution of bar strengths (deprojected maximum isophotal ellipticity) is different for Type III

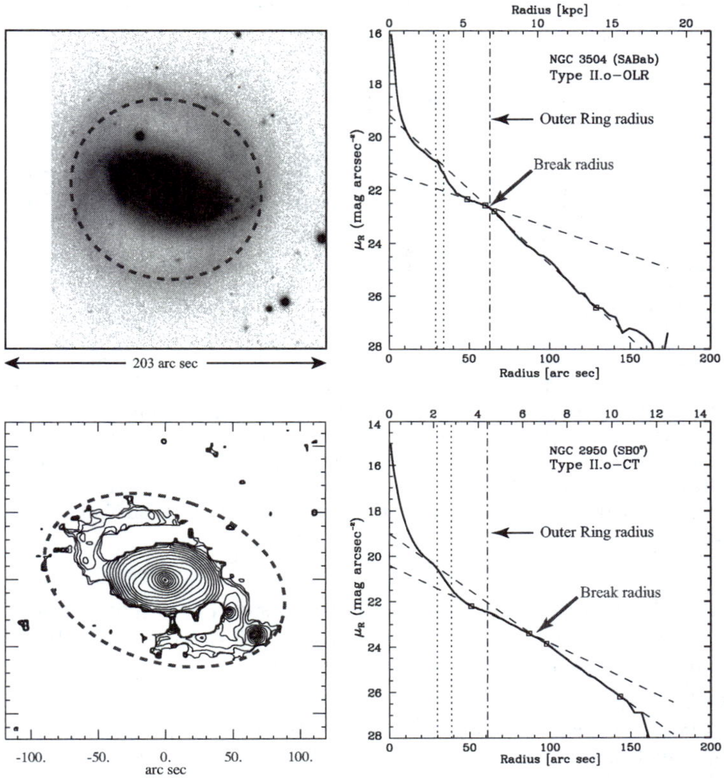


Figure 1. OLR breaks (II.o-OLR) versus classical truncations (II.o-CT). *Top:* A Type II.o profile where the break coincides with an outer ring. *Bottom:* A Type II.o profile where the break is well *outside* the outer ring. In both cases, the break radius is indicated by the dashed ellipse (*left-hand panels*) and the arrow (*right-hand panels*). For NGC 3504, we display an SDSS $r$-band image; for NGC 2950, we have subtracted a model of the outer disk from the SDSS image in order to bring out the (faint) outer pseudo-ring.

and non-type III profiles: bars in Type III galaxies have a median (deprojected) ellipticity of $\approx 0.35$, versus $\approx 0.50$ for Types I and II.

Type II profiles (of all kinds) are clearly more common in barred galaxies (49% of barred galaxies versus 27% of unbarred galaxies). However, they seem to be relatively unaffected by bar *strength*: for example, the Type II.o frequency is essentially identical for both SB and SAB galaxies, and there is no difference in bar ellipticity between Type I and II.o profiles.

What *does* appear to affect the presence of the Type II.o profiles is the *environment*. Specifically, barred galaxies in the Virgo cluster have a much lower frequency of Type II.o profiles (10%) than do barred galaxies in the field (49%). Most of the "missing" II.o profiles in the Virgo Cluster are Type I profiles instead. Although the numbers are relatively small, the result is significant at the

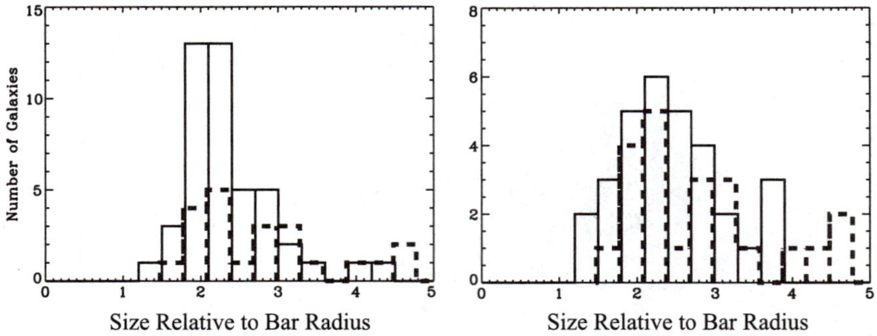


Figure 2. Comparisons of outer ring sizes and Type II.o break radii. *Left:* Outer-ring sizes from Buta & Crocker (1993), using inner-ring size as a proxy for bar radius, along with break radii from our sample (*dashed histogram*). *Right:* As before, except using outer-ring sizes from our sample (Erwin, Pohlen, & Beckman 2007). In both cases, II.o break radii share the same size distribution as outer rings (except for a tail at large radii), suggesting that these breaks are, like outer rings, related the outer Lindblad resonances of bars.

99.8% level. *Why* this should be so is hard to say, particularly since we don't yet know why some field galaxies have Type I profiles and others have Type II. One speculative possibility is that Type II.o-OLR profiles (which are the vast majority of the field II.o profiles) require several Gyr of interaction between the bar and gas in the outer disk. If the gas is removed rapidly enough, as might be the case for ram-pressure stripping of spirals which fell into the Virgo cluster several Gyr ago, then the OLR break formation process might be choked off before it can proceed to completion.

**Acknowledgments.** PE was supported by DFG Priority Program 1177; this work was also supported by grants No. AYA2004-08251-CO2-01 from the Spanish Ministry of Education and Science and P3/86 of the Instituto de Astrofísica de Canarias.

## References

Athanassoula, E., & Misiriotis, A. 2002, MNRAS, 330, 35
Buta, R., & Crocker, D. A. 1993, AJ, 105, 1344
Buta, R., & Combes, F. 1996, Fund. Cosmic Phys., 17, 95
Debattista, V. P., Mayer, L., Carollo, C. M., Moore, B., Wadsley, J., & Quinn, T. 2006, ApJ, 645, 209
Elmegreen, B. G., & Hunter, D. A. 2006, ApJ, 636, 712
Erwin, P., Beckman, J. E., & Pohlen, M. 2005, ApJ, 626, L81
Erwin, P., Pohlen, M., & Beckman, J. E. 2007, AJ, submitted
Freeman, K. C. 1970, ApJ, 160, 811
Hunter, D. A., & Elmegreen, B. G. 2005, ApJS, 162, 49
Pohlen, M., & Trujillo, I. 2006, A&A, 454, 759
Pohlen, M., Dettmar, R.-J., Lütticke, R., & Aronica, G. 2002, A&A, 392, 807
Schaye, J. 2004, ApJ, 609, 667
Valenzuela, O., & Klypin, A. 2003, MNRAS, 345, 406

*Pathways through an Eclectic Universe*
*ASP Conference Series, Vol. 390, © 2008*
*J. H. Knapen, T. J. Mahoney, and A. Vazdekis, eds.*

# Resolving the Age–Metallicity Degeneracy

Alexandre Vazdekis

*Instituto de Astrofísica de Canarias, E-38200 La Laguna, Spain*

**Abstract.** Despite the great developments that have been achieved in the last decade in disentangling most relevant stellar population parameters, in particular with the inclusion of absorption linestrength indices, the age–metallicity degeneracy still lurks in these studies. Here I summarize our effort for developing the stellar population synthesis method, which includes models predicting full spectral energy distributions of stellar populations at moderately high resolution. We show how these model spectral energy distributions open up new possibilities to overcome this degeneracy, such as optimized indices with enhanced sensitivities to age, metallicity, and abundance ratios of various key species. These tools are applied to well-selected samples of early-type galaxies embedded in various environments, allowing us to further our understanding of the origin of the color–magnitude relation and of galaxy formation timescales as a function of cluster properties.

## 1. Introduction

Studies of stellar populations are nowadays providing tight constraints on our picture of how galaxies form and evolve. Indeed, there are a number of results that appear to contradict the standard versions of semi-analytic models within a hierarchical galaxy formation framework (Kauffmann & Charlot 1998). It is, for example, difficult to reconcile these models with the tight color–magnitude relation (CMR) found at higher redshifts (e.g., Blakeslee et al. 2003). From spectroscopic linestrength studies it has been shown that larger galaxies have greater mean ages than their smaller counterparts (Kauffmann et al. 2003; Caldwell, Rose, & Concannon 2003). Furthermore, nearby giant elliptical galaxies show an [Mg/Fe] overabundance (e.g., Worthey, Faber, & González 1992; González 1993), suggesting that the bulk of their stars were formed at very early times.

To derive most relevant stellar population parameters, such as ages and metallicities, observed colors or absorption linestrengths are compared to the predictions of stellar population synthesis models. The most popular approach consists in comparing the data to single age, single metallicity stellar populations (SSPs). Therefore we obtain mean luminosity-weighted ages and metallicities, which are biased towards the most recent star formation episodes. Abundance ratios of key elements, such as [Mg/Fe], provide important clues to how fast the bulk of the stars formed at the earliest stages of galaxy formation (e.g., Worthey et al. 1992), as the magnesium, which is an $\alpha$-element, is ejected into the interstellar medium (ISM) by Type II supernovae (SNe) on very short timescales ($< 10\,$Myr), whereas the iron-peak elements are produced by SNe Ia, which occur on timescales of $\sim$1 Gyr. Thus by joining these two pieces of information we are able to constrain, via the SSP approach, the star formation history. This simple approach has the advantage that the analysis is performed on the basis of a significantly lower number of free parameters in comparison to the use of more complex stellar population models (e.g., Vazdekis et al. 1996)

However, the study of the integrated light of the stellar systems is affected by a number of problems, such as the degeneracy between the age of a burst and its strength (e.g., Leonardi & Rose 1996), the effects of the initial mass function (e.g., Vazdekis et al. 2003), or the well-known age–metallicity degeneracy (e.g., Worthey 1994; Arimoto 1996). The latter makes a galaxy look redder either because it is older or more metal-rich, although the inclusion in the analysis of low resolution absorption linestrengths, such as the widely used optical indices of the Lick/IDS system, has partially lifted this degeneracy as the effects of the ages and metallicities are selectively emphasized in various features (e.g., Trager et al. 1998 and references therein).

I will summarize here an effort to overcome the age–metallicity degeneracy that we started in the mid-'90s under the inspiration of John Beckman. In Sect. 2 the stellar population synthesis models that we have developed to tackle this problem are introduced. In Sect. 3 new indices, which include more age/metallicity sensitive linestrength index definitions are shown. In Sect. 4 we apply these newly developed methods to early-type galaxies in clusters of different properties.

## 2. Stellar Population Synthesis Models

The quality of stellar population synthesis models relies on the ingredients employed, mainly the stellar tracks and the stellar spectral libraries. Our model implements the scaled solar theoretical isochrones of the Padova group (Girardi et al. 2000), as these models cover a wide range of ages and metallicities and include the latest stages of the stellar evolution through the thermally pulsing AGB regime to the point of complete envelope ejection (employing a synthetic prescription). The isochrones, which are based on homogeneous input physics, were updated with an improved version of the equation of state, the opacities of Alexander & Ferguson (1994), and a mild convective overshoot scheme. A helium fraction was adopted according to the relation: $Y \approx 0.23 + 2.25Z$. The lower mass cut-off is $0.15\,M_{\odot}$. The highest metallicity covered is $Z = 0.03$. Note, however, that for the first versions of our models we employed the Bertelli et al. (1994) isochrones, which extend the metallicity range to $Z = 0.05$.

Unlike other models, which make use of theoretical stellar atmospheres, we use relations inferred on the basis of extensive empirical stellar photometric libraries to transform the theoretical parameters of the isochrones to the observational plane, i.e., colors and magnitudes. These relations allow us to obtain each color as a function of the temperature, metallicity, and gravity, and we generally use the metallicity-dependent empirical relations of Alonso, Arribas, & Martínez-Roger (1996, 1999) for dwarfs and giants, respectively.

These ingredients are required to predict fluxes, colors, surface-brightness fluctuations and mass-to-light ratios of stellar populations of different metallicities and ages (though our models are optimized for intermediate-age and old SSPs, i.e., older than 0.1 Gyr). To predict the absorption linestrength indices of the

Lick/IDS system at FWHM > 8 Å, our model incorporates the empirical fitting functions of Worthey et al. (1994) and Worthey & Ottaviani (1997), which relate the index strengths to the stellar atmospheric parameters ($T_{\rm eff}$, $\log g$, [Fe/H]) of the Lick library.

New empirical stellar spectral libraries, with flux-calibrated response curves and much improved atmospheric parameter coverage, have allowed us to predict full spectra of stellar populations at moderately high resolution (rather than predicting a number of linestrengths at low resolution). In Vazdekis (1999) we employed the library of Jones (1999), which is composed of ~700 stars, for predicting SSP spectra in two narrow ranges around 4000 Å and 5000 Å at resolution 1.8 Å (FWHM). We have developed new stellar libraries to increase the spectral coverage of our models. The library of Cenarro et al. (2001a,b), composed of ~700 stars, is incorporated to predict spectra in the Ca II triplet region around 8500 Å at resolution 1.5 Å (FWHM). We have recently published a new empirical stellar spectral library, called MILES, which covers the whole optical range (λλ3500–7500 Å) at 2.3 Å(FWHM) (Sánchez-Blázquez et al. 2006). The major feature of this library, which is composed of 985 stars, is its unprecedented stellar parameter coverage (Cenarro et al. 2007). There are alternative models, such as those of Bruzual & Charlot (2003), which predict spectra covering the range λλ3500–9000 Å at resolution 3 Å (FWHM).

To be able to apply the predictions based on the Lick/IDS system, computed via the empirical fitting functions, higher quality data must be transformed to match the lower spectral resolution (FWHM > 8 Å) and non-flux-calibrated response curve of this system (Worthey & Ottaviani 1997). Therefore, these model predictions are not optimal for exploiting galaxy spectra obtained with 8–10 m class telescopes. However, the SSP spectral energy distributions (SEDs) introduce a new methodology for analyzing spectra of galaxies and stellar clusters (Vazdekis 1999). First, the resolution and dispersion of the synthetic spectra are adjusted to match the instrumental configuration used for the observations. Second, the model spectra are smoothed to match the galaxy velocity dispersion. Finally, the synthetic spectra can either be compared to the whole spectrum (or part of it) or the comparison can be made on the basis of selected linestrengths, which are measured in both the observed and the model SSP spectra (e.g., Vazdekis et al. 2001b; Yamada et al. 2006). In addition, these model SEDs allow us to define alternative indices to those of the Lick system, as will be shown in Section 3.

## 3. Breaking the Age–Metallicity Degeneracy

The new models predicting full SEDs of stellar populations at moderately high resolution open up new possibilities for overcoming the age–metallicity degeneracy, as they allow us to define and optimize new indices with enhanced sensitivities to the ages and metallicities, including the abundance ratios of key species. This is now possible as we can shift and vary the widths of the pseudocontinua and feature passbands required to define an index and measure it directly on the model SSP spectral library. Therefore, the optimization of its sensitivity to, for example, age or metallcity can be done with no need to perform measurements on individual stellar spectra (Vazdekis & Arimoto 1999). In addition, the index definition can be tuned to minimize for example the signal-to-noise (S/N) requirements, to decrease its sensitivity to the spectral resolution and velocity dispersion, wavelength shifts due to rotation, redshift or wavelength calibration accuracy.

Particular effort has been devoted to defining new age indicators. We have mainly focused on two Balmer lines to develop these indices, namely H$\gamma$ and H$\beta$. Taking full advantage of the information available in the integrated SSP spectra, we defined a set of indices based on the H$\gamma$ feature, which provide unprecedented power at breaking the age–metallicity degeneracy, as they are virtually free from metallicity effects. These indices consist of pseudo-equivalent width measurement, relying on the pseudocontinuum peaks immediately longward and shortward of H$\gamma$. In addition, they are an improvement over earlier pseudo-equivalent width H$\gamma$ indices (Rose 1994; Jones & Worthey 1995) in being relatively insensitive to spectral resolution. For this purpose four indices, called H$\gamma_\sigma$, were defined to cover the velocity dispersion range of most galaxies. Table 1 lists the index definition passbands, and their range of applicability in $\sigma$. A strong limitation of the H$\gamma_\sigma$ indices is that a S/N (per Å) above $\sim$170, depending on the age and $\sigma$ ranges, is required. However, such high S/N values could be achieved by integrating over the spatial direction. A full description of the characteristics and properties of this set of H$\gamma$ age indicators can be found in Vazdekis & Arimoto (1999) and Vazdekis et al. (2001a).

An alternative age indicator based on the H$\beta$ feature has recently been proposed to overcome some of the limitations of the H$\gamma_\sigma$ set of indices (J. L. Cervantes-Rodríguez & A. Vazdekis, in preparation). This index, named H$\beta_0$, is virtually insensitive to metallicity in comparison to the standard H$\beta$ definition of the Lick/IDS system. When this index is plotted versus a metallicity indicator for SSPs of different ages and metallicities we obtain a nearly orthogonal model grid, allowing us to disentangle the age and the metallicity. The S/N required to obtain an accurate measurement of this index is very similar to that needed for the H$\beta$ Lick/IDS, i.e., S/N $> 40$. It is worth recalling that these values need not be transformed to the Lick/IDS system response curve nor to the Lick/IDS spectral resolution, as would have been the case if model predictions based on the Lick/IDS empirical fitting functions (e.g., Worthey 1994; Vazdekis et al. 1996; Thomas, Maraston, & Bender 2003) were to be applied.

One should note that Balmer lines, and particularly H$\beta$, might be filled in with nebular emission and therefore the estimated ages can be significantly older if not corrected for this effect (e.g., González 1993). Examples of such corrections can be found in de la Rosa et al. (2007). However, a detailed treatment, such as employing methods that simultaneously fit both the stellar spectrum and the emission lines to separate these two components, is highly recommended (e.g., Sarzi et al. 2006).

We are currently developing a non-spectroscopic approach to obtain relevant stellar population parameters with great accuracy. We are modeling a new set of indicators based on the tunable filters of the Day One instrument OSIRIS on the Spanish 10.4 m GTC. These filters, which reach high enough resolution to be able to measure the absorption features, can be positioned on a particular feature and on selected pseudocontinua to mimick spectroscopic index definitions. We have shown that these indicators provide a greater age/metallicity disentangling

Table 1. Age index definitions

| Index | Blue continuum Å | Index bandpass Å | Red continuum Å | $\sigma_{\rm total}$ [a] km s$^{-1}$ | S/N [b] per Å | Ref. [c] |
|---|---|---|---|---|---|---|
| H$\gamma_{\sigma\leq130}$ | 4329.00–4340.47 | 4333.25–4363.00 | 4352.50–4368.25 | 60–130 | 200 | 1 |
| H$\gamma_{125}$ | 4330.00–4340.49 | 4333.00–4352.74 | 4359.25–4368.75 | 100–175 | 200 | 2 |
| H$\gamma_{200}$ | 4331.00–4340.75 | 4332.00–4352.25 | 4359.25–4368.75 | 150–225 | 300 | 2 |
| H$\gamma_{275}$ | 4331.50–4341.00 | 4331.50–4351.88 | 4359.25–4368.75 | 225–300 | 400 | 2 |
| H$\beta_0$ | 4821.18–4838.40 | 4839.28–4877.10 | 4897.45–4915.85 | 0–400 | 50 | 3 |

[a] Velocity dispersion range of applicability ($\sigma^2_{\rm total} = \sigma^2_{\rm galaxy} + \sigma^2_{\rm instr}$).

[b] Required S/N Å$^{-1}$ to distinguish the ages of 12 Gyr and 17.5 Gyr.

[c] References: (1) Vazdekis et al. (2001a), (2) Vazdekis & Arimoto (1999), (3) J. L. Cervantes-Rodríguez & A. Vazdekis (in preparation).

power than their standard spectroscopic Lick/IDS index counterparts. Among other advantages, this approach is more efficient than standard spectroscopy if a limited number of key indicators is to be studied, as the acquisition of full galaxy spectra with standard spectroscopy is particularly expensive in terms of telescope time. The wide field of view of this instrument and the fact that we obtain two-dimensional images represents an additional advantage over long/multislit spectroscopic facilities. Furthermore, the current integral field units that are often used for obtaining galaxy spectra (e.g., SAURON) cover very small areas (usually smaller than 1 arcmin diameter). Therefore, this new approach will be useful for mapping the ages/metallicities of nearby unresolved galaxies (i.e., out to $\sim$3 effective radii of a galaxy), or to observe fainter targets that cannot be reached with spectroscopic methods (Cervantes-Rodríguez, Vazdekis, & Cepa 2006).

### 3.1. Estimating Abundance Ratios and Total Metallicity

Massive early-type galaxies do not follow scaled solar abundance ratios, as they show a [Mg/Fe] overabundance (e.g., Worthey et al. 1992). However, they are usually compared to predicted linestrength indices that rely on stellar libraries observed in the solar neighborhood, i.e., with a scaled solar element mixture (for stars with metallicities close to solar). It is becoming common practise to perform the analysis with the aid of models predicting Lick/IDS linestrength indices that are specifically computed for different abundance ratios (e.g., Tantalo, Chiosi, & Bressan 1998; Trager et al. 2000; Thomas et al. 2003). This is possible by including the sensitivities of these line indices to the abundance variations of the different elements from theoretical stellar atmospheric computations (e.g., Tripicco & Bell 1995). Another approach relies on isochrones which are specifically computed for different $\alpha$-enhancements (e.g., Salasnich et al. 2000; VandenBerg et al. 2000) as higher metallicity isochrones cannot be mimicked by scaled solar isochrones of the same total metallicity (Salaris & Weiss 1998). Linestrengths predicted on the basis of these isochrones are, however, based on the same empirical fitting functions used for the standard model predictions, i.e., with scaled solar ratios. Therefore none of these approaches is fully self-consistent as the same abundance pattern should apply to both the stellar tracks and the stellar spectra. At the time of writing this paper, the only self-consistent model predictions published were by Coelho et al. (2007), on the basis of a fully theoretical treatment, where both the stellar tracks and the stellar isochrones were computed with exactly the same element partition.

Inspired by the original studies that led to the [Mg/Fe] overabundance discovery we propose here an alternative approach, which employs only scaled solar model grids. We plot linestrength indices that are mostly dominated by specific elements, e.g., Mg*b* by Mg, $\langle Fe \rangle$ by Fe, versus a powerful age indicator, such as $H\beta_0$ or $H\gamma_\sigma$ (see Table 1). For a given object the resulting age estimates do not depend on the metallicity indicator in use, as the model grids are rather orthogonal. If the object were enhanced in, for example, Mg over Fe the metallicities obtained from the Mg-dominated index plot will be higher than the one from the Fe-dominated indicator. These metallicities can be used to compute abundance ratios, estimated in this way, which strongly correlate with the values obtained from models that incorporate these effects from stellar atmospheric computations (e.g., Thomas et al. 2003), as shown in de la Rosa et al. (2007). Note that this way of estimating the abundance ratios does not attempt to determine the true abundances as, among other reasons, the metallicity indicators employed do include contributions from other elements. However, this approach is fully appropriate to performing a qualitative analysis for disentangling relative abundance patterns among galaxies or stellar clusters. An advantage of applying this approach is that the results obtained are only mildly model-dependent, as only scaled solar models are employed.

To estimate the total metallicity of galaxies with [Mg/Fe] $> 0$ we plot one of these age-sensitive indices versus an $\alpha$-enhancement insensitive indicator, such as the combined indices [MgFe] (González 1993) or [MgFe]' (Thomas et al. 2003).

## 4. Results

This section shows various results obtained with these new models and methodology for analyzing galaxy spectra. These results might have important implications for our picture of how galaxies form and evolve, showing the relevance of performing detailed stellar population studies with improved tools of analysis.

### 4.1. Ages and Metallicities in Virgo Elliptical Galaxies

Elliptical galaxies yield a robust CMR at low and medium redshift (e.g., Bower, Lucey, & Ellis 1992). This relation could be either caused by a variation of the mean stellar metallicity along the sequence (e.g., Kodama & Arimoto 1997), or by a conspiracy between age and metallicity variations (e.g., Ferreras, Charlot, & Silk 1999). The first possibility would be supported by a dissipational collapse scenario (e.g., Chiosi & Carraro 2002). The second possibility would be supported within a hierarchical galaxy formation framework (e.g., Kauffmann & Charlot 1998) if galaxies had some recent star formation due to late mergers. However, an unambiguous analysis of the ages and metallicities along the CMR has been hampered by the age–metallicity degeneracy.

Spectra of extremely high quality (S/N $> 100$; obtained at the WHT and Subaru telescopes) of a well selected sample of ellipticals along the CMR of the Virgo cluster were analyzed with the $H\gamma_\sigma$ indices providing very accurate age and metallicity estimates, as described in Section 3. We find that these galaxies are all older than ~8 Gyr. Our analysis led to a positive correlation of the metallicity with galaxy luminosity, color, or velocity dispersion, but no clear age trend along the CMR for galaxies with $\sigma$ above $100\,\mathrm{km\,s^{-1}}$ (Vazdekis et al. 2001b). However, when smaller galaxies are taken into account we find a large age scatter (Yamada et al. 2006; Yamada et al., this volume, p. 326). Therefore, lower mass galaxies are on average younger than their more massive counterparts. This result, which is derived from a very detailed stellar population analysis, is in good agreement with the findings from more statistically meaningful studies (e.g., Kauffmann et al. 2003; Caldwell et al. 2003).

We also find that the metallicity, as measured by Fe or Ca lines, does not increase significantly with galaxy luminosity (or $\sigma$), but does increase when measured with Mg or CN indices. The [Mg/Fe] and [CN/Fe] abundance ratios increase with galaxy mass along the CMR of this cluster. This result suggests a shortening of the characteristic formation timescales for the bulk of the stellar populations of these ellipticals with increasing galaxy mass. Note, however, that not all the abundance ratios increasing with $\sigma$ are $\alpha$ elements (e.g., CN), and neither do all the $\alpha$-elements track Mg (e.g., Ca; see also Vazdekis et al. 1997).

### 4.2. Star Formation Timescales in Intermediate and Massive Elliptical Galaxies

A pivotal result from absorption linestrength studies has been the finding that giant ellipticals show [Mg/Fe] overabundances with respect to the scaled solar element ratios (e.g., Worthey et al. 1992). However, the study of chemical abundance patterns in early-type galaxies is still in its infancy. For example, some elements such as Ca do not show any enhancement with respect to Fe (Vazdekis et al. 1997; Cenarro et al. 2003). So far, only the contribution of type II and type Ia SNe, which are used to interpret the [Mg/Fe] ratios, have been considered for the formation timescales. However, there are other processes such as nucleosynthetic pollution from intermediate mass stars, including neutron-capture elements, which can give us complementary probes of star formation timescales. In fact, Sánchez-Blázquez et al. (2003) have discovered significant differences in the CN index between galaxies in the Virgo and Coma clusters.

In Carretero et al. (2004) we find a clear trend between the [CN/Fe] and [C/Fe] abundances measured in intermediate to massive early-type galaxies ($150 < \sigma < 250\,\mathrm{km\,s^{-1}}$) and cluster X-ray luminosity from higher S/N galaxy spectra of the various releases of the Sloan Digital Sky Survey. The selected galaxies are all morphologically confirmed early-type galaxies according to their light profiles following Bernardi et al. (2003). We find that galaxies in richer clusters (i.e., higher $L_X$ values) show lower [CN/Fe] and [C/Fe] abundance ratios. Note that our sample includes clusters richer than Coma and poorer than Virgo.

However, the [Mg/Fe] abundance ratio is constant and enhanced with respect to the solar element mixture. This is interpreted using the varying formation timescales for CN, Mg, and Fe, suggesting that early-type galaxies in more massive clusters are assembled on shorter timescales than those galaxies within less massive clusters. This is because the CN is ejected into the ISM on timescales larger than those for the Mg and shorter than the ones for the Fe. Although this result is in qualitative agreement with the standard semi-analytic models, there is a severe quantitative disagreement as the involved timescales ($< 1\,\mathrm{Gyr}$)

as derived from the [Mg/Fe] abundance ratios, which do not depend on the environment, are significantly shorter than those predicted by these models.

Although the CN abundance ratio seems to be an appropriate clock to measure differential timescales for galaxies of $\sigma \sim 200\,\mathrm{km\,s^{-1}}$ in clusters of different richness, it is important to note that there are hints that this might not be the case for more massive galaxies because their formation timescales should have been even faster. A detailed study of the stellar populations of 27 massive galaxies in four galaxy clusters richer than Coma at $z \sim 0.2$ has shown that the relative abundances of both [Mg/Fe] and [CN/Fe] increase with galaxy velocity dispersion. However, while [Mg/Fe] is independent of cluster richness, the slope of [CN/Fe] increases with increasing cluster richness (Carretero, Vazdekis, & Beckman 2007). These relations suggest that the formation timescales of intermediate mass galaxies ($\sigma \sim 200\,\mathrm{km\,s^{-1}}$) are shorter in rich clusters than those of their counterparts in poorer clusters. The formation timescales are even shorter for the most massive galaxies ($\sigma > 300\,\mathrm{km\,s^{-1}}$), although in this case a new "chemical clock" must be identified to resolve these shorter timescales.

**Acknowledgments.** I would like to thank J. Beckman, who brought me the opportunity to develop and further my career in astronomy, in particular during my PhD thesis. His ideas and the many discussions with him have inspired most of my work. I would like to acknowledge all the members of the MILES collaboration. I also wish to thank N. Arimoto and Y. Yamada, as well as my students, C. Carretero and J. L. Cervantes, for performing important bits of the work shown here. AV is a Ramón y Cajal Fellow of the Spanish Ministry of Education and Science. This work has been supported by the Spanish Ministry of Education and Science grant AYA2004-03059.

## References

Alexander, D. R., & Ferguson, J. W. 1994, ApJ, 437, 879
Alonso, A., Arribas, S., & Martínez–Roger, C. 1996, A&A, 117, 227
Alonso, A., Arribas, S., & Martínez–Roger, C. 1999, A&AS, 140, 261
Arimoto, N. 1996, in ASP Conf. Ser., 98, From Stars to Galaxies: The Impact of Stellar Physics on Galaxy Evolution ed. C. Leitherer, U. Fritze-von Alvensleben, & J. Huchra (San Francisco: ASP), p. 278
Bernardi, M. et al. 2003, AJ, 125, 1882
Bertelli, G., Bressan, A., Chiosi, C., Fagotto, F., & Nasi, E. 1994, A&AS, 106, 275
Blakeslee, J. P., et al. 2003, ApJ, 596, L143
Bower, R. G., Lucey, J. R., & Ellis, R. S. 1992, MNRAS, 254, 601
Bruzual, G., & Charlot, S. 2003, MNRAS, 344, 1000
Caldwell, N., Rose, J. A., & Concannon, K. D. 2003, AJ, 125, 2891
Carretero, C., Vazdekis, A., Beckman, J. E., Sánchez-Blázquez, P., & Gorgas, J. 2004, ApJ, 609, L45
Carretero, C., Vazdekis, A., & Beckman, J. E. 2007, MNRAS, 375, 1025
Cenarro, J., Cardiel, N., Gorgas, J., Peletier, R. F., Vazdekis, A., & Prada, F. 2001a, MNRAS, 326, 959
Cenarro, J., Gorgas, J., Cardiel, N., Pedraz, S., Peletier, R. F., & Vazdekis, A. 2001b, MNRAS, 326, 981
Cenarro, J., Gorgas, J., Vazdekis, A., Cardiel, N., & Peletier, R. F. 2003, MNRAS, 339, L12
Cenarro, J., et al. 2007, MNRAS, 374, 664
Chiosi, C., & Carraro, G. 2002, MNRAS, 335, 335

Cervantes-Rodríguez, J. L., Vazdekis, A., & Cepa, J. 2006, NewAR, 49, 670
Coelho, P., Bruzual, G., Charlot, S., Weiss, A., Barbuy, B., & Ferguson, J. 2007, MNRAS, 382, 498
de la Rosa, I. G., de Carvalho, R. R., Vazdekis, A., & Barbuy, B. 2007, AJ, 133, 330
Ferreras, I., Charlot, S., & Silk, J. 1999, ApJ, 521, 81
Girardi, L., Bressan, A., Bertelli, G., & Chiosi, C. 2000, A&AS, 141, 371
González, J. J. 1993, PhD Thesis, Univ. of California at Santa Cruz
Jones, L. A. 1999, PhD Thesis, Univ. North Carolina, Chapel Hill
Jones, L. A., & Worthey, G. 1995, ApJ, 446, L31
Kauffmann, G., & Charlot, S. 1998, MNRAS, 294, 705
Kauffmann, G. et al. 2003, MNRAS, 341, 54
Kodama, T., & Arimoto, Y. 1997, A&A, 320, 41
Leonardi, A., & Rose, J. A. 1996, AJ, 111, 182
Rose, J. A. 1994, AJ, 107, 206
Salasnich, B., Girardi, L., Weiss, A., & Chiosi, C. 2000, A&A, 361, 1023
Sánchez-Blázquez, P., Gorgas, J., Cardiel, N., Cenarro, J., & González, J. J. 2003, ApJ, 590, L91
Sánchez-Blázquez, P., et al. 2006, MNRAS, 371, 703
Sarzi, M., et al. 2006, MNRAS, 366, 1151
Salaris, M., & Weiss, A. 1998, A&A, 335, 943
Tantalo, R., Chiosi, C., & Bressan, A. 1998, A&A, 333, 419
Thomas, D., Maraston, C., & Bender, R. 2003, MNRAS, 339, 897
Trager, S., Worthey, G., Faber, S., Burstein, D., & González, J. 1998, ApJS, 116, 1
Trager, S, Faber, S. M., Worthey, G., & González, J. J. 2000, AJ, 119, 1645
Tripicco, M. J., & Bell, R. A. 1995, AJ, 110, 3035
VandenBerg, D. A., Swenson, F. J., Rogers, F. J., Iglesias, C. A., & Alexander, D. R. 2000, ApJ, 532, 430
Vazdekis, A., Peletier, R. F., Beckman, J. E., & Casuso, E. 1997, ApJS, 111, 203
Vazdekis, A., Cenarro, J., Gorgas, J., Cardiel, N., & Peletier, R. F. 2003, MNRAS, 340, 1317
Vazdekis, A., Kuntschner, H., Davies, R. L., Arimoto, N., Nakamura, O., & Peletier, R. 2001b, 551, L127
Vazdekis, A., Salaris, M., Arimoto, N., & Rose, J. 2001a, ApJ, 549, 274
Vazdekis, A., & Arimoto, N. 1999, ApJ, 525, 144
Vazdekis, A. 1999, ApJ, 513, 224
Vazdekis, A., Casuso, E., Peletier, R. F., & Beckman, J. E. 1996, ApJS, 106,307
Worthey, G. 1994, ApJS, 95, 107
Worthey, G., & Ottaviani, D. L. 1997, ApJS, 111, 377
Worthey, G., Faber, S. M., & González, J. J. 1992, ApJ, 398, 69
Yamada, Y., Arimoto, N., Vazdekis, A., & Peletier, R. F. 2006, ApJ, 637, 200


## Discussion

*Sarzi*: Tunable filters seem to be a wonderful idea; however, you will not have spectra to measure velocity dispersion and estimate the impact that this has on your indices. Presumably you can tune your filters to minimize the impact of velocity dispersion. Have you accounted for this effect?

*Vazdekis*: The new indicators based on the tunable filters have been tested and optimized to be nearly insensitive to velocity dispersion.

*Pathways through an Eclectic Universe*
*ASP Conference Series, Vol. 390, © 2008*
*J. H. Knapen, T. J. Mahoney, and A. Vazdekis, eds.*

# Disky Nuclei in Early- to Intermediate-type Disk Galaxies

Marc Balcells

*Instituto de Astrofísica de Canarias, 38200 La Laguna, Tenerife, Spain*

**Abstract.** Structural analysis of S0–Sbc galaxies imaged with *HST* at near-infrared wavelengths shows many nuclei to be flattened systems with surface brightness comparable to that of the spheroidal bulge component. While standard photometric decomposition would assign that light to the bulge component, its flattened shape and disky isophotes argue for including them as part of the (non-exponential) disk component of the galaxy.

## 1. Introduction

Many morphological components have been identified in galaxies, e.g., disk, bar, bulge, oval, nucleus (e.g., Block et al. 2000). Which particular components we emphasize in galaxy structural analysis is a choice that reveals our scientific preferences. It also conditions the outcome of the analysis. Since the mid-eighties, galaxy morphological (or "structural") analysis has emphasized the disk–bulge structural decomposition. Less attention was paid to spiral structure, ovals, rings, etc. One reason for the popularity of this particular decomposition may derive from the goal of reducing galaxy formation processes into two classes, slow-gentle (leading to disks), and fast-violent (leading to spheroids). In cold dark matter (CDM) cosmology, the fractional mass in bulges should be related to the relative importance of merging in galaxy formation.

Several definitions have been used over the past few decades (see recent reviews by Wyse, Gilmore, & Franx 1997; Kormendy & Kennicutt 2004). For Hubble and Sandage, "bulge" referred to the smooth light distribution that protrudes above the disk in edge-on views. Such a concept has generally been replaced by the "operational definition" that bulges are any central light in excess of the inward extrapolation of the exponential disk profile, to the point that many astronomers today believe the latter to be the true definition of bulge (Carollo, Ferguson, & Wyse 2000).

I prefer the definition from Hubble and Sandage, and have in the past attempted bulge–disk decompositions following this principle (Andredakis, Peletier, & Balcells 1995), as done by Kent (1985). In this paper, I present an analysis of bulge nuclear structures in which bulge–disk decomposition has been carried out following the second method listed above, i.e., by fitting a Sérsic-plus-exponential model to the surface brightness profile. The result I reach is that, for a large fraction of bulges, the densest central parts are flattened structures, i.e., they are alike to disks. To me this suggests that Sérsic-plus-exponential decompositions may mislead us as to the true fraction of disk vs spheroidal light in galaxies.

## 2. Data

Because galaxy nuclei are dusty and show rising profiles well inward of ground-based resolutions, we imaged our galaxy sample with *HST*, in bandpasses F450W, F814W (WFPC2), and F160W (NICMOS camera 2). Nineteen S0–Sbc galaxies from the Balcells & Peletier (1994) sample were imaged. The elliptically averaged near-IR profiles were linked smoothly with $K$-band profiles we had obtained from ground-based images (Peletier & Balcells 1997), yielding profiles that extend from typically 10 pc to a few kpc, and cover from the nuclear regions to the outer disk. These profiles were fitted with a number of fitting models that included a Sérsic function and an exponential function. In over 90% of the cases, the Sérsic+exponential fits showed strong non-trivial improvements when additional nuclear components were included; these consisted of either a point source or an inner exponential, or a combination of the two. We thus obtained parameters for the simplest model that fitted each galaxy. The results, and a full analysis of nuclear and global properties of bulges, are given in Balcells, Graham, & Peletier (2007a,b).

## 3. Extended Nuclear Components

Extended nuclear components were found in 60% of the galaxies. They show typical sizes of a few hundred parsecs, similar to those of inner disks, inner bars (Erwin 2004), and inner star-forming rings (Knapen et al. 2006). Scale lengths and central surface brightnesses are compared to those from the outer galaxy disks in Figure 1. Most noteworthy is that nuclear disks are brighter by about 5 mag arcsec$^{-2}$ than the outer disks. The central surface brightness is always comparable to that of the underlying Sérsic component; this is a consequence of the fitting process, which is not capable of detecting inner exponential components fainter than ~1 mag arcsec$^{-2}$ below the underlying Sérsic component.

Isophotal analysis of the *HST* F160W images shows a second important characteristic of the nuclear exponential components: all of them show increased ellipticity with respect to the outer bulge, and disky (positive $C4$) isophotes. I show a typical example in Figure 2. Given the inclined aspect of the present sample, the isophotes indicate that the extended nuclear components are flat structures.

The unavoidable conclusion is that the nuclear regions of early- to intermediate-type disk galaxies, when imaged at *HST* resolution, show positive deviations from the best-fit Sérsic model with flattened shapes. The existence of nuclei is of course not new. Galaxies have been reported to host *nuclei* for quite some time (Matthews et al. 1999). However, it appears that the goal alluded to in § 1. of simplifying galaxy structure made us ignore nuclei as fine-scale details. Such simplification appears to have reached its limits; when the nuclear light is bundled together with the bulge, one obtains Sérsic indices up to $n = 4 - 6$ (Andredakis et al. 1995), suggesting similar structures to elliptical galaxies and a prevalence of merging processes, whereas, when properly modeling nuclei, the shape, or concentration index $n$ of the Sérsic component becomes significantly lower ($n = 1.7 \pm 0.7$; Balcells et al. 2003), a concentration regime that can be attained by secular processes.

Because both the surface brightness profiles and the isophotal shapes show deviations with respect to the remainder of the bulge, it makes sense to speak of nuclear components as separate structural, and probably dynamical, entities. We associate them with nuclear disks such as those shown in various talks at this conference. Peletier et al. (1999) analyzed populations for these regions, concluding that, while extended bulges are old, the nuclei themselves often harbor ongoing star formation and are dusty, with typical $A_V \sim 0.6-1.0$, corresponding to $A_H = 0.1 - 0.2$.

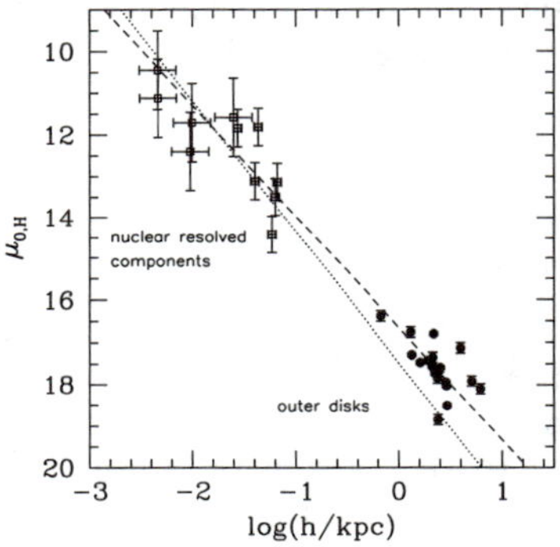


Figure 1. Disk scale-length in kpc vs extrapolated central surface brightness for inner extended components and for main galaxy disks. (*From Balcells et al. 2007a.*)

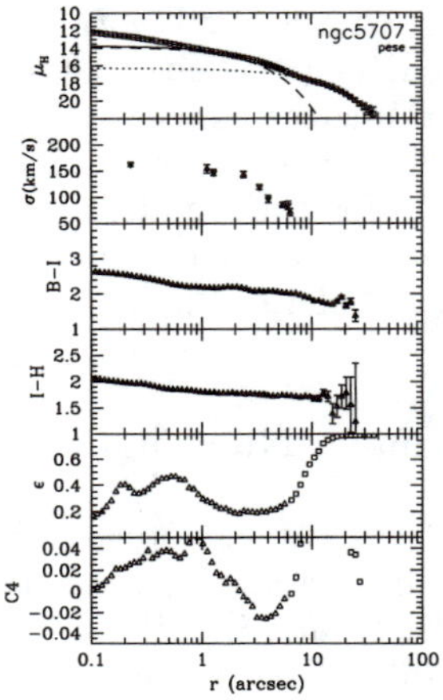


Figure 2. Profiles of surface brightness, velocity dispersion, colors, ellipticity, and $c4$ diskiness coefficient, for the Sab galaxy NGC 5707. In the region of the nuclear component, the ellipticity and diskiness profiles show an inward upturn.

## 4. Discussion

The extended nuclear components raise several open questions. First is their relation to nuclear structures in late-type spirals such as M33. Measurements of densities, colors, and kinematics can provide the relevant clues. Second is their structural characterization—are these disks, bars, or rings? Are they related to star-forming rings typically seen in barred galaxies? Does the presence of young stars indicate that these structures are young, or are the young stars the icing on top of an older, underlying nuclear disk? Color profiles, which trace a smooth outward bluing through the nucleus, bulge, and disk regions (Peletier et al. 1999) are easier to interpret if ages and metallicities do not strongly vary with radius, i.e., if nuclei are predominantly old; but further modeling is needed to disentangle age, metallicity, and dust effects. Many of these components show rapid rotation and velocity dispersion dips (Falcón-Barroso et al. 2006).

In the framework of this conference, I would like to emphasize the disky nature of these components. Disky isophotes are one of the properties used to define pseudo-bulges (Kormendy & Kennicutt 2004). One option would be to simply bundle these components into a generic class of pseudo-bulges. However, given their disky nature, these components may also be taken simply as part of the galaxy disk. Galaxy disks are generally assumed to have exponential surface brightness profiles. However, even galaxies without prominent bulges show deviations with respect to the exponential model, including bars (Knapen et al. 2000), truncations, and antitruncations (Erwin, Beckman, & Pohlen 2005). In galaxies with prominent bulges, information on the shape of the disk surface brightness profile in the region dominated by the bulge is nearly absent. Up to now it has been customary to simply extrapolate the outer disk's exponential profile inward. The presence of components with the structure, geometry, and kinematics of disks gives strong suggestions that we may want to move beyond the exponential disk paradigm; to bundle together all flat components of galaxies into the disk, and leave the term bulge for those components that are thicker than the disk: the so-called classical bulges, and the peanut-shaped bulges.

**Acknowledgments.** Many thanks to John Beckman for years of inspiring work, and to Johan Knapen for organizing this conference. Alister Graham, Reynier Peletier, and Lilian Domínguez contributed to the work presented here.

## References

Andredakis, Y. C., Peletier, R. F., & Balcells, M. 1995, MNRAS, 275, 874
Balcells, M., Graham, A. W., Domínguez-Palmero, L., & Peletier, R. F. 2003, ApJ, 582, L82
Balcells, M., Graham, A. W., & Peletier, R. F. 2007a, ApJ, 665, 1084
Balcells, M., Graham, A. W., & Peletier, R. F. 2007b, ApJ, 665, 1104
Balcells, M., & Peletier, R. F. 1994, AJ, 107, 135
Block, D., Puerari, I., Stockton, A., & Ferreira, D. (eds) 2000, Toward a New Millennium in Galaxy Morphology. From $z = 0$ to the Lyman Break (Dordrecht: Kluwer)
Carollo, C. M., Ferguson, H. C., & Wyse, R. F. G. (eds) 2000, The Formation of Galactic Bulges (Cambridge: Cambridge University Press)
Erwin, P. 2004, A&A, 415, 941
Erwin, P., Beckman, J. E., & Pohlen, M. 2005, ApJ, 626, L81
Falcón-Barroso, J., et al. 2006, MNRAS, 369, 529
Kent, S. M. 1985, ApJS, 59, 115
Knapen, J. H., Mazzuca, L. M., Böker, T., Shlosman, I., Colina, L., Combes, F., & Axon, D. J. 2006, A&A, 448, 489
Knapen, J. H., Shlosman, I., & Peletier, R. F. 2000, ApJ, 529, 93
Kormendy, J., & Kennicutt, R. C. Jr. 2004, ARA&A, 42, 603 (KK04)
Matthews, L. D., et al. 1999, AJ, 118, 208
Peletier, R. F. & Balcells, M. 1997, New Astron., 1, 349
Peletier, R. F., Balcells, M., Davies, R. L., Andredakis, Y., Vazdekis, A., Burkert, A., & Prada, F. 1999, MNRAS, 310, 703
Wyse, R. F. G., Gilmore, G., & Franx, M. 1997, ARA&A, 35, 637

*Pathways through an Eclectic Universe*
*ASP Conference Series, Vol. 390, © 2008*
*J. H. Knapen, T. J. Mahoney, and A. Vazdekis, eds.*

# The Origin of Star Formation since $z = 1$

F. Hammer,[1] M. Puech,[2] Y. Yang,[1] and H. Flores [1]

[1] *GEPI, Observatoire de Paris, CNRS, Université Paris Diderot, 92195 Meudon, France*

[2] *ESO, Karl-Schwarzschild-Str. 2, D-85748 Garching bei Muñchen, Germany*

**Abstract.** The use of multiple integral field units with FLAMES/GIRAFFE at the VLT has revolutionized investigations of distant galaxy kinematics. This facility can recover the velocity fields of almost all emission-line galaxies with $I$(AB)$< 22.5$ at $z < 0.8$. We have gathered a unique sample of 63 velocity fields at $z = 0.4$–$0.75$, which are representative of $M_{\rm stellar} \geq 1.5 \times 10^{10}\, M_\odot$ emission-line ($W_0$([O II])$\geq 15$ Å) galaxies and are unaffected by cosmic variance. Taking into account all galaxies—with or without emission lines—in that redshift range, we find that $42 \pm 7$% have anomalous kinematics, including $26 \pm 7$% with complex kinematics, i.e., not supported by either rotation or by dispersion. The large fraction of complex velocity fields suggests the high impact of merging in shaping the galaxies in the intermediate-mass range. We discuss how this can be accommodated within the framework of current scenarios of galaxy formation, including for the Milky Way and M31.

## 1. Introduction

The cosmic star formation rate has declined by a factor $\sim 10$ since $z \sim 1$ (Lilly et al. 1996; Flores et al. 1999). At $z = 1$, the cosmic star formation density is clearly dominated by luminous infrared galaxies (LIRGs, Flores et al. 1999; Le Floc'h et al. 2006), whose number density evolves extremely rapidly. There is overall agreement between the integrated star formation rate (derived from mid-IR observations from first *ISO* and then *Spitzer*) and the stellar mass evolution (derived from near-IR observations). One may focus on the evolving properties of intermediate-mass galaxies (or $M_*$ galaxies) with stellar masses from $3 \times 10^{10}$ to $3 \times 10^{11}\, M_\odot$: they include most (from 65% to 80%) of the present-day stellar mass. About half of their stellar mass has been formed since $z = 1$, mostly during strong star formation episodes during which galaxies take the appearance of LIRGs (Hammer et al. 2005).

What were the physical processes driving star formation at these redshifts? The internal kinematics of distant galaxies is a powerful tracer of the major processes governing star formation and galaxy evolution in the early Universe, such as merging, accretion, and hydrodynamic feedback related to star formation and AGN (e.g., Barnes & Hernquist 1996; Barton, Geller, & Kenyon 2000; Dressler 2004). Thus, robustly measuring the internal kinematics of distant galaxies is crucial for understanding how galaxies formed and evolved.

Flores et al. (2006) have presented the first study of a small sample of 35 $I(\mathrm{AB}) < 22.5$ galaxies at $z = 0.4 - 0.75$ using the integral field multi-object spectrograph GIRAFFE on the ESO VLT. They found a large fraction of galaxies with anomalous velocity fields, i.e., with velocity fields discrepant from those of rotationally or dispersion (pressure) supported galaxies. Interestingly, they found that the large dispersion of the Tully–Fisher relation at moderate $z$ (e.g., Conselice et al. 2005) is mainly caused by galaxies with anomalous kinematics.

Here, we present the kinematics of a representative sample of 63 $M_*$ galaxies at $z = 0.4 - 0.75$ (Section 2), we discuss their properties (Section 3) and relate them to local galaxies, including the Milky Way and M31 (Section 4).

## 2. Velocity Fields of 63 $M_*$ galaxies at $z = 0.4$–$0.75$

### 2.1. A Representative Sample of Distant Galaxies

A Large Program at the VLT entitled IMAGES (Intermediate MAss Galaxy Evolution Sequence) aims to derive velocity fields from VLT/GIRAFFE for galaxies in the Chandra Deep Field (CDFS) and combine these observations with deep and high quality images from *HST*/ACS as well as with deep mid-IR photometry from *Spitzer*/MIPS. It has now reached more than half its completion target and 39 velocity fields of galaxies have recently been derived (Yang et al. 2007). Combined with the previous observations of Flores et al. (2006), it leads to a sample of 74 galaxies. Observing time with VLT/GIRAFFE was ranging from 8 to 13 hours for each galaxy in order to provide high $S/N$ in the 20 GIRAFFE-IFU pixels. In most cases the [O II] emission doublet is well resolved at a resolution of $9000 \times (1 + z)$. It provides us with 68 robust velocity fields for which there are 6 to 18 GIRAFFE pixels with high $S/N$. The remaining six galaxies have been rejected from the sample because their velocity fields include less than four pixels with $S/N > 4$ (see, for example, Flores et al. 2006 and Yang et al. 2007 for a discussion).

Figure 1 shows the distribution of the absolute luminosity ($M_J(\mathrm{AB})$) of the combined sample. Because in IMAGES we have deliberately selected $M_J(\mathrm{AB}) < -20.3$ galaxies, we assume a similar limit for the combined sample. This yielded a sample of 63 galaxies, which is well representative of the luminosity function at $z = 0.4$–$0.75$. Notice that within this redshift range, GIRAFFE is able to recover the velocity field of almost all galaxies with $W_0([\mathrm{O\,II}]) \geq 15$ Å. Note also that the combined sample includes galaxies from four independent fields of view (see Yang et al. 2007) and is unlikely to be affected by cosmic variance effects. In the following, we will only discuss the properties of the representative sample of 63 galaxies with $M_J(\mathrm{AB}) < -20.3$.

### 2.2. Properties of Velocity Fields of Distant Galaxies

Flores et al. (2006) have generated a classification scheme for distant galaxy kinematics. Indeed, at large distances the spatial resolution is insufficient to resolve the central regions of the galaxies. This implies that the observed velocity dispersion ($\sigma$) of a rotating body is the convolution of the actual random motions with the rotation. For a rotationally supported galaxy, it leads unavoidably to a well-defined peak in the center (see Fig. 2). A classification of velocity fields

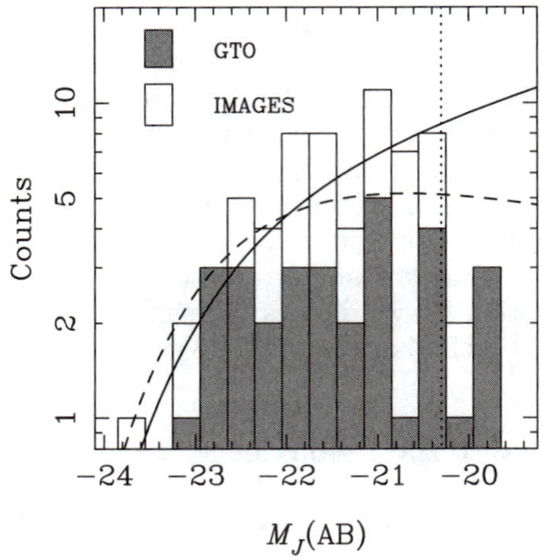


Figure 1. Number counts of selected galaxies versus AB absolute magnitude in the $J$-band. The GTO sample refers to Flores et al. (2006), the IMAGES sample to Yang et al. (2007). The vertical dotted line indicates the limit of the IMAGES program. Two luminosity functions derived from Pozzetti et al. (2003) are shown (*full line:* $z = 0.5$; *dashed line:* $z = 1$). Our sample of galaxies has redshifts ranging from $z = 0.4$ to $z = 0.75$. It turns out that the combined sample of 63 galaxies with $M_J(\mathrm{AB}) \leq -20.3$ is representative of galaxies with stellar masses greater than $1.5 \times 10^{10}\, M_\odot$ at $z \sim 0.6$.

and a description of their diagnostics may be found in Yang et al. (2007, see also Yang et al., this volume, p. 502). The robustness of the classification is provided by the high $S/N$ associated with GIRAFFE observations and by the fact that large scale motions are well resolved at $z = 0.4$–$0.75$, as evidenced by tests on redshifted velocity fields of various kinds of local galaxies.

Among the 63 velocity fields of the representative sample, Yang et al. (2007) find 20 rotating disks (32%), 16 rotating disks with perturbations (25%) and 27 galaxies with complex kinematics (43%, see Fig. 2). Within their classification, perturbed rotations correspond to a $\sigma$ peak discrepant from expectations for pure rotation. The large number of galaxies with complex kinematics is even more puzzling. Indeed, the complex kinematics class corresponds to objects for which the large scale motions are not aligned to the optical major axis, and which show a $\sigma$ map very discrepant from the expectations for rotation (see Fig. 2).

## 3. Nature of Velocity Fields of Distant Galaxies

We have gathered the largest existing sample of velocity fields for distant galaxies. Quite unexpectedly, only a small fraction of $z = 0.4$–$0.75$ galaxies are rotating disks. Let us now consider a representative sample of intermediate-mass galaxies at $z \sim 0.6$. At $z = 0.6$, Hammer et al. (1997) found that 60% of galaxies have $W_0([\mathrm{O\,II}]) \geq 15$Å. Let us assume that quiescent ($W_0([\mathrm{O\,II}]) < 15$Å)

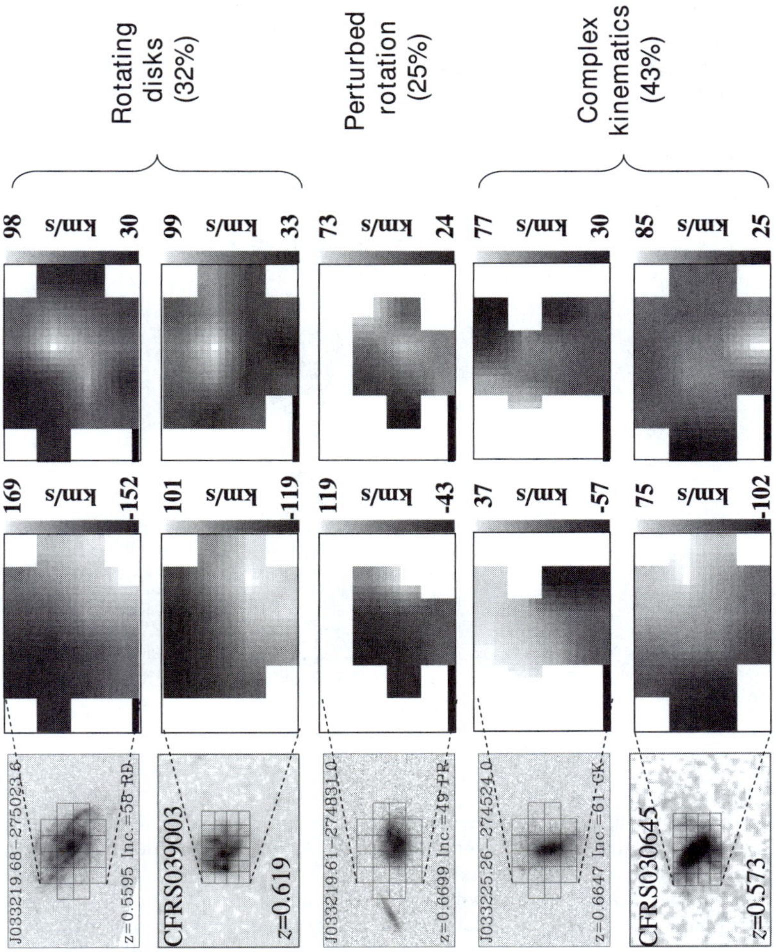


Figure 2. Four examples of kinematics of $z = 0.4 - 0.75$ galaxies. Each row corresponds to one galaxy, from *left* to *right*: *HST* ACS F755W/F814W image, observed velocity field and $\sigma$-map. The two top rows show regular rotating disks with a dynamical axis oriented along the main optical axis and a $\sigma$-map with a peak dominated by large scale rotational motions (see text); the three bottom rows show galaxies with anomalous kinematics, one with just a shift of the peak in the $\sigma$-map, e.g., a perturbed rotation that may be caused by the impact of a minor merger, the two others with dynamical axes misaligned relative to the main optical axis (the color version of this figure is available in the electronic version).

galaxy kinematics are either rotationally or dispersion supported, thus minimizing the fraction of galaxies with anomalous kinematics. Our results then imply that 42% ± 7% of $z = 0.4$–$0.75$ galaxies have anomalous kinematics, including 26% ± 7% possessing complex kinematics. In the local Universe, such a fraction for intermediate-mass galaxies has never been thoroughly estimated. Because up to 97% of intermediate mass galaxies are ellipticals, S0s, or spirals (see Hammer et al. 2005), it is likely that the fraction of galaxies with anomalous kinematics is close to a few percent today.

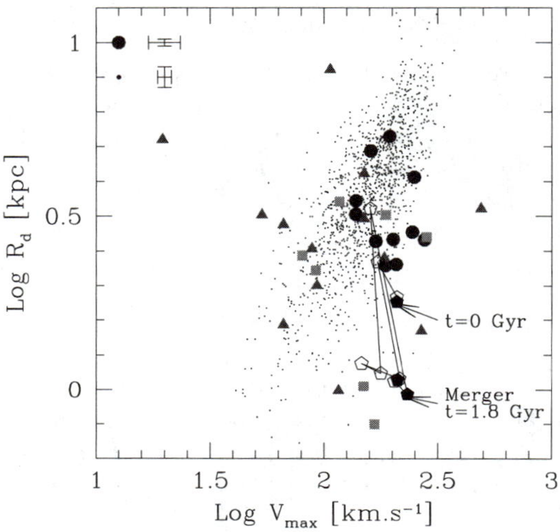


Figure 3. Disk scale length versus maximum rotational velocity, for the sample of 32 distant galaxies of Flores et al. (2006). Large dots (*blue in the on-line version*) represent galaxies classified as rotating disks, squares (*green*) those classified as perturbed rotators, and triangles (*red*) galaxies with complex kinematics. The small black dots represent the local sample of Courteau (1997) and Mathewson, Ford, & Buchhorn (1992). The median 1$\sigma$ uncertainties are indicated in the upper-left corner, for both distant rotating disks (*large blue dots*), and local disks (*small black dots*). Black pentagons represent simulated GIRAFFE observations using a hydrodynamical simulation of a major merger of two Sbc galaxies. The beginning and the end of the sequence, as well as the merger itself, are indicated by solid symbols.

There is a very rapid evolution of galaxy kinematics that is unlikely to be caused by artificial or instrumental factors. Indeed, the observed evolution of the Tully–Fisher relation provides us with a strong and independent confirmation. By observing galaxies with slit spectroscopy, Conselice et al. (2005) found a very scattered Tully–Fisher relation at moderate redshift such that one might wonder whether there is still a correlation between rotational velocity and stellar mass. Flores et al. (2006) and Puech et al. (2007) have convincingly shown that galaxies with non-relaxed kinematics are responsible for the large dispersions of both the Tully–Fisher and the $j_{\rm disk}$–$V_{\rm max}$ relationships. So it is beyond doubt that kinematics is among the most rapidly evolving properties of galaxies.

What physical process could explain such a dramatic evolution? Puech et al. (2007) have convincingly shown that merging may be responsible for the anomalous kinematics, as well as for the increase in the scatter of both the Tully–Fisher and the $j_{\rm disk}$–$V_{\rm max}$ relationships (see also Fig. 3). They have interpreted such a scatter as being due to a random walk process of galaxies, precisely consistent with encounters between galaxies during the hierarchical scenario of galaxy formation. Indeed, during a merger—especially a major merger—galaxies pass through various stages during which the disk may be destroyed, generating significant discrepancies with respect to the general behavior of isolated rotating disks. A scenario in which merging is the dominant physical process explains many evolutionary features since $z = 1$, including the number density evolution of LIRGs and the emergence of galaxies with complex morphologies and blue cores at $z > 0.4$ (see, for example, Hammer et al. 2005).

## 4. Relation to Local Galaxies and to Disk Formation

Whether the evolution of galaxy kinematics is associated with (major) mergers is still a pending question. If true, it means that many present-day galactic disks have been rebuilt at a relatively recent epoch. Taking into account the number fraction of both approaching mergers (those in pairs) and galaxies with complex velocity fields, Hammer et al. (2005, 2007) have concluded that from 50% to 70% of galaxies may have experienced a major merger and then a disk rebuilding phase since $z = 1$. Below, we discuss first how this can be consistent with properties of local galaxies, and then whether other alternatives may explain the evolution of galaxy kinematics.

Today, intermediate-mass galaxies are mostly spirals (70%, e.g., Hammer et al. 2005), which are dominated by rotating thin disks. It is beyond doubt that the Milky Way has not experienced any major interaction since the last 10–11 Gyr ($z = 3$). Is this true for other spiral galaxies, or, in other words, how is the Milky Way representative of local spirals? Hammer et al. (2007) have compared the global properties of the Milky Way to those of local spirals with similar rotational velocities, or $V_{\rm flat}$ which may be taken as a proxy for the total mass. They show that the Milky Way is quite discrepant (see Fig. 4), showing a disk radius and a stellar content (or mass) about half that of comparable disk galaxies. Only 7% of disk galaxies have properties similar to those of the Milky Way, while M31 is quite representative.

Moreover, the distant regions outside the disk of our Galaxy are also quite exceptional: compared to those of other galaxies, the stars inhabiting the outskirts of the Milky Way are exceptionally metal poor. The galactic outskirt properties may be affected particularly by past encounters with other galaxies. For example, when two galaxies merge, tidal debris and newly formed stars produced during the collision pollute their outskirts. Within this context, the outskirts of most spiral galaxies—including M31—show a significant fraction of metal-rich stars, evidencing that they have experienced more mergers in their past than has the Milky Way. All these pieces of evidence, taken together, support the hypothesis that the Milky Way is an exceptionally quiet galaxy.

In contrast to the Milky Way, M31 shows evidence for merging in the recent past (e.g., Ibata et al. 2004). Galaxy mergers generate additional material in galactic outskirts, including newly formed stars, which then enhance the stellar content and radius. This simply explains the peculiarities of the Milky Way, emphasizing that local spirals had a relatively active merger history in the recent past. It is indeed relevant to wonder whether merging may have shaped galaxies—including spirals—as we observe them today.

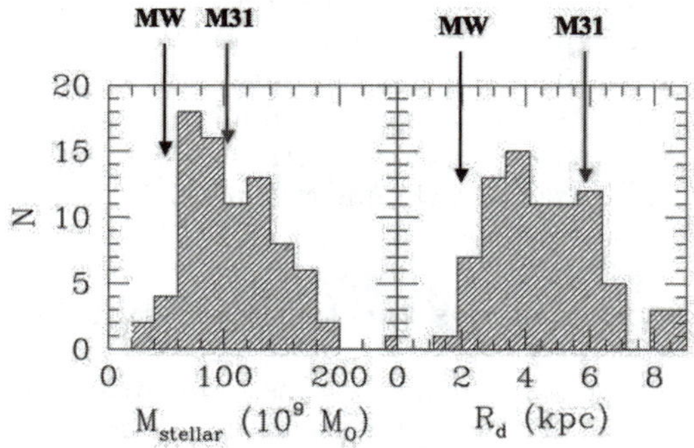


Figure 4. Stellar mass (in units of billions of solar masses) and disk radius (in units of kiloparsecs) distributions of spiral galaxies with rotational velocities similar to that of the Milky Way. Arrows (blue and red, respectively, in the online version) show the values for the Milky Way (and M31), indicating that its stellar content and disk radius are half the average values for external galaxies. To derive these quantities for the Milky Way, one needs to reconstruct or model the light distribution because we live within its disk, and extinction by dust may easily hide significant parts of it. Significant progress has been made recently (see Hammer et al. 2007), especially using infrared wavelengths to make such measurements, which are much less affected by dust extinction.

Although the merger alternative may explain many observations in the $z = 0.4-1$ redshift range (e.g., Hammer et al. 2005), could a less dramatic mechanism be at the origin of the peculiar kinematics found at $z = 0.4$–$0.75$? In fact, our observations take into account only the large-scale motions of the ionized gas and not those of the stellar component. While beyond the reach of present-day telescopes, observations of the latter are certainly crucial, although we still need to explain why the gas is so turbulent, while more abundant than it is today. Another possibility may be to advocate internal processes, such as bars, which may perturb the gaseous velocity fields. The presence of bars in the central region may create additional dispersion, although our spatial resolution is probably too poor to characterize bars kinematically. Moreover, the absence of number evolution of barred galaxies (e.g., Zheng et al. 2005 and references therein) does not support the hypothesis that bars could explain the strong evolution of galaxy kinematics.

## 5. Conclusions

We find a strong evolution in galaxy kinematics since $z = 0.6$, with a significant fraction of galaxies with complex velocity fields. Processes related to merging easily reproduce the observed properties of distant galaxies. Properties of their descendants—the local spiral galaxies—suggest that they have experienced more numerous and recent mergers than the Milky Way.

However, it has to be demonstrated in detail how present-day $M_*$ galaxies, mostly spiral galaxies, can be the end product of relatively recent mergers. In fact, galaxy simulations predict that in most cases, major merger end products are dispersion-supported galaxies, i.e., elliptical galaxies. Thus the question of angular momentum transfer is a key issue. There are indeed considerable uncertainties in the treatment of the baryonic matter in numerical simulations, which is illustrated by the so-called angular momentum problem (Steinmetz & Navarro 1999), and by the fact that no model has been able simultaneously to match the luminosity function and the Tully–Fisher zero point using standard ΛCDM parameters, as pointed out by Dutton et al. (2006). To reconcile these discrepancies, Dutton et al. need to assume an expanding halo, caused by a transfer of the angular momentum from a non-spherical and clumpy gas accretion when forming a primordial disk. This clearly leads to the question of forming disks by gaseous mergers, as illustrated by Robertson et al. (2006) and by Governato et al. (2007). The question of the required fraction of gas to form a new disk after a merger is still pending. It certainly needs some improvements in the treatment of the angular momentum transfer between baryonic and dark matter (see Kaufmann et al. 2007).

The observations presented here tell us that major mergers could have played an important role in shaping galaxies as we observe them today. Indeed many estimates of the merger rate found that a typical intermediate-mass galaxy should have experienced 0.5 to 0.75 major merger since $z = 1$ (see Rawat et al. 2007; Kartaltepe et al. 2007 and references therein), and more than one since $z = 2$–3. This illustrates further that the Milky Way has had an exceptionally quiet merger history (Hammer et al. 2007). It is unlikely that present-day $M_*$ galaxies—mostly spiral galaxies—have all escaped a major merging since $z = 2 - 3$. Alternatively, one may wonder whether the procedure used to simulate galaxy merger remnants is the correct one.

**Acknowledgments.** We are grateful to the technical team at GEPI for the unique achievement of GIRAFFE at the VLT.

### References

Barnes, J. E., & Hernquist, L. 1996, ApJ, 471, 115
Barton, E. J., Geller, M. J., & Kenyon, S. J. 2000, ApJ, 530, 660
Conselice, C. J., Bundy, K., Ellis, R. S., Brichmann, J., Vogt, N. P., & Phillips, A. C. 2005, ApJ, 628, 160
Courteau, S. 1997, AJ, 114, 2402
Dressler, A. 2004, Clusters of Galaxies: Probes of Cosmological Structure and Galaxy Evolution, ed. J. S. Mulchaey, A. Dressler, & A. Oemler (Cambridge: CUP), p. 206
Dutton, A. A., et al. 2007, ApJ, 654, 27

Flores, H., et al. 1999, ApJ, 517, 148
Flores, H., Hammer, F., Puech, M., Amram, P., & Balkowski, C. 2006, A&A, 455, 107
Governato, F., Willman, B., Mayer, L., Brooks, A., Stinson, G., Valenzuela, O., Wadsley, J., & Quinn, T. 2007, MNRAS, 374, 1479
Hammer, F., et al. 1997, ApJ, 481, 49
Hammer, F., Flores, H., Elbaz, D., Zheng, X. Z., Liang, Y. C., & Cesarsky, C. 2005, A&A, 430, 115
Hammer, F., Puech, M., Chemin, L., Flores, H., & Lehnert, M. 2007, ApJ, 662, 322
Ibata, R., Chapman, S., Ferguson, A. M. N., Irwin, M., Lewis, G., & McConnachie, A. 2004, MNRAS, 351, 117
Kartaltepe, J. S., et al. 2007, ApJS, 172, 320
Kaufmann, T., et al. 2007, MNRAS, 375, 53
Le Floc'h, E., et al. 2005, ApJ, 632, 169
Lilly, S. J., Le Fevre, O., Hammer, F., & Crampton, D. 1996, ApJ, 460, L1
Mathewson, D. S., Ford, V. L., Buchhorn, M. 1992, ApJS, 81, 413
Pozzetti, L., et al. 2003, A&A, 402, 837
Puech, M., Hammer, F., Lehnert, M. D., & Flores, H. 2007, A&A, 466, 83
Rawat, A., Hammer, F., Kembhavi, A. K., & Flores, H., 2007, ApJ, submitted
Robertson, B., Bullock, J. S., Cox, T. J., Di Matteo, T., Hernquist, L., Springel, V., & Yoshida, N. 2006, ApJ, 645, 986
Steinmetz, M., & Navarro, J. F. 1999, ApJ, 513, 555
Yang, Y. B., et al. 2007, A&A, submitted
Zheng, X. Z., Hammer, F., Flores, H., Assémat, F., & Rawat, A. 2005, A&A, 435, 507

## Discussion

*Combes*: You said that major mergers will solve the angular momentum problem. But the contrary is likely to be the case. In a merger the baryons lose their angular momentum against the dark matter halo by dynamical friction. Cold gas accretion is a way to keep angular momentum.

*Hammer*: That is indeed a possibility if the angular momentum of the baryonic matter is proportional to that of the total mass—the latter being increased after the major merger (conservation of angular momentum). Other assumptions on dynamical friction will necessarily lead to different conclusions. On the other hand observations (pair fractions and velocity fields) tell us that 50–70% of $M_*$ galaxies have probably experienced a major merger since $z \sim 1$. Most of these galaxies are indeed disks, and their angular momentum is larger than that of the Milky Way, suggesting a link between mergers (the bulk of $M_*$, including M31) and angular momentum (the Milky Way has not experienced any significant merger since $z \sim 3$, see Hammer et al. 2007).

*Kroupa*: The observed large number of major mergers for $L_*$ galaxies predicts the existence of large numbers of tidal dwarf galaxies (TDGs). Calculations have shown that all galaxies in the mass range $\sim 10^7$–$10^9\, M_\odot$ can be TDGs, posing a serious challenge for standard cosmology.

*Hammer*: This is an interesting suggestion that I have not considered yet. Indeed studies of the local group dwarfs show a large variation of star formation histories, which allow, for a significant fraction of them, the formation of TDGs. Whether those are evolving into present-day dwarfs or alternatively disrupted by other mechanisms (SN feedback, further accretion to the parent galaxy, etc.) might provide a solution to this important question.

*Beckman*: 1) Are the velocity fields derived from gas or from stars? 2) It is much easier to perturb gas than to perturb stars. For example gas–gas interaction due to high velocity gas accretion could possibly perturb gas in the disk but not lead to a heating of the stellar disk.

*Hammer*: 1) It is from gas. 2) Yes, and this has motivated us to observe the stellar kinematics of a few objects. However, these observations are quite challenging with the current generation of 8–10 m telescopes. Besides, our observed complex velocity fields derived from the gaseous component are truly unrelaxed, with large-scale motions decoupled from an expected rotation. Major mergers can reproduce this, but minor mergers apparently don't. It means that a very large fraction of gas should be accreted—asymmetrically—to reproduce our observations. There are no simulations to date showing such a complex velocity field by gas accretion while keeping the rotation for stars, all at large scales, which may require special conditions in already massive galaxies ($\sim M_*$).

*Silk*: 1) If major mergers are so common for $M_*$ galaxies at $z \sim 0.6$, how do you account for the thinness of the disks? 2) Could you comment on the zero-point for the Tully-Fisher relation that you find at $z \sim 0.6$?

*Hammer*: 1) Simulations by Robertson et al. (2006) may partly answer this question. Gas-rich mergers may produce thin disks with $v/\sigma \gg 1$ at least for the newly formed stars. Feedback might also help in quenching the star formation in the central region and expelling the high angular momentum gas to the galaxy outskirts, providing a further supply for forming a rotating disk later on. Interestingly, we do find that even the rotating disks at $z \sim 0.6$ have lower $v/\sigma$ ($\sim$3) than present-day disks (see Puech et al. 2007). So this mechanism (or another one) is necessary to explain the present-day thin disks. 2) We do find the same slope and zero-point for the stellar mass Tully–Fisher relation of rotating disks at $z \sim 0.6$ and at $z = 0$. However, we need more data to confirm it definitively. We interpret this finding by stating that since $z \sim 0.6$ the expected increase in stellar mass density is only 30%, or 0.1 dex in the Tully–Fisher relation.

From left to right: Gelys Trancho, François Hammer, and Myriam Rodrigues.

*Pathways through an Eclectic Universe*
*ASP Conference Series, Vol. 390, © 2008*
*J. H. Knapen, T. J. Mahoney, and A. Vazdekis, eds.*

# The ACS LCID Project: Old Main Sequence Turnoff Photometry of Isolated Local Group Dwarf Galaxies

Carme Gallart,[1] for the LCID team

[1]*Instituto de Astrofísica de Canarias, E-38200 La Laguna, Spain*

**Abstract.** The details of the early star formation history of isolated dwarf galaxies can shed light on the role of the cosmic re-ionization and supernova feedback in galaxy formation and evolution. In this paper, we present the project Local Cosmology from Isolated Dwarfs (LCID)[2], aimed at deriving detailed star formation histories for a sample of Local Group isolated dwarf galaxies. To accomplish this goal we have collected, using the ACS on board the *HST*, color–magnitude diagrams reaching their oldest main sequence turnoff ($V \simeq 29$ mag) with good photometric accuracy. Some results are briefly discussed.

## 1. Introduction

In current structure formation models, dwarf galaxies are the first to collapse, around 1 Gyr or so after the Big Bang, and thus their oldest populations probe the conditions of the very early Universe. In particular, the details of their early star formation histories (SFHs) can provide information on the effects on galaxy formation of the re-ionization which occurred at high redshift. They can also shed light on the role of other mechanisms, such as supernova feedback, in removing significant amounts of gas from small galaxy halos.

The majority of Local Group galaxies are dwarfs, and their SFHs can be studied very accurately from color–magnitude diagrams (CMDs) reaching the oldest main sequence turnoffs. They provide key details on the processes of formation and evolution of these galaxies and are therefore a complementary path to high redshift studies of galaxy formation and evolution.

The smallest *isolated* dwarf galaxies are the best probes for the above purposes since, unlike the Milky Way satellites, their evolution is not complicated by environmentally related mechanisms like tidal stirring (Mayer et al. 2006) or ram pressure stripping (Blitz & Robishaw 2000), which are expected to remove gas and strongly affect their SFHs. In addition, the radiation field of the early Milky Way is expected to dominate over the cosmological UV background (Mayer 2005) for the satellites.

Taking advantage of the great sensitivity and spatial resolution of the ACS on board the *HST*, we have undertaken a major project aimed at investigating the SFH of a selected sample of isolated Local Group dwarf galaxies. Our main goal is to obtain CMDs reaching the oldest main sequence turnoffs ($M_I \simeq +3$ mag) with good photometric accuracy, for six isolated dwarf galaxies (two dIrr galaxies: Leo A and IC 1613; the two isolated dSph galaxies in the Local Group (Cetus and Tucana); and two transition type dIrr/dSph: LGS3 and Phoenix).

---

[2]See http://www.iac.es/project/LCID.

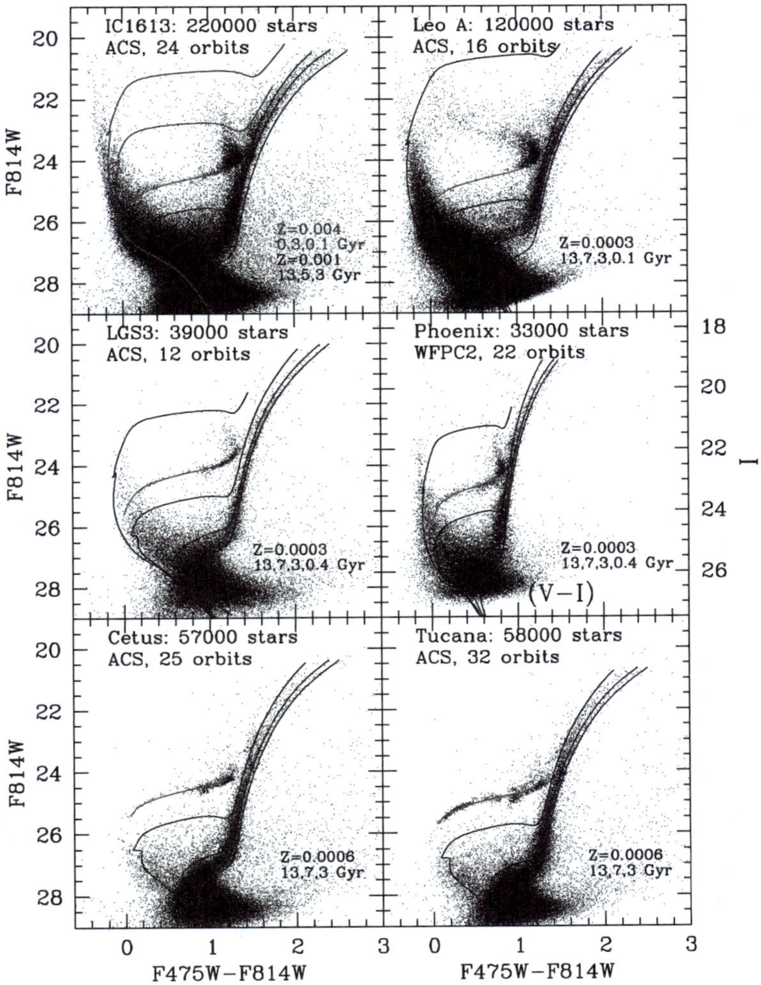


Figure 1. CMDs obtained for the six galaxies in our sample. In order to give a first indication of the range of ages and metallicities present in each galaxy, isochrones from Pietrinferni et al. (2004; overshooting set), with the ages and metallicities indicated in the labels, have been superimposed. The locus of the zero-age horizontal branch (ZAHB) for $Z = 0.0006$ or $Z = 0.0003$ (depending on the metallicity of the main population) is also shown. Distance moduli of $(m - M)_0 = 24.5$, 24.5, 24.1, 23.2, 24.5, and 24.8, and reddenings of $E(B - V) = 0.04$, 0.02, 0.05, 0.02, 0.03 and 0.03 for IC 1613, Leo A, LGS3, Phoenix, Cetus, and Tucana, respectively, have been adopted. Note the different band combination in the case of Phoenix. Determinations of the SFH of each system are under way through comparison with synthetic CMDs.

The data for this project have been obtained through two ACS programs (GO 10505, PI Gallart; GO 10590, PI Cole) with a total of 113 awarded orbits, and will be supplemented by data from a previous WFPC2 program for Phoenix, (GO 8706, PI Aparicio). The main aims of this project are the following:

- To obtain the SFH of each galaxy through comparison of the observed CMD with synthetic CMDs (see below).
- To study the radial gradients in the SFH for the galaxies in the sample in order to gain insight into the structural evolution within dwarf galaxies.
- To characterize the populations of variable stars in each galaxy, and to understand their relationships with the SFHs and stellar population gradients.

By comparing the results for isolated dwarfs with those obtained for the Milky Way and M31 satellites, we also expect to shed new light on the effects of environmental mechanisms, which are expected to dominate for the satellite sample. A detailed comparison between Cetus and Tucana, and the dSph satellites will be particularly useful in clarifying to what extent these two groups of objects may be related to each other.

## 2. The Star Formation History from Deep Color–Magnitude Diagrams

In this project, we have emphasized the need to obtain CMDs reaching the oldest main sequence turnoff with good photometric accuracy, in order to obtain reliable and detailed SFHs. In this case, the information is obtained directly from the main sequence, which is the best understood phase of stellar evolution from the theoretical point of view and also the one in which the location of stars in the CMD shows the highest sensitivity to age differences.

Figure 1 shows the CMDs of the galaxies in our sample. Note the variety of SFHs (though each pair of the same morphological type share important features), as evidenced by the comparison with selected isochrones. The similarity between the CMDs of Cetus and Tucana is particularly striking (and noteworthy—probably no two Milky Way dSph satellites are so similar!): both the metallicity and the age range of the main population are the same, and both have a similar *residual* star formation up to $\simeq 3$ Gyr ago. The horizontal branch (HB) morphology of both galaxies, however, is clearly different, with a well populated blue HB in Tucana and just a few blue HB stars in Cetus. Also, the characteristics of the variable star population are different (E. Bernard et al., in preparation). Preliminary SFHs for both galaxies using the detailed information in the old main sequence turnoff area (C. Gallart et al., in preparation; M. Monelli et al., in preparation) do show some differences which may account for the different HB characteristics.

The determination of the SFHs, through comparison of observed and synthetic CMDs, is under way using different approaches by different team members (e.g., Gallart et al. 1999; Bertelli & Nasi 2001; Skillman et al. 2003; Aparicio & Gallart 2004; and references therein, for a discussion of the different approaches; see also http://iac-star.iac.es/iac-star/). This will allow us to evaluate systematic differences in the results and to better understand the errors affecting the SFH determinations.

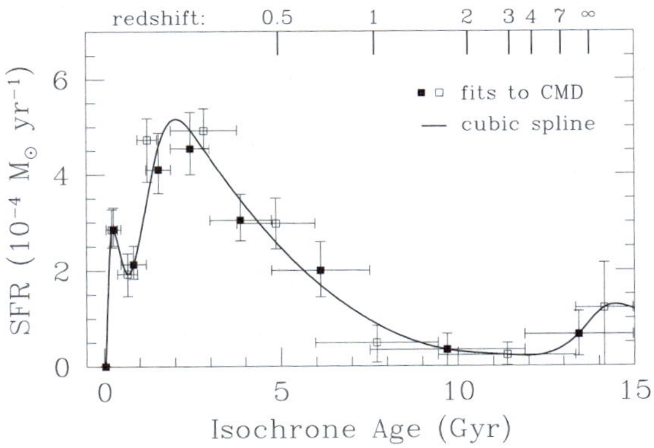


Figure 2. SFH for the central part of Leo A covered with the ACS. Data points with $1\sigma$ error bars show the best fit to the CMD for two age-binning schemes. The heavy line shows the results of a spline fit to the results, which represents our best estimate of the form of the Leo A SFH. The redshift derived using WMAP results and a flat $\Lambda$CDM cosmology, assuming lookback time $\equiv$ isochrone age, is given along the top axis.

The first results on a quantitative SFH have already been published for Leo A (Cole et al. 2007) and are shown in Figure 2. The bulk (over 90%) of the stars in the central part of this galaxy have formed within $\simeq 8$ Gyr from the present time. The apparent early peak of star formation above a subsequent minimum is not a high-confidence feature, as indicated by the error bars. However, some truly ancient stars are required to reproduce the observed CMD and provide the observed RR Lyrae stars (Dolphin et al. 2002).

**Acknowledgments.** I would like to thank the organizers for a wonderfully different (*eclectic!*) conference. Thanks Johan for the name tags for the kids. And Happy Scientific Birthday, John! I thank my parents, and Daniel's father, for joining and thus making it possible for me to attend conferences with Daniel's (and their!) company. Partial funding for this project is provided by the IAC (grant P3/94), the Spanish Ministry of Education and Science (grant AYA2004-06343), and NASA through grants GO-10505 and GO-10590 from the STScI.

## References

Aparicio, A., & Gallart, C. 2004, AJ, 128, 1465
Bertelli, G., & Nasi, E. 2001, AJ, 121, 1013
Blitz, L., & Robishaw, T. 2000, ApJ, 541, 675
Cole, A. A. et al. 2007, ApJ, 659, L17
Dolphin, A. E. et al. 2002, AJ, 123, 3154
Gallart, C., Freedman, W. L., Aparicio, A., Bertelli, G., & Chiosi, C. 1999, AJ, 118, 2245
Mayer, L. 2005, in IAU Col. 198, Near-field Cosmology with Dwarf Elliptical Galaxies, ed. H. Jerjen & B. Binggeli (Cambridge: Cambridge University Press), 220
Mayer, L., Mastropietro, C., Wadsley, J., Stadel, J., & Moore, B. 2006, MNRAS, 369, 1021

Pietrinferni, A., Cassisi, S., Salaris, M., & Castelli, F. 2004, ApJ, 612, 168
Skillman, E. D., Tolstoy, E., Cole, A. A., Dolphin, A. E., Saha, A., Gallagher, J. S., Dohm-Palmer, R. C., & Mateo, M. 2003, ApJ, 596, 253

## Discussion

*Díaz*: Can you get any constraint on the proportion of binaries from the comparison between synthetic and observed CMDs for your galaxy populations?

*Gallart*: The characteristics of the binary population (percentage of binaries and distribution of mass ratios) are free input parameters in the calculation of the synthetic CMDs, and in some cases it has been possible to constrain them through comparison with the observed CMD. In general, though, in order to reduce the parameter space to be tested, we assume some reasonable parameterization of the binary population according to, for example, what is observed in the solar neighborhood. Better fits are usually obtained if some reasonable fraction of binaries is assumed.

*Hammer*: Your result of various SFHs for dwarfs appears quite similar to those in other local dwarfs (e.g., Tolstoy and Hill). Have you plans to disentangle age/metallicity effects using spectroscopy (at least for luminous stars)?

*Gallart*: The age/metallicity degeneracy is less severe in the case of CMDs reaching the oldest main sequence turnoffs than in shallower data. In addition, these small galaxies have relatively low maximum metallicity, which limits the possible $Z(t)$ range of solutions. So even in the absence of spectroscopy it is possible to obtain reliable SFHs. However, members of our team have obtained calcium triplet spectra of red giant branch stars in several galaxies of our sample, and we plan to obtain these kinds of data for all of them. This will further constrain our solutions and in addition we will be able to check how accurate our solutions are in the absence of spectroscopy. In some cases SFR($t$) has been inferred from spectroscopic samples (since $Z$ is known, the age for each star can be derived from its position on the RGB). In our case we will also be able to compare SFR($t$) determinations from spectroscopic samples and from the full CMD.

*Beckman*: Have you extracted information about the spatial as well as the temporal distribution of the star formation in your galaxies?

*Gallart*: We plan to do both; there is work in progress for all of them except for Phoenix, for which S. L. Hidalgo, A. Aparicio, & D. Martínez-Delgado (2007, in preparation) have already derived both a global SFH and also the SFH in several elliptical annuli. What is observed (and qualitatively we observe the same in all the galaxies in our sample) is that the youngest stars age gradually as one moves outwards, but that the oldest population is everywhere equally old. This has been interpreted in terms of a "shrinking scenario" by Hidalgo, Marín-Franch, & Aparicio (2003, AJ, 125, 1247), who propose the hypothesis that in these small galaxies star formation moves inwards as gas is progressively exhausted in the outer parts.

*Pathways through an Eclectic Universe*
*ASP Conference Series, Vol. 390, © 2008*
*J. H. Knapen, T. J. Mahoney, and A. Vazdekis, eds.*

# Distinguishing Active Galactic Nuclei and Star Formation

Brent Groves[1] and Lisa Kewley[2]

[1]*Sterrewacht Leiden, Leiden University, PO Box 9513, 2300 RA Leiden, The Netherlands*

[2]*Institute for Astronomy, University of Hawaii, 2680 Woodlawn Drive, Honolulu, HI 96822, USA*

**Abstract.** Using the large emission-line galaxy sample from the Sloan Digital Sky Survey we show that star-forming galaxies, Seyferts, and low ionization nuclear emission-line regions (LINERs) form clearly separated branches on the standard optical diagnostic diagrams. We derive a new empirical classification scheme that cleanly separates these emission-line galaxies, using strong optical emission lines. Using this classification, we identify a few distinguishing host galaxy properties in each class, which, along with the emission-line analysis, suggest continuous evolution from one class to another. As a final note, we introduce models of both star-forming galaxies and AGN narrow-line regions that can explain the distribution of galaxies on standard emission-line ratio diagrams and possibly suggest new diagnostics across the emission spectrum.

## 1. Introduction

Since emission-line objects were first observed, ratios of strong lines have been used to distinguish the various classes of emission-line nebulae and ionization mechanisms. One of the first line ratio diagrams, and still one of the most powerful, was that of [N II]$\lambda$658.4/H$\alpha$ vs [O III]$\lambda$500.7/H$\beta$, proposed by Baldwin, Phillips, & Terlevich (1981) and commonly called the BPT diagram. This diagram uses strong optical lines of close proximity in the ratios, limiting reddening and spectrophotometric effects, and is able to clearly distinguish different classes of ionization. This and other diagnostic diagrams along these lines were later explored in detail by Veilleux & Osterbrock (1987, V&O). However, with the advent of the Sloan Digital Sky Survey (SDSS) we have been able to fill this diagram and explore details not seen before, enabling us to understand better the distribution and extent of the emission-line galaxies on this diagram.

## 2. SDSS Emission-Line Galaxies

In this work we used the SDSS Data Release 4 (DR4) spectroscopic galaxy sample (Adelman-McCarthy et al. 2006), which includes spectra of over 500 000 objects. The spectra are taken using 3 arcsec diameter fibers, positioned as close as possible to the centers of the target galaxies. The flux- and wavelength-calibrated spectra cover the range from 3800 to 9200 Å, with a resolution of $R \sim 1800$.

At the median redshift of the sample ($z \sim 0.1$) the spectroscopic fiber typically contains 20 to 40 percent of the total galaxy light and thus contains a component due to the host galaxy, as well as any nuclear AGN source. As described in Tremonti et al. (2004), we subtract the contribution of the stellar continuum from each spectrum, using the best fitting combination of template spectra from the population synthesis code of Bruzual & Charlot (2003). The best-fitting model can also be used to estimate stellar masses and star formation histories (Kauffmann et al. 2003a).

The sample criteria we use for the emission-line galaxies are described in detail in Kewley et al. (2006). The sample is limited to narrow emission-line galaxies with redshifts above 0.02. We also apply a signal-to-noise cut (S/N> 3) on the six dominant emission lines; H$\beta$ $\lambda$486.1 nm, [O III] $\lambda$500.7 nm, [O I] $\lambda$630.0 nm, H$\alpha$ $\lambda$656.3 nm, [N II] $\lambda$658.4 nm, and the [S II] $\lambda$671.6, 673.1 nm doublet. This gives a total sample of ~85 000 emission-line galaxies.

## 3. Emission-Line Galaxy Classification

The division of AGN and star-forming (SF) galaxies on the BPT diagram has been known for a while, and the work of Kewley & Dopita (2002) provided a theoretical basis for the distribution of SF galaxies in this and the V&O diagrams. This theoretical work provided a "maximal starburst" line and clear divisions between AGN and SF galaxies:

$$\begin{aligned}
\log([\mathrm{O\,III}]/\mathrm{H}\beta) &= 0.61/([\mathrm{N\,II}]/(\mathrm{H}\alpha - 0.47)) + 1.19, & (1)\\
\log([\mathrm{O\,III}]/\mathrm{H}\beta) &= 0.72/([\mathrm{S\,II}]/(\mathrm{H}\alpha - 0.32)) + 1.30, & (2)\\
\log([\mathrm{O\,III}]/\mathrm{H}\beta) &= 0.73/([\mathrm{O\,I}]/(\mathrm{H}\alpha - 0.59)) + 1.33, & (3)
\end{aligned}$$

with all AGN galaxies lying above this line.

After this work, a clearer delineation between AGN and SF galaxies in the BPT diagram was seen by Kauffmann et al. (2003b) when the large sample of SDSS galaxies was plotted. A dividing line was fitted to these, giving the empirical relation

$$\log([\mathrm{O\,III}]/\mathrm{H}\beta) = 0.61/([\mathrm{N\,II}]/(\mathrm{H}\alpha - 0.05)) + 1.3. \quad (4)$$

With these diagnostics, the AGN branch, composed of Seyfert galaxies, LINER-type galaxies, and composite AGN/SF galaxies, still remained mixed. However, when the V&O [S II]/H$\alpha$ and [O I]/H$\alpha$ vs [O III]/H$\beta$ diagrams are plotted with SDSS galaxies, a clear minimum is seen in the AGN branch (see fig. 3 of Kewley et al. 2006). When this minimum is fitted with a line in both diagrams, an empirical division for Seyferts and LINERs is obtained:

$$\begin{aligned}
\log([\mathrm{O\,III}]/\mathrm{H}\beta) &= 1.89\log([\mathrm{S\,II}]/\mathrm{H}\alpha) + 0.76, & (5)\\
\log([\mathrm{O\,III}]/\mathrm{H}\beta) &= 1.18\log([\mathrm{O\,I}]/\mathrm{H}\alpha) + 1.30. & (6)
\end{aligned}$$

Together, these equations define five different classes of emission-line galaxies, as shown in Figure 1.

SF galaxies, lying below the Kewley and Kauffmann curves, are the most numerous, making up ~64 000 (75%) of the SDSS sample.

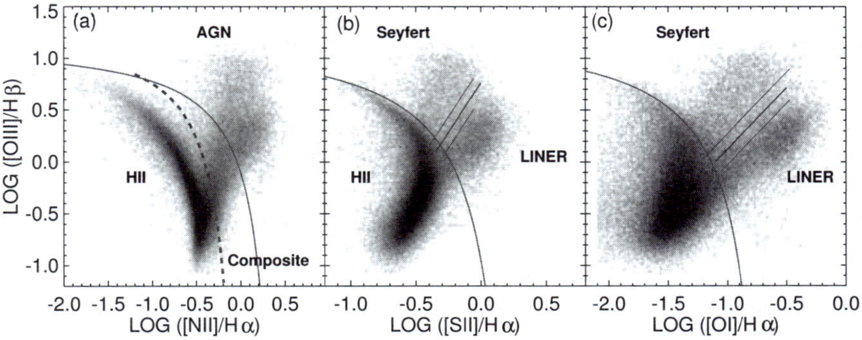


Figure 1. The three emission-line ratio diagnostic diagrams showing the six dividing lines. The solid curves in all three show the Kewley extreme starburst definition, the dashed curve in (a) shows the Kauffmann empirical relation, and the straight lines in (b) and (c) show the LINER/AGN divisions.

Composite AGN/SF galaxies, lying between the Kauffmann and Kewley curves on the BPT diagram, number only 5900, 7% of the sample.

Seyfert galaxies, lying above the Kewley curves and the [OI] and [SII] LINER lines, are only 3% of the SDSS sample (2400).

LINERs, lying on the other side of the LINER lines to Seyferts, are more numerous at 6000 galaxies (7%).

The remaining galaxies (8%) fall in different classifications in at least one of the diagrams and are classified as ambiguous.

These classifications allow the differences and similarities between SF galaxies and AGN to be considered. This was first done by Kauffmann et al. (2003b), who found that SF galaxies are predominantly found in younger, more metal-poor, later-type hosts than AGN. However, as they cover a much larger range in metallicity than AGN, their line properties are much more varied (Groves, Heckman, & Kauffmann 2006).

This was extended in detail in the work reported here (Kewley et al. 2006). The composite galaxies, as would be expected for an intermediate population, lie somewhere between AGN and metal-rich SF galaxies in their host properties. Seyferts tend to lie in the "green valley" in terms of their host properties, having intermediate properties between the blue, SF disk galaxies and the "red and dead" ellipticals. LINERs are older, more massive, less dusty, less concentrated, and they have higher velocity dispersions and lower [O III] luminosities than Seyfert galaxies have. Seyferts and LINERs are most strongly distinguished by their [O III] luminosities. One remarkable property of the AGN is that at fixed $L[\mathrm{OIII}]/\sigma_*^4$, which is an indicator of the black hole accretion rate relative to the Eddington rate, all differences between Seyfert and LINER host properties disappear. LINERs and Seyferts form a continuous sequence, with LINERs dominant at low $L/L_{\mathrm{Edd}}$ and Seyferts dominant at high $L/L_{\mathrm{Edd}}$. These results suggest that the majority of LINERs are AGN, and that the Seyfert/LINER dichotomy is analogous to the high/low-state models, and show that pure LINERs require a harder ionizing radiation field with lower ionization parameter than required by Seyfert galaxies, consistent with the low and high X-ray binary states.

## 4. A Model Understanding of AGN and SF Emission

In addition to using the emission-line diagrams to define *empirical* relationships and classifications, we can also use these diagrams to better define our *theoretical* models, which in turn help our understanding of emission-line objects. With the starburst models of Dopita et al. (2006), we can clearly see that the dominating horizontal extent of the SF galaxies in the BPT diagram is driven by metallicity variations, while the vertical spread is driven by a mix of stellar age and interstellar medium pressure.

Using the models of Groves, Dopita, & Sutherland (2004), we can see that the variation in the AGN branch is driven by a combination of effects, predominantly by ionization and metallicity in the horizontal, and ionization and SF/AGN mixing in the vertical [N II] direction.

Together with the observations, these models can help us correctly diagnose where exactly the ionizing photons are arising from in these galaxies, along with some important host galaxy parameters.

## 5. Conclusions

The SDSS has given us a large sample of $z \sim 0.07$ emission-line galaxies, truly filling out the BPT line ratio diagram. This has enabled clear, empirical classifications of emission-line galaxies to made, and an exploration of host properties of these different classes. These classifications and properties suggest some form of evolution, from the actively SF galaxies across the AGN Seyfert phase to the final, elliptical LINER phase. Along with new self-consistent theoretical models, the spread in this and other diagnostic diagrams is now being understood, allowing a full classification based on both observation and theoretical knowledge.

### References

Adelman-McCarthy, J. K., et al. 2006, ApJS, 162, 38
Baldwin, J. A., Phillips, M. M., & Terlevich, R. 1981, PASP, 93, 5 (BPT)
Bruzual, G., & Charlot, S. 2003, MNRAS, 344, 1000
Dopita, M. A., et al. 2006, ApJS, 167, 177
Groves, B. A., Dopita, M. A., & Sutherland, R. S. 2004, ApJS, 153, 75
Groves, B. A., Heckman, T. M., & Kauffmann, G. 2006, MNRAS, 371, 1559
Kauffmann, G., et al. 2003a, MNRAS, 341, 33
Kauffmann, G., et al. 2003b, MNRAS, 346, 1055
Kewley, L. J., & Dopita, M. A. 2002, ApJS, 142, 35
Kewley, L. J., Groves, B., Kauffmann, G., & Heckman, T. 2006, MNRAS, 372, 961
Tremonti, C. A., et al. 2004, ApJ, 613, 898
Veilleux, S., & Osterbrock, D. E. 1987, ApJS, 63, 295 (V&O)

**Discussion**

*Ercolano*: Do you have PDR physics in your code? If not your PAH predictions could contain uncertainties.

*Groves*: While my code (MAPPINGSIII) does include atomic and dust physics it does not include molecular species such as CO or $H_2$, which means that, at higher column depths, past the ionization front, our dust temperature could be incorrect. However, most of the PAH emission occurs at the inner edge of the PDR, where atomic physics dominates, and the PAH features should be accurate.

*Erwin*: How does your low Seyfert fraction (3%) square with the Ho et al. (1995, 1997) Palomar survey, which found a local Seyfert fraction of > 10%?

*Groves*: With a typical redshift of $z \sim 0.08$ and an SDSS fiber width of 3 arcsec most of our Seyferts include a large contribution of galactic light, which will move some of the these Seyferts down the AGN branch into the "transition" regime and likewise hide some of the transition objects.

*Díaz*: Do you have any information on line widths for your objects?

*Groves*: All type I objects have already been removed from the SDSS emission-line sample by principal component analysis. The remaining type II AGN may show faint indications of higher widths, but the typical S/N ratio means that this is very hard to detect.

Brent Groves (*left*) and Mark Westmoquette.

*Pathways through an Eclectic Universe*
*ASP Conference Series, Vol. 390, © 2008*
*J. H. Knapen, T. J. Mahoney, and A. Vazdekis, eds.*

# Classification of Galaxies by Their Radial Profiles: Unbarred Early Types

Rebeca Aladro,[1] L. Gutiérrez,[2,3] Peter Erwin,[4] and J. E. Beckman[2,5]

[1]*Universidad de La Laguna, La Laguna, Tenerife, Spain*
[2]*Instituto de Astrofísica de Canarias, E-38200 La Laguna, Tenerife, Spain*
[3]*Universidad Nacional Autónoma de México, Campus Ensenada, Mexico*
[4]*Max-Planck-Institut für extraterr. Physik, Garching, Germany*
[5]*Consejo Superior de Investigaciones Científicas, Spain.*

**Abstract.** We classify 46 broad band surface-brightness profiles in the visible from a sample of nearby early-type unbarred galaxies imaged at the ING (La Palma) and in the SDSS. A key result is that the fraction of these galaxies showing truncations in their outer disks is relatively small (of order 10%). The largest single category of galaxies show what Erwin et al. (2005) have termed "Type III" profiles, which have an exponential inner disk, and also an exponential outer disk but with a shallower slope. They show no sign of truncations out to beyond at least five scale-lengths. The second most frequent category is that of Freeman's (1970) "Type I" disks, which show a single exponential slope out to the radial limits of detection, with no sign of any truncation. Together, Types I and III account for over 75% of the galaxies in our sample. This is part of an ongoing program to use galaxy surface-brightness profiles to test models of galaxy formation.

## 1. The Sample and a Classification Scheme

Our sample is of 46 early-type galaxies chosen with the following criteria. They are catalogued as unbarred (although a few of them may in fact be weakly barred), nearby and well resolved, and with inclination angle less than 65°. For 18 of these we used SDSS images in the $r$ band to make the profiles, and the remaining 28 were observed by ourselves in the $r$ band with the Wide Field Camera on the 2.5 m Isaac Newton Telescope (La Palma). Very careful star masking and sky subtraction were followed by the extraction of radial surface brightness profiles down to a $5\sigma$ limiting brightness level.

Erwin et al. (2005) devised a classification scheme extending Freeman's classical scheme, in which Type I profiles are a single unbroken exponential and Type II profiles have outer slopes that are steeper than their inner slopes. The new class ("Type III") has an outer disk slope shallower than the inner disk slope. We note that some galaxies show the characteristics of Type II in their inner zones and Type III in their outer zones; we term these Type II/III (see also Beckman et al. 2007 and Pohlen et al., this volume, p. 247). The distribution of these disk types for our sample is shown in Figure 1.

## 2. Some Initial Inferences

As we have seen, Type III profiles have a shallower exponential at larger radii. It is probable that these are due to processes which modify the simple Type I single exponential profile, but there are two possibilities: either the stellar population in the outer disk is being somehow driven outwards, or the stellar population in the inner disk is being driven inwards.

Although our analysis in detail is not yet complete, in the sample of unbarred early-type galaxies presented here, we can find that, comparing this work with the barred galaxies studied by Erwin, Pohlen, & Beckman (2007), there is a much higher fraction of antitruncations (i.e., Type III profiles) than in barred early-type galaxies, and conversely a much lower proportion of Type II profiles. We are devising tests to see whether we can account for Type III profiles as Type Is with either an internal modification or an external modification, but so far with no clear outcome.

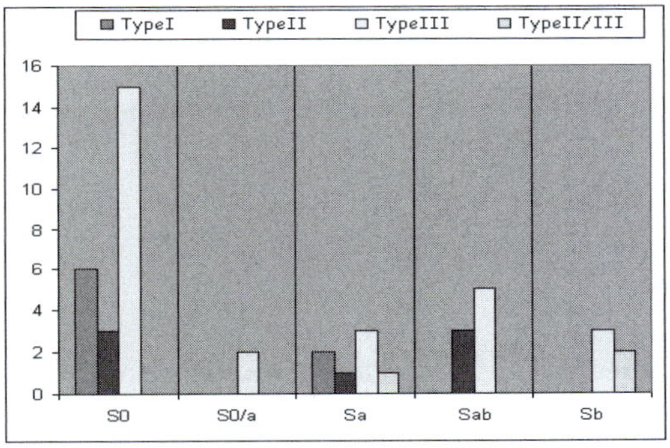


Figure 1. Histogram showing the frequency of the profiles according to their morphological type.

### References

Beckman, J. E., Erwin, P., Pohlen, M., Gutiérrez, L., Aladro, R., & Trujillo, I. 2007, in Mapping the Galaxy and Nearby Galaxies, ed. K. Wada & F. Combes (New York: Springer), p. 310

Erwin, P., Beckman, J. E., & Pohlen, M. 2005, ApJ, 626, L81

Erwin, P., Pohlen, M., & Beckman J. E. 2007, AJ, submitted

Freeman, K. C. 1970, ApJ, 160, 811

*Pathways through an Eclectic Universe*
*ASP Conference Series, Vol. 390, © 2008*
*J. H. Knapen, T. J. Mahoney, and A. Vazdekis, eds.*

# The Structural Properties of a Sample of Blue Compact Galaxies

R. O. Amorín, J. A. L. Aguerri, and C. Muñoz-Tuñón

*Instituto de Astrofísica de Canarias, E-38200 La Laguna, Spain*

**Abstract.** We study the structural properties of the low surface brightness stellar hosts in blue compact galaxies by fitting a two-dimensional Sérsic law to their optical images. The accurate isolation of the starburst emission is an important strength of the method. It allows us to obtain more reliable structural parameters, especially when a Sérsic law is fitted. Preliminary results on the structural properties of the host galaxy for a sample of 28 blue compact galaxies are presented here. Correlations between galaxy and starburst sizes, luminosities, central surface brightnesses, colors and the Sérsic index $n$, as well as with the neutral hydrogen mass and line width, $W_{20}$, are studied. Furthermore, we compare the structural properties of the sample with those of other types of dwarf galaxies in order to shed some light on possible evolutionary connections.

## 1. Introduction

Blue compact galaxies (BCGs) are gas-rich and metal-deficient objects, which display very intense and narrow emission lines in the optical, due to their intense star formation activity distributed in one or several star-forming bursts (Cairós et al. 2001a). The vast majority of BCGs present a low surface brightness (LSB) stellar host underlying the star-forming regions (Cairós et al. 2001a). This stellar component is an extended envelope that generally shows elliptical isophotes and displays red colors, indicative of an evolved stellar population. In order to establish the luminosities, structures and evolutionary status, as well as the star formation history of the BCGs, it is indispensable to characterize the LSB component. Their structure, kinematics and dynamical properties are important issues in dwarf galaxy research, as they help us to understand such processes as the regulation of the star-forming activity and the possible connections between different dwarf types.

In order to obtain accurate structural parameters of the host galaxies in BCGs we developed a 2D fitting technique, based on the GALFIT software (Peng et al. 2002). This technique allow us to obtain more confident models by using all the image information, taking special care in masking out the starburst emission. One of our main goals is to fit a Sérsic law in order to classify the BCG host galaxy into structures with low or high $n$ index with well-controlled uncertainties that could give us some clues about their dynamics. The structural parameters are also used to test possible scenarios that link different types of dwarfs and low mass star-forming galaxies by comparing their spectrophotometric properties. Furthermore, an accurate subtraction of the host galaxy is needed in order to reach a proper characterization of the starburst episode.

## 2. Modeling the Low Surface Brightness Stellar Host

The structural properties of the LSB stellar hosts in BCGs are often studied by fitting exponentials and/or $R^{1/n}$ models to the outer regions of their radial profiles. The limitations imposed by the presence of a large starburst overlapping the underlying component makes this kind of analysis a very difficult task. In this study, we carry out detailed 2D modeling of 28 star-forming galaxies previously presented in Cairós et al. (2001a,b). We follow the methodology developed in Amorín et al. (2007), where eight galaxies of this sample were fitted and the results compared with simulations and with their previously derived 1D fits (see Caon et al. 2005). The deep $BVR$ galaxy images ($\mu_{\rm B} \gtrsim 27$–$28$) were fitted by using a Sérsic law. Consistency checks were applied to estimate the reliability and accuracy of the fit. We explore the stability of the parameters with the region being fitted, i.e., their area, the consistency between parameters in different filters, and the uncertainties associated with sky subtraction, noise fluctuation, and seeing effects.

### 2.1. Ongoing work

We are now investigating correlations between the resulting structural parameters. Preliminary results show that the majority of the galaxies have Sérsic indexes $n \lesssim 2$, suggesting a disk-like structure. In this sample, larger and more luminous galaxies present larger and more luminous starbursts, as in previous studies, e.g., Papaderos et al. (1996). The colors of the host galaxies seem to be nearly constant with luminosity, with mean colors $\langle B-V \rangle = 0.52 \pm 0.25$ and $\langle B-R \rangle = 0.92 \pm 0.30$. We are also currently investigating how the Sérsic parameters relate to those of other types of galaxies, especially dwarf types. In this sample, BCG hosts seem to split into two groups, those with $M_{\rm B} \gtrsim -18$, which follow similar correlations in surface brightness and scale length with luminosity as dIs and dEs, and those with $M_{\rm B} \lesssim -18$, which follow the trend of spiral galaxies. Finally, possible correlations between the structure and the gas content of these galaxies are also being analyzed. The total luminosity and the size of both the host galaxies and the starburst correlates positively with their H I mass content. Low luminosity BCGs also clearly deviate from the correlation between $M_{\rm B}$ and the H I line width, $W_{20}$, followed by the spiral galaxies, showing larger velocities than expected for their luminosities.

**Acknowledgments.** Special thanks are given by the authors to N. Caon and L. M. Cairós for their collaboration in this project.

#### References

Amorín, R. O., Muñoz-Tuñón, C., Aguerri, J. A. L., Cairós, L. M., & Caon, N. 2007, A&A, 467, 541

Cairós, L. M., Caon, N., Vílchez, J. M., Gónzalez-Pérez, J. N., & Muñoz-Tuñón, C. 2001a, ApJS, 136, 2

Cairós, L. M., Vílchez, J. M., Gónzalez-Pérez, J. N., Iglesias-Páramo, J., & Caon, N. 2001b, ApJS, 133, 321

Caon, N., Cairós, L. M., Aguerri, J. A. L., & Muñoz-Tuñón, C. 2005, ApJS, 157, 218

Papaderos, P., Loose, H.-H., Fricke, K. J., & Thuan, T. X. 1996, A&A, 314, 59

Peng, C. Y., Ho, L. C., Impey, C. D., & Rix, H.-W. 2002, AJ, 124, 266

*Pathways through an Eclectic Universe*
*ASP Conference Series, Vol. 390, © 2008*
*J. H. Knapen, T. J. Mahoney, and A. Vazdekis, eds.*

# The Calcium Triplet Gradient of M32

A. J. Cenarro

*Instituto de Astrofísica de Canarias. E-38205 La Laguna, Tenerife, Spain*

N. Cardiel and J. Gorgas

*Dpto. Astrofísica y C.C. de la Atmósfera, UCM, Avda. Complutense s/n, 28040, Madrid, Spain*

**Abstract.** Based on 3D spectroscopic data of M32 in the range ~8400–8900 Å, we present the first reliable gradient of the calcium triplet (CaT) beyond one effective radius ($R_{\rm eff}$). Assuming the age–metallicity gradients inferred from optical spectra in previous work on M32, SSP models for the CaT predict a variation with radius that is qualitatively consistent with our observations. However, the CaT in M32 seems to be larger than predicted by the models, contrary to the low CaT values exhibited by massive ellipticals.

## 1. Introduction: The Puzzling Calcium Triplet in Ellipticals

Unlike any other metal line, the central CaT strengths of elliptical (E) galaxies anti-correlate with mass (Saglia et al. 2002; Cenarro et al. 2003, 2004). In addition, the absolute values of the CaT indices of massive Es are clearly lower than the model predictions, with dwarf Es being better reconciled by the models (Cenarro et al. 2003; D. Michielsen et al., in preparation). In fact, differences in age and metallicity exclusively do not explain the locus of E galaxies: calcium under-abundances and/or different dwarf-to-giant ratios have been proposed as plausible explanations for such a discrepancy.

To shed light on this issue, we performed 3D spectroscopy of the CaT region for M32, the best known low mass E in term of stellar population properties. Observations were carried out at the 3.5 m telescope in Calar Alto using PMAS, an integral field unit that in combination with the nod-and-shuffle technique allows an accurate removal of the dense sky line forest in the near infrared. Two different pointings of $16'' \times 16''$ (centered and off-centered by $36''$ along the major axis) allowed us to determine reliable gradients well beyond 1 $R_{\rm eff}$ ($\sim 35''$).

## 2. The CaT Gradient of M32

Radial velocity and velocity dispersion were determined for each spectrum in the data cube (a $4 \times 4$ binning was performed to increase the S/N of the off-centered pointing). The data cube spectra were then de-redshifted and broadened up to $\sigma = 110\,{\rm km\,s^{-1}}$ spectral resolution to avoid systematics among the CaT strengths.

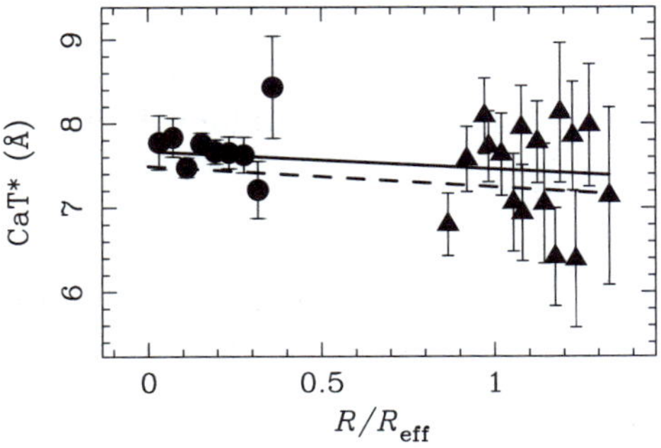


Figure 1. The CaT* gradient of M32 at $\sigma = 110\,\mathrm{km\,s^{-1}}$ spectral resolution. Dots and triangles correspond to the centered and off-centered 2D pointings respectively. The solid line illustrates an error-weighted linear fit to all data. The expected CaT* gradient, when assuming the age–metallicity trend along radius of M32 (see the text), is given by the dashed line.

Figure 1 shows a mild, negative gradient for the CaT* indices (Cenarro et al. 2001) of M32. For simplicity, single CaT* values from the center pointing were binned along radius (dots), with error bars being the larger of either the typical error of the mean or the CaT* rms within the bin. For the off-center indices (triangles), the initial $4 \times 4$ binning was kept. An error-weighted linear fit to all data is shown as a solid line.

On the basis of line-strength indices in the optical spectral range, Rose et al. (2005; see also Worthey 2004) reported that the mean age of M32 rises from 3 to 4 Gyr in the nucleus up to 6–7 Gyr at $30''$ radius, with metallicity dropping from [Fe/H] $\sim 0.0$ to $\sim -0.3$ dex. SSP model predictions by Vazdekis et al. (2003) for the CaT* index are over-plotted in Fig. 1 (dashed line) assuming such an age–metallicity trend: 3.5 Gyr and 0.0 dex at $R/R_{\mathrm{eff}} = 0$; 7 Gyr and $-0.38$ dex at $R/R_{\mathrm{eff}} = 1$. It is clear that the CaT* gradient of M32 can be qualitatively explained by just such age and metallicity effects. However, the absolute values—particularly in the center (CaT* $= 7.64 \pm 0.06$ within an aperture of radius $R_{\mathrm{eff}}/4$)—are significantly larger than predicted (CaT* $\sim 7.46$) and cannot be explained by any combination of age and metallicity. The large CaT* at the center of M32, also found by A. Praagman, R. F. Peletier & K. Ganda (priv. comm.), adds controversy to the intriguing puzzle of the CaT in Es.

## References

Cenarro, A. J., Cardiel, N., Gorgas, J., Peletier, R. F., Vazdekis, A., & Prada, F. 2001, MNRAS, 326, 959

Cenarro, A. J., Gorgas, J., Vazdekis, A., Cardiel, N., & Peletier, R. F. 2003, MNRAS, 339, L12

Cenarro, A. J.; Sánchez-Blázquez, P., Cardiel, N., & Gorgas, J. 2004, ApJ, 604, L101

Rose, J. A., Arimoto, N., Caldwell, N., Schiavon, R. P., Vazdekis, A., & Yamada, Y. 2005, AJ, 129, 712

Saglia, R. P., Maraston, C., Thomas, D., & Bender, R. 2002, ApJ, 579, L13

Vazdekis, A., Cenarro, A. J., Gorgas, J., Cardiel, N., & Peletier, R. F. 2003, MNRAS, 340, 1317

Worthey, G. 2004, AJ, 128, 2826

*Pathways through an Eclectic Universe*
*ASP Conference Series, Vol. 390, © 2008*
*J. H. Knapen, T. J. Mahoney, and A. Vazdekis, eds.*

# Dark Matter in Low Mass Surface Density Galaxies

Laurent Chemin

*Observatoire de Paris, section Meudon, GEPI, CNRS-UMR 8111 & Université Paris 7, 5 Pl. Janssen, 92195 Meudon, France*

Claude Carignan

*Laboratoire d'Astrophysique Expérimentale, Département de physique, Université de Montréal, Montréal, Québec, Canada*

Philippe Amram

*Observatoire Astronomique Marseille Provence, LAM, UMR 6110, 2 Pl. Le Verrier, 13248 Marseille, France*

**Abstract.** Low mass surface density spiral and irregular galaxies, like low surface brightness and dwarf galaxies, are unique laboratories to study the dynamical properties of dark matter halos because their mass is generally dominated by dark matter at all galactocentric radii. We present results from the largest sample ever assembled of high resolution H$\alpha$ velocity fields of low surface brightness and dwarf galaxies in order to study their mass distributions.

## 1. Observations

Observations of 39 galaxies have been obtained at the OmM 1.6 m, OHP 1.93 m, CFHT and ESO 3.6 m telescopes using Fabry–Perot (FP) interferometry with the FaNTOmM and Cigale instruments (Hernandez et al. 2003), allowing typical resolving powers at H$\alpha$ between 8 000–25 000, spectral samplings in the range 7–16 km s$^{-1}$, spatial samplings $\lesssim 1.5''$ and total scanning times of ~3 h per galaxy. The sample contains low surface brightness (LSB) and dwarf galaxies chosen in previous studies from Impey et al. (1996), McGaugh, Rubin, & de Blok (2001), de Blok & Bosma (2002), Swaters et al. (2003), or Monnier-Ragaigne et al. (2003). Galaxies for which dark matter (DM) halo shapes strongly disagree with the so-called "universal" cosmological halo (Navarro, Frenk, & White 1997) have been observed. Provisional results of the survey have been presented in Chemin et al. (2004) and Garrido et al. (2005). An adaptive binning is applied to the FP H$\alpha$ cubes in order to improve the signal detection in low signal-to-noise regions, as described in Chemin et al. (2006).

## 2. Results

An example is illustrated in Fig. 1 with the galaxy UGC 4325, which is supposed to exhibit a core-dominated halo according to de Blok & Bosma (2002). A hybrid rotation curve (RC) is derived from our H$\alpha$ velocity field and H I observations from Swaters et al. (2003). Figure 1 also shows the slit spectroscopy rotation curve from de Blok & Bosma (2002). The FP velocities are larger than the

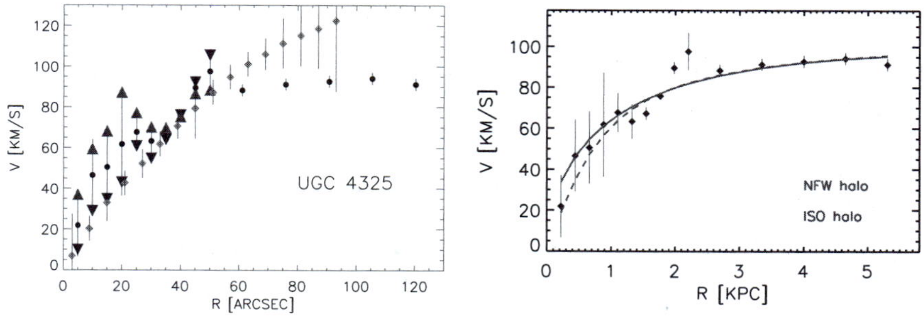


Figure 1. *Left panel:* Observed rotation curve of UGC 4325. Blue downward and red upward triangles are for the respective approaching and receding half of the disk, black circles for both sides fitted simultaneously and green squares are from slit spectroscopy data (de Blok & Bosma 2002). For $R > 50''$, black symbols are H I measurements (Swaters et al. 2003). *Right panel:* Mass models using the pseudo-isothermal sphere (ISO, green dashed curve) and the Navarro–Frenk–White formalism (NFW, magenta solid curve).

slit spectroscopy ones in the innermost regions of the galaxy. A significant asymmetry is detected between the approaching and receding halves of the disk. The ionized gas motions are thus non-axisymmetric in these regions. The FP and slit spectroscopy H$\alpha$ rotation curves are in good agreement in regions where the axisymmetry occurs ($\sim 30'' < R < 50''$).

Mass models of the rotation curve are shown in Fig. 1 for the minimum stellar disk hypothesis. Two halo models are fitted, a pseudo-isothermal sphere and the Navarro–Frenk–White formalism. Because of the asymmetry detected in the gas motions, the FP error bars are large in this region of the galaxy. As a result, the halo fits give equivalent results. This example illustrates that one clearly needs to model the gas motions taking into account perturbations of the potential if one wants to discriminate properly the DM halo shape within such low surface density galaxies. These results will be presented in a forthcoming article, with the dynamical analysis of the whole sample.

## References

Chemin, L., Amram, P., Carignan, C., Balkowski, C., van Driel, W., Cayatte, V., Hernandez, O., Boulesteix, J., & Marcelin, M. 2004, IAUS 220, ed. S. D. Ryder et al. (San Francisco: ASP), p. 333

Chemin, L., et al. 2006, MNRAS, 366, 812

de Blok, W. J. G., & Bosma, A. 2002, A&A, 385, 816

Garrido, O., Marcelin, M., Amram, P., Balkowski, C., Gach, J.-L., & Boulesteix, J. 2005, MNRAS, 362, 127

Hernandez, O., Gach, J.-L., Carignan, C., & Boulesteix, J. 2003, SPIE, 4841, 1472

Impey, C. D., Sprayberry, D., Irwin, M. J., & Bothun, G. D. 1996, ApJS, 105, 209

McGaugh, S. S., Rubin, V. C., & de Blok, W. J. G. 2001, AJ, 122, 2381

Monnier-Ragaigne, D., van Driel, W., Schneider, S. E., Balkowski, C., & Jarrett, T. H. 2003, A&A, 408, 465

Navarro, J. F., Frenk, C. S., & White, S. D. M. 1997, ApJ, 490, 493

Swaters, R. A., Madore, B. F., van den Bosch, F. C., & Balcells, M. 2003, ApJ, 583, 732

*Pathways through an Eclectic Universe*
*ASP Conference Series, Vol. 390, © 2008*
*J. H. Knapen, T. J. Mahoney, and A. Vazdekis, eds.*

# Low Luminosity Early-type Galaxies in the NGC 128 Group

I. Chilingarian,[1] O. Sil'chenko,[2] V. Afanasiev,[3] and P. Prugniel[4]

[1]*Obs. de Paris, LERMA, 61 Av. de l'Observatoire, Paris, F-75014*
[2]*Sternberg Astronomical Institute, Moscow, Russia*
[3]*Special Astrophysical Observatory RAS, N. Arkhyz, Russia*
[4]*CRAL Obs. de Lyon, 9 Av. Charles André, St.Genis Laval, F-69561*

**Abstract.** We present spatially resolved kinematics and stellar population parameters for three low luminosity galaxies in the NGC 128 group obtained by means of 3D spectroscopy. We briefly discuss their evolutionary scenarios.

## 1. Introduction

The NGC 128 group ($D = 57$ Mpc) includes (spectroscopically confirmed): the giant gas-rich S0 NGC 128 with a peanut-shaped bulge, the giant Sa NGC 125, three low-luminosity S0s closer than 100 kpc in projection to NGC 128: NGC 126 ($M_B = -18.5$, 61 kpc), NGC 127 ($M_B = -18.6$, 14 kpc), and NGC 130 ($M_B = -19.0$, 16 kpc), and several late-type spirals beyond 300 kpc. NGC 128 is known to have a gaseous disk counter-rotating with respect to the stars (Emsellem & Arsenault 1997).

## 2. Observations, Data Analysis, and Results

We obtained observations for NGC 126, NGC 127, and NGC 130 in 2005 and 2006 with the MPFS IFU spectrograph at the 6 m telescope of the SAO RAS. Using a novel stellar population fitting technique (Chilingarian et al. 2005, 2006, 2007a; Prugniel et al. 2005), we have derived maps of the stellar population parameters (SSP-equivalent age and metallicity), and the internal kinematics of the stars and the ionized gas (for NGC 127).

NGC 126 contains a prominent bar seen in direct images. It appears as an S-shaped structure in the velocity field. No emission lines are seen in the fitting residuals. NGC 127 is a gas-rich object with early-type morphology and ongoing star formation: strong emission lines are observed. H$\beta$ has been used to recover the velocity field of the ionized gas showing faster rotation ($\sim$100 km s$^{-1}$) than in the stars ($\sim$50 km s$^{-1}$), and an asymmetry on the SE part of NGC 127 (in the direction of NGC 128) that may be caused by perturbed motions of the gas near the region, where the accreting flow reaches NGC 127. NGC 130, located at nearly the same projected distance, is very different. Its relatively old stellar population, regular kinematics, and absence of ionized gas suggest it is located much farther from NGC 128. NGC 130 exhibits a spatially unresolved young

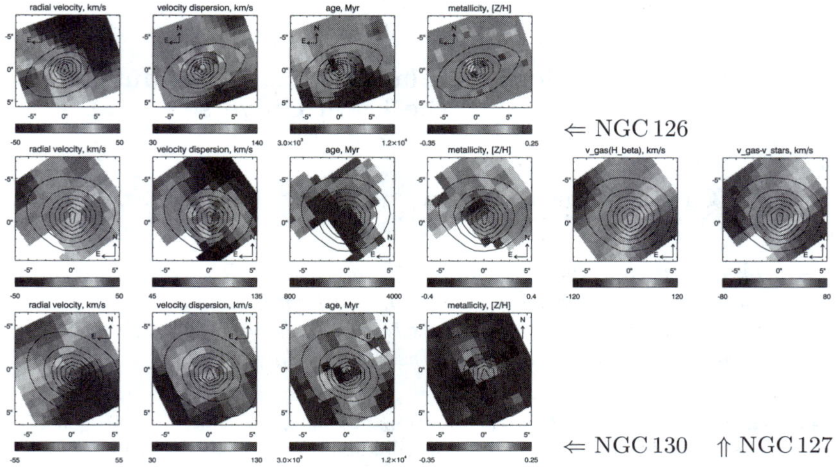


Figure 1. Internal kinematics and stellar populations of three galaxies. Radial velocity, velocity dispersion, age, and metallicity maps are shown for all three galaxies. The velocity field of the ionized gas (H$\beta$) and the difference between the velocities of the gas and stars are shown for NGC 127 (*middle row, right*).

nucleus ($\Delta t = 4\,\mathrm{Gyr}$) reminiscent of young central structures observed in bright Virgo cluster dE galaxies (Chilingarian et al. 2007b).

## 3. Discussion

While NGC 126 and NGC 130 look similar to cluster dS0/S0s, NGC 127 is a star-forming galaxy with unusual kinematics and a bridge connecting it to NGC 128. We propose the following scenario: NGC 127 has recently passed its pericenter, and we now observe an infall of gas from NGC 128 onto NGC 127. After gas removal and several Gyr of passive evolution, it will fade and will be indistinguishable from “normal” dE galaxies in terms of morphology and luminosity.

**Acknowledgments.** IC is grateful to the Organizing Committee of the conference for financial support.

### References

Chilingarian, I., et al. 2005, IAU Colloq. 198, Near-Fields Cosmology with Dwarf Elliptical Galaxies , ed. H. Jerjen & B. Binggeli (Cambridge: CUP), p. 105
Chilingarian, I. 2006, PhD thesis, MSU (Russia), UCBL (France), astro-ph/0611893
Chilingarian, I., et al. 2007a, MNRAS, 376, 1033
Chilingarian, I., et al. 2007b, AstL, 33, 292
Emsellem, E., & Arsenault, R. 1997, A&A, 318, L39
Prugniel et al. 2005, IAU Colloq. 198, Near-Fields Cosmology with Dwarf Elliptical Galaxies , ed. H. Jerjen & B. Binggeli (Cambridge: CUP), p. 73

*Pathways through an Eclectic Universe*
*ASP Conference Series, Vol. 390,* © *2008*
*J. H. Knapen, T. J. Mahoney, and A. Vazdekis, eds.*

# Properties of H I-selected Galaxies: The Contribution of Gas-rich, LSB Galaxies to the Local Universe

Diego A. García-Appadoo,[1,4] Andrew A. West,[2] Julianne J. Dalcanton,[3] and Mike J. Disney[4]

[1]*Argelander-Institut für Astronomie, Bonn Universität, Bonn, Germany*
[2]*Astronomy Department, University of California, Berkeley, USA*
[3]*Department of Astronomy, University of Washington, Seattle, USA*
[4]*Astronomy Department, Cardiff University, Cardiff, UK*

**Abstract.** Understanding the relationship between neutral hydrogen (H I) and stars in the interstellar medium of galaxies is integral to the study of galaxy evolution. Because stellar populations in galaxies form from collapsed clouds of hydrogen, we can naively claim that galaxies with few stars and large quantities of gas must be less "evolved" than those with little gas and many stars, and that galaxies evolve by converting their gas into stars. To first order this description is valid; we know that galaxies with large stellar populations must have once had fewer stars and more gas.

## 1. The Equatorial Strip

How a galaxy follows this evolutionary path is presumably much more complicated. Factors such as mass, metal content, gas fraction, internal motions, and other environmental conditions play an important role in determining how quickly H I gets converted into stars (e.g., van Zee et al. 1998; Kennicutt 1998).

The Equatorial Strip (Garcia-Appadoo 2005) is a unique sample which comprises 1077 galaxies selected purely by their H I signature from the Parkes radio telescope. This sample covers a large area of sky, is free from optical selection effects, and has complete, very high quality, optical data (Sloan Digital Sky Survey) for a subsample of 195 galaxies. The combination of the H I data with optical data allows us to make a comprehensive analysis of the properties of H I-selected extragalactic sources and investigate vital global correlations, all of which improves our understanding of gas-rich and low surface brightness (LSB) galaxies. The Equatorial Strip runs along the celestial equator from $-6^\circ < \delta < +10^\circ$ and through all RAs out to a distance of ~170 Mpc. The Equatorial Strip sample represents $\sim 14\%$ of the whole sky, covering an area of 5738 deg$^2$ and a total volume of $\sim 2.76 \times 10^6\,\mathrm{Mpc}^{-3}$.

LSB galaxies make up 12% of the sample, but no high luminosity or high H I mass LSB galaxies have been found. Consequently, LSB galaxies make up no more than 6% of the high luminosity, gas-rich population, and massive LSB galaxies contribute no more than 13% of the population at the 95% confidence level.

The bivariate brightness distribution (Fig. 1) and the luminosity function for the sample have been calculated and relationships found between surface

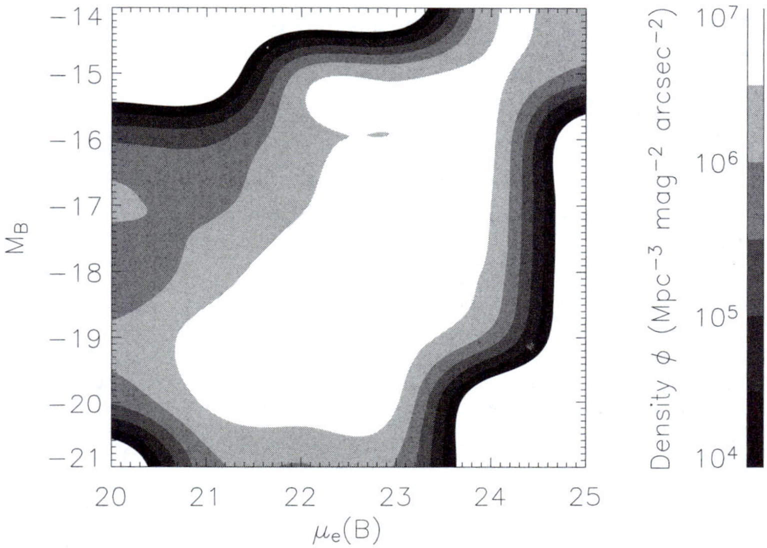


Figure 1. H I mass density–surface brightness distribution. The distribution is not peaked and the contribution is provided and shared by a wide variety of galaxies, from bright, high luminosity galaxies (*bottom left*) to low surface brightness, low luminosity galaxies (*top right*).

brightness and optical luminosity, H I, baryonic and dynamical mass. From these results we find that LSB galaxies contribute $35^{+29}_{-20}\%$ to the number density of gas-rich galaxies in the Universe but only $7^{+3}_{-2}\%$ to the luminosity density. They also contribute $21^{+7}_{-6}\%$ and $12 \pm 3\%$, respectively, to the neutral hydrogen (H I) and baryon density of gas-rich galaxies in the Universe.

The Equatorial Strip sample has unveiled many objects not found in optical surveys, ranging from very low surface brightness, very blue galaxies to extremely gas-rich galaxies.

**Acknowledgments.** Funding for the Sloan Digital Sky Survey (SDSS) has been provided by the Alfred P. Sloan Foundation, the participating institutions, the National Aeronautics and Space Administration, the National Science Foundation, the US Department of Energy, the Japanese Monbukagakusho, and the Max Planck Society.

## References

Garcia-Appadoo, D. A. 2005, PhD thesis, Cardiff University

Kennicutt, R. C. 1998, ARA&A, 36, 189

van Zee, L., Salzer, J. J., Haynes, M. P., ODonogue, A. A., & Balonek, T. J. 1998, AJ, 116, 2805

*Pathways through an Eclectic Universe*
*ASP Conference Series, Vol. 390, © 2008*
*J. H. Knapen, T. J. Mahoney, and A. Vazdekis, eds.*

# A Preliminary Budget for the Ionizing Photons in the H II Regions of M51

L. Gutiérrez[1,2] and J. E. Beckman[1,3]

[1]*Instituto de Astrofísica de Canarias, La Laguna, Tenerife, Spain*
[2]*Universidad Nacional Autónoma de México, Campus Ensenada, Mexico*
[3]*Consejo Superior de Investigaciones Científicas, Spain*

**Abstract.** Using *HST*/ACS broad- and narrow-band (H$\alpha$) images of M51 we have examined the ionizing radiation budget in a set of luminous H II regions in a spiral arm between 5 kpc and 11 kpc from the center. The extinctions within the regions were estimated, and a mean value for the differential extinction coefficient, $R_V$, of 3.3, similar to the Galactic value, was derived. The measured $A_V$ values for the regions was in the region 0.82–3.24, with a mean value of 1.57 magnitudes. The Lyman continuum (Lyc) luminosities of the ionizing clusters were derived from their absolute magnitudes, and thence the corresponding predicted H$\alpha$ luminosities from the H II regions; these exceeded the measured values by a factor of $\approx 7$. Even if this discrepancy is due entirely to Lyc extinction, the range of values for this would be from 0.8 to 3.7, too small to be explained except in a highly inhomogeneous model of gas and dust distribution. Assuming that in fact 50% of the Lyc escapes from a region, the need for a high degree of inhomogeneity is even greater.

## 1. The Method

The broad-band images showed point sources within the H II region boundaries (as defined in the H$\alpha$ image) whose absolute magnitudes showed them to be clusters. Plotting a $B - V$ vs $V - I$ color–color diagram, we estimated the internal visual extinction for a given region. We then inferred the Lyc luminosity of the cluster and derived a theoretical H$\alpha$ luminosity for the surrounding ionized gas. We took an average type for the ionizing stars as O7 V (Vacca 1994) and computed the equivalent O7 V number for each region; thence the total Lyc luminosity $Q_0$ (Vacca, Garmany, & Shull 1996). This can be used to compute the H$\alpha$ luminosity $L(\mathrm{H}\alpha)$ for "Case B" of Osterbrock (1989), assuming a dust-free H II region opaque to Lyc.

We used the $I-$band filter to continuum-subtract the H$\alpha$ image (Knapen et al. 2004) and derived absolute fluxes using the *HST* calibration factor and the distance to M51 of 8.4 Mpc (Feldmeier, Ciardullo, & Jacoby 1997).

The predicted luminosity is then compared with the extinction-corrected measured H$\alpha$ luminosity.

## 2. Results and Conclusions

The first results to note are extinction values: $A_V$ between 0.82 and 3.34 with a median value of 1.58, agreeing well with Calzetti et al. (2005). There is a slight correlation between reddening and galactocentric distance (Fig. 1, *left*) but our distance range is still rather too small.

A second result is the ratio of detected (extinction-corrected) H$\alpha$ to predicted H$\alpha$. This value is, strikingly, only 0.15; i.e., 85% of the predicted H$\alpha$ is not detected (see Fig. 1, *right*). Following Hill et al. (1997), defining the effective Lyc extinction $A_{\rm Lyc,eff} = 2.5\log[L(\mathrm{H}\alpha_{\rm pred})/L(\mathrm{H}\alpha_{\rm obs})]$, and assuming no Lyc photon escape, we find values of $A_{\rm Lyc,eff}$ between 0.8 and 3.7. Comparing directly with $A_V$ we can see that if the gas and dust distribution in H II regions were homogeneous, the two ranges of values are inconsistent (the values for $A_{\rm Lyc,eff}$ should be an order of magnitude greater). Assuming (following Oey & Kennicutt 1997, or Relaño, Peimbert, & Beckman 2002) that some 50% of Lyc photons escape from an H II region, the discrepancy with a homogeneous model is even greater. Only a highly clumpy scenario can account for these results. However, this is in good qualitative agreement with the now classical Osterbrock & Flather (1959) filling factor model for the gas in H II regions.

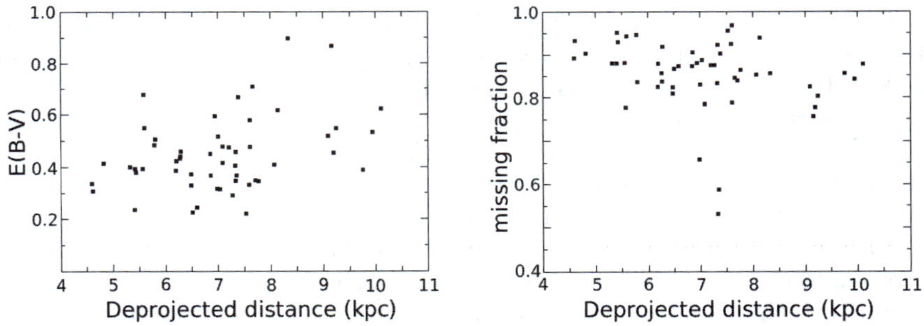


Figure 1. Plot of the reddening (*left*) and the missing fraction (*right*) vs the galactocentric distance.

### References

Calzetti, D., et al. 2005, ApJ, 633, 871
Feldmeier, J. J., Ciardullo, R., & Jacoby, G. H. 1997, ApJ, 479, 231
Hill, J. K., et al. 1997, ApJ, 477, 673
Knapen, J. H., Stedman, S., Bramich, D. M., Folkes, S. L., & Bradley, T. R. 2004, A&A, 426, 1135
Oey, M. S., & Kennicutt Jr., R. C. 1997, MNRAS, 291, 827
Osterbrock, D. 1989, Astrophysics of Gaseous Nebulae and Active Galactic Nuclei (Mill Valley: University Science Books)
Osterbrock, D., & Flather, E. 1959, ApJ, 129, 26
Relaño, M., Peimbert, M., & Beckman, J. 2002, ApJ, 564, 704
Vacca, W. D. 1994, ApJ, 421, 140
Vacca, W. D., Garmany, C. D., & Shull, J. M. 1996, ApJ, 460, 914

*Pathways through an Eclectic Universe*
*ASP Conference Series, Vol. 390, © 2008*
*J. H. Knapen, T. J. Mahoney, and A. Vazdekis, eds.*

# Age, Metallicity and $\alpha$-element Abundances in Stellar Populations

M. Koleva,[1,2] R. Gupta,[3] Ph. Prugniel,[1,4] and H. Singh[3]

[1] *Université de Lyon 1, CRAL, Observatoire de Lyon, CNRS, UMR5574; ENS de Lyon, France*
[2] *Department of Astronomy, University of Sofia, Bulgaria*
[3] *IUCAA, Pune, India*
[4] *GEPI, Observatoire de Paris, France*

**Abstract.** We build model spectra of stellar populations at high resolution and with variable $\alpha$-element enhancements. Analyzing spectra of galactic globular clusters we show that it is possible to derive [Mg/Fe] reliably and efficiently using spectra integrated along the line of sight, thus opening perspectives for investigating the enrichment process in galaxies and star clusters.

## 1. Flux-recalibrated CFLIB Library

The CFLIB stellar library (or IndoUS; Valdes et al. 2004) contains spectra of 1273 stars observed in the wavelength range 3460–9464 Å at a spectral resolution FWHM = 1 Å ($R \approx 5000$). The selection of the stars provides a good coverage of the $T_{\rm eff}$, $\log g$, [Fe/H] space. In the initial release, the spectra were flux-calibrated using the templates of Pickles (1998). This calibration was inaccurate for two main reasons: (1) the metallicity dependency of the spectral energy distribution (SED) was neglected, and (2) it relied on the precision of the atmospheric parameters. To generate high quality stellar population models, we corrected the flux calibration using the MILES library (Sánchez-Blázquez et al. 2006), which has a lower spectral resolution, but has almost the same spectral coverage and an accurate flux calibration. The re-calibrated library will be publicly available in a near future.

## 2. Modeled Spectra of Stellar Populations with $\alpha$-Enhancement

To build population models with parametric $\alpha$-elements we followed the recipe of Prugniel et al. (2007): the empirical library, supposed to have the $\alpha$/Fe pattern of the solar neighborhood, is differentially corrected for the effect of variable abundance using the Coelho et al. (2005) library. Two semi-empirical grids are produced with [Mg/Fe] = 0 and +0.4. Intermediate values are linearly interpolated as a function of the mass ratio of Mg to Fe.

Simple stellar population (SSP) models are built using Pegase.HR (Le Borgne et al. 2004) and the solar-scaled isochrones of the Padova group. The drawback of not using enhanced isochrones (Salasnich et al. 2000) may essentially affect the ages (the enhanced isochrones are bluer).

To determine the age, [Fe/H] and [Mg/Fe] of a population from its spectrum, we use the full spectrum-fitting method discussed in Koleva et al. (2006) implemented as a simple $\chi^2$ minimization where the free physical parameters of the model are those of the population mix, plus the internal kinematics. In addition to the free physical parameters, the model contains a multiplicative polynomial making the process insensitive to the shape of the continuum, just as the classical Lick indices are. This procedure has been found to be reliable and to give a three times better precision than Lick indices due to the optimized usage of the information. We performed Monte Carlo simulations and we have seen that there is no degeneracy between age–[Mg/Fe] or [Fe/H]–[Mg/Fe]. The absence of degeneracy between [Mg/Fe] and age allows us to have some confidence in the retrieved values of [Mg/Fe], since the uncertainties on the evolutionary tracks will mostly bias the age and not the other parameters.

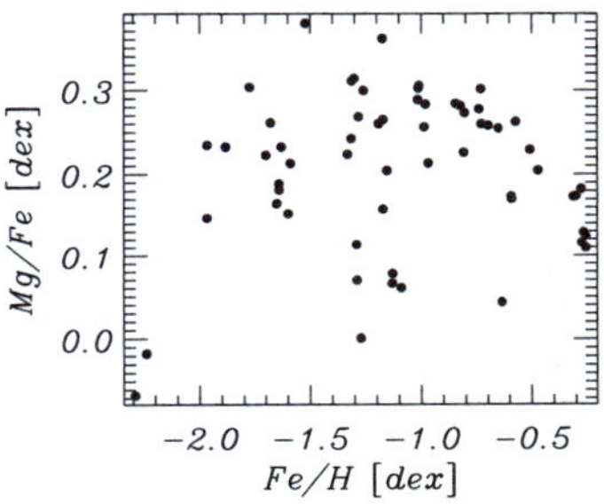


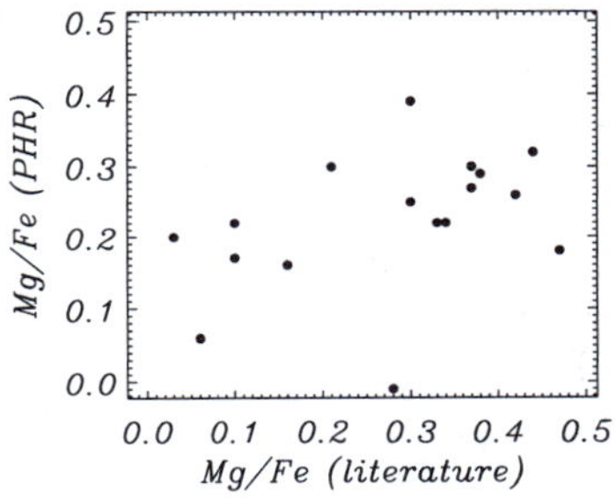


Figure 1. *Left:* Distribution of [Mg/Fe] as a function of [Fe/H] for GGCs. *Right:* Comparison between our values and those in the literature.

## 3. Application to Globular Clusters

We analyzed the spectra of galactic globular clusters (GGCs) from Schiavon et al. (2005). The distribution of [Mg/Fe] is shown in Fig. 1 by a comparison with the compilation of spectroscopic abundances (Pritzl, Venn, & Irwin 2005; Tantalo, Chiosi, & Piovan 2007). There is satisfactory agreement between the enhancement measured from stellar spectroscopy and our measurements from line-of-sight integrated spectra. The measurement of [Mg/Fe] from integrated spectra using full spectrum fitting is reliable and more precise than with the Lick indices.

### References


Coelho, P., Barbuy, B., Meléndez, J., Schiavon, R., & Castilho, B. 2005, A&A, 443, 735

Koleva, M., Bavouzet, N., Chilingarian, I., & Prugniel, Ph. 2006, arXiv:astro-ph/0602362v1

Le Borgne, D., Rocca-Volmerange, B., Prugniel, Ph., Lançon, A., Fioc, M., & Soubiran, C. 2004, A&A, 425, 881

Pickles, A. J. 1998, PASP, 110, 863

Prugniel, Ph., Koleva, M., Ocvirk, P., Le Borgne, D., & Soubiran, C. 2007, in Stellar Populations as Building Blocks of Galaxies, IAUS 241, ed. A. Vazdekis & R. F. Peletier (Cambridge: Cambridge University Press), p.68

Pritzl, B., Venn, K., & Irwin, M. 2005, IAUS, 228, 407

Salasnich, B., Girardi, L., Weiss, A., & Chiosi, C. 2000, A&A, 361, 1023

Sánchez-Blázquez, P., et al. 2006, MNRAS, 371, 703

Schiavon, R., Rose, J., Courteau, S., & MacArthur, L. 2005, ApJS, 160, 163

Tantalo, R., Chiosi, C., & Piovan, L. 2007, A&A, 462, 481

Valdes, F., Gupta, R., Rose, J. A., Singh, H. P., & Bell, D. J. 2004, ApJS, 152, 251

*Pathways through an Eclectic Universe*
*ASP Conference Series, Vol. 390, © 2008*
*J. H. Knapen, T. J. Mahoney, and A. Vazdekis, eds.*

# Ansae in Barred Galaxies

Inma Martinez-Valpuesta,[1,2] Johan H. Knapen,[1] and Ronald J. Buta[3]

[1]*Instituto de Astrofísica de Canarias, E-38200 La Laguna, Spain*
[2]*LAM, Observatoire Astronomique de Marseille Provence, 2 Place Le Verrier, F-13004 Marseille, France*
[3]*Department of Physics and Astronomy, Box 870324, University of Alabama, Tuscaloosa, AL 25487, USA*

**Abstract.** Many barred galaxies show a set of symmetric enhancements at the ends of the stellar bar, called the *ansae*, or "handles", of the bar. In this poster, we make a quantitative analysis of the occurrence of ansae in barred galaxies, making use of *The de Vaucouleurs Atlas of Galaxies* by Buta and collaborators. We find that ~40% of SB0s show ansae in their bars, thus confirming that ansae are common features in barred lenticulars. The ansae frequency decreases dramatically with later types, and hardly any ansae are found in galaxies of type Sb or later.

## 1. Introduction

Bars in disk galaxies are both common and long-lived (Martinez-Valpuesta, Shlosman, & Heller 2006, and references therein). In this poster we focus on a class of morphological feature at the outer region of the bar: the ansae.

Ansae may be described as a pair of density enhancements at the ends of a bar (Fig. 1), sometimes also referred to as "condensations" in older work (e.g., Danby 1965) or symmetric density knots (see Martinez-Valpuesta, Knapen, & Buta 2007). We show clear examples of an ansa bar and a non-ansa bar in Figure 1. More examples of ansae can be found in the literature (e.g., NGC 4262, NGC 2859, and NGC 2950).

We present a statistical analysis of the presence of ansae in barred galaxies based on *The de Vaucouleurs Atlas of Galaxies* (hereafter the Atlas; Buta, Corwin, & Odewahn 2007).

## 2. Observational Data and Statistics

We select all the barred galaxies from the Atlas. We assume that the sample of galaxies collected in the Atlas is random in the sense that the presence or absence of ansae did not influence the selection process of galaxies in the catalog. Based on the detailed description for each galaxy as given in the Atlas, we then separated all galaxies with and without ansae, and collected the statistics as a function of morphological type.

We estimate the uncertainties in all our derived fractions by using Poissonian statistics. Our sample consists of 267 barred galaxies, of which just 36 are of

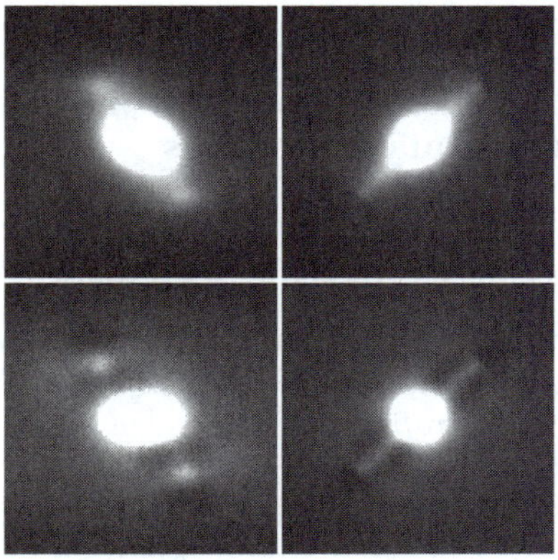

Figure 1. Two strongly-barred early-type galaxies, one having ansae (NGC 2983, *left panels*) and the other without ansae (NGC 4643, *right panels*). The lower panels show the same images after a smooth Ferrers bar model has been subtracted. The bar-subtracted images reveal very clearly that NGC 2983 has density enhancements at the ends of its bar while NGC 4643 shows no trace of such features. (*From Martinez-Valpuesta et al. 2007.*)

ansa type ($14\% \pm 2\%$). There is, however, a very strong dependence of the ansa fraction on morphological type, and 32 out of 36 ansa galaxies occur in the 120 galaxies of type Sab or earlier ($26\% \pm 4\%$). Even within this group of early-type galaxies we see that the highest fraction of ansae occurs in S0 galaxies (14 out of 39, or ($36\% \pm 8\%$). And among the lenticular galaxies, ansae are especially prevalent among those classified as "SB", or strongly barred, with $42\% \pm 10\%$ of the SB0s having ansae. Only five of the 35 Sab galaxies ($14\% \pm 6\%$), three of the 39 Sb, one of the 31 Sbc galaxies, and none of the 77 barred galaxies with later morphological types in the Atlas have ansae.

## 3. Conclusions

We have assessed the frequency of occurrence of ansae and our conclusion, based on a sample of 267 barred galaxies, is that ansae are very much concentrated in early-type galaxies. The highest fraction of ansae is found in strongly barred lenticulars, with $\sim 40\%$ of SB0 galaxies showing ansae. The overall fraction among S0 and Sa galaxies is around a third.

**Acknowledgments.** This work has been partially supported by the Peter Gruber Foundation Fellowship. RB acknowledges the support of NSF grant AST050-7140.

### References

Buta, R., Corwin, H., & Odewahn, S. 2007, The de Vaucouleurs Atlas of Galaxies (Cambridge: Cambridge University Press) (the Atlas)

Danby, J. M. A. 1965, AJ, 70, 501

Martinez-Valpuesta, I., Shlosman, I., & Heller, C. 2006, ApJ, 637, 214

Martinez-Valpuesta, I., Knapen, J. H., & Buta, R. 2007, AJ, 134, 1863

*Pathways through an Eclectic Universe*
*ASP Conference Series, Vol. 390, © 2008*
*J. H. Knapen, T. J. Mahoney, and A. Vazdekis, eds.*

# The Cepheid Variable Population in the Local Group Dwarf Irregular Galaxy Pegasus

I. Meschin,[1] C. Gallart,[1] S. Cassisi,[2] A. Aparicio,[1,3] and A. Rosenberg[1]

[1]*Instituto de Astrofísica de Canarias, E-38200 La Laguna, Spain*
[2]*INAF - Osservatorio Astronomico di Teramo, Teramo, Italy*
[3]*Departamento de Astrofísica, Universidad de La Laguna, Spain*

**Abstract.** We present the result of a search for Cepheids in the Local Group dwarf irregular galaxy Pegasus. We identified 20 Cepheid stars with periods $0^{\mathrm{d}}.6 < P < 15^{\mathrm{d}}.3$, compared them with previously published Cepheids, and carried out an accurate completeness analysis in order to obtain the fraction lost. Using their period–luminosity relation, we obtain a distance for Pegasus of $950 \pm 30$ kpc.

## 1. Method and Analysis

The search for Cepheids was made from a ground-based collection of 79 images (55 in $V$ and 24 in $I$, seeing $< 1''.35$) compiled over several years (1993–2004). Data reduction was performed in the IRAF environment. The photometric reduction of the frames was performed using the DAOPHOT-ALLSTAR-ALLFRAME suite of programs (Stetson 1987, 1994) for PSF-fitting photometry in crowded fields. Due to the inhomogeneity of the images, matching was more difficult than usual.

We created a "master" image by interpolating the individual frames and applying a median filter (using MONTAGE2). Variable star searching was performed in this master image. We made a first selection of variable candidates based in the variability index "*var*" defined as the ratio of the external $\sigma_{\mathrm{ext}}$ to the internal standard error $\sigma_{\mathrm{int}}$ as calculated by DAOMASTER from the individual magnitudes of the 55 $V$ images. We selected as variable candidates the stars having $var > 3$. In all, 319 stars fulfill this criterion.

## 2. Results

We constructed the color–magnitude diagram on the basis of our photometry. Only 204 of our candidates are in the instability strip. The light variations of these candidate pulsating variables were searched for periodicities using a phase dispersion minimization algorithm (Stellingwerf 1978). We visually inspected the light curves that each of the most likely periods would produce for each variable and we found periods for 43 candidates, but only some of them fulfill certain requirements besides being located next to the instability strip: a good period–luminosity relation and a correct location in the galaxy. In this way we discarded 54% of the candidates, and 20 bona fide Cepheids were retained. The $V$-band light curves for some of these stars are displayed in Figure 1.

We carried out an accurate completeness analysis by adding artificial Cepheid stars with the DAOPHOT routine ADDSTAR to the original frames in order to check the fraction of Cepheids lost. Cepheid magnitudes were simulated using the IAC-Star (Aparicio & Gallart 2004) and VARFINDER codes; we also used the prototype $\delta$ Cephei to introduce the pattern of variability. A completeness factor as a function of light curve amplitude was found. Briefly, we are finding only 64% of the Cepheids that exist in Pegasus.

The distance to Pegasus was obtained using our 20 Cepheid stars. We derived $950 \pm 30$ kpc. This measurement agrees with the earlier work by Aparicio (1994), who obtained from the absolute magnitude of the RGB tip (TRGB) $d = 955 \pm 44$ kpc and with the more recent study by McConnachie et al. (2005), $d = 919 \pm 30$ kpc, also obtained from the TRGB.

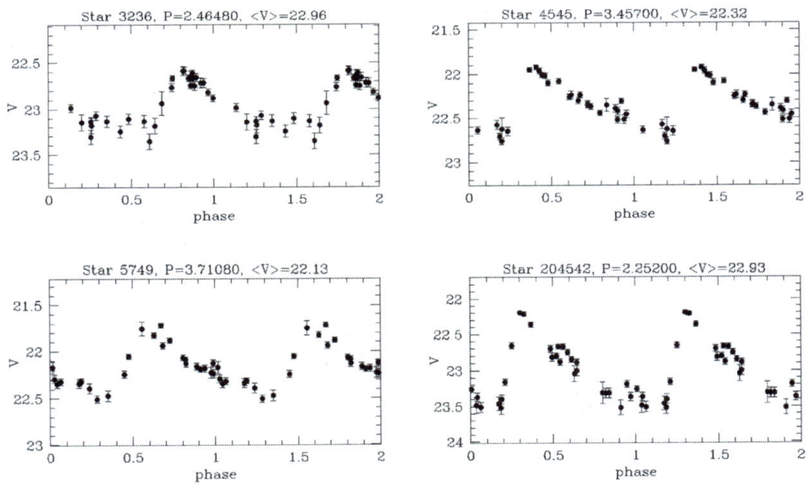


Figure 1. Some of the $V$-band light curves for some of the Pegasus Cepheids. Data are repeated over a second cycle for clarity.

## References

Aparicio, A. 1994, ApJ, 437, L27
Aparicio, A., & Gallart, C. 2004, AJ, 128, 1465
McConnachie, A. W., Irwin, M. J., Ferguson, A. M. N., Ibata, R. A., Lewis, G. F., & Tanvir, N. 2005, MNRAS, 356, 979
Stellingwerf, R. F. 1978, ApJ, 224, 953
Stetson, P. B. 1987, PASP, 99, 191
Stetson, P. B. 1994, AR&AA, 106, 250

*Pathways through an Eclectic Universe*
*ASP Conference Series, Vol. 390, © 2008*
*J. H. Knapen, T. J. Mahoney, and A. Vazdekis, eds.*

# Stellar Populations in Dwarf Elliptical Galaxies

Dolf Michielsen,[1] Mina Koleva,[2,3] Sven De Rijcke,[4] Werner W. Zeilinger,[5] Philippe Prugniel,[2,6] Herwig Dejonghe,[4] Anna Pasquali,[7] Ignacio Ferreras,[8] and Victor P. Debattista[9]

[1]*School of Physics and Astronomy, University of Nottingham, UK*
[2]*Université Lyon 1, Observatoire de Lyon, Av. Ch. André 9, F-69230 St. Genis Laval, France*
[3]*Department of Astronomy, St. Kl. Ohridski University of Sofia, J. Bourchier Blvd. 5, BG-1164 Sofia, Bulgaria*
[4]*Sterrenkundig Observatorium, Ghent University, Krijgslaan 281, S9, B-9000 Gent, Belgium*
[5]*Institut für Astronomie, Universität Wien, Türkenschanzstrasse 17, A-1180 Wien, Austria*
[6]*Observatoire de Paris, GEPI, Av. de l'Observatoire 61, F-75014 Paris, France*
[7]*Max-Planck-Institut für Astronomie, Koenigstuhl 17, D-69117 Heidelberg, Germany*
[8]*Physics Department, King's College London, London, UK*
[9]*Centre for Astrophysics, University of Central Lancashire, Preston, UK*

**Abstract.** Dwarf ellipticals (dEs) are the most numerous galaxy population in nearby dense environments. These low mass systems are very sensitive to both internal and external processes and are therefore ideal to test theories of galaxy formation and evolution. However, their low surface brightness makes detailed observations challenging. Here we present the first results based on FORS/VLT optical spectroscopy of fifteen dEs. We compare age and metallicity estimates from full spectrum fitting with those derived using Lick/IDS indices, and find very good agreement. The dEs have solar [$\alpha$/Fe] ratios and mean ages and metallicities in the range of 1–10 Gyr, and $-0.7 < [Z/\mathrm{H}] < 0$. These values suggest that present-day dEs in clusters formed stars until a recent epoch and were self-enriched by a long star formation history.

To date, stellar population studies of dEs are still scarce. To alleviate this situation, we obtained new optical spectroscopy of fifteen dEs in the Fornax cluster and in Southern groups. This sample is drawn from a VLT Large Programme in which we obtained NIR Ca II spectroscopy for the same dEs (Michielsen et al. 2003; De Rijcke et al. 2005). The present analysis focuses on the central stellar populations (inner 4″) where we have S/N/Å > 75.

To estimate mean, luminosity-weighted, ages and metallicities we carry out both the classical Lick/IDS analysis and a full spectrum fit (see Fig. 1). For the latter technique we perform a $\chi^2$ fit of the full observed spectrum against simple stellar population (SSP) models computed using Pegase-HR (Le Borgne et al. 2004), as discussed in Koleva et al. (2006). Several galaxies in our sample are known to contain emission lines and cannot easily be studied using in-

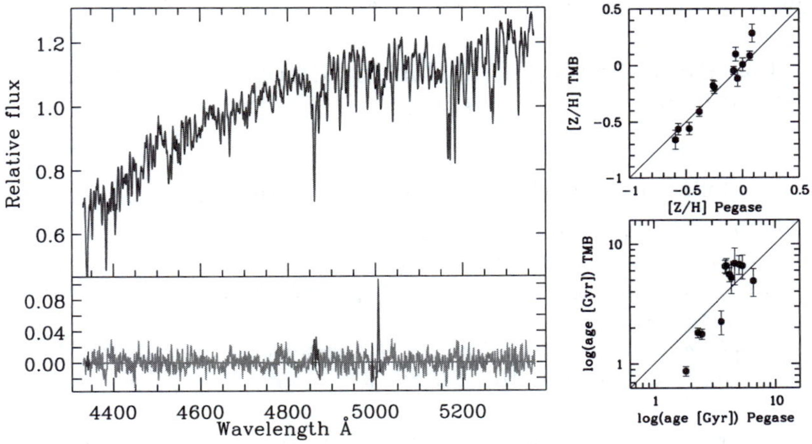


Figure 1. *Left:* An example of the fitting procedure where faint emission is detected in FCC 150. Failing to take this emission into account changes the age estimate from 3.8 to 4.5 Gyr. *Right:* Comparison of ages and metallicities estimated using two different analysis techniques. The dEs span a range of ages, from 1 to 9 Gyr, and metallicities, from −0.6 to +0.1 dex.

dices. Using the fitting technique, we can mask out emission lines from the fit. For emission-free dEs, we compare measured Lick/IDS indices with SSP predictions by Thomas, Maraston, & Bender (2003). The dEs are mostly compatible with solar [$\alpha$/Fe] ratios, justifying the use of the Pegase-HR models in the full-spectrum fitting. The mean, luminosity-weighted, ages and metallicities obtained using the two methods agree extremely well, bearing in mind that (i) the SSP models are different, and (ii) the method to compare the observations with the models is different.

To conclude, the age, metallicity and [$\alpha$/Fe] ratio of these dEs suggest they formed stars until recently and were self-enriched by a long star formation history. This fits in with an infall scenario of small disk or irregular galaxies that are transformed into dEs by the cluster environment.

**Acknowledgments.** MK acknowledges financial support from the LOC. Based on ESO-VLT data (programmes 075.B-0196 and 076.B-0179).

## References

De Rijcke, S., Michielsen, D., Dejonghe, H., Zeilinger, W. W., & Hau, G. K. T. 2005, A&A, 438, 491

Koleva, M., Bavouzet, N., Chilingarian, I., & Prugniel, P. 2006, in ESO Astrophysics Symposia, Science Perspectives for 3D Spectroscopy, ed. M. Kissler-Patig, M. M. Roth and J. R. Walsh

Le Borgne, D., Rocca-Volmerange, B., Prugniel, P., Lançon, A., Fioc, M., & Soubiran, C. 2004, A&A, 425, 881

Michielsen, D., De Rijcke, S., Dejonghe, H., Zeilinger, W. W., & Hau, G. K. T. 2003, ApJ, 597, L21

Thomas, D., Maraston, C., & Bender, R. 2003, MNRAS, 339, 897

*Pathways through an Eclectic Universe*
*ASP Conference Series, Vol. 390, © 2008*
*J. H. Knapen, T. J. Mahoney, and A. Vazdekis, eds.*

# The Role of Star Formation in Nitrogen Abundance Evolution

M. Mollá,[1] J. M. Vílchez,[2] Á. I. Díaz,[3] and M. Gavilán[3]

[1]*CIEMAT, Departamento de Investigación Básica, Avda. Complutense-22, E-28040 Madrid, Spain*
[2]*Instituto de Astrofísica de Andalucía, CSIC, Apdo. 3004, E-18080 Granada, Spain*
[3]*Departamento de Física Teórica, Universidad Autónoma de Madrid, E-28049 Cantoblanco, Madrid, Spain*

**Abstract.** We analyze the evolution of nitrogen resulting from a set of spiral and irregular galaxy models computed for a large number of input mass radial distributions and with various star formation efficiencies. We show that our models produce a nitrogen abundance evolution in good agreement with the observational data. Differences in the star formation histories of the regions and galaxies modeled are essential to reproduce the observational data in the N/O–O/H plane and the corresponding dispersion.

## 1. The Primary Nitrogen

When observations, especially metal-poor galaxy data, are plotted in the plane N/O vs O/H, it is evident that a primary N contribution must exist. Applying the classical Closed Box Model, the points may be limited by three curves, as plotted in Fig. 1a: 1) the one called NS, which shows the evolutionary tracks of N/O when N needs a seed of O to be created: $N/O = p_N(O)/p_O \propto O$; 2) the one called NP that appear when N is created directly from H or He: $N/O = p_N/p_O =$ constant; 3) NS + NP, when both contributions exist, which seems to be the apparent behavior reproduced by the data.

It is, however, necessary to take into account three factors not included in that simple scenario: 1) the mean lifetimes of stars, 2) the star formation efficiency, or the existence of different star formation histories or star formation rates in different galaxies or objects, and 3) the dependence of stellar yields. If the primary nitrogen is ejected by low and intermediate mass (LIM) stars, it will appear in the interstellar medium very abruptly after a time delay, which means when O/H has already reached a certain value. Obviously, this level will be higher or lower depending on the mean lifetime of these stars. But it also depends on the efficiency in forming stars: for a high efficiency the NP will appear later, or at higher O/H, than for a low one. Both facts produce a certain dispersion in the resulting abundances. Besides that, the metallicity-dependent stellar yields produce tracks that are elongated in comparison with that produced when only one yield value is used (for details see Mollá et al. 2006).

## 2. The Multiphase Chemical Evolution Model Grid

We have computed a grid of models (Mollá & Díaz 2005) with 44 theoretical galaxies of different total masses and ten possible efficiencies to form stars in each one. We have used the yields from Woosley & Weaver (1995) for massive stars and those from Gavilán, Buell & Mollá (2005) for LIM stars, which give results for our Galaxy (Gavilán, Mollá & Buell 2006) in excellent agreement with halo stars data (Israelian et al. 2004; Spite et al. 2005).

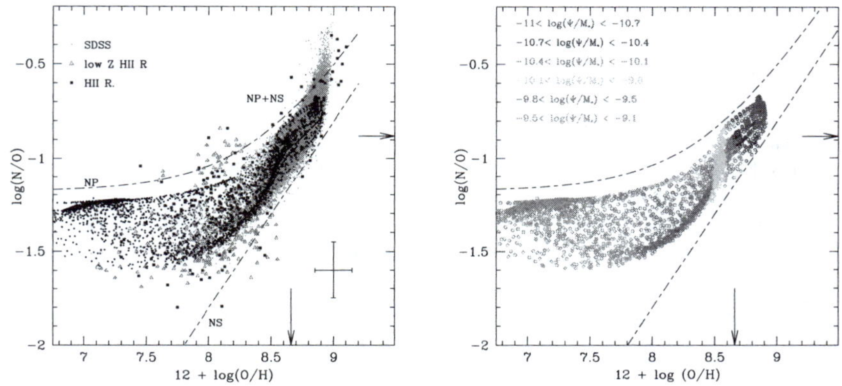


Figure 1. The relative abundance log(N/O) vs the oxygen abundance 12 + log(O/H). *Left:* Our model results for the present time as small black dots compared with data (colors as labeled) from authors given in Table 1 of Mollá et al. (2006). *Right:* The same results represented according to their present specific star formation rate, as labeled.

The results obtained (Mollá et al. 2006) for the present time are compared with observations in Figure 1 (*left*). The position of a galaxy in the N/O–O/H plane depends on the star formation histories and/or on the actual star formation rate as we may see in Fig. 1 (*right*): if the star formation occurred as a strong and early burst, with a low rate at the present time compared with the past maximum (red dots), the evolutionary track in that plane appears as very secondary and O/H and N/O are high. When the star formation occurs quietly with a rate higher now than in the past (cyan dots), the track is almost flat with low O/H and N/O abundances. This way we conclude that our grid of models reproduces and explains very well most of observational data in the N/O–O/H plane.

### References

Gavilán, M., Buell, J. F., & Mollá, M. 2005, A&A, 432, 861
Gavilán, M., Mollá, M., & Buell, J. F. 2006, A&A, 450, 509
Israelian, G., et al. 2004, A&A, 421, 649
Mollá, M., & Díaz, A. I. 2005, MNRAS, 358, 521
Mollá, M., Vílchez, J. M., Gavilán, M., & Díaz, A. I. 2006, MNRAS, 372, 1069
Spite, M., et al. 2005, A&A, 430, 655
Woosley, S. E., & Weaver, T. A. 1995, ApJS, 101, 181

*Pathways through an Eclectic Universe*
*ASP Conference Series, Vol. 390, © 2008*
*J. H. Knapen, T. J. Mahoney, and A. Vazdekis, eds.*

# A New Grid of Evolutionary Synthesis Models

M. Mollá,[1] M. L. García-Vargas,[2] A. Bressan,[3] and P. Gómez-Álvarez[4]

[1]*CIEMAT, Avda. Complutense 22, E-28040 Madrid, Spain*
[2]*FRACTAL, Castillo de Belmonte 1, E-28232 Las Rozas de Madrid, Spain*
[3]*INAF Osservatorio Astronomico di Padova, Viccolo dell' Osservatorio 5, I-35122 Padova, Italy*
[4]*ESAC, Villafranca del Castillo, PO Box 50727, E-28080 Madrid, Spain*

**Abstract.** We present new evolutionary synthesis models for simple stellar populations covering a wide range in age and metallicity. The most important difference with existing models is the use of NLTE atmosphere models for the hot stars (O, B, WR, post-AGB stars, and planetary nebulae) that have an important impact on the stellar cluster's ionizing spectra.

## 1. Synthesis Code Description

We have used the synthesis code by García-Vargas, Mollá, & Bressan (1998), updated by Mollá & García-Vargas (2000) and now newly revised. The basic grid is composed of simple stellar populations (SSPs) for five different initial mass functions (IMFs). The first one is a Salpeter (1955) power law with masses between 0.85 and 120 $M_{\odot}$. The other IMFs are Salpeter (1955), Ferrini, Penco, & Palla (1990), Kroupa (2002) and Chabrier (2003) functions, with masses between 0.15 and 100 $M_{\odot}$.

The isochrones are those from Bressan, Granato, & Silva (1998) for six different metallicities: $Z = 0.0004$, 0.001, 0.004, 0.008, 0.02, and 0.05. The age coverage is from $\log t = 5.00$ to 10.30 with a variable time resolution $\Delta(\log t) = 0.01$ in the youngest stellar ages. The WC and WN stars are identified in the isochrones according to their surface abundances.

The atmosphere models are from Lejeune, Cuisinier, & Buser (1997) with an excellent coverage in effective temperature, gravity and metallicity, for stars with $T_{\rm eff} \leq 25\,000$ K. For O, B, and WR stars we have taken the NLTE blanketed models by Smith, Norris, & Crowther (2002) at $Z = 0.001$, 0.004, 0.008, 0.02, and 0.04. There are 110 for O–B stars, with $25\,000\,{\rm K} < T_{\rm eff} \leq 51\,500\,{\rm K}$ and $2.95 \leq \log g \leq 4.00$, and 120 for WR stars (60 WN + 60 WC), with $30\,000\,{\rm K} \leq T^{*} \leq 120\,000\,{\rm K}$ and $1.3\,R_{\odot} \leq R^{*} \leq 20.3\,R_{\odot}$ for WN, and with $40\,000\,{\rm K} \leq T^{*} \leq 140\,000\,{\rm K}$ and $0.8\,R_{\odot} \leq R^{*} \leq 9.3\,R_{\odot}$ for WC. $T^{*}$ and $R^{*}$ are the temperature and the radius at a Roseland optical depth of 10.

To assign a model to each WR star, we use the relationships among opacity, mass loss, and wind velocity: $d\tau = -\kappa(r)\rho(r)dr$, where $\kappa(r) = -0.2(1 + X_{\rm S})$ and $X_{\rm S}$, the H surface abundance, is taken as 0.2 for WN and 0 for WC. The mass loss is: $dM/dt = 4\pi r^2 \rho(r) v(r) dr$ with $v(r) = v_{\infty}(1 - R_{\rm S}/r)^{\beta}$, taking $\beta = 1$. Integrating these equations, we find $R^{*}$ and then we select the closer atmosphere model.

For post-AGB stars and planetary nebulae with $T_{\rm eff}$ from 50 000 to 220 000 K we take the NLTE models by Rauch (2003). For higher temperatures we use black bodies.

## 2. Results

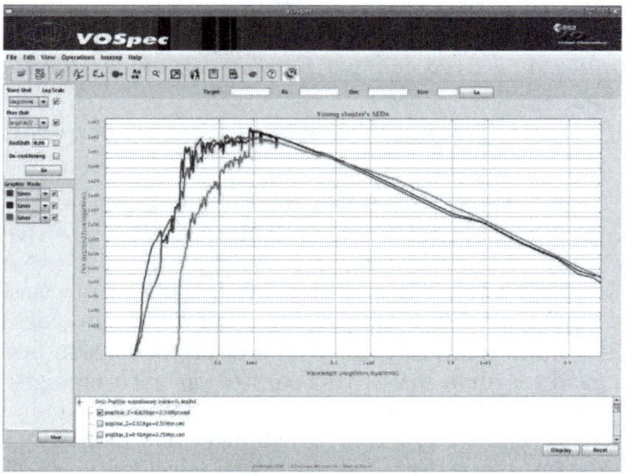


Figure 1. SEDs for some young clusters as they are seen in VOspec.

The resulting SEDs and HR diagrams[1] are available in the Virtual Observatory. The use of NLTE blanketed models produces fewer hard ionizing photons than old models (e.g., García-Vargas, Bressan, & Díaz 1995) which explains in a natural way the emission line ratios in low excitation high metallicity H II regions. Previous work needed a steeper IMF or mass segregation in small clusters, in disagreement with evidence from *HST* for the existence of very massive stars even in small clusters.

### References

Bressan, A., Granato, G. L., & Silva, L. 1998, A&A, 332, 135
Chabrier, G. 2003, ApJ, 586, L133
Ferrini, F., Penco, U., & Palla, F. 1990, A&A, 231, 391
García-Vargas, M. L., Bressan, A., & Díaz, A. I. 1995, A&AS, 112, 13
García-Vargas, M. L., Mollá, M., & Bressan, A. 1998, A&AS, 130, 513
Kroupa, P. 2002, Science, 295, 82
Lejeune, Th., Cuisinier, F., & Buser, R. 1997, A&AS, 125, 229
Mollá, M., & García-Vargas, M. L. 2000, A&A, 359, 18
Rauch, T. 2003, A&A 403, 709
Salpeter, E. E. 1955, ApJ, 121, 161
Smith, L., Norris, R., & Crowther, P. 2002, MNRAS, 337, 1309

[1]Models in http://esavo.esa.int/vospec/ and http://www.fractal-es.com/SEDmod.htm.

*Pathways through an Eclectic Universe*
*ASP Conference Series, Vol. 390, © 2008*
*J. H. Knapen, T. J. Mahoney, and A. Vazdekis, eds.*

# Stellar Line-strength Index Gradients along Bars

Isabel Pérez,[1,2] Patricia Sánchez-Blázquez,[3] and Almudena Zurita[2]

[1]*Kapteyn Astronomical Institute, University of Groningen, the Netherlands*
[2]*Departamento de Física teórica y del Cosmos, Universidad de Granada, Spain*
[3]*Centre for Astrophysics, University of Central Lancashire, UK*

**Abstract.** We present a detailed study of stellar line indices along the bar region for a sample of six early-type galaxies. We find positive gradients within the bar region in the metal indices in four of the six galaxies, and opposite trends in the other two. These latter two galaxies are classified as SAB and they present exponential bar light profiles. For all the galaxies we find a positive gradient in the Balmer indices. There is a clear correlation between the position of morphological features inside the bar region with changes in the slope and value of the indices that indicate, using stellar population analysis, changes in the stellar population, changing radially to younger and more metal rich populations for all the galaxies except for those two with exponential profiles.

## 1. Methodology

We have obtained high S/N long-slit spectra for a sample of six barred galaxies with the Double Beam Spectrograph (DBS) on the 2.3 m telescope at Siding Spring Observatory. The slit was placed along the bar major axis. We have derived radial profiles of the line-strength indices for the sample galaxies by measuring spectral indices of the Lick/IDS system Faber et al. (1985). All the details about the observations, data reduction, and analysis are presented in Pérez, Sánchez-Blázquez, & Zurita (2007). The results are presented in Fig. 1, which clearly shows the different trends of the radial distribution of the indices.

**Acknowledgments.** We thank John Beckman for sharing with all of us his eclectic 40 years of active research and for inspiring the celebration of this great conference. IP acknowledges financial support from the Netherlands Organisation for Scientific Research (NWO) and the Plan Nacional del Espacio of the Spanish Ministerio de Educación y Ciencia. AZ acknowledges support from the Consejería de Educación y Ciencia of the Junta de Andalucía, Spain.

### References

Faber, S. M., Friel, E. D., Burstein, D., & Gaskell, C. M. 1985, ApJ, 57, 711
Perez, I., Sánchez-Blázquez, P., & Zurita, A. 2007, A&A, 465, 9
Sarzi, M., et al. 2006, MNRAS, 366, 1151

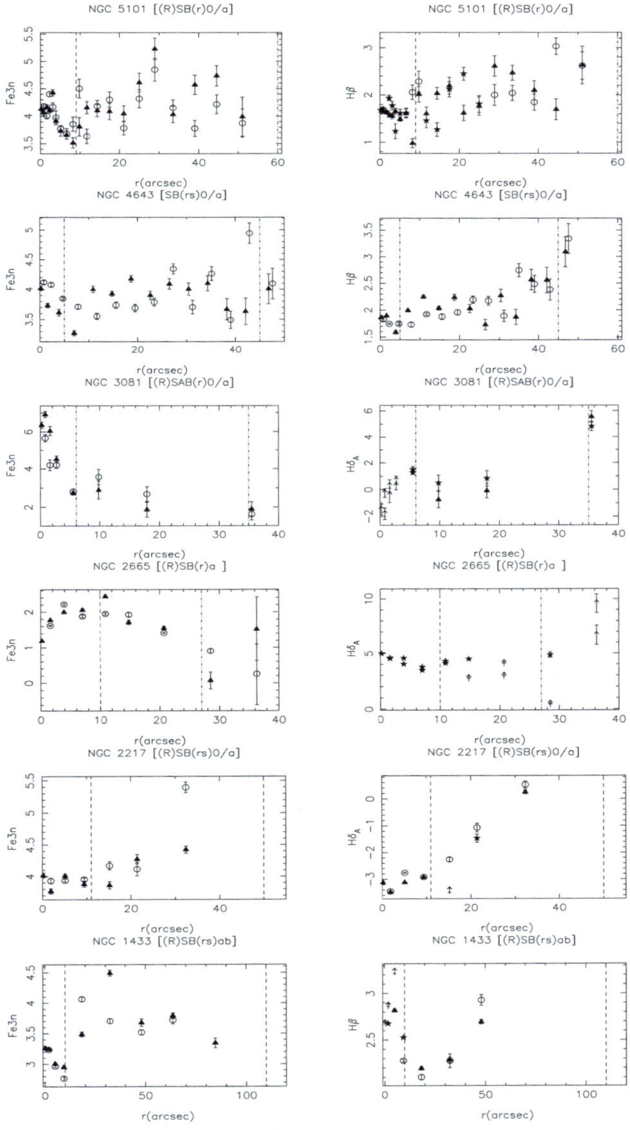


Figure 1. Radial distribution of two indices for each of the galaxies (a combined Fe index and the H$\beta$ index or the H$\delta$ image depending on the emission in the bin, Pérez et al. 2007). The open circles and the filled triangles represent both sides of the bar. We have marked all the points with detected emission with a star, where the detection threshold has been calculated with the prescriptions given in Sarzi et al. (2006). When the emission completely fills all the absorption Balmer lines, the correction is more uncertain. For those cases, we have marked the points with an arrow. The vertical dashed lines represent the end-point of morphological features, meaning the inner bar or nuclear disk and the primary bar. Notice the general change in the indices for the two regions and the gradients in both.

*Pathways through an Eclectic Universe*
*ASP Conference Series, Vol. 390, © 2008*
*J. H. Knapen, T. J. Mahoney, and A. Vazdekis, eds.*

# What Hydrodynamical Simulations Tell Us about the Radial Properties of the Stellar Populations in Ellipticals

Antonio Pipino,[1,2] Annibale D'Ercole,[3] and Francesca Matteucci[2]

[1]*Astrophysics, University of Oxford, Denys Wilkinson Building, Keble Road, Oxford OX1 3RH, UK*
[2]*Dipartimento di Astronomia, Universita di Trieste, Via G.B. Tiepolo 11, I-34100 Trieste, Italy*
[3]*INAF-Osservatorio Astronomico di Bologna, Via Ranzani 1, I-40127 Bologna, Italy*

**Abstract.** Elliptical galaxies probably host the most metal rich stellar populations in the Universe. The processes leading to both the formation and the evolution of such stars are discussed by means of a new gas dynamical model which implements detailed chemical evolution prescriptions. Moreover, the radial variations in the metallicity distribution of these stars are investigated by means of G-dwarf-like diagrams.

By comparing model predictions with observations, we derive a picture of galaxy formation in which the higher the mass of the galaxy is, the shorter are the infall and the star formation timescales. The galaxies seem to have formed from the outside in, namely the most external regions accrete gas, form stars and develop a galactic wind very quickly (a few Myr) compared to the central core, where the star formation can last up to 1 Gyr.

We show for the first time a model able to reproduce the mass–metallicity and the color–magnitude relations, as well as the radial metallicity gradient, and, at the same time, the observed either positive or negative slopes in the [α/Fe] abundance ratio gradient in stars.

## 1. The Model

We present preliminary results from a new class of hydrodynamical models for the formation of single elliptical galaxies (A. Pipino, A. D'Ercole, & F. Matteucci, in preparation) in which we implement detailed prescriptions for the chemical evolution of H, He, O, and Fe. Our aims are: i) to test and improve our previous predictions of an outside-in formation for the majority of ellipticals in the context of the supernovae-driven wind scenario, by means of a careful study of gas inflows/outflows; ii) to explain the observed slopes, either positive or negative, in the radial gradient of the mean stellar [α/Fe] and their apparent lack of any correlation with all the other observables (Melhert et al. 2003; Annibali et al. 2006; Sánchez-Blázquez et al. 2007).

We adopted a one-dimensional hydrodynamical model that follows the time evolution of the density of mass, momentum, and internal energy of a galaxy, under the assumption of spherical symmetry. In order to solve the equation of hydrodynamics with source term we made use of an improved version of the Bedogni & D'Ercole (1986) Eulerian, second-order, upwind integration scheme.

Our main novelty, in fact, is that we follow the chemical evolution of several elements, namely H, He, O, and Fe. This set of elements is good enough to characterize our simulated elliptical galaxy from the chemical evolution point of view. In fact, as shown by the time-delay model (Matteucci & Greggio 1986), the [$\alpha$/Fe] ratio is a powerful estimator of the duration of the star formation.

## 2. Results

All the models run undergo an outside-in formation as suggested by Pipino, Matteucci, & Chiappini (2006), in the sense that star formation stops earlier in the outermost than in the innermost regions, owing to the onset of a galactic wind. We find that the predicted variety of the gradients in the [$\alpha$/Fe] ratio can be explained by physical processes, generally not taken into account in simple chemical evolution models, such as metal-enhanced radial flows coupled with different initial conditions.

We find [Fe/H] gradient slopes in the range $-0.5$ to $-0.2$ dex per decade in radius and $-0.3$ dex per decade in radius for [$Z$/H], in agreement with the observations (e.g., Annibali et al. 2007).

Once transformed into predictions on the line-strength indices, these gradients in the abundances lead to $d\mathrm{Mg}_2/\log(R_{\mathrm{eff},*}/R_{\mathrm{core},*}) \sim -0.06$ mag per decade in radius, again in agreement with the typical mean values measured for ellipticals by several authors and confirming the Pipino et al. (2006) best-model predictions. The remarkable exception of some model with a steeper gradient seems to go in the direction of a few massive objects in the Ogando et al. (2005) sample.

By analyzing typical massive ellipticals, we find that all the models that show values for their chemical properties, including the [$\langle \mathrm{Fe/H} \rangle_V$] and the total metallicity gradients, within the observed ranges, show a variety of gradients in the [$\alpha$/Fe] ratio, either positive or negative, and one has no gradient at all.

The build-up of the gradients is very fast and we predict non-significant evolution after the first 0.5–1 Gyr.

### References

Annibali, F., Bressan, A., Rampazzo, R., Zeilinger, W. W., & Danese, L. 2007, A&A, 463, 455

Bedogni, R., & D'Ercole, A. 1986, A&A, 157, 101

Matteucci, F. & Greggio, L. 1986, A&A, 154, 279

Mehlert, D., Thomas, D., Saglia, R. P., Bender, R., & Wegner, G. 2003, A&A, 407, 423

Ogando, R. L. C., Maia, M. A. G., Chiappini, C., Pellegrini, P. S., Schiavon, R. P., & da Costa, L. N. 2005, ApJ, 632, 61

Pipino, A., Matteucci, F., & Chiappini, C. 2006, ApJ, 638, 739

Sánchez-Blázquez, P., Forbes, D. A., Strader, J., Brodie, J., & Proctor, R. 2007, MNRAS, 377, 759

*Pathways through an Eclectic Universe*
*ASP Conference Series, Vol. 390, © 2008*
*J. H. Knapen, T. J. Mahoney, and A. Vazdekis, eds.*

# Evolution of the Mass–Metallicity Relation from IMAGES

M. Rodrigues, H. Flores, and F. Hammer

*GEPI, Observatoire de Paris, CNRS, Université Paris Diderot, 5 Place Jules Janssen, F-92190 Meudon, France*

Y. C. Liang

*National Astronomical Observatories, Chinese Academy of Sciences, 20A Datun Road, Chaoyang District, Beijing 100012, China*

**Abstract.** The evolution of the stellar mass–metallicity relation is still a matter of intense debate ( Savaglio et al. 2005; Liang, Hammer, & Flores 2006). We present here preliminary results using a sample of intermediate mass galaxies with redshift between 0.4 and 0.8. We take special care of the methodology used to measure metallicity in order to avoid any systematic errors coming from bad flux measurement or unreliable extinction determination. We confirm the results of Liang et al. (2006) that at $z \sim 0.6$, galaxies have gas phases a half less enriched than in the local Universe.

## 1. Introduction

The IMAGES survey proposes to relate stellar masses to dynamical masses for a complete and representative sample of $z \leq 1$ galaxies in the *Chandra* Deep Field South, i.e., a complete study of galaxy kinematics and velocity fields, gas chemistry, and stellar populations at $z \leq 1$. We are currently working on the FLAMES/GIRAFFE data (F. Hammer et al., and Y. Yang et al., in preparation) and the FORS2 (600RI+600z) data, complemented with *Chandra*, *HST*/ACS, and *Spitzer* observations.

Integrated optical spectra will be used to understand how the metal content in galaxies evolves, which is one of the outstanding problems of modern astrophysics. We are currently estimating the gas abundances using extinction-corrected line ratios as provided by FORS2/MXU from a complete sample of distant galaxies. From these estimates, we will be able to constrain the evolution of the mass–metallicity relation with the redshift.

## 2. Observations and Data Reduction

We use the observations from the IMAGES survey with two different FORS2 VLT setups (the RI600 and z600 grisms, 3 hours each). We represent here a subsample of nine high signal-to-noise galaxies and $R = 1200$, with [O II]$\lambda$3727,[O III] $\lambda\lambda$4959,5007, and Balmer lines; we cover a wide range of stellar masses, selected from the 70 emission galaxies of the final sample.

For each galaxy, the spectra were flux-calibrated using standard stars. Broadband optical images were used to compute the spectroscopic aperture corrections and check the slope of each spectrum. The underlying Balmer absorption is another source of systematic error in flux measurement. We fit the stellar continuum with a linear combination of five star types. For a subsample of galaxies detected in the mid-IR (Le Floc'h et al. 2005), the extinction was measured by two independent methods: the Balmer decrement and the IR flux (Liang et al. 2006). Finally, [O/H + 12] was calculated from the $R23$ calibration of Tremonti et al. (2004).

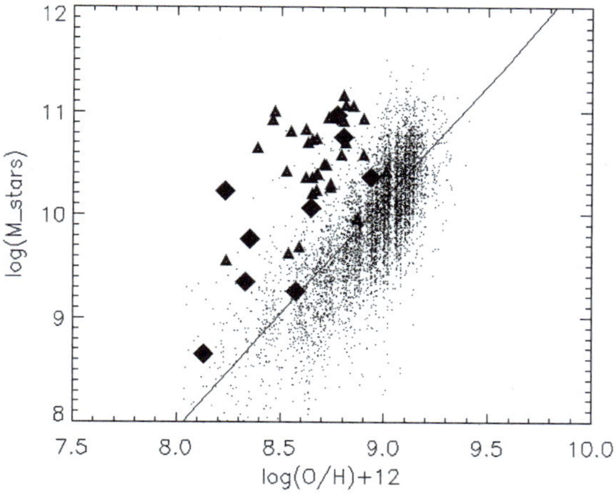


Figure 1. Mass–metallicity relation with the SDSS galaxies in the local sample (black dots and black line for fit), our sample (big symbols) and the sample $z \sim 0.65$ of Liang et al. (2006) (small symbols).

## 3. Results

We compare this preliminary sample to local galaxies from the Nearby Field Galaxy Survey, the KPNO International Spectroscopy Survey and the Sloan Digital Sky Survey. Metallicity and stellar masses have been determined by the same method in our sample and in the local one.

We confirm the previous results of Liang et al. (2006): distant galaxies present a metal content less than half of that of the local counterparts.

## References

Liang, Y. C., Hammer, F., & Flores, H. 2006, A&A, 447, 113
Savaglio, S., et al. 2005, ApJ, 635, 260
Tremonti, C., et al. 2004, ApJ, 613, 898
Le Floc'h, E., et al. 2005, ApJ, 632, 169

*Pathways through an Eclectic Universe*
*ASP Conference Series, Vol. 390, © 2008*
*J. H. Knapen, T. J. Mahoney, and A. Vazdekis, eds.*

# Comparison of Bar Strengths in the Optical and Near-infrared

Tom Speltincx, Eija Laurikainen, and Heikki Salo

*Univ. Oulu, Dept. Phys. Science, Div. Astronomy, Finland*

Ronald J. Buta

*Department of Physics and Astronomy, University of Alabama, USA*

Johan H. Knapen

*Instituto de Astrofísica de Canarias, E-38200 La Laguna, Spain*

**Abstract.** Bars are expected to play an important role in galactic dynamics. Since the rest-frame wavelength of near-infrared images of high redshift galaxies lies in the optical, a reliable comparison of low and high redshift bars requires an understanding of how the properties of a bar depend on the observing wavelength. For this purpose, we compare bar strengths for galaxies from the OSUBSGS sample in the $B$- and $H$-bands, using the gravitational bar torque method. We find that the ratio of bar torques in the $B$- and $H$-bands is fairly close to unity ($Q_{B/H} \sim 1.25$) and independent of the morphological type. Hence, we can make a good estimate of the "real" bar strength for galaxies at high redshift from the observed one, even if the morphological classification were uncertain. Also, we show that the gravitational bar torque method using a polar grid gives satisfactory results when applied to degraded images that mimic high redshift images. Typical errors in bar strength remain below 10% as long as the total bar length is at least ~10 pixels.

## 1. Comparing Bar Strengths

Bars are important drivers of galaxy evolution, which highlights the importance of studying the properties of bars at a large range of redshifts. Near-infrared (NIR) images of galaxies at $z \sim 1$ have their rest-frame wavelength in the optical. Therefore, in this study we use a subsample (Buta et al. 2004) of the Ohio State University Bright Spiral Galaxy Survey (OSUBSGS; Eskridge et al. 2002) to compare bar strengths in the optical ($B$-band) and NIR ($H$-band).

We use a gravitational bar torque method (Laurikainen & Salo 2002) in which we define the maximum of $Q_{\rm T}(r) = F_{\rm T}^{\rm max}(r)/\langle F_{\rm R}(r)\rangle$ as a single measure of bar strength, with $F_{\rm T}^{\rm max}(r)$ the maximum tangential force and $\langle F_{\rm R}(r)\rangle$ the azimuthally averaged radial force at radius $r$. We denote this maximum by $Q_{\rm g}$. Further, the method uses an exponential model for the vertical density profile and an empirical relation between the morphological type and the radial to vertical scale parameter ratio, $z_0/h_{\rm R}$ (de Grijs 1998).

In order to compare bar strengths in the $B$- and $H$-bands, we use four approaches, all based on the $Q_{\rm g}$ method. Independent of the used approach, we find that through morphological types $T = 0$–$6$ the bar strength in the $B$-band is about 1.25 times higher than that in the $H$-band (see Fig. 1). Hence, this number can be used to convert between bar strengths at high redshift and local ones. The fact that the bar strengths are higher in the optical can be attributed to bulge dilution. The bulge appears less prominent in the optical, causing the radial forces to be smaller and hence the bar torques to be larger.

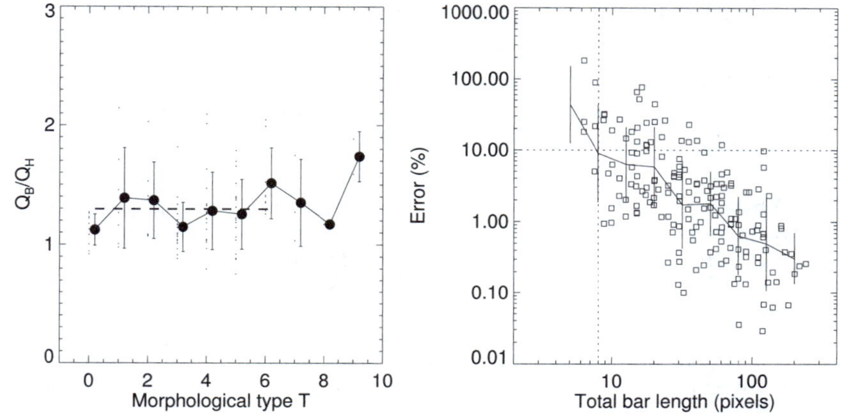


Figure 1. *Left panel:* Bar strength ratios were calculated by taking the average $Q_{\mathrm{T}}$ value in the bar region. Through types $T$ = 0–6 the $Q_B/Q_H$ ratio is nearly constant around 1.25. *Right panel:* Error in $Q_{\mathrm{g}}$ versus the total bar length in pixels. Bar strengths are within 10% of their original value as long as the bar, after degradation, stretches over at least 10 pixels.

The $Q_{\mathrm{g}}$ method makes explicit use of $h_{\mathrm{R}}$ to obtain $z_0$. At high redshift, $h_{\mathrm{R}}$ may be very difficult to estimate, and we therefore estimate $z_0$ through $z_0 \approx \langle z_0/r_{K_{20}} \rangle \, r_{K_{20}}$. Calculating the bar strengths again with this approximation, we find that there is no more than a 10–15% error, with an overestimate of the strength for early types and an underestimate for late types.

Finally, to test whether the $Q_{\mathrm{g}}$ method can provide good results when applied to low resolution images, we degrade our $B$-band images using the nearest-neighbor algorithm. As long as the total bar length is at least ten pixels, the bar torque estimates remain within 10% of the original value (see Fig. 1). As a comparison, we note that Jogee et al. (2004) find bars (at $z \leq 1$) with lengths of 6–88 pixels with the Advanced Camera for Surveys on the *Hubble Space Telescope*. We can therefore say that the $Q_{\mathrm{g}}$ method would reliably estimate bar torques also at high redshifts.

The authors are thankful for the financial support from the Academy of Finland, CIMO, and Emil Aaltonen Foundation.

## References

Buta, R., Laurikainen, E., & Salo, H. 2004, AJ, 127, 279
de Grijs, R. 1998, MNRAS, 299, 255
Eskridge, P., et al. 2002, ApJS, 143, 73
Jogee, S., et al. 2004, ApJ, 615, L105
Laurikainen, E., & Salo, H. 2002, MNRAS, 337, 1118

*Pathways through an Eclectic Universe*
*ASP Conference Series, Vol. 390, © 2008*
*J. H. Knapen, T. J. Mahoney, and A. Vazdekis, eds.*

# The Pattern Speeds of NGC 6946

Silvia Toonen,[1] Kambiz Fathi,[2] Jesús Falcón-Barroso,[1,3] John E. Beckman,[2] and P. Tim de Zeeuw[1]

[1]*Leiden Observatory, Leiden University, The Netherlands*
[2]*Instituto de Astrofísica de Canarias, E-38200 La Laguna, Spain*
[3]*ESTEC, Noordwijk, The Netherlands*

**Abstract.** We study the kinematics of the barred spiral galaxy NGC 6946 by investigating the velocity field from H$\alpha$ Fabry–Perot observations, determine the pattern speed of the bar by using the Tremaine–Weinberg method, and find a main pattern speed of $21.7^{+4.0}_{-0.8}$ km$^{-1}$ s$^{-1}$ kpc$^{-1}$. Our data clearly suggest the presence of an additional pattern with a pattern speed more than twice that of the large pattern in this galaxy. We use the epicycle approximation to deduce the location of the resonance radii and subsequently determine the pattern speed between the radii, and find that inside the inner inner Lindblad resonance radius a bar-like system has evolved.

## 1. Introduction

The study of the ISM and its kinematics enables us to address one of the more interesting questions in galactic evolution: by what mechanism or mechanisms is the optical appearance of a galaxy related to the evolution of its spiral or barred structure in response to the underlying dynamics of its stellar and gaseous components, and what is the role of such evolution on star formation in the host galaxy? At a distance of 5.5 Mpc, NGC 6946 provides us with a good opportunity to study and characterize the properties of interstellar gas and dust in detail. The observations were taken with the Fabry–Perot interferometer FaNTOmM (Hernandez et al. 2003) at the 1.6 m telescope at the Observatoire du Mont Mégantic, Québec, Canada. The resulting data cube of 40 planes separated by 0.21 Å was used to derive flux, line-of-sight velocity and velocity dispersion for each position. The radial velocity map shows the presence of a small, asymmetrically extended, inner bar-like system embedded in a large scale rotating structure.

## 2. Two Pattern Speeds

The Tremaine & Weinberg (1984) method makes it possible to measure the pattern speed of a barred disk galaxy without adopting a particular dynamical model. It is based on the ratio of measurements of the distribution of intensity-weighted velocity and position of a component that reacts to the density wave. We use this method with the refinement of Merrifield & Kuijken (1995). The distribution of the measurements along two lines in Fig. 1a suggests not one, but

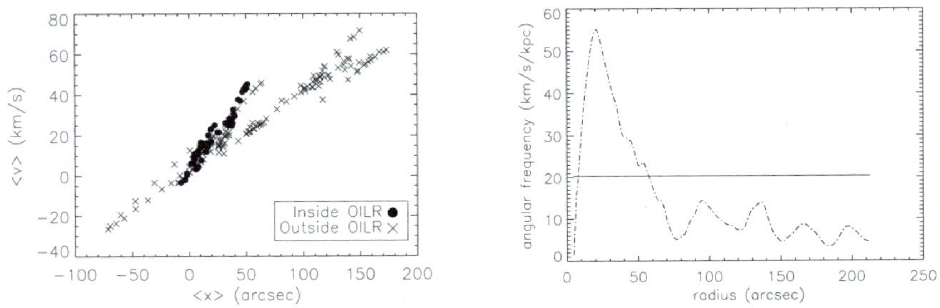


*(a)* *(b)*

Figure 1. *(a)* Pattern speed determination with the Tremaine & Weinberg method. The pattern speed is given by the ratio of $\langle v \rangle$ and $\langle x \rangle$. Measurements through the center are distributed along the line of higher slope, where the measurements solely through the exterior of the NGC 6946 are accumulated along the line of lower slope, indicating two distinct pattern speeds. *(b)* The $(\Omega - \kappa/2)$ curve, with $\Omega$ the angular frequency and the $\kappa$ the epicyclic frequency, crosses the horizontal line representing the pattern speed of 21.7 $\mathrm{km^{-1}\,s^{-1}\,kpc^{-1}}$, indicating two ILRs.

*two* distinct pattern speeds for measurements including and excluding the central regions. This observational result suggests the presence of an additional, inner, pattern that rotates at a different pattern speed roughly inside the outer inner Lindblad resonance (ILR). The outer pattern rotates at $21.7 \pm 0.5\,\mathrm{km^{-1}\,s^{-1}\,kpc^{-1}}$ and the inner pattern rotates at $47.0 \pm 1.4\,\mathrm{km^{-1}\,s^{-1}\,kpc^{-1}}$.

## 3. Resonance Radii

We use the epicycle approximation to deduce the location of the ILR radii and the corotation radius. Figure 1b shows the $(\Omega - \kappa/2)$ curve. This indicates an inner and an outer ILR. As described by Maciejewski (2004), the curve can be interpreted as representing a galaxy without a central mass concentration or density cusp. Figure 1b also indicates that a bar-like system with solid body rotation has evolved inside the inner ILR, as confirmed by the steeply rising rotation curve in this galaxy.

In a forthcoming paper, we will present an extended analysis of the pattern speeds and a diagnosis of the resonance radii, within the framework of epicycle approximation, in NGC 6946.

## References


Hernandez, O., Gach, J., Carignan, C., & Boulesteix, J. 2003, SPIE, 4841, 1472
Maciejewski, W. 2004, MNRAS, 354, 892
Merrifield, M. R., & Kuijken, K. 1995, MNRAS, 274, 933
Tremaine, S., & Weinberg, M.D. 1984, ApJ, 282, L5

*Pathways through an Eclectic Universe*
*ASP Conference Series, Vol. 390, © 2008*
*J. H. Knapen, T. J. Mahoney, and A. Vazdekis, eds.*

# The Gemini Library of Late Spectral Templates for Stellar Kinematics Analysis in the CO 2.3 $\mu$m Region

Cláudia Winge

*Gemini Observatory, Casilla 603, La Serena, Chile*

Rogemar A. Riffel and Thaisa Storchi-Bergmann

*Depto. de Astronomia, IF/UFRGS, Av. Bento Gonçalves, 9500, CEP91501-970, Porto Alegre, RS, Brazil*

**Abstract.** We present a new spectroscopic library of late spectral type stellar templates in the near-IR range 2.15–2.42 $\mu$m, at $R = 5900$ resolution, oriented to support stellar kinematics studies in external galaxies, such as the direct determination of the masses of supermassive black holes in nearby active (and non-active) galaxies. The combination of high spectral resolution and state-of-the-art instrumentation available on 8 m-class telescopes has made analysis of circumnuclear stellar kinematics using the near-IR CO bandheads one of the most frequently used techniques for such studies, and this library aims to provide the supporting datasets required by the higher spectral resolution and spectral coverage achieved with modern near-IR spectrographs. The data are particularly well suited for two Gemini instruments—GNIRS and NIFS—but can be easily adjusted for use with any other near-IR spectrograph at similar or lower resolution.

## 1. Introduction

The internal kinematic information from a galaxy spectrum can be obtained through different techniques. The most commonly used one employs a cross-correlation method to extract the velocity dispersion from the galaxy spectrum using a single star spectrum as a template.

The CO bandheads from evolved stars are the strongest absorption features in the 1–3 $\mu$m range of stellar systems. The features are sharp and deep, and at least the first two overtones are located in regions of the IR spectrum relatively clean of telluric lines. The equivalent width of the CO bands is a function of effective temperature and surface gravity of the star: decreasing the surface gravity or the effective temperature increases the equivalent width.

The fitting technique, however, is very sensitive to template mismatch, as the characteristics of the stellar spectrum used as template affect the fitting result. The net effect is that the velocity dispersion measured from the galaxy spectrum will increase as the equivalent width of the template star CO bandhead decreases. To obtain a reliable measurement, one should provide the fitting program with a variety of template stellar spectra, which will be attributed different weights to obtain the best fit, but in the near-IR at $R \sim 6000$ there are very few, if any, stellar template libraries available for this kind of study.

## 2. Observations and Description of the Library

Observations were carried out in queue mode between 2006 August and 2007 January as a "poor weather" (IQ=ANY, CC=90%) Director's Discretionary Program (GS-2006B-DD-3). The configuration used was the GNIRS IFU with the 111 l/mm grating, yielding a resolution of 3.42 Å (44 $\mathrm{km\,s^{-1}}$ FWHM) at 2.293 $\mu$m.

Data reduction was done using the `gemini.gnirs` IRAF package, and followed standard procedures (flat-fielding, sky subtraction, wavelength calibration, 1D spectral extraction from all the slices with reasonable signal, coadding, telluric calibration). The proprietary period has been waived on all data. The raw datasets are available to the general community and can be retrieved from the Gemini Science Archive (GSA) at the CADC.

The library currently contains 29 stars observed in the range 2.24–2.42 $\mu$m, ranging from spectral types F7 III to M3 III. Of these, 23 objects were also observed on a second setting, extending the wavelength coverage down to 2.15 $\mu$m. The sample spans a range of EW(CO 2.32 $\mu$m) from 3–22 Å, and the spectra have S/N $>$ 50 at the blue side of the first CO overtone. We plan to expand the number of stars by retrieving from the GSA observations with GNIRS and NIFS obtained for other programs as the proprietary period expires.

The reduced spectra can be downloaded (in standard FITS format) from the library pages in the Gemini website at:
http://www.gemini.edu/sciops/instruments/nir/spectemp/index.html

**Acknowledgments.** The authors would like to thank the Gemini Deputy Director and Head of Science, J.-R. Roy, and the Gemini South Head of Science Operations, Michael West, for the time allocation for this project. The data were obtained under program ID GS-2006B-DD-3. This paper is based on data obtained at the Gemini Observatory, which is operated by AURA, Inc., under a cooperative agreement with the NSF on behalf of the Gemini partnership: NSF (United States), PPARC (United Kingdom), NRC (Canada), CONICYT (Chile), ARC (Australia), CNPq (Brazil) and CONICET (Argentina).

*Pathways through an Eclectic Universe*
*ASP Conference Series, Vol. 390, © 2008*
*J. H. Knapen, T. J. Mahoney, and A. Vazdekis, eds.*

# Stellar Populations of Nearby Elliptical Galaxies

Y. Yamada,[1] N. Arimoto,[1] A. Vazdekis,[2] and R. F. Peletier[3]

[1]*National Astronomical Observatory of Japan, 2-21-1 Osawa, Mitaka, Tokyo 181-8588, Japan*
[2]*Instituto de Astrofísica de Canarias, E-38200 La Laguna, Spain*
[3]*Kapteyn Astronomical Institute, University of Groningen, Postbus 800, NL-9700 AV Groningen, The Netherlands*

## 1. Introduction

Since elliptical galaxies are common objects in the Universe, many people have studied their formation and evolution. Elliptical galaxies have been traditionally considered as old stellar populations from several lines of evidence, such as the tightness of the color–magnitude relations and the fundamental plane. On the other hand, spectroscopic observations have shown the presence of young stellar populations in early-type galaxies. There is controversy over whether elliptical galaxies are old or not, but it has been difficult to measure ages due to age–metallicity degeneracy on spectral features (color, line indices); i.e., colors and line indices can be reproduced in several ways as a combination of ages and metallicities (Arimoto 1986; Worthey 1994).

## 2. Observations and Method

Spectral data were taken with the Subaru Telescope, the ESO-NTT, and the WHT. The data are moderately high resolution spectra (2.4–3.1 Å) and with an extremely high signal-to-noise ratio (115–490/Å). We used the $H\gamma_\sigma$ method (Vazdekis & Arimoto 1999) to disentangle the age–metallicity degeneracy. The $H\gamma_\sigma$ method is quite a different route to age determination from previous studies, which used the Lick/IDS system (González 1993; Kuntschner 2000; Sánchez-Blázquez et al. 2006); i.e., we can directly compare an observational SED to a simple stellar population (SSP) model SED without any complex conversions to the Lick/IDS system. Thanks to the capability of $H\gamma_\sigma$ indices, we can also derive the metallicity for each element from metallic indices.

## 3. Results and Discussion

We measured luminosity-weighted ages and metallicities of 14 elliptical galaxies in the Virgo cluster, whose luminosity spans 5.8 mag. Less massive galaxies show a large age dispersion (3–15 Gyr), while massive galaxies are uniformly old ($> 10$ Gyr). We also found that $[Z_{\rm Mg}/Z_{\rm Fe}]$ increases with the velocity dispersion $\sigma$ of galaxies (Fig. 1). This suggests that less massive galaxies have longer star formation periods than massive ones. These results are not clear in Vazdekis et al. (2001), who used the same method, since the magnitude range of their sample is smaller (3.3 mag). We measured ages and metallicities of 13 elliptical galaxies in a low density environment (field and groups) as well.

We found that massive galaxies show an age spread of 6–15 Gyr, and that the less massive ones, as well as those in the Virgo cluster, showed a spread of 3–10 Gyr (Fig. 2).

The age and metallicity trends which we found were not clear in previous studies due to the age–metallicity degeneracy of the indices, errors coming from the complex conversion into the Lick/IDS system and from emission correction. Since the ages and metallicities which we derived from $H\gamma_\sigma$ indices are consistent with those from fitting with SSP models, those ages and metallicities are reliable measurements. We conclude that although galaxies have variety in their formation process, in principle, massive galaxies tend to finish star formation earlier than less massive ones, and galaxies in denser regions tend to finish star formation earlier than galaxies in lower density regions.

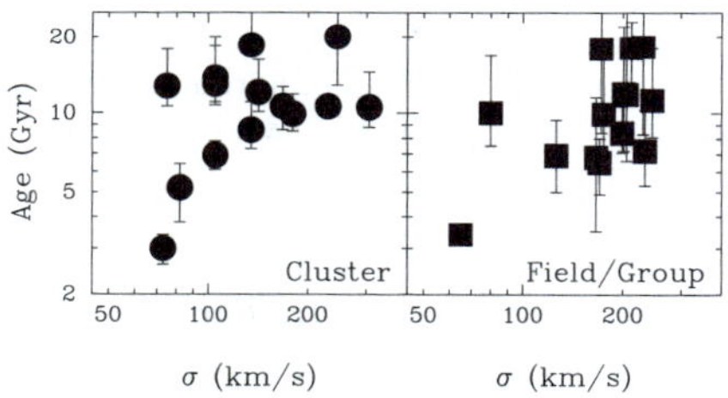


Figure 1. Velocity dispersion vs age derived from the $H\gamma_\sigma$–[MgFe] index–index diagram for cluster (*left*) and field/group galaxies (*right*).

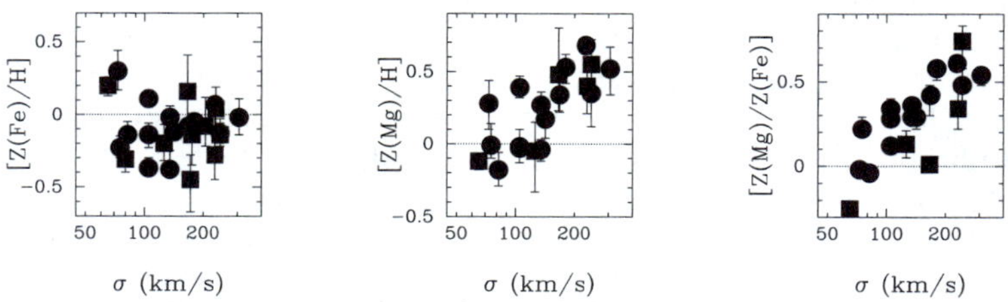


Figure 2. Velocity dispersion vs $[Z_{\rm Fe}/{\rm H}]$, $[Z_{\rm Mg}/{\rm H}]$, and $[Z_{\rm Mg}/Z_{\rm Fe}]$ derived from Fe3 and Mg *b* indices. Circles and squares represent Virgo cluster ellipticals and field/group ellipticals, respectively.

## References


Arimoto, N. 1996, in ASP Conf. Ser., 98, From Stars to Galaxies: The Impact of Stellar Physics on Galaxy Evolution, ed. C. Leitherer, U. Fritze-von Alvensleben, & J. Huchra (San Francisco: ASP), p. 287

González, J. J. 1993, PhD Thesis, University of California, Santa Cruz

Kuntschner, H. 2000, MNRAS, 315, 184

Sánchez-Blázquez, P., Gorgas, J., Cardiel, N., & González, J. J. 2006, A&A, 457, 787

Vazdekis, A., & Arimoto, N. 1999, ApJ, 525, 144

Vazdekis, A., Salaris, M., Arimoto, N., & Rose, J. A. 2001, ApJ, 549, 274

Worthey, G. 1994, ApJS, 95, 107

*Pathways through an Eclectic Universe*
*ASP Conference Series, Vol. 390, © 2008*
*J. H. Knapen, T. J. Mahoney, and A. Vazdekis, eds.*

# New Clues to Unveiling the Recent Massive Star Formation History in the Bar of NGC 1530

Almudena Zurita,[1] Isabel Pérez,[1,2] and Mónica Relaño[1]

[1]*Dpto. de Física Teórica y del Cosmos, Universidad de Granada, 18071 Granada, Spain*
[2]*Kapteyn Astronomical Institute, University of Groningen, The Netherlands*

**Abstract.** We have carried out $N$-body/SPH modeling and obtained $BVRK_s$ and H$\alpha$ photometry of the barred galaxy NGC 1530 to investigate the conditions under which H II regions form and their spatial relation to the principal dynamical features of the bar. We find differences between the H$\alpha$ equivalent widths of the H II regions of the trailing and leading zones of the bar. This result is interpreted in terms of an age effect, which constrains models of massive star formation in bar environments.

## 1. Where Do Massive Stars Form in Bars?

The formation and evolution of bars is a subject of recognized importance for the evolution of galaxies. It has led to the development of sophisticated modeling of the gas dynamics in barred potentials that explains considerably well the main morphological features observed in barred galaxies (e.g., Athanassoula 1992). However, how star formation (SF) occurs in bars is not yet well understood. Although it has traditionally and qualitatively been associated with zones in which a balance between the presence of gas shocks (high gas density) and low shear (allowing the collapse of molecular clouds) exists, there is no consensus yet on the parameters that determine where and how SF occurs.

The bar of NGC 1530 is an excellent laboratory to study the link between gas dynamics and SF: it has one of the largest bars ever observed (length ~24 kpc), high quality ionized gas kinematic data exist (Zurita et al. 2004), and it shows evidence for recent SF along the bar.

## 2. Results

$N$-body/SPH modeling (Pérez, Fux, & Freeman 2004) has been carried out using the gravitational potential obtained from the light distribution of the galaxy. The best-fit pattern speed is $\sim 10\,\mathrm{km\,s^{-1}\,Mpc^{-1}}$ (implying a fast bar). The model reproduces well the main observed morphological and kinematic features, namely the presence and amplitude of projected velocity gradients perpendicular to the bar, which are trailing with respect to the predicted higher gas density zones along the bar.

H$\alpha$ equivalent widths (EW$_{\mathrm{H}\alpha}$) and $B$-, $V$-, $R$-, and $K_s$-band fluxes have been measured for the H II regions of the bar. Our most important result comes from the EW$_{\mathrm{H}\alpha}$ measurements: the H II regions located further from the bar dust

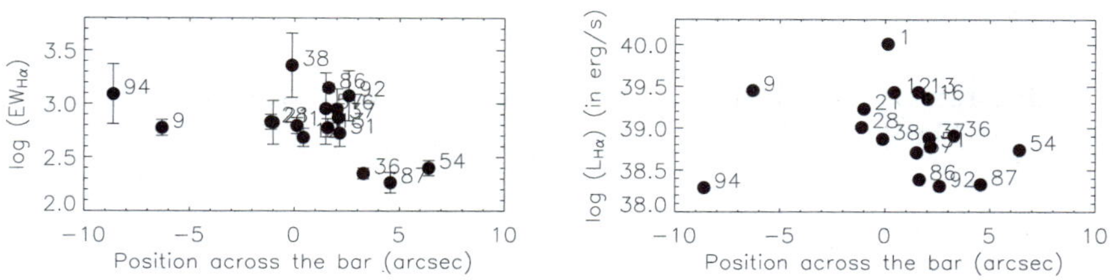


Figure 1. $EW_{H\alpha}$ (*left*) and H$\alpha$ luminosity (*right*) of the H II regions of the bar of NGC 1530 as a function of their projected distance to the bar dust lane. The numbers indicate the corresponding H II region identification numbers as given in the NGC 1530 H II region catalog (Relaño et al. 2005). Negative and positive distances indicate H II regions located on the trailing and leading sides of the bar dust lane, respectively.

lane on the leading side of the bar have lower $EW_{H\alpha}$ (by a factor ~4–5) than H II regions located on the trailing side and those closer to the bar dust lane (see Fig. 1). In terms of broad band colors and H$\alpha$ luminosities there is no obvious difference (within the error bars) between the H II regions located on the leading and trailing sides of the bar dust lane.

Among the most important factors affecting the low $EW_{H\alpha}$ of H II regions are: ionizing photon leakage, dust extinction of ionizing radiation, and age effects (for ages $\lesssim$ 3 Myr, the initial mass function has little effect on the $EW_{H\alpha}$ vs age relation). Dust extinction and/or photon leakage alone do not seem to give a plausible explanation. If they were the cause, then approximately 80% of the ionizing photons would be leaking or absorbed by dust for the regions of lower $EW_{H\alpha}$, while this percentage would be only ~20% for the regions located on the trailing side or closer to the bar dust lane. The most likely explanation, then, seems to be an age difference between both groups of H II regions. This would imply that the H II regions of lower $EW_{H\alpha}$ are ~3 Myr younger than the rest. Together with the presence of H II regions on the trailing side of the bar and its spatial association with dust spurs (A. Zurita & I. Pérez, in preparation), this would support the scenario proposed by Sheth et al. (2000), in which massive stars form in the dust spurs of the trailing side and continue their elliptical orbits and travel towards the leading side (passing through the bar dust lane).

**Acknowledgments.** AZ acknowledges the Consejería de Educación y Ciencia de la Junta de Andalucía for support, and IP the Netherlands Organisation for Scientific Research (NWO) and the Plan Nacional del Espacio del Ministerio de Educación y Ciencia.

## References

Athanassoula, E. 1992, MNRAS, 259, 345
Pérez, I., Fux, R., & Freeman, K. 2004, A&A, 424, 799
Relaño, M., Beckman, J. E., Zurita, A., Rozas, M., & Giammanco, C. 2005, A&A, 431, 235
Sheth, K, Regan, M. W., Vogel, S. N., & Teuben, P. J. 2000, ApJ, 532, 221
Zurita, A., Relaño, M., Beckman, J. E., & Knapen, J. H. 2004, A&A, 413, 73

*Pathways through an Eclectic Universe*
*ASP Conference Series, Vol. 390, © 2008*
*J. H. Knapen, T. J. Mahoney, and A. Vazdekis, eds.*

# Panel Discussion I. Star Formation in Galaxies: How Do We Continue?

Johan H. Knapen (Reporter)

*Instituto de Astrofísica de Canarias, E-38200 La Laguna, Spain*

**Abstract.** This is the written account of the first of two panel discussions, on *Star formation in galaxies: how do we continue?* The chair of the panel was Phil James, and panel members were John Beckman, Torsten Böker, Daniela Calzetti, Angeles Díaz, and Rob Kennicutt. The panel and audience discussed the following four questions: 1) What are the most critically needed techniques to give accurate measurements of total rates and efficiencies of star formation? 2) Do we understand the form of the initial mass function and its variation as a function of redshift and environment? 3) Are there multiple modes of star formation in galaxies (bulge vs disk, burst vs continuous) or does the Schmidt law explain everything? 4) How do we bring together our understanding of star formation in our Galaxy and in external systems?

JAMES: [Introduced the panel members, then introduced the four questions, and asked John Beckman to give his opinions on the first: What are the most critically needed techniques to give accurate measurements of total rates and efficiencies of star formation (SF)? ]

BECKMAN: Rob [Kennicutt] and Daniela [Calzetti] will have more to say on this topic but I will make some remarks to start the discussion. One question is on missing photons: which fraction of the SF rate (SFR) might be missed due to escaping photons when we observe a whole galaxy (assuming that the initial mass function [IMF] issue is covered). We know that photons can easily leak out of H II regions, but most will get soaked up somewhere in the galaxy so the global SFR should take them into account anyway. But there is a caveat. In grand design galaxies in particular, a large fraction, up to 50%, of the emission may be diffuse. This should not be missed but the diffuse emission can easily fall below the detection threshold. Locally this is most probably not a problem, but it might well be more significant at higher redshifts.

DÍAZ: I would like to stress the importance of high accuracy in measurements. A second point is that we then need similar accuracy in the tools we subsequently use for analysis. We need to know and quantify the influence of, for example, leaking photons, but also second-order effects which we must take into account as the observations become better. We also need to analyze the scatter which may well carry physical information.

KENNICUTT: To me, the key is the issue of efficiencies. Thanks to the multiwavelength revolution in observations over the last few years we now have a lot of information on many different issues. But I worry about our knowledge of the interstellar medium, about molecular masses involved, and about cloud cores. New facilities like CARMA and ALMA are coming, but we need experts to say where improvement is most needed.

BÖKER: A main outstanding question is that of recent versus current SF. Even H$\alpha$ gives information on the "past" SF, and to observe the current SF we need submillimeter observations. This is not yet being incorporated into the discussion on SFRs.

CALZETTI: Among the advances that new millimeter telescopes and instruments (ALMA, LMT [Large Millimeter Telescope], CARMA) will bring is the possibility of detecting CO emission in the interarm regions of galaxies, another under-explored region of the parameter space of galactic environments.

JAMES: Thank you panel members. I now invite comments from the audience.

KRUMHOLZ: I would like to comment on questions (1) and (4), and stress that we need to worry most about, and clarify, the efficiency of SF. In a Galactic context, the term efficiency is used to describe the fraction of mass which is transformed into stars over the lifetime of a cloud. But in an extragalactic context, efficiency is used to denote the SFR over gas mass, which is an instantaneous measure and not one over a lifetime. My comment, then, is that we should rather *not* use the term efficiency, or define carefully whether we are talking about instantaneous efficiency or over a lifetime.

HENSLER: I agree with Mark [Krumholz] but let me add that a galaxy is not a closed box, and so in addition we need to consider how, e.g., infall influences the efficiency.

SHETH: The new CARMA telescope is now working, and early work on M51 shows that it works very well. But rather than concentrating on HCN, we should use 1 mm continuum which is where the improvements in CARMA are the most significant compared to OVRO/BIMA and other observatories. The 1 mm continuum would trace the dense dusty regions in galaxies.

EVANS: A worry is whether even with ALMA we will be able to sort out observational issues to obtain molecular gas masses to a precision better than a factor of three. In particular the calibration of CO measurements is difficult and we need simulations of molecular clouds in galaxies to understand what the observations mean and what their real uncertainties are.

CHURCHWELL: IR observations are sensitive to the most massive young stellar objects (YSOs), whereas X-ray observations are sensitive to magnetic activity in lower mass YSOs, so together one has a chance to determine the IMF for the whole stellar mass range within a given cluster. It's the lower mass range that has historically been the most difficult to establish and since most of the mass is in lower mass stars, this is the most important part of the IMF that needs to be determined with good precision. My student is trying to use the GLIMPSE survey to determine the current global SF rate in the Galaxy, but to do this he has to determine the IMF, especially for lower mass YSOs, for various environments in the disk of our Galaxy. Our hope is that X-ray data from *Chandra* and other X-ray satellites will provide the tool to do this.

LEHNERT: I would like to address the differences that we heard on Monday between the results on SF modeling, from Mark [Krumholz] and Ian [Bonnell]. We need to know how SF happens in different environments within and outside the Galaxy, and the key point is how do we approach this.

KRUMHOLZ: The difference is mainly due to a limitation in the CPU power we can use. Ian [Bonnell] can model larger volumes because he doesn't include the effects of radiation.

LEHNERT: ...but are you sure you know the physics?

KRUMHOLZ: The limitation is in the code and in the CPU, not in the physics.

KROUPA: We've heard earlier in the week (in the talk by Alves) that the stellar IMF shape is already found in the mass distribution of the cloud cores. But theoretical work seems to be wrong, for instance I quote the failure of Klessen, Spaans, & Jappsen (2007) to reproduce the mass function of the Arches cluster. This is very worrying indeed, and we need more work on the theoretical side.

BONNELL: The Klessen et al. result of a top-heavy IMF depends primarily on their input assumptions of the dust physics and background radiation field. I would not misconstrue their result as implying we do not understand the dynamics of the SF process, but rather that there remain many unknowns that are still input into the numerical simulations.

ALVES: I propose the use of a combination of the extinction technique, which is easy and works well in local clouds, with the study of clouds at larger distances as described by Churchwell on Monday. The prospects of this combination are good, but it won't be easy.

EVANS: I totally agree. The cores we see in local clouds go up to 2–3 $M_{\odot}$ in mass and have nothing to do with massive SF. Up to now we cannot resolve the much larger cores, and even with ALMA this will be very tough.

JAMES: I propose that we move on to point (3): Are there multiple modes of SF in galaxies (bulge vs disk, burst vs continuous), or does the Schmidt law explain everything? Let's hear from our panel members.

BECKMAN: Before going to point (3) I would like to add something to the discussion on the previous point, namely that the correspondence between the IMFs as derived from cloud cores and stars is almost too good to be true—and let's hope it is indeed true. This makes the theorists' life much easier, all they will have to do is scale... I think the measured turnover in the stellar IMF to low masses, plus the turnover in the mass function of the cloud cores in Jõao [Alves]'s work is clearly one of the key novelties so far in this meeting.

JAMES: On to point (3) then—we should discuss whether the Schmidt–Kennicutt law is really the last word, or whether it may in some sense be holding back new developments in this area, and whether we should thus look to move on to a more physically grounded approach (while still citing Rob [Kennicutt] appropriately)?

BECKMAN: Rob?

KENNICUTT: I'm thinking on my feet here... In yesterday's talks we heard about SF in nuclei, we know now that this occurs over scales of hundreds of parsecs, and is involved in the formation of bulges. We have heard of central SFRs of up to $1000\,M_\odot\,\mathrm{yr}^{-1}$. It is remarkable that one law can cover all this range of SF, from the very low mass regime to the extremes just mentioned. I wonder whether we have any evidence that there is more than one mode, or more than one phenomenon. Does the single Schmidt law reflect a single universal mode of SF, or could there be two or more modes that happen to follow the same dependence of SFR and gas surface densities? I guess the answer in this respect is very much like we have discussed already with regards to the IMF: we don't see any evidence for a change, but we should be astounded if indeed there weren't any.

BECKMAN: Suggestions on whether there is evidence from abundance ratios for more than one mode of SF come and go—does anyone know what the current state of affairs is?

BÖKER: The SF mode that produced bulges in the early Universe must be different from the currently dominant one: a stellar mass in the order $10^9$–$10^{10}\,M_\odot$ was formed in a very short time.

KENNICUTT: Or, for example, how did a globular cluster like M13 or Omega Cen form? Have we seen anything at this conference that is relevant to SF in that context?

DÍAZ: Considering clusters of $1000\,M_\odot$ (not $1000\,M_\odot\,\mathrm{yr}^{-1}$!), do we have there a violent mode of SF, such as the one that Rosa [González-Delgado] talked about? It cannot simply be adding up lots of little SF events, this doesn't seem to work so we must have another mode of SF.

CALZETTI: I would like to present a concrete example of how observations can sometime be misleading in the interpretation of the modes of SF. Back in the mid-90s, analysis of UV images of nearby starburst galaxies led to the discovery that only a small fraction (about 20% or less) of the UV light was associated with clearly identifiable stellar clusters; most of the UV light (~80%) was associated with a diffuse stellar component. During that period, the hypothesis was advanced that SF had two "flavors" (modes): a cluster mode and a "diffuse" mode. Subsequent studies, over the course of the following ten years, have shown that the diffuse UV component is likely the "product" of the dissolution of the stellar clusters over timescales of about 10–20 Myr; once the clusters have evaporated, the UV light will appear to be originating from a diffuse stellar population. Hence, there appears to be one single mode of SF in starbursts, the "cluster" mode.

JAMES: Let me restate the question posed by John [Beckman]: is there any evidence from stellar abundance ratios for the existence of different modes of SF?

SILK: In this respect, we remain biased by the local Universe. Even if we can cope with objects like M13, we continue to have big problems at high redshifts.

BASTIAN: Looking at smaller scales, we know from a decade of star cluster research with the *HST* that there are *not* different modes. The relations between the SFR and the number of clusters present, and similarly the relation between the SFR and the mass of the most massive cluster in the galaxy argue strongly that "starbursts" and quiescent galaxies lie on the same continuum. In addition, work done by myself and also by Bruce Elmegreen has shown that star clusters themselves are simply part of a continuous distribution of SF in galaxies, from sub-parsec to kpc scales.

BÖKER: This is a valid point, but it refers to the very local Universe, and to the current SF. The past may have been very different.

NESVADBA: We looked at the SF in detail in one galaxy at $z \sim 2.6$, as close as we can at present at high redshift (Nesvadba et al. 2007). We found an SFR of $\sim 500\,M_\odot\,\mathrm{yr}^{-1}$, and no evidence for an AGN contribution (which includes the X-ray). The SF efficiency we found is very similar to that found at low redshifts, which implies that the SF properties are very similar. The same is in fact true for the SF-related feedback. There are obviously limitations working at these redshifts, but empirically, we have not found alarmingly large discrepancies.

KRUMHOLZ: A comment on the differences in the IMF. If the expected number of low mass stars were present in the ring around the Galactic Center (given the observed number of B-stars) we should have seen X-ray emission from them, but haven't.

HAMMER: We don't even know if there is a Schmidt law at $z \sim 0.6$. Indeed the gas distribution has a big effect on this law, and we know that there is much more gas at a redshift of unity. How is this gas distributed? I am not sure that question (3) is relevant.

CAPUZZO-DOLCETTA: May I stress that even after the talk by Kennicutt it is not clear what the Schmidt law actually means physically. As it stands, it is a phenomenological law.

EDMUNDS: I wonder why we must go on about variations in the IMF. The point is that it is fair enough that the IMF changes in strange places like the Galactic Center or the metal-poor early Universe. But we don't have to change the overall IMF—this is great science!

LEHNERT: The Schmidt law has been alluded to as an empirical one, but the question is what keeps the SF inefficient. It may be related to the effect of shear (cf. Mark Krumholz) due to differential rotation in the large scale velocity fields of galaxies, which will be much lower in the central parts with solid body rotation. So, what will control SF in these regions? And why do spirals look young, but disks look old?

BALCELLS: Let me put on a theorist's hat. I would then want the SF to depend on dynamical parameters, such as the velocity dispersion. Gas clouds should collapse in different ways, and I would like to have a parameter that prescribes when the gas can become molecular. For instance, in ellipticals the gas is too hot so it won't form stars. Similarly, there is gas at the centers of spirals but it cannot collapse. All this should really go into the Schmidt law, but in that case will it still work?

KENNICUTT: There are examples of that, for example, the Silk law (which relates the SFR density to the ratio of gas density and dynamical time, and thus identifies the number of "compression events", or the number of times the gas goes through the spiral arms). We have tested this, but there is no clear signature. For instance, in disks the SF declines exponentially with radius, but the velocity dispersion profile is flat with radius. So, yes, we do need to move to a physical basis of the Schmidt law, but Occam's razor tells us that density alone does the trick.

SARZI: About SF in early-type galaxies, it is interesting to notice that the objects that for sure experienced recent SF all have gas emission characterized by very small velocity dispersion, independent of the apparent ionization of the gas, as traced by the [O III]/H$\beta$ line ratio.

KRUMHOLZ: In response to Marc Balcells, there is no lack of theoretical ideas, quite the contrary, there are perhaps more theories than theorists. There are papers by, e.g., Klessen, Silk, and myself on how the Schmidt law may be related to the underlying physics.

SILK: I agree that the Schmidt law is a very good fit and is a great result, but the more interesting physics is in the scatter.

DÍAZ: Indeed, we should perhaps concentrate on those objects that do not fit the Schmidt law and then try to find the physical drivers that made them deviate.

CALZETTI: We studied M83 in detail to test this possibility. If the interesting environments to study are those where SF deviates from the Schmidt–Kennicutt law, then the outer regions of galaxies need to be excluded from those studies. The combination of recent data in H I (from the VLA, data obtained by Fabian Walter), mid-infrared (from *Spitzer*), and UV (from *GALEX*) of the outer regions of M83 suggests that the UV emission in those extreme environments is associated with SF that is compatible with the Schmidt–Kennicutt law. Those regions are located at a distance from the center of about four times the H$\alpha$ cut-off radius of M83.

BECKMAN: Partly in answer to Mike Edmunds' point, the interest lies in trying to identify the objects that do not fit the relation, and then to find the physics drivers for them.

JAMES: I wonder if there are any further comments, because we need to finish in five minutes. Would anyone from our panel perhaps wish to comment on the fourth question: How do we bring together our understanding of SF in our Galaxy and in external systems?

CALZETTI: When I formulated question number four last night, this is what I had in mind. We have seen during the course of this conference that there is a direct (linear) correlation between SFR and dense cores. I believe this implies that we are moving the problem from having to explain physically the Schmidt–Kennicutt law in its original formulation, to having to explain the correlation between the density of dense cores and the total gas density.

KENNICUTT: If I refer back to the introduction of Neal [Evans]'s talk, I think we need to come together in our studies of the small and the large scales. Perhaps the key to tying the Schmidt law to the underlying physics lies in detailed studies of SF at small scales. So Neal, let me put you on the spot: what should we observe?

EVANS: We need to speak, first of all, a common language, use a common terminology, especially on issues of efficiency and timescales. There is a difference between whether, and how fast, you make stars, and on large scales SF tends to be slow. Remember that the actual process of SF occurs in dense cores, which are two steps removed from the scales where the Kennicutt–Schmidt relations operate. First, you need to understand what controls the formation of molecular clouds, which is probably best studied with very high resolution and sensitivity (ALMA and CARMA) in nearby galaxies. Then you need to understand what controls the fraction of a molecular cloud that turns into dense cores (a few percent in local molecular clouds, but maybe very high in extreme starbursts). This is probably best studied in our own Galaxy, but it requires a careful combination of observations with models. Those two approaches together may bridge the gap.

SHETH: May I take the opportunity to advertise a conference organized by the Spitzer team in Pasadena this autumn, where we plan to bring together these two communities, and where we hope to start answering these questions.

JAMES: Our time has run out, so may I ask John Beckman to make a closing comment?

BECKMAN: It is great to see more people getting together to discuss these themes, the more people together, the more we think we know. There are many new observational results, for instance multiwavelength determinations of SFRs using *GALEX*, *Spitzer*, and other facilities, and also on gas and its detailed structure. So the observations are getting better. There still seem to be too many theories so what we must perhaps think of are observational results that cannot be reproduced by any theories.

JAMES: With those words we finish this panel discussion. Let me thank our panel members and the members of the audience.

**References**

Klessen, R. S., Spaans, M., & Jappsen, A.-K. 2007, MNRAS, 374, L29
Nesvadba, N. P. H., et al. 2007, ApJ, 657, 725

# Session IV

# Feedback and Infall

*Pathways through an Eclectic Universe*
*ASP Conference Series, Vol. 390, © 2008*
*J. H. Knapen, T. J. Mahoney, and A. Vazdekis, eds.*

# Feedback in Galaxy Formation

Joseph Silk

*Department of Physics, University of Oxford, Oxford OX1 3RH, UK*

**Abstract.** The physics of forming galaxies constrains their evolution in a highly nonlinear way. Observational phenomenology motivates the need for negative feedback to operate via supernova explosions in disk galaxies and via AGN outflows for quenching star formation in massive early-type galaxies. AGN-triggered and supernova-enhanced positive feedback may also be required to account both for ultra-luminous infrared galaxies and the presence of extremely red galaxies at high redshift.

## 1. Introduction

We lack any fundamental understanding of either star or galaxy formation. Our models are largely based on phenomenology. Local observations suggest that the Kennicutt law, which relates star formation rate to gas content, may be universal, encompassing quiescent star-forming galaxies, starbursts, and even individual star-forming complexes. However, whether phenomenological modeling of the star formation rate, efficiency, and initial mass function from studies of the nearby Universe has any relevance to the extreme conditions of the early Universe is far from certain. Two modes of star formation most likely are required to explain young, gas-rich, star-forming disks that are currently forming stars with low efficiency and old, gas-poor massive spheroids that formed stars with high efficiency but long ago. These two modes of star formation refer, respectively, to formation of disks along with so-called pseudo-bulges, and to elliptical and massive spheroid formation. I will describe some simple analytical prescriptions for galaxy formation that potentially provide templates for numerical simulations.

## 2. Disk Galaxy Formation

The characteristic timescale for star formation is several gigayears for an Sb galaxy such as the Milky Way, an order of magnitude longer than the dynamical timescale. The explanation comes from studying smaller scales. The processing of gas into stars in molecular clouds is decelerated by feedback associated with both protostellar outflows and supernova explosions of massive stars. The lifetime of a molecular cloud is up to an order of magnitude longer than its free-fall time. While this ratio of timescales is empirically uncertain, it is clear that the observed gas turbulent motions dominate the gas pressure and hence counterbalance gravity. Since the turbulence most likely is driven by magnetically regulated protostellar feedback in combination with the self-gravitating disk instabilities that drive cloud formation, cloud star formation rates are expected to be self-regulating.

On a global scale, self-regulation is important via maintenance of the disk cold gas surface density above the Toomre threshold for gravitational instability. The regulation is controlled by the star formation timescale and especially by the gas supply. Star formation depletes the gas supply, and dynamical interactions between stars and molecular clouds heat up the young stellar disk. Maintenance of global disk instability requires a cold disk, and the gas supply to maintain this is provided via gas accretion by continuous infall from the halo as well as by minor mergers. The disk gas will consequently be turbulent, and this is reflected in the cloud internal velocity structures. More generally, supersonic turbulence is believed to play an important role in accounting for the longevity of star formation in disks. However, the global driving of the turbulence, which normally would dissipate on a short timescale, is not fully understood. Numerical simulations of entire disks generally do not resolve individual star formation and stellar explosions on dynamically relevant scales and hence have difficulty in following the dissipation of the turbulence. It has nevertheless been argued that the Schmidt–Kennicutt law is quantitatively reproduced by disk gravitational instability-driven turbulence (Kravtsov 2003; Li, Mac Low, & Klesse 2005). However, higher resolution and more realistic modeling of the interstellar medium is needed in order to study the efficiency of star formation. While the Toomre instability accounts for the location of cloud (and star) formation, it cannot account for the low observed efficiency in a realistic interstellar medium. Supernova feedback in a multiphase interstellar medium is the dominant effect that regulates the rate of star formation (Slyz et al. 2005; Tasker & Bryan 2006).

One can gain some insight into the feedback by simple analytic considerations. The low efficiency of disk star formation, with a characteristic gas conversion timescale of $M_{\rm gas}/\dot{M}_* \sim 3\,{\rm Gyr}$ is primarily due to supernova feedback. One can quantify this from a simple interpretation of the fraction of gas converted into stars per disk rotation time, observed globally to be about 2%. The fraction of initial supernova energy available for stirring up the interstellar gas, after allowing for radiative losses, is the ratio of the interstellar gas cloud velocity dispersion, about $11\,{\rm km\,s^{-1}}$, to the velocity at which an expanding supernova remnant enters the approximately momentum-conserving phase, around $400\,{\rm km\,s^{-1}}$. The coincidence between this ratio and the observed inefficiency of gas conversion suggests that cloud turbulent velocities are indeed maintained by supernovae. This motivates a self-regulation model for disk star formation, controlled by the interplay between star formation and the negative feedback associated with massive star death. One further complication is that the disk is not a closed environment. Supernova explosions trigger venting of disk gas into the halo. The disk gas cannot escape, at least for disks similar in mass to the Milky Way galaxy, but erupts into the halo in galactic fountains. These fountains eventually cool, and provide part of the reservoir of cold gas that maintains disk instability.

The cold dark matter (CDM) formation ansatz asserts that there is continuing accretion of cold gas. How the gas heats up on interacting with the halo is debated, but the current gas content is expected to have augmented the total baryon content by up to 30% over the past Hubble time. This means that a typical $L_*$ galaxy should contain on the order of $10^{10.5}\,M_\odot$ in diffuse gas. This is a factor of five or so larger than the typical atomic/molecular gas content in the inner galaxy, and so most of this gas must be in the outer regions and ionized, virialization requiring it to be hot ( $\gtrsim 10^7$ K) and so visible in X-rays. X-ray observations of massive disk galaxies such as the Sombrero (Li, Wang, & Hameed 2007) find only of order $10^9\,M_\odot$ in hot gas in the bulge region out to ~20 kpc. The best explanation seems to be that supernova heating has driven most of the gas into the outer halo or beyond. Superwinds indeed are found around some disk galaxies (Strickland 2007). This is puzzling from the theoretical perspective because supernovae are considered to be incapable of driving a strong wind from the gravitational potential well of a massive galaxy (Mac Low & Ferrara 1999). Also, at least one example is known of an extended X-ray halo around a normal massive spiral galaxy (Pedersen et al. 2006). In this case, gravitational accretion and shock heating provides the most plausible source for heating the gas. It seems unlikely that this is the norm, however, otherwise there would be excessive accumulation of gas in the halo.

The Milky Way Galaxy (MWG) is one of the best-studied examples in terms of baryon content and star formation history. Nevertheless, there is no understanding of how it and other late-type disk galaxies acquired thin disks and small bulges. The angular momentum transfer problem is still unresolved. This may be connected to the baryon content of the MWG. It is clear that known baryon reservoirs only account for at most two-thirds of the initial baryon content. Where are these baryons today? The situation for the MWG is complicated, because the Local Group provides a potential buffer against gas infall. An explanation of the paucity of halo gas in the MWG may lie in a complex history for the gas. One possibility is that substantial amounts of gas were ejected into the Local Group intergalactic medium (IGM). There are indeed indications that groups and clusters of galaxies are baryonically closed systems (Gonzalez, Zaritsky, & Zabludoff 2007). Moreover, the intergalactic gas is enriched. Presumably enriched debris was ejected, most likely in the early stages of galaxy formation.

Ejection could not have happened recently, given the depth of the MWG gravitational potential well, but could have happened during the assembly phase of the MWG. Winds were easily driven from the star-forming dwarf galaxies that merged into the MWG, some of these still remaining in the halo as low surface brightness fossil relics. To complicate matters further, we know that the merging history of our Galaxy itself involved dwarfs that were still gas-rich. Star formation and enrichment occurred during the merging process, as the chemical abundances of the nearby dwarfs show similar trends with metallicity to MWG halo stars, but are generally offset to lower metallicity. Presumably much of the enriched gas eventually formed the disk, but up to ~50% could plausibly have been ejected by supernova-driven winds into the IGM. In nearby dwarf starbursts, we indeed find that the wind mass flux is comparable to the star formation rate. The halo structure of our galaxy reveals a complex pattern of tidal substructure from disrupted satellites.

## 3. Early-Type Galaxies

The bright galaxy end of the luminosity function is dominated by early-type galaxies and displays an exponential cut-off with a characteristic luminosity (fit to the Schechter function) of $L_* \approx 3\times10^{10}\,L_\odot$. In the $\Lambda$CDM cosmological model, gas infall continues until the present epoch. Approximately 90% of the baryons are presently in the IGM and constitute a huge gas reservoir that potentially is available for accretion onto galaxies and eventual star formation. While massive galaxies are generally old, some 30% of nearby early-type galaxies reveal residual levels of recent star formation. Of order 10% of their stellar mass has formed within the past Gyr. However, massive early-type galaxies are generally gas-poor and predominantly consist of very old stars.

An emerging paradigm asserts that massive spheroid formation is terminated by the effects of AGN outflows on the halo gas reservoir (Bower et al. 2006; Croton et al. 2006). Massive outflows quench infall, strip the halo of gas, and lead to old (and red) ellipticals. Hot gas accretion does not feed star formation until the heat supply is turned off. Heating by continuing infall may ensure that this never happens completely. However, if gravitational infall and the associated drag heating are responsible for the bulk of late quenching, it is difficult to form thin disks, compounding the problem of disk galaxy formation. Indeed, the so-called over-merging problem continues to pose a challenge to theory in view of increasing evidence that stellar mass assembly occurred early.

Early suppression of star formation can be accomplished by AGN outflows, which heat and expel the gas reservoir. Numerous low luminosity AGN may play a role in inhibiting late epoch cooling flows in galaxy clusters. At the current epoch, there is still substantial gas accretion, and low luminosity AGN may heat the intracluster gas on a timescale of a Gyr. The entropy of the hot gas can be accounted for by AGN heating, since the required preheating of the gas occurred in the early stages of cluster formation.

AGN outflows are driven by infall onto supermassive black holes (SMBHs). SMBH growth is in phase with galaxy formation, or at least with the global star formation rate. AGN are fed during the gas-rich phase of galaxy formation. Correlation of SMBH mass with the spheroid potential suggests formation of both is contemporaneous. The starburst association with spheroid formation occurs during the SMBH growth phase. In the gas-rich bulge formation phase, the momentum input from AGN is dominant by an order of magnitude or more relative to input from supernovae. Indeed, supernovae provide negative feedback in galactic disks and drive winds from potential wells of low mass galaxies. In protospheroids, blowout occurs and star formation terminates when the SMBH–$\sigma$ relation is saturated. The SMBH-quenching hypothesis leads to a SMBH–spheroid velocity dispersion correlation that fits the observed normalization and slope (Silk & Rees 1998; King 2003).

Quenching is likely to be associated with massive momentum-driven winds. Neutral (Na I-absorbing) winds are ubiquitous in ultra-luminous infrared galaxies (ULIRGs, Martin 2005). The outflow rate is on the order of the star formation rate, suggesting that the mass of ejected baryons indeed is comparable to that remaining in stars. Supernovae cannot account for outflows from massive galaxies. The AGN outflows play an essential role in quenching star formation. Without these early outflows and gas deprivation, too many ellipticals would be too luminous and too blue. Quenching has a memory even at low redshift: once the residual gas is ejected and heated, its new adiabat means that relatively little of the gas will cool sufficiently to produce excessively blue early-type galaxies. Some will, of course, and this may suffice to account both for the residual star formation found in 30% of nearby ellipticals (Yi et al. 2005) as well as the occurrence of some nearby morphologically identified early-type galaxies that are blue (Schawinski et al. 2007).

### 3.1. Positive Feedback

AGN outflows can also trigger star formation before quenching it. The phenomena are most likely causally connected but we do not know the associated arrow of time. That is, does AGN growth precede and trigger spheroid formation, or is AGN growth a consequence and eventual terminator of spheroid formation?

The star formation timescale decreases for luminous starbursts at high redshift. One almost certainly needs to enhance star formation, and this means one must seek a mechanism for positive feedback. What is less clear is whether the high efficiency needed for massive spheroid formation, already in place by $z \sim 2$, requires a distinct mechanism from disk star formation. Supernova feedback plays an important role in disks. Star formation ages ($\sim M_*/\dot{M}_*$) as short as 0.2 Gyr are inferred at $z = 2$–3 for massive early-type galaxies from spectrophotometric population synthesis modeling (Maraston et al. 2006) and enhanced [$\alpha$/Fe] ratios (Thomas et al. 2005). The universal connection between gas content and star formation rate, applicable to ULIRGs as well as quiescent disks, suggests supernovae must continue to play a feedback role, but an additional ingredient may be needed to accelerate the rate of supernovae. The downsizing phenomenon seen for massive galaxy formation argues for a role for AGN, which display similar downsizing, and have a SMBH mass function that is weighted towards massive galaxies and correlates with the old spheroid component. AGN triggering may provide an explanation both for the short timescale and for downsizing of star formation.

AGN are present in at least 50% of local ULIRGs (Veilleux 2006). Supermassive black holes can induce outflows that overpressure the most massive interstellar clouds and provide positive feedback, especially for protospheroids. In a hierarchically forming protogalaxy, the outflow occurs in an inhomogeneous interstellar medium. Twin jets propagate, developing bow shocks that become unstable. A cocoon-like bipolar outflow develops (Saxton et al. 2005). Because of the overpressure, this initially propagates over a timescale considerably shorter than the dynamical timescale. The star formation rate is of order the product of the usual efficiency factor for supernova momentum input and the ratio of $M_{\rm gas}$ to $t_{\rm jet}$. The condition $t_{\rm jet} \ll t_{\rm dyn}$ means that cloud overpressuring results in triggered star formation and an enhanced supernova rate compared to the case when AGN-induced star formation is insignificant (Silk 2005). Triggered star formation results in accelerated spheroid growth. The multiplicative factor of AGN-triggered supernovae is expected to be most effective in deep potential wells. Globally launched, as opposed to nuclear, winds testify to the multiplicative action of the AGN trigger. Spatially resolved kinematics in some nearby AGN provides direct evidence for global launching (Everett & Murray 2007).

The optimal conditions for gas accretion into the central core and central black hole growth require molecular hydrogen cooling to be suppressed in the first generation of protogalactic clouds and dwarf galaxies (Spaans & Silk 2006; Lodato & Natarajan 2006). A short cooling time leads to premature fragmentation and (inefficient) disk-mode star formation. A long cooling time $t_{\rm cool} \sim t_{\rm dyn}$ promotes radial gas flow into the protobulge, which grows by accretion and merging until intermediate-mass black holes (IMBHs) grow into SMBHs. This differentiation via $t_{\rm cool}/t_{\rm dyn}$ tracks halo mass, with the ratio being shortest in the least massive halos. Suppressed feedback in early-forming dwarfs is also due to the predominance of Type II supernovae, which are far less efficient at feedback than Type Ia supernovae both because cooling is suppressed and because successive massive star supernovae explode from the same (and unique) massive star cluster, with the consequence that their debris and momentum tend to be channeled into a low density porous interstellar medium, leaving the cold, denser gas in place. Evidence for this interpretation is found from the steepening of the faint end of the galaxy luminosity function towards high redshifts.

A more or less monolithic build-up of gas occurs until the SMBH is massive enough, at 0.1% of the protobulge mass, to give significant feedback and trigger the efficient mode of star formation. This leads to a contemporaneous and top-down formation scenario for both SMBH and spheroid growth. Downsizing is further favored if Bondi accretion powers jets, as suggested by observations of X-ray luminous ellipticals (Allen et al. 2006). The mass accretion rate is proportional to $(GM/\sigma^2)^2\rho\sigma \overset{\propto}{\sim} M^2$ for adiabatic compression, if the specific entropy in the accreting gas is assumed to be scattered about a universal value (corresponding to preheating and enrichment of the IGM). A similar dependence (of mass accretion rate on protostellar mass) is observed for accretion flows onto T Tauri stars. The associated jets and outflows are magnetically accelerated from the rotating accretion disk, suggesting that a universal jet mechanism may be responsible (Banerjee & Pudritz 2006). Similar scalings are found over a wide range of black hole masses, from stellar to galactic scales (Gallo et al. 2006). The resulting nonlinear mass dependence, if applicable to SMBHs, motivates downsizing for both the spheroid and the SMBH if the associated outflows trigger star formation.

### 3.2. The Observational Case

There are increasing indications of a connection between starburst and AGN activity. Both SMBHs and galaxies display anti-hierarchical growth rates, the more massive systems forming earlier. Both show a rapid decline at low redshift. Radio observations provide a way of tracing high redshift luminous starbursts, and at redshift beyond unity at least a quarter of these host obscured X-ray luminous AGN (Richards et al. 2006). The AGN fraction increases from 20% at $z < 1$ to 60% at $z \sim 2.5$ (Seymour et al. 2006). Extended NUV emission is associated with high redshift radio galaxy lobes on scales of tens of kpc (Labiano et al. 2006), characteristic of recent star formation.

A young age is independently inferred from the radio characteristics of the compact AGN components. Optical spectroscopy demonstrates that the photoionization, which produces the extended Lyman alpha emission associated with luminous radio galaxies, especially at high $z$, most likely is due to an enhanced young star population (Villar-Martín et al. 2007). Further evidence cited by these authors for the role of young stars rather than scattered nuclear light comes from the reduced UV rest-frame polarization and enhanced rate of submillimeter detections for $z > 2.5$ radio galaxies, and the fact that the smaller high $z$ radio sources show evidence for the strongest jet–gas interactions and the most luminous young stellar populations.

However, the causal nature of the interaction is not yet demonstrated. The lack of resolution at high redshifts means that one cannot exclude either a simultaneous external trigger of both phenomena, such as a merger event, or star formation triggering of the AGN activity by runaway stellar collisions and explosions. At least one local example, Minkowski's object (Croft et al. 2006), demonstrates that starbursts can be triggered by radio jets. At high redshift, alignments are found between radio lobes and star formation on large scales (Feain et al. 2007), and it is even possible that star formation-generated enrichment can account for the jet-aligned CO structures as well as the unpolarized jet-associated UV continuum (Klamer et al. 2004) found at high redshift. This is suggestive but circumstantial evidence that provides a crucial missing ingredient for the triggering hypothesis: triggering requires a gas reservoir, whereas quenching would anticorrelate molecular gas and outflows.

Another crucial clue to the nature of the AGN–spheroid interaction is embedded in the correlation between black hole mass and spheroid velocity dispersion. There are indications that massive SMBHs are in place before spheroids are fully assembled, at the highest redshifts (Bertoldi et al. 2003; Maiolino et al. 2007; Ho 2007). At lower redshift, a study of radio galaxies finds a factor of $\sim$4 systematic enhancement in SMBH mass, inferred by assuming that the broad emission line region gas is confined gravitationally in the vicinity of the SMBH, in the redshift range $1 \lesssim z \lesssim 2$ (McLure et al. 2006). If the SMBHs are in place first, the case for positive feedback is self-evident. However, not all data support this view (Borys et al. 2005); in particular, obscured AGN at $z \sim 2$ may be undergoing a major growth spurt and hence be undermassive compared to their optically luminous counterparts (Alexander et al. 2007).

## 4. Summary

A strong case can be made for feedback in galaxy formation. Supernovae and AGN outflows play essential roles. In disk galaxies, supernova remnants self-regulate the interstellar medium, driving enriched gas fountains out of the disk. The gas cools in the halo and eventually accretes onto the disk as high velocity clouds. During the spheroid formation phase, the central SMBH growth results in outflows that induce accelerated star formation. Induced supernovae can drive winds even at early stages of spheroid formation. Since the rate of momentum input from triggered supernovae soon dominates that from the AGN; this phenomenon can occur even during the phase that the SMBHs are subdominant. Eventually when the SMBH attains its final mass, the AGN wind itself is strong enough to thoroughly clean out the spheroid and thereby terminate the feeding of any ongoing star formation.

Unfortunately, there is no robust prescription for implementing feedback, or indeed star formation, into simulations of galaxy formation. The situation is especially grave for forming spheroids where one lacks the framework of the global non-axisymmetric disk instability, which sets the scene for disk star formation via spiral density wave generation of molecular cloud complexes. Even here, the subsequent steps, involving Jeans instability and turbulence generation, culminating in cloud core collapse and star formation, involve as much phenomenology as theory. For example, while supersonic turbulence is observed to dominate on all scales of the interstellar medium, the turbulence driver is disputed. Large-scale gravitational instabilities, magnetohydrodynamic Parker instabilities, supernova explosions, and protostellar outflows are all conjectured to play a role. Most likely, all provide input at some level to multiscale turbulence. Hence, our modeling results depend entirely on the phenomenological input. One may hope that the output is insensitive to the details of the input. Perhaps the most serious issue is that, whatever mix of physics is relevant in the nearby Universe, there is little reason to be confident that a similar prescription holds at early epochs in the Universe. All predictions for the high redshift Universe should be regarded as qualitative.

**References**

Alexander, D., et al. 2007, preprint
Allen, S. W., Dunn, R. J. H., Fabian, A. C., Taylor, G. B., & Reynolds, C. S. 2006, MNRAS, 372, 21
Banerjee, R., & Pudritz, R. 2006, ApJ, 641, 949
Bertoldi, F., et al. 2003, A&A, 409, L47
Borys, C., Smail, I., Chapman, S. C., Blain, A. W., Alexander, D. M., & Ivison, R. J. 2005, ApJ, 635, 853
Bower, R. G., Benson, A. J., Malbon, R., Helly, J. C., Frenk, C. S., Baugh, C. M., Cole, S., & Lacey, C. G. 2006, MNRAS, 370, 645
Croton, D., et al. 2006, MNRAS, 365, 11
Croft, S., et al. 2006, ApJ, 647, 1040
Everett, J., & Murray, N. 2007, ApJ, 656, 93
Feain, I. J., Papadopoulos, P. P., Ekers, R. D., & Middelberg, E. 2007, ApJ, 662, 872
Gallo, E., Fender, R. P., Miller-Jones, J. C. A., Merloni, A., Jonker, P. G., Heinz, S., Maccarone, T. J., & van der Klis, M. 2006, MNRAS, 370, 1351
Gonzalez, A., Zaritsky, D., & Zabludoff, A. 2007, ApJ, 666, 147
Ho, L. 2007, ApJ, 669, 821
King, A. 2003, ApJ, 596, 27
Klamer, I. J., Ekers, R. D., Sadler, E. M., & Hunstead, R. W. 2004, ApJ, 612, L97
Kravtsov, A. 2003, ApJ, 590, L1
Labiano, A., O'Dea, C. P., Barthel, P. D., & Vermeulen, R. C. 2006, in Highlights of Spanish Astronomy IV, ed. A. Figueras et al., in press (astro-ph/0611738)
Lodato, G., & Natarajan, P. 2006, MNRAS, 371, 1813
Li, Y., Mac Low, M., & Klesse, R. S. 2005, ApJ, 626, 823
Li, Z., Wang, D., & Hameed, S. 2007, MNRAS, 376, 960
McLure, R. J., Jarvis, M. J., Targett, T. A., Dunlop, J. S., & Best, P. N. 2006, MNRAS, 368, 1395
Maraston, C., Daddi, E., Renzini, A., Cimatti, A., Dickinson, M., Papovich, C., Pasquali, A., & Pirzkal, N. 2006, ApJ, 652, 85 (Erratum 2007, ApJ, 656, 1241)
Martin, C. 2005, ApJ, 621, 227
Mac Low, M., & Ferrara, A. 1999, ApJ, 513, 142
Maiolino, R., et al. 2007, A&A, 472, L33
Naab, T., Khochfar, S., & Burkert, A. 2006, ApJ, 636, 81

Pedersen, K., Rasmussen, J., Sommer-Larsen, J., Toft, S., Benson, A. J., & Bower, R. G. 2006, New Astron., 11, 465
Richards, A., et al. 2006, in Latest Results from the Deepest Astronomical Surveys, (San Francisco: ASP), in press (astro-ph/0612075)
Saxton, C., et al. 2005, MNRAS, 359, 781
Schawinski, K., et al. 2007, preprint
Seymour, N., et al. 2006 in Latest Results from the Deepest Astronomical Surveys, (San Francisco: ASP), in press (astro-ph/0612134)
Silk, J. 2005, MNRAS, 364, 133
Silk, J., & Rees, M. 1992, A&A, 331, 1
Slyz, A. D., Devriendt, J. E. G., Bryan, G., & Silk, J. 2005, MNRAS, 356, 737
Spaans, M., & Silk, J. 2006, ApJ, 652, 902
Strickland, D. K. 2007, MNRAS, 376, 523
Tasker, E., & Bryan, G. 2006, ApJ, 641, 878
Thomas, D., et al. 2005, ApJ, 621, 673
Veilleux, S. 2006, in Extreme Starbursts: Near and Far, in press (astro-ph/0603028)
Villar-Martín, M., Humphrey, A., De Breuck, C., Fosbury, R., Binette, L., & Vernet, J. 2007, MNRAS, 375, 1299
Yi, S., et al. 2005, ApJ, 619, L111

## Discussion

*Hammer*: 1) Don't you think that "downsizing" might be naturally explained in the frame of the ΛCDM? It assumes that high masses are preferentially formed earlier in the higher density regions (e.g., Neistein et al. 2006, MNRAS, 372, 933; Mouri & Taniguchi 2007, ApJ, 659, 1752). 2) Which physical mechanism can explain AGN feedback? How is merging related to this?

*Silk*: 1) Downsizing is seen in large-area deep surveys, which are dominated by field galaxies. More significantly, both low mass and massive galaxies are made hierarchically, massive objects forming later, at any given overdensity. 2) AGN feedback is due to outflows whose energy comes from accretion onto supermassive black holes. These impact the protogalaxies if the AGN activity coincides with galaxy formation. Mergers may be responsible for driving the accretion event that feeds the black hole.

*Balcells*: 1) You mentioned tidal disruption as a mechanism to reconcile predictions of dwarf galaxy abundance with observations. Can tidal disruption bring down the numbers of dark matter satellites as well? 2) For many years, we understood that the effect of adding baryons to the DM evolution was adiabatic contraction. I agree that the opposite process, i.e., the softening of DM cores when dense baryon clouds/clusters spiral in due to dynamical friction, needs to be considered as well. Do we know which of the two processes dominates?

*Silk*: 1) Tidal disruption is important for the smaller dark matter clumps and satellites. We see similar physics in action via the production of stellar tidal tails, such as those connected with the Sagittarius dwarf and other, even darker, relics. 2) Adiabatic contraction plays a role initially as the baryons dissipate. Once star formation occurs and supermassive black holes form, a consequence is injection of energy into the gas via supernova explosions and AGN outflows, driving bulk gas motions. Another effect is that the initial baryon collapse is highly asymmetric, resulting in a tumbling, short-lived gaseous bar. The heating effects on the dark matter are dynamical friction, which must help soften the central cusps, and dark matter concentration. However, no realistic models that include both the adiabatic contraction steepening and subsequent dynamical heating have yet been produced.

*Beckman*: 1) The Roche-Pinto et al. measurements of the star formation rate variations occurring on timescales of $\sim 10^9$ yr must be giving us information about scales in the galaxy considerably in excess of 1 kpc, which rather suggests that the star clusters feedback mechanism cannot easily explain them. 2) You quoted D. Thomas et al. (2006) as evidence for the more rapid massive star formation in massive galaxies. You might have quoted Carretero et al. (2004, ApJ, 609, L45; 2006, ApJ, 648, L115), who showed the same result.

*Silk*: 1) The linear shear of cold disk gravitational instability applied to the Milky Way gives a maximum growth rate for a scale of around 1 kpc, I would expect that this scale characterizes the gas reservoir via giant molecular cloud formation for "local" star formation. The rise and fall of the minibursts of star formation in the disk model that I described are controlled by the supply and eventual exhaustion of the gas. I would expect that of order 1 kpc would be the natural scale for any disk star formation rate variations. 2) The recent Thomas et al. (2006) reference is to an SDSS DSR4 analysis. Early work led by D. Thomas on 2004 gave the first quantitative results on the [$\alpha$/Fe] variations with spheroid velocity dispersion (and mass).

From left to right: Joe Silk, S. N. P. Gupta, and Gerhard Hensler.

*Pathways through an Eclectic Universe*
*ASP Conference Series, Vol. 390, © 2008*
*J. H. Knapen, T. J. Mahoney, and A. Vazdekis, eds.*

# The Earliest Star Formation History: Learning from Globular Cluster Systems

Markus Kissler-Patig

*ESO, Karl-Schwarzschild-Str. 2, 85748 Garching, Germany*

**Abstract.** The earliest star formation history is well traced by metal-poor globular clusters. These are the oldest dated objects in the Universe and are present in all giant (and many dwarf) galaxies. What do they tell us about the early phases of the star formation in the Universe? Did this early phase stop before metal-rich populations formed? And did the latter extend in all environments over significant time? How are these phases linked to structure formation? These are the questions that we intend to address in various studies, e.g., by including globular clusters in cosmological simulations; by investigating old metal-poor populations in low surface brightness galaxies, or by studying the element ratios in metal-rich populations.

## 1. Star Clusters as Tracers of the Star Formation History

Let us start with the usual reminder: most, if not all, stars, formed in clusters. Thus, studying the formation history of star clusters will also reveal facts about the star formation history (with some caveats, of course). This in turn sheds light on the galaxy formation and evolution history.

This pathway to understanding galaxy formation and evolution is surely eclectic. Star clusters, though, present many advantages, bearing in mind some caveats. Their age and chemical composition is more easily determined than those of diffuse stellar populations. Thus, they often reveal distinct star formation episodes not seen in the diffuse light of galaxies. Further, globular clusters are the oldest known objects in the Universe, providing insight into the earliest epochs of star formation—the topic of this contribution.

## 2. Star Cluster vs Star Formation: Some Puzzles...

The vast majority of galaxies host many sub-populations of star clusters. But most seem to show two main, distinct populations: an old metal-poor one and an (on average old) metal-rich one. This dichotomy is present in pretty much all galaxies studied to date, hinting at two universal epochs and/or mechanisms of star cluster formation. What are these two epochs/mechanisms? Do they reflect the star formation history of the host galaxy?

In my view, the two main puzzles when connecting the formation history of star clusters with the one of stars are evident in Fig. 1, taken from Harris, Harris, & Poole (1999, their figure 11). It shows, for NGC 5128, the metallicity distributions of the stars (top panel) and the star clusters (bottom panel). The two puzzles are:

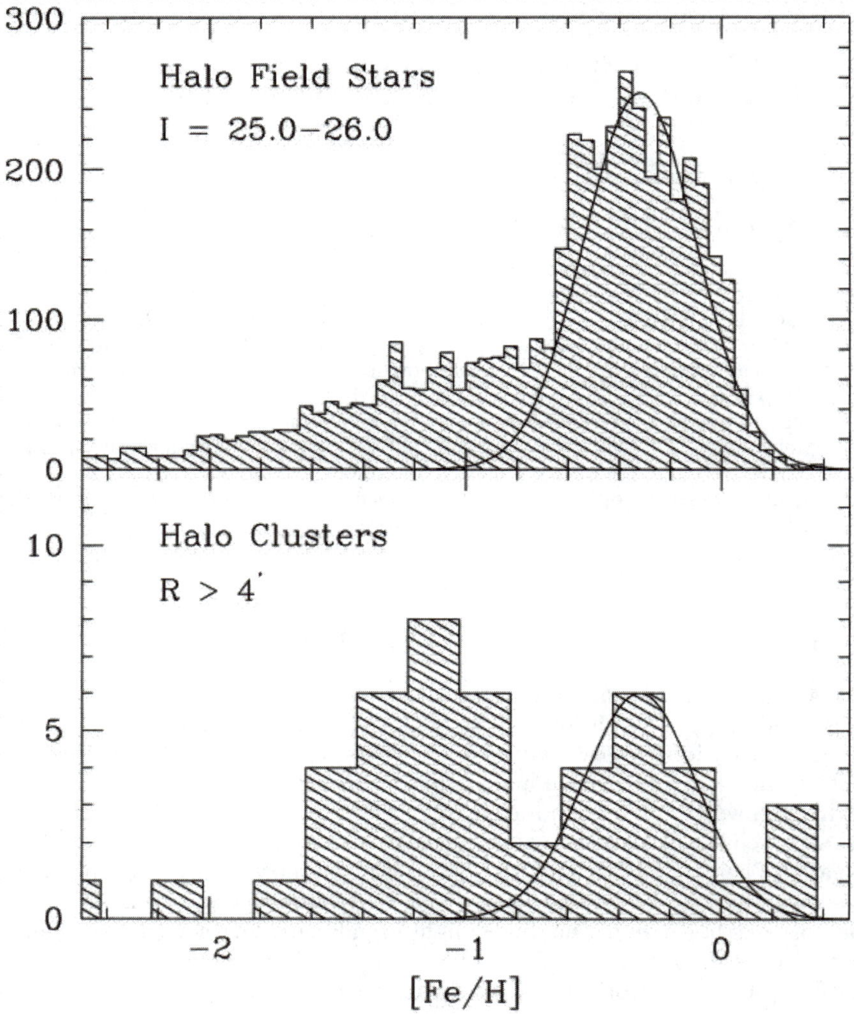


Figure 1. Puzzles... why do these metallicity distributions for stars (*top*) and star clusters (*bottom*) in NGC 5128 look so different, when their formation histories should be coupled?

1. Assuming that stars and star clusters of similar metallicity form at the same epoch, why was the efficiency of star cluster formation with respect to star formation so much higher in the metal-poor regime than in the metal-rich regime? and

2. Why does the metallicity distribution of star clusters show a universal gap, while that for stars does not, but rather a continuous increase?

The answer to these two question will certainly tell us a lot about the formation of stars and star clusters, as well as the earliest epochs of star formation. In order to answer these questions, we are currently following several approaches, outlined below.

## 3. Star Clusters in Cosmological Simulations

Together with Qi Guo and Simon D. M. White, we are attempting to include the formation of star clusters in cosmological simulations. Our starting point is the millennium simulation (Springel et al. 2005). Crudely, star cluster formation occurs in two ways in the simulations: cold gas turns into star clusters (following a Schmidt–Kennicutt type of law), and mergers induce the formation of star clusters. Both mechanisms act with different efficiencies at different epochs. We compare the results of the simulations at redshift $z = 0$ with observational results in the local Universe to set constraints on the mechanisms, epochs, etc.

We have so far not been able to reproduce in a satisfactory manner the properties of local globular cluster systems. Are we missing a special mode of star/star cluster formation at high redshift? Is star cluster formation too poorly understood?

## 4. Star Clusters in Low Surface Brightness Galaxies

Together with Daniela Villegas et al., we were wondering whether the early phase of metal-poor globular cluster formation is indeed universal. In particular, we were curious as to whether old, massive, low surface brightness galaxies also hosted metal-poor (presumably halo) globular clusters, i.e., whether low surface brightness galaxies, which are thought to have followed a very different star formation history than high surface brightness galaxies, shared at least the earliest star formation phases with "normal" galaxies.

We did not expect to find star clusters in galaxies that seemed never to have experienced a violent enough star formation episode to create star clusters. Yet, to our surprise, in our sample of massive galaxies spanning a range of surface brightness, all galaxies have very similar old metal-poor globular cluster populations. We conclude that the properties (numbers, luminosities, colors) of globular cluster systems in low and in high surface brightness galaxies are essentially indistinguishable (details to be found in Villegas et al. 2008).

## 5. Star Clusters with Extreme Chemical Abundances

Finally, we attempt to explore the earliest epochs of cluster formation through their chemical abundances. In particular, we investigated the $\alpha$-element-to-iron ratios in globular clusters together with Thomas H. Puzia and Paul Goudfrooij. These ratios can be used as clocks for the enrichment of the interstellar medium. They reflect the mix of enrichment from Type II supernovae (SNe II, fast evolution) and SNe I (slow evolution). Surprisingly, we have found (exclusively in elliptical galaxies) a metal-rich sub-population with extreme $\alpha$/Fe ratios. The observed ratios essentially forbid any enrichment by later SNe I, and thus point at an extremely early formation—potentially earlier than the formation of metal-poor, halo globular clusters . . . (details to be found in Puzia, Kissler-Patig, & Goudfrooij 2006).

## 6. Some Conclusions . . .

The above studies show that the earliest epochs of star and star cluster formation are still poorly understood, also with respect to the re-ionization epoch of the Universe.

The little that we have learned from the above, as so often, has raised more questions than it has answered. For now we conclude that:

- We seem to be missing in cosmological simulations a crucial point in the earliest phases of star formation. We cannot satisfactorily reproduce the formation of star clusters in the current framework.
- Low and high surface brightness galaxies appear, surprisingly, to share very similar formation epochs/mechanisms of building their old halos; even though their later star formation histories appear to be very different.
- Extreme abundance ratios indicate that the first metal-rich star clusters must have formed within a few 100 Myr of the first stars. Could metal-rich globular clusters have preceded the metal-poor ones?!

Star clusters have not ceased to reveal facts about the earliest epoch of star formation in the Universe. The coming years promise exciting results, both observationally and from simulations.

**Acknowledgments.** I wish to thank all my collaborators on the various parts of the presented work: Qi Guo and Simon White who are introducing me to the secrets of semi-analytic models; Daniela Villegas, Andrés Jórdan, Paul Goudfrooij, and Martin Zwaan who are exploring the low surface brightness Universe with me; and Thomas Puzia and Paul Goudfrooij, with whom I reflect on the meaning of mysterious abundance ratios. Further, warmest thanks to the organizers for such a pleasant conference—and to John Beckman for enlightening our pathways through an eclectic Universe.

### References

Harris, G. H. L., Harris, W. E., & Poole, G. B. 1999, AJ, 117, 855
Puzia, T. H., Kissler-Patig, M., & Goudfrooij, P. 2006, ApJ, 648, 383
Springel, V., et al. 2005, Nature, 435, 629
Villegas, D., Kissler-Patig, M., Jórdan, A., Goudfrooij, P., & Zwaan, M. 2008, AJ, 135, 467

### Discussion

*Yang*: Clusters are formed early in galaxies. And galaxies, as we know, will experience the merging processes. So in this case clusters will be destroyed.

Can we know what fraction of clusters will be destroyed during galaxy mergers and what fraction survive till today?

*Kissler-Patig*: Existing globular clusters are not destroyed in great numbers in mergers, as far as we observe. Thus, mergers will have a small, not a major, influence on existing globular cluster systems. Another question of course is whether mergers form a significant number of star clusters. The answer is yes, but the star/star cluster ratio does not seem significantly unbalanced by mergers.

*Bastian*: Can you comment on the "turn-over" in magnitude (probably also in mass) in the cluster population in low surface brightness galaxies? In particular, the "turn-over" is often interpreted as the preferential disruption of low mass clusters due to the gravitational potential of the host galaxy. But the potential of the low surface brightness galaxy is presumably much lower than in Galactic type objects.

*Kissler-Patig*: Our study (Villegas et al. 2007) shows that low surface brightness galaxies show the "universal" globular cluster luminosity function with high surface brightness galaxies including a peak (i.e., a break in the mass function), as seen in all spirals and ellipticals. If the shape/peak of the cluster luminosity/mass function was heavily influenced by cluster disruption, low surface brightness galaxies would pose a puzzle.

From left to right: Eduardo Battaner, Mónica Relaño, Johan Knapen, Jaime Beckman, John Beckman, and Almudena Zurita.

*Pathways through an Eclectic Universe*
*ASP Conference Series, Vol. 390, © 2008*
*J. H. Knapen, T. J. Mahoney, and A. Vazdekis, eds.*

# On the Origin of Early-type Galactic Nuclei

R. Capuzzo-Dolcetta and P. Miocchi

*Dept. of Physics, Universitá di Roma "La Sapienza", Rome, Italy*

**Abstract.** The ACS Virgo Cluster Survey by Côté and collaborators shows the presence of compact nuclei at the photocenters of many early-type galaxies. It is argued that they are the low mass counterparts of nuclei hosting super-massive black holes detected in bright galaxies. If this view is correct, then one should think in terms of central massive objects, either super-massive black holes or compact stellar clusters that accompany the formation of almost all early-type galaxies. In this observational frame, the hypothesis that galactic nuclei may be the remains of globular clusters driven inward to the galactic center by dynamical friction and there merged, finds an exciting possible confirmation. In this short paper we report on our recent results on globular cluster mergers obtained by means of detailed $N$-body simulations.

## 1. Globular Cluster Merging in Galactic Centers

The Virgo Cluster Survey performed by Côté et al. (2006), using the ACS on the *Hubble Space Telescope*, has provided evidence of the presence of compact nuclei at the photocenters of many early-type galaxies in the cluster, whose luminosity distribution is much better fitted by an extended (King) profile than by a point source (see Böker et al. 2002 for a similar finding in late-type spirals). The half mass radii of the nuclei, $r_{\rm h}$, are in the range $2 < r_{\rm h}\,[{\rm pc}] < 62$, with $\langle r_{\rm h} \rangle = 4.2$ pc, and correlate with the nuclear luminosity: $r_{\rm h} \propto L_{\rm n}^{0.5\pm0.03}$. The mean of the frequency function for the nucleus-to-galaxy luminosity ratio in nucleated galaxies, $\log \eta = -2.49 \pm 0.09$, is indistinguishable from that of the SBH (super-massive black hole)-to-bulge mass ratio, $\log(M_{\rm BH}/M_{\rm gal}) = -2.61 \pm 0.07$, calculated in 23 early-type galaxies.

On these bases, Côté et al. (2006) argue that resolved stellar nuclei are the low mass counterparts of nuclei hosting SBHs detected in bright galaxies. If this view is correct, then one should think in terms of central massive objects, either SBHs or compact stellar clusters (CSCs), that accompany the formation and/or early evolution of almost all early-type galaxies.

It is clear that these characteristics of the nuclei of early-type galaxies fit well into a scenario of multiple globular clusters (GCs) merging in the inner galactic regions, corresponding to a "dissipationless" scenario as an alternative to the "dissipational" scenario, the latter being based mainly on speculative hypotheses (see, for example, Babul & Rees 1992) supported by just a few quantitative results (Mihos & Hernquist 1994). The validity of the "dissipationless" hypothesis first requires a detailed $N$-body simulation of the interaction of stellar clusters in the inner regions of their parent galaxy, taking into account both cluster–cluster and cluster–galaxy dynamical interactions. The latter depend on tidal distortion, acting on the cluster internal motion, and on dynamical friction, acting on the cluster orbital motion. Not many simulations have been presented in the literature dealing with the possible formation scenarios for nuclear clusters (NCs). Among these, we highlight those by Fellhauer & Kroupa (2005), who find that super-massive star clusters, such as W3 in NGC 7252, are very likely the merging remnants of objects spanning a range of masses from the $\omega$ Cen scale to that of the smallest dwarf ellipticals. Interestingly, the resulting super cluster is a stable and bound object, whose density profile is well fitted by a King profile, even if mass loss of the merger product occurs at every perigalacticon passage. Bekki et al. (2004) examined a merger formation scenario in which NCs are the product of multiple mergings of star clusters in the central regions of galaxies. The simulations of Bekki et al. are done in a simplified way, neglecting the role of the galactic external potential and with a relatively low resolution. Therefore, it is not clear how realistic their interesting results are that show (i) the full merging of the interacting clusters in a significantly short time ($\sim 10^7$ yr), (ii) the triaxiality in the merger configuration, and (iii) the existence of scaling relations among velocity dispersion, luminosity, effective radius, etc.

## 2. The Simulations

Here, we report briefly on some of the results of R. Capuzzo-Dolcetta & P. Miocchi (in preparation), where we study whether and how the merging of various massive GCs decayed by dynamical friction in the inner galactic region may occur.

The main questions to answer are: (i) given some (realistic) initial conditions for a set of GCs which experienced a significant orbital decay, whether they are undergoing a full merger; (ii) if so, how much time is needed; (iii) what the final structure of the merged supercluster is; and (iv) whether it attains a quasi-steady state.

The answers to these questions are of overwhelming importance for lending substance to the interpretation of the formation of early-type galaxy nuclei via merging of decayed GCs, a hypothesis first raised by Tremaine, Ostriker, & Spitzer (1975) and subsequently studied extensively by Capuzzo-Dolcetta and collaborators (see Capuzzo-Dolcetta & Vicari 2005, and references therein). We consider GCs as $N$-body systems moving within a triaxial galaxy represented by an analytical potential, subject also to the deceleration due to dynamical friction. We studied the merging process occurring among four GCs already decayed in the inner region of the galaxy. The galaxy where the GCs move is represented by a self-consistent triaxial potential. In this potential, the dynamical friction yields decay times significantly shorter than the Hubble time (Capuzzo-Dolcetta & Vicari 2005). We have simulated four massive GCs having an initial King profile with total mass, central velocity dispersion, and core radius in the range $M = 42 - 54 \times 10^6\, M_\odot$, $\sigma_0 = 25 - 36\,\mathrm{km\,s^{-1}}$, and $r_\mathrm{c} = 2.2$–$3.8\,\mathrm{pc}$. They are initially located within 100 pc from the galactic center (Fig. 1). Simulations are done with the parallel "ATD" $N$-body code (Miocchi & Capuzzo-Dolcetta 2002), with a total of $10^6$ particles.

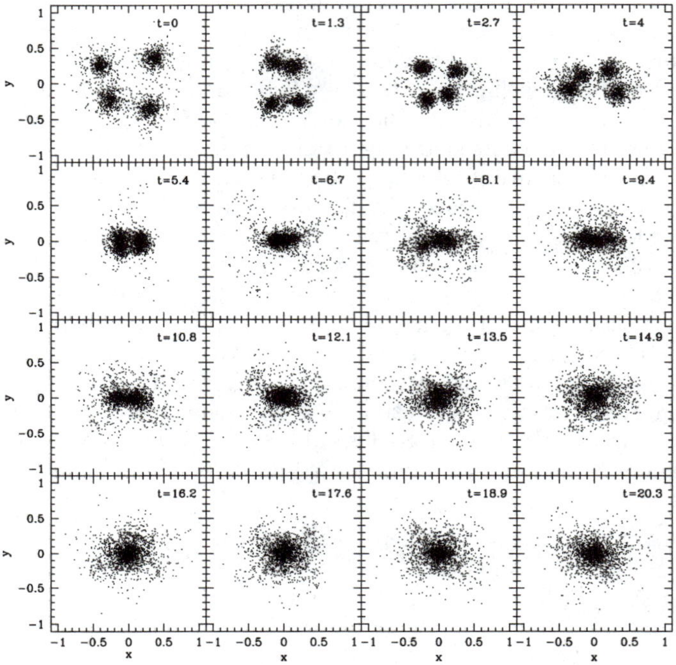


Figure 1. Snapshots of the cluster evolution (projected on the $x$–$y$ plane) from $t = 0$ to $t = 20.3$ Myr. Lengths are in units of 200 pc and time is in Myr.

## 3. Results

The merging occurs rather quickly: after ~15 Myr, corresponding to ~20 galactic core crossing times ($t_{\rm cross}$), the merging is completed and the resulting system attains a quasi-equilibrium configuration (see Fig. 1). The total projected density profile in the nuclear region is remarkably similar to those recently observed in the central regions of nucleated early-type galaxies in the Virgo cluster (Fig. 2). The final CSC morphology is that of an axisymmetric ellipsoid (axial ratios 1.4:1.4:1, ellipticity ~0.3) without figure rotation. In the velocity dispersion–mass plane, the CSC is located closer to the scaling relation followed by GCs than to that of elliptical galaxies. This is due to our system being located deep inside the galaxy potential.

## 4. Conclusions

We have studied the modes of interaction of four globular clusters which, due to their large initial masses, have experienced a significant dissipation of their orbital energy by dynamical friction caused by the galactic triaxial field. The GC motion in the inner galactic region is characterized by a further barycenter braking due to the galactic tidal torque, followed by a full merger that is com-

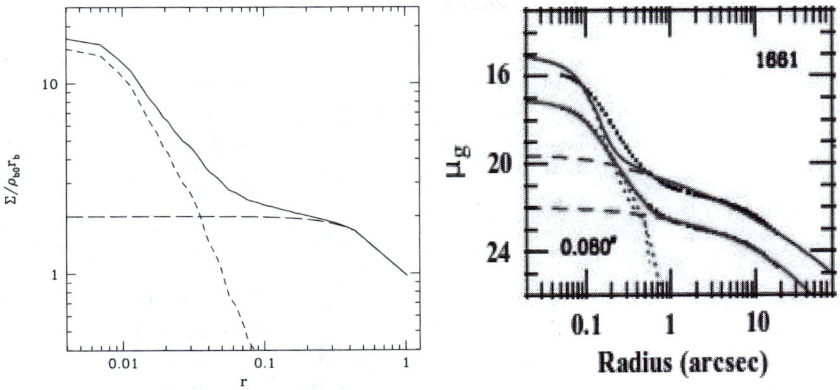


Figure 2. *Left:* Surface density profile (in proper units) for our last computed CSC configuration (*short dashed line*), overlapping the galaxy profile (*long dashed*) so as to give the total density (*solid line*). *Right panel:* The $g$-band surface brightness profile of NGC 1661, where the lower solid line is the King model fit to the data (*black dots*) and the upper solid line is the point source fit. The dashed line is the galaxy contribution. All the curves are translated for display convenience. The median half light radius (0.080″) is reported. (*From Côté et al. 2006.*)

pleted in $\sim 20\, t_{\rm cross}$. The merger result retains some of the characteristics of the pre-existing objects, leading to a single cluster with a King density profile plus a halo which is diffusing in the external field. Because of this, we can speak with some confidence of a secularly stable CSC. Even in the limits of our simulations (both on the number of merging objects and on the integration time, which covers $\sim 45\, t_{\rm cross}$), we may extrapolate from our results that the nuclei of many early-type elliptical galaxies may have formed by the merging of a few tens of GCs massive enough to have decayed into the inner galactic regions in a time short compared to the Hubble time and compact enough to survive to the tidal stresses.

## References

Babul, A., & Rees, M. J. 1992, MNRAS, 255, 346
Bekki, K., Couch, W. J., Drinkwater, M. J., & Shioya, Y. 2004, ApJ, 610, L13
Böker, T., Laine, S., van der Marel, R. P., Sarzi, M., Rix, H.-W., Ho, L. C., & Shields, J. C. 2002, AJ, 123, 1389
Capuzzo-Dolcetta, R., & Vicari, A. 2005, MNRAS, 356, 899
Côté, P., et al. 2006, ApJS, 165, 57
Fellhauer, M., & Kroupa, P. 2005, MNRAS, 359, 223
Mihos, J. C., & Hernquist, L. 1994, ApJ, 437, L47
Miocchi, P., & Capuzzo-Dolcetta, R. 2002, A&A, 382, 758
Tremaine, S., Ostriker, J. P., & Spitzer, L. Jr. 1975, ApJ, 196, 407

## Discussion

*Böker*: 1) A comment: the nuclear star clusters of late-type spirals are not young. Many of them do have a young stellar population, but only in addition to a much more massive, old (~10 Gyr) population. This implies that the clusters have been in place for a long time but get "rejuvenated" regularly. 2) A question: it seems plausible that the very compact ($r_e \sim$ 2–5 pc) stellar nuclei can be produced by a series of mergers of GCs, each of which has a similar size. Can you comment on this?

*Capuzzo-Dolcetta*: 1) Thanks for your comment, which actually reduces the difference between nuclear star clusters in late-type spirals and those in many early-type galaxies. 2) The answer to your question is that the $N$-body simulations of the merging of massive GCs show that after the merging event in the inner galaxy potential well the central density is just slightly less (a factor of 1.7) than the sum of the individual GC central densities, and the half mass radius is correspondingly slightly larger (a factor of five). This makes it plausible to form not-too-compact nuclei by merging GCs; of course everything should be tested by more extended simulations (in time and in number of merging objects).

Roberto Capuzzo-Dolcetta.

*Pathways through an Eclectic Universe*
*ASP Conference Series, Vol. 390, © 2008*
*J. H. Knapen, T. J. Mahoney, and A. Vazdekis, eds.*

# Galactic Disk Warps due to Intergalactic Accretion Flows onto the Disk

M. López-Corredoira,[1] J. Betancort-Rijo,[1,2] and J. E. Beckman[1,3]

[1] *Instituto de Astrofísica de Canarias, E-38200 La Laguna, Spain*
[2] *Departamento de Astrofísica, University of La Laguna, Spain*
[3] *Consejo Superior de Investigaciones Científicas (CSIC), Spain*

**Abstract.** The accretion of the intergalactic medium onto the gaseous disk is used to explain the generation of galactic warps. A cup-shaped distortion is expected, due to the transmission of the linear momentum, but this effect is small for most incident inflow angles and the predominant effect turns out to be the transmission of angular momentum, i.e., a torque giving an integral-sign shaped warp. The torque produced by a flow of velocity $\sim 100\,\mathrm{km\,s^{-1}}$ and baryon density $\sim 10^{-25}\,\mathrm{kg\,m^{-3}}$, which is within the possible values for the intergalactic medium, is enough to generate the observed warps and this mechanism offers quite a plausible explanation. The inferred rate of infall of matter, $\sim 1\,M_{\odot}\,\mathrm{yr}^{-1}$, to the Galactic disk that this theory predicts agrees with the quantitative predictions of chemical evolution resolving key issues, notably the G-dwarf problem.

Sánchez-Salcedo (2006) suggests that this mechanism is not plausible because it would produce a dependence of the scale height of the disk with the galactocentric azimuth in the outer disk, but rather than being an objection, this is another argument in favor of the mechanism because this dependence is actually observed in our Galaxy.

## 1. The Mechanism

In López-Corredoira et al. (2002, hereafter LBB), we proposed a mechanism to explain the formation of warps (S- or U-shaped, or a combination of both) in spiral galaxies: the accretion of the intergalactic medium onto the disk. In a Milky Way-like galaxy, the mean density of baryonic matter in the intergalactic medium needed to produce the observed warp is around $10^{-25}\,\mathrm{kg\,m^{-3}}$ when the infall velocity at large distance is $\sim 100\,\mathrm{km\,s^{-1}}$ (LBB). These numbers were confirmed independently by Sánchez-Salcedo (2006). This hypothetical low density net flow is a very reasonable physical assumption and would explain why most spiral galaxies are warped. No massive halo is necessary, neither are high values of magnetic fields or satellite companions, and the presence of these elements would not modify the present conclusions qualitatively.

## 2. Why Accretion onto the Disk Rather than into the Halo?

If the halo axis were misaligned with respect to the disk axis due to the accretion of intergalactic matter onto the disk, the disk would be torqued by the halo and this would also give rise to warps (Ostriker & Binney 1989). However, our

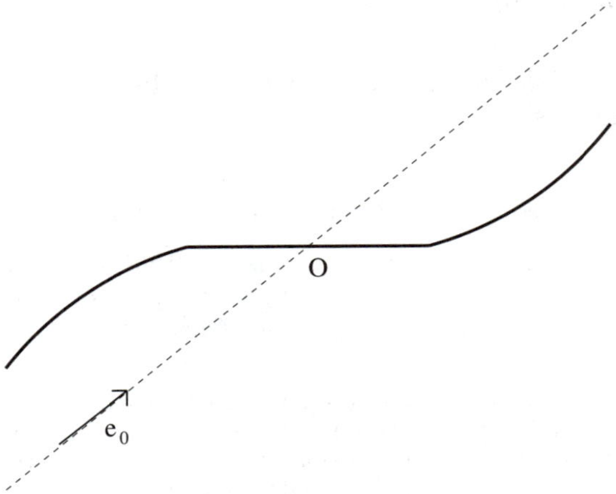


Figure 1. Graphical representation of the orientation of the S-warp with respect to the velocity of incoming flow. The azimuthal angle, $\phi_W$, with maximum amplitude of the warp, which is not plotted, is the same as the azimuthal angle of $\vec{e_0}$.

opinion is that our LBB mechanism is preferable. The direct accretion of matter by the disk is more plausible than accretion by the halo because the low density of baryonic matter in the halo produces a very small amount friction to trap intergalactic flows. Moreover, there is no correlation between the amplitude of the warps and the mass of the halo (derived from rotation curves; Castro-Rodríguez et al. 2002). Lenticular galaxies, which have disks with no gas, do not show warps (Sánchez-Saavedra et al. 2003). If the mechanism to produce warped disks were a purely gravitational interaction, such as the halo–disk interaction proposed by Ostriker & Binney (1989), it would not distinguish between gas and stars, and lenticulars might also have a warp. However, it seems that the gas in the disk plays an important role, in favor of our mechanism, which requires the friction with the disk gas to trap the intergalactic matter, although other explanations in terms of lower mass of the halo in lenticulars are in fact possible.

In any case, even if the accretion into the halo were a possible mechanism, the continuous accretion of low metallicity gas onto the disk is also a fact (Tinsley 1980), so the mechanism proposed here can always act.

## 3. Different Types of Observations Successfully Explained by This Hypothesis

The accretion of intergalactic flows with these characteristics gives rise to:

- Good agreement with the observations of the chemical evolution of the Milky Way by contributing $\sim 1\,M_\odot\,\mathrm{yr}^{-1}$ of low metallicity gas to the disk (LBB). This turns out to be of the order of the accretion rate required to resolve the G-dwarf problem in our Galaxy (Tinsley 1980), as well as explaining a number of phenomena of chemical evolution that require the long term infall of low metallicity gas.

- The frequency of warps and its amplitude is dependent on environment (García-Ruiz, Sancisi, & Kuijken 2002). The most isolated galaxies are more frequently warped, although with less amplitude. It seems clear that warping is due to something related to the environment rather than to the intrinsic properties of the galaxies, to something which is not related to the proximity of other galaxies. The accretion of intergalactic matter onto the disk seems a good candidate to explain these observational facts.

- An older version of the present mechanism was proposed by Kahn & Woltjer (1959), but this was rejected because it predicted only cup-shaped (U-shape) warps. This is no longer a valid criticism to the present model. Integral-shape (S-shape) warps with the amplitude and shape observed for instance in the Milky Way are reproduced by this theory as a consequence of the transmission of angular momentum (LBB). Even if the global angular momentum of the intergalactic medium is nil, the redistribution of momentum in different rings of the galaxy due to gravity produces a net torque in each ring (LBB). Figure 1 illustrates the relationship between the wind velocity and the S-shape warp.

- Rather than being a criticism, the fact that this scenario produces U-shaped warps is now an argument in its favor. U-shaped warps are explained by this mechanism as being due to a transmission of linear momentum (LBB), while the S-shaped warps are due to the transmission of angular momentum. LBB predict that the frequency of U-shaped warps should be lower than that of S-warps, only occurring when the flow is accreted from nearly polar directions (low angle with the rotational axis of the disk). This was corroborated with a different method of calculation devised by Sánchez-Salcedo (2006). This lower frequency of U-warps over S-warps is observed (García-Ruiz et al. 2002; Sánchez-Saavedra et al. 2003). The asymmetric cases can also be explained by the present theory as a combination of S-warps and U-warps (LBB; Saha & Jog 2006). Up to now, there has been no alternative explanation for the U-warps or the asymmetric warps.

- Sánchez-Salcedo (2006) raises the criticism that this mechanism is not plausible because it would produce a dependence of the scale height of the disk with galactocentric azimuth ($\phi$; defined to be zero in the Sun–galactic center line) in the outer disk. Rather than being an objection, however, it is another argument in favor of our model because this dependence is actually observed in our Galaxy: Voskes & Burton (2006, their figure 15) and Levine, Blitz, & Heiles (2006, their figure 5) have shown that the scale height of the outer disk ($R > 20\,\mathrm{kpc}$) is 2–3 times higher on average for $0 < \phi < 180°$ than for $180° < \phi < 360°$. This is in agreement with the expectations of our model: the wind comes from the direction of the southern warp, $\phi \approx 270°$ [to produce a U-warp northwards which causes a smaller amplitude of the (S+U)-warp of the southern part], so a higher pressure is expected for the region around the southern warp and consequently a lower thickness therein. This effect is expected to appear in the outer disk because here the pressure dominates over the self-gravitation of the disk. Sánchez-Salcedo (2006) derived roughly a factor of 6 with our LBB mechanism between the minimum and maximum scale height at $R = 22\,\mathrm{kpc}$. There are differences as high as this in Levine et al. (2006, their figure 4), but this factor might be somewhat lower for other reasons. For instance, if we take into account the response time ($t_{\mathrm{response}}$) of the disk and the rotation ($\omega(R)$): the scale height corresponds to the average pressure between $[\phi - \omega(R)t_{\mathrm{response}}]$ and $\phi$. The fact that the intergalactic medium might not be continuous but with some clouds would change the factor. Also the variation of some free and unknown parameters such as the angle of the intergalactic flow with respect to the rotation axis of the galaxy ($\theta_0$ in LBB) produces variations in the factor.

## 4. Conclusions

Several mechanisms can generate warps: intergalactic magnetic fields (Battaner, Florido, & Sánchez-Saavedra 1990), gravitational interaction with satellites, or with the halo misaligned by an accretion of intergalactic matter into it (Ostriker & Binney 1989); or our proposed mechanism in LBB of accretion of intergalactic flows onto the disk. This last option seems to offer a very plausible scenario: it is quantitatively consistent with many observations and works, independently of other ingredients of galaxies and their structure. A key advantage of the LBB theory is that it explains facts like U-shaped warps or variations in scale height with azimuth that other theories do not explain.

### References

Battaner, E., Florido, E., Sánchez-Saavedra, M. L. 1990, A&A, 236, 1
Castro-Rodríguez, N., López-Corredoira, M., Sánchez-Saavedra, M. L., & Battaner, E. 2002, A&A, 391, 519
García-Ruiz, I., Sancisi, R., & Kuijken, K. 2002, A&A, 394, 769
Kahn, F. D., & Woltjer, L. 1959, ApJ, 130, 705
Levine, E. S., Blitz, L., & Heiles, C. 2006, ApJ, 643, 881
López-Corredoira, M., Betancort-Rijo, J., & Beckman, J. E. 2002, A&A, 386, 169 [LBB]
Ostriker, E. C., & Binney, J. J. 1989, MNRAS 237, 785
Saha, K., & Jog, C. J. 2006, A&A, 446, 897
Sánchez-Saavedra, M. L., Battaner, E., Guijarro, A., López-Corredoira, M., & Castro-Rodríguez, N. 2003, A&A 399, 457
Sánchez-Salcedo, F. J. 2006, MNRAS, 365, 555
Tinsley, B. M. 1980, Fund. Cosmic Phys., 5, 287
Voskes, T., & Burton, W. B. 2006 (astro-ph/0601653)

## Discussion

*Pagel*: Another chemical evolution-related reason for postulating inflow may arise from the high (near primordial) deuterium abundance found in parts of the interstellar medium.

*López-Corredoira*: Yes, indeed this kind of evidence inspired the elaboration of our theory.

*Trujillo*: There is increasing evidence that galaxies are not aligned randomly but oriented with the spin vectors preferentially parallel to the surface of the large scale structures. Would you expect to find galaxies in these structures having a similar warp pattern?

*López-Corredoira*: Yes, if there is a common direction of the average velocity in the accretion for each galaxy. A common direction of the extragalactic magnetic field, for the explanation of warps generated by magnetic fields, would also lead us to expect it.

*James*: Does this analysis require all the accreted material, $1\,M_\odot\,\mathrm{yr}^{-1}$, to approach the disk from the same direction?

*López-Corredoira*: $1\,M_\odot\,\mathrm{yr}^{-1}$ is the total amount accreted by all the rings, yes, and in our rough approach we have considered there to be an "average" direction from which the whole flow comes. This is only the mean direction; there may be flows from other directions due to turbulence, but we calculate the average effect of the accretion.

Eduardo Battaner (*left*) and Martín López-Corredoira.

*Pathways through an Eclectic Universe*
*ASP Conference Series, Vol. 390, © 2008*
*J. H. Knapen, T. J. Mahoney, and A. Vazdekis, eds.*

# Total Contribution of the Warm/Hot Intergalactic Medium to the X-ray Background

Andrzej M. Sołtan

*Nicolaus Copernicus Astronomical Center, Bartycka 18, 00-716 Warsaw, Poland*

**Abstract.** The interaction of the infalling and escaping matter around galaxies generates areas of the warm/hot intergalactic medium with temperatures of $10^5$–$10^7$ K. These high temperature regions emit X-rays that enhance the fluctuating component of the background. Variations in the background at angular scales of 0.5–8 arcmin are analyzed using *XMM–Newton* observations. The amplitude of the autocorrelation function generated by a non-uniform distribution of sources contributing to the background is discussed. An upper limit for the fluctuations due to the warm/hot component is obtained.

## 1. Introduction

The density of the luminous baryonic matter in the local Universe amounts roughly to 0.02 of the critical matter density, $\Omega_{\rm cr}$, whereas the total baryonic matter exceeds $0.04\,\Omega_{\rm cr}$. A resolution of this *local baryon deficit* was given by Cen & Ostriker (1999), who investigated the evolution of the primordial intergalactic matter which at moderate redshifts produces the Lyman-$\alpha$ forest. According to their simulations, a diffuse matter component slowly flows towards large concentrations of matter, such as galaxies. The rate at which the matter accumulates in galaxies is defined by the structure of the gravitational potential and the interaction of the infalling gas with the galactic wind. Although the details of this process are not well recognized, a coherent general picture emerges from hydrodynamic simulations carried out by several groups (e.g., Davé et al. 2001; Bryan & Voit 2001; Croft et al. 2001). In the areas surrounding galaxies the infalling material is heated by shocks and enriched with heavy elements. In effect, galaxies in the local Universe are submerged in bubbles of hot plasma with temperatures of $10^5$–$10^7$ K, the so called warm/hot intergalactic medium (WHIM).

Theoretical considerations are unable to give the detailed description of the spatial distribution of the WHIM. Neither are plasma-physical parameters (density, temperature and metal abundances) well constrained by simulations. In this paper, observational methods effective in searches for WHIM emission are discussed. At temperatures above $\sim 10^6$ K the WHIM is a source of soft X-ray emission. Spatial variations of the WHIM temperature and density make this emission highly non-uniform. I give a brief summary of the analysis of the soft X-ray background (XRB) fluctuations using the correlation functions. Results of the correlation analysis of the WHIM component are then presented and first quantitative estimates of the WHIM contribution to the XRB are obtained.

## 2. X-Ray Background Fluctuation Measures

Fluctuations in the XRB surface brightness have been investigated using the *ROSAT* All-Sky Survey (Sołtan et al. 1999) and the PSPC pointing observations (Sołtan, Freyberg, & Trümper 2001). The autocorrelation function (ACF) was determined for separations between 2 arcmin and ~10 deg. The XRB is produced predominantly by extragalactic point-like sources, mostly various classes of AGN. A non-negligible fraction of the XRB is generated by extended sources, viz. clusters of galaxies. Thus, at angular scales exceeding the width of the instrumental point spread function, the ACF signal results from the clustering of the point sources and from clusters. The ACF based on the *ROSAT* observations was estimated at energies above ~1 keV. Since the fluctuations of the XRB produced by the WHIM are expected to appear mostly below this energy threshold, the *ROSAT* data were only marginally sensitive to the WHIM.

The cross-correlation function (CCF) of the XRB and the galaxy distribution was measured by Sołtan et al. (1997) and Sołtan, Freyberg, & Hasinger (2002) using the *ROSAT* observations and by Sołtan, Freyberg, & Hasinger (2005) and Sołtan (2006) using *XMM–Newton* EPIC MOS data. The correlation signal of fluctuations generated by the WHIM is expected to decay rapidly at separations above ~2 arcmin and the *ROSAT* observations gave only indications of the emission by the WHIM component. The better angular and energy resolution of the X-ray telescope/detector combination of the *XMM–Newton* satellite has allowed quantitative estimates to be made of the WHIM emission correlated with the galaxy population (see below). The question of whether the WHIM signal clearly visible in the CCF generates the XRB fluctuations detectable also in the ACF is discussed in the last section of the paper.

## 3. WHIM Emission Correlated with Galaxy Distribution

The distribution of counts in a large number of *XMM–Newton* pointings has been correlated with the distribution of galaxies with $17 < m_b < 20$. The galaxy sample was selected from the APM Galaxy Survey (Maddox et al. 1990) supplemented with galaxies extracted from the Minnesota Automated Plate Scanner (MAPS).[1] The average redshift of the galaxies used in the investigation is $z = 0.12$.

To facilitate the separation of the expected soft signal of the WHIM from the much harder contribution from AGN and clusters, the X-ray data have been split into five energy bands: 0.3–0.5 keV, 0.7–1.0 keV, 1.0–1.35 keV, 1.9–3.0 keV, and 3.0–4.5 keV. In all the energy bands an extended excess emission surrounding the galaxies is visible. The typical soft X-ray halo around an individual galaxy has a radius of ~300 kpc. The average observed fluxes and luminosities corresponding to this emission, assuming the redshift of 0.12, are given in Table 1.

Although the CCF shows clear enhancements of the XRB surface brightness around galaxies, the WHIM signal is low in comparison to the total XRB flux in all the energy bands. The surface brightness of the individual galaxy halo

[1]The MAPS database is supported by the University of Minnesota and is available at `http://aps.umn.edu/`.

Table 1. Average extended emission around the sample galaxies

| Energy band [keV] | Excess flux [erg s$^{-1}$ cm$^{-2}$] | Luminosity [erg s$^{-1}$ keV$^{-1}$] |
|---|---|---|
| 0.3–0.5 | $8.7 \times 10^{-16}$ | $1.6 \times 10^{41}$ |
| 0.7–1.0 | $7.4 \times 10^{-16}$ | $9.2 \times 10^{40}$ |
| 1.0–1.35 | $3.4 \times 10^{-16}$ | $3.7 \times 10^{40}$ |
| 1.9–3.0 | $6.1 \times 10^{-16}$ | $2.1 \times 10^{40}$ |
| 3.0–4.5 | $6.8 \times 10^{-16}$ | $1.7 \times 10^{40}$ |

amounts just to 7% of the XRB in the 0.7–1.0 keV energy band and is below even this value in the remaining bands. Consequently, the expected fluctuations in the integral XRB induced by the WHIM component correlated with galaxies are small.

## 4. Autocorrelation Function of the XRB on Small Scales

The ACF of the XRB has been estimated using a large set of *XMM–Newton* EPIC MOS observations. The data suitable for the analysis have been carefully selected from the *XMM* archives. One should note that the available X-ray observations are strongly affected by instrumental effects and contaminated by a particle background. In order to ensure high homogeneity of the data, only the central CCD detectors of the MOS cameras have been used. Consequently, the calculations have been performed for separations below 8 arcmin. The detailed procedures applied in the analysis to isolate the ACF signal will be described elsewhere.

Because the contribution to the XRB by different classes of sources is a function of energy, also the fluctuation amplitude is expected to depend on the energy band. In particular, at energies below ~1 keV, emission by the WHIM and the hot plasma in the Galaxy generate an increasing fraction of the XRB, and the ACF amplitude is expected to show this effect. Above ~1 keV, however, the entire ACF amplitude is undoubtedly generated mostly by clustering of AGN and the ACF signal should be constant.

### 4.1. AGN Contribution to the XRB Fluctuations

The correlation analysis in three hard energy bands gave identical results. The ACF amplitude is also consistent with the amplitude extrapolated from the larger separations determined using the *ROSAT* data band R6 centered at 1.1 keV (Sołtan et al. 2001). The ACF based on the combined data is shown in Figure 1. The apparent constant relative amplitude of the XRB fluctuations at energies above ~1 keV supports a straightforward interpretation, namely that these fluctuations arise from inhomogeneities in the spatial distribution of the sources generating the XRB.

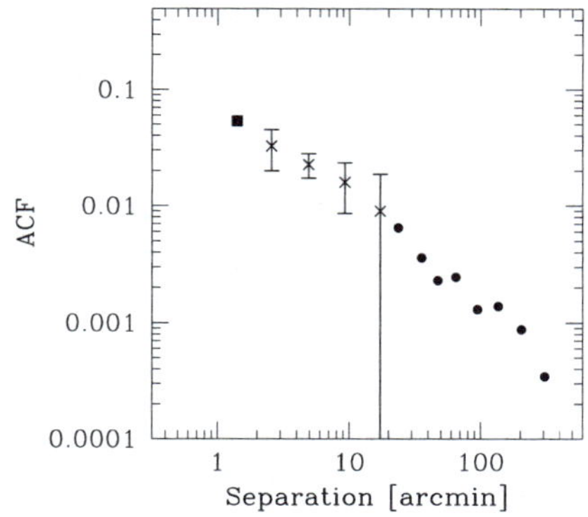


Figure 1. The ACF of the combined *ROSAT* and *XMM–Newton* observations: *ROSAT* All-Sky Survey—dots, *ROSAT* pointings—crosses, *XMM* pointings—square.

### 4.2. WHIM Contribution to the XRB Fluctuations

The relative amplitude of fluctuations *below* 1 keV is *smaller* than that at higher energies. This is consistent with the conjecture that an increasing fraction of the XRB is produced by hot gas in the Galaxy. To assess the WHIM contribution to the observed fluctuations, one needs to estimate what fraction of the XRB at these low energies is produced by AGN. The present data are insufficient to give a satisfactory answer. Simple extrapolation of the power law which approximates the average AGN spectrum above 1 keV generates fluctuations smaller than actually observed. In this case, the fraction of the ACF amplitude attributed to the WHIM would be substantially greater than the fluctuations found using the CCF analysis. However, it is well established that most AGN exhibit a softening of their spectra at low energies. Calculations show that a moderate steepening of the average spectrum of sources contributing to the XRB would reproduce the observed amplitude of the ACF. Consequently, no excess WHIM emission above that detected in the cross-correlation analysis would be required.

**Acknowledgments.** This work has been partially supported by the Polish KBN grant 1 P03D 003 27.

### References

Bryan, G. L., & Voit, G. M. 2001, ApJ, 556, 590
Cen, R., & Ostriker, J. P. 1999, ApJ, 514, 1
Croft, R. A. C., Di Matteo, T., Davé, R., Hernquist, L., Katz, N., Fardal, M. A., & Weinberg, D. H. 2001, ApJ, 557, 67
Davé, R. et al. 2001, ApJ, 552, 473
Maddox, S. J., Sutherland, W. J., Efstathiou, G., & Loveday, J. 1990, MNRAS, 243, 692
Sołtan, A. M. 2006, A&A, 460, 59
Sołtan, A. M., Freyberg, M. J., & Hasinger, G. 2002, A&A, 395, 475

Sołtan, A. M., Freyberg, M. J., & Hasinger, G. 2005, A&A, 436, 67
Sołtan, A. M., Hasinger, G., Egger, R., Snowden, S., & Trümper, J. 1997, A&A, 320, 705
Sołtan, A. M., Freyberg, M., Hasinger, G., Miyaji, T., Treyer, M., & Trümper, J. 1999, A&A, 349, 354
Sołtan, A., Freyberg, M. J., & Trümper, J. 2001, A&A, 378, 735

## Discussion

*Beckman*: 1) Can you give us an estimate of the X-ray luminosity of a single galaxy as observed in its surrounding WHIM (you gave an estimate of the mass in the WHIM at $3.7 \times 10^{11}\ M_{\odot}$)? 2) What is thought to be the physical mechanism producing this emission?

*Sołtan*: 1) An individual WHIM halo produces, in the soft band, less than 2 counts in a 10 ks observation. At the median redshift of the galaxies in the sample of 0.12, it gives a luminosity of $\sim 10^{40}\,\mathrm{erg\,s^{-1}}$. 2) The WHIM interacts with the matter expelled from galaxies. It is non-gravitationally heated by the feedback processes. Simulations show that these interactions generate shocks which heat the intergalactic material and increase its metallicity.

*Sarzi*: $10^6$ K gas also powers coronal line emission in the FUV, such as O VI$\lambda\lambda$ 1032, 1038. Have you considered whether such emission from the WHIM could be detected as FUV galactic halos, or if it can be important for the FUV background?

*Sołtan*: The O VI line has been detected in the absorption spectrum of a nearby AGN, and it was attributed to a presence of a foreground WHIM cloud. I cannot give you an immediate answer as to whether the WHIM emission in this line would be measurable.

*Silk*: How do you exclude a contribution from soft X-ray emission intrinsic to galaxies, such as X-ray sources or hot ISM, that could contaminate your WHIM signal?

*Sołtan*: The cross-correlation amplitude has in fact a very strong peak at small separations (in all the energy bands), indicating sources within galaxies. However, the angular profile of this emission is narrow and practically indistinguishable from the PSF. These point-like sources had no major effect on the analysis of the extended emission.

*Pathways through an Eclectic Universe*
*ASP Conference Series, Vol. 390, © 2008*
*J. H. Knapen, T. J. Mahoney, and A. Vazdekis, eds.*

# Infall and Accretion

F. Combes
*LERMA, Observatoire de Paris, 61 Av. de l'Observatoire, F-75014, Paris, France*

**Abstract.** Gas infall and accretion play a fundamental role in galaxy formation, and several processes of accretion are reviewed. In particular cold accretion may solve to some extent the angular momentum problem in disk formation, while it is aggravated by mergers. Gas accretion is one of the main actors in secular evolution: it is required to account for recurrent bar formation, and to explain the feedback cycles of formation of bulges and black holes, with correlated masses. Infall is also required to fuel a regular and almost stationary star formation history. Star formation is quenched for galaxies in clusters when gas accretion is suppressed through stripping. The brighter central galaxy can benefit, however, from gas accretion through cooling flows moderated by AGN feedback. Hot and cold feedback scenarios can be considered, to account for a stationary cooling flow, and to explain the observed filamentary CO and H$\alpha$ structures.

## 1. Cold Gas Accretion on Galaxies

According to the conventional scenario of $\Lambda$CDM hierarchical galaxy formation, mass is essentially assembled by mergers, and the baryonic gas is shock-heated to the virial temperature of the total structure ($10^6$ K for a Milky Way-type galaxy), before being accreted, until the hot gas cools down. This scenario ignores the possibility of dense gas being present in contrasted structures, which can cool much more quickly and radiate its gravitational energy while remaining cold. Numerical simulations with enough resolution now clearly show two modes of accretion (Keres et al. 2005): hot gas accretion, mainly around massive structures, where shocks can be violent, and cold gas falling in along filaments, the fraction of cold gas being larger in low mass halos ($M_{\rm CDM} < 3 \times 10^{11}\, M_\odot$). Cold gas accretion dominates at high redshift. In addition, the star formation rate history intimately follows the total gas accretion, as shown in Figure 1.

The observation of a clear bimodality between galaxies—a blue sequence still accreting gas and forming stars, while a red sequence consists of massive red galaxies, in general in a rich environment, where gas and new stars are absent—has been interpreted by the existence of these two modes of gas accretion, by Dekel & Birnboim (2006). Below a certain halo mass of $\sim 10^{12}\, M_\odot$, the accretion shock is not sufficient to heat the gas to the virial temperature, and galaxy disks can be formed by cold gas flows, yielding efficient star formation. In contrast, in more massive structures the gas heated by shocks is diluted and sensitive to feedback due to active nuclei, quenching star formation and producing "red and dead" galaxies, which today do not evolve any further, their observed properties changing only as expected for passive evolution.

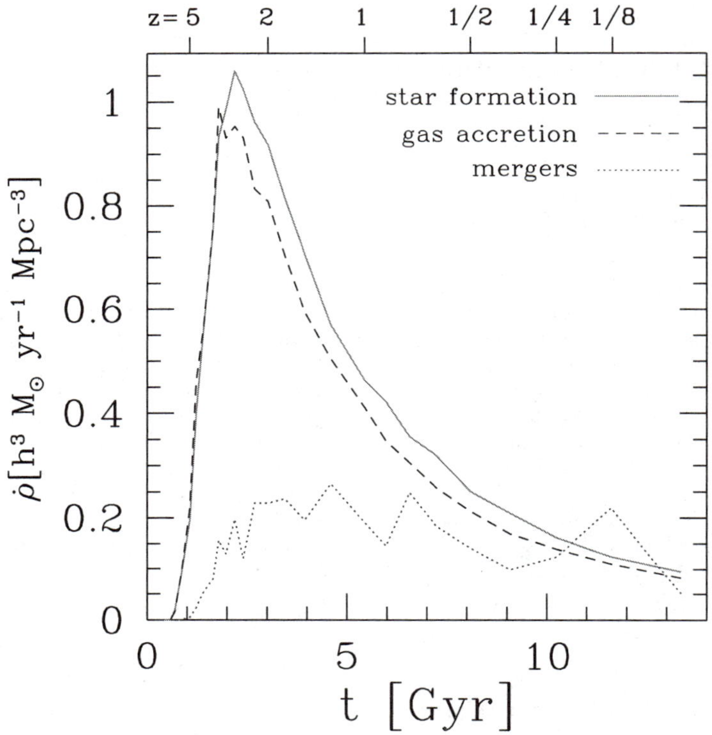


Figure 1. Comparison as a function of time of the star formation rate (*solid line, green in the electronic version*), the smooth gas accretion rate onto simulated galaxies (*dashed line, blue*), and mass gained in mergers (dotted line, red), by Keres et al. (2005).

The bimodality is observed in big samples of galaxies (SDSS, 2dF, MGC, . . .), and clearly separates the early-type "red" galaxies from the late-type "blue" ones, according to their stellar distribution (Sersic index, for example), or their surface density at the effective radius. The total galaxy mass is related to this separation, the red sequence being more massive (e.g., Driver et al. 2006). This observation is also reflecting the downsizing phenomenon: early-type galaxies appear to have formed the majority of their stars very early in the Universe, while star formation is going on now in smaller spirals and dwarfs. Is this incompatible with the hierarchical scenario?

## 2. Star Formation History and Surface Density

In fact, the two sequences of galaxies are distinguished by their stellar surface densities (Kauffmann et al. 2003). Low surface brightness (LSB) galaxies are small spirals or dwarfs, with high gas content, high and young star formation, while high surface brightness (HSB) galaxies are massive early-type galaxies, with concentrated mass, characterized by an old stellar population. The transition occurs at the visible mass of $3 \times 10^{10}\, M_\odot$, or stellar surface density of $3 \times 10^8\, M_\odot\, \mathrm{kpc}^{-2}$. But star formation histories are more strongly correlated with surface mass density than with stellar mass.

This observation could be interpreted as the consequence of star formation feedback. At low mass concentration, the potential well is not deep enough to retain the gas after supernova explosions, and star formation is quenched, which leads to a lower star formation efficiency in LSB galaxies. Following Dekel & Silk (1986), the energy provided by supernovae can be estimated as $E_{\mathrm{SN}} \sim \epsilon\nu\,(dM^*/dt)\, t_{\mathrm{rad}}$, where $\epsilon = 10^{51}\,\mathrm{erg}$ is the energy released by one supernova, the frequency of supernovae per unit mass is $\nu$ =1 for $100\, M_\odot$ stars formed, and the characteristic time $t_{\mathrm{rad}}$ marks the end of the adiabatic phase. This amount of energy from supernovae is able to disperse the gas (of mass $M_{\mathrm{g}}$), when it equals its kinetic energy $1/2\, M_{\mathrm{g}} V^2$, where $V$ is the rotational velocity of the galaxy. There is a transition in galaxy rotational velocity (or potential well depth) where the gas begins to flow out, at the critical velocity $V_{\mathrm{SN}} \sim 100\,\mathrm{km\,s^{-1}}$, which corresponds in order of magnitude to the transition found between LSB and HSB galaxies.

### 2.1. Quenching of Star Formation

As shown in Fig. 2, the main parameter determining the star formation efficiency is the stellar surface density. Below the critical value of $3 \times 10^8\, M_\odot\, \mathrm{kpc}^{-2}$, the average star formation rate per unit mass does not depend on mass or surface density, or concentration. The scatter in star formation rate increases with surface density. The star formation rate is then more bursty, and depends essentially on the external gas accretion. Above the threshold in surface density, galaxy growth through star formation shuts down. The timescale to consume the gas is inversely proportional to the stellar surface density.

### 2.2. Origin of the Bimodality

There are therefore two interpretations for the origin of the bimodality. The total mass could be the determining factor, and the transition could occur at the critical halo mass of $3 \times 10^{11}\, M_\odot$: above this halo mass, the external gas is not accreted cold, but is heated in shocks and then has no time to cool (subject in addition to AGN feedback, Dekel & Birnboim 2006). Alternatively, above the critical surface density of stars of $3 \times 10^8\, M_\odot\, \mathrm{kpc}^{-2}$, the gas could be quickly transformed into stars, and the time spent in the "blue" star-forming sequence could be short. According to this view, galaxies pass from the red sequence to the blue sequence and back, but the blue phase is much shorter at high mass and concentration than at low mass (Kauffmann et al. 2006a).

## 3. Gaseous Filaments around Galaxies

Numerical techniques are progressively refined to take into account colder gas, and simulate the cold gas infall. While most simulations still consider the "cold gas" to be at $10^4$ K only, infalling multi-phase gas has been studied by Kaufmann

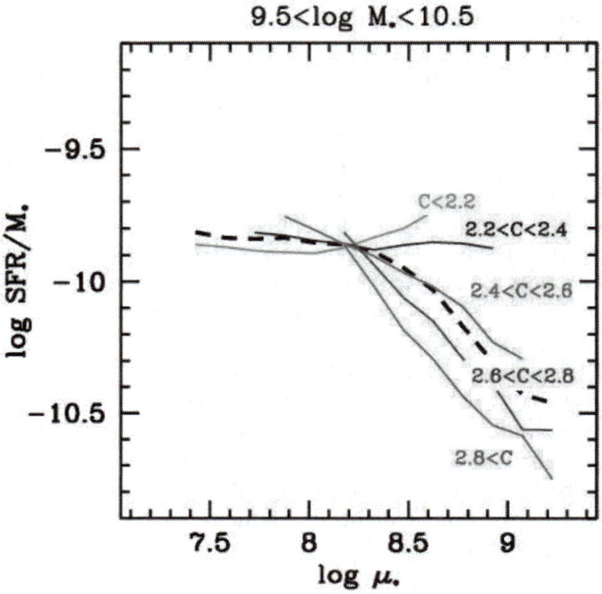


Figure 2. The star formation rate per unit stellar mass depends essentially on surface density, and on the concentration. Here, this specific SFR is plotted versus surface density for masses between $3\times10^9$ and $3\times10^{10}$ $M_\odot$, and different concentrations in different colors (the black dashed line is the total). (*From Kauffmann et al. 2006.*)

et al. (2006b), and the cooling/heating of dense clumps is now introduced in the SPH scheme through considerations of $H_2$ and HD molecules (Maio et al. 2007).

The gaseous filaments are hard to observe because of their low density and dilution. However, with the high sensitivity of present-day deep searches, it is possible to see them in their connection to galaxies. The example of the edge-on galaxy NGC 891 is spectacular (Fraternali et al. 2007). Not only is a thick gaseous envelope seen at a thickness of 8 kpc, but H I gas filaments are also obvious up to 20 kpc above the galaxy plane. The rotational velocity of this gas decreases with $z$, and its origin cannot be due to a galactic fountain phenomenon, coming from the star formation feedback, which would conserve angular momentum.

In some galaxies, the fountain effect is detected in addition (for example the gas outflows in NGC 2403), while in general inflow is observed as high or intermediate velocity clouds (HVCs, IVCs) in the Milky Way. Gaseous halos are only explained by accretion of external gas (Fraternali & Binney 2006).

## 4. Gas Angular Momentum Lost in Mergers

External gas infall could help to solve the problem of the angular momentum catastrophe and the formation of too small galaxy disks in the conventional scenario (Navarro & Steinmetz 2000). The gas loses its angular momentum essentially through mergers, but not during cold accretion (D'Onghia et al. 2006). Tidal torques and the corresponding dynamical friction during galaxy interactions transfer the gas orbital angular momentum to the more extended dark matter (e.g., Barnes & Hernquist 1992), but also to the extended gas in the outer filaments (see Fig. 4). After a violent merger, the gas has condensed in a compact disk of radius lower than 3.5 kpc. Only diffuse cold accretion after the last merger can reform an extended disk, but only a small fraction of the baryonic mass is then in the extended disk (larger than 3.5 kpc). This gas accretion can be in the form of either smooth gas or substructures less than 1/10 in mass, which are easily stripped.

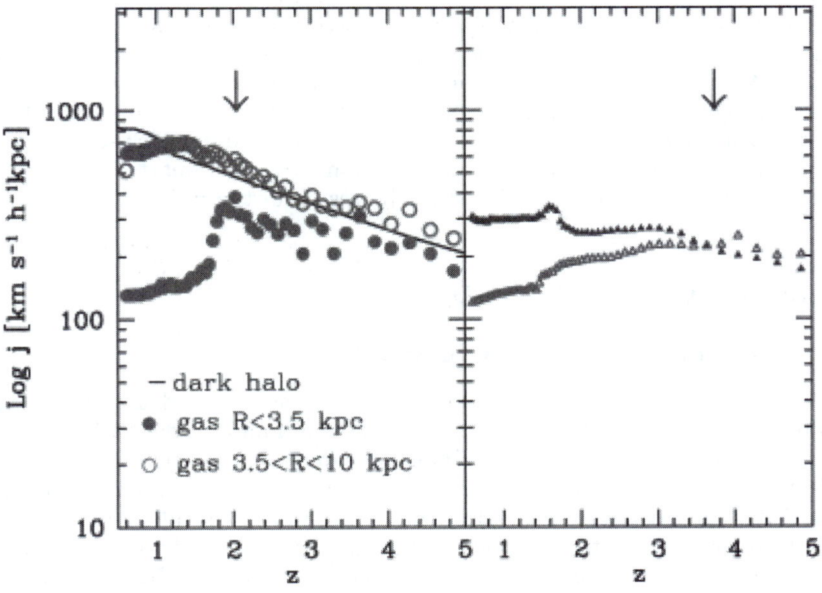


Figure 3. How galaxies lose their angular momentum during ΛCDM simulations. *Left:* The specific angular momentum as a function of redshift of the dark matter (*solid line*) and the gas in the compact central disk (*filled circles*) or the gas in the envelope (*empty circles*). *Right:* The evolution of the angular momentum of the dark matter (*upper curve*) or the gas (*lower curve*) of the massive progenitor that has already formed a compact disk at $z = 2$. (*From D'Onghia et al. 2006*.)

Since each major merger exchanges angular momentum between the baryons and the dark matter, the latter could be expected to possess more specific angular momentum in halos having experienced more mergers. In fact, this does not appear to be the case in numerical simulations, since after a certain period of relaxation, the excess of angular momentum is expelled from the halo through the virial radius (D'Onghia & Navarro 2007). The specific angular momentum is a function of radius, and its excess yields the expansion of the particles, and inflation of the halos.

### 4.1. Why the Angular Momentum Problem Is Still not Solved

To keep a significant angular momentum in baryonic disks, there should be cold gas accretion along cosmic filaments after the last major merger, where the angular momentum loss occurs. To form substantial disks, there should be no major mergers from $z = 1$ until the present time, but this is not likely in the hierarchical scenario and according to cosmological simulations. Statistically, there exist dark halos with no merger after $z = 1$, and subsequent gas accretion will then explain extended gas disks around big bulges. But the main problem is to form spiral disks with no bulge at $z = 0$.

The gas inflow from filaments has an angular momentum direction generally offset from that of the galaxy disks. This skewed accretion will in general create warps, and progressively re-orient the spins of galaxies (e.g., Jiang & Binney 1999). When the offset is extreme and the accretion is perpendicular, then polar rings could be formed (Maccio, Moore, & Stadel 2006). This is an efficient way to form such systems, which reveal young disks around early-type systems. In polar ring galaxies formed in such accretions, the dark matter component is predicted to be quite round, at least in the region of the visible galaxy.

## 5. Infall of Gas and Galaxy Dynamics

### 5.1. Star Formation History in Spirals

In order to reproduce the observed abundances of metals, models of the chemical evolution of the Milky Way require a continuous infall of gas with metallicity about 0.1 times the solar value. Infall is required in particular to solve the well-known G-dwarf problem, i.e., the observation of a very narrow range in metallicity for most long-lived stars near the Sun. The infall of gas is also supported by the constant or increasing star formation rate scenario inferred from the local distribution of stars (e.g., Haywood, Robin, & Crézé 1997). Other abundance problems require also an infall rate integrated over the entire disk of the Milky Way of a few solar masses per year at least (Casuso & Beckman 2001). This infall dilutes the enrichment arising from the production of heavy elements in stars and thereby prevents the metallicity of the interstellar medium from increasing steadily with time. Some of this gas could come from the HVCs infalling onto our Galaxy disk (Wakker et al. 1999).

In M31, *HST* color–magnitude diagrams reveal that there exists a similar G-dwarf problem, even more severe than in the Milky Way (Worthey & Espana 2004). Here also, closed-box models cannot explain the stellar distributions. The star formation rate in spirals in the middle of the Hubble sequence has also kept of the same order of magnitude over the Hubble time, instead of the expected exponential decrease expected from closed box models (e.g., Kennicutt, Tamblyn, & Congdon 1994), and favors gas accretion.

### 5.2. Secular Evolution

Gas accretion is an essential ingredient for secular evolution driven by bars and spirals. Without gas, stellar disks are heated by spiral waves and become featureless, or barred only after several dynamical times. Gas accretion can maintain spiral structure, and can also reform bars, which are weakened or destroyed in the evolution. Bar/spiral gravity torques produce radial gas inflow, and nuclear starbursts. Dynamical instabilities then regulate themselves, since the gas inflow itself can destroy the bar. The bar destruction involves two mechanisms: first, the central mass concentration, built after the gas inflow, destroys the orbital structure sustaining the bar, scatters particles, and pushes them into chaotic orbits. Second, the gas inflow itself weakens the bar, by exchanging its angular momentum to the stars forming the bar (Bournaud, Combes, & Semelin 2005).

Secular evolution then includes several bar episodes in a galaxy lifetime. A spiral galaxy rich in gas (at least 5% of the disk mass) is unstable with respect to bar formation. Gravity torques are then efficient to drive the matter inwards. The galaxy morphological type evolves towards early types, the mass is concentrated, the bulge is developed through horizontal and vertical resonances. This weakens the bar, and when the galaxy again becomes axisymmetric, gas can be accreted from the outer parts by viscosity (Bournaud & Combes 2002). The gas accretion, if significant with respect to the disk mass, can reduce the bulge-to-disk ratio (by replenishing the disk), and make the galaxy disk unstable again to a new bar.

### 5.3. AGN Fueling

Gas flows due to bar gravity torques are an attractive explanation to explain AGN fueling. However, low luminosity AGN, such as Seyfert galaxies, are not observed in good correlation with bar strength. This could be due to the widely different timescales between the gas flows from the outer parts of galaxies to the nuclear region (100 pc–1 kpc) and the short nuclear timescales. Also, the fact that gas flows destroy the bar could play a large role in this absence of a correlation.

More details on the evolutionary scenario have been searched through gas mapping at high resolution in a dozen Seyfert galaxies by the NUGA project (e.g., García-Burillo et al. 2005). In most galaxies, gas has been found concentrated in rings at inner Lindblad resonances (ILRs), and the nuclear torques were measured positive, meaning that no fueling was occurring.

The accumulation of the gas into ILR rings is the first step of the secular evolution scenario, in galaxies where mass is concentrated in bulges, which is likely for spiral galaxies with massive central black holes. Through bar torques, the gas is driven inwards from corotation to the ILR and accumulates in the ring, usually the site of active star formation. Since the sign of the torques changes at a resonance, positive gravity torques then prevent the gas to flow further in: under the bar forcing, gas inside the ring is evacuated outwards. Only in the central 10 pc around the black hole is the potential mostly axisymmetric, and viscosity could help to fuel the AGN. A more consequent feeding has to wait for the weakening of the bar through the gas flows. The gas of the nuclear ring is then liberated to form a smoother disk through viscosity, which has become competitive against gravity torques.

The AGN fueling in early-type spirals can be viewed as a two step process: first, gravity torques bring the gas of the large-scale disk to the nuclear ring, and when the dynamical feedback has weakened the bar, the viscous torques smooth out the ring and bring gas to the central 10 pc, where it is under the influence of the Keplerian potential of the black hole. The disk becomes axisymmetric and the cycle can be restarted at its first step. In particular, the disk will be prone to a new bar instability if gas is accreted from the outer parts of the disk. In any case, the AGN phase is always correlated with a weak bar phase.

Only in one galaxy, NGC 2782 (L. K. Hunt et al., in preparation) have the nuclear torques been measured negative and the fueling caught in action (see Fig 4). This means that the fueling phase is very short, and only a minor fraction of active galaxies is observed in this phase.

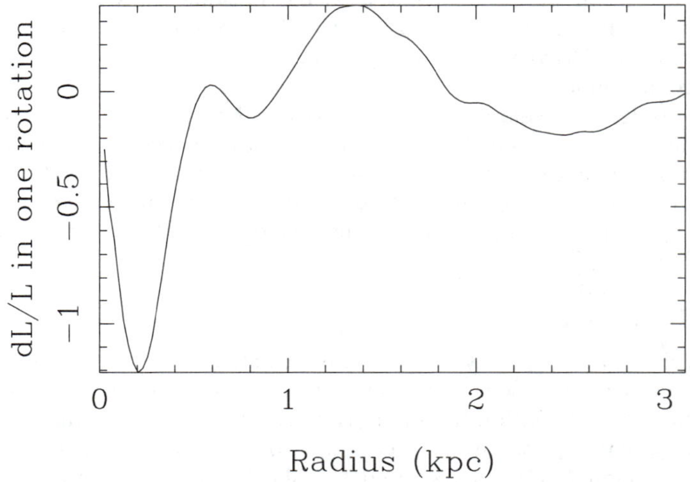


Figure 4. Estimation of the gravity torques and the relative angular momentum lost by the gas in a rotation period ($L$ is the angular momentum perpendicular to the disk). The torque is negative towards the center, which means the gas is fueling the AGN in NGC 2782 (L. K. Hunt et al., in preparation).

## 6. Cooling Flows in Galaxy Clusters

Galaxy clusters are interesting to study in the context of gas accretion, since they are certainly the prototype model of "hot accretion" in massive dark matter halos. Since the hot gas is observed and contains most of the baryons, it can be regarded as certain that cold accretion is restricted to a minimum. The cooling time of the gas at the center of clusters becomes smaller than the Hubble time, so an important gas flow has been predicted for many years. The cold gas has now been observed (23 detected galaxies in CO at 2.6 mm, Edge 2001; Salomé, & Combes 2003), but the amplitude of the flow is ten times lower than expected in the early times. Moreover, X-ray lines reveal gas down to a temperature of 2 keV but not below (Peterson & Fabian 2006).

In the last decade, results from the *Chandra* and *XMM* satellites have established that the cooling flow phenomenon is self-regulated. The simple view of unimpeded spherical flow is now completely revised to take into account the re-heating processes, and the feedback due to the active nucleus or black hole: shocks, radio jets, acoustic waves, bubbles . . . The cooling gas is not only observed in CO emission, but also directly from the pure rotational lines of the $H_2$ molecule. For instance, large molecular masses are infalling onto the central galaxy in Perseus (Johnstone et al. 2007). The $H_2$ lines have been surprisingly strong in many cooling flow clusters, as discovered by *Spitzer* (e.g., Egami et al. 2006).

### 6.1. Amount of Cooling Gas

One of the puzzles in cooling flows was that the gas with temperatures $10^4$ and $10^7$ K was not seen in the expected quantity. Gas at $T = 10^{5.5}$ K is now seen in O VI emission (with *FUSE*) in many clusters, for instance in Abell 1795 and Perseus (Bregman et al. 2006), giving cooling rates of $\sim 30\, M_\odot\, \mathrm{yr}^{-1}$, about three times lower than previously expected.

The cold gas masses detected through CO emission are compatible with these cooling rates. The masses derived are conservative, since they are obtained with the same CO-to-$H_2$ conversion factor as solar metallicity spiral galaxies. Masses are between $3 \times 10^8$ and $4 \times 10^{10}\, M_\odot$ (Edge 2001; Salomé & Combes 2003). In some cases, high resolution maps have been obtained with millimeter interferometers, and it has been clearly established that the CO emission is not correlated with any central galaxy, but originates from the cooling flow. The CO gas kinematics is related to the cluster and is not rotation around the central galaxy. This is the case for Abell 1795, where a cooling wake is observed in X-rays (Fabian et al. 2001). The cooling rate has been estimated to $200\, M_\odot\, \mathrm{yr}^{-1}$ inside a radius of 200 kpc (Ettori et al. 2002). The CO emission has been shown to follow the cooling wake (Salomé & Combes 2004), and also the 60 kpc H$\alpha$ filament mapped by Cowie et al. (1983). The main CO velocity is centered on the cluster velocity. In Perseus, the CO emission follows the conspicuous H$\alpha$ filamentary structure, mapped by Conselice, Gallagher, & Wyse (2001). Again the molecular component is not in rotation in NGC 1275 (Salomé et al. 2006). The radio jets from 3C84 create holes and bubbles in the hot gas, which is compressed at the boundaries, and cools faster. This means that cold clumps could be found far from the central cooling activity.

Two main models are therefore advanced to account for cooling and heating processes: hot and cold feedback. From the point of view of hot feedback, the hot intercluster medium (ICM) cools and is accreted close to the AGN. The observed cold gas far from the center should then come from galaxies. In the cold feedback scenario, cold gas is accreted on large scales. Cooling can occur at large distance because of non-linear over-dense blobs, created in wide regions due to past AGN activity, which decreases the cooling timescale for these perturbed regions (e.g., Pizzolato & Soker 2005). According to this point of view, there is a co-existence of multiphase (cold and hot) gas.

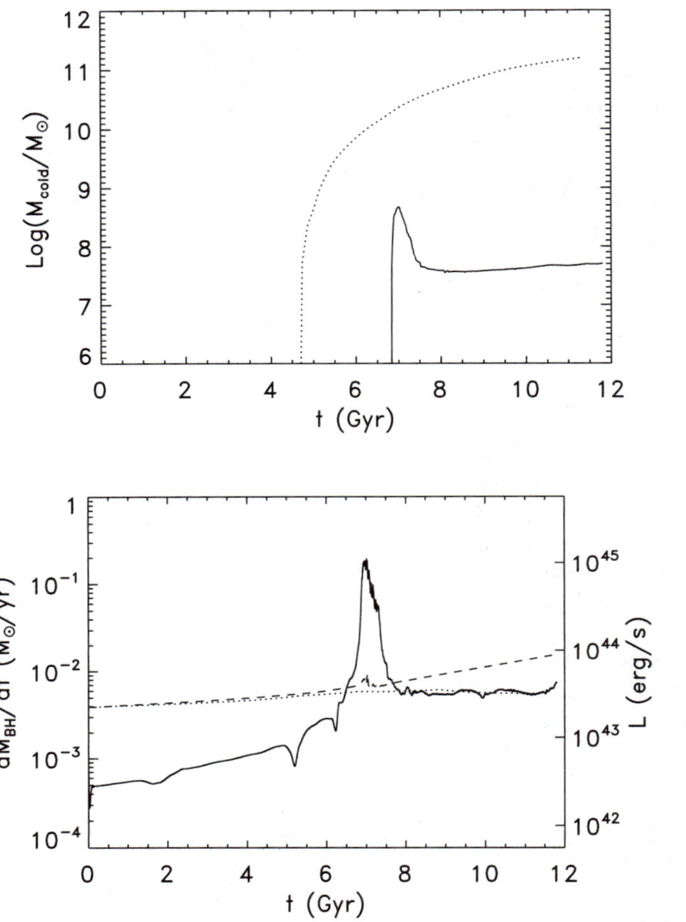


Figure 5. AGN feedback in cooling flows. The *left panel* shows the mass of the cold gas as a function of time, in the cooling flow simulation (*dotted line*), and with AGN feedback (*solid line*); the *solid line* in the *right panel* shows the accretion rate by the black hole (*left axis*) or the corresponding jet mechanical luminosity (*right axis*) in the simulation with AGN feedback. The *dashed line* shows the X-rays emitted by the ICM in the same simulation, compared with the *dotted line* without feedback. (*From Cattaneo & Teyssier 2007.*)

## 6.2. Numerical Simulations

The cold feedback scenario has been considered in recent multi-phase simulations of the cluster environment (Revaz, Combes & Salomé 2007). The AGN feedback is schematized by following hot bubbles during their buoyant rise across the cluster. Very cold gas is spontaneously formed by the perturbation of the hot ICM, and forms a filamentary structure in the wake and in the rim of the bubbles. The amount of cold gas formed and its kinematics are in agreement with observations in CO and H$\alpha$.

The AGN feedback is therefore not only negative in re-heating the cooling gas but is also positive in allowing some further gas to cool.

Alternatively, the hot feedback scenario has also been simulated by Cattaneo & Teyssier (2007), who show that the catastrophic cooling is regulated by the AGN jets, and that a stationary state can be reached to reproduce observations (see Fig. 5). The cooling gas is regulated to just the right amount required to fuel the AGN activity.

## 7. Conclusions

Cold gas accretion along cosmic filaments is an important way to form and grow galactic disks. In particular, cold gas accretion dominates at high redshift. Simulations with enough resolution reveal that not all the gas is shock heated at the virial radius, as in the usually adopted paradigm.

Cold accretion is a way to solve the angular momentum problem in the formation of galaxy disks. In the hierarchical scenario, the angular momentum of baryons is lost in mergers to the dark matter halos. The gas accreted after the last merger does not lose angular momentum and is able to form large disks, although around already massive bulges. The baryons will keep more specific angular momentum globally if galaxy formation does rely more on smooth matter accretion than mergers.

Gas accretion is also required for disk evolution, being one of the main actors in secular evolution. It is needed to maintain spiral structure, reform bars, asymmetries, warps . . .

Cooling flows in galaxy clusters are a prototypical example of hot accretion. Cold gas observed in CO and $H_2$ line emission is now compatible with the revised expected cooling rates. The cooling process is self-regulated through AGN feedback. In the hot feedback scenario, self-regulation can occur close to the central AGN, while cold feedback involves a much wider region, where dense clumps of gas can cool far from the AGN. This scenario could explain the large-scale filamentary structure observed in CO and H$\alpha$ in nearby cooling flows.

**Acknowledgments.** I am very grateful to the organizers for a very instructive and pleasant meeting that reflected so vividly John Beckman's eclectic career.

### References

Barnes, J. E., & Hernquist, L. 1992, ARAA, 30, 705
Bournaud, F., & Combes, F. 2002, A&A, 392, 83
Bournaud, F., Combes, F., & Semelin, B. 2005, MNRAS, 364, L18
Bregman, J. N., Fabian, A. C., Miller, E. D., & Irwin, J. A. 2006, ApJ, 642, 746
Casuso, E., & Beckman, J. E. 2001, ApJ, 557, 681
Cattaneo, A., & Teyssier, R. 2007, MNRAS, 376, 1547
Conselice, C. J., Gallagher III, J. S., & Wyse, R. F. G. 2001, AJ, 122, 2281

Cowie, L. L., Hu, E. M., Jenkins, E. B., & York, D. G. 1983, ApJ, 272, 29
Dekel, A., & Birnboim, Y. 2006, MNRAS, 368, 2
Dekel, A., & Silk, J. 1986, ApJ, 303, 39
D'Onghia, E., Burkert, A., Murante, G., & Khochfar, S. 2006, MNRAS, 372, 1525
D'Onghia, E., & Navarro, J. F. 2007, MNRAS, 380, L58
Driver, S. P., et al. 2006, MNRAS, 368, 414
Edge, A. 2001, MNRAS, 328, 762
Egami, E., Rieke, G. H., Fadda, D., & Hines, D. C. 2006, ApJ, 652, L21
Ettori, S., Fabian, A. C., Allen, S. W., & Johnstone, R. M. 2002, MNRAS, 331, 635
Fabian, A. C., Sanders, J. S., Ettori, S., Taylor, G. B., Allen, S. W., Crawford, C. S., Iwasawa, K., & Johnstone, R. M. 2001, MNRAS, 321, L33
Fraternali, F., & Binney, J. 2006, MNRAS, 366, 449
Fraternali, F., Binney, J., Oosterloo, T., & Sancisi, R. 2007, New Astron. Rev., 51, 95
Garcia-Burillo, S., Combes, F., Schinnerer, E., Boone, F., & Hunt, L. K. 2005, A&A, 441, 1011
Haywood, M., Robin, A. C., & Crézé, M. 1997, A&A, 320, 440
Jiang, I.-G., & Binney, J. 1999, MNRAS, 303, L7
Johnstone, R., Hatch, N., Ferland, G., Fabian, A., Crawford, C., & Wilman R. 2007, MNRAS, in press
Kauffmann, G., et al. 2003, MNRAS, 341, 54
Kauffmann, G., Heckman, T. M., De Lucia, G., Brinchmann, J., Charlot, S., Tremonti, C., White, S. D. M., & Brinkmann, J. 2006a, MNRAS, 367, 1394
Kaufmann, T., Mayer, L., Wadsley, J., Stadel, J., & Moore, B. 2006b, MNRAS, 370, 1612
Kennicutt, R. C., Tamblyn, P., & Congdon, C. E. 1994, ApJ, 435, 22
Keres, D., Katz, N., Weinberg, D. H., & Davé, R. 2005, MNRAS, 363, 2
Maccio, A. V., Moore, B., & Stadel, J. 2006, ApJ, 636, L25
Maio, U., Dolag, K., Ciardi, B., & Tornatore, L. 2007, MNRAS, 379, 963
Navarro, J. F., & Steinmetz, M. 2000, ApJ, 538, 477
Peterson, J. R., & Fabian, A. C. 2006, Phys. Rep., 427, 1
Pizzolato, F., & Soker, N. 2005, ApJ, 632, 821
Revaz, Y., Combes, F., & Salomé, P. 2007, A&A, submitted
Salomé, P., & Combes, F. 2003, A&A, 412, 657
Salomé, P., & Combes, F. 2004, A&A, 415, L1
Salomé, P., Combes, F., Edge, A. C., Crawford, C. , Erlund, M., Fabian, A. C., Hatch, N. A., Johnstone, R. M., Sanders, J. S., & Wilman, R. J. 2006, A&A, 454, 437
Wakker, B. P., et al. 1999, Nature, 402, 388
Worthey, G., & Espana, A. L. 2004, in Origin and Evolution of the Elements, from the Carnegie Observatories Centennial Symposia, ed. A. McWilliam and M. Rauch, http://www.ociw.edu/ociw/symposia/symposium4/proceedings.html

## Discussion

*Silk*: Cold gas accretion along filaments must be clumpy, leading to cloud–cloud collisions at the virial velocity and therefore to gas heating. Without high enough resolution this may not be effectively traced in the simulations. If so, there should be similar implications for star formation between hot and cold gas accretion.

*Combes*: Certainly the cold gas accretion comes in clumps, subject to collisions, but these are mild, similar to what happens in the ISM of galaxies. There is no violent head-on collision here, the external gas smoothly settles in a large "accretion disk" in the outer parts of galaxies, as we see in H I extended disks.

*Hammer*: How may cold gas accretion explain the numerous populations of LIRGs (20–200 $M_\odot\,\mathrm{yr}^{-1}$) at $z \sim 0.5$–1, i.e., representing 15% of $M_*$ galaxies? Is this also consistent with the observations of major mergers? A final concern is the angular momentum problem that you consider decreasing after merging. I wonder if accounting for dark and baryonic matter is done properly (not only in simulations but also in comparing simulations to observations . . .).

*Combes*: Some of the LIRGs could be due to cold gas accretion. The bar in galaxies self-regulates the gas infall and can produce a starburst, i.e., of the order of 30–100 $M_\odot\,\mathrm{yr}^{-1}$. Of course, mergers can also explain some of them, and most of the ULIRGs. Concerning the angular momentum of the gas in the merger, this does not depend much on the gas physics or the techniques of the simulations, it is essentially due to gravity torques during the merger that exchange angular momentum between the gas and the dark matter.

*Hensler*: The coexistence of hot and cold matter in cooling flows should lead to heat conduction that evaporates the cool gas within ~100 Myr. Do filaments, then, only survive if shock fronts continue compression of the gas or if magnetic fields prevent heat conduction?

*Combes*: Indeed, there is some evidence from the comparison between observations and simulations that heat conduction must be prevented from occurring at a normal rate, and authors are invoking magnetic fields to achieve that (although it is difficult to measure it in the filaments).

*Kroupa*: Gilmore et al. (2007, ApJ, 663, 948) claim to have found very strong unambiguous evidence that we live in a warm dark matter (WDM) universe—how would this change the notion of cold/hot infall and accretion?

*Combes*: Indeed, the WDM universe will have less small-scale structure, so it is a solution to the missing dwarf satellites problem. But as for the hot/cold gas accretion, this will not change much for galaxy formation.

Françoise Combes.

*Pathways through an Eclectic Universe*
*ASP Conference Series, Vol. 390, © 2008*
*J. H. Knapen, T. J. Mahoney, and A. Vazdekis, eds.*

# Gas Feedback and Angular Momentum Redistribution in Barred Galaxies within Dark Matter Halos

I. Berentzen,[1,2] I. Shlosman,[2] I. Martínez-Valpuesta,[2,3] and C. H. Heller[4]

[1] *Astronomisches Rechen-Institut, Zentrum für Astronomie, University of Heidelberg, Germany*
[2] *Department of Physics & Astronomy, University of Kentucky, Lexington, USA*
[3] *Instituto de Astrofísica de Canarias, La Laguna, Spain*
[4] *Department of Physics, Georgia Southern University, Statesboro, USA*

**Abstract.** We have analyzed the effects of gas on the disk–halo evolution in barred galaxies using $N$-body/SPH simulations. Our main focus has been the angular momentum ($J$) redistribution mediated by the stellar bar and the subsequent evolution of the galactic disk. We find that the stellar bars in disks with 0–8 percent gas fraction are primarily affected by the growth of a central mass concentration owing to the bar-driven gas inflow to the center. The almost abrupt weakening of the bars observed in all models is triggered either by vertical buckling (in gas-poor disks) or by resonance heating (in gas-rich disks). The dark matter halo serves as the angular momentum sink for the stellar bar, thus compensating it for the angular momentum received from the gas.

Galactic stellar bars are of paramount importance to short and long term galaxy evolution by breaking the axial symmetry of disks and efficiently redistributing mass and angular momentum, $J$, within the galaxy. Although the gas in the disk typically amounts to only a very small fraction of the galaxy mass, it can strongly alter the disk evolution due to its dissipative nature. Here, we analyze the evolution of such disk–halo systems in the presence of various fractions, $f_{\rm gas}$ = 0–8 percent, of gas by means of $N$-body/SPH simulations. Details of the initial conditions and the code used for the simulations are given in Berentzen et al. (2007).

The stellar bars in our models form and evolve initially in a qualitatively similar way, reaching a maximum bar strength after some 1–1.5 Gyr (see Fig. 1a). Past this maximum, the more gas-poor models exhibit a plateau, followed by a sharp drop in the bar strength. This plateau gradually disappears with increasing gas fractions. Following the substantial weakening of the bar, the subsequent evolution also differs: gas-poor disks exhibit a growing bar, while gas-rich models show a weakly declining bar. The borderline between our gas-poor and -rich disks lies at gas fractions of about 3%, but is clearly model-dependent and will be affected by various processes, such as the degree of clumpiness in the gas and feedback from stellar evolution.

We find that, beyond the bar-driven gas inflow and the resulting gradual buildup of a central mass concentration (CMC), the angular momentum exchange between the gas and the stars (see Fig. 1b) has no direct impact on the bar strength in our models. Specifically, the $J$-transfer from the gas to the stel-

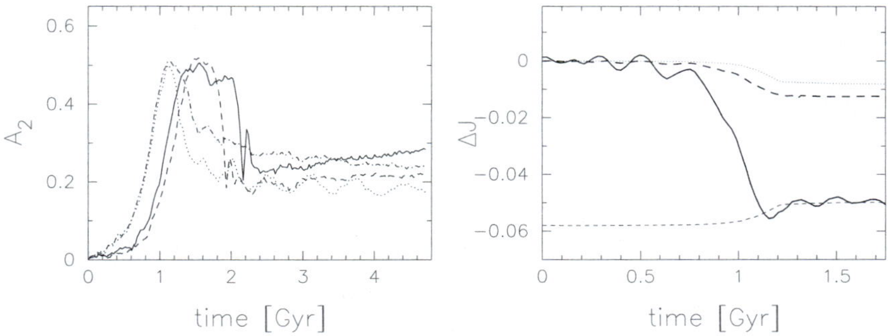


Figure 1. *Left:* Evolution of the bar strength. Shown are models with gas fractions of 8% (solid), 4% (dashed), 2% (dotted-dashed) and 0% (dotted). *Right:* Change in angular momentum, $\Delta J$, in the 8% gas model. Shown is $\Delta J$ of the stars (dashed lines) and gas (solid line). The thin lines show the $J$ flow from the gas to the stars (dashed) and vice versa (dotted), calculated from integration of the corresponding gravitational torques in the simulation. The thin dotted line has been shifted in $y$ to match the solid one.

lar bar does not play a role in the substantial weakening of the bar, as reported in the literature (Bournaud, Combes, & Semelin 2005). The reason for this is that the excess of $J$ in the stellar bar is drained by the *live* dark matter (DM) halo (see Fig. 1b). Instead, our models show that the weakening of the bar is caused by vertical buckling instability in the gas-poor disks, and by a vigorous heating of the stellar velocity dispersion by the CMC in the gas-rich case. Both of these processes are accompanied by a vertical swelling of the stellar bar: in the gas-poor models the stellar bar buckles out of the disk plane (with a transient break in the vertical symmetry) and leads to the well-known formation of a boxy/peanut-shaped bulge. With an increasing gas fraction, however, the CMC heats up the central kpc of the stellar disk on a dynamical timescale, followed by a slow (secular) stage of bar thickening, with no visible signs of buckling. This action of the CMC leads to the formation of progressively more elliptical bulges.

We have tested the above results by removing the gravitational torques exerted by the gas on the stars in the disk—models with torques reduced by 25–100 percent exhibited the same bar evolution. Neither is the observed behavior of the stellar bar in our models altered if a supermassive black hole (SBH) is allowed to grow in the center—models with and without a SBH behave very similarly. It appears that the DM halo shapes have by far larger effects on the bar fate than the disk gas, as shown by Berentzen & Shlosman (2006).

**Acknowledgments.** We gratefully acknowledge support from NASA, NSF, the Volkswagen Foundation and by the Gruber Foundation.

## References

Berentzen, I., & Shlosman, I. 2006, ApJ, 648, 807
Berentzen, I., Shlosman, I., Martínez-Valpuesta, I., & Heller, C. H. 2007, ApJ, 666, 189
Bournaud, F., Combes, F., & Semelin, B. 2005, MNRAS, 364, L18

*Pathways through an Eclectic Universe*
*ASP Conference Series, Vol. 390, © 2008*
*J. H. Knapen, T. J. Mahoney, and A. Vazdekis, eds.*

# Observations of Complex H, a High Velocity Cloud Close to the Galactic Plane

V. Buenrostro,[1] J. Lockman,[2] and J. E. Beckman[1,3]

[1]*Instituto de Astrofísica de Canarias, La Laguna, Tenerife, Spain*
[2]*National Radio Astronomy Observatory, Green Bank, WV, USA*
[3]*Consejo Superior de Investigaciones Científicas, Spain*

**Abstract.** Using 21 cm observations of Complex H with the Green Bank Radio Telescope, we have derived its LSR velocity of $\sim 200\,\mathrm{km\,s^{-1}}$ and an H I mass of $7.5 \times 10^{7}\,M_{\odot}$. This gives useful support to the scenario in which the Galactic disk is being built up by the continual accretion of clouds of this type from the intracluster medium of the Local Group. Complex H is detected as shedding a gas tail, and has a shock front at its leading surface, both phenomena attributable to its interaction with the halo gas of the Milky Way.

## 1. Observations and Data Reduction

Complex H is one of the most interesting high velocity clouds in our Galaxy, since it is one of the biggest clouds of H I known not to have any stars in it. It was discovered by Hulsbosch (1971) and Dieter (1971), covering around $478\,\mathrm{deg}^2$ at a very low galactic latitude, with a velocity with respect to the Local Standard of Rest of $-200\,\mathrm{km\,s^{-1}}$, an orbital inclination of $i = 230°$, and a distance from the Galactic Center of $R = 33\,\mathrm{kpc}$. In 2002, a small fraction of the complex was observed by Lockman (2003), and by 2003 he observed a much bigger fraction of the complex, at $108° \leq l \leq 138°$ and $-11° \leq b \leq 11°$, and here we show the results of reducing and analyzing the data.

Observations were done with the 100 m "Robert C. Byrd Green Bank Telescope" (GBT), at the National Radio Astronomy Observatory in the year 2003. The angular resolution of the telescope at 21 cm is $9'.1$ (FWHM). Reduction and calibration were done during the Summer of 2005 and 2006 with the software package called GBTIDL, created for data obtained with the GBT. The spectra were assembled in a data cube using AIPS, and the data cube was visualized using AIPS++. It should be noted that a set of data is missing in the data cube, so the cube is not entirely complete for the time being. The reason is that there is a velocity shift in this data set that has not been fixed yet.

## 2. Results and Conclusions

The distance was calculated by Lockman (2003) using a model in which Complex H moves in a circular retrograde orbit around the Galactic Center. Assuming that the rotation curve is almost flat over tens of kiloparsec and assuming a constant circular velocity of $220\,\mathrm{km\,s^{-1}}$, we find that its distance from the Galactic Center is $R \sim 33\,\mathrm{kpc}$, and that the orbital inclination is about $i = 230°$. The orbit crosses the Galactic plane at $\theta = 38° \pm 5$ from the radius vector joining the Galactic Center to the Sun in the direction of Galactic rotation at a distance from the Sun of $r \sim 27$ kpc. With the new data cube, the mass was calculated to be $\sim 7.5 \times 10^7\, M_\odot$. In reality, this is a lower limit. The mass will be higher when we can take into account the data missing in the data cube. It is of great utility to generate the moment maps of a data cube in order to be able to derive the integrated emission and the velocity distribution from the observed spectra.

Moment 0 gives us the integrated value of the spectrum, from which the cloud mass can be derived. Moment 1 is the coordinate weighted by intensity used to find the velocity fields. An interesting feature of the moment maps obtained is that we can see how the cloud is accelerating in the direction of the Galactic plane, leaving behind it a big tail of gas with velocities lower than that of the center of the cloud, and close to the Galactic rotation velocity, perhaps because of interaction with the Milky Way. We can also see that it is being decelerated at the front, probably because of the shock front it forms with the Galactic medium. With these results, it is evident, as Lockman (2003) had stated first, that the complex is moving at a mean speed of around $-200\,\mathrm{km\,s^{-1}}$ in a southeasterly direction, leaving behind it a big tail of H I gas, because of a probable interaction with the Milky Way. The frontal part is being decelerated by a possible shock front. Although the estimated mass of Complex H is around $7.5 \times 10^7\, M_\odot$ if we consider a distance $r$ from the Sun to the Complex of $\sim 27$ kpc, the mass must be higher because there are data missing from the data cube.

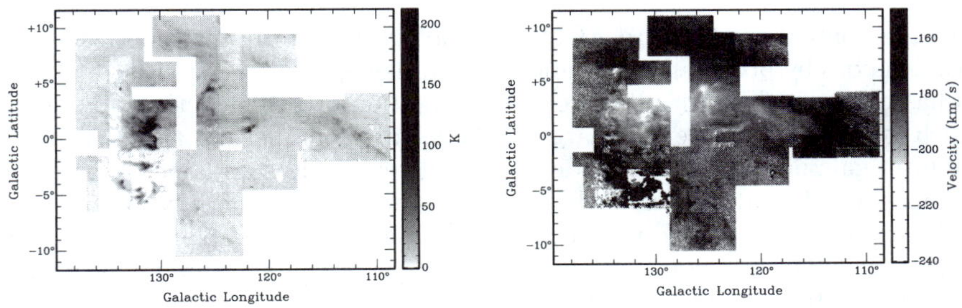


Figure 1. Moments 0 (*left*) and 1 (*right*) of Complex H.

## References

Dieter, N. H. 1971, A&A, 12, 59
Hulsbosch, A. M. 1971, A&A, 14, 489
Lockman, F. J. 2003, ApJ, 591, L33

*Pathways through an Eclectic Universe*
*ASP Conference Series, Vol. 390, © 2008*
*J. H. Knapen, T. J. Mahoney, and A. Vazdekis, eds.*

# Triggering of Cloud Collapse in a Galactic Disk by Infall of a High Velocity Cloud

Emilio Casuso,[1] John E. Beckman,[1,2] and Valeria Buenrostro[1]

[1]*Instituto de Astrofisica de Canarias, E-38200 La Laguna, Tenerife, Spain*
[2]*Consejo Superior de Investigaciones Científicas, Madrid, Spain*

**Abstract.** We express in linearized form the five classical coupled equations representing the rotating gas disk of a standard disk galaxy. We solve them in three limiting cases for the gas disk: 1) infinite thickness and uniform rotation, 2) finite thickness and uniform rotation, and 3) infinite thickness and shearing box approximating differential rotation. We then test the effect of a giant high velocity cloud (HVC) colliding with the disk at velocities in excess of $100\,\mathrm{km\,s^{-1}}$. We find that the usual Jeans criterion for the limit of cloud stability is modified by the additional term arising from the effect of the HVC collision with the gas disk. This term, which contains the velocity of the incoming cloud, and a characteristic scale for the shock front, is closely comparable in magnitude to the original term (the square of the product of the sound speed and the characteristic wave number) in the static Jeans equation and also similar in magnitude to the global effect of shear. This result shows that an HVC falling onto a disk that contains clouds close to Jeans equilibrium will generally be effective in triggering cloud collapse and subsequent star formation.

In the context of standard $\Lambda$CDM cosmology the accumulation of matter to form galaxies by progressive accretion is nowadays considered standard. Emphasis has consistently been placed on the formation of sub-galactic units of stars which go on to coalesce into larger galaxies and on the absorption of minor, satellite, galaxies by large galaxies. For entirely understandable observational reasons much less attention has been paid to the evolution of galaxies via secular gas accretion. There is no reason to assume, however, the absence of relatively large numbers of gas clouds at the low mass end of the mass function of galaxy scale clouds, too small to convert a significant fraction of their masses into stars, which would be available within groups of galaxies to fuel the secular evolution of larger galaxies by accretion.

We first consider the energy balance of the interaction of an HVC with a gas disk. Mac Low & Klessen (2004) showed that for interstellar gas clouds in general, support against gravitational collapse and maintenance of the observed internal velocity field depend on continued driving of the internal turbulence. The ratio $R$ between the dissipation rate for isothermal supersonic turbulent energy and that supplied by the shock front caused when the HVC impinges on the interstellar medium (ISM) of the Galactic disk is between $10^{-3}$ and 3. For values of $R \leq 1$, the impact of the HVC will be sufficient to break the equilibrium of gas clouds, leading to gravitational collapse and in turn to possible star formation so that we can already see that in general terms the star forming process will tend to be accelerated by the impact of a large HVC on the gas disk.

A more quantitative look at the effect of the infall of a large HVC onto the plane of a galaxy follows Chandrasekhar (1954). We will do this by examining the way the dynamical input affects the Jeans condition for the collapse of a gas cloud in the plane, introducing the force of the shock front induced by the HVC, which leads to the (modified) Jeans condition. We first write down, in a simplified way, the dynamical equations. These include the equation of mass conservation, and that describing the generation of the gravitational field. To enable us to obtain analytically solvable equations we make the highly simplifying assumption that the Galactic disk is an extended homogeneous gaseous disk with an angular velocity $\vec{\Omega}$. Then the fluctuations in the velocity field $\vec{u}$, density $\delta\rho$, pressure $\delta p$, and gravitational potential $\delta V$ are governed by the linearized equations (Casuso, Beckman, & Buenrostro 2006), where we assumed that the HVC falls perpendicularly to the plane of the disk (in the $z$ direction), and that $\rho_{\rm HVC} \sim 10^{-4}\rho_{\rm ISM}$. During the interaction of the HVC with the disk the shock front destabilizes some fraction of the gas favoring gravitational collapse, while after a relaxation time those clouds which have not collapsed to form stars will incorporate the shock energy as internal turbulent energy (the stability of these clouds is thereby increased), until the energy is dissipated and equilibrium is re-established.

This simple but direct approximate treatment, using a range of HVC physical properties taken from observations, shows us how the Jeans criterion for stability against collapse is modified by the dynamical effect of the infalling gas cloud. The modification is expressed in a single term added to the left hand side of the stability condition:

$$c^2k^2 \mp \frac{10^{-4}V_{\rm HVC}^2 k}{3d} \leq 4\pi G\rho. \qquad (1)$$

Using typical values for the parameters we show that for the three cases this term is similar in amplitude to the original term found in the simplest standard Jeans case, but if we also introduce the differential rotation through the shearing-box approximation we find that the shear term is of similar magnitude, and operates in the opposite sense to the effect of the HVC, as would be expected intuitively. This result shows that in a variety of circumstances the impact of an HVC with gas density much less than that of the disk gas will be to trigger those clouds which were close to equilibrium into a state of collapse, and hence by inference such impacts will tend to trigger star formation in any galactic disk, where the previous conditions prevailing are those close to equilibrium against gravitational collapse. This strongly suggests that HVCs should be effective in triggering star formation under these circumstances, a conclusion which may have rather wide implications for galaxy evolution.

## References

Casuso, E., Beckman, J. E., & Buenrostro, V. 2006, PASP, 118, 833
Chandrasekhar, S. 1954, ApJ, 119, 7
Mac Low, M. M., & Klessen, R. S. 2004, Rev. Mod. Phys., 76, 125

*Pathways through an Eclectic Universe*
*ASP Conference Series, Vol. 390, © 2008*
*J. H. Knapen, T. J. Mahoney, and A. Vazdekis, eds.*

# Supernova Feedback in Galaxy Formation

Y. Dubois and R. Teyssier

*Laboratoire AIM, CEA/DSM–CNRS–Université Paris Diderot, DAPNIA/SAp, 91191 Gif sur Yvette, France*

**Abstract.** The hierarchical model of galaxy formation is known to suffer from the "over-cooling" problem: the high efficiency of radiative cooling results in too much baryonic matter in a condensed phase (namely, cold gas or stars) when compared to observations. A solution proposed by many authors (see Springel & Hernquist 2003; Fujita et al. 2004; Rasera & Teyssier 2005) is feedback due to supernova (SN) driven winds or active galactic nuclei. Modeling SN feedback by direct injection of thermal energy usually turns out to be inefficient in galaxy-scale simulations, due to the quasi-instantaneous radiation of the SN energy. To avoid this effect, we have developed a new method to incorporate SN feedback in cosmological simulations: using temporary test particles, we reproduce explicitly a local Sedov blast wave solution in the gas distribution. We have performed several self-consistent runs of isolated Navarro, Frenk, & White (1996, hereafter NFW) halos with radiative cooling, star formation, SN feedback and metal enrichment using the adaptive mesh refinement code RAMSES (Teyssier 2002). We have explored the influence of SN feedback on the formation and the evolution of galaxies with different masses. We have studied the efficiency of the resulting galactic winds, as a function of the mass of the parent halo.

## 1. Numerical Ingredients

We have implemented a solution in a kinetic form for SNe in the adaptive mesh refinement (AMR) code RAMSES using typically 100 test particles per SN event, called "debris particles", to mimic a Sedov blast wave propagating up to two cells in radius. Particles are produced two by two with exactly opposite velocity vectors in the SN rest frame in order to conserve the global momentum of the simulation. When the particles have sufficiently propagated into the gas, they return all their mass and momentum to the gas, creating locally a Sedov blast wave flow. They also enrich the interstellar medium with metals. We also use the following ingredients in our galaxy formation simulation: metallicity-dependent atomic cooling and UV heating (no molecular cooling), star formation following a standard Schmidt law with a local characteristic timescale between $t_0 = 3$ and 8 Gyr. Following Springel & Hernquist (2003), we also model the effect of thermal energy injection due to SN feedback as a polytropic equation of state for the cold gas, where the polytropic index depends on the density of the gas (between 1.3 and 2).

Our simulations were performed using self-gravity with isolated boundary conditions for the gas only. The dark matter is considered as a static NFW halo. Initially, the gaseous halo also follows an NFW profile. The total baryonic mass fraction in the halo is fixed at $f_{\rm b} = 15\%$. It is initially in hydrostatic equilibrium with a Bullock et al. (2001) rotation profile in order to obtain a centrifugally supported disk. The halo spin parameter was chosen to be $\lambda = 0.04$ or $0.1$. We have also imposed a density cut-off at twice the virial radius. Boundary conditions for the hydrodynamics were taken as zero gradient or "outflow boundaries".

In all our simulations, we start with a coarse grid of $128^3$ cells covering a cubic region of three virial radii in length. We have adaptively refined our grid up to four more levels of refinement for the $10^{10}\,M_\odot$ dark matter halo and up to five more levels for the $10^{11}\,M_\odot$ one. The refinement strategy is based on a quasi-Lagrangian approach (the mass in each cell is roughly constant; see Teyssier 2002).

## 2. Conditions for the Apparition of a Galactic Wind

We obtained different results depending on the mass of the parent halo. In the first case of a dwarf galaxy ($10^{10}\,M_\odot$), an early wind appears after 1–3 Gyr (depending on $t_0$ and $\lambda$). For a medium mass galaxy ($10^{11}\,M_\odot$), a galactic fountain appears expelling gas from the disk but then falls back, unable to escape freely.

We can explain this behavior by studying the competition between the ram-pressure of the infalling gas and the SN luminosity using a very simple analytic model to compute the epoch at which the wind develops. We consider an NFW density profile for the halo and compute the accretion rate assuming the halo to be cooled instantaneously. We then compute the resulting star formation rate. By equating the SN luminosity and the accretion luminosity, we can deduce a characteristic time for the apparition of the galactic wind. We have one free parameter, $\chi$, the conversion efficiency of SN energy into galactic wind energy. Comparing our simulation results with our analytical model, we can deduce an approximate value of $\chi$ = 1–2%. For the most massive galaxy, the wind luminosity is not efficient enough to break out the ram-pressure of the infalling gas.

As simple as this analytical model might seem to be, it could shed light on the criterion for the apparition of a galactic wind. The virial mass of the parent halo is the main parameter that determines whether or not a wind can appear. We therefore arrive at a conclusion similar to that of Fujita et al. (2004) but using a more realistic cosmological setting: for efficient feedback, the main challenge is for the wind to be able to overcome the ram-pressure of infalling material.

### References

Bullock, J. S., et al. 2001, ApJ, 555, 240
Fujita, A., Mac Low, M.-M., Ferrara, A., & Meiksin, A. 2004, ApJ, 613, 159
Navarro, J. F., Frenk, C. S., & White, S. D. M. 1996, ApJ, 462, 563
Rasera, Y., & Teyssier, R. 2005, A&A, 445, 1
Springel, V., & Hernquist, L. 2003, MNRAS, 333, 649
Teyssier, R. 2002, A&A, 385, 337

*Pathways through an Eclectic Universe*
*ASP Conference Series, Vol. 390, © 2008*
*J. H. Knapen, T. J. Mahoney, and A. Vazdekis, eds.*

# *FUSE* and *Chandra* Observations of VV 114 and Haro 11: Feedback in Local Lyman Break Analogs

John Grimes, Tim Heckman, and Dave Strickland

*Center for Astrophysical Sciences, Johns Hopkins University, 3400 N. Charles St, Baltimore, MD 21218, USA*

**Abstract.** Recently, a population of local ultraviolet luminous galaxies have been identified as having many properties similar to those of high redshift Lyman break galaxies. We present *FUSE* and *CHANDRA* observations of the two nearest analogs, VV 144, and Haro 11. Both galaxies have galactic outflows with velocities of a few hundred km s$^{-1}$. Hot gas ($kT \sim 0.6$–$0.7$ keV) surrounds both galaxies and has properties similar to those observed in other local star-forming galaxies.

Understanding starburst-driven galactic winds in Lyman break galaxies (LBGs) is an essential step in understanding how star formation affects later galaxy formation. In the FUV, absorption line studies can examine the velocity, composition, and phase of the cool and warm material being accelerated by the winds. X-ray observations trace the hot gas ($T \sim 10^7$ K) from the collision of the high speed outflow with the cool, ambient interstellar medium (ISM) material. Unfortunately, LBGs are difficult to observe in the FUV and impossible in the soft X-ray bands. We must therefore observe local LBG analogs to extend our knowledge of LBGs into the FUV and X-ray wavelength ranges.

Both VV 114 and Haro 11 have been identified through SDSS and *GALEX* observations as local LBG analogs (Hoopes et al. 2003). They are similar to LBGs in $L_{\rm UV}$, UV surface brightness, star formation rate (SFR), galaxy size, and galaxy mass. Both galaxies also show morphological evidence of merger activity.

*FUSE* observations of VV 114 and Haro 11 show strong ISM absorption lines similar to those seen in LBGs. Blueshifted wings of the absorption lines suggest outflows of entrained material with velocities of ~350 and ~240 km s$^{-1}$ in VV 114 and Haro 11, respectively. Unlike some high redshift LBGs (Shapley et al. 2006), neither galaxy shows convincing evidence of Lyman continuum emission.

O VI emission is the dominant cooling mechanism for coronal ($10^{5.5}$ K) gas. O VI cooling could inhibit an outflow's growth if a significant fraction of the available energy was radiated away. O VI$\lambda\lambda$1032,1038 emission is observed in Haro 11 but not in VV 114. If associated with the outflow of Haro 11 < 15% of the available energy is lost to cooling, suggesting that energy loss is not significant. The complicated O VI line profiles are not consistent with the ISM absorption features for either Haro 11 or VV 114, indicating that the O VI is produced in different physical regions of the galaxy. In VV 114, the O VI absorption feature is significantly blueshifted relative to the other lines. The O VI is not entrained within the outflow but instead could be produced at the interface between the outrushing gas and cooler, ambient material. The Haro 11 O VI feature is consistent with a P Cygni profile of an expanding shell of gas with an outflow velocity of $\sim 120\,\mathrm{km\,s^{-1}}$.

Hot (0.68 and 0.62 keV, respectively) optically thin, thermalized gas surrounds both Haro 11 and VV 114 (see Fig. 1). Both galaxies have elevated $\alpha$-element/Fe ratios relative to solar, suggesting enrichment from SN II ejecta.

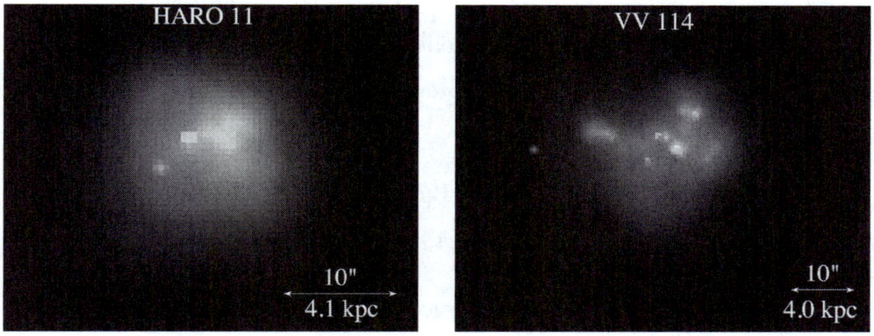


Figure 1. *Chandra* images (0.3–8.0 keV) of Haro 11 (*left*) and VV 114 (*right*).

We have also compared the X-ray, $K$-band, and FIR properties of VV 114 and Haro 11 to those of a sample of 23 local star-forming galaxies. VV 114 and Haro 11 follow the same relationship between FIR emission and thermal X-ray emission (suggesting a near constant efficiency in turning kinetic energy from the winds into X-ray emission). The scale of the soft X-ray emission is more strongly correlated with the $K$-band (a proxy for stellar mass) than the FIR (SFR) luminosity. There is a mild correlation between the SFR per unit area and the temperature of the hot gas.

Both VV 114 and Haro 11 have strong, blueshifted absorption lines that appear to be similar in equivalent width and velocity dispersion to those seen in LBGs, suggesting that they have similar galactic scale outflows. Significant amounts of enriched gas and energy are contained within the hot, thermalized X-ray gas. As this material most closely traces the hot wind fluid ($T \sim 10^8$ K), this suggests that star formation driven winds can energize and enrich the intergalactic medium in the early Universe. O VI radiative cooling does not appear to be a significant source of energy loss in galactic winds and should not significantly inhibit the growth of the outflow. We find that LBG analogs are excellent local laboratories for detailed high energy studies of high redshift LBGs.

**References**

Hoopes, C. G., Heckman, T. M., Strickland, D. K., & Howk, J. C. 2003, ApJ, 596, L175
Shapley, A. E., Steidel, C. C., Pettini, M., Adelberger, K. L., & Erb, D. K. 2006, ApJ, 651, 688

*Pathways through an Eclectic Universe*
*ASP Conference Series, Vol. 390, © 2008*
*J. H. Knapen, T. J. Mahoney, and A. Vazdekis, eds.*

# Feeding Active Galaxies: Gas Dynamics from the Outskirts to the Very Center

S. Haan,[1] E. Schinnerer,[1] C. G. Mundell,[2] F. Combes,[3]
S. García-Burillo,[4] and E. Emsellem[5]

[1]*Max-Planck-Institut für Astronomie, Königstuhl, 17, 69117-Heidelberg, Germany*

[2]*Astrophysics Research Institute, Liverpool John Moores University, Twelve Quays House, Egerton Wharf, Birkenhead, CH41 1LD, UK*

[3]*Observatoire de Paris, LERMA, 61 Av. de l'Observatoire, 75014-Paris, France*

[4]*Observatorio Astronómico Nacional (OAN)-Observatorio de Madrid, Alfonso XII, 3, 28014-Madrid, Spain*

[5]*Université de Lyon 1, CRAL, CNRS, UMR5574, Observatoire de Lyon; ENS de Lyon, France*

**Abstract.** We analyze the gas dynamics of 16 nearby spiral galaxies. The H I gas distributions and kinematics indicate that some properties of the atomic gas correlate with the type of the AGN present. Furthermore, gas inflow rates are derived as a function of radius and location within the disks using combined H I and CO data.

## 1. Introduction

Study of the process of fueling the centers of active galaxies with gas is paramount for understanding galaxy formation, in particular secular evolution and the origin of nuclear activity. The majority of AGN show low ionization narrow emission-line regions (LINERs) with luminosities lower than Seyfert galaxies and quasars. Although Seyferts and LINERs are powered by the same basic process, i.e., accretion of matter by a supermassive black hole, the Seyfert/LINER distribution is bimodal (Groves et al. 2006), implying different accretion modes (for a general review see Ulrich, Maraschi, & Urry 1997; Combes 2001). Despite numerous studies, the origin of this difference is still under debate. Furthermore, an understanding of AGN feeding, i.e., how to remove the angular momentum from gas at larger radii, is still missing. Observations over the last decade suggest that a hierarchy of mechanisms might combine to transport gas from large (kpc) scales down to the inner (pc) scales.

**The NUGA Project:** To study the gas transport from the outskirts of galaxies to the central AGN we complemented the IRAM key project NUclei of GAlaxies (NUGA, PIs: García-Burillo and Combes) with VLA observations (HI-NUGA, PI: Schinnerer).

*CO Observations:* The millimeter CO (2→1) and (1→0) lines were observed with the IRAM Plateau de Bure Interferometer to study the molecular gas in galaxy centers. The NUGA CO survey observed 12 (core sample; total:~30) nearby low luminosity AGN host galaxies at high spectral (3–6 km s$^{-1}$) and angular (~0.5″) resolution.

*H I Observations:* For our detailed H I study presented here, 16 galaxies from the total NUGA sample were selected for observations of their 21 cm line emission of neutral H I gas using the NRAO Very Large Array (VLA) interferometer in its C- and D-array configuration with ~20″ angular and ~5.2 km s$^{-1}$ spectral resolution. Our AGN classification (Seyfert, LINER, and starburst galaxies) is based on the method of Kewley et al. (2006). All galaxies are spirals ranging in Hubble type from Sa to Sbc.

## 2. Results

The atomic gas distributions and kinematics for a sample of 16 galaxies are studied and possible relations with the large environment and the AGN type present (Seyfert, LINER) are investigated. Four galaxies (25%) are surrounded by companions with associated H I emission and 56% of our sample has dynamically disturbed H I disks. A comparison between H I properties and AGN type suggests several correlations: LINER host galaxies seem to have a larger fraction of disturbed H I disks, a significantly higher percentage of H I ring geometry, and a slightly smaller H I extent than Seyfert host galaxies. No correlation between the presence of companions and the AGN type was found.

The gas inflow rates using CO and H I data are studied via gravity torque maps extracted from NIR/optical images and the gas distributions (see García-Burillo et al. 2005). A pilot study with the combined molecular and atomic gas distribution was performed on NGC 6951. Preliminary results show that 1) the directions of torques are similar for the large scale (20 kpc) and the central scale (1 kpc) and 2) the torque changes sign in different quadrants of the gas disk, which can be attributed to the action of the outer bar and the nuclear oval, respectively. In combination with recently observed VLA B-array data (giving 7″ resolution) we will apply the gravity torque method to seven galaxies of our sample in order to derive gas inflow rates and to assess the efficiency of the feeding of the central black hole.

**Acknowledgments.** SH is supported by the German DFG under grant number SCHI 536/2-1.

### References

Combes, F. 2001, in Advanced Lectures on the Starburst-AGN Connection, ed. I. Aretxaga, D. Kunth, & R. Mújica (Singapore: World Scientific), p. 223

García-Burillo, S., Combes, F., Schinnerer, E., Boone, F., & Hunt, L. K. 2005, A&A, 441, 1011

Groves, B., Kewley, L., Kauffmann, G., & Heckman, T. 2006, New Astron. Rev., 50, 743

Kewley, L. J., Groves, B., Kauffmann, G., & Heckman, T. 2006, MNRAS, 372, 961

Ulrich, M.-H., Maraschi, L., & Urry, C. M. 1997, ARA&A, 35, 445

*Pathways through an Eclectic Universe*
*ASP Conference Series, Vol. 390, © 2008*
*J. H. Knapen, T. J. Mahoney, and A. Vazdekis, eds.*

# Tidal Tails around Globular Clusters

M. Montuori,[1] R. Capuzzo-Dolcetta,[2] P. Di Matteo,[3] and P. Miocchi[2]

(1) *CNR-INFM, Roma, Italy*
(2) *Dept. Physics, Universitá di Roma "La Sapienza", Roma, Italy*
(3) *Observatoire de Paris, LERMA, Paris, France*

**Abstract.** We present the results of detailed $N$-body simulations of clusters moving in a realistic Milky Way potential. The strong interaction with the bulge and the disk of the Galaxy leads to the formation of tidal tails, emanating from opposite sides of the cluster. Their orientation and morphology may be interpreted easily in terms of a comoving frame of coordinates.

Tidal tails are often observed around dwarf galaxies and globular clusters (GCs) (see, for example, Odenkirchen et al. 2002). Figure 1 shows the formation and evolution of tidal tails around a GC, simulated as an $N$-body system moving in a Milky Way potential (Montuori et al. 2007). Tidal tails clearly show up after less than one orbital crossing. Their extension and orientation depend on the velocity and acceleration of the cluster along the orbit. The outer part (at distance $>$ 7–8 tidal radii) of the tails is aligned with the cluster orbit, while the inner part shows a shape which varies with time. When the cluster approaches its apogalacticon, tails divide into multiple "arms", as observed in NGC 288 and Willman 1.

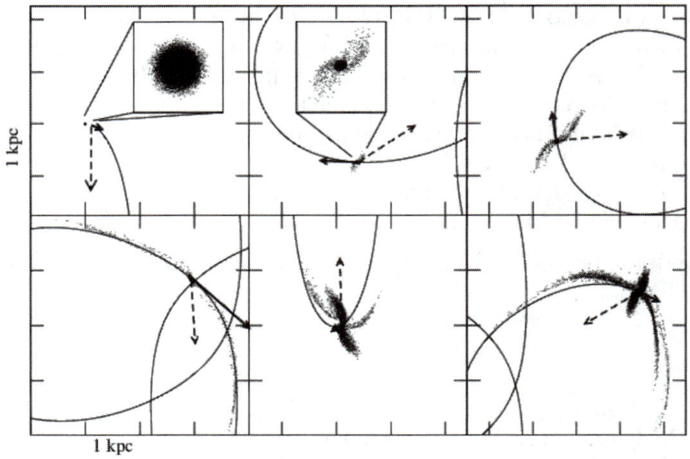


Figure 1. The solid arrow is the GC velocity vector; the dashed arrow points to the Galactic Center; the curve is the GC orbit.

Figure 2 shows the behavior of some orbital quantities. Because the orientation of the inner tails is highly correlated to the cluster orbital phase and with the local orbital angular acceleration, the orbital path cannot be deduced directly from the orientation of the tails, unless a sufficiently large field around the cluster is observed.

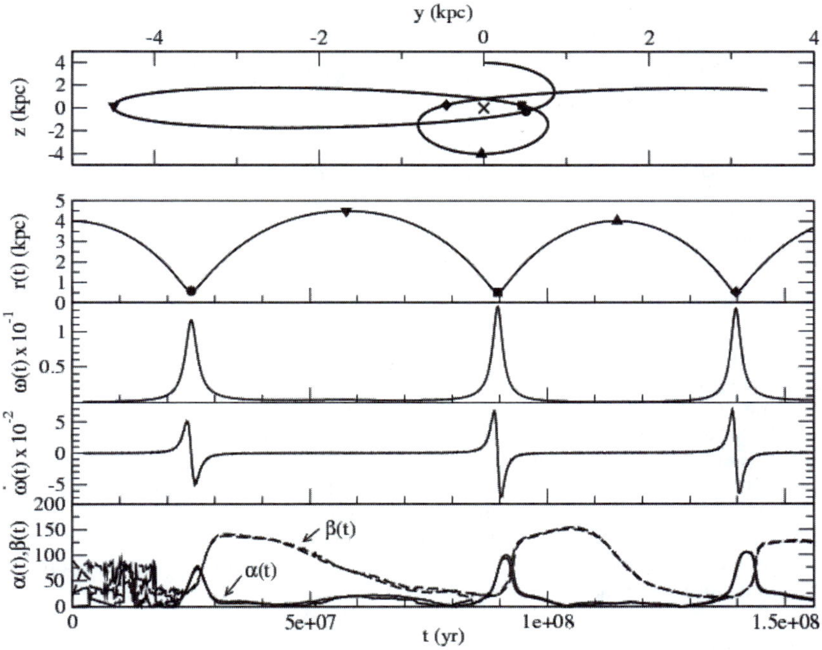


Figure 2. From the top down: GC orbit (*1st panel*); GC distance to the Galactic Center (*2nd panel*); GC angular velocity (*3rd panel*); GC angular acceleration (*4th panel*); angles $\alpha(t)$ and $\beta(t)$ between inner tail and radius vector $r$ and angular velocity, respectively (*5th panel*). All the quantities shown in panels 2 to 5 are vs time.

**References**

Odenkirchen, M., et al. 2002, AJ, 124, 1497

Montuori, M., Capuzzo Dolcetta, R., Di Matteo, P., Lepinette, A., & Miocchi, P. 2007, ApJ, 659, 1212

*Pathways through an Eclectic Universe*
*ASP Conference Series, Vol. 390, © 2008*
*J. H. Knapen, T. J. Mahoney, and A. Vazdekis, eds.*

# High Redshift Radio Galaxies: AGN Feedback in Extreme Starbursts

Nicole P. H. Nesvadba,[1,2] M. D. Lehnert,[1] C. De Breuck,[3] A. Gilbert, [4] and W. van Breugel[4]

[1] *GEPI, Observatoire de Paris, CNRS, Université Denis Diderot, 5 Place Jules Janssen, F-92190 Meudon, France*
[2] *Marie Curie Fellow*
[3] *European Southern Observatory*
[4] *Institute of Geophysics and Planetary Physics, Lawrence Livermore National Laboratory*

**Abstract.** We present the initial results of an ongoing study to identify the fingerprints of powerful AGN feedback during the formation epoch of massive galaxies at high redshift by tracing the dynamics of their optical emission line gas through integral field spectroscopy. In a small sample of radio galaxies at $z = 2$–3.5, we find energetic outflows of $\sim 10^{9-10}\,M_{\odot}$ of ionized gas being driven out with velocities of up to $1000\,\mathrm{km\,s^{-1}}$. Geometry, timescale, and energy arguments indicate the outflows are related to the AGN and are probably driven through the mechanical energy of the radio jet. With kinetic energies of $\geq 10^{60}$ erg, sufficient to unbind significant gas masses, these outflows will potentially escape even the most massive halos.

## 1. Introduction

AGN feedback is a crucial component of state-of-the-art models of galaxy formation, tailored to solve some of the outstanding mysteries in the evolution of massive galaxies (Croton et al. 2006). By not postulating a phase of powerful AGN feedback terminating the formation epoch of massive galaxies, theorists fail to explain the slope of the mass function at its upper end (Benson et al. 2003), as well as the apparent "anti-hierarchical" properties of massive galaxies at low redshift (old, evolved stellar populations, low gas content, and little ongoing star formation). However, direct observational constraints are scarce.

## 2. The "Smoking Gun" of AGN Feedback

Using the near-infrared integral field spectrograph SINFONI on the VLT, we have observed a small sample of six high redshift radio galaxies (HzRGs) at redshifts $z = 2.2$–3.5 in the $H$ and $K$ bands (Nesvadba et al. 2006, 2007a, 2008). Four galaxies in our sample have extended radio lobes ($R > 50\,\mathrm{kpc}$), and two galaxies are compact, with radii $< 10\,\mathrm{kpc}$ in the radio. Measuring the morphologies and gas kinematics of the restframe optical line emission, we find that all galaxies with extended radio sources have extended nebulosities with 20–30 kpc radius that are aligned with the radio jet axis (Fig. 1). The velocity fields resemble inflating "bubbles" with uniform internal motion but large relative velocities between the two regions ($> 1000\,\mathrm{km\,s^{-1}}$, Fig. 1). Three

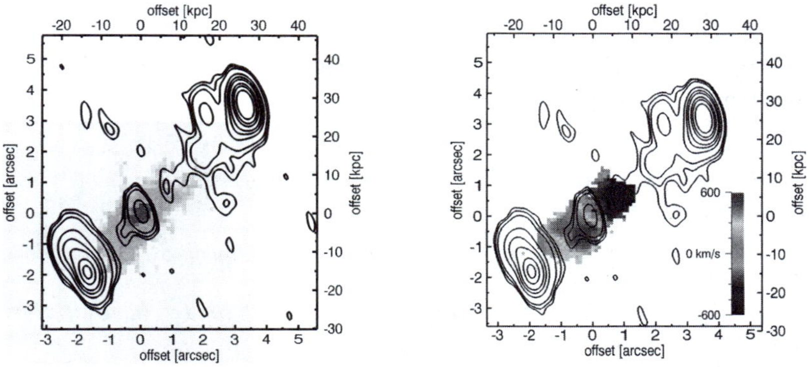


Figure 1. *See electronic version for a color version of this figure.* [O III]λ5007 line image (*left*) and [O III]λ5007 relative velocities (right) in MRC 0406-244 at $z = 2.4$. Contours show the VLA 1.4 GHz morphologies.

arguments indicate that the gas kinematics are most likely related to the radio jet (Nesvadba et al. 2006, 2007a):

- **Geometry** In all extended radio galaxies, the outflows are aligned with the axis of the radio jets, while the radius of the radio lobes exceeds the extent of the gas. The two galaxies with compact radio sources have compact emission line morphologies (Nesvadba et al. 2008).
- **Timescales** Estimating the age of the outflows, $\tau$, from their spatial extent, $R$, and velocities, $V$, as $\tau \approx R/V \approx$ a few $\times 10^7$ yr, we find that the age of the outflows appears very similar to typical ages of radio jets (a few $\times 10^{7-8}$ yr).
- **Kinetic Energies** The observed outflow rates of $10^{2-3}\,M_\odot\,\mathrm{yr}^{-1}$, correspond to a total mass loss of a few $\times 10^{10} M_\odot$ of gas in $\sim 10^{7-8}$ yr. Moreover, kinetic energies of $\sim 10^{44-45}\,\mathrm{erg\,s^{-1}}$ must be injected to power the observed gas dynamics, compared to $10^{45-46}\,\mathrm{erg\,s^{-1}}$ provided by the jets, implying efficient coupling. Over $10^7$ yr, this means that $10^{60}$ erg are injected into the gas, about the binding energy of the dark matter halo of a massive galaxy.

These outflows appear significantly stronger than starburst-driven winds at similar redshifts (Nesvadba et al. 2007b), and, intriguingly, they meet all the necessary requirements of being the "smoking gun" of powerful AGN feedback. Thus, unlike for other proposed AGN feedback mechanisms, we do have direct evidence of jet-driven outflows at high redshift.

## References

Benson, A., Bower, R., Frenk, C., Lacey, C., Baugh, C., & Cole, S. 2003, ApJ, 599, 38
Croton, D. J., et al. 2006, MNRAS, 365, 11
Nesvadba, N. P. H., Lehnert, M., Eisenhauer, F., Gilbert, A., Tecza, M., & Abuter, R. 2006, ApJ, 650, 693
Nesvadba, N. P. H., Lehnert, M., De Breuck, C., Gilbert, A., & van Breugel, W. 2007a, A&A, 475, 145
Nesvadba, N. P. H., et al. 2007b, ApJ, 657, 725
Nesvadba, N. P. H., Lehnert, M., De Breuck, C., Gilbert, A., & van Breugel, W. 2008, A&A, submitted

*Pathways through an Eclectic Universe*
*ASP Conference Series, Vol. 390, © 2008*
*J. H. Knapen, T. J. Mahoney, and A. Vazdekis, eds.*

# Mapping the Roots of the Galactic Wind in NGC 1569 with the Gemini GMOS IFU

M. S. Westmoquette,[1] K. M. Exter,[2] L. J. Smith,[1,2] and J. S. Gallagher[3]

[1]*Department of Physics and Astronomy, UCL, Gower Street, London, WC1E 6BT, UK*
[2]*Space Telescope Science Institute, 3700 San Martin Drive, Baltimore, MD 21218, USA*
[3]*Department of Astronomy, University of Wisconsin-Madison, 5534 Sterling, 475 North Charter St., Madison WI 53706, USA*

## 1. Introduction

NGC 1569 is classified as a post-starburst dwarf irregular galaxy and, at a distance of $2.2 \pm 0.6$ Mpc, is the closest example of a starburst. It has undergone a recent episode of star formation that resulted in a chaotic interstellar medium (ISM; Hunter, Hawley, & Gallagher 1993; Heckman et al. 1995; Stil & Israel 2002). NGC 1569 has a galaxy-wide outflow observed in the optical and X-rays that appears (Martin, Kobulnicky, & Heckman 2002) to emanate from the full extent of the H$\alpha$ disk. The most conspicuous star-forming region is located 90 pc west of the famous super star cluster (SSC) A, and exhibits evidence for much on-going activity.

To investigate the roots of the wind outflow, the interaction of individual winds from clusters with their environments must be probed at high angular scales. However, spectral resolution is equally important for determining properties such as gas dynamics and excitation mechanisms.

## 2. Observations and Emission Line Fitting

In November 2004, queue-mode observations using the Gemini-North Multi-Object Spectrograph (GMOS) integral field unit were obtained covering four regions near the center of NGC 1569. Only results from one position are discussed here. Through 500 object fibers arranged in a rectangular array (each with a diameter of $0\farcs2$), we obtained a spectral coverage of 4740–6860 Å with a dispersion of 0.34 Å/pix. The data were reduced using the standard Gemini pipeline.

We used an interactive curve-fitting program written in IDL (PAN; Dimeo 2005) to automate the fitting of multiple Gaussian components to each of the emission lines observed. Decomposing each compound line profile means we can analyze the properties of each component separately, allowing us to disentangle flux, velocity, and line width information.

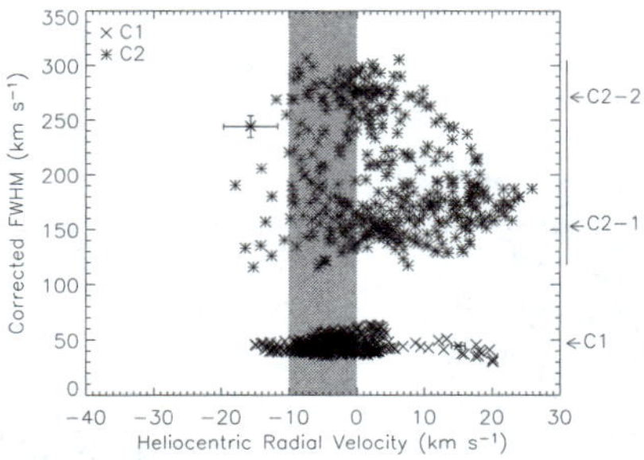


Figure 1. Hα line width (FWHM) vs radial velocity (relative to $v_{sys}$). The shaded region represents the velocity of the star cluster within the field.

## 3. Line Components and Interpretation

In general, we find the line profiles to be composed of a bright, narrow component (hereafter C1) overlaid on fainter, broad emission (hereafter C2). We find bright C1 emission over the whole field, well characterized by a Gaussian function with a distinct lower limit to its FWHM of $\sim$35 km s$^{-1}$. We also measure a fainter line component (C2) over the whole field with line widths between FWHM $\sim$ 100–300 km s$^{-1}$ (see Fig. 1). Through mapping out the properties of the components, we find that the width of the broad underlying emission (C2) is both highly correlated with the flux of C1 and anti-correlated with its own flux.

We interpret C1 to represent the turbulent ISM, where its motions (width) result from a convolution of the general stirring effects of the starburst and gravitational virial motions (thus giving rise to the clear FWHM lower limit). Additional minor contributions may result from unresolved expanding shell components along the line of sight.

C2 cannot result from any of the effects described above since none is able to produce such broad line widths, or such strong correlations between the emission line maps described above. However, we know that C2 must somehow be linked to effects produced by the starburst. What happens when cluster winds impact the surfaces of cool ISM gas clouds? Shearing between the high velocity flow and the cloud surface can set up a turbulent mixing layer (Slavin, Shull, & Begelman 1993) that can drive gas evaporation and mechanical ablation. We therefore conclude that C2 represents emission from tenuous gas within a turbulent mixing layer set up on the surface of the observed gas knot.

### References

Dimeo, R. 2005, PAN User Guide
Heckman, T. M., Dahlem, M., Lehnert, M. D., Fabbiano, G., Gilmore, D., & Waller, W. H. 1995, ApJ, 448, 98
Hunter, D. A., Hawley, W. N., & Gallagher, J. S. 1993, AJ, 106, 1797
Martin, C. L. 1998, ApJ, 506, 222
Martin, C. L., Kobulnicky, H. A., & Heckman, T. M. 2002, ApJ, 574, 663
Slavin, J. D., Shull, J. M., & Begelman, M. C. 1993, ApJ, 407, 83
Stil, J. M., & Israel, F. P. 2002, A&A, 392, 473

Bernard Pagel and David Lambert (*above*), and Pavel Kroupa (*below*).

# Session V

# Galaxy Evolution

*Pathways through an Eclectic Universe*
*ASP Conference Series, Vol. 390, © 2008*
*J. H. Knapen, T. J. Mahoney, and A. Vazdekis, eds.*

# The Assembly History of Massive Galaxies: What Do We Know?

Christopher J. Conselice

*School of Physics & Astronomy, University of Nottingham, University Park, Nottingham, NG7 2RD, UK*

**Abstract.** Understanding the formation history of massive galaxies is one of the most popular and longstanding problems in astronomy, with observations and theory addressing how and when these systems assembled. Since the most massive galaxies in today's Universe, with $M_* > 10^{11} M_\odot$, are nearly all elliptical with uniform old stellar populations, we must probe higher redshifts to discover their full origins. A recent consensus has developed that nearly all $M_* > 10^{11} M_\odot$ galaxies we see today were established by $z \sim 1$, with at most a factor of two growth in stellar mass and number densities at lower redshifts. We review the evidence for this and discuss how recent observations of star formation rates, colors, and morphologies of massive galaxies at $z < 1$ with $M_* > 10^{11} M_\odot$ show that these systems are still experiencing some evolution. Massive galaxies undergo on average a single major merger at $z < 1.5$, and roughly half are experiencing star formation at the same redshifts. The highest mass galaxies, with $M_* > 10^{11.5} M_\odot$, appear in similar abundance at $z < 2$, suggesting that extremely massive galaxies are mostly formed very early in the Universe. Observations at $z > 1.5$ demonstrate that major galaxy mergers are the primary method for assembling these massive galaxies, with nearly all of this merging occurring at $z > 2$, with on average four to five major mergers taking place at $z = 1.5$–3.

## 1. Introduction

Determining when and how galaxies in the Universe formed is one of the most outstanding problems in cosmology and galaxy formation. Galaxies are predicted in cold dark matter based models of structure formation to assemble gradually with time through the merging of smaller systems (e.g., White & Rees 1978). While there is some evidence for this process, at least in terms of galaxies (e.g., Le Fevre et al. 2000; Patton et al. 2002; Conselice et al. 2003a; Conselice, Chapman, & Windhorst 2003b; Bridge et al. 2007), many details are still lacking. Alternatively, massive galaxies, which are mostly ellipticals in today's Universe (e.g., Conselice 2006a), may have formed in a very rapid collapse of gas (e.g., Larson 1974).

As such, massive galaxies are largely the test-bed for galaxy models. Understanding their evolution observationally is therefore an important test of the physics behind galaxy formation. As star formation and merging activity has been seen in ellipticals from $z \sim 0$ to $z \sim 1$ (Stanford et al. 2004; Lin et al. 2004; Teplitz et al. 2006), it is not clear when or how the most massive galaxies finally assembled. If it were possible to date every star in nearby massive galaxies, we could in principle determine the formation epoch and timescales of these systems by examining their individual stars. We cannot, however, resolve stars in any but the nearest galaxies, and their integrated stellar properties, such as colors, become degenerate after about 5 Gyr (e.g., Worthey 1994). Stellar ages also do not necessarily correlate with the assembly of mass through, for example, merging activity (Conselice 2006b; De Lucia et al. 2006; Trujillo et al. 2006). An alternative approach towards understanding massive galaxies and their evolution is empirically measuring the number densities, morphologies, star formation rates, and stellar masses of massive systems at some fiducial time, and to compare these to similar quantities at different times (redshifts), and with models.

Observational evidence suggests that passively evolving massive galaxies exist at $z \sim 1$, and likely at even earlier times, at $z > 2$ (Fontana et al. 2004; Daddi et al. 2004; Glazebrook et al. 2004; Saracco et al. 2005). Recent claims also exist for the establishment of the full massive galaxy population by $z \sim 1$ (e.g., Drory et al. 2005; Bundy, Ellis, & Conselice 2005, 2006; Borch et al. 2006; Cimatti, Daddi, & Renzini 2006). However, what is not yet clear is if number densities measured in these surveys are able to rule out evolution at $z < 1$ due to uncertainties in measuring stellar masses, number densities, and cosmic variance.

On the other hand, at $z > 1.5$ it appears that there are significantly fewer massive galaxies than at $z < 1.5$ (e.g., Fontana et al. 2004). Observationally, a large fraction of the most massive galaxies at $z > 1.5$ are undergoing major mergers, which are able to construct the stellar masses of these galaxies rapidly (e.g., Conselice 2006b). The merger history at $z < 1.5$ is not as clear, with observations inconsistent on whether there is evolution in the massive galaxy population at $z < 1$ (e.g., Bell et al. 2004; Brown et al. 2007; Scarlata et al. 2007). We argue in this review, after taking into account all sources of error and by examining the physical properties of massive galaxies at $z < 1.5$, that at least one major merger is occurring within these systems. A significant fraction of massive galaxies at $z < 1.4$ also have not yet acquired a smooth elliptical structure, and have ongoing star formation. Thus, while the bulk of the stellar mass in massive galaxies is present by $z \sim$ 1–1.5, there is still observable evolution. Throughout this review we use a standard cosmology of $H_0 = 70\,\mathrm{km\,s^{-1}\,Mpc^{-1}}$, $\Omega_\mathrm{m} = 1 - \Omega_\lambda = 0.3$, and a Chabrier initial mass function (IMF) for stellar mass calculations.

## 2. Number and Mass Densities

The most basic method for understanding the evolution of massive galaxies is measuring how their number densities and integrated mass densities change as a function of time. This is typically done through the use of stellar masses (e.g., Brinchmann & Ellis 2000; Conselice, Blackburne, & Papovich 2005a; Conselice et al. 2005b; Bundy et al. 2005, 2006). Alternatively, it has remained popular to determine the number densities for luminous, red, or elliptical galaxies, although these selection methods will produce biases when trying to understand the evolution of massive galaxies, as star formation and morphological evolution are occurring in "early-type" galaxies at $z \sim 1$ (e.g., Stanford et al. 2004; Teplitz et al. 2006). Recent work on measuring densities suggests that within the uncertainties galaxies with large stellar masses, with $M_* > 10^{11}\,M_\odot$, are largely in place at $z \sim 1$ (Glazebrook et al. 2004; Bundy et al. 2005, 2006; Cimatti et al. 2006).

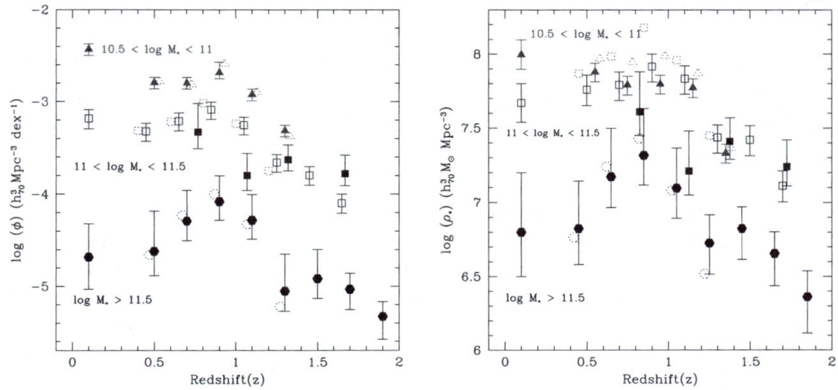


Figure 1. *Left:* The evolution in the number densities for galaxies of various stellar masses in the redshift range $z \sim 0.4$–1.4. *Right:* The stellar mass density evolution as a function of stellar mass in the same redshift intervals. The points at $z \sim 0$ are taken from Cole et al. (2001). The error bars listed on both the numbers and stellar mass densities reflect uncertainties from stellar mass errors, as well as cosmic variance, and counting statistics. The dashed symbols near each data point show how these values would change if just using photometric redshifts. For comparison, we show the stellar mass densities for systems with $10^{11}\,M_{\odot} < M_* < 10^{11.5}\,M_{\odot}$ from Glazebrook et al. (2004) plotted as solid boxes (blue in the electronic version). Note that shifts of $\pm 0.05$ in redshift have been applied so that the data points and error bars do not overlap.

Figure 1 shows an updated version of how the number and mass densities of galaxies with stellar masses $M_* > 10^{11.5}\,M_{\odot}$ and $10^{11}\,M_{\odot} < M_* < 10^{11.5}\,M_{\odot}$ evolve out to $z \sim 2$, as seen in the large $1.5\,\text{deg}^2$ Palomar Observatory Wide-Field Infrared Survey (Conselice et al. 2007a) covering the DEEP2 fields (Davis et al. 2003, 2006) from Conselice et al. (2007b). Figure 1 also shows the number densities of galaxies within these mass ranges measured in the nearby Universe to $z \sim 0.2$ by the 2MASS/2dF galaxy surveys (Cole et al. 2001), normalized using the same Chabrier IMF used for the higher redshift stellar masses. The number density evolution of these massive galaxies demonstrates that statistically there is very little to no evolution at $z < 1$ for the $M_* > 10^{11}\,M_{\odot}$ systems. This appears to support the idea that massive galaxies are present by $z \sim 1$ (e.g., Cimatti et al. 2006; Bundy et al. 2006; Brown et al. 2007).

However, as can be seen by eye in Fig. 1, within the observational errors, there is some evolution in number densities for $M_* > 10^{11}\,M_{\odot}$, and perhaps $M_* > 10^{11.5}\,M_{\odot}$ selected galaxies between $z \sim 1$–1.5. The evolution in the number and mass densities can be examined quantitatively in a number of ways. When considering evolution just within this sample from $z = 1.5$ to $z = 1$, there are significant increases at masses $10^{11}\,M_{\odot} < M_* < 10^{11.5}\,M_{\odot}$, both in terms of number and mass densities. This is also the case when considering evolution between $z \sim 1.5$ and $z \sim 2$. However, galaxies with $M_* > 10^{11.5}\,M_{\odot}$ show an increase by a factor of $2.7^{+1.8}_{-1.7}$ in number densities between $z = 1.5$ to 0.4. This is significant only at the $< 2\sigma$ level, considering all the uncertainties. In fact, all of this evolution occurs at $z > 1$. Furthermore, there is a factor of $1.3^{+0.74}_{-0.53}$ increase in the mass density associated with $M_* > 10^{11.5}\,M_\odot$ galaxies in the same redshift range, although this is also at less than $3\sigma$ significance. There is an increase of $11.2^{+8.7}_{-4.9}$ in number densities, and a factor of $5.5^{+4.3}_{-1.7}$ increase in mass densities for systems with $M_* > 10^{11.5}\,M_\odot$ from $z \sim 2$ to $z \sim 1$. This is also an insignificant increase, and it is impossible to rule out that massive galaxies with $\mathrm{M}_* > 10^{11.5}\,M_\odot$ are all in place at $z < 2$.

An analysis of Fig. 1 shows that the number density of systems with $10^{11}\,M_\odot < M_* < 10^{11.5}\,M_\odot$ increases by a factor of $2.2^{+0.57}_{-0.41}$ between $z = 1.4$ and $z = 0.4$, a result significant at $> 4\sigma$. Just as for the most massive systems, this evolution occurs completely at $z > 1$. Similarly, there is a factor of $2.1^{+0.6}_{-0.35}$ increase in the integrated mass density for systems with $10^{11}\,M_\odot < M_* < 10^{11.5}\,M_\odot$ within the same redshift range, also at $> 4\sigma$ confidence. After correcting for incompleteness there is a factor of $14.5^{+4.1}_{-2.8}$ evolution in the number densities, and a factor of $10.7^{+3.1}_{-2.0}$ in mass densities between $z \sim 2$ and $z \sim 1$ for galaxies with $10^{11}\,M_\odot < M_* < 10^{11.5}\,M_\odot$ (Conselice et al. 2007b).

The observed evolution is such that the most massive systems with $M_* > 10^{11}\,M_\odot$ increase in number and mass densities by factors $> 2$–$3$ at a significance $> 3\sigma$. Taken as a whole, we calculate that the scenario whereby the stellar mass and number densities of massive galaxies do not evolve between $z \sim 1.5$ to $z \sim 0.4$ can be rejected at $> 8\sigma$ confidence. Therefore, it does not appear that high mass galaxy formation, with the possible exception of $M_* > 10^{11.5}\,M_\odot$ systems, is complete by $z \sim 1.4$, yet it is largely completed by $z \sim 1$. Therefore, the redshift range $z \sim 1 - 1.5$ is the final epoch for the build-up of the majority of the mass in massive galaxies. However, there could easily be a factor of two or three evolution in the mass and number densities for massive galaxies at $z < 1$, and we would not be able to measure this based on the current uncertainties in the measurements of these quantities. The best way to approach this problem is to determine if massive galaxies have any ongoing evolution based on physical features, through structure and star formation.

## 2.1. Structures and Morphologies

Investigating the structures and morphologies of galaxies is becoming recognized as one of the most important methods for understanding galaxies (e.g., Conselice 2003; Cassata et al. 2005; Trujillo et al. 2006), and for tracing the merger history at higher redshifts (e.g., Conselice et al. 2003a; Bridge et al. 2007).

Early work showed that massive galaxies at $z < 1$ are generally early types or disks (Brinchmann & Ellis 2000; Bundy et al. 2005). The overlap of Palomar NIR imaging and *Hubble* ACS imaging in the Extended Groth Strip (Davis et al. 2006) allows us to study in detail $> 500$ galaxies with stellar masses $M_* > 10^{11}\,M_\odot$ at $z < 1.4$. For nearly all of these systems, their magnitudes are bright enough such that effects due to redshift do not affect the ability to classify these systems either by eye or through quantitative methods (e.g., Conselice et al. 2000c; Windhorst et al. 2002; Papovich et al. 2005; Taylor-Mager et al. 2006; Conselice et al. 2007b).

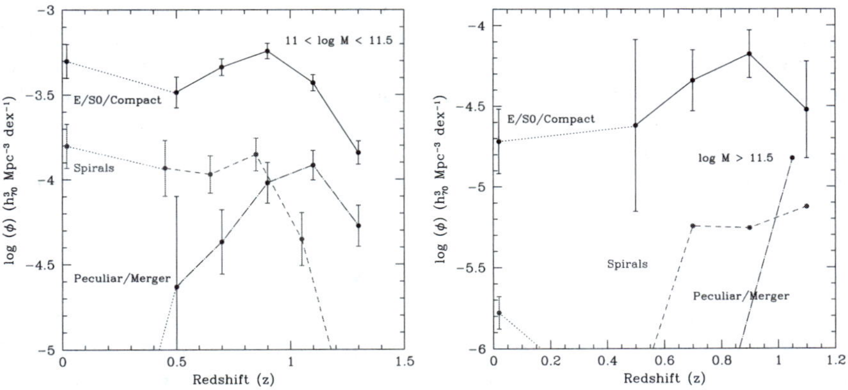


Figure 2. Visual estimates of the morphological evolution for massive galaxies with $M_* > 10^{11}\,M_\odot$ taken from Conselice et al. (2007b) using ACS data. This figure is divided into two mass ranges, and shows the evolution of galaxy number densities at $z < 1.4$. *Left:* The morphological breakdown up to $z \sim 1.4$ for systems with $10^{11}\,M_\odot < M_* < 10^{11.5}\,M_\odot$. *Right:* The corresponding evolution for galaxies with $M_* > 10^{11.5}\,M_\odot$. The morphological fractions and densities for the $z = 0$ systems are taken from Conselice (2006a).

Conselice et al. (2007b) find a significant amount of morphological diversity among the $M_* > 10^{11}\,M_\odot$ galaxies (Fig. 2). At $z < 1$, 69% of $M_* > 10^{11}\,M_\odot$ systems are early types (elliptical, S0, compact), while 10% are disks, and 18% are peculiars. This is perhaps a surprisingly high fraction of peculiars within a massive galaxy selected sample and suggests that some of these systems are still undergoing some type of mass assembly, possibly through merging or star formation. This changes slightly when examining only the most massive systems with $M_* > 10^{11.5}\,M_\odot$ . These galaxies are ~90% early type over all redshifts, with a roughly similar number of mergers and disks making up the remainder. These results remain essentially the same, to within 5%, after considering how the Eddington bias may bring lower mass galaxies into our mass cuts due to observational uncertainty (see also Brinchmann & Ellis 2000; Bundy et al. 2005).

It is clear, however, that ~30% of galaxies with $M_* > 10^{11}\,M_\odot$ at $z < 1.5$ are not early types, which suggests that there is evolution in the massive galaxy population that cannot be seen simply through changes in number and mass densities. The disk galaxies show that there is some star formation occurring, and the peculiars reveal merger activity within this population.

Although early types (classified E/S0/compact) dominate the massive galaxy population at both the $M_* > 10^{11}\,M_\odot$ and $M_* > 10^{11.5}\,M_\odot$ selection limits, these galaxies often contain evidence for morphological peculiarities. Usually, these are in the form of outer low surface brightness features, or multiple cores. A total of 68 out of 263 (26±3%) ellipticals with $M_* > 10^{11}\,M_\odot$ show some internal substructure visible by eye (Conselice et al. 2007b). These objects are perhaps seen in other ways, such as through color gradients and color structures in ellipticals (e.g., Menanteau et al. 2005; Stanford et al. 2004; Teplitz et al.

2006) resulting from star formation, and which may be related to these features. Previous studies have generally found that it is the lower mass ellipticals that contain these star formation signatures. These morphological disturbances, however, do not appear more common in the lower mass ellipticals, and in fact, 36% of the $M_* > 10^{11.5}\,M_\odot$ ellipticals show this signature—a higher fraction than in the $M_* > 10^{11}\,M_\odot$ population. These peculiarities are likely the result of recent merging activity in these systems within the past 1–2 Gyr before we observe them.

## 2.2. Star-Forming Properties of Massive Galaxies

*General Trends in the Color–Magnitude Diagram* A major question concerning high mass galaxies is whether or not these systems have ongoing star formation at high redshift. While it is commonly thought that massive and early-type galaxies have largely finished their assembly and star formation by $z \sim 1$, detailed investigations suggest otherwise (e.g., Stanford et al. 2004; Teplitz et al. 2006; Conselice et al. 2007b).

One way to understand the star formation history of massive galaxies is to examine their position in color–magnitude diagrams. We show in Fig. 3 the $M_B$ vs. $(U-B)_0$ diagram for $M_* > 10^{11}\,M_\odot$ galaxies at $z < 1.4$ taken from Conselice et al. (2007b). Galaxies appear to separate into a red sequence and a blue cloud in this parameter space (e.g., Strateva et al. 2001; Baldry et al. 2004; Bell et al. 2004; Faber et al. 2005). At the highest redshift bin shown, $1.2 < z < 1.4$, there is a significant number of massive galaxies that are not on the red sequence. The fraction of massive galaxies on the red sequence, however, increases at lower redshifts. This shows that massive systems with $M_* > 10^{11.5}\,M_\odot$ generally fall in the red sequence region at all redshifts, but with a significant number of systems in the blue cloud region at $z > 0.8$. The fraction of $M_* > 10^{11}\,M_\odot$ galaxies on the red sequence increases with time at all masses. Galaxies with lower masses show a similar pattern, yet lower mass galaxies always have a lower fraction of galaxies on the red sequence at all redshifts, up to $z \sim 1.4$.

This leads to a very important conclusion regarding the red sequence and high mass galaxies. Previous studies have examined the increase in the amount of stellar mass on the red sequence, finding as much as a factor of two increase since $z \sim 1$ (Bell et al. 2004; Faber et al. 2005; Brown et al. 2007). However, this increase is due to galaxies appearing on the red sequence which were previously blue, and not due to in situ growth on the red sequence itself. This can be clearly seen by massive galaxies gradually moving onto the red sequence with time. This effect is also revealed in the decline in the number of blue massive galaxies found in the Universe since $z \sim 1$ (e.g., Bundy et al. 2006). This is not consistent with the idea that the red sequence grows solely through the so-called "dry mergers". Although merging may be present within the red sequence, and within our massive galaxy sample, it does not appear to be the dominant method by which the red sequence grows.

*Star Formation Rates* Quantifying star formation in massive galaxy samples can be done in several ways, including rest-frame UV emission, emission line fluxes, and *Spitzer* MIPS 24 $\mu$m data. Perhaps surprisingly, about half of all massive galaxies at $z \sim 1$ are detected at 24 $\mu$m, after removing AGN contam-

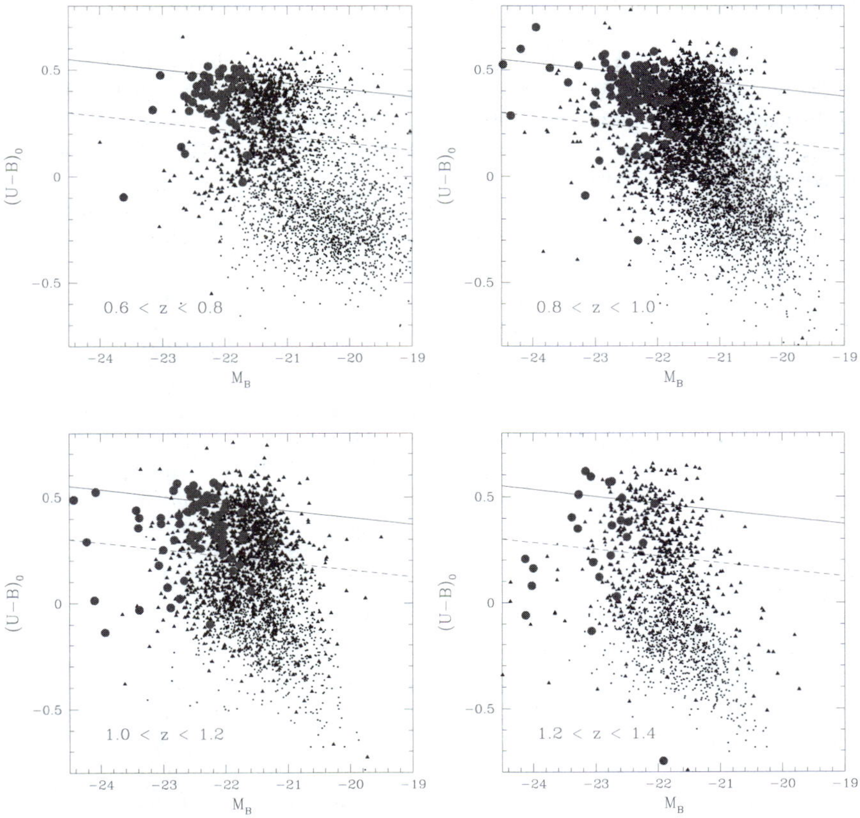


Figure 3. The $(U-B)_0$ vs. $M_B$ diagram for galaxies with $M_* > 10^{10.5}\,M_\odot$ from $z = 0.6$ to $z = 1.4$. The large red points on each panel are for galaxies with $M_* > 10^{11.5}\,M_\odot$. The blue triangles show the location of systems with $10^{11}\,M_\odot < M_* < 10^{11.5}\,M_\odot$. The solid line in each diagram shows the location of the red sequence, as defined in Faber et al. (2005), and the dashed line is the demarcation between red and blue galaxies.

ination based on *Chandra* detections (Conselice et al. 2007b). After matching the MIPS and [O II] star formation indicators, Conselice et al. (2007b) find that ~40% of the $M_* > 10^{11}\,M_\odot$ systems at $0.4 < z < 1.4$ are detected at $24\,\mu$m. A total of $37 \pm 5\%$ of the systems with $M_* > 10^{11.5}\,M_\odot$ within this redshift range are detected at $24\,\mu$m, with an average star formation rate of $70\,M_\odot\ \mathrm{yr}^{-1}$.

For galaxies with stellar masses $10^{11}\,M_\odot < M_* < 10^{11.5}\,M_\odot$, Conselice et al. (2007b) find that the fraction of galaxies undergoing star formation remains roughly similar at all redshifts. Circumstantially, this is consistent with the fact that the fraction of spirals+peculiars in this mass cut remains roughly constant throughout this redshift range. Interestingly, the fraction of systems which are undergoing star formation is higher than the non-elliptical fraction, showing that some morphologically classified massive ellipticals must be undergoing star formation (Stanford et al. 2004; Teplitz et al. 2006). The fraction of $M_* > 10^{11.5}\,M_\odot$ galaxies with a significant $24\,\mu$m detection declines slightly at lower redshift, from 33% at $1.2 < z < 1.4$ to 14% at $0.4 < z < 0.6$, consistent with a drop in the morphological fraction of non-ellipticals. This, however, is certainly a lower limit to the evolution, as the number of galaxies detectable at $24\,\mu$m declines at higher redshifts.

Conselice et al. (2007b) find, similar to previous studies utilizing IR star formation indicators (e.g., Le Floc'h et al. 2005), a decline with redshift for the massive galaxy population. After fitting these star formation histories up to their plateau (i.e., at $z \sim 1$) as a power law $\sim (1+z)^\alpha$ we can quantify the star formation history differences between $M_* > 10^{11}\,M_\odot$ galaxies and $M_* > 10^{11.5}\,M_\odot$ galaxies. For systems with $M_* > 10^{11.5}\,M_\odot$, Conselice et al. (2007b) find that the star formation rate declines as $\alpha = 6 \pm 2.2$, and for systems with $10^{11}\,M_\odot < M_* < 10^{11.5}\,M_\odot$ the slope is fit as $\alpha = 4.1 \pm 0.64$. The overall decline in the entire galaxy population's star formation history can be parameterized as $\alpha = 3 - 4$ (Hopkins 2004; Le Floc'h et al. 2005). It appears that while the $10^{11}\,M_\odot < M_* < 10^{11.5}\,M_\odot$ galaxies have a similar decline as the overall field, the highest mass galaxies show a faster decline.

## 3. Galaxy Merging

A major question concerning massive galaxies is the role of mergers in their formation. Galaxy mergers are occurring in the Universe, and it is likely that they play some role in the formation of massive galaxies, but the details are still debated. If mergers are occurring at $z < 1$, there are very few of them and they might be nearly all dissipationless "dry" mergers, without star formation. On the other hand, major mergers appear to be the dominate method for forming massive galaxies at $z > 2$.

*CAS Structural Analysis and the Merger Rate* The CAS (concentration, asymmetry, clumpiness) parameters allow us to probe the structures of galaxies quantitatively and are a major method for determining mergers in a galaxy population (e.g., Conselice, Bershady, & Jangren 2000a; Conselice, Bershady, & Gallagher 2000b; Bershady, Jangren, & Conselice 2000; Conselice 2003; Conselice et al. 2004, 2005; Casatta et al. 2005; Bridge et al. 2007). The CAS system can also be used to identify relaxed massive ellipticals. The basic idea is that galaxies have light distributions that reveal their past and present formation modes (Conselice 2003). One benefit in using the CAS system for finding mergers is that it allows us to quantify the merger rate and the number of mergers occurring in a galaxy population (Conselice et al. 2003a; Conselice 2006b).

Massive galaxies with $M_* > 10^{11}\,M_\odot$ at $z < 1.4$ are generally found in CAS space, where they are expected according to their visual morphologies. One important exception is that many of the visually classified undistorted early types are not located in the corresponding $z \sim 0$ part of CAS diagrams, being slightly too asymmetric (Conselice et al. 2007b).

Using CAS values for massive galaxies with $M_* > 10^{11}\,M_\odot$ Conselice et al. (2007b) have determined the evolution of the non-dry merger fraction for massive galaxies (cf. Hernandez-Toledo et al. 2006 for understanding dry mergers within CAS). By using the criteria, outlined in Conselice (2006b), of

$$A > 0.35, \;\; A > S, \tag{1}$$

Conselice et al. (2007b) have determined the merger fraction for the $M_* > 10^{11}\, M_\odot$ galaxies out to $z \sim 1.4$. There is a slight decrease with redshift in the merger fraction such that it declines as $(1+z)^{1.3}$, similar to the evolution seen in lower mass galaxies at $z < 1$ (Conselice et al. 2003; Bridge et al. 2007).

Using the number densities of massive systems, and timescales for CAS mergers, we can calculate the merger rate for $M_* > 10^{11}\, M_\odot$ galaxies based on the merger fraction and the timescales for merging derived from $N$-body models. From this, a major merger timescale of $\tau = 0.43 \pm 0.05$ Gyr for a galaxy with a mass of $3 \times 10^{11}\, M_\odot$ is calculated (Conselice et al. 2007b).

The merger rate of massive galaxies at $z < 1$ can then be calculated through the merger rate equation,

$$\Re(z) = f_{\rm m}(z) \cdot \tau_{\rm m}^{-1} n_{\rm m}(z), \tag{2}$$

where $n_{\rm m}(z)$ is the number density of objects, and $f_{\rm m}(z)$ is the merger fraction.[1] We find that, statistically, the merger rate for these $M_* > 10^{11}\, M_\odot$ galaxies is constant from $z \sim 0.4$–1.4, and is on average $\log\langle\Re\rangle$ (Gyr$^{-1}$ Gpc$^{-3}$) $= 4.3^{+0.4}_{-0.7}$.

We can furthermore calculate the total number of major mergers a galaxy with $M_* > 10^{11}\, M_\odot$ undergoes from $z \sim 1.4$ to $z \sim 0.4$ using equation (11) in Conselice (2006b). We calculate that the average number of mergers a massive galaxy with $M_* > 10^{11}\, M_\odot$ undergoes from $z \sim 1.4$ to 0.4 is $N_{\rm m} = 0.9^{+0.7}_{-0.5}$. Thus, on average, a massive galaxy will experience about one major merger from $z \sim 1.4$ to 0.4, roughly consistent with other results (Conselice 2006b; Bell et al. 2006).

*Dry Merging at $z < 1$* One issue that is not clear is how much of the merger and star formation process, and especially the controversial and hard to find dry mergers, are responsible for the addition of mass in massive galaxies at $z < 1$. We can address this question using mass functions and the measured star forming histories of galaxies with $M_* > 10^{11}\, M_\odot$ and $10^{11}\, M_\odot < M_* < 10^{11.5}\, M_\odot$. While the star formation history matches the increase in stellar mass, in massive galaxies, to within $< 3\sigma$ at any one redshift, the fact that the star formation history is consistently lower implies that star formation cannot account statistically for the total increase in stellar mass. This implies that part of the mass growth in these systems must be accounted for by mergers, or galaxies with masses lower than each stellar mass limit evolving into the higher mass bin due to star formation and/or merging.

Between the $M_* > 10^{11.5}\, M_\odot$ and $10^{11}\, M_\odot < M_* < 10^{11.5}\, M_\odot$ bins, the amount of stellar mass added to the higher mass bin can be measured partially through the star formation rate. The star formation rate in a bin will increase the amount of mass in that bin, and star formation in a lower mass

[1] Note that this is not the galaxy merger fraction, which is the fraction of galaxies merging and is roughly double the merger fraction (Conselice 2006b).

bin will increase the mass and number densities in higher mass bins by bringing up galaxies. When comparing the changes in the mass function to the amount of new mass from star formation, it is clear that changes in mass and number densities cannot be totally accounted for only by star formation. The remainder of the excess must be produced through merging. In Conselice et al. (2007b) it is argued that the amount of merging is such that at least one major merger is occurring for $M_* > 10^{11.5}\, M_\odot$ galaxies at $z < 1.5$. Since these galaxies are largely early types, it is likely that many of these mergers are dry (see also Bell et al. 2005). This roughly agrees with the number of mergers calculated through the CAS parameters.

*High Redshift $z > 2$ Mergers* The situation at higher redshifts, however, appears to be different, and it is likely that the greatest part of the stellar mass in modern ellipticals was put into place through major mergers at $z > 1.5$. Massive galaxies at $z > 2$ are not smooth ellipticals, but appear peculiar, even in the optical rest-frame (Conselice et al. 2005). In fact, the CAS values for these galaxies reveals a merger fraction of roughly 40–50% (Conselice et al. 2003).

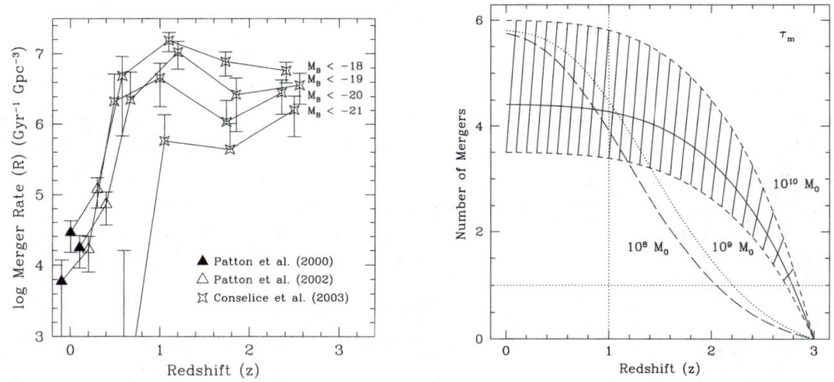


Figure 4. Evolution of the merger rate, in units of Gyr and co-moving Gpc$^3$, as a function of redshift and observed magnitude (*left*) and empirically determined integrated number of major mergers since $z \sim 3$ (*right*). The right panel shows the cumulative number of mergers for galaxies at three different masses, with the typical error range shown on the $M_* > 10^{10}\ M_\odot$ evolution. These merger fraction data are taken from previously published work (see Conselice 2006b and references therein).

By integrating the merger rate since $z \sim 3$, Conselice (2006b) finds that a typical massive galaxy with $M_* > 10^{10}\, M_\odot$ undergoes $4.4^{+1.6}_{-0.9}$ mergers from $z = 3$ to $z = 0$ (Fig. 4). Most of these mergers are at $z > 1.5$. An additional feature of the $N$-body models analyzed in Conselice (2006b) is the ability to determine the merging galaxy mass ratios that can produce high asymmetries. The result of this is that the CAS method is only sensitive to major mergers; that is, mergers with a mass ratio of 1:3 or lower (see also Hernandez-Toledo et al. 2005). This also allows us to determine how much mass is likely added to galaxies due to the merger process since $z \sim 3$. The result is that a galaxy which undergoes on average four to five major mergers will increase its total mass by a factor of $\sim$10. This is consistent with direct observations that show that the most massive galaxies are generally in place by $z \sim 1.5$–2, but are significantly depleted in number at $z > 2$ (Fontana et al. 2004).

## 4. Discussion and Summary

The last few years have seen a number of studies designed to determine when massive galaxies in the Universe formed. These studies have generally concluded that massive galaxies in the Universe, typically those with $M_* > 10^{11}\,M_\odot$, are largely formed by $z \sim 1$, but with considerable uncertainty. This is found to be the case statistically in nearly all studies in terms of mass and number densities for systems with $M_* > 10^{11}\,M_\odot$ from $z \sim 1$, although there is measurable evolution from $z \sim 1.4$ (Conselice et al. 2007b).

While studies have found that the number and mass densities of massive galaxies are similar at $z < 1$, this does not necessarily imply that there is no evolution in this population. Using results from the wide and deep Palomar Observatory Wide-Field Infrared Survey, combined with DEEP2 spectroscopy, we can directly select and study the properties and evolution of $M_* > 10^{11}\,M_\odot$ galaxies at $0.4 < z < 1.4$. Based on the findings of this survey (Bundy et al. 2006; Conselice et al. 2007b), it appears that the stellar mass and number densities of $M_* > 10^{11}\,M_\odot$ galaxies do not change significantly at $z < 1$. We cannot, however, rule out factors of two to three in number and mass density evolution for these systems, based solely on densities, due to the uncertainties in these measurements.

Other methods besides densities are therefore needed to determine conclusively whether massive galaxies have finished forming by $z \sim 1$. The fact that a high fraction of massive galaxies are forming stars (40%), and are non-elliptical (30%), at $z \sim 1$ suggests that there is active evolution. We find that in addition to star formation activity, typical massive galaxies with $M_* > 10^{11}\,M_\odot$ will undergo, on average, a single merger at $z < 1.4$. Most of the formation for these massive galaxies occurs at higher redshifts. Observationally, there is a significant decrease in the number densities of massive galaxies at $z > 2$. These systems are also observed to be undergoing a significant amount of merging which is likely how they build up most of their mass by $z \sim 2$.

Furthermore, these results show that the study of "early-types", defined through luminosity, color, morphology or stellar mass, at high redshift must be carefully done, and results of studies will vary significantly, depending on selection. It is clear, particularly at high redshift, that red galaxies are not the equivalent of massive galaxies, or elliptical galaxies, and each of these populations must be studied individually.

**Acknowledgments.** I thank the members of the Palomar, DEEP2, AEGIS, and *Spitzer* teams for their contribution to this work. Support from a NSF Astronomy and Astrophysics Fellowship and PPARC are gratefully acknowledged.

### References

Baldry, I. K., et al. 2004, ApJ, 600, 681

Bell, E., et al. 2004, ApJ, 608, 752
Bell, E., et al. 2006, ApJ, 640, 241
Bershady, M. A., Jangren, J. A., & Conselice, C. J. 2000, AJ, 119, 2645
Borch, A., et al. 2006, A&A, 453, 869
Bridge, C., et al. 2007, ApJ, 659, 931
Brinchmann, J., & Ellis, R. S. 2000, ApJ, 536, L77
Brown, M. J. I., et al. 2007, ApJ, 654, 858
Bundy, K., et al. 2006, ApJ, 651, 120
Bundy, K., Ellis, R. S., & Conselice, C. J. 2005, ApJ, 625, 621
Cassata, P., et al. 2005, MNRAS, 357, 903
Cimatti, A., Daddi, E., & Renzini, A. 2006, A&A, 453, L29
Cole, S., et al. 2001, MNRAS, 326, 255
Conselice, C. J., Bershady, M. A., & Jangren, A. 2000a, ApJ, 529, 886
Conselice, C. J., Bershady, M. A., & Gallagher, J. S. 2000b, A&A, 354, L21
Conselice, C. J., Gallagher, J. S., Calzetti, D., Homeier, N., & Kinney, A. 2000c, AJ, 119, 79
Conselice, C. J. 2003, ApJS, 147, 1
Conselice, C. J., Bershady, M. A., Dickinson, M., & Papovich, C. 2003a, AJ, 126, 1183
Conselice, C. J., Chapman, S. C., & Windhorst, R. A. 2003b, ApJ, 596, L5
Conselice, C. J., et al. 2004, ApJ, 600, L139
Conselice, C. J., Blackburne, J., & Papovich, C. 2005a, ApJ, 620, 564
Conselice, C. J., Bundy, K., Ellis, R., Brichmann, J., Vogt, N., & Phillips, A. 2005b, ApJ, 628, 160
Conselice, C. J. 2006a, MNRAS, 373, 1389
Conselice, C. J. 2006b, ApJ, 638, 686
Conselice, C. J., et al. 2007a, ApJ, 660, L55
Conselice, C. J., et al. 2007b, MNRAS, submitted
Daddi, E., et al. 2004, ApJ, 600, L127
Davis, M., et al. 2003, SPIE, 4834, 161
Davis, M., et al. 2007, ApJ, 660, L1
De Lucia, G., Springel, V., White, S. D. M., Croton, D., & Kauffmann, G. 2006, MNRAS, 366, 499
Drory, N., Salvato, M., Gabasch, A., Bender, R., Hopp, U., Feulner, G., & Pannella, M. 2005, ApJ, 619, L131
Faber, S., et al. 2005, astro-ph/0506044
Fontana, A., et al. 2004, A&A, 424, 23
Glazebrook, K., et al. 2004, Nature, 430, 181
Hernandez-Toledo, H. M., et al. 2005, AJ, 129, 682
Hernandez-Toledo, H. M., et al. 2006, AJ, 132, 71
Hopkins, A. 2004, ApJ, 615, 209
Larson, R. B. 1974, MNRAS, 166, L585
Le Fevre, O., et al. 2000, MNRAS, 311, 565
Le Floc'h, E., et al. 2005, ApJ, 632, 169
Lin, L., et al. 2004, ApJ, 617, L9
Menanteau, F., et al. 2005, ApJ, 620, 697
Papovich, C., Dickinson, M., Giavalisco, M., Conselice, C. J., & Ferguson, H. C. 2005, ApJ, 631, 101
Patton, D. R., et al. 2002, ApJ, 565, 208
Saracco, P., et al. 2005, MNRAS, 357, L40
Scarlata, C. et al. 2007, astro-ph/0701746
Stanford, S. A., Dickinson, M. E., Postman, M., Ferguson, H. C., Lucas, R. A., Conselice, C. J., Budavari, T., & Somerville, R. 2004, AJ, 127, 131
Strateva, I., et al. 2001, AJ, 122, 1861
Taylor-Mager, V., Conselice, C., Windhorst, R., & Jansen, R. 2007, ApJ, 659, 162
Teplitz, H. I., et al. 2006, AJ, 132, 853

Trujillo, I., et al. 2006, MNRAS, 373, L36
White, S. D. M., & Rees, M. J. 1978, MNRAS, 183, 341
Windhorst, R., et al. 2002, ApJS, 143, 113
Worthey, G. 1994, ApJS, 95, 107

## Discussion

*Böker*: Without kinematic information, how certain can you be about the classification in early- and late-type galaxies? At least some of the "asymmetric" early types look similar to disks with spiral structure.

*Conselice*: These galaxies are very bright, and it is clear that if they had a disk or spiral arms they would be much brighter than the features we see in the distorted ellipticals. These features are also not made of young stars, are not symmetric, and have a very low surface brightness.

*Silk*: 1) Do the numerical simulations provide enough major mergers to support your inference that more than 50% of massive galaxies undergo major mergers by $z \sim 2$? 2) Can minor mergers account for some or even all of the mass assembly of massive galaxies?

*Conselice*: Some semi-analytic simulations underpredict the merger fraction, but this may not be a general feature of models and simulations. Much more work needs to be done to investigate this within models. It is unlikely that many of the asymmetric galaxies are minor mergers, as the galaxies we see are too asymmetric to be accounted for by minor mergers. While a minor merger could produce significant amounts of star formation in gas rich young galaxies, these young star-forming regions would be localized in small areas. What we are seeing at $z \sim 2$, using NICMOS on *HST*, is that these asymmetric galaxies are asymmetric due to large-scale ($> 1\,\mathrm{kpc}$) features that are elongated and smooth. This is evidence that these asymmetries are not star-forming, but are older large-scale features, likely from merging.

Corrado Giammanco (*left*) and Mónica Relaño.

*Pathways through an Eclectic Universe*
*ASP Conference Series, Vol. 390, © 2008*
*J. H. Knapen, T. J. Mahoney, and A. Vazdekis, eds.*

# Strong Size Evolution of the Most Massive Galaxies since $z \sim 2$

Ignacio Trujillo[1] and Christopher J. Conselice[2]

[1]*Instituto de Astrofísica de Canarias, E-38200 La Laguna, Spain*

[2]*School of Physics and Astronomy, University of Nottingham, University Park, Nottingham NG7 2RD, UK*

**Abstract.** We explore the size evolution of 831 very massive galaxies ($M_\star \geq 10^{11}\, h_{70}^{-2}\, M_\odot$) since $z \sim 2$ using the combined capabilities of the large near infrared Palomar/DEEP-2 survey, and the superb resolution of the *HST* ACS camera. We split our sample according to their light concentration using the Sérsic index $n$. At a given stellar mass, both low ($n < 2.5$) and highly ($n > 2.5$) concentrated objects were much smaller in the past than their local massive counterparts. The evolution is particularly strong for the highly concentrated (spheroid-like) objects. At $z \sim 1.5$, massive spheroid-like objects were a factor of $4(\pm0.4)$ smaller (i.e., almost two orders of magnitude denser) than those we see today. These small sized, high mass galaxies do not exist in the nearby Universe, suggesting that this population merged with other galaxies over several billion years to form the largest galaxies we see today.

## 1. Introduction

In the nearby Universe, the population of galaxies with stellar masses greater than $10^{11}\, M_\odot$ is dominated by large early-type galaxies with correspondingly large sizes. These nearby systems contain old and metal-rich stellar populations that formed quickly in the early Universe. However, it has remained unknown whether the stars in these galaxies were all assembled in the same system, or formed in lower mass galaxies that later merged.

Two major formation models have been proposed in order to explain the properties of these galaxies: the so-called monolithic collapse model (Eggen, Bell, & Sandage 1962) and the hierarchical merging scenario (White & Frenk 1991). These two mechanisms have both observational and theoretical successes and drawbacks. For instance, evidence in favor of a fast and dissipative monolithic collapse is the fact that the bulk of stars in massive ellipticals are old (Mannucci et al. 2001) and have high [$\alpha$/Fe] ratios (i.e., short star formation timescales; Worthey, Faber, & Gonzalez 1992). On the other hand, supporting a hierarchical merger scenario, observations find a decline in the number of massive galaxies at high $z$. This decline is moderate since $z \sim 1$ and much stronger at higher $z$ (Daddi, Cimatti, & Renzini 2000).

Exploring the assembly of massive galaxies with cosmic time through number density analysis or merger rate estimations is difficult to conduct since, first, very massive galaxies are scarce and, second, their clustering properties make them strongly affected by field-to-field variance associated with limited observed volumes.

A much more straightforward approach to the assembly problem of massive galaxies is to explore the size evolution of these systems at a given stellar mass. In a "monolithic-like" scenario where a galaxy is fully assembled after the formation of its stars, the stellar mass–size relation should remain unchanged as cosmic time evolves. In the hierarchical merging scenario, however, the stellar mass–size relation will evolve as a result of the increase in size after each galaxy merger. The goal of this work is to shed some light on this issue by exploring the sizes of a large sample of very massive galaxies since $z < 2$. This will help to clarify whether the evolutionary scenario for these objects is hierarchical or monolithic-like. In what follows, we adopt the following cosmology: $\Omega_{\rm m} = 0.3$, $\Omega_\Lambda = 0.7$, and $H_0 = 70\,{\rm km\,s^{-1}\,Mpc^{-1}}$.

## 2. Results and Discussion

We use the Palomar/DEEP-2 survey (Davis et al. 2003; Bundy et al. 2006) to define a sample of 831 galaxies with masses larger than $10^{11}\,h_{70}^{-2}\,M_\odot$ located over $\sim$710 arcmin$^2$ in the Extended Groth Strip (EGS). This field was imaged with the Advanced Camera for Survey (ACS) in $V$ (F606W, 2660 s) and $I$ (F814W, 2100 s). In this region, the Palomar near-infrared $K_{\rm s}$-band imaging has a typical depth of greater than $K_{\rm AB} = 22.5$ mag ($5\sigma$). About 70% of the galaxies brighter than $R_{\rm AB} = 24.1$ mag in this field were targeted by the DEEP-2 Galaxy Redshift survey. In our case, 410 objects have a spectroscopic redshift ($\sim$50% of the full dataset). We supplemented our spectroscopic redshift catalogue with photometric redshifts with an accuracy of $\delta z/(1+z) \approx 0.07$ in the range $0.2 < z < 1.4$ and $\delta z/(1+z) < 0.22$ for $1.4 < z < 2$ (Conselice et al. 2007).

The determination of stellar masses for each of our galaxies follows a standard multicolor ($BRIJK$) stellar population fitting technique producing uncertainties of $\approx$0.2 dex. The largest systematic source of error comes from the assumed initial mass function (IMF); in this paper we have used the Chabrier IMF. The structural parameters used in this paper were measured using the ACS $I$-band images. Sizes (as parameterized by the half-light or effective radius) were estimated using the GALFIT code (Peng et al. 2002).

To quantify the size evolution, we measure the ratio between the observed size, and the expected size at a given stellar mass by comparing with the SDSS distribution (Shen et al. 2003) at different redshifts. The evolution of the median and the dispersion of the above ratio are shown in Figure 1. The observed size evolution of $M_\star > 10^{11}\,M_\odot$ galaxies is stronger than the one found in previous work using less massive galaxies.

To illustrate this size decrease at progressively higher $z$, Fig. 2 shows the appearance of six of our $n > 2.5$ (i.e., spheroid-like) galaxies with the same stellar mass but at different lookback times. To allow a fair comparison between the sizes of objects at very different redshifts, the limiting surface brightness in each panel is changed according to the cosmological surface brightness dimming $\sim (1+z)^4$. For this reason the surface brightness of the object at $z = 0.35$ is shown down to $\sim$3.25 mag brighter than that of the object at $z = 1.85$.

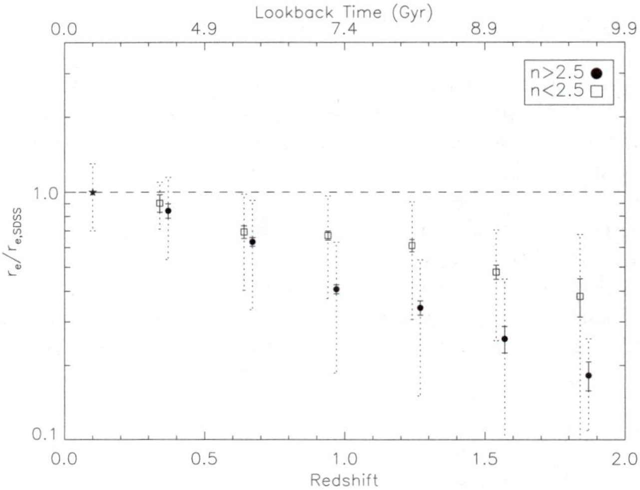


Figure 1. Size evolution of the most massive galaxies with lookback time. The evolution with redshift of the median ratio between the sizes of the galaxies in our sample and the galaxies of the same stellar mass in the SDSS local comparison sample is shown. Solid points indicate the size evolution of spheroid-like galaxies. Open squares show the evolution for disk-like galaxies. The small error bars indicate the uncertainty ($1\sigma$) at the median position. The large error bars represent the dispersion in the distributions.

The size evolution of the most massive galaxies found here disagrees with a scenario where massive galaxies were fully assembled in the early Universe and have subsequently evolved passively until today (i.e., a pure "monolithic" scenario). In fact, our findings agree with a scenario where a relevant mass fraction of the most massive galaxies possibly formed in a short "monolithic-like" collapse and then evolve through some major gas-rich or gas-poor merging. In this sense, the two scenarios would be just different phases of galaxy formation and evolution. Our results consequently point to a scenario in which the stellar populations of the most massive galaxies we observe today were located in different "primordial" massive galaxy pieces in the early Universe.

## References

Bundy, K. et al. 2006, ApJ, 651, 120
Conselice, C. J., et al. 2007, ApJ, 660, L55
Daddi, E., Cimatti, A., & Renzini, A. 2000, A&A, 362, L45
Davis, M., et al. 2003, SPIE, 4834, 161
Eggen, O. J., Lynden-Bell, D., & Sandage, A. R. 1962, ApJ, 136, 748
Mannucci, F., Basile, F., Poggianti, B. M., Cimatti, A., Daddi, E., Pozzetti, L., & Vanzi, L. 2001, MNRAS, 326, 745
Peng, C. Y., Ho, L. C., Impey, C. D., & Rix, H. W. 2002, AJ, 124, 266
Shen, S., Mo, H. J., White, S. D. M., Blanton, M. R., Kauffmann, G., Voges, W., Brinkmann, J., & Csabai, I. 2003, MNRAS, 343, 978
White, S. D. M., & Frenk, C. S. 1991, ApJ, 379, 52
Worthey, G., Faber, S. M., & Gonzalez, J. J. 1992, ApJ, 398, 69

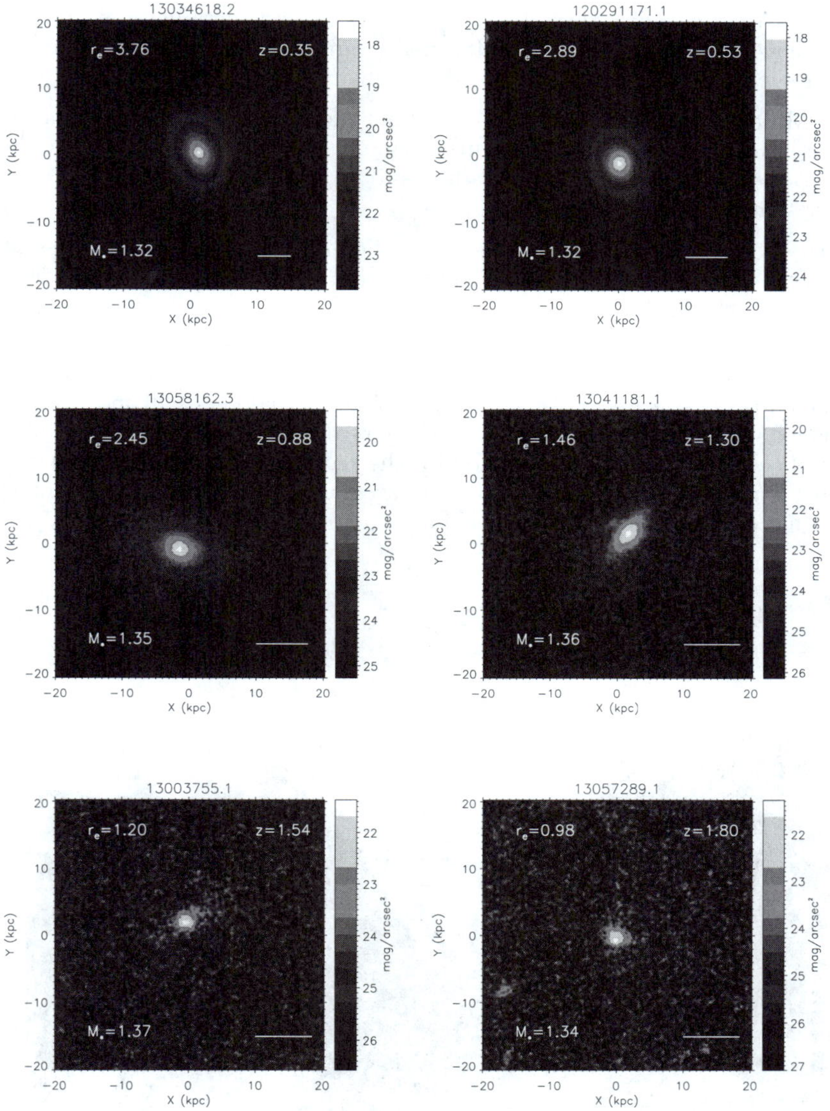


Figure 2. Size evolution of equal stellar mass objects at different lookback times. The panel shows six concentrated ($n > 2.5$, spheroid-like) representative galaxies in our sample at different redshifts. Effective radii are given in kpc and stellar masses in units of $10^{11}\ M_{\odot}$. Galaxies are shown with different surface brightness limits to account for the cosmological surface brightness dimming. The solid line indicates one arcsec angular size.

## Discussion

*Pipino*: Observations show that elliptical galaxies exhibit metallicity gradients. How can they fit into your scenario of dry mergers given the fact that simulations show that such a process is actually washing out the gradients?

*Trujillo*: Most of the information we have about metallicity gradients in nearby ellipticals is constrained to the very inner regions ($r < r_e$). I think we should wait for studies showing metallicity gradients to further galactocentric distances to answer that question properly.

*Edmunds*: 1) Are you and Conselice saying different things? Does the mass of these objects (the most massive ones) change from $z \sim 2$ to now? 2) How does surface density change?

*Trujillo*: Conselice finds a mild evolution in the number density of these objects since $z \sim 2$. Our results are not in contradiction with that. If our objects grow efficiently in size ($r_e \propto M_*$) a couple of major mergers will move them to the local stellar mass–size relation. Mergers decrease the number density but this effect can be compensated for by the fact that less massive systems ($M < 10^{11} M_\odot$) will move to the more massive bins maintaining the number density almost constant, as Conselice is finding.

From left to right: Michael Pohlen, Ignacio Trujillo, Peter Erwin, and Jesús Falcón-Barroso.

*Pathways through an Eclectic Universe*
*ASP Conference Series, Vol. 390, © 2008*
*J. H. Knapen, T. J. Mahoney, and A. Vazdekis, eds.*

# Evolution of Dust Attenuation and Star Formation Activity from $z = 0.7$ to $z = 0.2$ for a Sample of UV-selected Galaxies

Jorge Iglesias-Páramo and Jonatan Hernández-Fernández

*Instituto de Astrofísica de Andalucía, Camino Bajo de Huétor 50, 18008 Granada, Spain*

Véronique Buat, Denis Burgarella and Alessandro Boselli

*Observatoire Astronomique Marseille Provence, Laboratoire d'Astrophysique de Marseille, BP 8, 133761 Marseille Cedex 12, France*

Cong K. Xu

*California Institute of Technology, MC 405-47, 1200 East California Boulevard, Pasadena, CA 91125, USA*

Tsutomu T. Takeuchi

*Institute for Advanced Research, Nagoya University, Furo-cho, Chikusa-ku, Nagoya 464-8601, Japan*

**Abstract.** We study the ultraviolet to far-infrared spectral energy distributions of a sample of intermediate-redshift ($0.2 \leq z \leq 0.7$) UV-selected galaxies from the ELAIS-N1 and ELAIS-N2 fields by fitting a multiwavelength dataset to a library of GRASIL templates. The dust attenuation of low mass galaxies increases as redshift decreases, and this trend seems to disappear for galaxies with $M_* \geq 10^{11} M_\odot$. This result is consistent with a mass-dependent evolution of the dust to gas ratio, which could be driven by a mass-dependent efficiency of star formation in star-forming galaxies. The specific star formation rates decrease with increasing stellar mass at all redshifts, and for a given stellar mass the specific star formation rate decreases with decreasing redshift.

## 1. The Dataset

Our dataset contains multiwavelength (UV to far infrared) photometric data of galaxies in the ELAIS fields. The UV data were taken by *GALEX* and consist of two-band photometry at 1530 Å (FUV) and 2315 Å (NUV). The optical to FIR data correspond to the *Spitzer* Wide-area InfraRed Extragalactic (SWIRE) survey and consist of 12-band photometry: $U, g', r', i', Z$ taken at the 2.5 m Isaac Newton Telescope with the WFC; 3.6 $\mu$m, 4.5 $\mu$m, 5.8 $\mu$m, and 8.0 $\mu$m from *Spitzer*/IRAC; and 24 $\mu$m, 70 $\mu$m, and 160 $\mu$m from *Spitzer*/MIPS. In addition, photometric redshifts are provided for all the galaxies on the list. Details of the photometry of the SWIRE sources and of the quality of the photometric redshifts are given in Rowan-Robinson et al. (2005). Our UV-selected sample consists of a total of 6980 galaxies with $m_{\rm NUV} \leq 22.5\,\rm mag$ and a redshift of $0.2 < z < 0.7$.

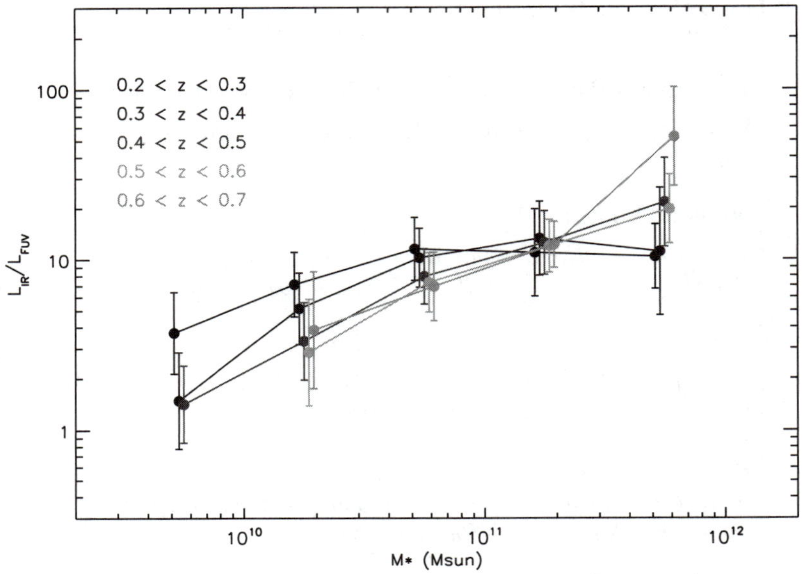


Figure 1. $M_*$ vs volume-weighted average $L_{\rm IR}/L_{\rm FUV}$ for the different redshifts. The average and standard deviation values (corresponding to the error bars in the figure) have been estimated as in Buat et al. (2007).

## 2. Synthetic Spectral Energy Distributions

A set of synthetic templates of spectral energy distributions (SEDs) were computed with the code GRASIL.[1] This code treats with particular care the effects of a dusty interstellar medium for different geometries of the dust and stars, and is coupled with a chemical evolution code (CHE_EVO) that takes into account in a consistent way the chemical evolution of the galaxy and the star formation history. This code describes one-zone open models depending only on time, including the infall of primordial gas. A more detailed description of this code can be found in Silva et al. (1998). Details on the grid of models can be found in Iglesias-Páramo et al. (2007). The physical parameters associated with a given galaxy are derived from fits of the observed data to the synthetic templates using a Bayesian approach.

---

[1]Full details concerning the use of the GRASIL code can be found at the website address http://web.pd.astro.it/granato/grasil/grasil.html. A web interface to use GRASIL can be found at http://web.pd.astro.it/galsynth/index.php.

## 3. Dust Attenuation versus Stellar Mass

Figure 1 shows the volume-weighted average $M_*$ vs $L_{\rm IR}/L_{\rm FUV}$ at different redshifts. It can be seen that $L_{\rm IR}/L_{\rm FUV}$ increases with stellar mass at all redshifts for galaxies with $M_* \leq 10^{11}\,M_\odot$. A galaxy with $M_* \approx 2 \times 10^{10}\,M_\odot$ shows $L_{\rm IR}/L_{\rm FUV}$ a factor of $\simeq$3 higher at $z = 0.65$ than at $z = 0.25$. This difference is lower for a galaxy with $M_* \approx 7 \times 10^{10}\,M_\odot$, a factor of $\simeq$2. For galaxies with $M_* \geq 10^{11}\,M_\odot$ the trend seems to flatten and $L_{\rm IR}/L_{\rm FUV}$ reaches an almost constant value with redshift. The plot also shows that for galaxies more massive than $\log M_* = 11.5$ the evolutionary trend of $L_{\rm IR}/L_{\rm FUV}$ seems to invert in the sense that it decreases with redshift. Whether this effect is real or not is still to be confirmed since the statistics are scarce for this last mass bin.

A tentative explanation of this evolution of $L_{IR}/L_{FUV}$ with stellar mass could be associated with the total dust content of star-forming galaxies, which is linked to the mass–metallicity relation reported at different redshifts (Zaritsky, Kennicutt, & Huchra 1994; Tremonti et al. 2004; Savaglio et al. 2005; Erb et al. 2006). This result suggests a mass-dependent rate of chemical enrichment for galaxies that can be explained by invoking a mass-dependent star formation history. In fact, $N$-body simulations by Brooks et al. (2007) claim that low star formation efficiencies are required for low mass galaxies in order to explain the mass–metallicity relation.

## 4. The Specific Star Formation Rate

Figure 2 shows the specific star formation rate (SSFR) as a function of the stellar mass. For a given $M_*$, the SSFR is found to decrease with redshift. The slope of the variation of the SSFR with $M_*$ and the rate of decline of SSFR with $z$ for different stellar masses are clues to understanding the origin of the downsizing process. We find a shallower variation of SSFR versus $M_*$ (i.e., a larger SSFR for massive galaxies) as compared to other works aimed at studying the evolution of the SSFR–$M_*$ variation and covering (at least) the same redshift range (e.g., Feulner et al. 2005; Bauer et al. 2005; Zheng et al. 2007). The SSFR is also found to decrease by a larger factor for massive galaxies than for low mass ones from $z = 0.7$ to $z = 0$. As an example, the SSFR lowers by a factor of $\simeq$8 from $z \approx 0.65$ to 0.25 for galaxies more massive than $10^{11}\,M_\odot$, whereas this factor is of the order of 4 for galaxies with $M* \approx 10^{10}\,M_\odot$. This result differs from the recent findings of Zheng et al. (2007), who conclude a similar decrease of SSFR with redshift for all stellar masses from $z = 0.9$ to the present time.

**Acknowledgments.** *GALEX* (*Galaxy Evolution Explorer*) is a NASA Small Explorer, launched in April 2003. We gratefully acknowledge NASA's support for the construction, operation, and science analysis for the *GALEX* mission, developed in cooperation with the Centre National d'Etudes Spatiales of France and the Korean Ministry of Science and Technology. Support for this work, part of the *Spitzer Space Telescope* Legacy Science Program, was provided by NASA through an award issued by JPL under NASA contract 1407. This article is based on observations made with the Isaac Newton Telescope, operated on the island of La Palma by the Isaac Newton Group at the Spanish Observatorio del Roque de los Muchachos of the Instituto de Astrofísica de Canarias. This work

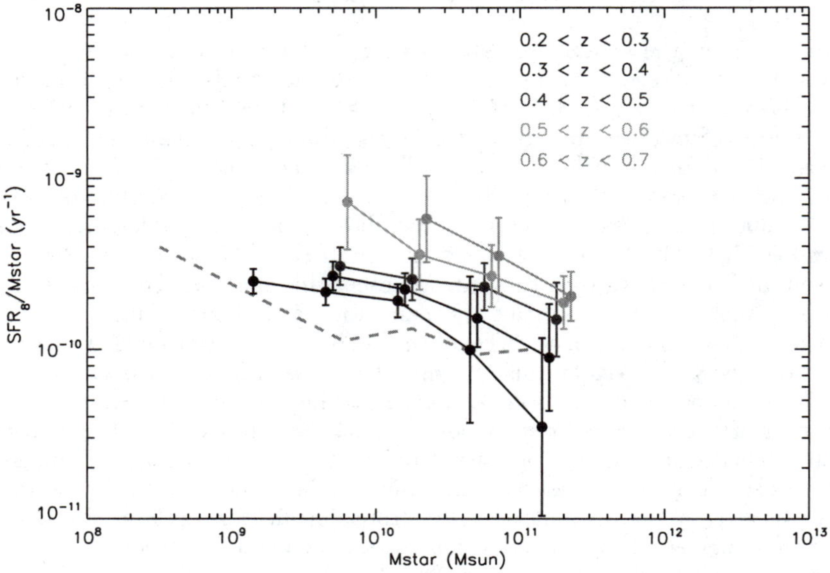


Figure 2. $M_*$ vs volume-weighted average SSFR for the different redshift bins. The dashed cyan line corresponds to the relation for UV selected local galaxies reported by Buat et al. (2007).

has been partially funded by the projects AYA 2004-08260-C03-02 of the Spanish PNAYA and TIC114 of the Junta de Andalucía. TTT has been supported by the Program for Improvement of Research Environment for Young Researchers from Special Coordination Funds for Promoting Science and Technology (SCF) commissioned by the Ministry of Education, Culture, Sports, Science and Technology (MEXT) of Japan.

## References

Bauer, A. E., Drory, N., Hill, G. J., & Feulner, G. 2005, ApJ, 621, L89
Brooks, A. M., Governato, F., Booth, C. M., Willman, B., Gardner, J. P., Wadsley, J., Stinson, G., & Quinn, T. 2007, ApJ, 655, L17
Buat, V., et al. 2007, ApJS, 173, 404
Erb, D. K., Shapley, A. E., Pettini, M., Steidel, C. C., Reddy, N. A., & Adelberger, K. L. 2006, ApJ, 644, 813
Feulner, G., Gabasch, A., Salvato, M., Drory, N., Hopp, U., & Bender, R. 2005, ApJ, 633, L9
Iglesias-Páramo, J., et al. 2007, ApJ, 670, 279
Rowan-Robinson, M., et al. 2005, AJ, 129, 1183
Savaglio, S., et al. 2005, ApJ, 635, 260
Silva, L., Granato, G. L., Bressan, A., & Danese, L. 1998, ApJ, 509, 103
Tremonti, C. A., et al. 2004, ApJ, 613, 898
Zaritsky, D., Kennicutt, R. C., Jr., & Huchra, J. P. 1994, ApJ, 420, 87

Zheng, X. Z., Bell, E. F., Papovich, C., Wolf, C., Meisenheimer, K., Rix, H.-W., Rieke, G. H., & Somerville, R. 2007, ApJ, 661, L41

## Discussion

*Churchwell*: In order to achieve the very nice model fits to observed SEDs for very different types of galaxies you must have assumed different morphologies from galaxy to galaxy. Could you give us some indication of your assumptions about scale height (in spirals like M51) for cold gas, dust and massive stars? Also, local extinction vs interstellar medium extinction around massive stars? It seems like a large number of parameters would be needed to obtain the precise fits you obtain, correct?

*Iglesias*: The geometrical parameters involved in the models are the scale height (assumed to be 0.4 kpc for stars and diffuse gas) and the scale length (assumed to be 4 kpc for stars and diffuse gas). Around massive stars the extinction is higher than around other stars because they are assumed to spend most of their life within molecular clouds where the extinction is larger than in the diffuse ISM. In total I allow just six free parameters to build the grid of models.

*James*: What can you say about the ratio of extinction towards SF regions to that towards the old stellar population in typical galaxies?

*Iglesias*: Taking M51 as an example, the model that best fits the SED of this galaxy (see Silva et al. 1998) shows a total dust attenuation at $\lambda \sim 1500$ Å of 1.6 mag, out of which 1.2 mag correspond to molecular clouds (SF regions). At optical wavelengths ($\lambda \sim 4500$ Å), the model gives a total attenuation of 0.32 mag (0.15 mag corresponding to molecular clouds).

Ruyman Azzollini, the LOC driver.

*Pathways through an Eclectic Universe*
*ASP Conference Series, Vol. 390, © 2008*
*J. H. Knapen, T. J. Mahoney, and A. Vazdekis, eds.*

# The Redshift Evolution of Bulges and Disks of Spiral Galaxies in COSMOS

Kartik Sheth, Lori Spalsbury, Nick Scoville, and the COSMOS Collaboration

*California Institute of Technology, USA*

**Abstract.** We present a preliminary analysis of the bulge and disk properties for a sample of over 4000 $L^*$ spiral galaxies at $z < 0.84$ from the COSMOS 2 square degree survey. We find that for early Hubble type spiral galaxies (Sa–Sb), the bulge-to-disk ratio is roughly constant over the last 7 Gyr of lookback time. This suggests that bulges of early type spirals were in place early on, consistent with other downsizing signatures. There is a monotonic increase in the bulge-to-disk ratios of late type spirals but that likely reflects the well-known decline in the star formation rate from $z \sim 1$ to the present. For this sample of $L^*$ spirals, we also find that the median exponential scale length of disks remains unchanged at 3.1 kpc from $z = 0.0$ to $z = 0.84$.

## 1. Introduction

The redshift evolution of galaxy properties provides important constraints on the evolutionary history of the Hubble sequence. In this contribution, we present a preliminary analysis from the COSMOS 2 square degree survey of the disk size and bulge prominence in a sample of $L^*$ spiral galaxies. With this same sample, we have found in a parallel study a dramatic evolution in the bar fraction from $z \sim 0.8$ to the present, but the majority of the evolution occurs in low mass, low luminosity spiral galaxies. The high luminosity, high mass and red spiral galaxies seem to have matured early, not only in their star formation rate, but also dynamically in developing galactic structures such as bars (Sheth et al. 2007). We explore how the evolution in the bar fraction relates to the evolution of bulges and disk sizes.

## 2. Sample Selection

An overview of COSMOS is given in Scoville et al. (2007a) and details of the *HST* observations are described in Scoville et al. (2007b). The photometric catalog and redshift measurements used for this study are in Mobasher et al. (2007) and Capak et al. (2007). We choose all spiral (Sa–Sd) galaxies (spectral type $T_{\rm phot} > 2$, details on $T_{\rm phot}$ in Mobasher et al. 2007) at $z < 0.84$ that are $L^*$ and brighter (we assume one magnitude of evolution: $M_V^* = -20.7$ at $z = 0$ to $-21.7$ at $z = 0.9$, Capak et al. 2003). The redshift limit restricts the analysis to rest-frame $g$-band observations (see detailed discussion in Sheth et al. 2007). With these criteria we are left with a sample of 4409 $L^*$ spirals which are then fitted with GALFIT (Peng et al. 2002).

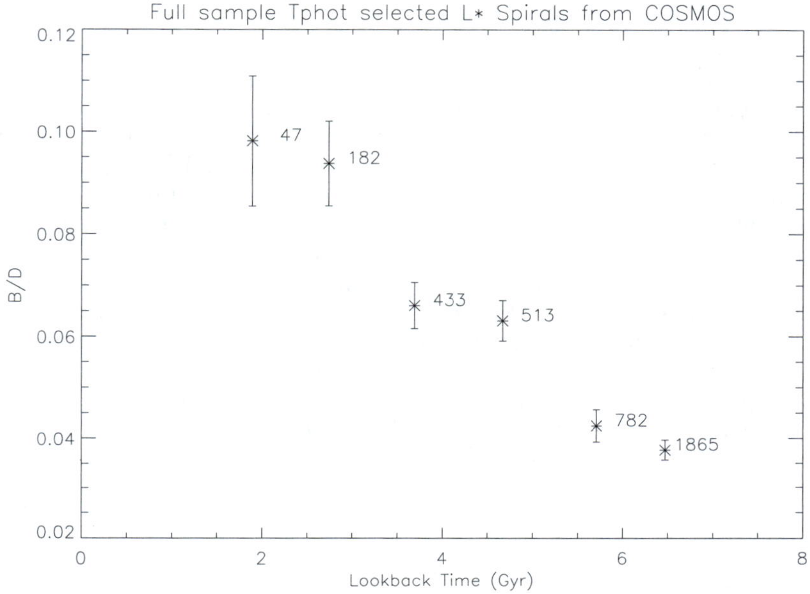


Figure 1. The median bulge-to-disk ratio for the entire COSMOS sample as a function of lookback time. The error in the median is computed by Monte Carlo simulations, taking observed values for $B/D$ and superposing random errors with a dispersion equal to the estimated errors in measurement of the luminosities. The numbers next to the data points reflect the galaxies in each bin. We see a monotonic increase in $B/D$ from $z = 0.84$ to the present.

We fitted each galaxy iteratively with a two-component (Sersic + exponential) model until the residuals were minimized. Each galaxy and residuals were individually examined by eye for robustness. We were able to obtain good results for 3822 galaxies where we measured a slew of parameters such as the luminosity of the disk and bulge, the Sersic index, the exponential disk scale length ($R_e$), etc. The bulge-to-disk ($B/D$) ratio is calculated by dividing the measured luminosity for the Sersic component by that of the exponential component. We did not try to deconvolve the contribution from the bars or AGN in these measurements because we assume that the median value is representative of the bulk properties of the sample.

## 3. Evolution of the Bulge-to-Disk Ratio

The median bulge-to-disk ratio for all the spirals in our sample is shown in Figure 1. There is a monotonic increase in the $B/D$ ratio over the last 7 Gyr.[1] However, the known fading of disks relative to the bulge can also mimic this trend. Since star formation is known to decrease dramatically from $z \sim 1$ to the present, this result should be interpreted with caution.

[1]We assume a $\Lambda$-dominated cosmology ($H_0 = 70\,\mathrm{km\,s^{-1}\,Mpc^{-1}}$, $\Omega_M = 0.3$, $\Omega_\Lambda = 0.7$).

However, the result is completely different for early Hubble type galaxies (Sa–Sb). For these systems the median $B/D$ is roughly constant, and even decreases slightly from $z \sim 0.84$ to the present! Note that our selection of early types is based on the SED (color) of the system and not on its structural properties such as bulge properties, etc. The result in Fig. 2 implies that in the red, early Hubble type systems, the bulge prominence stays constant or decreases to the present day. The expected selection effect of $k$-correction would have been smaller $B/D$ ratios at higher redshifts because we observe increasingly bluer rest-frame light, which probes the star-forming disk;[2] but the observed trend is exactly the opposite. Figure 2 is thus a robust result which indicates that the bulges of early Hubble type systems were already in place at $z = 0.84$, consistent with other downsizing signatures. This does *not* mean that all early type bulges were present by $z = 0.84$, it simply means that a bulge in an early type galaxy, once formed, does not change significantly over time.

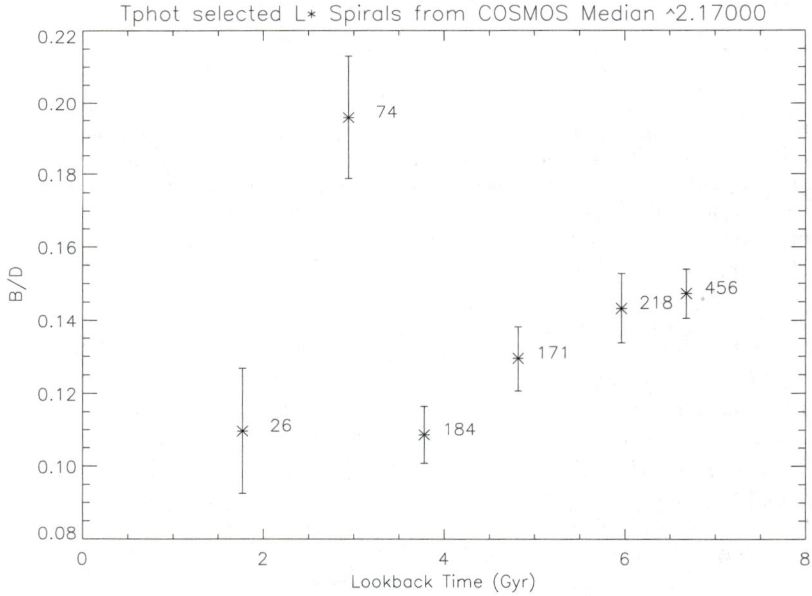


Figure 2. The same $B/D$ redshift evolution as in Fig. 1 but for early Hubble-type spirals only ($2 < T_{\text{phot}} < 3$). The $B/D$ is roughly constant. This implies that bulges of early Hubble-type spirals were already in place at a lookback time of 7 Gyr, consistent with other signatures of downsizing. The figure should *not* be misinterpreted as showing that all early type bulges were in place at that time. All this figure shows is that the relative prominence of bulges in early Hubble-type spirals does not change significantly over time.

In the late type systems we see the same monotonic increase in the $B/D$ ratio as we move from high to low redshifts, but this cannot be disentangled from the

[2]The COSMOS field was observed only in the F814W band with the ACS.

effects of $k$-correction. If real, it suggests that the bulges in late type systems are growing with time.

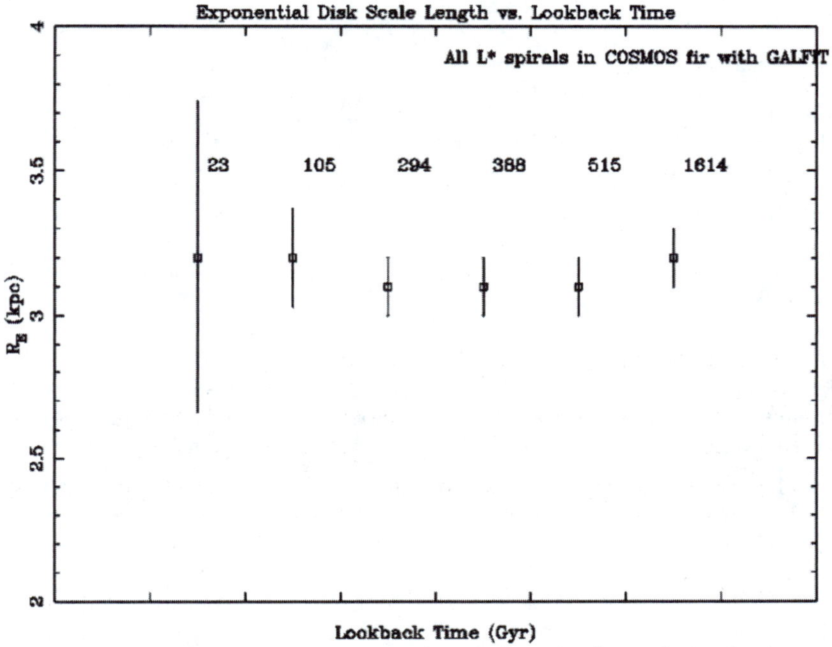


Figure 3. The median exponential scale length of spirals is shown as a function of redshift. We find no change in the disk scale length for this $L^*$ sample, consistent with previous studies.

## 4. Evolution of the Disk Scale Length

Finally, we studied the redshift evolution of the exponential disk scale length of the sample. As shown in Fig. 3, we see no evolution in the median exponential scale length of the disk. Since we specifically choose to explore the same part of the luminosity function by allowing $L^*$ to vary, the constant scale length suggests a size–magnitude relationship like the one seen in the local Universe.

## 5. Conclusions

• Analysis of $L^*$ spirals in COSMOS shows that the median bulge-to-disk ratio in early Hubble-type spirals (Sa–Sb) is roughly constant from $z = 0.84$ to the present. Bulges of early type spirals were in place at a look back time of 7 Gyr and bulges in these systems show little evolution.
• There is a monotonic increase in $B/D$ ratio for late type systems but the change is hard to disentangle from $k$-correction effects and may simply reflect a fading of the stellar disk rather than an increase in the bulge luminosity.

• The median exponential scale length of spirals remains constant at 3.1 kpc over the last 7 Gyr of lookback time.

## References

Capak, P. 2003, PhD Thesis, University of Hawaii
Capak, P. et al. 2007, ApJ, submitted
Mobasher, B. et al. 2007, ApJS, 172, 117
Peng, C. Y., Ho, L. C., Impey, C. D., & Rix, H.-W. 2002, ApJ, 124, 266
Scoville, N. Z. et al. 2007a, ApJS, 172, 1
Scoville, N. Z. et al. 2007b, ApJS, 172, 38
Sheth, K. et al. 2007, ApJ, submitted

## Discussion

*Trujillo*: You say that at given morphological type $B/D$ has not evolved since $z \sim 0.8$. Since morphological types are based on visual classification where the prominence of the bulge over the disk is one of the keys to specifying the type, I wonder whether the lack of evolution is not just a definition issue?

*Sheth*: The morphological type here is *not* determined from a visual classification. It is based on galaxy colors (SEDs matched to spectral templates for different spirals). So the $B/D$ measurement is an independent measure from type. The robustness of the $B/D$ invariance at redshift for early types (spectral) means that bulges in these galaxies were in place as early as 7 Gyr ago. The change of $B/D$ in late types could be evolution of the bulge but could also be the fading of the disk.

Eduardo Battaner.

*Pathways through an Eclectic Universe*
*ASP Conference Series, Vol. 390, © 2008*
*J. H. Knapen, T. J. Mahoney, and A. Vazdekis, eds.*

# The Physical Properties of Lyman Break Galaxies at $z > 5$: Outflows and the "Pre-enrichment Problem"

M. D. Lehnert,[1] M. Bremer,[2] A. Verma,[3] L. Douglas,[2] and N. Förster Schreiber[4]

[1]*GEPI, Observatoire de Paris, France*
[2]*Physics Department, University of Bristol, UK*
[3]*Physics Department, University of Oxford, UK*
[4]*MPE, Germany*

**Abstract.** We discuss the properties of Lyman break galaxies at $z > 5$ as determined from disparate fields covering approximately 500 arcmin$^2$. While the broad characteristics of the Lyman break galaxy population, such as luminosity functions and clustering amplitude, has been discussed extensively in the literature, we focus on the detailed physical properties of the sources in this large survey ($\gtrsim$ 100 with spectroscopic redshifts). Specifically, we discuss ensemble mass estimates, stellar mass surface densities, core phase space densities, star formation intensities, characteristics of their stellar populations, etc., as obtained from multi-wavelength data (rest-frame UV through optical) for a subsample of these galaxies. In particular, we focus on evidence that these galaxies drive vigorous outflows and speculate that this population may solve the so-called "pre-enrichment problem". The general picture that emerges from these studies is that these galaxies, observed about 1 Gyr after the Big Bang, have properties consistent with being the progenitors of the densest stellar systems in the local Universe—the centers of old bulges and early-type galaxies.

## 1. High Redshift Galaxies and Cosmology

Galaxies at the highest redshifts, $z > 5$, are vital objects in observational cosmology. They formed during an era of one of the most dramatic phase changes in the history of the Universe—the epoch of re-ionization. Remarkably, these earliest galaxies were able to dictate the phase of the general medium in which all galaxies are imbedded. Because of their enhanced [$\alpha$/Fe] ratios, which are the product of core collapse supernova enrichment, and old ages, the densest (i.e., globular clusters) and most massive stellar systems (i.e., giant ellipticals) are thought to have formed at such high redshifts in spectacular bursts of star formation of relatively short duration (< 100 Myr to 1 Gyr). The short duration of the star formation but the large accumulated mass must mean that these galaxies had very high star formation intensities—the rate of star formation per unit surface area. It is well known that galaxies with high star formation intensities drive vigorous outflows of metals and energy. Such outflows from high redshift galaxies could have both cleared material from the galaxian surroundings allowing ionizing photons to escape and enrich the intergalactic medium (IGM). Through this mechanism, problems of how the IGM was re-ionized and enriched in metals even at the highest redshifts yet observed (this is the so-called "pre-enrichment problem") could be solved. But can we find direct evidence for this scenario in the physical properties of high redshift galaxies? This is what we will attempt to address here.

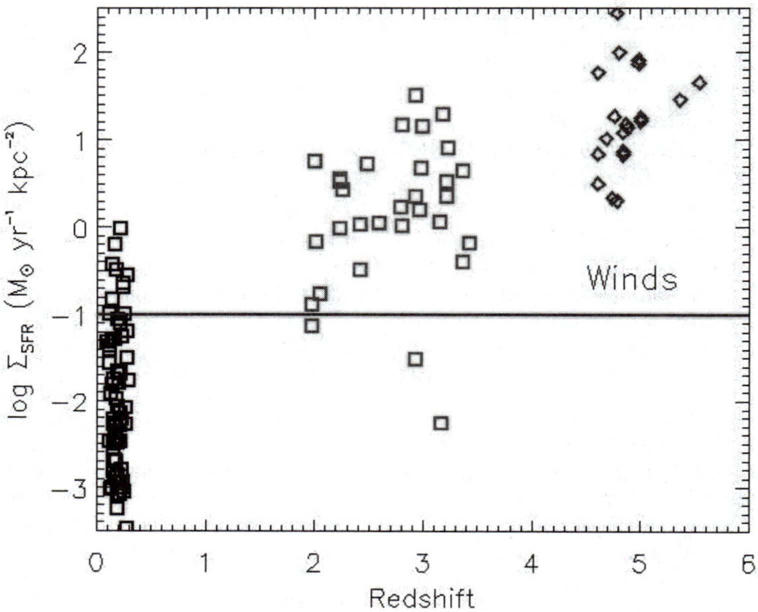


Figure 1. We compare the star formation rate intensities—the rate of star formation per unit projected area—of the $z \approx 5$ LBGs (*diamonds*; Verma et al. 2007), $z \approx 3$ LBGs (*squares*; Papovich, Dickinson, & Ferguson 2001), and a local sample of UV-selected galaxies (*squares*; Heckman et al. 2005). The horizontal line indicates the threshold above which local starburst exhibit large scale outflows (Heckman 2001; Lehnert & Heckman 1996). All of the galaxies studied here lie above this threshold and thus by analogy, must be driving outflows. We note that the evidence is overwhelming that virtually all $z \approx 3$ LBGs are driving outflows (Adelberger et al. 2003).

## 2. Our Comprehensive Study of $z > 5$ Galaxies

At $z \sim 5$, there have been a number of significant spectroscopic surveys investigating the nature of Lyman break galaxies (LBGs, or "$V$- or $R$-band dropouts"), two of which have been led by some of the authors. There are now over 100 confirmed redshifts in total at $z \gtrsim 5$ through the work of our ESO Large Programme (the ESO Distant Galaxy Survey—ERGS; L. Douglas et al., in preparation) and study of the field we call "BDFs" (e.g., Lehnert & Bremer 2003; M. D. Lehnert et al., in preparation). The ESO LP is based on the EDIScS fields, which are ten widely separated intermediate-redshift cluster fields, while the "BDFs" are four adjacent fields observable from Paranal in winter. Our total sample constitutes the largest sample of $z > 5$ LBGs ("$R$-band drop outs") with spectroscopically confirmed redshifts. In all of these fields, we have deep $R$-, $I$-, $z$-band, and IRAC data (3.6, 4.5, 5.8, and 8.0 $\mu$m). Some of the fields also have observations in other optical/infrared bands, including MIPS 24 $\mu$m data and *Hubble Space Telescope* ACS imaging. In addition, we have been studying a similar population of galaxies in the GOODS-S, taking advantage of the deep multi-wavelength data set (Verma et al. 2007).

## 3. What Have we Concluded?

From our analysis of some of these extensive data sets (Verma et al. 2007), we conclude:

• The LBGs at $z \approx 5$ are young, with typical ages less than 100 Myr, compact, with typical half-light radii of about 1 kpc, and are rapidly forming stars, 10 to 200 $M_{\odot}\,\mathrm{yr}^{-1}$. These ages are very young suggesting that they have been intensely forming stars for only a few dynamical times (estimated from stellar mass estimates and sizes). They therefore may likely represent a population of "primordial galaxies".

• These sources are generally so young that they likely did not substantially contribute to re-ionization. The estimates of their ages from the spectral energy distribution (SED) suggests that they formed at redshifts of about 6–7. Their contribution to re-ionization would likely be after the Universe was already substantially ionized.

• From SED fitting from the rest-frame UV through optical, the LBGs at $z \approx 5$ are typically about a factor of 10 less massive than similarly selected galaxies at $z \approx 3$ (a few $10^9\,M_{\odot}$ compared to a few $10^{10}\,M_{\odot}$; Shapley et al. 2001). These galaxies represent about 1% of the local mass density (Verma et al. 2007).

• The LBGs at $z \approx 5$ have high stellar mass surface densities, $\mu_{\mathrm{stars}} \approx 2$–$6 \times 10^8\,M_{\odot}\,\mathrm{kpc}^{-2}$, and have very roughly approximated core phase densities of $10^{-6}\,\mathrm{pc}^3\,\mathrm{km}^3\,\mathrm{s}^{-3}$. Both of these are similar to that of bulges and spheroids of $M^{\star}$ galaxies at low redshift.

• The LBGs at $z \approx 5$ have high star formation intensities, well above those needed to drive winds at low redshift, $\Sigma_{\mathrm{SFR}} \gg 0.1\,M_{\odot}\,\mathrm{kpc}^{-2}\,\mathrm{yr}^{-1}$ (Fig. 1; Heckman 2001; Lehnert & Heckman 1996).

• If the $z \approx 5$ LBGs drive strong outflows then they may be able to solve the "pre-enrichment problem"—whereby the metallicity of the IGM does not appear to evolve strongly with redshift at least out to almost $z = 6$ (Songaila 2001; Ryan-Weber et al. 2006). Quantitatively (crudely) estimating the likely metal contribution of these galaxies suggests that they are able to solve the pre-enrichment problem (see Fig. 2).

**Acknowledgments.** MDL wishes to thank Johan Knapen for organizing such a delightful meeting in honor of the contributions of John Beckman and for his extreme patience in waiting for this contribution.

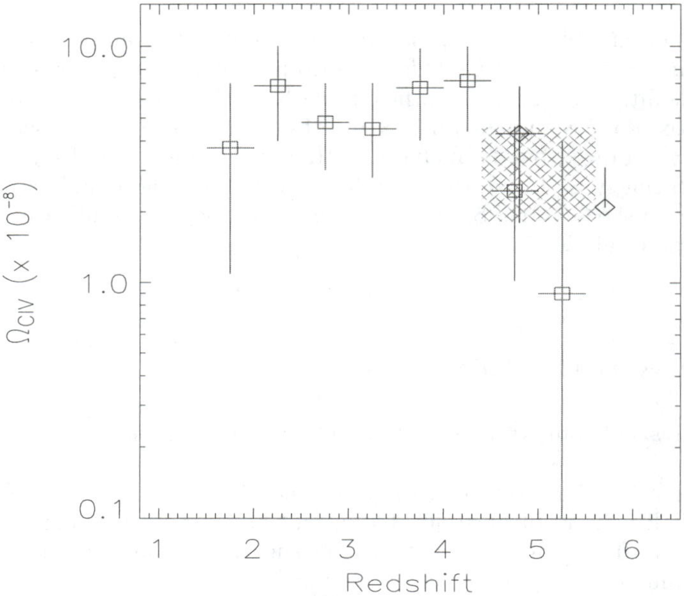


Figure 2. The co-moving density of C IV absorbers does not appear to evolve strongly with redshift (Songaila 2001). This suggests that a large fraction of the metals in the IGM may have been in place within 1 Gyr after the Big Bang. Comparing the data of the contribution to the closure density due to C IV absorbers, $\Omega_{\mathrm{C\,IV}}$ (Songaila 2001; *squares*, and Pettini et al. 2003; Ryan-Weber et al. 2006; the highest redshift diamond, from Ryan-Weber et al., is a lower limit), we find that it is plausible that the $z \approx 5$ LBGs could eject sufficient metals to explain the IGM absorption line observations. Our range of estimates was derived assuming that $\dot{M}_{\mathrm{outflow}} = \dot{M}_{\mathrm{star\,formation}}$ (Lehnert & Heckman 1996), the closure density, that the ionization fraction of $C^{3+} = 0.5$, the metallicity of the outflow is $0.2\,Z_{\odot}$, $\Omega_{\mathrm{b}}h^2 = 0.023$, and the range of star formation rates derived in Verma et al. (2007).

## References

Adelberger, K. L., Steidel, C. C., Shapley, A. E., & Pettini, M. 2003, ApJ, 584, 45
Heckman, T. M., et al. 2005, ApJ, 619, 35
Heckman, T. M., 2001, in ASP Conf. Ser., 240, Gas and Galaxy Evolution, ed. J. E. Hibbard, M. Rupen, & J. H. Gorkom (San Francisco: ASP), p. 345
Lehnert, M. D., & Bremer, M. 2003, ApJ, 593, 630
Lehnert, M. D., & Heckman, M. 1996, ApJ, 472, 546
Papovich, C., Dickinson, M., & Ferguson, H. C. 2001, ApJ, 559, 620
Pettini, M., Madau, P., Bolte, M., Prochaska, J. X., Ellison, S. L., & Fan, X. 2003, ApJ, 594, 695
Ryan-Weber, E. V., Pettini, M., & Madau, P. 2006, MNRAS, 371, L78
Shapley, A. E., Steidel, C. C., Adelberger, K. L., Dickinson, M., Giavalisco, M., & Pettini, M. 2001, ApJ, 562, 95
Songaila, A. 2001, ApJ, 561, L153
Verma, A., Lehnert, M. D., Förster Schreiber, N. M., Bremer, M. N., & Douglas, L. 2007, MNRAS, 377, 1024

*Pathways through an Eclectic Universe*
*ASP Conference Series, Vol. 390, © 2008*
*J. H. Knapen, T. J. Mahoney, and A. Vazdekis, eds.*

# The FIR–Radio Relationship at High and Low Redshifts

Catherine Vlahakis

*Argelander-Institut für Astronomie, Universität Bonn, Auf dem Hügel 71, D-53121 Bonn, Germany*

Stephen Eales

*School of Physics and Astronomy, Cardiff University, 5 The Parade, Cardiff CF24 3AA, UK*

Loretta Dunne

*School of Physics and Astronomy, University of Nottingham, University Park, Nottingham, NG7 2RD, UK*

**Abstract.** We use the results of the SCUBA Local Universe Galaxy Survey, a submillimeter survey of galaxies in the nearby Universe, to investigate the relationship between the far-infrared–submillimeter and radio emission of galaxies at both low and high redshift. At low redshift we show that the correlation between radio and far-infrared emission is much stronger than the correlation between radio and submillimeter emission, which is evidence that massive stars are the source of both the far-infrared and radio emission. At high redshift we show that the submillimeter sources detected by SCUBA are brighter sources of radio emission than predicted from the properties of galaxies in the local Universe. We discuss possible reasons for the cosmic evolution of the relationship between radio and far-infrared emission.

## 1. Introduction

One of the most intriguing relationships in astronomy is that between the non-thermal radio and far-infrared (FIR) emission from galaxies. It is one of the strongest correlations in astronomy, with a tight correlation over five decades of luminosity (e.g., Yun, Reddy, & Condon 2001). Yet, although the relationship has been known for over two decades, the cause of the relationship is still unclear. The standard explanation is that high mass stars both heat the dust, which then emits FIR emission, and provide the source of radio emission (e.g., Condon 1992, and references therein). Alternative explanations have been proposed but, despite surprisingly little direct evidence, most workers in the field have assumed that the standard explanation (i.e., ongoing star formation) is the correct one. Whatever its cause, the FIR–radio relation in nearby galaxies provided a valuable tool for estimating the redshifts of the galaxies discovered in the first deep submillimeter (submm) surveys (e.g., Smail, Ivison, & Blain 1997). Carilli & Yun (1999) suggested that an alternative to optical spectroscopy would be to estimate redshifts from the ratio of submm to radio flux densities, assuming that the FIR–radio relation is the same at all redshifts.

In this work we explore the FIR–radio relationship at high and low redshifts, using the results of the SCUBA Local Universe Galaxy Survey (SLUGS), a submm survey of ~200 galaxies in the local Universe, about half FIR-selected (Dunne et al. 2000b; IRS sample) and half optically selected (Vlahakis, Dunne, & Eales 2005; OS sample). Our submm measurements allow us to test a basic prediction of the standard theory: since the intensity of the interstellar radiation field is greater in regions containing large numbers of OB stars, the dust in these regions will be hotter than in the general interstellar medium, and thus we would expect the correlation between FIR and radio emission to be tighter than that between submm and radio emission. We then consider the FIR–radio relation at high redshift by predicting the ratios of radio to submm emission one would see at high redshift if there is no cosmic evolution in the FIR–radio relationship. Full details of this work can be found in Vlahakis, Eales, & Dunne (2007).

## 2. The FIR–Radio Relationship in the Low Redshift Universe

The relationship between the 1.4 GHz luminosity and 60 $\mu$m luminosity is shown in the left-hand panel of Figure 1. For the OS sample there is a very tight correlation ($r_s = 0.97$, probability $< 0.001$). The less tight correlation for the IRS sample could be related to the fact that luminous IRAS galaxies tend to be mergers or more active systems, but we note that the apparent difference in scatter may also be explained by the large number of upper limits for the OS galaxies, which were not considered for this analysis. There are seven OS galaxies which have "excess" 1.4 GHz emission relative to the correlation, all of which are early types, likely indicating the presence of an active galactic nucleus (AGN).

The relationship between the 1.4 GHz luminosity and 850$\mu$m luminosity is shown in the right-hand panel of Figure 1. The correlation for the OS sample ($r_s$=0.73, probability $< 0.001$) is clearly much less tight than that between radio and 60 $\mu$m luminosity (the rms scatter is 0.13 for 60 $\mu$m–radio and 0.40 for 850 $\mu$m–radio). This is exactly the behavior we would expect if the standard explanation (i.e., ongoing star formation) for the FIR–radio relation is correct.

## 3. The FIR–Radio Relationship in the High Redshift Universe

Carilli & Yun (2000) predicted the evolution of the radio–submm spectral index between 1.4 GHz and 850 $\mu$m ($\alpha_{1.4}^{850}$) with redshift for 17 low redshift galaxies. The ratio of 850 $\mu$m to 1.4 GHz flux density is related to $\alpha_{1.4}^{850}$ by the equation

$$\alpha_{1.4}^{850} = 0.42 \times \log\left(\frac{S_{850}}{S_{1.4}}\right)$$

(Carilli & Yun 1999). Similarly, Dunne, Clements, & Eales (2000a) presented predictions for the IRS SLUGS sample. These previous studies were selected in the FIR, but our OS SLUGS sample contains a population of cold ($T \sim 20$ K) dusty galaxies which is missing from *IRAS*-selected samples (Vlahakis et al. 2005). Thus such galaxies have so far been unrepresented in any analysis of the effectiveness of the Carilli & Yun method as a redshift indicator.

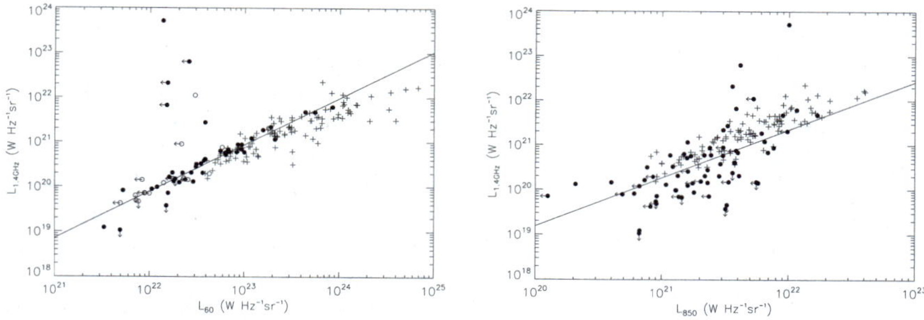


Figure 1. 1.4 GHz luminosity versus 60 $\mu$m luminosity (*left*) and 850 $\mu$m luminosity (*right*) for the OS and IRS samples (circles and crosses respectively; filled circles show OS galaxies detected at 850 $\mu$m, open circles non-detections; for clarity, those with upper limits in both wavebands are not shown).

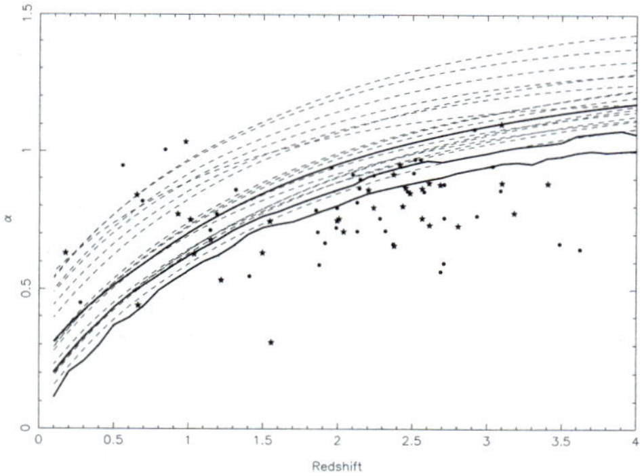


Figure 2. The $\alpha_{1.4}^{850}$–z relationship for the 17 OS SLUGS galaxies detected at both 850 and 450 $\mu$m (dashed lines). Also shown are the median of the predictions (and $\pm 1\sigma$ spread) for the 25 IRS SLUGS galaxies with the highest 60 $\mu$m luminosities (thick solid lines), and the deep SCUBA sources with both spectroscopic redshifts and radio measurements from Chapman et al. (2005) (points; stars indicate those with the most robustly determined redshifts).

We use the fitted spectral energy distributions for the 17 OS galaxies detected at both 850 $\mu$m and 450 $\mu$m (see Vlahakis et al. 2005) to make a new analysis of the $\alpha_{1.4}^{850}$–redshift relation. The predicted change in $\alpha_{1.4}^{850}$ with redshift for each OS galaxy is plotted in Fig. 2, along with the median of the predictions for the 25 most luminous IRS galaxies, and the deep SCUBA sources from Chapman et al. (2005). Clearly, many of the OS galaxies predict a very different $\alpha_{1.4}^{850}$–redshift relationship to the IRS galaxies, with about two thirds of the OS galaxies lying outside the $1\sigma$ variation of the IRS predictions and having higher $\alpha_{1.4}^{850}$ at a given redshift. Therefore, without an independent measurement of redshift, a given source could be either IRS-like and at higher redshift or OS-like and at lower redshift, which is essentially the degeneracy noted by Blain (1999). When the OS galaxies are included, the uncertainty in the Carilli & Yun method is much greater than would be evident from the IRS curves alone.

For the high redshift galaxies it is clear that these sources are significantly brighter at radio wavelengths (or alternatively fainter at FIR wavelengths) than the predictions for either SLUGS sample. Evidence that the ratio of radio to FIR emission may be higher at high redshift than at low redshift has also been found by other authors (e.g., Kovács et al. 2006). Investigating possible explanations, we find it unlikely that either a correlation between $\alpha_{1.4}^{850}$ and luminosity at zero redshift or significant AGN contribution to the 1.4 GHz emission of the Chapman et al. sources is the explanation. We find that using only those sources with the most robustly determined redshifts (following Aretxaga et al. 2007) does not affect the result we see in Fig. 2, and thus it seems unlikely that the redshifts are unreliable. There is evidence that the radio sources associated with high redshift dust sources are quite different from those today (Chapman et al. 2004), suggesting that physical conditions may be different for low and high redshift sources. In this case one might expect to see some evolution of the FIR–radio relation with redshift, and we find this the most likely explanation.

## 4. Conclusions

We find a very tight correlation between 1.4 GHz and 60 $\mu$m luminosity for the OS SLUGS galaxies. There is much more scatter in the correlation between radio and 850 $\mu$m luminosity, thus providing evidence that the standard explanation for the FIR–radio relation (ongoing star formation) is correct.

There is much more scatter in the $\alpha_{1.4}^{850}$-redshift relation for OS galaxies than for IRS galaxies, suggesting that the temperature of the bulk of the dust significantly affects the position of a galaxy on the $\alpha_{1.4}^{850}$–$z$ diagram. In order for the Carilli & Yun method to be reliable as a redshift estimator for deep submm sources we conclude one would need measurements of their dust temperatures.

The majority of the deep SCUBA sources from Chapman et al. (2005) are brighter at radio wavelengths than any of our OS $\alpha_{1.4}^{850}$–redshift predictions. We cannot exclude the possibility that this simply reflects that our low redshift galaxies are not suitable templates for the high redshift sources, but we think it more likely that this shows that the FIR–radio relation is evolving with redshift.

## References

Aretxaga, I., et al. 2007, MNRAS, 379, 1571

Blain, A. W. 1999, MNRAS, 309, 955

Carilli, C. L., & Yun, M. S. 1999, ApJ, 513, L13

Carilli, C. L., & Yun, M. S. 2000, ApJ, 530, 618

Chapman, S. C., Smail, I., Windhorst, R., Muxlow, T., & Ivison, R. J. 2004, ApJ, 611, 732

Chapman, S. C., Blain, A. W., Smail, I., & Ivison, R. J. 2005, AJ, 622, 772

Condon, J. J. 1992, ARA&A 30, 575

Dunne, L., Clements, D. L., & Eales, S. A. 2000a, MNRAS, 319, 813

Dunne, L., Eales, S., Edmunds, M., Ivison, R., Alexander, P., & Clements, D. L. 2000b, MNRAS, 315, 115

Kovács, A., Chapman, S. C., Dowell, C. D., Blain, A. W., Ivison, R. J., Smail, I., & Phillips, T. G. 2006, ApJ, 650, 592

Smail, I., Ivison, R. J., & Blain, A. W. 1997, ApJ, 490, L5

Thompson, T. A., Quataert, E., Waxman, E., Murray, N., & Martin, C. L. 2006, ApJ, 645, 186
Vlahakis, C., Dunne, L., & Eales, S. 2005, MNRAS, 364, 1253
Vlahakis, C., Eales, S., & Dunne, L. 2007, MNRAS, 379, 1042
Yun, M. S., Reddy, N. A., & Condon, J. J. 2001, ApJ, 554, 803

## Discussion

*Sarzi*: In your sub-mm/radio relation it would seem that $\sim$2/3 of the optically selected galaxies lie systematically below the relation. Are these upper limits? If so, is their position dictated by the sensitivity of your instrument? And if they are not upper limits, are these objects indeed radio-faint or do they have larger amounts of very cold dust?

*Vlahakis*: The "cluster" of objects below the relation which you are referring to are all upper limits. Those with 850 $\mu$m limits are distributed both above and below the relation, those with radio limits were assigned the same radio flux (the flux limit of the survey), but it is those with both submm and radio limits which produce this apparent "cluster" of points. These consist of a variety of Hubble types, including ellipticals, which may explain why they are not detected. The "clustering" is likely simply due to the way upper limits have been attributed in both the submm and radio. It is probably not a real feature.

*Krumholz*: Could the change you appear to see in the FIR–radio correlation at high redshift be due to the evolution of galactic magnetic fields? If so, how would the fields have to be different to explain your results?

*Vlahakis*: Yes, this is a possibility. Estimates of the magnetic field using the "minimum energy" argument yield values a factor of $\sim$100 higher for vigorously star-forming galaxies such as Arp 220 compared to normal galaxies like the Milky Way. The radio emission will thus also be much higher, which would agree with results we see in the $\alpha$–redshift diagram if the relativistic electrons do escape from their galaxy before they radiate all their energy. Arp 220 is one of the objects in the *IRAS*-selected SLUGS sample, so that the sources which lie below the local predictions would have to have quite different properties. Thompson et al. 2006) have investigated the possibility that magnetic fields are actually much higher than "minimum energy" estimates.

Fumi Egusa (*left*) and Isaura Fuentes.

*Pathways through an Eclectic Universe*
*ASP Conference Series, Vol. 390, © 2008*
*J. H. Knapen, T. J. Mahoney, and A. Vazdekis, eds.*

# Disk–Halo Interplay in Galaxy Evolution

Isaac Shlosman
*Department of Physics & Astronomy, University of Kentucky, Lexington, KY 40506, USA*

**Abstract.** Some aspects of disk–halo interactions for models of disk galaxies in and out of equilibrium are reviewed. Specifically, we focus on disk–halo resonant interaction without and in the presence of a gas component. Another issue is disk growth within an assembling triaxial dark matter halo. We argue that while the triaxiality is the result of the merger process and the radial orbit instability, it is the developing chaos that damps the first generation of bars and washes out the halo prolateness. This chaos is triggered by the gravitational quadrupole interaction(s) in the system and supported by a number of other processes characteristic of baryons.

## 1. Introduction

The current paradigm of galaxy formation requires that galactic disks grow and evolve within dark matter (DM) halos. As such, baryonic disks serve as a testbed for studying the DM properties, its dynamics and morphology, as well as mass, momentum and energy exchange within the disk–halo system. Interaction between the galactic disks and DM halos can in principle involve an exchange of these quantities. For example, observations of the low angular momentum H I gas deep in the halo of NGC 891 require cold extragalactic gas influx (e.g., Fraternali et al. 2007), maybe through filaments (Dekel & Birnboim 2006). On the other hand, the X-ray halos of starbursts and some normal galaxies can be explained by the supernova-heated gas driven from the disk or by the shock-compressed gas in the halo (e.g., Strickland et al. 2004).

The situation is much more straightforward with the angular momentum ($J$) flow in the system. Galactic disks are rotationally supported, while the DM halos have low $J$ and therefore a low spin parameter $\lambda$ (e.g., Barnes & Efstathiou 1987; Frenk, White, & Efstathiou 1988). Most disk galaxies are barred, especially in the NIR (e.g., Knapen, Shlosman, & Peletier 2000), with bar properties that remain steady up to redshift ~1 at least (Jogee et al. 2004; Elmegreen, Elmegreen, & Hirst 2004), and except for the very early Hubble types, they exhibit a spiral structure. This prevailing *disk asymmetry* is extremely important for enhancing the $J$ transfer between different morphological components—the alternative way can be achieved, e.g., by dynamical friction of baryons against the DM. The current understanding of galactic bar formation relies on the spontaneous breakup of the axial symmetry in the disk, the so-called classical bar instability (e.g., Hohl 1971), originally applied to isolated disks. Paradoxically, while Lynden-Bell & Kalnajs (1972) showed that bars and spirals facilitate the $J$ transfer, DM halos have been considered as the main stabilizers against the

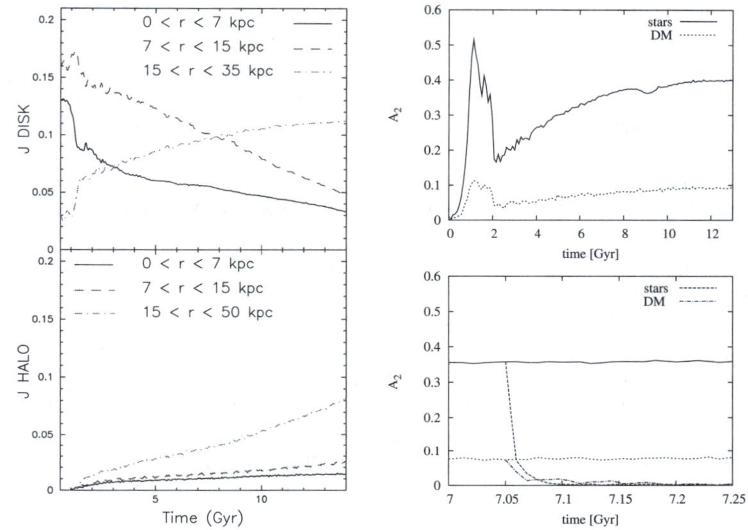


Figure 1. *Left:* Evolution of angular momentum, $J$, in the disk (*upper panel*) and the halo (*lower panel*). Note that $J$ saturates in the disk outside the bar CR. (*From Martinez-Valpuesta, Shlosman, & Heller 2006.*) *Right:* The $m = 2$ amplitude, $A_2$, of the stellar and DM ghost bars (*upper panel*) evolution. The fast dissolution of the ghost bar (*lower panel*) after its stellar counterpart is axisymmetrized. Note the very short time period for the lower frame. (*From I. Berentzen & I. Shlosman, unpublished.*)

bar instability for years (Ostriker & Peebles 1973), though they appear rather to be sinks of the disk momentum (e.g., Athanassoula & Misiriotis 2002).

Within the framework of the classical bar instability, a stellar bar can develop only if the unstable region loses its $J$. Gravitational torques serve as the mechanism that drives this process (Lynden-Bell & Pringle 1974). Their action can be described in terms of a non-local viscosity (Lin & Pringle 1987; Shlosman 1991) that shortens the characteristic timescale of the $J$ transfer dramatically, making this a dynamical rather than a secular process. In principle, $J$ can flow across the corotation radius (CR) to the outer disk, to the DM halo, or to a flyby galaxy in case of a tidal galaxy interaction. The first option limits the $J$ flow because the mass of the outer disk is typically ~20% of the disk mass and its ability to absorb $J$ quickly saturates (Fig. 1, *left*). On the other hand, the halo is massive; its inner part has a mass comparable to that of the inner disk, and its overall $J$ is small—the halo is not supported by rotation. Figure 1 shows also that the outer halo, beyond the disk radius, is fully susceptible to the $J$ transfer, after the inner halo efficiency has decreased.

The relatively local density response in the DM halo to the growing bar in the disk is to generate a shadow "ghost" bar in the DM (Athanassoula 2005, 2007; Berentzen & Shlosman 2006). The ghost has the pattern speed of the stellar bar but its mass distribution differs and is more centrally concentrated. The particle orbits in the DM ghost, therefore, are in resonance with the stellar orbits. The principal difference between these two bars is that the DM bar is not strictly speaking a bar at all—it is not self-gravitating *per se* and represents a gravitational wake induced in the DM by its stellar counterpart. When the stellar bar is axisymmetrized, the DM bar dissolves in a fraction of a crossing time (Fig. 1, *right*).

## 2. Angular Momentum Transfer in Collisionless Disk–Halo Systems

The draining of $J$ from the inner disk appears to proceed selectively—some of the orbits lose much more than their neighbors (Athanassoula 2004; Martinez-Valpuesta et al. 2006). Figure 2 reveals this phenomenon—stellar/DM particles populating orbits with specific frequencies $\nu \equiv (\Omega - \Omega_{\rm b})/\kappa$ that correspond to lower resonances, 0, $\pm$ 1:2, 1:3, 1:4, etc., dominating the $J$ exchange. Here $\Omega$ is the angular velocity, $\Omega_{\rm b}$ is the bar pattern speed, and $\kappa$, the radial epicyclic frequency. More precisely, the particles that are more affected are those ultimately trapped by the above resonances. The dominant resonance in the disk that emits $J$ is the inner Lindblad resonance (ILR), and the dominant resonance in the halo that absorbs it is the CR. So the direction of the $J$ transfer is confirmed to be from the inner disk to the halo. Figure 10 of Martinez-Valpuesta et al. (2006) shows a "forest" of lower resonances in the halo outside the CR which appear to be active in absorbing $J$ from the disk.

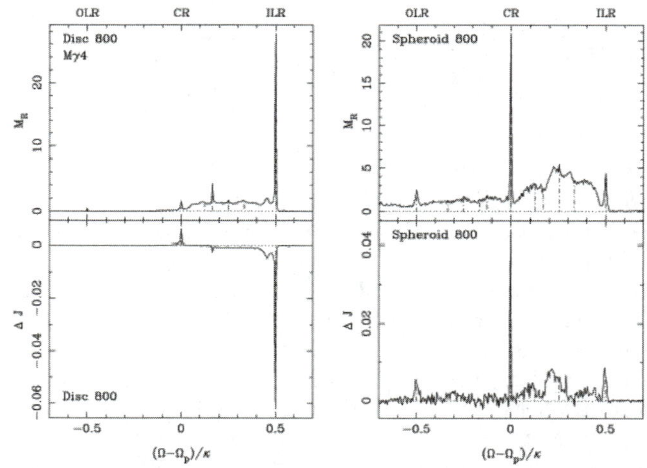


Figure 2. Resonances and $J$ exchange. *Upper panels:* Disk or halo mass per unit normalized frequency $\nu \equiv (\Omega - \Omega_{\rm b})/\kappa$ as a function of $\nu$. *Lower panels:* $\Delta J$ by the particles as a function of $\nu$. *Left panels:* Disk. *Right panels:* DM halo. The vertical dash-dotted lines give the positions of the main resonances (ILR $\rightarrow$ 0.5, CR $\rightarrow$ 0, OLR $\rightarrow$ $-0.5$, etc). (*From Athanassoula 2004.*)

The bar growth is closely related to the $J$ loss from the disk within the CR. All orbits within the bar are near-resonant, i.e., trapped by the resonances, mostly by the ILR. As the resonances sweep across the phase and configuration spaces, because the bar slows down, disk particles can lose their $J$ abruptly, being added (mostly) to the outer bar in this process (Martinez-Valpuesta 2006). Most of the untrapped particles by the bar can be found between its end and the CR. Bars that do not slow down, also do not grow in length. For the trapped particles, much of the $J$ evolution has ceased.

While Athanassoula (2004) and Martinez-Valpuesta et al. (2006) froze the gravitational potential in order to integrate the particle orbits, Ceverino & Klypin (2007) used a live potential for the relatively short time period of 1 Gyr. It is not clear whether the latter method provides any advantage because the short integration time results in broader and less defined resonances.

Different halos are expected to exhibit various efficiencies in their resonance interaction with the disk. So far, we are aware of two factors that govern this process: the velocity dispersion in the halo and its mass density near the lower resonances (Athanassoula 2003); however, additional dependencies are expected.

Cosmological simulations of DM structure show that the halos tend to acquire a universal density profile, $\rho(r)$, which is characterized by a central cusp with $\rho(r) \sim r^{-1}$ (e.g., Navarro, Frenk, & White 1997, hereafter NFW). This trend is challenged by observations of galactic rotation curves that typically indicate the presence of a central flattening, i.e., core, instead of the cusp (e.g., Bosma 2004), and by a theoretical estimate of dynamical friction of a stellar bar against the DM background, which requires a density core to explain why bars are still spinning (e.g., Sellwood 2006, 2007). But what is the possible cause of the DM core?

A number of alternatives have been explored for this purpose. Weinberg & Katz (2007) have advanced the idea that stellar bars transfer $J$ to the DM cusp and flatten it. However, the use of an analytical bar and other conditions that prevent the system from responding in a self-consistent way make their arguments unconvincing, especially the need for a large number of particles, $N > 10^8$, to resolve the bar–cusp interaction (e.g., Sellwood 2007). Live bars do not dissolve the DM cusps in the axisymmetric halos even with an effective $N \sim 10^{10}$ (J. Dubinski, I. Shlosman, & I. Berentzen, in preparation), but see Holley-Bockelmann, Weinberg, & Katz (2005). A possibility that the supernova feedback-injected energy flattens the cusp has been proposed (Mashchenko, Couchman, & Wadsley 2006). The other option is that the DM cusp has been flattened prior to or during the disk formation, when massive baryonic clumps descended to the center via dynamical friction (El-Zant, Shlosman, & Hoffman 2001; Tonini, Lapi, & Salucci 2006). It should be pointed out that the cusp existence requires a "temperature" inversion, when solving the Jeans equations without the central velocity anisotropy. Such an anisotropy was analyzed by Dehnen & McLaughlin (2005). In the absence of the DM velocity anisotropy, the cusp is thermodynamically unstable, and only the lack of an energy flow in the collisionless matter allows for its existence—a condition that is resolved by the dynamical friction. Similarly, the DM cusps in galaxy clusters can be erased by massive baryonic clumps—which are played out by individual galaxies (El-Zant et al. 2004).

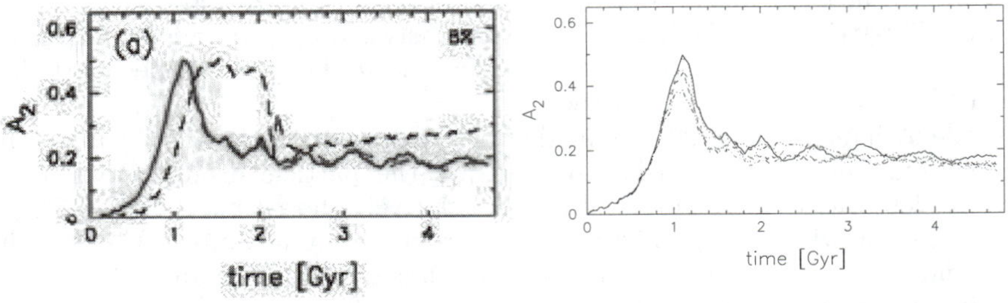


Figure 3. *Left:* Evolution of the stellar bar ($m = 2$ mode amplitude $A_2$) in models with $f_g = 0\%$ (solid line) and 8% (dashed) over the first 5 Gyr. *Right:* Evolution of $f_g = 8\%$ model (solid line) and the test models with gradually subtracted gravitational torques from the gas onto the stars—25% subtracted (dotted), 50% (dash-dotted), 75% (dashed) and 100% (dash-dot-dot-dotted). (*From Berentzen et al. 2007.*)

## 3. Angular Momentum Transfer in the Presence of Gas

While the $J$ redistribution in the collisionless disk–halo system has many additional caveats, it is interesting to explore the effects of an additional component: the gas. Currently, the gas fraction in disk galaxies probably does not exceed 10%, but gas is known to influence the galaxy evolution well beyond its fractional mass. Moreover, the gas fraction in disks at redshifts above unity can be substantially higher; left alone without the dissipative effects in the gas, the disk would not form in the first place.

Figure 3 (*left*) shows the disk–halo evolution in the presence of various gas fractions, $f_g \sim 0\% - 8\%$ (Berentzen et al. 2007). Similar bars develop in all models and experience a vertical buckling that "heats up" the stellar "fluid" along the rotation axis, weakening the bar. Although $f_g$ differs dramatically among these models, the evolution of the bar strength, $A_2$, shows the same maximum and nearly identical decrease in $A_2$. This decrease has been mistakenly attributed to the $J$ transfer from the gas to the stars in the bar by Bournaud, Combes, & Semelin (2005). On the other hand, the extent of the plateau in $A_2$ depends strongly on $f_g$ and clearly anti-correlates with it.

To show explicitly that the $J$ transfer from the gas does not alter the bar evolution over the first 5 Gyr, Berentzen et al. (2007) have run a number of test models, specifically removing the gravitational torques from the gas onto the stars and DM (Fig. 3, *right*). The resulting $A_2$ curves are very similar. This test has been supplemented by the direct analysis of $J$ redistribution in the disk halo system. The inner disk (inside the CR) loses its $J$ to the outer disk and the DM halo. The inner halo absorbs the disk $J$ and forms the ghost bar. Figure 4 (*left*) shows that the gas inputs into the stars $\sim 13\%$ of $J$ lost by the disk stars to the halo. The overall $J$ balance in the system is hardly affected by the gas presence. Surprisingly, *the halo is capable of absorbing* J *from the disk on a relatively short timescale.* This timescale is defined by the gas, but the bar evolution is not. Bournaud et al. (2005) concluded that the input of $J$ by the gas torques contributes to the bar dissolution—the halo used in this work was

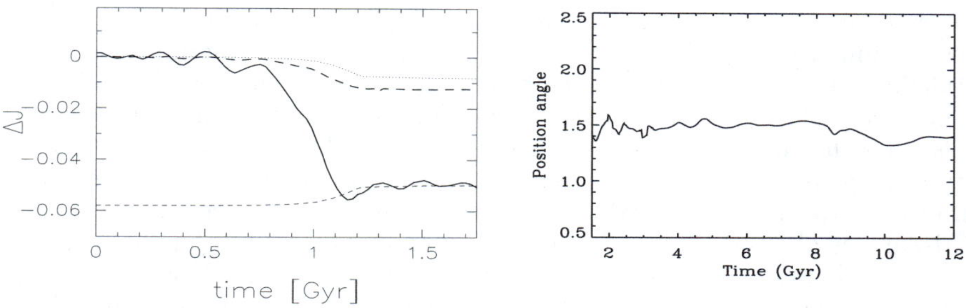


Figure 4. *Left:* Change of angular momentum, $\Delta J$, within in the bar CR in the 8% gas model from Fig. 3 (*left*). The solid line shows $J$ lost by the stars, calculated directly from the model evolution. The thin dashed line shows $J$ acquired by the stars from the gas, calculated from integration of the corresponding gravitational torques in the simulation (it was shifted down in $y$ to match the solid line), and vice versa, from the stars to the gas (dotted line). The thick dashed line provides the $J$ flow in the gas component, obtained from the model evolution. (*From Berentzen et al. 2007.*) *Right:* Position angle (in radians) of the major axis of the DM halo in a cosmological simulation of a disk/halo formation in the ΛCDM Universe with the WMAP3 parameters. (*From E. Romano-Diaz et al., in preparation.*)

frozen and therefore not allowed to participate in the overall $J$ exchange. Note that the sharp decrease in $A_2$ of the bar is not related to the angular momentum at all—it results from the buckling instability.

## 4. DM Halo Shapes and Disk Morphology: The Source of Chaotic Dynamics

The angular momentum redistribution in the disk-halo system is facilitated by the departure from axial symmetry. It is of paramount importance, therefore, that cosmological simulations universally lead to the formation of triaxial DM halos, i.e., both prolate and flattened (e.g., Bullock 2002; Allgood et al. 2006). Hence we can expect high-$z$ halos to follow this trend. On the other hand, observations indicate that halos in the nearby Universe are rather oblate and somewhat flattened, based on a number of arguments, e.g., the coherence of the Sagittarius tidal stream (Ibata et al. 2001), polar rings (Sparke 1986; although their shape can be affected by their self-gravity), diffuse X-ray emission around ellipticals (Buote et al. 2002), H I isophotes of galactic disks (Merrifield 2002), weak lensing (Hoekstra, Yee, & Gladders 2004), etc. The main issue, therefore, is *how the prolate DM halos at high redshifts have been transformed into oblate ones in the local galaxies.* Mass, angular momentum and energy exchange between the disk and the halo are expected to contribute to this process.

While triaxial collisionless systems are not exactly in equilibrium and are expected to evolve secularly, the addition of baryons accelerates this process dramatically and reduces the asymmetry (Dubinski 1994; Kazantzidis et al. 2004). However, these studies did not address the reason(s) for the shape evolution.

Why do baryons reduce the halo triaxiality so dramatically and how is this effect related to the internal dynamics of the system? Even in pure DM halos, within the hierarchical merging framework, rich substructure forms in the ΛCDM cosmology. These DM clumps join the main halo mostly along filaments. The "thermalization" of such streamers and the accompanying dynamical friction that the clumps experience when deep inside the main halo are themselves irreversible processes that randomize the DM trajectories. The clumps also transfer $J$ to the smooth DM component—again changing the DM orbit shapes (El-Zant et al. 2001, 2004; Tonini et al. 2006). The resulting inhomogeneities inside the halo serve as multiple isotropic scattering centers. This process, while being inherently related to the disk formation, in fact, precedes it. In elliptical galaxies it may dominate the reshaping of the DM figure. Baryons only enhance this trend. Dissipation leads to their accumulation in the inner halo, thus increasing the baryon-to-DM mass ratio there and making them more dynamically important. If the baryons are allowed to collapse to a disk, they become rotationally supported. But within the prolate background potential, the disk acquires a substantial quadrupole moment, as axisymmetric trajectories do not exist under these conditions.

A new and interesting phenomenon occurs because the halo figure tumbles slowly while the baryons in the disk are rotationally supported (Shlosman 2007; Heller, Shlosman, & Athanassoula 2007b; E. Romano-Diaz et al., in preparation). Under these conditions, the inner ILR (if it exists) approaches the center, while the outer ILR moves to large radii outside the halo. This assures that the baryons in the disk respond out of phase with the DM potential—the required response of trajectories between the ILRs. Support for the halo potential shape weakens and its prolateness is diluted. *The near absence of halo figure rotation has important consequences for the dynamics of the galactic disks.*

Even more important is the specifics of the baryon response to halo prolateness —large-scale shocks induced in the growing and gas-rich disk by the gravitational torques—which act as a finite perturbation from the halo onto the disk. As a result, the disk will be strongly asymmetric as long as the halo maintains its equatorial prolateness within the disk region. The halo and disk quadrupoles will acquire different pattern speeds (Heller, Shlosman, & Athanassoula 2007a,b). Their mutual interaction will lead to chaos and dissolution of the least massive quadrupole, whether this is a spontaneous or induced bar (El-Zant & Shlosman 2002). This last work found that the presence of a large-scale stellar bar in the disk immersed in a mildly prolate halo triggers chaotic behavior in the least massive component: the bar. Using the Liapunov exponents as a measure of chaos in the bar, El-Zant & Shlosman found that the stellar trajectories quickly become chaotic, dissolving the bar. When the slowly tumbling nonaxisymmetric perturbation in the *potential* has reached 10%, almost all trajectories integrated in time were found to be chaotic and have large Liapunov exponents. More centrally concentrated halos appear to be more dominated by chaos.

Figure 5 quantifies the developing chaos in terms of the Liapunov exponents in three specific cases of 5 kpc flat density core DM halos. Interestingly, a pure DM triaxial halo, with no bar, is nearly regular (Fig. 5, *left*), and no chaos is associated with this potential over the time period exceeding the Hubble time. An axisymmetric halo that hosts a bar exhibits a healthy amount of chaos around

Figure 5. Comparison of developing chaos in various systems given by gray-scale diagrams showing the maximal Liapunov exponents at 50 Gyr. The $x$-axis gives the cylindrical distance from the rotation axis and extends to 6 kpc. The $y$-axis gives the particle velocities normalized by the local circular velocity, $v/v_c$, and extends to $v/v_c = 1.2$. The frames display the Liapunov exponents for the particle trajectories in the system of (*left*) a pure triaxial DM halo with a core radius of 5 kpc and equatorial potential ellipticity of 0.1; (*middle*) an axisymmetric halo with a core radius of 5 kpc, and with a stellar bar of ~5 kpc; (*right*) a triaxial halo with a core radius of 5 kpc and equatorial potential ellipticity of 0.1, and with a 5 kpc stellar bar. The gray-scale is logarithmic with the white background corresponding to a characteristic time of $10^6$ yr or less, and the darkest background to $10^{10}$ yr or more. (*From El-Zant & Shlosman 2002.*)

the stellar (and ghost) bar CR, as expected (Fig. 5, *middle*). It is the addition of the fast tumbling bar within a prolate halo that generates *continuous* chaos within the CR and at most of the velocities (Fig. 5, *right*). The characteristic timescale for the trajectories' divergence is much less than the Hubble time. The halo core potential can then be approximated as a quadratic form where the motion is separable in Cartesian coordinates. Centrally concentrated potentials do not have such symmetries—the oscillations in the different degrees of freedom are, in general, coupled, and no global integrals of motion exist. This makes plausible a situation where most trajectories in the central regions conserve only energy, in which case self-consistent solutions become impossible.

Galaxies that are not initially centrally concentrated may acquire a central mass concentration (CMC) by accreting gas via spontaneous or tidally induced bars. The resulting coupling between the degrees of freedom will destroy the integrals of motion for a large enough fraction of orbits and dissolve the bar, which is replaced by, for example, a CMC. This scenario requires massive CMCs, e.g., supermassive black holes (SBHs) in excess of $10^9\,M_\odot$, which are not confirmed by observations (El-Zant et al. 2003).

Heller et al. (2007b) have compared the *shape* evolution of an assembling DM halo with and without the baryons. Figure 6 (*left*) shows the time evolution of the pure DM shape from 2 kpc to 300 kpc. Figure 6 (*right*) exhibits the same evolution when the disk grows by converting the gas to stars during the collapse of an isolated density perturbation embedded in the Hubble flow. The halo triaxiality is reasonably stable with time for the pure DM models at all radii—only a mild secular decrease was found over the Hubble time. Models with baryons show different behavior within and outside the disk radius (roughly

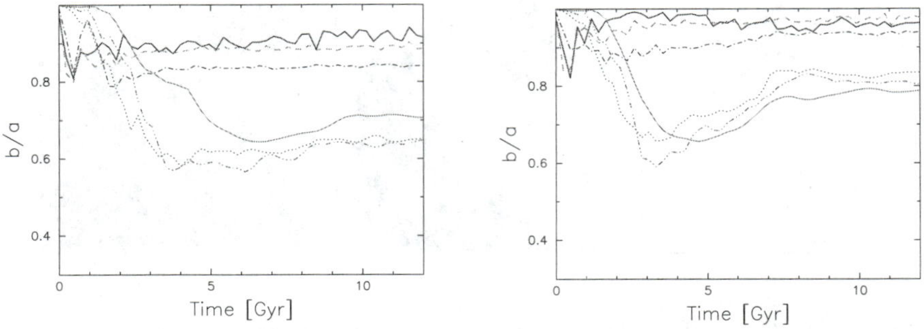


Figure 6. Evolution of DM halo intermediate-to-major axial ratio $b/a$ for a pure DM model (*left*) and a DM with baryons model (*right*). The curves correspond to 2 kpc (thick solid), 5 kpc (dashed), 10 kpc (dot-dashed), 50 kpc (thin solid), 100 kpc (dotted), and 300 kpc (dash-dot-dot-dotted). (*From Heller et al. 2007b.*)

15 kpc). The inner halo prolateness is washed out by the time a disk has formed, ~2–3 Gyr. The outer halo prolateness and flatness go through an adjustment phase and then stabilize at some value, remaining substantial.

Berentzen, Shlosman, & Jogee (2006) have verified in live $N$-body simulations that the interaction of two or more quadrupoles leads to chaotic dynamics both in the disk and in the inner halo. A typical decay time of stellar bars and disk ovalness was found to lie in the range of a few Gyr, with a caveat that the halo itself suffers a reduction of its quadrupole in the feedback, and, in extreme cases, the halo prolateness can be damped ahead of that in the disk. More cuspy halos have been losing their prolateness within the radius faster, allowing the bar to survive. This effect was further quantified in simulations with a tenfold growing stellar disk embedded in the halo which assembles under cosmological initial conditions within a 4 Mpc computational box (Berentzen & Shlosman 2006). More massive disks have erased the halo prolateness faster and more efficiently —hence, maximal disks are particularly apt to axisymmetrizing the background halos. Moreover, the loss of prolateness by the halo allowed the maximal disk to develop a next generation bar via classical bar instability, albeit at much later times. On the other hand, disks dominated by the DM at all radii will have a much smaller effect, if at all, on the shapes of the inner halo.

The emerging conclusion from the above works underlines the importance of self-consistent models in understanding DM halo shape evolution. If the disk is not allowed to generate one or more substantial quadrupoles, or, in simple terms, is prevented from responding to the halo prolateness, the basic ingredients for its feedback onto the halo are virtually eliminated. There are three basic ways the disk responds to the halo: (1) its orbits are out-of-phase with the halo potential (the disk is elongated normal to the halo major axis), (2) halo-induced and spontaneous stellar bars develop and tumble fast, and (3) extensive stellar arms in the outer disk are triggered by the halo shape. Recent interesting work by Debattista et al. (2007) has reached a different conclusion, based on simulations with a *frozen* disk which, therefore, cannot respond in any of the above fashions. Furthermore, the disk is grown adiabatically, thus only reversible processes are induced in the halo, making its orbits rounder. In fact, there is no basic difference between slowly growing a rigid disk or a Plummer sphere within a responsive halo—the latter experiment was performed by Dubinski (1994). It is natural that Debattista et al. find that chaos plays no role in reshaping the halo, and that only the (DM) substructure has an irreversible effect on the halo shape.

To summarize, *the origin of the halo triaxiality lies to a certain degree in the major merger process and in the radial orbit instability, while the reduction in halo triaxiality is governed by developing chaos and transfer of the angular momentum to the DM.* These processes can be facilitated through the basic mechanism—the separation of baryons from the DM that is driven by dissipation. Specifically, (1) the cooling baryons enhance the substructure in the overall mass distribution during the halo assembly. The resulting clumps deposit energy and $J$ in the background DM via dynamical friction—this includes minor mergers and quiescent accretion and is an irreversible process. (2) The ability of the baryons to cool allows them to become more centrally concentrated (in the global and not only in the local sense)—the centrifugal barrier delays this process, by converting it from a dynamical to a secular one. However, even the presence of this barrier does not assure the dynamical-to-secular transition because the baryonic disk forming in the center will not be axisymmetric in the presence of a prolate DM potential, and even if it did form axisymmetrically, the classical bar instability would spontaneously break this symmetry. (3) The resulting interaction of two or more quadrupoles that produce incommensurable perturbing forces both on baryons and the DM will generate chaos in the disk and the triaxial halo and act to reduce the triaxiality. (4) The central cusp, whether baryonic or DM, can induce chaos as well, although this is not assured in the most general sense.

## 5. Evolving Disk Morphology: Bars, Bulges, and All That

The origin of galactic disks has emerged as one of the sticking issues in our overall understanding of the galaxy formation process. Large gaps permeate our knowledge of the origin of disk structural components, e.g., bars, bulges, SBHs, etc. The path to the observed tight correlation between the properties of the central SBH and the surrounding bulge is also unknown, although suggestions abound. Some of the components require a certain degree of dissipation to accompany their formation, such as the disk itself, certain types of bulges, and the SBH—this erases the memory of the initial conditions and the link to cosmology. In particular, the origin timescale for current disk morphology and the amount of evolution that occurred since is only vaguely understood. A number of controversies have emerged: First, the angular momentum distribution of the observed (baryonic) disks does not match that of the disks obtained in cosmological numerical simulations—the latter appear to have radial scalelengths that are too small (e.g., Sommer-Larsen, Gotz, & Portinari 2002)—the so-called angular momentum "catastrophe". Moreover, the central bulges form too massive and there are difficulties in explaining the widespread bulgeless disks. Second, the numerical simulations of DM halo formation show that they display a universal density profile characterized by a steep central density cusp (e.g., NFW), while observations point rather to a flat density core (e.g., de Blok & Bosma

2002)—the core "catastrophe". The central difficulty appears to be the conservation of baryonic $J$ during the disk formation and major merger events. The standard picture of slow adiabatic collapse of a smooth baryonic component only aggravates the problem by dragging the DM inward in an adiabatic compression.

These and other discrepancies between observations and modeling cannot be fully resolved by the current theory—additional high-resolution numerical simulations are required, focusing on the effect of various parameters of star formation, energy feedback from the stellar evolution and growing SBHs. Indeed, recent numerical modeling has made progress in understanding the complexity of disk formation and evolution. Models with *frozen* spherically symmetric DM halos, e.g., by Samland & Gerhard (2003) and Immeli et al. (2004), have used a chemo-dynamical code to investigate the bulge stellar population, as well as a possible bulge formation from massive spiraling-in gas clumps, accompanied by a strong central starburst (see also Shlosman & Noguchi 1993). Okamoto et al. (2005) have assumed that the star formation proceeds in two modes—in a high density environment or shock-triggered, leading to elliptical or disk galaxies, respectively. Hence, the star formation processes can affect the final $J$ of the galaxy. Governato et al. (2004, 2007) have claimed to produce disks with realistic exponential scalelengths to fit the $I$-band and baryonic Tully–Fisher relations, and to reduce the number of visible satellites around the Milky Way-type halo, by combining the UV background and SN feedback.

Heller et al. (2007a,b) have focused on the formation of the disk and its structural components within the context of the collapse of an isolated density perturbation embedded in the Hubble flow. In agreement with Fig. 4b, $J$ was channeled into the internal circulation, leaving the DM halo figure tumbling $\sim \pi\,\mathrm{rad}$ in a Hubble time. At $z = 0$, a range of nearly bulgeless to bulge-dominated disks were obtained, with flat rotation curves which account for the observed disk/halo contributions. A relatively large number of models was run to investigate the effect of star formation and stellar evolution feedback onto the disk properties, normally unresolved in cosmological simulations. Several single-varied parameter evolutionary sequences have been constructed, such as varying the energy thermalization parameter, which characterizes the feedback from stellar evolution, the star formation threshold parameter, the collapse time for the star-forming clouds, and the gravitational softening in the gas. A number of correlations along these sequences have been found; e.g., the thermalization parameter appears to correlate with the bulge prominence, and a lower density threshold leads to smaller, thicker, disks.

While the bar origin lies in the $J$ loss by the inner disk, exponential bulges appear to be largely by-products of the bar evolution. The vertical buckling instability in a bar leads to the formation of boxy/peanut-shaped bulges (e.g., Combes & Sanders 1981; Combes et al. 1990). Although the bar weakens as a result (Raha et al. 1991; Pfenniger & Friedli 1991) and shortens, in principle it cannot be destroyed (Martinez-Valpuesta & Shlosman 2004). Multiple bucklings of the bar add up to the growth of its bulge, irrespective of the gas inflow (Martinez-Valpuesta et al. 2006).

We discussed earlier that the halo shape speeds up the $J$-exchange in the system. This is especially pronounced in the early stages of disk growth, when its rotation curve is probably dominated by the DM. The initial disk response to a finite perturbation from the halo will naturally lead to a bar. The resulting mass cascade to the center, e.g., in the form of nested bars (Shlosman, Frank, & Begelman 1989), can create the environment for the formation and initial growth of a SBH (Begelman, Volonteri, & Rees 2006).

## 6. Conclusions

We have summarized some aspects of disk–halo interactions in equilibrium and forming disk galaxies. Because the overall problem is of a high complexity, progress in this direction has been rather slow and dependent on numerical simulations. Nevertheless, certain trends have emerged. The baryonic disk and DM halo exchange mass, angular momentum and energy during the initial formation phase as well as during the subsequent quiescent evolution. A number of processes, like star formation as well as formation of the central SBH, require a much deeper understanding before they can be modeled beyond the phenomenological approach. On the other hand, nonlinear dynamics and the role of chaos in the evolution of disks and halos can be quantified already at present.

**Acknowledgments.** I acknowledge illuminating discussions with my colleagues, too numerous to list here. I am grateful to my collaborators Ingo Berentzen, Clayton Heller, Yehuda Hoffman, Inma Martinez-Valpuesta and Lia Athanassoula on these issues. The relevant grants from NASA and NSF are acknowledged. I also thank the conference organizers for generous support.

### References

Allgood, B., Flores, R. A., Primack, J. R., Kravtsov, A. V., Wechsler, R. H., Faltenbacher, A., & Bullock, J. S. 2006, MNRAS, 367, 1781
Athanassoula, E., & Misiriotis, A. 2002, MNRAS, 330, 35
Athanassoula, E. 2003, MNRAS, 341, 1179
Athanassoula, E. 2004, in IAU Symp. 220, Dark Matter in Galaxies, ed. S. D. Ryder, D. J. Pisano, M. A. Walker, & K. C. Freeman (San Francisco: ASP), p. 255
Athanassoula, E. 2005, in Nonlinear Dynamics in Astronomy and Physics, ed. S. Gottesman & J. R. Buchler, Annals N. Y. Acad. Sci., 1045, p. 168
Athanassoula, E. 2007, MNRAS, 377, 1569
Barnes, J., & Efstathiou, G. 1987, ApJ, 319, 575
Begelman, M. C., Volonteri, M., & Rees, M. J. 2006, MNRAS, 370, 289
Berentzen, I., Shlosman, I., & Jogee, S. 2006, ApJ, 637, 582
Berentzen, I., & Shlosman, I. 2006, ApJ, 648, 807
Berentzen, I., Shlosman, I., Martinez-Valpuesta, I., & Heller, C. H. 2007, ApJ, 666, 189
Bosma, A. 2004, in IAU Symp. 220, Dark Matter in Galaxies, ed. S. D. Ryder, D. J. Pisano, M. A. Walker, & K. C. Freeman (San Francisco: ASP), p. 39
Bournaud, F., Combes, F., & Semelin, B. 2005, MNRAS, 364, L18
Bullock, J. S. 2002, in The Shapes of Galaxies and Their Dark Halos, ed. P. Natarajan (Singapore: World Scientific), p. 109
Buote, D. A., Jeltema, T. E., Canizares, C. R., & Garmire, G. P. 2002, ApJ, 577, 183
Ceverino, D., & Klypin, A. 2007, MNRAS, 379, 1155
Combes, F., & Sanders, R. H. 1981, A&A, 96, 164
Combes, F., Debbash, F., Friedli, D., & Pfenniger, D. 1990, A&A, 233, 92
Debattista, V. P., Moore, B., Quinn, T., Kazantzidis, S., Maas, R., Mayer, L., Read, J., & Stadel, J. 2007, astro-ph/07070737
de Blok, W. J. G., & Bosma, A. 2002, A&A, 385, 816

Dehnen, W., & McLaughlin, D. 2005, MNRAS, 363, 1057
Dekel, A., & Birnboim, Y. 2006, MNRAS, 368, 2
Dubinski, J. 1994, ApJ, 431, 617
Elmegreen, B. G., Elmegreen, D. M., & Hirst, A. C. 2004, ApJ, 612, 191
El-Zant, A., Shlosman, I., & Hoffman, Y. 2001, ApJ, 560, 636
El-Zant, A., & Shlosman, I. 2002, ApJ, 577, 626
El-Zant, A., Shlosman, I., Begelman, M. C., & Frank, J. 2003, ApJ, 590, 641
El-Zant, A., Hoffman, Y., Primack, J., Combes, F., & Shlosman, I. 2004, ApJ, 607, L75
Fraternali, F., Binney, J., Oosterloo, T., & Sancisi, R. 2007, New Astron., 51, 95
Frenk, C. S., White, S. D. M., & Efstathiou, G. 1988, ApJ, 327, 507
Governato, F., Mayer, L., Wadsley, J., Gardner, J. P., Willman, B., Hayashi, E., Quinn, T., Stadel, J., & Lake, G. 2004, ApJ, 607, 688
Governato, F., Willman, B., Mayer, L., Brooks, A., Stinson, G., Valenzuela, O., Wadsley, J., & Quinn, T. 2007, MNRAS, 374, 1479
Heller, C. H., Shlosman, I., & Athanassoula, E. 2007a, ApJ, 657, L65
Heller, C. H., Shlosman, I., & Athanassoula, E. 2007b, ApJ, 671, 226
Hoekstra, H., Yee, H. K. C., & Gladders, M. D. 2004, ApJ, 606, 67
Hohl, F. 1971, ApJ, 168, 343
Holley-Bockelmann, K., Weinberg, M., & Katz, N. 2005, MNRAS, 363,991
Ibata, R., Irwin, M., Lewis, G. F., & Stolte, A. 2001, ApJ, 547, L133
Immeli, A., Samland, M. O., Gerhard, O., & Westera, O. 2004, A&A, 413, 547
Jogee, S., et al. 2004, ApJ, 615, L105
Kazandzidis, S., Kravtsov, A. V., Zentner, A. R., Allgood, B., Nagai, D., & Moore, B. 2004, ApJ, 611, L73
Knapen, J. H., Shlosman, I., & Peletier, R. F. 2000, ApJ, 529, 93
Lin, D. N. C., & Pringle, J. E. 1987, MNRAS, 225, 607
Lynden-Bell, D., & Kalnajs, A. G. 1972, MNRAS, 157, 1
Lynden-Bell, D. & Pringle, J. E. 1974, MNRAS, 168, 603
Martinez-Valpuesta, I., & Shlosman, I. 2004, ApJ, 613, L29
Martinez-Valpuesta, I. 2006, PhD Thesis, University of Hertfordshire
Martinez-Valpuesta, I., Shlosman, I., & Heller, C. H. 2006, ApJ, 637, 214
Mashchenko, S., Couchman, H. M. P., & Wadsley, J. 2006, Nature, 442, 7102
Merrifield, M. R. 2002, in Shapes of Galaxies and Their Dark Matter Halos, ed. P. Natarajan (Singapore: World Scientific), p. 170
Navarro, J. F., Frenk, C. S., & White, S. D. M. 1997, ApJ, 490, 493 (NFW)
Okamoto, T., Eke, V. R., Frenk, C. S., & Jenkins, A. 2005, MNRAS, 363, 1299
Ostriker, J. P., & Peebles, P. J. E. 1973, ApJ, 186, 467
Pfenniger, D., & Friedli, D. 1991, A&A, 252, 75
Raha, N., Sellwood, J. A., James, R. A., & Kahn, F. D. 1991, Nature, 352, 411
Samland, M., & Gerhard, O. 2003, A&A, 399, 961
Sellwood, J. A. 2006, astro-ph/0610468
Sellwood, J. A. 2007, astro-ph/07041047
Shlosman, I., Frank, J., & Begelman, M. C. 1989, Nature 338, 45
Shlosman, I. 1991, in IAU Colloq. 124, Paired and Interacting Galaxies, ed. J. Sulentic & W. Keel (Dordrecht: Kluwer), p. 689
Shlosman, I., & Noguchi, M. 1993, ApJ, 414, 473
Shlosman, I. 2007, in IAU Symp. 235, Galaxy Evolution Across the Hubble Time, ed. F. Combes & J. Palous (Cambridge: CUP), p. 24
Sommer-Larsen, J., Gotz, M., & Portinari, L. 2003, ApJ, 596, 47
Sparke, L. S. 1986, MNRAS, 219, 657
Strickland, D. K., Heckman, T. M., Colbert, E. J. M., Hoopes, C. G., & Weaver, K. A. 2004, ApJ, 606, 829
Tonini, C., Lapi, A., & Salucci, P. 2006, ApJ, 649, 591
Weinberg, M. D., & Katz, N. 2007, MNRAS, 375, 460

## Discussion

*Beckman*: 1) Why are the DM halos prolate? 2) As an observer what can you suggest to me as useful tests for the presence of DM halos which can be derived from your work?

*Shlosman*: 1) Two reasons exist (and contribute to) inducing the triaxiality in halos: major mergers and radial orbit instability. 2) The existence of DM manifests itself in numerous ways in galactic dynamics. For example, the formation of stronger and larger bars depends on the sink of angular momentum from the disk; the flatness of Galactic rotation curves comes from the DM contribution; the shapes of peanut/boxy bulges are affected by the DM mass distribution, etc.

Françoise Combes and Isaac Shlosman.

*Pathways through an Eclectic Universe*
*ASP Conference Series, Vol. 390, © 2008*
*J. H. Knapen, T. J. Mahoney, and A. Vazdekis, eds.*

# Formation and Evolution of Bulges

E. Athanassoula[1] and I. Martínez-Valpuesta[1,2]

[1]*LAM, OAMP, 2 place Le Verrier, 13248 Marseille cedex 04, France*
[2]*Instituto de Astrofísica de Canarias, E-38200 La Laguna, Tenerife, Spain*

**Abstract.** After presenting three ways of defining a bulge component in disk galaxies, we introduce the various types of bulges, namely the classical bulges, the boxy/peanut bulges and the disk-like bulges. We then discuss three specific topics linked to bulge formation and evolution, namely the coupled time evolution of the bar, buckling, and peanut strengths; the effect of velocity anisotropy on peanut formation; and bulge formation via bar destruction.

## 1. What Is a Bulge?

Three ways of defining a bulge have been used so far, one morphological, the second photometric and the third kinematic. Based on morphology, a bulge is a component of a disk galaxy that has a smooth light distribution that swells out of the central part of a disk viewed edge on. This definition has the disadvantage of being applicable only to edge-on systems and the advantage of necessitating only an image of the galaxy. The second definition is based on photometry and defines a bulge as the extra light in the central part of the galaxy, above the exponential profile fitting the remaining (non-central) part of the disk. In earlier papers this component was fitted with an $r^{1/4}$ law, while more recent ones use its generalization to an $r^{1/n}$ law (Sérsic 1968). This definition has the advantage of being applicable to disk galaxies independent of their inclination. It also has the advantage of leading to quantitative results about the light distribution but has the disadvantage of assigning to the bulge any extra central luminosity of the disk, independent of its origin. The third definition is based on kinematics, and in particular on the value of $V/\sigma$, or more specifically on the location of the object on the ($V/\sigma$, ellipticity) plot, which is often referred to as the Binney diagram (Binney 1978, 2005). This definition, potentially quite powerful, has unfortunately been very little used so far, due to the small number of galaxies for which the necessary data are available, a situation which is rapidly improving with large surveys, such as SAURON (Bacon et al. 2001; de Zeeuw et al. 2002; Peletier, this volume p. 232).

The lack of a single, clear-cut definition of a bulge is due to the fact that disk galaxies are viewed in different orientations and also to the fact that not all types of data are available for all objects. Nevertheless, it has led to considerable confusion and to the fact that bulges are an inhomogeneous class of objects. Indeed, many different types of objects, with very different properties and formation histories are included in the general term "bulge". To remedy this, Kormendy (1993) and Kormendy & Kennicutt (2004) distinguish classical bulges from pseudo-bulges, the latter category encompassing all bulges that are not classical. Athanassoula (2005a) argues that pseudo-bulges are also an inhomogeneous class of objects, and thus distinguishes three types of bulges.

*Classical bulges* are formed by gravitational collapse or hierarchical merging of smaller objects and corresponding dissipative gas processes. The material forming this bulge could be externally accreted, or could come from clumps in the protodisk. In general, bulges of this type are formed before the actual disk (e.g., Steinmetz & Müller 1995; Noguchi 1998; Immeli et al. 2004). Nevertheless, a bulge can also form from material externally accreted at much later stages (e.g., Pfenniger 1991; Athanassoula 1999; Aguerri, Balcells, & Peletier 2001; Fu, Huang, & Deng 2003). Their morphological, photometric and kinematic properties are similar to those of ellipticals.

*Box/peanut bulges* (B/Ps) form from a vertical instability of the disk material. This has often been observed in $N$-body simulations of bar-unstable disks, where the initial stage of bar formation is followed by a puffing up of the inner parts of the bar (e.g., Combes & Sanders 1981; Combes et al. 1990; Raha et al. 1991; Athanassoula & Misiriotis 2002; Athanassoula 2003, 2005a; O'Neil & Dubinski 2003; Martínez-Valpuesta & Shlosman 2004; Debattista et al. 2004, 2006; Martínez-Valpuesta, Shlosman, & Heller 2006). Viewed side on (i.e., edge on with the line of sight along the bar minor axis), this structure protrudes from the disk and has a characteristic boxy or peanut shape whose size is of the order of a few disk scale-lengths. Thus, a box/peanut bulge is just *part* of a bar seen side on.

Finally *disk-like bulges* form from inflow of (mainly) gas material to the center of the galaxy and subsequent star formation (e.g., Athanassoula 1992; Friedli & Benz 1993; Heller & Shlosman 1994; Wada & Habe 1995) . The torque exerted by the bar pushes gas, and to a lesser extent also stars, to the inner parts of the disk where they form an inner disk. Star formation can be very high there, due to the increased gas density. Thus the result of this process should be a central disk, or disk-like object, whose stellar component should include a sizable fraction of young stars, and whose size should be less than—or of the order of—a kpc. Such a component could harbor sub-structures such as spirals, or bars, as is indeed sometimes observed (Kormendy & Kennicutt 2004, and references therein). It is thus clear that disk-like bulges are very different objects from boxy/peanut bulges, since they are much smaller, have a different shape, different kinematics and provide a different type of excess on the radial photometric profiles. They also have different formation histories.

The different formation histories of these three types of objects lead to different properties—morphological, photometric, and kinematic—which in turn help classify observed bulges into one of the three above-mentioned types. Nevertheless, as stressed by Athanassoula (2005a), different types of bulges often co-exist and it is possible to find all three types of bulges in the same simulation, or galaxy.

Realizing the non-homogeneity of the objects classified as bulges and attempting to classify them is only the first step. Much more work is now necessary, particularly on two issues. The first one is understanding the formation and evolution of these types of objects. The second is to predict the properties of these objects, starting from their formation scenarios. The latter is particularly important in order to bridge the gap between classification schemes based on formation histories and classification schemes based on observed properties. Here, we make small contributions to both these issues by using $N$-body simulations. In Sect. 2 we present the time evolution of the bar, the buckling and the peanut strengths and their interplay. Section 3 discusses the velocity anisotropy and its link to the above strengths. Finally, in Sect. 4 we discuss the photometric properties of a destroyed bar and boxy/peanut bulge.

## 2. Time Evolution of the Bar, Buckling and Peanut Strengths

The formation of boxy/peanut bulges has been witnessed in a large number of simulations, starting with Combes & Sanders (1981). It is now well understood in terms of orbital structure and particularly in terms of the instability of the periodic orbits that constitute the backbone of the bar (Pfenniger 1984; Skokos, Patsis, & Athanassoula 2002; Patsis, Skokos, & Athanassoula 2002).

The time evolution of the bar, buckling, and peanut strengths are plotted in Fig. 1 for a simulation which develops a strong bar. The time is given in computer units and, for a reasonable calibration, 100 computer units correspond to 1.4 Gyr (Athanassoula & Misiriotis 2002). The initially unbarred disk forms a bar roughly between times 150 and 250 (lower panel). We define as bar formation time the time at which the bar growth is maximum (i.e., when the slope of the bar strength as a function of time curve is maximum) and indicate it by the first vertical line in Figure 1. The bar strength reaches a maximum at a time noted by the second vertical line, and then drops, at a time which is denoted by the third vertical line. It reaches a minimum, at a time given by the fourth vertical line and then starts increasing again. The upper panel shows the buckling strength, i.e., the vertical asymmetry as a function of time. Before the bar forms, the disk is vertically symmetric, with the first indications of asymmetry occurring after bar formation. Then the asymmetry grows very abruptly to a strong, clear peak and then drops equally abruptly. The time of the buckling is given by the peak of this curve and is very clearly defined. It is important to note that, to within the measuring errors, it coincides with the time of bar decay (third vertical line). This is not accidental. We verified this in a very large number of simulations and thus can establish the link between the buckling episode and the decay of the bar strength (Raha et al. 1991; Martínez-Valpuesta & Shlosman 2004). The middle panel shows the strength of the peanut as a function of time. This quantity grows abruptly with time after the bar has reached its maximum amplitude and during the time of the buckling. This abrupt growth is followed by a much slower increase over a long period of time. Taken together, these figures show that the bar forms vertically thin, and only after it has reached a maximum strength does the buckling phase occur. The time intervals during which bar formation, or peanut formation, or buckling occur are all three rather short, of the order of a Gyr or less, and they are followed by much longer times of slower bar and B/P evolution.

This particular simulation has a second, much weaker, buckling episode around time 700. This occurs very often in simulations developing strong bars. It is seen clearly in the peanut strength development, as a second abrupt increase

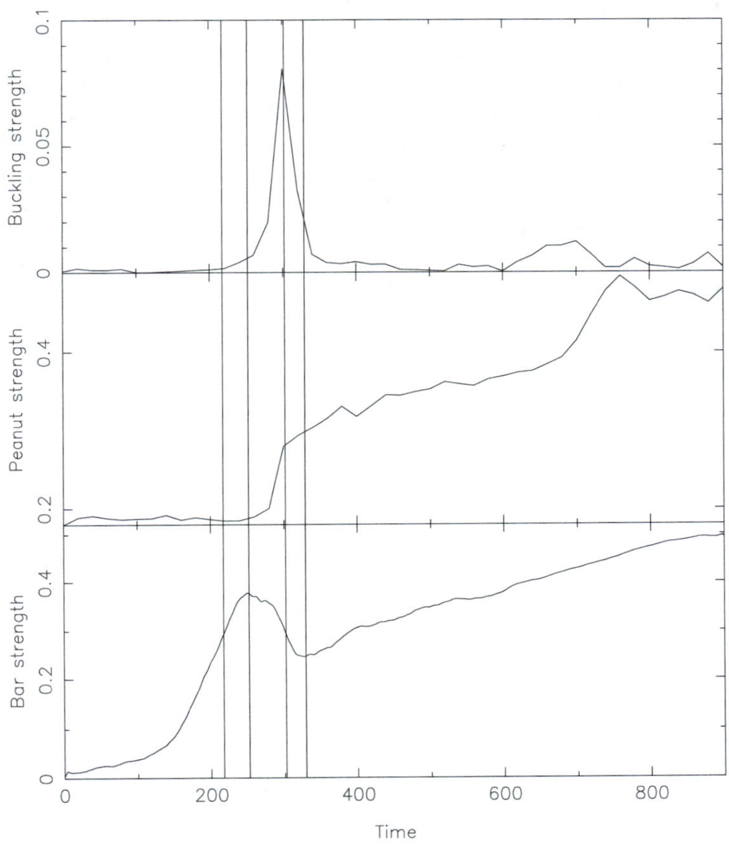


Figure 1. Time evolution of three peanut- or bar-related quantities, namely the buckling strength (asymmetry; *upper panel*), peanut strength (*middle panel*) and bar strength (*lower panel*). The thin vertical lines mark characteristic times linked to bar formation and evolution. From left to right, these are the bar formation time, the maximum amplitude time, the bar decay time, and the bar minimum amplitude time (see text).

of the peanut strength (Athanassoula 2005b; Martínez-Valpuesta et al. 2006). There are indications for it in the buckling strength, but not in the bar strength. As already mentioned, there are several possible definitions of the bar and the buckling strengths (E. Athanassoula & I. Martínez-Valpuesta, in preparation; Martínez-Valpuesta & Athanassoula, this volume, p. 463), and exactly how clearly the buckling episodes are seen in time evolution plots depends somewhat on the definition adopted. For example, since the first buckling concerns more inner than outer parts (Martínez-Valpuesta et al. 2006), one can calculate the quantities in Fig. 1 using, instead of the whole disk, only the radial range mainly concerned in the first or in the second buckling, which allows the buckling in question and its effects to be seen better.

## 3. Velocity Anisotropy and Peanut Formation

As already mentioned, peanut formation is linked to the vertical instability of parts of the main family of periodic orbits constituting the bar. This family is widely known as the $x_1$ family. Its stability can be followed from the stability diagram (see, for example, figures 3 and 4 of Skokos et al. 2002), which shows that, at the positions where $x_1$ becomes unstable, other families bifurcate. These are linked to the $n$:1 vertical resonances and extend well outside the disk equatorial plane. As shown by Patsis et al. (2002), some of them are very good building blocks for the formation of peanuts, because they are stable and because their orbits have the right shape and extent. Thus, orbital structure theory can go a long way towards explaining the formation and properties of the peanut.

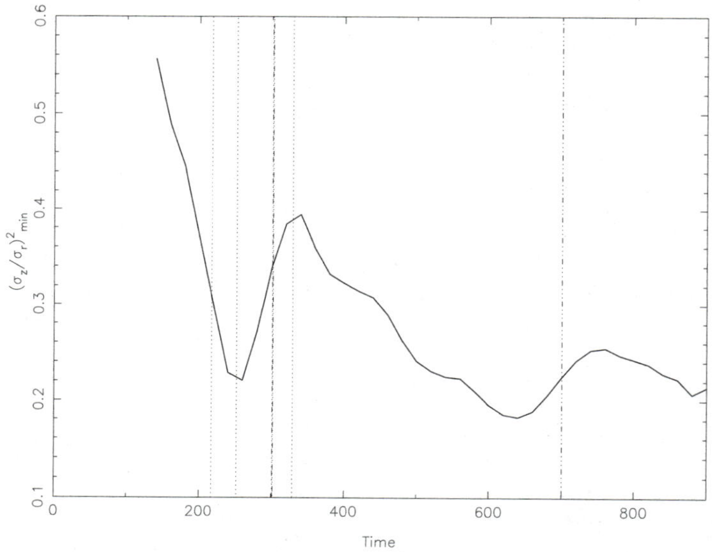


Figure 2. Time evolution of the $\sigma_z^2/\sigma_r^2$ ratio. The thin vertical lines mark characteristic times linked to bar formation and evolution. From left to right, the first four are the bar formation time, the maximum amplitude time, the bar decay time and the bar minimum amplitude time, all corresponding to the first buckling episode (see text and Fig. 1). The fifth and last vertical line (*dash-dot-dot-dot*) marks the time of the second buckling.

An alternative approach explains the buckling and peanut formation as due to the bending, or fire-hose, instability, studied analytically in the linear regime (Toomre 1966; Araki 1985). These studies assign a critical value of $R_\sigma = \sigma_z^2/\sigma_r^2$, igniting the onset of the instability, which is around 0.1. A number of simulations, however, have shown that this instability sets in already at much larger values (e.g., Merritt & Sellwood 1994; Sotnikova & Rodionov 2003).

To test this hypothesis, we calculate the radial and $z$ components of the disk velocity dispersion as a function of radius (averaging over azimuth and height). We then find the minimum value of their $R_\sigma$ ratio and plot its time evolution in Figure 2. The thin vertical lines in this figure are at the locations found from

Fig. 1 and mark the characteristic times linked to bar formation and evolution; namely, the bar formation time, the maximum amplitude time, the bar decay time, and the bar minimum amplitude time. Their location is clearly linked to the various evolutionary phases of $R_\sigma$. This, however, need not necessarily be seen as the cause of the buckling, but can also be seen as its result. Indeed, as the bar forms $\sigma_r$ increases drastically, so that $R_\sigma$ decreases. The bar amplitude then reaches its maximum and starts decreasing, while the peanut starts forming. During this time, $\sigma_r$ decreases, while $\sigma_z$ increases. As a result, between these two time intervals the $R_\sigma$ ratio should reach a minimum and then increase again, as is indeed seen in Figure 2. The bar amplitude then reaches a minimum, which corresponds to a minimum of $\sigma_r$ and therefore to a maximum of $R_\sigma$. This is followed by a slow decrease of $R_\sigma$, which is stopped by the second buckling episode. The value of $R_\sigma$ at which this instability sets in is much less extreme than that predicted by the above-mentioned analytical works, but is in good agreement with other $N$-body simulations.

More work is necessary before we fully understand the respective roles of the orbital structure results and of the velocity anisotropy effects on the formation and evolution of B/P structures. Both explain part of the story, but many aspects of their interplay are still unclear. Orbital structure results can tell us whether the appropriate building blocks are available or not, and this is essential since the lack of the appropriate building blocks can prohibit the formation of a given structure. Furthermore, studies of the properties of the building-block orbits are essential for understanding the properties of the B/P structures. On the other hand, it is necessary to group all these building blocks into one coherent unit and here collective effects are essential. They can also place limits on the formation of B/P structures, as well as give information on their properties. We will discuss the respective input from the two methods further elsewhere.

## 4. Bar and Peanut Destruction

The effect of a central mass concentration (CMC) on bar evolution has been well studied with the help of purely stellar $N$-body simulations, placing lower limits on the mass and central concentration necessary for the bar to be destroyed. Comparison with observations, however, shows that these limits are much higher than the measured values (Shen & Sellwood 2004; Athanassoula, Lambert, & Dehnen 2005 and references therein), thus casting doubts as to whether it is possible to destroy bars in this way. On the other hand, simulations with an SPH, or sticky particle representation of the gas come to a different conclusion (e.g., Friedli & Benz 1993; Berentzen et al. 1998; Bournaud & Combes 2002). Whether the difference between the two types of simulations could be due to the role of the gas in the angular momentum exchange is still debated (Bournaud, Combes, & Semelin 2005; Athanassoula et al. 2005; Berentzen et al. 2007). We will not attempt to settle the issue here. We will just study the photometric properties after the bar is destroyed, without worrying about whether the necessary CMC is compatible with observations, or not, or what the role of the gaseous component is. Our aim here is to see whether these photometric properties are compatible with those of a bulge, and, if so, of what type of bulge.

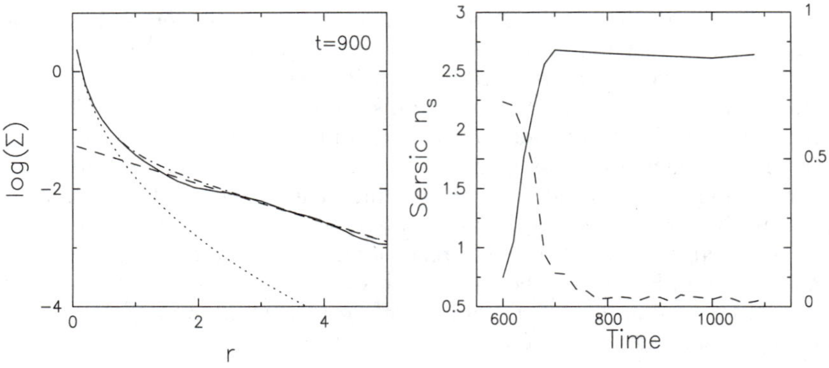


Figure 3. Bulge formation via bar destruction (see text). *Left panel:* Radial density profile of the luminous material (*solid line*) and its decomposition in a disk (*dashed*) and bulge (*dotted*) component. The sum of the two is given by a dash-dotted line. *Right panel:* Time evolution of the Sérsic index (*solid line*) and of the bar strength (*dashed line*). All quantities are in computer units as given by Athanassoula et al. (2005).

The simulations used in this analysis are those of Athanassoula et al. (2005). For several times and in every simulation, we obtained the radial density profile and then decomposed it into a disk and a bulge component, fitted by an exponential disk and a Sérsic profile, respectively. An example for a case with initially a strong bar with a clear peanut, in which we gradually grew a massive CMC, is shown in the left panel of Figure 3. The corresponding value for the Sérsic index is $n_s = 2.6$, i.e., approaching the range of values found for classical bulges (e.g., Kormendy & Kennicutt 2004). We have performed this decomposition for several times during the simulation and plot the values of $n_s$ as a function of time in the right panel of Fig. 3, where we also plot the bar strength as a function of time. For the simulation we are analyzing, the CMC has been grown gradually in 100 computer units (600–700), or, equivalently, 1.4 Gyr. We note that, during that time, the bar strength drops and $n_s$ increases, both drastically, while after the CMC has reached its final mass, both values stay nearly constant. The final value of the Sérsic index depends on the mass of the CMC.

The above shows that, at least as far as the photometrical definition presented in Sect. 1 is concerned, a decayed bar can give rise to an object with a Sérsic profile approaching those of classical bulges. Its shape when the snapshot is viewed edge on is compatible with that of a classical bulge. Thus, by two of the three definitions presented in Sect. 1, it is tempting to classify this object as a classical bulge. Work to study its kinematics is under way in order to complete the classification and see whether such objects are compatible with classical bulges (E. Athanassoula, J. A. L. Aguerri, & I. Martínez-Valpuesta, in preparation).

**Acknowledgments.** It is a pleasure to thank our collaborator, Alfonso Aguerri, for many interesting discussions and a fruitful collaboration, and A. Bosma, M. Balcells, P. Erwin, C. Heller, R. Peletier, M. Pohlen, and I. Shlosman for stimulating discussions and encouragement. This work has been partially supported by a Peter Gruber Foundation Fellowship to IMV and by grant ANR 06 BLAN 0172.

## References

Aguerri, J. A. L., Balcells, M., & Peletier, R. F. 2001, A&A, 367, 428
Araki, S. 1985, PhD Thesis, MIT
Athanassoula, E. 1992, MNRAS, 259, 345
Athanassoula, E. 1999, in ASP Conf. Ser., 160, Astrophysical Discs, ed. J. A. Sellwood & J. Goodman (San Francisco: ASP), 351
Athanassoula, E. 2003, MNRAS, 341, 1179
Athanassoula, E. 2005a, MNRAS, 358, 1477
Athanassoula, E. 2005b, in Planetary Nebulae as Astronomical Tools, ed. R. Szczerba, G. Stasińska & S. K. Górny (New York: AIP), 333
Athanassoula, E., Misiriotis, A. 2002, MNRAS, 330, 35
Athanassoula, E., Lambert, J. C., & Dehnen, W. 2005, MNRAS, 363, 496
Bacon, R., et al. 2001, MNRAS, 326, 23
Berentzen, I., Heller, C. H., Shlosman, I., & Fricke, K. J. 1998, MNRAS, 300, 49
Berentzen, I., Shlosman, I., Martínez-Valpuesta, I., & Heller, C. H. 2007, ApJ, 666, 189
Binney, J. 1978, MNRAS, 183, 501
Binney, J. 2005, MNRAS, 363, 937
Bournaud, F., & Combes, F. 2002, A&A, 392, 83
Bournaud, F., Combes, F., & Semelin, B. 2005, MNRAS, 364, L18
Combes, F., Sanders, R. H. 1981, A&A, 96, 164
Combes, F., Debbasch, F., Friedli, D., & Pfenniger, D. 1990, A&A, 233, 82
Debattista, V. P., Carollo, M., Mayer, L., & Moore, B. 2004, ApJ, 604, L93
Debattista, V. P., Carollo, M., Mayer, L. Moore, B., Wadsley, J., & Quinn, T. 2006, ApJ, 645, 209
de Zeeuw, P. T. et al. 2002, MNRAS, 329, 513
Friedli, D., & Benz, W. 1993, A&A, 268, 65
Fu, Y. N., Huang, J. H., & Deng, Z. G. 2003, MNRAS, 339, 442
Heller, C. H., & Shlosman, I. 1994, ApJ, 424, 84
Immeli, A., Samland, M., Gerhard, O., & Westera, P. 2004, A&A, 413, 547
Kormendy, J. 1993, in IAU Symposium 153, Galactic Bulges, ed. H. Dejonghe & H. J. Habing (Dordrecht: Kluwer), 209
Kormendy, J., & Kennicutt, R. C. 2004, ARA&A, 42, 603
Martínez-Valpuesta, I., & Shlosman, I. 2004, ApJ, 613, 29
Martínez-Valpuesta, I., Shlosman, I., & Heller, C. 2006, ApJ, 637, 214
Merritt, D., & Sellwood J. A. 1994, ApJ, 425, 551
Noguchi, M. 1998, Nature, 392, 253
O'Neill, J. K., & Dubinski, J. 2003, MNRAS, 346, 251
Patsis, P., Skokos, Ch., & Athanassoula, E. 2002, MNRAS, 337, 578
Pfenniger, D. 1984, A&A, 134, 373
Pfenniger, D. 1991, in Dynamics of Disc Galaxies, ed. B. Sundelius (Göteborg: Göteborg University Press), p. 191
Raha, N., Sellwood, J. A., James, R. A., & Kahn, F. D. 1991, Nature, 352, 411
Sérsic, J. 1968, Atlas de Galaxias Australes, (Córdoba: Obs. Astron. Córdoba)
Shen, J., & Sellwood, J. A. 2004, ApJ, 604, 614
Skokos, H., Patsis, P., & Athanassoula, E. 2002, MNRAS, 333, 847
Sotnikova, N. Ya., & Rodionov, S. A. 2003, AstL, 29, 321
Steinmetz, M., & Müller, E. 1995, MNRAS, 276, 549
Toomre, A. 1966, Geophys. Fluid Dyn., N66-46, 111
Wada, K., & Habe, A. 1995, MNRAS, 277, 433

## Discussion

*Trujillo*: You said that to destroy bars you need CMCs with masses of 5% or larger. Actual black holes have 1% or less of the mass of the bulge. How are they going to destroy bars?

*Athanassoula*: Indeed it is doubtful whether bars can be destroyed in this way. They can, however, get destroyed in simulations with gas, although it is still unclear whether this is due to the role of the gas in the angular momentum exchange, or not. It is, therefore, worth studying the photometric properties of destroyed bars and we can do this without having to solve first the problem of what the role of the gas is, or should be. The results I showed were made in that spirit.

Isaac Shlosman and Lia Athanassoula. In the background: Preben Grosbøl (*left*) and Reynier Peletier.

*Pathways through an Eclectic Universe*
*ASP Conference Series, Vol. 390, © 2008*
*J. H. Knapen, T. J. Mahoney, and A. Vazdekis, eds.*

# Boxy/Peanut Bulges and Stellar Bars

Inma Martinez-Valpuesta[1,2] and E. Athanassoula[2]

[1]*Instituto de Astrofísica de Canarias, E-38200 La Laguna, Spain*
[2]*LAM, Observatoire Astronomique de Marseille Provence, 2 Place Le Verrier, F-13004 Marseille, France*

**Abstract.** Boxy/peanut bulges in disk galaxies have been associated with stellar bars. We analyze their properties in a large sample of $N$-body simulations, using different methods to measure their strength, shape, and possible asymmetry, and then to intercompare the results. Some of these methods can be applied to both simulations and observations. In particular, we seek correlations between bar and peanut properties, which, when applied to real galaxies, will give information on bars in edge-on galaxies and on peanuts in face-on galaxies.

## 1. Introduction

Both simulations and theoretical studies have shown that bars are not vertically thin morphological features but have a considerable vertical extent and a vertical structure, known as boxy/peanut (B/P) bulges (Combes & Sanders 1981; Combes et al. 1990). Comparisons between observations and $N$-body simulations have more firmly established this direct connection (Athanassoula 2005, and references therein). Observations have shown that both bars and B/P bulges are quite predominant in disk galaxies, and that the corresponding frequencies are in good agreement with the link between the two structures (Lütticke, Dettmar, & Pohlen 2000).

We measure the peanut properties in a large sample of several hundred $N$-body simulations ran by one of us (EA) for different purposes. More information on these simulations and on their properties can be found in Athanassoula & Misiriotis (2002) and Athanassoula (2003, 2007). In particular, we seek correlations between the properties of the bar and the properties of the B/P bulge.

## 2. Methods for Measuring Bar and Peanut Strength and Correlations

In order to measure the bar strength, we use standard Fourier decomposition and take the amplitude of the second ($m = 2$) component:

$$C_{m,r}(R) = \left| \sum_{j=1}^{N_s} m_j e^{im\theta_j} \right|, \quad m = 2, \tag{1}$$

where $m_j$ and $\theta_j$ are the mass and azimuthal angle of particle $j$. This can be done either globally or as a function of radius. In the former case the summation

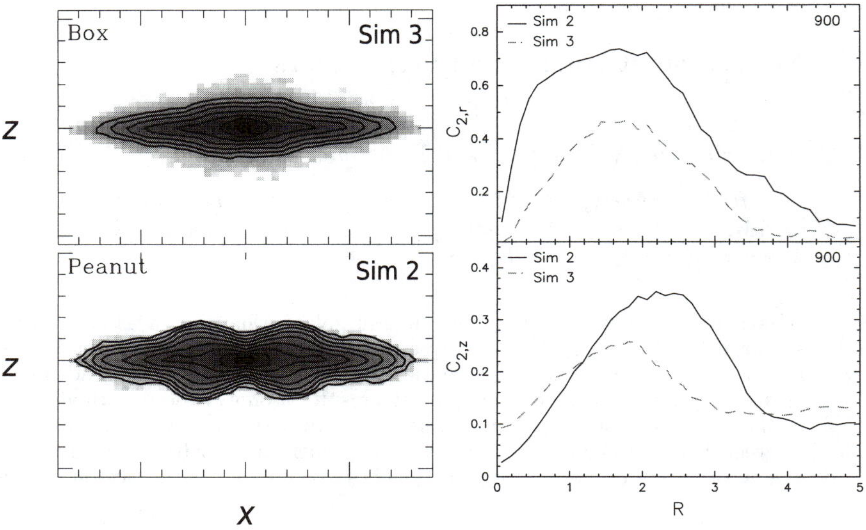


Figure 1. *Right:* Strength of the bar (*top panel*) and of the B/P bulge (*lower panel*) vs radius, for both snapshots. *Left:* Edge-on, side-on view of the boxy (*top panel*) and the strong peanut (*lower panel*) bulges used in the right panels.

is carried out over all particles in the disk, while in the latter $N_s$ is the number of particles in a given cylindrical shell of radius $R$ and $C_{m,r} = C_{m,r}(R)$.

The vertical asymmetry and the strength of the B/P can be measured in a similar way, by

$$C_{m,z} = \left| \sum_{j=1}^{N_s} z_j e^{imz_j/(5z_0)} \right|, \qquad m = 1, 2, \tag{2}$$

where $N_s$ can be either the number of particles in the disk component, or the number of particles per vertical column or cut (in which case $C_{m,z} = C_{m,z}(R)$) and $z_0$ is the scale height of the initial exponential disk. In Fig. 1 we apply these two definitions to two simulation snapshots, one with a boxy bulge and the other with a peanut, or X-shaped bulge.

We have also applied statistical methods, based on the distribution of the $z$ coordinates of the particles in a given vertical cut (perpendicular to the equatorial plane) of a snapshot seen edge on, with the bar viewed side on. We use the median of the absolute values and the standard deviation ($\sigma_z$) to measure the strength of the peanut, the kurtosis to measure its shape, and the skewness and mean to measure the strength of the buckling event, i.e., the asymmetry during the vertical instability.

Our methods to calculate the strength of the B/P bulge correlate nicely (Fig. 2, *right panels*) and can also be applied to real, highly inclined galaxies. Even more interesting, we find correlations between the strength and shape of the B/P bulge and the strength of the bar (Fig. 3). Note in this figure how some simulations

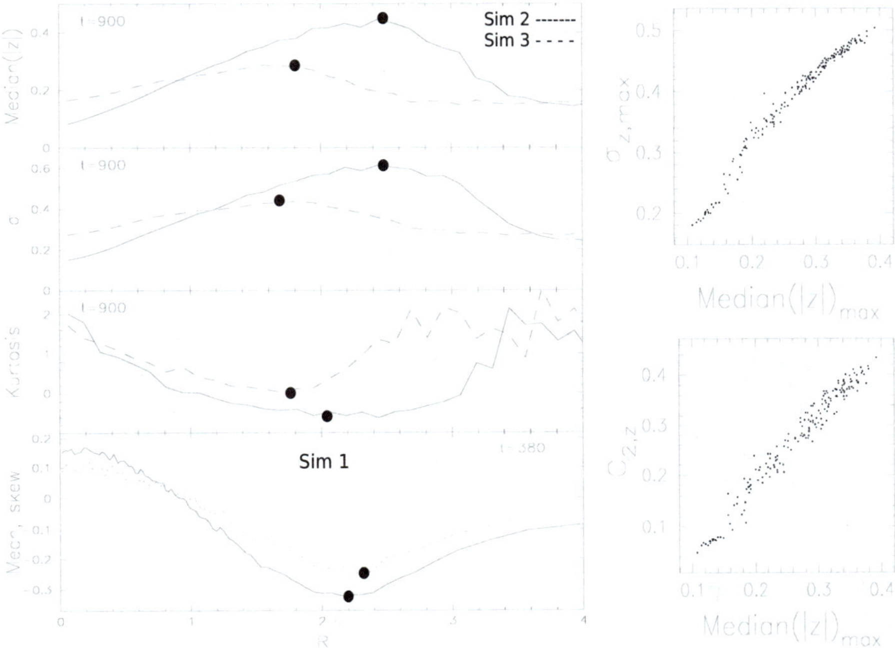


Figure 2. *Left panels*: Results of the statistical methods vs radius. The upper three panels are for the boxy and peanut snapshots shown in Fig. 1 and the lower one for a snapshot caught at the time of the buckling. *Right panels*: Two correlations showing how the statistical moments method agrees with the Fourier analysis method.

undergo several buckling events, as previously noted by Martinez-Valpuesta, Shlosman, & Heller (2006).

## 3. Conclusions

We have presented several methods to calculate the strength of the bar and the strength, shape, and asymmetry of the B/P bulge and have found strong correlations between their results. The most important correlation relates the strength of the bar with the strength of the B/P bulge, the strongest bars having the strongest peanuts. We also find that the strength of the peanut depends on the number of buckling episodes it undergoes, the strongest bars undergoing more buckling episodes (Fig. 3). Finally, we find a very interesting result about $C_{2,z}(R)$, i.e., about the shape of the radial density profiles along cuts perpendicular to the equatorial plane. For strong bars, having a strong peanut or X-shaped bulge, this profile is more flat-topped, while for weaker bars, with more boxy-like bulges, it is more peaked. All the results summarized here are discussed at length by E. Athanassoula & I. Martinez-Valpuesta (2008, in preparation).

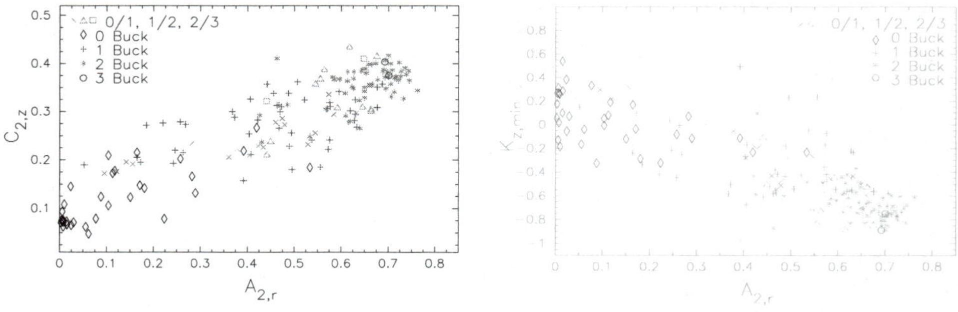


Figure 3. Correlations between bar and peanut properties. Each symbol corresponds to one simulation. The type of symbol is related to the number of buckling events suffered by the bar during its evolution. *Left panel*: Strength of the B/P bulge measured with our Fourier based method vs the strength of the bar. *Right panel*: Shape of the B/P bulge (i.e., shape of the radial density profiles along cuts perpendicular to the equatorial plane, measured by the minimum of the kurtosis) plotted as a function of bar strength.

**Acknowledgments.** This work has been partially supported by the Peter Gruber Foundation Fellowship to IMV, and by grant ANR 06 BLAN 0172 to EA.

## References

Athanassoula, E., 2003, MNRAS, 341, 1179
Athanassoula, E., 2005, MNRAS, 358, 1477
Athanassoula, E., 2007, MNRAS, 377, 1569
Athanassoula, E., & Misiriotis, A. 2002, MNRAS, 330, 35
Combes, F., & Sanders, R. H. 1981, A&A, 96, 164
Combes, F., Debbasch, F., Friedli, D., & Pfenniger, D. 1990, A&A, 233, 82
Lütticke, R., Dettmar, R.-J., & Pohlen, M. 2000, A&AS, 145, 405
Martinez-Valpuesta, I., Shlosman, I., & Heller, C. 2006, ApJ, 637, 214

## Discussion

*Cenarro*: What is the typical timescale of the evolution of the bar morphology that your models predict? Is it long enough to expect a systematic variation or evolution with redshift?

*Martínez-Valpuesta*: We work with a typical timescale of around 10 Gyr. The bar has a noticeable evolution in length, strength, and thickness that can be followed for around 6 or 7 Gyr. So it should possible to detect a variation in bar properties with redshift. This has already been checked by Jogee et al. (2004, ApJ, 615, L105), who found good agreement in terms of length evolution. But still higher resolution observations should be used.

*Sarzi*: What is the fraction of time in your simulation where you see the actual buckling of the bar when, in particular, the bar is asymmetric with respect to the disk plane? These kinds of objects appear to be exceedingly rare. Are the observations consistent with your simulations? The census of B/P bulges may be incomplete, but could you account for possible effects (the presence of gas, non-perfectly edge-on perspective) that might reduce the strength of the asymmetry in your simulations?

*Martínez-Valpuesta*: In general, we can distinguish two types of buckling. One is a fast buckling, which is the first to occur and lasts for less than a Gyr. The other one happens slowly, usually at later time in the evolution and lasts for around 3 Gyr. It should be possible to observed this slow buckling structure, but in general, any observational sample avoids the weird (e.g., asymmetric) galaxies. It could also affect the detection of the inclination and the dust lanes—although in simulations we can detect the asymmetry even at lower inclinations, sometimes even up to 70°.

*Otí*: How do you arrive to the expressions of the magnitudes you use to characterize the shape of the galaxy bulges?

*Martínez-Valpuesta*: In general, a good method of detecting symmetries and asymmetries is to use Fourier analysis. With a simple decomposition you can get the amplitude of the symmetric and asymmetric modes, i.e., the bar and the buckling in our case. In the case of the strength of the peanut, we slightly modified the Fourier decomposition to account not just for the symmetry but also for the thickness.

From left to right: Inma Martínez-Valpuesta, Françoise Combes, Ingo Berentzen.

*Pathways through an Eclectic Universe*
*ASP Conference Series, Vol. 390,* © *2008*
*J. H. Knapen, T. J. Mahoney, and A. Vazdekis, eds.*

# Downsizing in Elliptical Galaxies: Some Clues from Self-consistent Hydrodynamical Simulations

R. Domínguez-Tenreiro,[1] J. Oñorbe,[1] A. Serna,[2] and A. C. González-García[1]

[1] *Universidad Autónoma de Madrid, Spain*
[2] *Universidad Miguel Hernández, Elche, Spain*

**Abstract.** We report on the links among the dissipation rate history, the star formation rate history, and the history of mass assembly in a sample of elliptical-like objects (ELOs) identified, at $z = 0$, in a set of self-consistent hydrodynamical simulations operating in a cosmological context. These simulations provide clues on the physical processes involved in elliptical formation. They indicate that most of the dissipation involved in the mass assembly of a given ELO occurs in a violent early phase at high $z$ and on very short timescales, as a consequence of ELO assembly out of gaseous material and its transformation into stars. It turns out that the dissipation rate history is reflected by the star formation rate history (very high in this early phase). During the later slower phase of mass assembly, ELO stellar mass growth essentially occurs without any dissipation through mergers, and the star formation rate substantially decreases. We briefly discuss some observational implications of this scenario for ELO formation unveiled by our simulations, and in particular the downsizing effect.

## 1. Motivation

Different observational results point to apparently paradoxical results concerning the formation of elliptical galaxies (Es). None of the formation scenarios proposed so far (monolithic collapse or hierarchical mass assembly, see discussion in Matteucci 2003; Faber et al. 2005; Domínguez-Tenreiro et al. 2006) can recover all the observational information on Es we have now at our disposal.

Downsizing (the fact that more massive galaxies, and in particular massive Es, have their star formation [SF] stopped earlier on than less massive ones, Cowie et al. 1996) is one important example of such paradoxes. A related issue is the tilt of the fundamental plane of Es relative to the virial plane and its preservation since, at least, $z \sim 1.5$.

It is very convenient to study such issues in connection with the global cosmological model. Hydrodynamical simulations in a cosmological context provide a very suitable method for exploring the underlying physical processes, once one gets realistic galaxy-like objects in the simulations. In this contribution we report on some results from such simulations providing a scenario for E formation.

## 2. ELO Formation: Some Clues from Cosmological Simulations

We simulated the evolution of a 10 Mpc side periodic box within a concordance model using the DEVA code (for details see Serna, Domínguez, & Sáiz 2003; Oñorbe et al. 2006). Virtual galaxies of different morphological types form in the simulations. Elliptical-like objects (ELOs) have been identified as those having a prominent, dynamically relaxed stellar spheroidal component with no extended disks and very low gas content. Our simulations indicate (see Sierra González de Buitrago et al. 2003; Domínguez-Tenreiro et al. 2006), that ELOs are assembled out of the mass elements that at high $z$ are enclosed by those overdense regions $R$ whose local coalescence length $L_c(t, R)$ (Vergassola et al. 1994) grows much faster than average, and whose mass scale (total mass enclosed by $R$), is of the order of an E galaxy virial mass. These overdense regions act as flow convergence regions (FCRs), whose baryon content defines the particles that will end up in a bound configuration forming an ELO. FCRs contain a hierarchy of attraction basins toward which a fraction of the matter flows, feeding the clumps they host. Another fraction of the matter remains diffuse.

Analytical models, as well as $N$-body simulations, indicate that two different phases operate during halo mass assembly: first, a violent fast one, where the mass aggregation rates are high, and then, a slower one, with lower mass aggregation rates (Wechsler et al. 2002; Zhao et al. 2003; Salvador-Solé, Manrique, & Solanes 2005). Is is found that, on a given scale, overdense regions first expand more slowly than average, then they turn around and collapse through fast global compressions involving the cellular structure elements they enclose and in particular nodes connected by filaments, which experience fast head-on fusions (i.e., multiclump collapse, see Thomas, Greggio, & Bender 1999).

Our hydrodynamical simulations indicate that these fast head-on mergers result in strong shocks and high cooling rates of their gaseous component and, consequently, in strong and very fast SF bursts that transform most of the available cold gas at the FCR into stars (Fig. 1). For the massive ELOs this happens between $z = 6$ and $z = 2.5$ and mainly corresponds to a cold mode of gas aggregation, as in Keres et al. (2005). Consequently, most of the dissipation involved in the mass assembly of a given ELO occurs in this violent early phase at high $z$; moreover, its rate history is reflected by the SF rate history (Fig. 1). At the end of this phase most ELO stars are already formed and the proto-ELOs are virialized systems. After that, our simulations indicate that mergers in the slow phase imply only a modest amount of energy dissipation or star formation (i.e., dry mergers). A strong SF burst and dissipation follow a major merger only if enough gas is still available at the FCR after the early violent phase. This is unlikely in any case for those FCRs out of which ELOs form, and it becomes more and more unlikely as the ELO mass scale grows. The average number of mergers (major or minor) for ELOs in the slow phase is about one, and about 50% of ELOs in the sample have experienced a major merger event at $2 < z < 0$.

## 3. Dissipation, Star Formation, and Mass Assembly

An illustration of how the two phases operate is given in Figure 1. The cooling rate history (*dotted line*) results from collapse-induced mergers at high $z$; they take place at separated subunits within the FCR, and their rate is very high during this fast phase (*left*). During the slow phase (*right*) the rate decreases considerably. These merging patterns are also reflected by the SF rate history (*thin continuous line*) so that, as explained above, most stars form during the fast phase, but within separate clumps. In this figure the mass-aggregation tracks along the main branches of the merger tree are also shown, both for

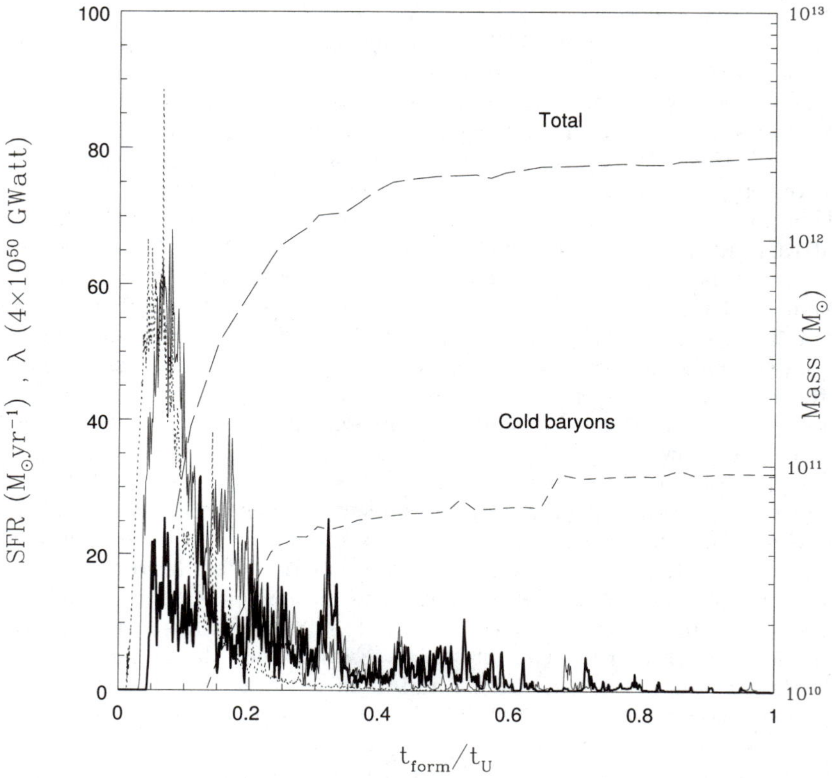


Figure 1. *Thin lines:* The cooling and SF rate histories (*dotted* and *continuous lines*, respectively), and the mass-aggregation tracks (*dashed lines*) for a typical massive ELO. *Thick line:* The SF rate history for a low mass ELO.

total mass (*long-dashed line*) and for cold baryon mass (i.e., stars and cold gas, *short-dashed line*). A merger event results in a discontinuity in the track. Note that the last major merger at $t_{\rm form}/t_{\rm U} \sim 0.7$ results in a rather modest SF burst, because no more gas is available to form stars. However, the ELO mass increases by a factor of ~2 due to merging activity during the late slow phase, joining previously formed stars belonging to different clumps.

## 4. Effects on Stellar Populations, Mass Distribution, and Kinematics

The physical laws responsible for clustering at high $z$ (Vergassola et al. 1994; Domínguez-Tenreiro, Sáiz, & Serna 2004) imply that clustering processes begin earlier on and go faster in FCRs enclosing higher masses. Consequently, stars form earlier on and on shorter timescales as the ELO mass scale increases (Fig.

1). Indeed, in this figure the continuous thick line is the SF rate history for a low mass ELO; we can see that the age distribution of the stars has, in this case, a younger mean and a broader width than in the case of a more massive ELO, that is, the continuous thin line. This is just a basic dynamical effect (i.e., *dynamical* downsizing). Other physical processes could also have contributed to the observed downsizing and stellar age effects (Kawata & Gibson 2005; Cattaneo et al. 2006).

Moreover, as a consequence of dissipation, mass homology breaking appears in the early, violent phase of mass assembly and is essentially preserved during the later, slower phase. A consequence is that a dynamical plane, tilted with respect to the virial plane and whose observational counterpart is the fundamental plane, appears in the violent phase (Oñorbe et al. 2005, 2006) and is roughly preserved along the slower phase, consistent with observations (see discussion in Domínguez-Tenreiro et al. 2006, and references therein). Note, however, that as the tracks in Fig. 1 indicate, merging during the slow phase causes the ELO mass (and consequently, the size and the velocity dispersion) to increase (Oñorbe, Domínguez-Tenreiro, & Serna, this volume, p. 498).

Other effects of dry merging are shape and kinematic transformations. ELO concentration, ellipticity and rotational support decrease (A. C. González-García et al., in preparation). Otherwise, the physical processes involved in ELO formation unveiled by our simulations might also explain other properties of local ellipticals recently inferred from observations (for example, the appearance of blue cores), as well as the increase in the stellar mass contributed by the elliptical population since higher $z$, or some QSO properties.

**Acknowledgments.** We are indebted to the CCCFC (UAM) for computing facilities and to the MCyT (Spain) and the regional government of Madrid for financial support through grants AYA-154292-C03-01 and the ASTROCAM Astrophysics network (S-0505/ESP-0237), respectively.

## References

Cattaneo, A., et al. 2006, MNRAS, 370, 1651
Cowie, L. L., et al. 1996, AJ, 112, 839
Domínguez-Tenreiro, R., Sáiz, A., & Serna, A. 2004, ApJ, 611, L5
Domínguez-Tenreiro, R. et al. 2006, ApJ, 636, L77
Faber, S. M., et al. 2005, astro-ph/0506044
Kawata, D., & Gibson, B. K. 2005, MNRAS, 358, L16
Keres, D., Katz, N., Weinberg, D. H., & Dave, R. 2005, MNRAS, 363, 2
Matteucci, F. 2003, ApSS, 284, 539
Oñorbe, J., et al. 2005, ApJ, 632, L57
Oñorbe, J., et al. 2006, MNRAS, 373, 503
Salvador-Solé, E., Manrique, A., & Solanes, J. M. 2005, MNRAS, 358, 901
Serna, A., Domínguez-Tenreiro, R., & Sáiz, A. 2003, ApJ, 597, 878
Sierra-González de Buitrago, M. M., Domínguez-Tenreiro, R., & Serna, A. 2003, in Highlights of Spanish Astronomy III, ed. J. Gallego, J. Zamorano, & N. Cardiel (Dordrecht: Kluwer), p. 171
Thomas, D., Greggio, L., & Bender, R. 1999, MNRAS, 302, 537
Vergassola, M., Dubrulle, B., Frisch, U., & Noullez, A. 1994, A&A, 289, 325
Wechsler, R. H. et al. 2002, ApJ, 568, 52
Zhao, D. H., Mo, H. J., Jing, Y. P., & Borner, G. 2003, MNRAS, 339, 12

*Pathways through an Eclectic Universe*
*ASP Conference Series, Vol. 390, © 2008*
*J. H. Knapen, T. J. Mahoney, and A. Vazdekis, eds.*

# Primordial Light Element Abundances

Paolo Molaro

*INAF Osservatorio Astronomico di Trieste, I-34143 Trieste, Italy*

**Abstract.** After the first few minutes of its existence, the Universe evolved through conditions of temperature and density that permitted the first synthesis of astrophysically interesting abundances of D, $^3$He, $^4$He, and $^7$Li. Relic abundances are sensitive probes of the nucleon density, as are acoustic oscillations in the cosmic microwave background, some 400 000 years after the Big Bang; both allow a stringent cross check. The high precision estimate of the baryon density by *WMAP* is currently used as input parameter for standard Big Bang nucleosynthesis to interpret primordial abundances, rather than these being directly derived from the observation of light elements, as was common before. New atomic physics and the identification of systematics lead to an upward revision of the $^4$He primordial abundance of $Y_{\rm p} = 0.2477 \pm 0.0029$, removing a major source of tension between standard Big Bang nucleosynthesis and *WMAP*. The D/H ratio as measured in QSO high redshift absorbing clouds shows an excess of scatter but the mean value is found to be in spectacular agreement with the *WMAP* $\Omega_{\rm b}$ prediction. The Li/H ratio as recently redetermined in halo dwarfs is more than a factor four lower than expected. We argue that the difference reduces to a factor two when the infrared flux method $T_{\rm eff}$ scale is adopted. Diffusion has been suggested to have depleted Li in halo dwarfs by the required amount to remove the gap; however, this would imply an implausibly high abundance of the more fragile $^6$Li detected in some halo dwarfs, thus leaving the puzzle open.

## 1. John Beckman and the Light Elements

The organizers asked me to review the status of the light elements, which is one of J. E. Beckman's interests, although a bit aside the main theme of this unique and enjoyable conference. John's interest in light elements dates back to the '80s, the epoch of his definitive move to the Instituto de Astrofísica de Canarias (IAC). The start was a search for Be II lines in HD 76932 by means of *IUE* spectra, a star belonging to the first sample of halo stars in which François and Monique Spite observed primordial Li. The upper limit provided for Be showed that spallation processes of high energy cosmic rays were not an alternative to primordial nucleosynthesis to make up for all or a fraction of the Li observed in the star, thus supporting its primordial origin (Molaro & Beckman 1984). This work was followed by the first Be detections in halo stars by means of the Image Photon Counting System at the Isaac Newton Telescope (INT) on La Palma and reported in Rebolo et al. (1988a). In the same years, an observational Li campaign was conducted from La Palma with the INT, and the first results were presented at the second IAP workshop in Paris (Beckman, Rebolo, & Molaro 1986), and then in a more definitive shape in Rebolo, Beckman, & Molaro (1988b). This was the first confirmation of the primordial Spite and Spite halo plateau made by an independent group, significantly enlarging the sample of halo stars in number and physical properties, in particular with a significant extension in metallicity down to [Fe/H] = −3.5, which is not very far from where we are today. After these pioneering works at the IAC, R. García-López and E. Martín, among others, extended the work to other light elements (B) and to other environments such as stellar clusters, cool X-ray binary companions, and brown dwarfs—a field still very active today at the IAC.

## 2. The *WMAP*–$\Omega_b h^2$ Primordial Abundances

The number of baryons is set once for all at baryogenesis and is constant in a comoving Universe. When the Universe was a few minutes old it evolved through conditions of temperature and density which permitted the first synthesis of astrophysically interesting abundances of D, $^3$He, $^4$He, and $^7$Li. While the $^4$He abundance is determined primarily by the universal expansion rate and is sensitive to the number of relativistic particles at the nucleosynthesis epoch, D and Li abundances depend strongly on the available nuclei. In standard Big Bang nucleosynthesis (SBBN) with three neutrino flavors the only free parameter is the relative number of baryons to photons, namely $\eta = n_b/n_\gamma$. In the range of interest $3 < 10^{10}\eta < 8$, D/H and Li/H show the strongest dependence on $\eta$ and therefore on the particle density; these are $\propto \eta^{-1/6}$ and $\propto \eta^2$, respectively.

The value of the baryon density can be measured from the acoustic oscillations in the cosmic microwave background (CMB) since the baryonic component determines the inertia of the mass–photon fluid. The ratio of the compressible peaks of the CMB power spectrum are sensitive to the baryonic density and allow a very precise measure of $\Omega_b$. Considering the *WMAP* only three-year data, Spergel et al. (2007) derived $\Omega_b h^2 = 0.02233^{+0.00072}_{-0.00091}$.

The number of photons is also constant in a comoving Universe after $e^\pm$ annihilation and therefore $\eta = 10^{10}(n_b/n_\gamma)_0 = (273.9 \pm 0.3)\Omega_b h^2$, where the small error enters in the conversion from number to mass densities (Steigman 2006). The *WMAP* value for $\Omega_b h^2$ yields $\eta = 10^{10}(n_b/n_\gamma)_0 = 6.116^{+0.197}_{-0.249}$, which fixes the primordial abundances of the light elements in the framework of SBBN. By using Kneller & Steigman's (2004) approximations of primordial yields in the range $3 < 10^{10}\eta < 8$ we obtain the predicted primordial abundances reported in Table 1, where the primordial $^4$He, $Y_p$, is given in mass fraction and the other light elements in number fractions with respect to hydrogen.

Table 1. WMAP primordial abundances

| Element | SBBN + *WMAP* |
|---|---|
| $Y_p$ | $0.2482\ ^{+0.0004}_{-0.0003}$ |
| $^3$He/H | $(10.5 \pm 0.6) \times 10^{-6}$ |
| D/H | $(25.7\ ^{+1.7}_{-1.3}) \times 10^{-6}$ |
| Li/H | $(4.41^{+0.3}_{-0.4}) \times 10^{-10}$ |

However, some degeneracies are present in the derivation of the cosmological parameters from the CMB power spectrum. In particular, $\Omega_b h^2$ depends on the index of the power law of the primordial fluctuations and on the opacity at the re-ionization epochs. For instance, the *WMAP* three-year combined with all other relevant datasets with a running spectral index is providing $\Omega_b h^2 = 0.02065^{+0.00083}_{-0.00082}$. This value gives $\eta_{10} = 5.656^{+0.227}_{-0.225}$ and the predicted primordial abundances become: $Y_p = 0.2474^{+0.0004}_{-0.0003}$, D/H= $(29.1^{+1.9}_{-1.8}) \times 10^{-6}$, and Li/H $= (3.8 \pm 0.4)^{-10}$ with a shift upwards of about 11% for D/H abundance, and one of 14% downwards for Li/H. Thus, uncertainties regarding priors are a significant source of systematic error that slightly exceed the statistical error in the prediction of $\Omega_b h^2$. Nevertheless, the CMB precision estimate of the baryon density remains high and should be used as input parameter to interpret primordial abundances rather then being directly derived from the observations of light elements, as was usual before *WMAP*. Only an unlikely knowledge of a primordial D or Li abundance with an accuracy greater than 10–15% could reverse the logical path.

## 3. Helium

Primordial helium is best determined through recombination of He II and H II lines in extragalactic, chemically almost unevolved H II regions, with an extrapolation to zero metallicity to account for stellar production. Figure 1 shows the behavior of the $^4$He determinations over the last few decades. After many years with rather low values we have now approached the *WMAP* values, thus removing a major source of tension between *WMAP* and SBBN. Olive & Skillman (2004) stressed the importance of systematic errors and derived $Y_p = 0.2495 \pm 0.0092$, while Peimbert, Luridiana, & Peimbert (2007) make a list of 13 different types of errors, several of systematic origin, entering in the primordial $^4$He determinations. Fukugita & Kawasaki (2006) carefully considered the underlying $^4$He stellar absorption and derived $Y_p = 0.250 \pm 0.004$. Peimbert et al. (2007) derived $Y_p = 0.2477 \pm 0.0029$ by using the new recombination coefficients of He I from Porter et al. (2005) and excitation collisional coefficients for H from Anderson et al. (2002). This new physics alone determines a 0.008 dex increment in the $^4$He determination. By using the Porter et al. emissivities Izotov, Thuan, & Stasińska (2007) obtained $Y_p = 0.2516 \pm 0.0011$, which exceeds by $2\sigma$ the predicted *WMAP* value, implying a slight deviation, but of a different sign, from SBBN.

$^3$He has a complex post-SBBN evolution with unclear stellar production. Bania, Rood, & Balser (2002), after 20 years of heroic observations of the most distant Galactic H II region, suggest $11 \pm 2$ ppm (parts-per-million) as the best upper limit on the primordial $^3$He abundance, essentially derived from the observation of a single H II region. This is in excellent agreement with the *WMAP* value of $10.5 \pm 0.6$ ppm, but considering the large uncertainties involved, $^3$He does not impose strong constraints on SBBN.

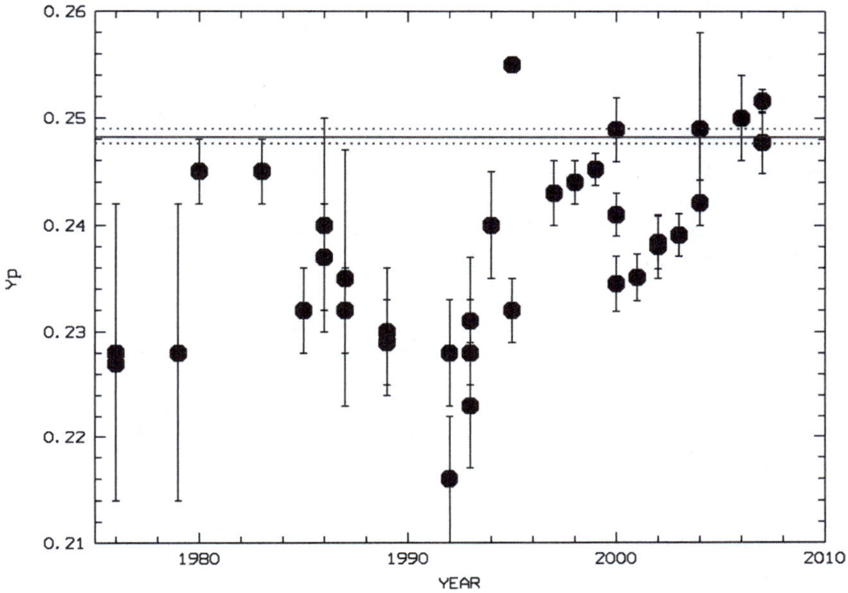


Figure 1. Historical record of $^4$He determinations. The *WMAP*–SBBN predicted primordial value is also shown with the (red) line and $1\sigma$ error.

## 4. Deuterium

Deuterium has a very simple chemical evolution history. It cannot be produced by other sources than the SBBN and whenever it is cycled in stars it is entirely burned out. The possibility of observing primordial D/H in distant chemically unevolved clouds had already been predicted by Adams (1976) but the first observations were those of Tytler, Fan, & Burles (1996), and after a decade of 8–10 m large telescope observations we remain with only eight measurements. The observations are shown in Fig. 2, where they are plotted with respect to the neutral hydrogen column densities, which imply different types of absorbers. The sub-DLA cases are those which better match the *WMAP* value. Kirkman et al. (2003) claimed the presence of a correlation between D/H and H I, but the new measurements of O'Meara et al. (2006) of $33.1 \pm 4.3$ ppm at $\log N(\mathrm{H\,I}) = 20.67\,\mathrm{cm}^{-2}$ and that of Crighton et al. (2004) of $16.0 \pm 2.5$ ppm at $\log N(\mathrm{H\,I}) = 18.25\ \mathrm{cm}^{-2}$ showed that the supposed correlation with H I was an artifact produced by poor statistics. The systems show low metallicities in the range $-3 < [\mathrm{Si/H}] < -1$ and no hint of a correlation between D/H and metallicity or redshift. At these metallicities, no significant astration is foreseen according to computations by Romano et al. (2003), and the measurements should reflect essentially pristine gas composition. From Fig. 2 it is evident that the dispersion of the measures exceeds the reported errors, suggesting at face value either the presence of scatter in D/H or an underestimation of the errors in some or all measurements. The unweighted average of the D/H gives $28.1 \pm 9$ ppm, with the central value in excellent agreement with the 25.7 $^{+1.7}_{-1.3}$ ppm of *WMAP*. It is really very unfortunate that we do not understand the cause of the observed scatter since it prevents the potential use of D/H for the identification of the priors in the CMB analysis.

A matter of concern are the Galactic D/H abundances, which turn out to be more complex than the high redshift Universe. *FUSE* has provided a lot of new data confirming the dispersion in the Galactic disk values, with a bimodal distribution of high (D/H $= 23.1 \pm 2.4$ ppm) and low values ($10.1 \pm 3.3$ ppm), once cleaned for the local bubble values, which cluster around a third value of $15.8 \pm 2.1$ ppm (Savage et al. 2007; Linsky et al. 2006). The measurements show a correlation with Fe abundances, suggesting a depletion of D into dust grains. A new measurement of the deuterium hyperfine transition towards the Galactic anticenter region of $21 \pm 2.3$ ppm is also consistent with the high D/H values in the solar neighborhood (Rogers, Dudevoir, & Bania 2007). The high values should therefore be more representative of the gas phase abundances but they impose tight constraints on the amount of minimum recycling foreseen in the Galactic chemical evolution and require infall of primordial D (Steigman, Romano, & Tosi 2007).

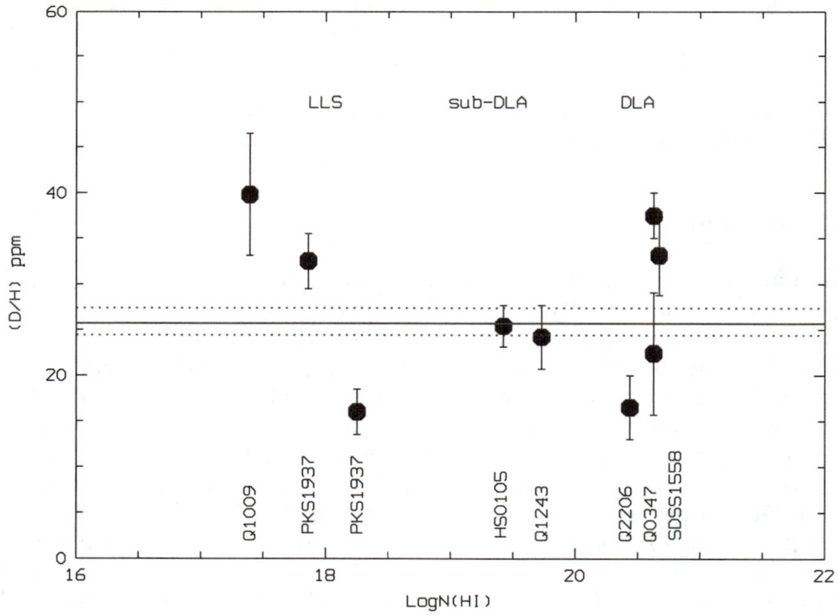


Figure 2. D/H versus hydrogen column density. The labels identify the QSO where the D/H has been measured: Q1009+299 is from Tytler et al. (1996), PKS1937-1009 are from Burles & Tytler (1998) and from Crighton et al. (2004), HS0105+1619 is from O'Meara et al. (2001); Q1243+3047 is from Kirkman et al. (2003), Q2206-199 is from Pettini & Bowen (2001); for Q0347-383 there are two determinations of the same system by D' Odorico, Dessauges-Zavadsky, & Molaro (2001) and by Levshakov et al. (2002); SDSS 1558-0031 is from O'Meara et al. (2006)

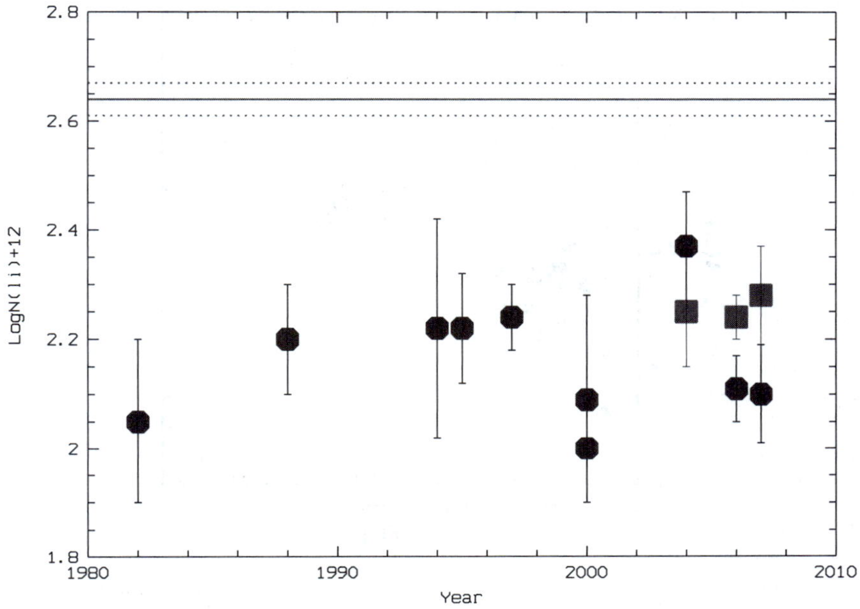


Figure 3. Blue circles are the historical record of primordial Li/H determinations in field halo stars since Spite & Spite (1982). Red squares are re-determinations after correction for some systematics, as explained in the text. $A(\mathrm{Li}) = (\log\,(\mathrm{Li/H}) + 12)$.

## 5. Lithium

### 5.1. Off, but by How Much?

The more significant Li measurements in halo stars since Spite & Spite (1982) are shown in Fig. 3, where $A(\mathrm{Li}) = (\log\,(\mathrm{Li/H}) + 12)$. At variance with the other elements, there is a clear gap between the measured values and the *WMAP* predictions that seems to increase with time instead of converging to the *WMAP* value. The most recent works are those by Asplund et al. (2006), characterized by data with very high signal to noise, and Bonifacio et al. (2007), who increase the data at low metallicity significantly. Their results, recomputed on common scales, are shown in Fig. 4, taken from Bonifacio et al. (2007). Figure 4 shows an increase in the dispersion at low metallicity, the presence of a slope with metallicity, or more precisely of a discontinuity in the Li abundance at [Fe/H] $\approx -2.5$ and even a possible drop in Li abundance at about [Fe/H] $= -3.5$. Both works conclude a rather low Li primordial value, somewhere around $A(\mathrm{Li}) = 2.1$, or even 2.0 if the observed slope is extrapolated to metallicities of Fe/H $= -3.5$. Thus, we note that the most recent values are very close to the initial value of 2.05 proposed by the Spites more than 20 years ago.

Li is mostly ionized and remains very sensitive to the $T_{\mathrm{eff}}$. At the temperatures of interest an error of 100 K in $T_{\mathrm{eff}}$ translates into about 0.08 dex in Li

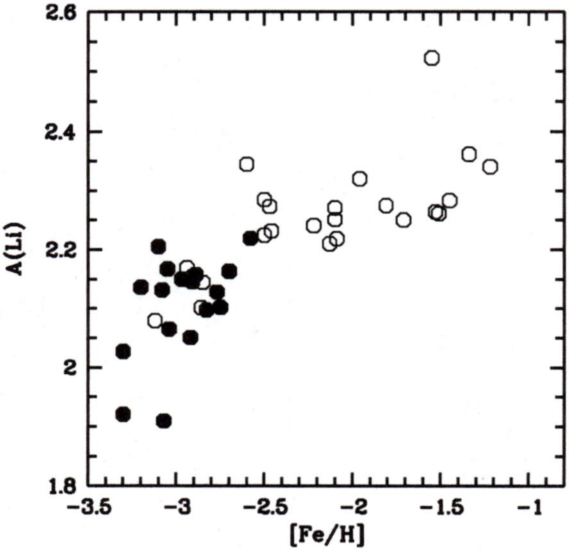


Figure 4. Li/H from Asplund et al. (2006; open circles) and Bonifacio et al. (2007; filled circles) plotted on a common scale. (*From Bonifacio et al. 2007.*)

abundance. Both Asplund et al. (2006) and Bonifacio et al. (2007) use $T_{\rm eff}$ obtained through H$\alpha$ fitting and adopting the theory of collisional broadening by Barklem, Piskunov, & O'Meara (2000) and Barklem et al. (2002). This new treatment provides Li values lower by $\approx$0.1 dex at lower metallicities in comparison to the Ali & Griem (1966) theory generally adopted in the preceding works. Thus, these $T_{\rm eff}$ are likely responsible for the abrupt change in the Li abundances observed at [Fe/H] $\approx -2.5$, and for the low values obtained at lower metallicities. For instance, keeping only the sample of 19 stars with metallicity [Fe/H] $> -2.5$ from Asplund et al. (2006), we obtain $A$(Li) $= 2.24 \pm 0.04$. Moreover, the data of Bonifacio et al. (2007) but with different $T_{\rm eff}$ obtained by using the Ali & Griem (1966) theory, provide the value of $A$(Li) $= 2.28 \pm 0.09$, significantly enhancing by $\approx$0.2 dex the average value (Bonifacio et al. 2003).

As pointed out in Molaro et al. (1995) the adoption of $T_{\rm eff}$ obtained with the infrared flux method (IRFM) levels out any slope of Li with metallicity. This has also been confirmed by Bonifacio & Molaro (1997) and Meléndez & Ramírez (2004), who make use of the IRFM $T_{\rm eff}$ and, while sharing several stars in common with Asplund et al. (2006), do not find any evidence of a Li slope with metallicity. Figure 5 of Asplund et al. compares the $T_{\rm eff}$ obtained from H$\alpha$ fitting plus Barklem theory for collisional broadening with those obtained with the IRFM method and reveals a divergence between the two scales at low metallicities, i.e., [Fe/H] $< -2.5$. The IRFM method encounters problems with reddening correction and with the difficulty of having accurate $K$-magnitudes, but is less model-dependent and therefore less sensitive to our ability to model the stellar atmosphere. Moreover, Barklem (2007) points out that NLTE effects could be important in the wings of the Balmer lines, thus calling into question the $T_{\rm eff}$ obtained in this way.

Meléndez and Ramírez (2004), by means of the IRFM value, obtained $A(\mathrm{Li}) = 2.37 \pm 0.06$. However, this value cannot be taken at face value since it is obtained by means of stellar atmospheric models, which provide systematically higher Li abundance than other codes. They use the Kurucz models, which adopt an approximate treatment for overshooting, which has been shown to provide an unsatisfactory color match at low metallicity and should not be used (Castelli, Gratton, & Kurucz 1997). Considering this effect, their value should be lowered by about 0.08 dex, plus an additional correction for the NLTE correction not taken into account, to reach a final value of $\sim 2.25$.

Thus, eliminating the data values obtained with $T_{\rm eff}$ from the Barklem theory at low metallicities, which is likely affected by systematics, it is possible to find a convergence of the various measures and methods around a common range of values in the range $A(\mathrm{Li}) = 2.24$–$2.28$. This still remains significantly below the *WMAP* predicted value, but not as much as the factor four implied by the measures by Asplund et al. (2006) and Bonifacio et al. (2007).

The hypothetical presence of a drop in Li at very low metallicities may find support also in the absence of Li in HE 1327-2326 a dwarf with $[\mathrm{Fe/H}] = -5.45$ (Aoki et al. 2006). However, HE 1327-2326 is chemically rather peculiar with $[\mathrm{C/Fe}] = +4$ and the chemical composition may reflect single supernova contamination where most of the Li is burned out. On the other hand, the more "normal" metal poor dwarf CS 22876-32 with $[\mathrm{Fe/H}] = -3.7$ shows $A(\mathrm{Li}) = 2.2$ in the main component of the binary system suggesting a normal behavior of Li at very low metallicities (González Hernández et al. 2008)

## 5.2. Is the Li Puzzle Solved?

Stellar depletion, either by rotationally induced mixing or diffusion, has been suggested as a possible mechanism to reduce the Li abundance in halo stars. These models predict more dispersion in the Li abundances than is observed in the halo stars and a pronounced downturn in the Li abundances at the hot end of the Li plateau. Richard, Michaud, & Richer (2005) invoke some extra turbulence to limit the diffusion in the hotter stars and to restore uniform Li abundance along all the plateau. Recently, Korn et al. (2006) claimed the detection of a diffusion "signature" in the globular cluster NGC 6397. According to their analysis the turnoff (TO) stars in this cluster exhibit lower iron and lithium abundances than the slightly more evolved stars. These observations may be interpreted in the framework of the diffusive models of Richard et al. (2005) and this has been regarded as the solution of the Li problem (Charbonnel 2006).

The Korn et al. result is very suggestive; however, it again relies heavily on the adopted temperature scale, and an increase of only 100 K in the effective temperature assigned to the TO stars would remove the abundance differences between these and subgiant stars. Previous analyses of the same cluster by Castilho et al. (2000) and Gratton et al. (2001), with different assumptions on the effective temperature scale, failed to find any difference in [Fe/H] among TO, subgiant, and giant stars. Bonifacio et al. (2002), studying the Li in this cluster, obtained a slightly hotter temperature for the TO stars, and $A(\mathrm{Li}) = 2.34$, which suffices to eliminate any Li difference between the TO stars and subgiants. On close inspection, globular clusters show Li scatter among the mean sequence stars which cannot be related to diffusive processes since they show Li–O and Li–Na anticorrelations. These have been observed in NGC 6752 (Pasquini et al. 2005) and 47 Tuc (Bonifacio et al. 2007) and suggest that the formation of globular clusters could be a complex process where it may be difficult to disentangle small diffusive signatures.

The diffusion hypothesis encounters an additional problem with the $^6$Li detection in some halo stars by Asplund et al. (2006). In the presence of diffusion of the sort required to reproduce the $^7$Li expected for *WMAP*, $^6$Li is depleted by the same processes by an even larger amount. When pre-main sequence destruction is also considered, the resulting original $^6$Li value is almost comparable to the $^7$Li level. As noted by Asplund et al. (2006), the Richard et al. (2005) T6.25 model, when applied to the star LP 815-43, implies an implausibly high original $^6$Li value of $\sim$2.6. This level of $^6$Li would pose serious problems for $^6$Li synthesis, so that the presence of the Li isotope challenges the diffusion hypothesis (cf. Prantzos 2007)

The reason for the disagreement of the observations with the *WMAP* value is not readily identified. Atmospheric modeling problems are unlikely but not completely ruled out. Observations of the Li subordinate, which forms in a different layer in the atmosphere, provide results consistent with those of the main Li line (Bonifacio & Molaro 1998; Asplund et al. 2006). First results from 3D atmospheric models show very small departures from 1D NLTE Li abundances with small but negative corrections, thus eventually exacerbating the disagreement with the *WMAP*-based prediction (Asplund, Carlsson, & Botner 2003; Barklem, Belyaev, & Asplund 2003).

A nuclear fix to the Li problem is also unlikely (Cyburt, Field, & Olive 2004). Coc et al. (2004) suggested that the cross-section of $^7\mathrm{Be}(d,p)2^4\mathrm{He}$, and $^7\mathrm{Be}(d,\alpha)^5\mathrm{Li}$ could be larger, but recent measures of this cross-section at energies appropriate to the Big Bang environment found it to be ten times *smaller*, leaving little room for a Li change due to nuclear cross-sections (Angulo et al. 2005).

Piau et al. (2006) suggested that significant matter processing in pregalactic Population III stars could lower the Li abundance and also explain the presence of Li dispersion among the metal-poor sample as well as the absence of Li in HE 1327-2326. However, as pointed out by Prantzos (2007), the considerable amount of matter recycling in Population III stars should produce significant amounts of heavy elements, which are not seen.

Although a solution of the Li problem may still reside in our ability to reconstruct the complexity of the stellar atmosphere, systematic errors in $T_{\mathrm{eff}}$ or 3D effects able to increase the Li value to match the *WMAP* value seem unlikely. It is thus possible that the disagreement may be real, revealing the presence of unaccounted processes in the SBBN. Neutral decaying particles (Jedamzik 2006, but see Ellis et al. 2005) or a variation of the deuteron binding energy (Dent, Stern, & Wetternich 2007) have already been considered and are among the more intriguing scenarios for the solution of the puzzle.

**Acknowledgments.** It is a pleasure to thank the organizers for their kind invitation to this unique conference and Piercarlo Bonifacio, Valentina Luridiana, and Bernard Pagel for enjoyable discussions on the light elements. Special thanks to John Beckman for his advice and inspiration.

## References

Adams, T. F. 1976, A&A, 50, 461
Ali, A. W., & Griem, H. R. 1966, Phys. Rev., 144, 366
Anderson, H., Ballance, C. P., Badnell, N. R., & Summers, H. P. 2002, J. Phys. B Atom. Molec. Phys., 35, 1613
Angulo, C., et al. 2005, ApJ, 630, L105
Aoki, W., et al. 2006, ApJ, 639, 897
Asplund, M., Carlsson, M., Botnen, A. V. 2003, A&A, 399, L31
Asplund, M., Lambert, D. L., Nissen, P. E., Primas, F., & Smith, V. 2006, ApJ, 644, 229
Bania, T. M., Rood, R. T., & Balser, D. S. 2002, Nature, 415, 54.
Barklem, P. S., Piskunov, N., & O'Mara, B. J. 2000, A&A, 363, 1091
Barklem, P. S., Stempels, H. C., Allende Prieto, C., Kochukhov, O. P., Piskunov, N., & O'Mara, B. J. 2002, A&A, 385, 951
Barklem, P. S., Belyaev, A. K., & Asplund, M. 2003, A&A, 409, L1
Barklem, P. S. 2007, A&A, 466, 327
Beckman, J. E., Rebolo, R., & Molaro, P. 1986, in Advances in Nuclear Astrophysics, Proc. Second IAP Workshop (Gif-sur-Yvette: Editions Frontieres), p. 29
Bonifacio, P., & Molaro, P. 1997, MNRAS, 285, 847
Bonifacio, P., & Molaro, P. 1998, ApJ, 500, L175
Bonifacio, P., et al. 2002, A&A, 395, 515
Bonifacio, P., et al. 2003, IAUJD, 15, 39
Bonifacio, P., et al. 2007, A&A, 462, 851
Bonifacio, P., et al. 2007, A&A, 470, 153
Burles, S., & Tytler, D. 1998, ApJ, 507, 732
Castelli, F., Gratton, R. G., & Kurucz, R. L. 1997, A&A, 318, 841
Castilho, B. V., Pasquini, L., Allen, D. M., Barbuy, B., & Molaro, P. 2000, A&A, 361, 92
Charbonnel, C. 2006, Nature, 442, 636
Coc, A., Vangioni-Flam, E., Descouvemont, P., Adahchour, A., & Angulo, C. 2004, ApJ, 600, 544
Crighton, N. H. M., Webb, J. K., Ortiz-Gil, A., & Fernández-Soto, A. 2004, MNRAS, 355, 1042
Cyburt, R. H., Fields, B. D., & Olive, K. A. 2004, Phys. Rev. D, 69, 123519
Dent, T., Stern, S., & Wetterich, C. 2007, Phys. Rev. D, 76, 063513
D'Odorico, S., Dessauges-Zavadsky, M., & Molaro, P. 2001, A&A, 368, L21
Ellis, J., Olive, K. A., & Vangioni, E. 2005, Phys. Lett. B, 619, 30
Fukugita, M., & Kawasaki, M. 2006, ApJ, 646, 691
González Hernández, J. I., et al. 2008, A&A, in press (arXiv:0712.2949)
Gratton, R., et al. 2001, A&A, 369, 87
Izotov, Y. I., Thuan, T. X., & Stasińska, G. 2007, ApJ, 662, 15
Jedamzik, K., Choi, K.-Y., Roszkowski, L., & Ruiz de Austri, R. 2006, J. Cosmol. Astroparticle Phys., 7, 7
Kirkman, D., Tytler, D., Suzuki, N., O'Meara, J. M., & Lubin, D. 2003, ApJS, 149, 1
Kneller, J. P., & Steigman, G. 2004, New J. Phys., 6, 117
Korn, A. J., et al. 2006, Nature, 442, 657
Levshakov, S. A., Dessauges-Zavadsky, M., D'Odorico, S., & Molaro, P. 2002, ApJ, 565, 696
Linsky, J. L., et al. 2006, ApJ, 647, 1106
Melendéz, J., & Ramírez, I. 2004, ApJ, 615, L33

Molaro, P., & Beckman, J. E. 1984, A&A, 218, 197
Molaro, P., Primas, F., & Bonifacio, P. 1995, A&A, 295, L47
Olive, K. A., & Skillman, E. D. 2004, ApJ, 617, 29
O'Meara, J. M., Tytler, D., Kirkman, D., Suzuki, N., Prochaska, J. X., Lubin, D., & Wolfe, A. M. 2001, ApJ, 552, 718
O'Meara, J. M., Burles, S., Prochaska, J. X., Prochter, G. E., Bernstein, R. A., & Burgess, K. M. 2006, ApJ, 649, L61
Pasquini, L., et al. 2005, A&A, 441, 549
Peimbert, M., Luridiana, V., & Peimbert, A. 2007, ApJ, 666, 636
Pettini, M., & Bowen, D. V. 2001, ApJ, 560, 41
Piau, L., Beers, T. C., Balsara, D. S., Sivarani, T., Truran, J. W., & Ferguson, J. W. 2006, ApJ, 653, 300
Porter, R. L., Bauman, R. P., Ferland, G. J., & MacAdam, K. B. 2005, ApJ, 622, L73
Prantzos, N. 2007, Space Science Rev., 130, 27
Rebolo, R., Abia, C., Beckman, J. E., & Molaro, P. 1988a, A&A, 193, 193
Rebolo, R., Beckman, J. E., & Molaro, P. 1988b, A&A, 192, 192
Richard, O., Michaud, G., & Richer, J. 2005, ApJ, 619, 538
Rogers, A. E. E., Dudevoir, K. A., & Bania, T. M. 2007, AJ, 133, 162
Romano, D., Tosi, M., Matteucci, F., & Chiappini, C. 2003, MNRAS, 346, 295
Savage, B. D., Lehner, N., Fox, A., Wakker, B., & Sembach, K. 2007, ApJ, 659, 1222
Spergel, D. N., et al. 2007, ApJS, 170, 377
Spite, M., & Spite, F. 1982, A&A, 297, 483
Steigman, G. 2006, J. Cosmol. Astroparticle Phys., 10, 016
Steigman, G., Romano, D., & Tosi, M. 2007, MNRAS, 378, 576
Tytler, D., Fan, X.-M., & Burles, S. 1996, Nature, 381, 207

## Discussion

*Silk*: What are the implications of the $^6$Li detection in halo stars for a possible slope in $^7$Li at the lowest metallicities?

*Molaro*: The observed $^6$Li plateau, at face value, implies a corresponding $^7$Li plateau. However, $^6$Li suffers from pre-main sequence depletion resulting in a tilted plateau with more $^6$Li at higher metallicities. Thus, a slope in $^7$Li should be produced, but detectable only at relatively higher metallicities, let's say higher than $[Fe/H] = -2$. If we consider that diffusion occurs, then the amount of original $^6$Li is very high and it may show a complicated behavior with metallicity, so that it is not easy to tell what the implications are for $^7$Li.

*Luridiana*: Is there any independent observational evidence, apart from the abundances of the Li isotopes, that might prove or disprove the efficiency of diffusion?

*Molaro*: Yes, the various elements should show a specific pattern revealing the presence of diffusion. For instance, Korn et al. (2006) claimed Fe variations along with those in Li. However, the effects are very small and a small change in the effective temperature would also produce the same change.

*Pathways through an Eclectic Universe*
*ASP Conference Series, Vol. 390, © 2008*
*J. H. Knapen, T. J. Mahoney, and A. Vazdekis, eds.*

# Where Are the Missing Baryons and Metals?

Bernard E. J. Pagel

*Astronomy Centre, University of Sussex, Brighton BN1 9QH, UK*

**Abstract.** The question of missing baryons arises because the smoothed-out density of baryons in the Universe is known from *WMAP* and primordial deuterium to be close to $\Omega_{\rm B} = 0.04$, whereas the corresponding figure for stars, based on the luminosity density from galaxy redshift surveys, is only 1/10 of this, or less, giving rise to the concept of "baryonic dark matter".

A clue to the nature of the missing baryons comes from the statistics of the Lyman-$\alpha$ forest at redshifts of 2 to 3, where the abundance of the corresponding ionized gas is estimated to be virtually equal to $\Omega_{\rm B}$. Most of that gas is probably still there, partly in the form of remaining Lyman-$\alpha$ forest and partly in the form of still more highly ionized intra-cluster and warm–hot intergalactic gas (WHIM). Most of the "metals" seem to be about equally shared between the WHIM and stars.

## 1. Introduction

Only about 28 percent of the energy density of the Universe is matter, and only 4 percent (i.e., 1/7 of the matter) is baryonic, as we know from primordial deuterium and the angular fluctuation spectrum of the microwave background (Spergel et al. 2003). Of this 4 percent, only about 1/10 is in the form of known stars and interstellar gas in galaxies, which gives rise to the question of missing baryons or baryonic dark matter. A related question is that of the metal budget in the Universe, which is related through stellar yields to star formation, past and present. The cosmic densities of various forms of matter and energy have been very thoroughly discussed in the "Domesday Book" by Fukugita & Peebles (2004; hereinafter FP), following earlier work by Persic & Salucci (1992), Fukugita, Hogan, & Peebles (1998) and Finoguenov, Burkert, & Böhringer (2003), among others.

## 2. Stars and the Interstellar Medium

Cosmic luminosity densities have been estimated from various galaxy redshift surveys, e.g., SDSS, 2DF, Millennium survey (Cole et al. 2001; Driver et al. 2007; FP) giving

$$L_B = (1.9 \pm 0.2) \times 10^8 \, L_\odot \, \mathrm{Mpc}^{-3} \quad (1)$$
$$L_R = (2.3 \pm 0.2) \times 10^8 \, L_\odot \, \mathrm{Mpc}^{-3} \quad (2)$$
$$L_z = (3.6 \pm 0.4) \times 10^8 \, L_\odot \, \mathrm{Mpc}^{-3}. \quad (3)$$

These luminosity densities may be compared to the cosmic critical density, through an assumed mass:light ratio,

$$\rho_{\rm crit} = 1.4 \times 10^{11}\, M_\odot\, {\rm Mpc}^{-3}. \tag{4}$$

FP use stellar population synthesis models designed for star formation histories characteristic of different kinds of galaxies to derive

$$\langle M_{\rm stars}/L_z \rangle \simeq 1.23 \pm 0.33, \tag{5}$$

assuming an initial mass function (IMF) with the Salpeter slope above $1\,M_\odot$, but flattened below; with the conventional Salpeter slope extended to $0.1\,M_\odot$, the ratio would be 1.5 times larger. Equation (1) then leads to

$$\Omega_{\rm stars} = 0.0024 \pm 0.0007. \tag{6}$$

A very similar value is derived by FP from the multicolor photometry of 2MASS combined with 2DF data by Cole et al. (2001), so they adopt a mean:

$$\Omega_{\rm stars} = 0.0027 \pm 0.0005, \tag{7}$$

with 74 percent of stellar mass in spheroids and 26 percent in disks.[1] This estimate is shown by FP to be consistent with less certain estimates of past star formation rates as a function of redshift and current supernova rates.

FP use the recent blind H I survey HIPASS (Zwaan et al. 2003) to give the mass of atomic gas in galaxies:

$$\Omega_{\rm HI} = (4.2 \pm 0.7) \times 10^{-4}, \tag{8}$$

and the CO survey of Keres, Yun, & Young (2003) giving

$$\Omega_{\rm H_2} = (1.6 \pm 0.6) \times 10^{-4}, \tag{9}$$

to give a total cold gas density (including an allowance for helium)

$$\Omega_{\rm HI+H_2} = (7.8 \pm 0.9) \times 10^{-4}. \tag{10}$$

García-Appadoo et al. (this conference, p. 298) have made a blind H I survey designed to pick up gas-rich low surface brightness (LSB) galaxies, estimating that these contribute 12 percent of the baryon density of gas-rich galaxies in general, i.e.,

$$\Omega_{\rm gas-rich\ LSB} \simeq 1.5 \times 10^{-4} \tag{11}$$

only. Gas-poor LSBs remain a significant unknown.

Another significant "known unknown" is the contribution of MACHOs in galactic halos. Alcock et al. (2000) estimated a mass of MACHOs in our own

---

[1] On the other hand, Driver et al. (2007) find 58 percent currently in the form of disks, 39 percent in red spheroids and 3 percent in blue spheroids and pseudo-bulges.

Galaxy comparable to or even greater than that of stars, which if typical of galaxies in general would at least double the stellar density from Equation 7. However, Tisserand et al. (2007), using more stringent criteria in the EROS project, have found that estimate to be a strict upper limit. Furthermore, there are severe nucleosynthesis and galactic-dynamical constraints on the white-dwarf sized MACHOs postulated by Alcock et al. (Fields, Freeze, & Graff 2000; Zhao 2002), to which I might add that there are additional constraints from extragalactic background light (EBL; Aharonian et al. 2006), since a substantial proportion of the EBL itself comes from the formation of white dwarfs (Pagel 2002).

## 3. Hot Intergalactic Gas

### 3.1. Intra-cluster Plasma

From total masses of clusters of galaxies (Reiprich & Böhringer 2002), FP use $\Omega_{\rm B}$, $\Omega_{\rm stars}$ and $\Omega_{\rm m}$ to deduce the cosmic density of their hot-gas content

$$\Omega_{\rm b,cl} = 0.0018 \pm 0.0007, \tag{12}$$

which is about 1/2 of stars + cold gas. This still leaves a substantial gap to be bridged, but offers a hint as to where the solution may lie.

### 3.2. Warm–Hot Intergalactic Medium (WHIM)

It has been known for some years that at redshifts between 2 and 3, there must have been enough ionized gas associated with the Lyman-$\alpha$ forest clouds to account for $\Omega_{\rm B}$ with ease (Rauch et al. 1998). Numerical cosmological simulations suggest that, of this gas, some is still there as photo-ionized intergalactic gas producing a Lyman-$\alpha$ forest at zero redshift, while the rest left over from star formation has been heated by shocks and feedback processes to temperatures above $10^5$K forming the web-like warm–hot intergalactic medium (WHIM; Cen & Ostriker 1999, 2006; Cen & Fang 2006). How much of this expected gas has actually been detected?

The observations come from *FUSE* and *HST* detections of Lyman-$\alpha$ and O VI at modest redshifts in the FUV in absorption on lines of sight to quasars. X-ray emission lines give more limited information, as discussed by Soltan (this conference, p. 364). Lehner et al. (2007) have given the latest picture of the low redshift Lyman-$\alpha$ forest, which presents a considerable range of line widths (10–100 km s$^{-1}$). Above a limiting column density of log $N_{\rm HI} = 13.2$, they estimate

$$\Omega_{\rm Ly\,forest,\mathit{z}=0} = 0.013\ (b < 40\,{\rm kms}^{-1}) \tag{13}$$
$$= 0.009\ (b > 40\,{\rm kms}^{-1}), \tag{14}$$

totaling about 50 percent of the baryons, and this is presumably a lower limit because the narrow-line column-density distribution function is $N^{-1.76}$. The narrow lines and part of the broad lines sample the cool, photo-ionized intergalactic medium (IGM), while the rest of the broad lines sample the WHIM, the contribution of which increases towards lower redshifts as expected from the simulations. O VI had been previously detected in some of the Lyman-$\alpha$ forest systems (Danforth & Shull 2005) and the distribution of equivalent widths per unit redshift is well fitted by the simulations of Cen & Fang (2006) involving galactic superwinds. A claimed detection of O VII absorptions at redshifts of 0.011 and 0.027 in front of Mrk 421 using *Chandra* during a bright phase of the quasar (Nicastro et al. 2005) aroused considerable interest, but was later found from summed *XMM–Newton* spectra with better signal-to-noise ratio to be spurious (Kaastra et al. 2006). Cen & Fang (2006) compared Nicastro's claimed distribution function of numbers per unit redshift at a given equivalent width with their simulations, which predicted nearly an order of magnitude less and were thus triumphantly vindicated by the *XMM–Newton* result. The observations are therefore quite consistent with the simulations, from which Cen & Ostriker (2006) have deduced that the fraction of baryons in the WHIM is 40–50 percent, i.e.,

$$\Omega_{\rm WHIM} = 0.020 \pm 0.002. \tag{15}$$

This presumably includes the broad Lyman-$\alpha$ clouds of Eq. 14, but may have only limited overlap with the narrow-line clouds of Eq. 13, so that WHIM + Lyman-$\alpha$ forest between them could account for 3/4 of the baryons. Estimates of the various components to the baryon density today are shown in the first column of the top half of Table 1. Lyman-$\alpha$ forest and WHIM between them account for most of the baryons, while most of the "metals" (see below) are in stars or WHIM.

## 4. Where Are the "Metals"?

Having decided where the baryons may be, it is of interest to say something about the distribution of "metals", i.e., heavy elements expelled into the interstellar medium as a result of nuclear processes in stars. Following Edmunds & Phillips (1997), we take the average stellar abundance to be solar, leading to the entries in the second and third columns of the second line of Table 1. There is also a substantial contribution to the cosmic metal budget from the X-ray gas in clusters of galaxies, but this is exceeded or even dwarfed by the WHIM, following Cen & Ostriker's prediction of a median metallicity of 0.18 $Z_\odot$.

The metallicity of the Lyman-$\alpha$ forest at low redshift is not known, but it is presumably quite low; according to simulations by Cen & Ostriker (1999), the lowest density regions in the Universe now have

$$Z_{\rm low\,dens.} \simeq 0.03\,Z_\odot = 0.0005, \tag{16}$$

which is what is assumed in the table. Thus the Lyman-$\alpha$ forest, while important for the baryon budget, contributes insignificantly to the metal budget.

## 5. Where Were the Baryons and Metals at $z = 2.5$?

The lower half of the table gives corresponding estimates for the Universe at a substantial redshift, which involve still larger uncertainties. Pettini (1999) pointed out several years ago that, while from the "Madau plot" about 1/4 of present-day stars are estimated to have been in place by $z = 2.5$, far less of the corresponding 1/4 of present-day metals had been detected at that time. This gave rise to numerous suggestions including: $\Omega_Z \simeq 2 \times 10^{-5}$ as dust in submm-detected galaxies (Dunne, Eales, & Edmunds 2003), $\Omega_Z \simeq 1.4 \times 10^{-5}$ as hot gas associated with proto-clusters of galaxies (Finoguenov, Burkert, & Böhringer 2003) and a combination of faint galaxies and IGM including Lyman-$\alpha$ forest and sub-DLAs (Bouché et al. 2005, 2006, 2007). Bouché et al. (2005) argued that the dust contribution had been overestimated. Thus, while most of the baryons are undoubtedly in the Lyman-$\alpha$ forest, it is difficult to give a clear answer to the question of where the metals are; they could be spread around pretty evenly, the largest single location being in stars or the hypothetical proto-cluster hot gas.

Anyway, lines 2–5 of the lower half of the table are based on the survey by Bouché et al. (2007), the double entry for DLAs accounting for the possibility of some high metallicity systems being obscured by dust.

Table 1. Baryon and metal budgets

| | | $z = 0$ | | | |
|---|---|---|---|---|---|
| Component | $\Omega_B$, $10^{-2}$ | | $Z, 10^{-2}$ | $\Omega_Z$, $10^{-5}$ | |
| X-r gas, clusters | 0.2 | | 0.7 | 1.4 | |
| Stars | 0.27 | | 1.7 | 4.6 | ←Most |
| WHIM (BLA, O VI) | 0.9–2.0 | ←Most | 0.2: | 1.8–4: | ← met. |
| Ly-$\alpha$ forest (NLA, O VI) | 1.0–1.5 | ←bar. | .05: | 0.5–0.8 | |
| H I + $H_2$ | 0.06 | | | | |
| Gas-rich LSB | 0.02: | | | | |
| Total | 2.4 –4: | | | 8 – 11: | |
| Predicted | 4.4 | | | | |
| | | $z = 2.5$ | | | |
| Protocl./hot gas? | 0.0–0.2 | | 0.7 | 0.0–1.4 | |
| Stars (LBG, sbm.) | .05–0.1 | | 1.0 | 0.5 – 1.0 | |
| ISM, dust | .02–.04 | Most | 0.8 | 0.2 – 0.4 | |
| Ly-$\alpha$ forest | 3 – 5 | ←bar. | .005 | .15 – .25 | |
| subDLAs | .04–.4 | | 0.1 | .05 – 0.5 | |
| DLAs | .15 | | 0.1 | .15(+.3?) | |
| Total | 3.4–5.4 | | | 2.4 – 4 | |
| Predicted | 4.4 | | | 2 – 4: | |

Mean metallicity of the Universe $\langle Z \rangle \equiv \Omega_Z/\Omega_B \simeq 3 \times 10^{-3}$ or about 0.2 solar.

Yield $p \equiv \Omega_Z/\Omega_s \simeq 0.03–0.04 \simeq 2.5\, Z_\odot \simeq$ Salpeter (0.1).

## References

Aharonian, F., et al. 2006, Nature, 440, 1018
Alcock, C., et al. 2000, ApJ, 542, 281
Bouché, N., Lehnert, M. D., Aguirre, A., Péroux, C., & Bergeron, J. 2007, MNRAS, 378, 525
Bouché, N., Lehnert, M. D., & Péroux, C. 2005, MNRAS, 364, 319

Bouché, N., Lehnert, M. D., & Péroux, C. 2006, MNRAS, 367, 16
Cen, R., & Fang, T. 2006, ApJ, 650, 573
Cen, R., & Ostriker, J. P. 1999, ApJ, 514, 1
Cen, R., & Ostriker, J. P. 2006, ApJ, 650, 560
Cole, S., Norberg, I. R. P., Baugh, C. M., & the 2dFGRS team 2001, MNRAS, 326, 255
Danforth, C. W., & Shull, J. M. 2005, ApJ, 624, 555
Driver, S. P., Allen, P. D., Liske, J., & Graham, A. W. 2007, ApJ, 657, 85
Dunne, L., Eales, S. A., & Edmunds, M. G. 2003, MNRAS, 341, 589
Edmunds, M. G., & Phillipps, S. 1997, MNRAS, 292, 733
Fields, B. D., Freeze, K., & Graff, D. S. 2000, ApJ, 534, 265
Finoguenov, A., Burkert, A, & Böhringer, H. 2003, ApJ, 594, 136
Fukugita, M., Hogan, C. J., & Peebles, P. J. E. 1998, ApJ, 503, 518
Fukugita, M., & Peebles, P. J. E. 2004, ApJ, 616, 643 (FP)
Kaastra, J. S., et al. 2006, ApJ, 652, 189
Keres, D., Yun, M. S., & Young, J. S. 2003, ApJ, 582, 659
Lehner, N., et al. 2007, ApJ, 658, 680
Nicastro, N., et al. 2005, ApJ, 629, 700
Pagel, B. E. J. 2002, in ASP Conf. Ser., 253, Chemical Enrichment of Intracluster and Intergalactic Medium, ed. R. Fusco Femiano & F. Matteucci (San Francisco: ASP), p. 489
Pettini, M. 1999, in ESO Workshop Proc., Chemical Evolution from Zero to High Redshift, ed. J. Walsh & M. Rosa (Berlin: Springer), p. 233
Persic, M., & Salucci, P. 1992, MNRAS, 258, 14
Rauch, M., et al. 1998, ApJ, 489, 1
Reiprich, T. H., & Böhringer, H. 2002, ApJ, 567, 716
Spergel, D. M., et al. 2003, ApJS, 148, 175
Tisserand, P., et al. 2007, A&A, 469, 387
Zhao, H. 2002, MNRAS, 336, 159
Zwaan, M. A. et al. 2003, AJ, 125, 2842

## Discussion

*García-Appadoo*: What are your views on the possible existence of "dark galaxies", i.e., baryonic matter with no stellar content, and how much do you think they could contribute to the total baryonic budget?

*Pagel*: There is some wiggle room if you have reasons for believing that such objects exist.

*Edmunds*: If $\langle Z \rangle > 0.1$, it implies lots of low metallicity gas. Where is it?

*Pagel*: In Cen & Ostriker's model, the Ly$\alpha$ forest gas and WHIM dominate the baryonic mass at present.

*Pathways through an Eclectic Universe*
*ASP Conference Series, Vol. 390, © 2008*
*J. H. Knapen, T. J. Mahoney, and A. Vazdekis, eds.*

# A Quantitative Analysis of the Morphology of Star Formation in Galaxies at Different Redshifts

Ruymán Azzollini[1] and John E. Beckman[1,2]

[1] *Instituto de Astrofísica de Canarias, E-38200 La Laguna, Spain*
[2] *Consejo Superior de Investigaciones Científicas, Serrano, 117, E-28006 Madrid, Spain*

**Abstract.** Here, we present briefly the methodology we have implemented to study the spatial distribution of massive star formation in galaxies. To do this, we analyze a sample of 759 galaxies in a range of redshifts extending up to $z \sim 1$. We use data from different surveys, which span in wavelength from FUV to NIR (*GALEX*, COMBO17, and GOODS-*HST*). The aim is to develop methodology which can be used to help decide between different basic models of galaxy formation using the spatial distribution of star formation within galaxies as the test parameter. In situ knowledge of the star formation distribution as a function of redshift will be a valuable counterpart to population synthesis studies, which at best integrate this kind of information over long periods.

## 1. Morphological Parameters as a Function of Redshift and Physical Parameters of Galaxies

In this study we have used four parameters which describe the morphology of a distribution of light, and which have been applied before to the study of galaxy morphology: Concentration ("C", Bershady, Jangren, & Conselice 2000), Asymmetry ("A", Abraham et al. 1996), Clumpiness ("S", Conselice 2003) and "M20/90" (a modification of the M20 parameter by Lotz, Primack, & Madau 2004). We performed photometry on the GOODS-*HST*/ACS (Giavalisco et al. 2004) images, in all four bands: B435W, V606W, I775W, and Z850LP, and measured the four morphological parameters in this photometric sample. This catalog of objects was matched to the "Master Compilation of GOODS/CDFS-S Spectroscopy".[1] This is our parent sample, with 986 galaxies, of which 759 (77%) have redshifts < 1.1.

We have also used data from COMBO17 (Wolf et al. 2004) and *GALEX* (Martin et al. 2005), and the KCORRECT software (Blanton et al. 2003), to obtain rest-frame absolute magnitudes in several bands from 1500 to 6000 Å and ultimately to estimate stellar mass ($M+$), mean surface mass density ($\mu^*$), star formation rate (SFR), and specific SFR (sSFR) for all objects under study.

In Fig. 1 we show how the morphological parameters under study change with redshift, as measured in different bands. It is a key point for our work that the B435W band falls below 4000 Å in the rest-frame spectra of galaxies if

[1] http://www.eso.org/science/goods/spectroscopy/CDFS_Mastercat/ .

$z > 0.2$, so it is progressively more dominated by young stars (it can be deemed as a proxy star formation tracer). We see how in this band objects appear less concentrated, more asymmetric, and clumpy, and have larger values of M20/90 in B435W than in other bands. Though these graphs could be interpreted in terms of how galaxies could be assembling their stellar content in the last 8 Gyr, we have to make further analyses. The effects of decreasing $S/N$ and lower spatial resolution, and the shift of bands along the rest-frame SED with redshift should be taken into account in an unbiased way to avoid serious errors in the interpretation of these results.

We also studied the four morphological parameters as a function of $M^*$, $\mu^*$, SFR and sSFR. No clear correlation between the morphology (both of young and old stars) and the physical properties under study could be seen in those graphs. There is only a mild tendency of smaller objects with higher $\mu^*$ to be also less clumpy, and objects with larger $M^*$ and $\mu^*$ to have stronger cores relative to total mass.

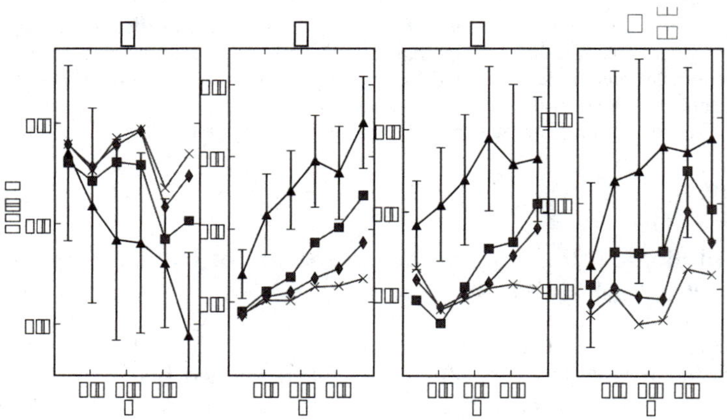

Figure 1. Morphological parameters vs redshift. Values have been binned in redshift ($\Delta z = 0.2$). Code: crosses : Z850, diamonds : I775, squares: V606 and triangles: B435. Error bars represent the standard deviation of the values at each bin, though only the errors in $B$ are shown for clarity (the other are similar).

## References


Abraham, R. G., Tanvir, N. R., Santiago, B. X., Ellis, R. S., Glazebrook, K., & van den Bergh, S. 1996, MNRAS, 279, L47

Bershady, M., Jangren, A., & Conselice C. 2000, AJ, 119, 2645

Blanton, M. R., et al. 2003, AJ, 125, 2348

Conselice, C. 2003, ApJS, 147, 1

Giavalisco, M., et al. 2004, ApJ, 600, 93

Lotz, J. M., Primack, J. & Madau, P. 2004, AJ, 128, 163

Martin, D. C., et al. 2005, ApJ, 619, 1

Wolf, C., et al. 2004, A&A, 421, 913

*Pathways through an Eclectic Universe*
*ASP Conference Series, Vol. 390, © 2008*
*J. H. Knapen, T. J. Mahoney, and A. Vazdekis, eds.*

# Self-consistent Models of Triaxial Cuspy Galaxies with Dark Matter Halos

R. Capuzzo-Dolcetta,[1] L. Leccese,[1] D. Merritt,[2] and A. Vicari[1]

[1] *Dept. of Physics, Universitá di Roma "La Sapienza", Rome, Italy*
[2] *Rochester Institute of Technology, Rochester, USA*

**Abstract.** We have constructed realistic, self-consistent models of triaxial elliptical galaxies embedded in triaxial dark matter halos. Self-consistent solutions by means of the standard orbital superposition technique introduced by Schwarzschild were found in each of the three cases studied. Chaotic orbits were found to be important in all of the models, and their presence was shown to imply a possible slow evolution of the shapes of the halos. The equilibrium velocity distribution is reproduced by a Lorentzian function better than by a Gaussian. Our results demonstrate for the first time that triaxial dark matter halos can co-exist with triaxial galaxies.

## Models and Methods

We report briefly here some of the results extensively presented and discussed in Capuzzo-Dolcetta et al. (2007). We considered three different mass models for the galaxy+halo system. In all the three cases considered we adopted triaxial shapes for both components, but with a different radial distribution for the dark (DM) and luminous matter (LM). In the first model the DM was assumed to have the same axial ratios as the LM (0.7:0.86:1), while the second and third models were characterized by halos that were more prolate (0.5:0.66:1) and more oblate (0.7:0.93:1), respectively, than the LM.

In the absence of analytic expressions for the orbital integrals, one must resort to numerical methods to construct self-consistent models of galaxies or DM halos. For this purpose, we followed the Schwarzschild (1979) scheme, which is also often referred to as the *orbital superposition* technique. We generalized Schwarzschild's method to the construction of two-component (LM and DM), self-consistent models as follows. First, we constructed a catalog of orbits integrating the equations of motion for a particle moving in the potential generated by the sum of the LM and DM density distributions. Then, two different grids of cells, one for the LM and one for the DM components, were built.

## Results

We were able to successfully construct self-consistent models in all of the three cases considered for the shape of the DM halo. A large fraction of orbits in all the models is chaotic; nevertheless, the evolution of the models' shapes over a Hubble time is likely to be small. This result suggests that DM halos can coexist with triaxial galaxies. The velocity distributions are non-Gaussian in shape; the distribution is more peaked around the maximum and with broader tails, and is hence better represented with a Lorentzian profile (see Fig. 1). This finding suggests that care be taken about the unquestioned use of Gaussian distributions as the initial conditions for simulations.

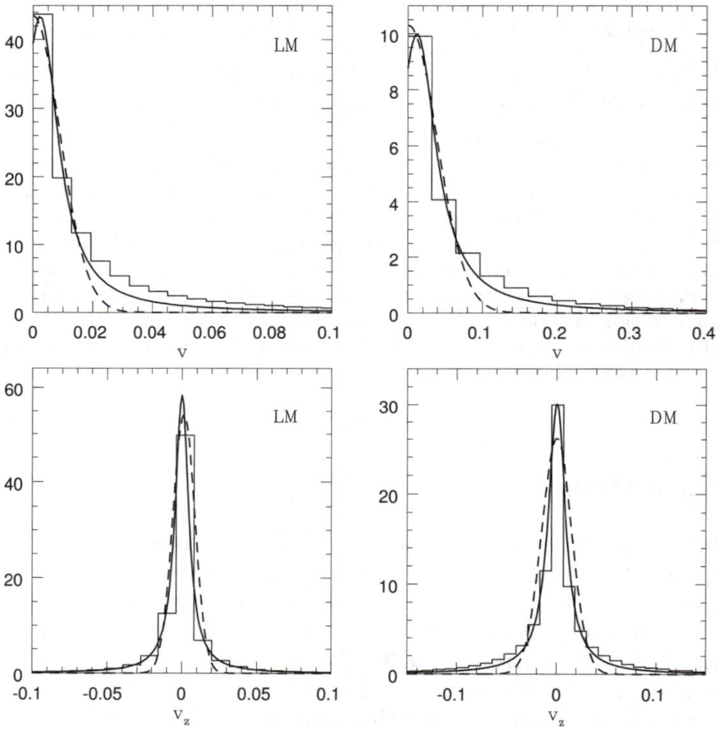


Figure 1. Histograms of the velocity distributions of LM (*left column*) and DM (*right column*). *Upper panels:* The distribution of the absolute value of the velocity. *Lower panels:* The distribution of the $z$-component of the velocity. *Solid* and *dashed* curves are the best Lorentzian and Gaussian fits, respectively.

## References

Capuzzo-Dolcetta, R., Leccese, L., Merritt, D., & Vicari, A. 2007, ApJ, 666, 165
Schwarzschild, M. 1979, ApJ, 232, 236

*Pathways through an Eclectic Universe*
*ASP Conference Series, Vol. 390, © 2008*
*J. H. Knapen, T. J. Mahoney, and A. Vazdekis, eds.*

# Hot GRB-selected Submillimeter Galaxies

M. J. Michałowski, J. Hjorth, J. M. Castro Cerón, and D. Watson

*Dark Cosmology Centre, Niels Bohr Institute, University of Copenhagen, Juliane Maries vej 30, 2100 Copenhagen Ø, Denmark*

**Abstract.** Using detailed spectral energy distribution fits we present evidence that submillimeter- and radio-bright gamma-ray burst host galaxies are hotter counterparts to submillimeter galaxies. This hypothesis makes them of special interest since hotter submillimeter galaxies are difficult to find and are believed to contribute significantly to the star formation history of the Universe.

## 1. Introduction

It is widely accepted that cold and luminous submillimeter (submm) galaxies (SMGs) are dominant contributors to the star formation history of the Universe at redshifts $z \sim 2$–3 (Chapman et al. 2005). On the other hand, the host galaxies of long duration gamma-ray bursts (GRBs, originating in the collapses of very massive stars, e.g., Hjorth et al. 2003) are found to be subluminous (Fruchter et al. 2006) and low mass (Castro Cerón et al. 2006). Four of them (GRBs 980703, 000210, 000418, and 010222) have been firmly detected in submm and/or radio (Berger, Kulkarni, & Frail 2001; Berger et al. 2003; Tanvir et al. 2004). In this paper we discuss the possibility that these submm-bright GRB hosts may represent the hotter counterparts of SMGs. For details see Michałowski et al. (2008) and Michałowski & Hjorth (2007).

## 2. SED Modeling and Results

In order to model the spectral energy distributions (SEDs) of GRB hosts we used the GRASIL software (Silva et al. 1998), a numerical code that calculates the spectrum of a galaxy by means of a radiative transfer method applied to photons produced by a stellar population and reprocessed by dust.

In Figure 1 we compare the total infrared luminosities and dust temperatures of GRB hosts (derived from the SED fits) with well-studied galaxies both local and at high-$z$. It is apparent that GRB hosts are hotter than SMGs with the same luminosity. This gives a hint that GRB events may pinpoint a population of ultraluminous infrared galaxies (ULIRGs) at high redshifts with dust hotter than in typical SMGs. The search for such galaxies is important because they likely contribute to the star formation history at the same level as SMGs.

We note that the majority of the galaxies shown in Fig. 1 also have higher dust temperatures compared to SMGs. However, all of them are local galaxies, so cannot be considered as counterparts of high redshift SMGs and their submm emission has been detected only because of their proximity.

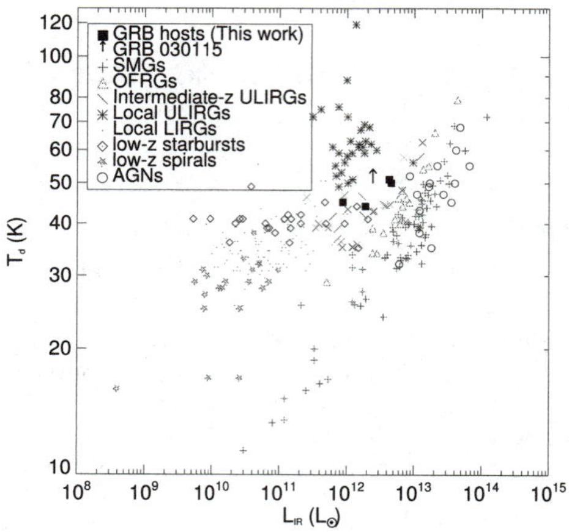


Figure 1. Dust temperature as a function of infrared luminosity. The GRB hosts discussed here and GRB 030115 are compared with different galaxy samples (see Michałowski et al. 2008, and references therein). Note that for GRB 030115 and the OFRGs lower limits to dust temperature are shown. GRB hosts seem to reflect the properties of intermediate-$z$ ULIRGs, the bright end of starburst galaxies and OFRGs—the candidates for hotter counterparts of SMGs.

GRB hosts may be consistent with a population of optically faint radio galaxies (OFRGs) having similar infrared luminosities and (likely) temperatures. Although the majority of OFRGs lie at $z \sim 2$ (Chapman et al. 2004), some of them are within the redshift range of the GRB hosts discussed here ($z = 0.85$–$1.48$). OFRGs have been suggested to be hotter counterparts of SMGs (Chapman et al. 2004), so the same may be true for GRB hosts.

**Acknowledgments.** The Dark Cosmology Centre is funded by the Danish National Research Foundation.

## References

Berger, E., Kulkarni, S. R., & Frail, D. A. 2001, ApJ, 560, 652
Berger, E., Cowie, L. L., Kulkarni, S. R., Frail, D. A., Aussel, H., & Barger, A. J. 2003, ApJ, 588, 99
Castro Cerón, J. M., Michałowski, M. J., Hjorth, J., Watson, D. J., Fynbo, J. P. U., & Gorosabel, J. 2006, ApJ, 653, L85
Chapman, S. C., Smail, I., Blain, A. W., & Ivison, R. J. 2004, ApJ, 614, 671
Chapman, S. C., Blain, A. W., Smail, I., & Ivison, R. J. 2005, ApJ, 622, 772
Fruchter, A. S., et al. 2006, Nat, 441, 463
Hjorth, J., et al. 2003, Nat, 423, 847
Michałowski, M. J., & Hjorth, J. 2007, AIPC, in press
Michałowski, M. J., Hjorth, J., Castro Cerón, J. M., & Watson, D. 2008, ApJ, 672, 817
Silva, L., Granato, G. L., Bressan, A., & Danese, L. 1998, ApJ, 509, 103
Tanvir, N. R., et al. 2004, MNRAS, 352, 1073

*Pathways through an Eclectic Universe*
*ASP Conference Series, Vol. 390, © 2008*
*J. H. Knapen, T. J. Mahoney, and A. Vazdekis, eds.*

# The ALHAMBRA Survey

M. Moles,[1] J. A. L. Aguerri,[2] E. J. Alfaro,[1] N. Benítez,[3] T. Broadhurst,[4] J. Cabrera-Caño,[5] F. J. Castander,[6] J. Cepa,[2] M. Cerviño,[1] A. Fernández Soto,[7] R. M. González Delgado,[1] L. Infante,[8] V. J. Martínez,[7] J. Masegosa,[1] I. Márquez,[1] A. del Olmo,[1] J. Perea,[1] F. Prada,[1] J. M. Quintana,[1] and S. F. Sánchez[9]

[1]*Instituto de Astrofísica de Andalucía (CSIC), E-18008 Granada, Spain*
[2]*Instituto de Astrofísica de Canarias, E-38200 La Laguna, Spain*
[3]*IMAFF (CSIC), Madrid, Spain*
[4]*School of Physics and Astronomy, Tel Aviv University, Israel*
[5]*Facultad de Física, Universidad de Sevilla, Sevilla, Spain*
[6]*Institut d'Estudis Espacials de Catalunya, CSIC, Barcelona, Spain*
[7]*Department d'Astronomia i Astrofisica, Universitat de Valencia, Valencia, Spain*
[8]*Departamento de Astronomía, Pontificia Universidad Católica, Santiago, Chile*
[9]*Centro Astronómico Hispano Alemán, Almería, Spain*

**Abstract.** ALHAMBRA is a survey of eight separate regions of the sky with a total area of 4 sq. deg. The fields are observed with the LAICA (visible) and OMEGA-2000 (near infrared) cameras mounted on the 3.5 m telescope of CAHA observatory. The survey is done with a set of 20 adjacent medium-band filters that cover the visible range. These filters, together with three standard near-infrared ones, allow for a detailed measurement of the spectral energy distribution and redshift of a multitude of objects, stretching over an unprecedented range of distances for such a large-area survey. In particular, the basic motivation of the whole ALHAMBRA project is the cosmic evolution of galaxies. ALHAMBRA has been designed to cover the niche between deep pencil-beam surveys, which are necessarily restricted to small areas, and large-area shallow surveys that only probe the local Universe. The project enjoys Spanish guaranteed observing time status at the Calar Alto facilities. In this poster, we present the project and the first results obtained.

## 1. Goals of the Project

ALHAMBRA (Advanced Large, Homogeneous Area Medium-Band Redshift Astronomical; Moles et al. 2005) will produce accurate photometric redshifts for a large number of objects, enough to track cosmic evolution in a kind of cosmic tomography. ALHAMBRA is imaging 4 sq. deg, with 20 contiguous, equal-width (310 Å), medium-band filters covering from 3500 to 9700 Å, plus the standard $JHK_s$ near-infrared (NIR) bands. The optical photometric system has been designed to maximize the number of objects with accurate classification by spectral energy distribution and redshift, and to be sensitive to relatively faint emission features in the spectrum. The observations are being carried out with the Calar Alto 3.5 m telescope using the wide field cameras in the optical, LAICA, and in the NIR, OMEGA-2000. The magnitude limit, for a total of 100 ks integration time per pointing, is $AB = 25$ mag (for an unresolved object, $S/N = 5$) in the optical filters from the blue to 8300 Å, and ranges from $AB = 24.7$ to 23.4 mag for the redder ones. The limit in the NIR, for a total of 15 ks exposure time per pointing, is $K_s = 20$, $H = 21$, $J = 22$.

We will obtain accurate redshift values ($\Delta z/(1+z) \leq 0.03$) for about $6.6 \times 10^5$ galaxies with $I \leq 25$ mag (60% completeness level), and $z_{\rm med} = 0.74$. This accuracy, together with the homogeneity of the selection function, will allow the study of the large scale structure evolution, the galaxy luminosity function and its evolution, the identification of clusters of galaxies, and many other studies, without the need for any further follow-up. It will also provide exciting targets for detailed studies with 10 m-class telescopes. Given its area, spectral coverage, and depth, the ALHAMBRA survey will also produce valuable data for Galactic studies.

The survey will provide a 50% completeness limit of $AB \sim 24.2$ mag for objects with secure redshift. For objects with high quality redshifts the equivalent limit will be $AB \sim 25.4$ mag. An extra advantage of the ALHAMBRA data is that the redshift determination is based on the general shape of the spectrum and not on particular emission and/or absorption features. It is precisely because of this absence of biases that the survey is more valuable, and because of the testability associated with the photometric redshift measurements that the survey is especially adapted to studies of cosmic evolution. The median redshift of ALHAMBRA (90% completeness) will be ~0.63, with approximately 2000 galaxies with secure redshifts $z \geq 5$.

Because of its very nature ALHAMBRA occupies a niche which lies halfway between traditional photometric and spectroscopic surveys. Over half a million galaxies down to $AB \sim 25$ mag will be observed, with high quality redshifts being measured for them. This represents a number of objects comparable to that of the widest spectroscopic surveys, but covering considerably smaller areas. On the other hand, ALHAMBRA will reach deeper than most photometric surveys over a relatively large area.

Further details about the project are available at http://alhambra.iaa.es:8080/ and a recent press release in http://www.caha.es/alhambra-the-history-of-the-universe-at-sight.html or http://www.uv.es/obsast/alhambra/pressrelease.html.

## References

Moles, M., et al. 2005, astro-ph/0504545

Figure 1. Image formed by combining the individual optical frames (only 15 filters in this case) corresponding to a 15 arcmin area of the Alhambra-8 field.

*Pathways through an Eclectic Universe*
*ASP Conference Series, Vol. 390, © 2008*
*J. H. Knapen, T. J. Mahoney, and A. Vazdekis, eds.*

# The Luminous Matter Profiles in Ellipticals since $z \sim 1.5$: The Photometric Plane and its Evolution

J. Oñorbe,[1] R. Domínguez-Tenreiro,[1] and A. Serna[2]

[1] *Universidad Autónoma de Madrid, Spain*
[2] *Universidad Miguel Hernández, Elche, Spain*

**Abstract.** We present results of an analysis of the stellar mass distributions of relaxed elliptical-like objects (ELOs) identified, at $z = 0$, $z = 0.5$, $z = 1$, and $z = 1.5$, in a set of self-consistent hydrodynamical simulations operating in the context of a concordance cosmological model, confirming the existence of a photometric plane, as observations suggest.

Some recent observational and numerical results on elliptical galaxies (Es) strongly suggest that their stellar populations formed early on through dissipative processes, and that they lack significant structural and dynamical sample evolution at intermediate redshifts (La Barbera et al. 2005; Domínguez-Tenreiro et al. 2006; De Lucia et al. 2006), except that the objects became on average more compact, at fixed stellar mass, with increasing $z$ (Trujillo et al. 2006). In the present work we study the structural properties of elliptical galaxies using hydrodynamical simulations operating in a cosmological context. The code used in our simulations is DEVA. We have run (Serna, Domínguez-Tenreiro, & Sáiz 2003) five simulations where elliptical-like objects (ELOs) have been identified at redshifts $z = 0$, $z = 0.5$, $z = 1$, and $z = 1.5$. We have found that the projected mass density profiles of ELOs at different redshifts can be well fitted by a Sérsic law (Oñorbe et al. 2006), and that the scaling correlations agree with observational data (Vazdekis, Trujillo & Yamada 2004; D'Onofrio 2001), in particular the Fundamental Plane relation.

We have analyzed an analogous relation of the Photometric Plane (i.e., a relation among effective radius, stellar mass and Sérsic index, Khosroshahi et al. 2000) that we define as the Structural Fundamental Plane (SFP): $\log R^{\rm star}_{\rm e,bo} = A \log M^{\rm star}_{\rm bo} + B \log n + C$. We have calculated the orthogonal least-squares fit of this equation for the ELO sample at $z = 0$ obtaining: $A_{z0} = 0.0501$; $B_{z0} = 0.83251$; $C_{z0} = -8.5141$ $\sigma_{\rm nRM,z0} = 0.0578$. In Fig. 1 we plot the edge-on projection of this plane for the $z = 0$ sample, as well as for samples at other redshifts in the same direction.

We have confirmed observational results (La Barbera et al. 2005) that indicate that the logarithms of $n$, $R^{\rm star}_{\rm e,bo}$, and $M^{\rm star}_{\rm bo}$ populate, at $z = 0$ and $z = 0.5$, a flattened ellipsoid close to a two-dimensional plane (the SFP). The slight deviation of this plane at higher redshifts can be explained by a mild evolution of the $R^{\rm star}_{\rm e,bo} - M^{\rm star}_{\rm bo}$ relation. Present results show that the Photometric Plane could be an interesting alternative tool for the study of elliptical galaxies at high redshifts instead of the Fundamental Plane, which requires a heavy amount of time for measuring velocity dispersions.

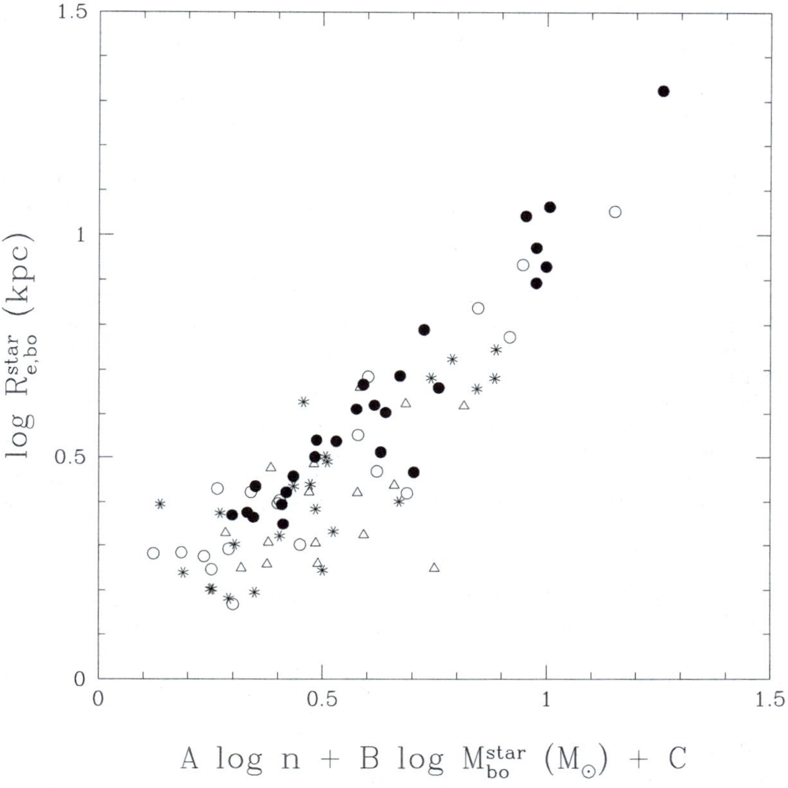


Figure 1. The edge-on projection of the SFP at $z = 0$ (full circles). Empty circles, crosses, and empty triangles stand for the projection of the $z = 0.5$, $z = 1$, and $z = 1.5$ samples, respectively, in the same direction.

**Acknowledgments.** We are indebted to the CCCFC (UAM) for computing facilities and to the MCyT (Spain) and the regional government of Madrid for financial support through grants AYA-154292-C03-01 and the ASTROCAM Astrophysics network (S-0505/ESP-0237), respectively.

## References

De Lucia, G. et al. 2006, MNRAS, 366, 499
D'Onofrio, M. 2001, MNRAS, 326, 1517
Domínguez-Tenreiro, R., et al. 2006, ApJ, 636, L77
Khosroshahi, H. G., et al. 2000, ApJ, 531, L103
La Barbera, F., et al. 2005, MNRAS, 358, 111
Oñorbe, J., et al. 2006, MNRAS, 373, 5030
Serna, A., Domínguez-Tenreiro, R., & Sáiz, A. 2003, ApJ, 597, 878
Trujillo, I., et al. 2006, MNRAS, 373, L36
Vazdekis, A., Trujillo, I., & Yamada, Y. 2004, ApJ, 601, L33

*Pathways through an Eclectic Universe*
*ASP Conference Series, Vol. 390, © 2008*
*J. H. Knapen, T. J. Mahoney, and A. Vazdekis, eds.*

# Modeling the Formation and Evolution of Disk Galaxies

M. J. Stringer[1,2] and A. J. Benson[1]

[1]*Department of Astrophysics, Keble Rd., Oxford, OX1 3RH, UK*
[2]*Theoretical Astrophysics, Caltech, MC130-33, 1200 E. California Blvd., Pasadena, CA 91125, USA*

**Abstract.** Inspired by recent work on feedback in disk galaxies (Efstathiou 2000; Silk 2003) and on the angular momentum distribution in simulated gas halos (Sharma & Steinmetz 2005), a fully dynamic model of disk galaxy formation and evolution has been developed. This is used to demonstrate how observed galactic systems could have formed from halos similar to those found in simulations and applies physically motivated models of star formation and feedback to explore whether the true nature of these processes would be manifest from local and cosmological observables. This is made possible by computational integration with the galaxy formation model developed originally by the group at Durham University (Cole et al. 2000).

Galaxies within the GALFORM hierarchical model can now be decomposed into $\sim$10 pc regions so that their formation can be followed without any prior assumption of disk profile. Their subsequent evolution can be studied across the entire disk, as described in Stringer & Benson (2007).

Three initial findings are presented here. Firstly, the predicted stellar surface density profiles, governed predominantly by the inital angular momentum distribution of the halos, are qualitatively similar to those observed, exhibiting the same key features of an exponential decrease and a central peak. Secondly, the predicted circular velocities are in poor agreement with both observed rotation curves (Fig. 1) and the Tully–Fisher relation (Fig. 2).

Finally, the model suggests that a constant *local* star formation efficiency per orbit may produce regions which significantly deviate from a Schmidt–Kennicutt law whilst the *global* star formation rates could still follow a trend with comparable slope and scatter to the observations.

**Acknowledgments.** MJS acknowledges a PPARC studentship at the University of Oxford and the hospitality of the CTCP at Caltech. AJB acknowledges support from the Gordon and Betty Moore Foundation.

## References

Cole, S., Lacey, C. G., Baugh, C. M., & Frenk, C. S. 2000, MNRAS, 319, 168
Kassin, S., de Jong, R. S., & Pogge, R. W. 2006, ApJ, 643, 804
Kennicutt, R. C. 1998, ARA&A, 36, 189
Efstathiou, G. 2000, MNRAS, 317, 697
Silk, J. 2003, MNRAS, 343, 249
Sharma, S., & Steinmetz, M. 2005, ApJ, 628, 21
Stringer, M. J., & Benson, A. J. 2007, MNRAS, 382, 641

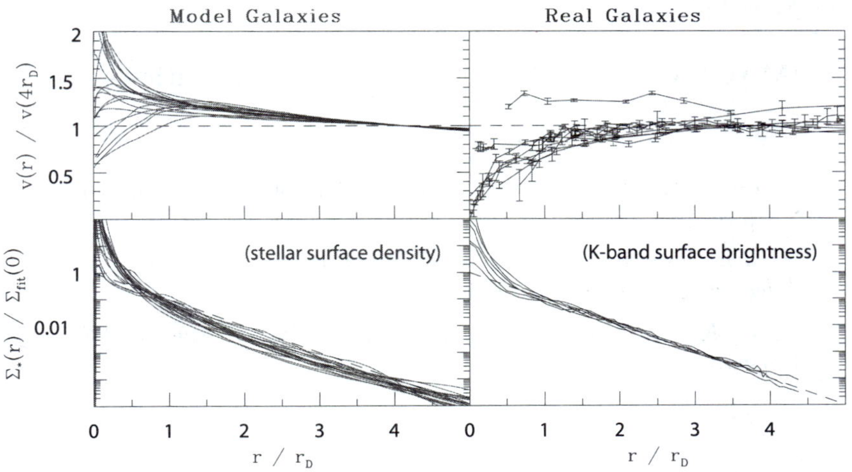


Figure 1. The radial profiles of real and modeled galaxies, each in radial units of its own particular scalelength, $r_D$. Rotation curves, in units of their value at four scalelengths, are plotted in the top row. Stellar surface density and surface brightness are plotted in the second row in units of the central fitted value. The 11 real galaxies are from Kassin, de Jong, & Pogge (2006), and have bulge fractions of 20% or less. The modeled systems are taken from a realization of hundreds of hierarchically forming halos, of which the 20 here are those central galaxies which have the same range of mass and bulge fraction as their real counterparts.

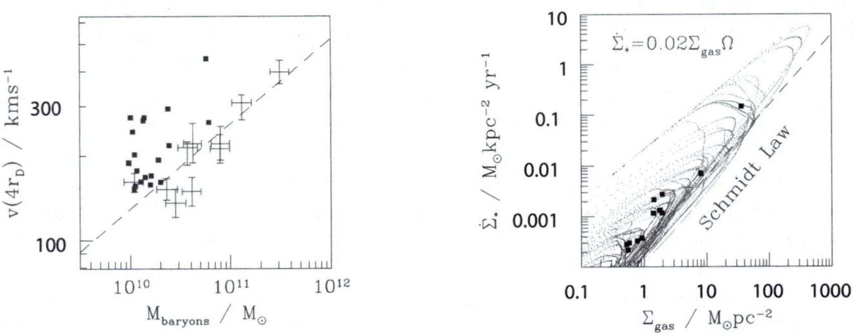


Figure 2. Global properties of the galaxies from Figure 1. *Left:* The circular velocity, at four scalelengths, for the real systems (error bars, with a dotted line of best fit) and those from the model (squares). *Right:* Star formation rates from the modeled galaxies, which applied a "Silk law" of star formation, as given in the equation in the figure. Values are plotted both for every individual annular region (dots) and for each disk as a whole (squares). The dotted (blue) line shows the traditional observational relationship from Kennicutt (1998), $\dot{\Sigma}_\star = 2.5 \times 10^{-4} \left[\Sigma_{\rm gas}/M_\odot\,{\rm pc}^{-2}\right]^{1.4} M_\odot\,{\rm kpc}^{-2}$.

*Pathways through an Eclectic Universe*
*ASP Conference Series, Vol. 390, © 2008*
*J. H. Knapen, T. J. Mahoney, and A. Vazdekis, eds.*

# IMAGES I. Classification of Galaxy Kinematics since $z = 1$

Y. Yang,[1] H. Flores,[1] F. Hammer,[1] B. Neichel,[1] M. Puech,[2] and N. Nesvadba[1]

[1] *GEPI, Observatoire de Paris, CNRS, University Paris Diderot, 5 Place Jules Janssen, Meudon, France*

[2] *ESO, Karl-Schwarzschild-Strasse 2, D-85748 Garching bei München Germany*

The multi-IFU mode of FLAMES/GIRAFFE at the VLT is used to study the 2D-kinematics to understand the evolution of galaxies at $z < 0.8$. A full description of the sample is given in Yang et al. (2007) and Hammer et al. (this volume, p. 268). We have classified the kinematics of 63 galaxies at $z = 0.4$–$0.75$ into three classes (see Fig. 1 for examples): rotating disks (RD), perturbed rotations (PR), complex kinematics (CK). Flores et al. (2006) have developed a diagnostic method to quantitatively test how a given velocity field differs from a pure rotating disk model. Indeed, at large distances the spatial resolution is not sufficient to resolve the central regions of the galaxies. This implies that the observed velocity dispersion ($\sigma$) of a rotational body is the convolution of the actual random motions with the rotation. For a rotationally supported galaxy, it unavoidably leads to a well defined peak in the center (see Figure 1). Flores et al.'s scheme allows us to compare any velocity field to what it could be if it were a rotational disk. For each velocity field, one can generate first a rotation curve assuming that the largest scale motion is indeed due to rotation, and then a pseudo-$\sigma$ map. The deviation from a rotational body can be measured from differences in amplitude and in position of the $\sigma$ peak between the pseudo rotational $\sigma$ map and the observed one. Yang et al. (2007) have calculated error bars for these differences using Monte Carlo simulations, leading to a very robust diagnostic diagram for the 63 observed galaxies (Figure 2). This robustness is provided by the large S/N of the GIRAFFE observations and by the fact that large scale motions are well resolved at $z = 0.4$–$0.75$, as evidenced by tests on redshifted velocity fields of various kinds of local galaxies.

We confirm and consolidate the results of Flores et al. (2006), i.e., that a significant fraction of intermediate mass galaxies had perturbed or complex kinematics 5 Gyr ago. Even if we assume that all quiescent galaxies at $z = 0.4 - 0.75$ had well ordered velocity fields, we find that $(41 \pm 7)\%$ of the galaxies are not kinematically relaxed, including $(26 \pm 7)\%$ of galaxies that show complex kinematics. Undoubtedly, galaxy kinematics is evolving very rapidly, since most present-day galaxies in the same mass range are likely to have ordered velocity fields. If major mergers are the main mechanism responsible for the large fraction of complex velocity fields, it implies that, since $z = 1$, from 50% to 70% of intermediate mass galaxies have experienced a major merger. This supports the spiral rebuilding scenario as proposed by Hammer et al. (2005).

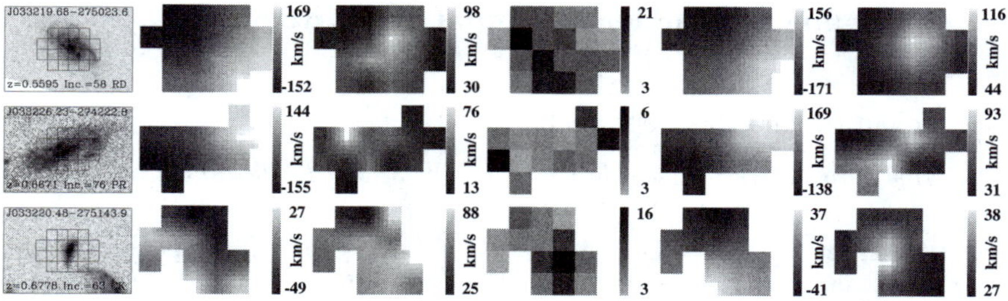

Figure 1. Examples for each kinematical class, from top to bottom, RD, PR, CK. For each galaxy we show the *HST*/ACS F755W images, the observed velocity fields, $\sigma$ maps and S/N maps, and the model velocity fields and $\sigma$-maps. We have applied a $5 \times 5$ linear interpolation to the velocity fields and $\sigma$ maps for visualization.

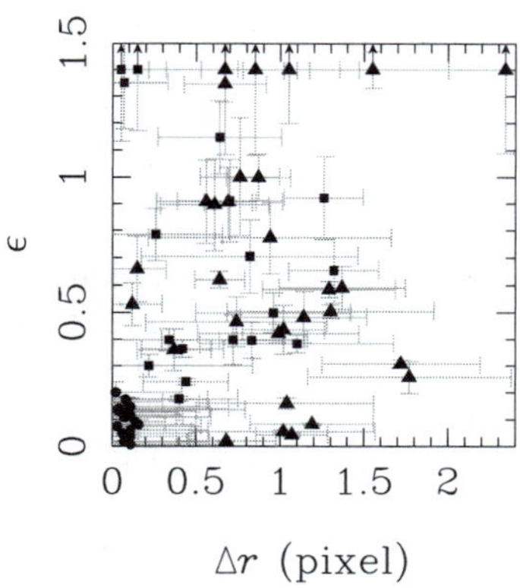


Figure 2. Diagnostic diagram for the kinematics of galaxies. $\Delta r$ is the spatial separation between the peaks of the observed and the modeled $\sigma$-maps. The $\epsilon$ gives the relative difference between the $\sigma$ of the model and the observation at the reference location of the peak of the model $\sigma$-map. The (blue) dots, (green) squares and (red) triangles represent the RD, PR, and CK galaxies respectively. The typical error for $\Delta r$ is 0.3 GIRAFFE pixels and 15% for $\epsilon$.

**Acknowledgments.** We thank the entire GIRAFFE team at the Observatories in Paris and Geneva, and ESO for this unique instrument.

## References

Flores, H., Hammer, F., Puech, M., Amram, P., & Balkowski, C. 2006, A&A, 455, 107

Hammer, F., Flores, H., Elbaz, D., Zheng, X. Z., Liang, Y. C., & Cesarsky, C. 2005, A&A, 430, 115

Yang, Y. B., Flores, H., Hammer, F., Neichel, B., Puech, M., & Nesvadba N. 2007, A&A, submitted

*Pathways through an Eclectic Universe*
*ASP Conference Series, Vol. 390, © 2008*
*J. H. Knapen, T. J. Mahoney, and A. Vazdekis, eds.*

# Panel Discussion II. Reconciling Observations and Modeling of Star Formation at High Redshifts

Johan H. Knapen (Reporter)

*Instituto de Astrofísica de Canarias, E-38200 La Laguna, Spain*

**Abstract.** This is the written account of the second of two panel discussions, on *Reconciling observations and modeling of star formation at high redshifts.* The chair of the panel was Pavel Kroupa, and panel members were Marc Balcells, John Beckman, Christopher Conselice, and Joseph Silk. After a short introduction by each of the panelists, panel and audience entered into a lively discussion, centered around the following six themes: the mass function of pre-stellar gas clouds; a possible top-heavy initial mass function at high redshifts versus mini-quasars as the first sources of ionization; the integrated galactic initial mass function; possible differences in specific star formation rates in disks and in massive galaxies; whether merging rates yield a wrong prediction for massive galaxies, and what is the physics behind the onset of the red sequence of galaxies; and the case of dark matter-dominated dwarf galaxies versus tidal dwarf galaxies.

KROUPA: Welcome to this second panel discussion, focused on observations and modeling of star formation (SF) in the high redshift Universe. Each of our panelists will start by giving a brief introduction to an aspect of this general topic that they find particularly interesting, and which they propose for further discussion. Let me ask John Beckman to start.

BECKMAN: The point I would like to make is not (at any rate not yet) on SF at high redshifts but on the initial mass function (IMF). The question is whether there is a universal IMF wherever and whenever stars have formed. It would imply, for example, that metallicities in stellar populations could be derived more precisely using that IMF. We have already heard during this week that nobody has found overwhelming evidence for variations in the IMF. But as absence of evidence is not evidence of absence can we conclude that the IMF is in fact universal? The reasons for suspecting that it might not be are that physical considerations seem to tell us that the IMF ought to depend on metallicity, gravity and interstellar medium (ISM) pressure, as examples. What I am about to say is detailed in a poster by Casuso, Beckman, & Buenrostro (these proceedings, p. 386). In that work we were, using very simple assumptions, able to make an initial uniform spectrum of cloud core masses which is brought about by considering motion within a closed box, e.g., a spiral arm, and letting the system evolve. Using an algorithm in which the tendency of clouds in collision to merge or to fission is determined by their masses and their mutual velocities, we find results which agree well with what is measured locally in the ISM, but are also stable over time. Our model uses an arbitrary initial mass distribution of the clouds, and the output results depend only slightly on the input parameters, and very little on epoch after an initial relaxation time of order a few Myr. This model is metallicity-independent and virtually independent of dynamics. Its simplicity and robustness may enable us to gain some insight into the possible existence of a universal IMF.

BALCELLS: The confirmation that the color distribution of galaxies is bimodal has far-reaching consequences for our understanding of galaxy evolution. It has provided a simple, but inspiring, unifying theme for galaxy evolution. Galaxies are born and grow in the *blue cloud*, a sort of nursery for galaxies. Eventually, galaxies come of age, and migrate to the *red sequence*, a sort of retirement paradise, where massive galaxies live forever, growing only moderately by aggregation.

Irrespective of how much of this picture will eventually be shown to correspond with reality, one obvious question arises: what physical processes drive galaxies away from the blue cloud into the red sequence?

One basic hypothesis is that SF stops as a result of gas exhaustion, due to the ongoing SF. A second hypothesis is that gas may get consumed, removed and/or heated to virial temperatures during mergers. A third hypothesis, developed by Dekel and collaborators, exploits the presence of the most massive galaxies in the red sequence: the gas accreting onto dark matter halos along the filaments of the cosmic web remains cool when falling onto low mass halos, while it shock-heats to virial temperatures in galaxies with masses above about $\log(M/M_\odot) > 11.5$. This model amounts to a phase transition in the external interactions of galaxies. A final hypothesis would be that, as the potential of the galaxy deepens as a result of SF, eventually a phase transition occurs to the galaxy's ISM, and a new equilibrium, or quasi-equilibrium, is found at higher temperatures.

What makes the red sequence interesting is that it presents us with a lot of physics to analyze, in a territory where theorists and observers can contribute, compete, and, hopefully, lead us to a better understanding of the workings of galaxies.

CONSELICE: Figure 1 shows the mass and number densities for massive galaxies as a function of redshift. As can be seen, the densities for the most massive galaxies with $\log M > 11.5$ are nearly all in place, statistically, at $z < 2$. The two solid lines show the Millennium simulation results for the same quantities. As can be seen, the Millennium results underpredict the number of massive galaxies by up to two orders of magnitude. Although there is good agreement between the data and the models at $z = 0$, there is a large disagreement at higher redshifts, which continues to grow. This shows that the processes responsible for the assembly of massive galaxies in the Millennium simulation, which are generally mergers, occur much later in the simulation than for real galaxies. The formation of massive galaxies is therefore occurring much earlier than predicted. It is also observed that galaxy mergers are common at $z > 2$, and it is likely that galaxy merging

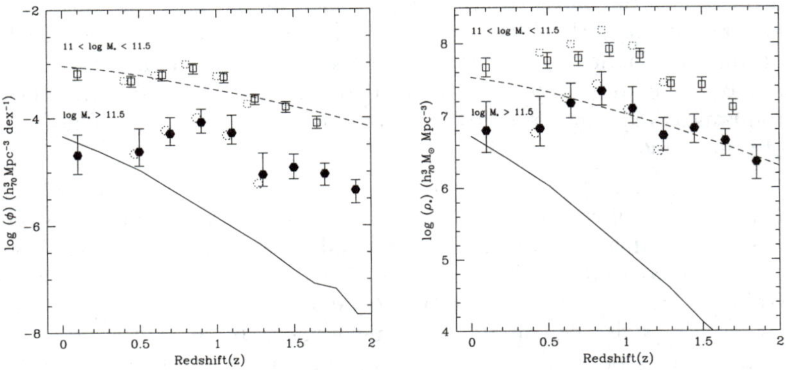


Figure 1. A comparison between Palomar NIR survey data (Conselice et al. 2007), and models from the Millennium simulation (e.g., De Lucia et al. 2006). The dashed line shows the predicted evolution in the number and stellar mass densities for $10^{11}\,M_\odot < M^* < 10^{11.5}\,M_\odot$ systems, while the solid line shows the same predicts for galaxies with stellar masses $M^* > 10^{11.5}\,M_\odot$. As can be seen, for the most part the simulated massive galaxies do not assemble quickly enough to match the observations.

is not occurring early enough in the simulations to produce the number of distant massive galaxies.

SILK: I wish to address the limitations of numerical simulations of galaxy formation and evolution. The problem is one of inadequate resolution. If one wishes to address SF one must go to very small scales and an enormous dynamical range—beyond reach currently. New rules are needed for sub-grid physics. Our current knowledge is based on observational phenomenology. However, detailed observations made locally are too complicated to implement, so assumptions must be made. These are then fine-tuned as more detailed knowledge becomes available, but it is a fundamental question as to how robust the predictions are. This is a general statement, but here are a few examples:

- There are three main problems with cold dark matter theory (CDM). These are the frequency of dwarf galaxies, of massive galaxies, and the ubiquity of halo cusps. The solutions to these problems may be tracers of early SF. Why is it that the cusps in galaxy centers cannot be erased by, e.g., a bar? Models look at the effect of the bar on its surroundings, but a bar is accompanied by inflow, and by continuous re-formation. So one must study the frequencies of dwarf and massive galaxies in different environments and metallicities, as well as the cores of halos. Any correlations obtained may lead to potential insights on fossil SF.
- Disks formed stars inefficiently but massive galaxies formed stars very efficiently. These insights are due to the derivation of specific SF rates (SFRs) from spectrophotometric modeling, and from [$\alpha$/Fe] abundance ratio studies. One resolution may lie in invoking two modes of

SF, with the inefficient disk mode being associated with supernova feedback and the efficient massive spheroid mode being at least partly regulated by AGN outflows.

- The first ionizing sources at high redshift may have been massive stars, possibly with a top-heavy IMF, or mini-quasars. Are intermediate mass black holes required to account for re-ionization if the IMF is not top-heavy, and if so, what are their signatures?

My bottom line is that one should not believe current models, especially as far as their predictive power is concerned.

KROUPA: I'd like to bring up two additional problems which may well be associated with a currently wrong view of structure formation and the build-up of stellar content of galaxies:

1. Joe [Silk] already mentioned that CDM theory predicts the wrong number of dark matter dominated satellites of $L^*$ galaxies. In this context I'd like to emphasize this problem further: we know that when late-type galaxies interact, tidal arms take away the angular momentum and energy. When these are gas rich, galaxy-sized objects are often observed to form in the arms; sometimes a dozen star-forming dIrr-type knots are evident per interacting galaxy pair. Theoretical work shows the number of self-gravitating knots to scale with the gas fraction. Okazaki & Taniguchi (2000) apply a standard hierarchical structure formation prescription for the build-up of galactic mass, and show that if only one or two long-lived tidal dwarf galaxies (TDGs) survive per encounter involving late-type galaxies then *all* dE galaxies can be accounted for. In particular, their work shows that the observed morphology–density relations for both dwarf and giant galaxies in the field, groups of galaxies, and clusters of galaxies are readily explained. Okazaki & Taniguchi made very conservative assumptions on the number of TDGs formed, and the actual number of TDGs is likely to be much higher because the early galaxies had a larger fraction of their mass in gas.
   My own work (Kroupa 1997; Metz, & Kroupa 2007) shows that the interpretation that the Milky Way dSph satellites are dark matter dominated is probably wrong and that they are most probably ancient TDGs. Their spatial distribution about the Milky Way supports this conclusion (Metz, Kroupa, & Jerjen 2007).
   Taken together, this would suggest a major logical problem with CDM theory because there would not be any room in the observed cosmos for any type of dark matter dominated satellite galaxy; CDM theory would predict the formation of more than enough TDGs to explain all known dwarf galaxies. The problem is that these TDGs cannot be explained away—they are an inherent consequence of fundamental conservation laws.
2. In cosmological applications the stellar IMF is often taken to be of an invariant Salpeter form. Sometimes top-heavy forms are considered. Recently it has emerged, though, that the stellar IMF of a

whole galaxy, the "integrated galactic IMF" (IGIMF) is a result of adding up all the stellar IMFs in all the star clusters that are forming. The IGIMF turns out to be steeper than the invariant stellar IMF and to depend on the SFR of a galaxy. Low mass galaxies have steeper IGIMFs than massive galaxies and also a larger variation of the IGIMF. This has important implications for extragalactic astrophysics, as, for example, the mass–metallicity relation of galaxies would be a natural result of such a variation of the IGIMF (Köppen, Weidner, & Kroupa 2007). Also, the H$\alpha$-SFR calibration would be wrong for low mass galaxies; the observed H$\alpha$ fluxes would imply significantly larger true SFRs than currently thought (Pflamm-Altenburg, Weidner, & Kroupa 2007). Clearly, this needs to be taken into account in future stellar population studies.

I would now like to propose the following six points to discuss with the audience:

1. What is the mass function of pre-stellar gas clouds?
2. Can we distinguish between a top-heavy IMF at high redshifts, and mini-quasars as the first ionizing sources?
3. Is there a galaxy-wide IGIMF, along the lines of the description given by Weidner & Kroupa (2005)? We know the Salpeter description does not work.
4. Are there differences in the specific SFRs and in the [$\alpha$/Fe] relative abundances between disks and massive galaxies?
5. Do merging rates yield a wrong prediction for massive galaxies, and how does this relate to the red sequence?
6. Are apparently dark matter-dominated dwarf galaxies really TDGs, or is theory trying to explain something that isn't there?

Can I invite comments from the audience please?

PAGEL: According to Kashlinsky et al. (2007), a small contribution to the extragalactic background light could come from supermassive Population III objects for which they have measured an angular correlation function resembling that of galaxies.

KROUPA: And we should not just consider the dark matter paradigm, but also different gravitational laws to study where our understanding breaks down.

ALVES: If the characteristic mass function of the IMF is the critical Bonnor–Ebert mass (cf. Alves, Lombardi, & Lada 2007), then I would naively expect, at least, a bottom-deficient IMF for the first stars, as $M_{\rm BE} \propto c^2/p^{0.5}$.

SILK: The back-of-the-envelope calculation, which uses a sound speed approximation to $H_2$ cooling, indeed gives about $1000\,M_\odot$ for the mass of the first stars to form in a metal-free environment. High dynamic range simulations of the first stars initially confirmed the naive predictions for the *minimum* mass. More recent simulations, which are now able to study more than one realization of collapse over a wide dynamic range, find a dispersion in the central temperature and density gradients that controls the accretion and angular momentum transfer rates. A range in first star mass is found, down to as low as $10\,M_\odot$, and up to $100\,M_\odot$. These stars are in the normal stellar mass range.

KRUMHOLZ: Theoretical models and simulations seem to suggest that the Arches cluster should have a top-heavy IMF, but observations have shown that the Arches IMF is basically the same as everywhere else in the Galaxy. Given the failure of our models for the Arches, we should be wary of applying theoretical models that predict varying IMFs to the high redshift universe.

DOMÍNGUEZ-TENREIRO: I would like to comment on the issue of the top-heavy IMF versus mini-quasars as the first ionizing source. Cosmic flows can be described on a mathematical level by the Burgers equation, a generic equation in the framework of non-linear physics. This equation predicts that flows unavoidably develop singularities (i.e., black holes in the current context) under generic conditions. So, it is very likely that mini-quasars appear very early in the evolution of the Universe, and they would act as sources of the first ionization. Of course, this does not exclude a top-heavy IMF at high redshift, too.

PELETIER: I would like to reiterate an up to now unsolved problem: the low Ca II IR triplet values in giant ellipticals. This was published by Saglia et al. (2002), Cenarro et al. (2003) and Falcón-Barroso et al. (2003). Assuming a normal Salpeter or Kroupa IMF, an average old age, and a high metallicity, necessary to fit absorption lines in the optical, the derived Ca abundance is lower than solar, unless the IMF is top-heavy at the low metallicity end. Below solar [Ca/Fe] ratios are in principle unattractive, since Ca is an alpha-element, and alpha-elements are generally over-abundant with respect to Fe in giant ellipticals. The observations are beyond question, and the community is still waiting for a satisfactory solution for a line which is relatively well understood.

SHLOSMAN: I would like to respond to Joe [Silk]'s apparent criticism of numerical simulations of dark matter cusp destruction by stellar bars. I fully agree that results of simulations should always be taken with a healthy dose of skepticism, *unless* they are supported by compelling theoretical arguments and common sense. In the particular case of the dark matter cusps, all numerical simulations with live bars (including our still unpublished results of $N \sim 10^{10}$ particles by Dubinski, Shlosman, & Berentzen) show that the cusps do *not* dissolve (except in simulations by Holley–Bockelmann). Moreover, there are no compelling theoretical arguments as to why this should occur. The models now encompass pure collisionless cases as well as gaseous disks.

SILK: Proto-galactic bars in a gas-rich environment have not yet been adequately studied. I do not think the dust has settled yet on the issue of cusp softening.

VAZDEKIS: $[\alpha/\mathrm{Fe}]$ overabundance ratios are usually assumed to be a result of highly efficient SF. But if the prediction by, for example, Schneider et al. (2003) of a top-heavy IMF for metallicities lower than $10^{-5}$ is true, these overabundance ratios can be explained in part without invoking such high SF efficiency. Furthermore, a combination of the two scenarios might be appropriate for explaining the observed overabundance ratios.

KROUPA: But if one were to have a top-heavy IMF, could one make a higher $[\alpha/\mathrm{Fe}]$ without the need for a higher SF efficiency?

SILK: As you must also produce the Fe, you need type Ia supernovae, and so a top-heavy IMF will not be enough.

PIPINO: I would like to make two comments on points that were previously raised. Firstly, on Vazdekis's quest for Population III stars, we showed (Matteucci & Pipino 2005) that if you take Population III stars into account as a real first (and single) generation of stars, you wouldn't notice their effect after $10^{11}\,M_\odot$ of stars have been created with a standard Salpeter IMF, over a period of at least half a Gyr. Secondly, on the Ca depletion in ellipticals: it is a false problem because Ca is also produced by Type I supernovae in a non-negligible way (up to half of the total mass can be synthesized by supernovae Ia for a standard IMF), as is Si, therefore a ratio of $[\mathrm{Ca/Mg}] \leq 0$ is naturally predicted by chemical evolution models (as described in Pipino & Matteucci 2004).

PELETIER: This paper still claims that Ca, as an alpha-element, is overabundant with respect to Fe, while in these ellipticals [Ca/Fe] lower than solar seems to be needed. I agree that we have to look for the solution also in the direction of the chemical enrichment models.

HENSLER: As one learns from changes in element production in massive stars by refinements of the stellar models over the last years, a word of caution should be emphasized not to rely too much on abundances and their relations and take them too seriously as tracers, e.g., of changes in the IMF and SF episodes. So I agree that we should be more careful with interpretations of models until we have made sure that all the relevant physical processes are properly included, and until their influences on galaxies within the complex network of matter cycles are intensively explored.

HAMMER: I would just like to question the assumption that considering their specific SFRs, disk SF is not efficient. This is certainly true for local disks but what about the period six to eight Gyr ago? Some 15% of the $M^*$ galaxy population is made with luminous infrared galaxies (LIRGs) which can double their mass in a few times $10^8$ yr. And almost half of them look to be disks, some even with a normal-looking rotation curve.

BALCELLS: Maybe a way of addressing the question is to study whether LIRGs can be understood in terms of quiescent or burst-mode SF, or whether there is some continuum between the extremes.

HAMMER: Well, spectroscopy shows that in general LIRGs are the same as, or very similar to, spirals, which is a purely observational result.

KRUMHOLZ: I object, as I did in the first panel discussion and on other occasions, to the use of the term efficiency. Unless we have ironclad evidence that high redshift SF does *not* follow the same rules as local SF, e.g., the universal IMF and the Schmidt–Kennicutt and Gao-Solomon correlations, we should not hypothesize the existence of additional "modes" of SF. It is significant that local LIRGs and ULIRGs follow the same correlations as normal galaxies, and, from what we can tell, high redshift ones do too.

HAMMER: How do you know this ? There may be no different physics at high $z$ but certainly different conditions (gas, merging occurrence, etc.).

KRUMHOLZ: I repeat, the basic point is that we should only introduce new physics if we really have to.

SILK: So ULIRGs lie on the Kennicutt law?

KRUMHOLZ: Yes indeed, local ULIRGs do lie on both the Kennicutt law and the Gao–Solomon law.

BALCELLS: From the point of view of a galaxy observer, there are indeed different modes, for instance, ellipticals have an $[\alpha/\mathrm{Fe}]$ ratio that is not found in disks, which implies the existence of a burst versus a quiescent mode. So I think we can talk about two modes. The two modes must reflect different physical conditions—though not necessarily new physical processes.

STRINGER: Just a short comment: do different modes of SF necessarily need to correspond to different IMFs?

BECKMAN: Not necessarily.

SILK: Could be the IMF, or it could be due to AGN.

KROUPA: This is a question of interpretation—some panel members would suggest no. Personally, I think that under extreme star-forming rates ($> 10^2\,M_\odot$ yr$^{-1}$) a top-heavy IMF does emerge. One indication of this is that the metallicity distribution of the Milky Way bulge and of the M31 bulge are easily reproduced with a top-heavy IMF, while a universal IMF does not match the observations (Ballero, Kroupa, & Matteucci 2007). In Bonn we are working on constraining the IMF for extreme starbursts. It is too early to state definite results, but a top-heavy IMF appears to be necessary.

KROUPA: I need to start wrapping up, and was wondering whether anyone wants to comment on our fifth point, on merging rates and the red sequence.

CONSELICE: The merger rate or fraction is a very difficult parameter to determine, and observations presently disagree with theory. Much more work needs to be done before we can make statements on this.

TRUJILLO: The small sizes of galaxies at a redshift of around 1.5 are not in contradiction with current $N$-body and semi-analytical simulations. A couple of major mergers will locate these objects within the local stellar mass-size relation (Boylan-Kolchin, Ma, & Quataert 2006; Khochfar & Silk 2006).

CONSELICE: The smaller sizes of these galaxies also demonstrate that merging is occurring, but how the observed merger rate, itself, agrees or not with the models still needs to be determined as there is no clear cut answer at the moment.

KROUPA: Let's hear a final comment from the audience before we turn back to the panel?

HAMMER: There is a lot of emphasis on the reproduction of the merger rate by simulations, but there is also a question to observers. In 1999 we made the first observation of pairs and of the merger rate up to a redshift of unity in the framework of the Canada–France Redshift Survey (CFRS). Since that time a very large number of papers has appeared on the subject, with different assumptions and methods (evolution-corrected or not, using blue light, red light, etc.). The overall result, also on the observers' side, is real confusion. I suggest that we should resolve this confusion before going to different simulations which also have various predictions. Otherwise, this "problem" will never end!

BALCELLS: The observational results can also usually be changed with sample selection. In work from our group, $B$-band vs $K$-band selection yields different $z$-evolution of the merger fraction, even when $I$-band asymmetries are used for both diagnostics. Pixellation and cosmological dimming need to be carefully calibrated when using asymmetries or pair fractions to infer the $z$-evolution of the merger fractions. Going beyond $z \sim 1$ becomes especially tricky as CCD-based imaging surveys start to sample the rest-frame UV continuum. Given these difficulties, disagreement between different observers is almost unavoidable.

CONSELICE: Merger fractions can be calculated in various ways and need to be determined quantitatively before comparing them with observations.

SILK: On the one hand the merger rate does not agree well with the models. But on the other, I am sure that modelers are very clever and so will come up with new models which will fit much better. But the question is whether we will learn any new physics from it?

KROUPA: We have run out of time, so before finalizing this panel discussion I would like to ask John [Beckman] to make a final statement.

BECKMAN: It is refreshing to see so much discord about the IMF. There are good reasons for uncertainty about the question of whether under extreme star-forming conditions or under conditions of very low metallicity the IMF is non-standard. These have been mentioned in the discussion and are as always a question of how to interpret the observations, notably of element abundance ratios. The question of whether Population III occurred or not is closely linked with this and is also up for grabs for the same sort of reasons. It is also refreshing to find that there are no clear answers to the problem of the lack of cuspiness in galaxy centers, or indeed to the question of why the semi-analytical models, with mergers at their heart, give such a relatively poor account of the evidence from stellar populations of the way and the rates at which galaxies have evolved. All this bodes well for the next generation of researchers.

**References**


Alves, J., Lombardi, M., & Lada, C. J. 2007, A&A, 462, L17
Ballero, S. K., Kroupa, P., & Matteucci, F. 2007, A&A, 467, 117
Boylan-Kolchin, M., Ma, C.-P., & Quataert, E. 2006, MNRAS, 369, 1081
Cenarro, A. J., Gorgas, J., Vazdekis, A., Cardiel, N., & Peletier, R. F. 2003, MNRAS, 339, L12
Conselice, C. J., et al. 2007, ApJ, 660, L55
De Lucia, G., Springel, V., White, S. D. M., Croton, D., & Kauffmann, G. 2006, MNRAS, 366, 499
Falcón-Barroso, J., Peletier, R. F., Vazdekis, A., & Balcells, M. 2003, ApJ, 588, L17
Kashlinsky, A., Arendt, R. G., Mather, J., & Moseley, S. H. 2007, ApJ, 654, L5
Khochfar, S., & Silk, J. 2006, ApJ, 648, L21
Köppen, J., Weidner, C., & Kroupa, P. 2007, MNRAS, 375, 673
Kroupa, P. 1997, New Astron., 2, 139
Matteucci, F., & Pipino, A. 2005, MNRAS, 357, 489
Metz, M., & Kroupa, P. 2007, MNRAS, 376, 387
Metz, M., Kroupa, P., & Jerjen, H. 2007, MNRAS, 374, 1125
Okazaki, T., & Taniguchi, Y. 2000, ApJ, 543, 149
Pipino, A., & Matteucci, F. 2004, MNRAS, 347, 968
Pflamm-Altenburg J., Weidner C., & Kroupa P. 2007, ApJ, 671, 1550
Saglia, R. P., Maraston, C., Thomas, D., Bender, R., & Colless, M. 2002, ApJ, 579, L13
Schneider, R., Ferrara, A., Salvaterra, R., Omukai, K., & Bromm, V. 2003, Nat, 422, 869
Weidner, C., & Kroupa, P. 2005, ApJ, 625, 754

From left to right: Alexandre Vazdekis, Rosa Domínguez-Tenreiro, and Ignacio Trujillo (*above*), and (again from left to right) John Beckman, Ángeles Díaz, Rob Kennicutt, Torsten Böker, and Daniela Calzetti during the first panel discussion (*below*).

# Session VI

# Eclectics

*Pathways through an Eclectic Universe*
*ASP Conference Series, Vol. 390, © 2008*
*J. H. Knapen, T. J. Mahoney, and A. Vazdekis, eds.*

# An Eclectic Look at Stellar Compositions

David L. Lambert
*The W. J. McDonald Observatory, The University of Texas at Austin, Austin, Texas, USA*

**Abstract.** Much effort in stellar spectroscopy is devoted to abundance analyses of stars which satisfy exactly or broadly the hypothesis that the composition of the stellar atmosphere is that of the star's natal cloud. Studies of Galactic chemical evolution, for example, exploit this hypothesis. This essay, on the contrary, discusses stars for which the hypothesis demonstrably fails because either the star since birth has gained products of nucleosynthesis or other than nuclear processes have operated to change the composition of the atmosphere. This eclectic look comments on diffusion in HgMn stars and in residents of the Spite plateau, the H-deficient cool carbon (HdC) stars, and diverse stars considered to be affected by dust–gas winnowing.

## 1. Introduction

In composing my contribution to the meeting, I decided to attempt an eclectic look at stellar compositions. My starting point was to look up "eclectic" in the compact Oxford dictionary that has been my companion since grammar school. There, I found that as an adjective eclectic is defined as 'deriving ideas or style from a varied range of sources'. My forty years of residence in the US have taught me that English and American uses of words may differ. So, I turned to Webster's dictionary where the definition is subtly different—'selecting what seems best of various styles or ideas'. I was reminded at this point of a remark attributed to George Bernard Shaw—'England and America are two countries divided by a common language'. Lia Athanassoula explained to me that Webster's definition with the insertion of the word 'best' is faithful to the Greek origin of the word. In this contribution, I shall highlight a selection—an eclectic selection—of beguiling differences in stellar compositions.

At this meeting where the focus is decidedly on "star formation", I shall begin by introducing the restricted hypothesis about the composition of a stellar atmosphere—the composition of a stellar atmosphere is exactly that of the star's natal interstellar cloud. This hypothesis is demonstrably false for many stars, and, indeed, in the strictest sense of applying to all elements, may be false for all stars. When the hypothesis is invalid, there are frequently multiple causes operating in parallel or in tandem to cause the abundance anomalies. This is fortunate because it provides a phenomenon-rich playground for observers and theoreticians alike.

If a Nelsonian blind eye is turned to minor failures (e.g., the depletion of lithium in some stars), the narrow hypothesis is broadly applicable to main sequence stars and even giants in the pursuit of the chemical evolution of the

Galaxy and external stellar systems. Galactic chemical evolution, a major practice of observers, is not discussed here although it is one with a direct link to questions associated with star formation; i.e., what is the initial mass function? What is the star formation rate?

## 2. Failure Modes

The failure modes for the narrow hypothesis are classifiable in a variety of ways. Perhaps, the simplest top level classification would have two groups: (A) those for which observers and/or theoreticians invoke nucleosynthesis as the principal agent and (B) those for which processes other than nucleosynthesis are identified or suspected as the key agent. Construction of a full classification scheme for the failure modes would call for mixed modes, such as A(B) and B(A).

In group A, products of nucleosynthesis from events occurring after the birth of a star have contaminated the stellar atmosphere. A group A process may be resolved into a site for nucleosynthesis and a mechanism for transferring the products of nucleosynthesis from the site to the stellar atmosphere. The site may be within the star with the contaminated atmosphere or in a companion star.

A leading entry for group A must be the sequence of dredge-up events occurring in red giants. The full sequence of first, second, and third dredge-up is not enjoyed by all stars, as is well known. Severe changes of surface composition occur in the third dredge-up experienced by low mass (i.e., initial $M_* < 8\,M_\odot$) stars in which products from H-burning and He-burning layers deep within the star are brought by convection to the atmosphere. This observer cannot help but recall Merrill's (1952) discovery of Tc, an element with no stable isotopes, in the spectra of S stars. Later, production of Tc was linked to the neutron capture *s*-process. And later still, the site of the *s*-process was identified with the He shell sandwiched between the red giant's extended H-rich envelope and the C-O electron degenerate core. The Tc and other products are convected to the surface as part of a third dredge-up event. Operational details of the *s*-process continue to bemuse theoreticians.

Severe mass loss may expose interior layers in which nucleosynthesis occurred previously. Here, examples include the Wolf–Rayet stars: massive stars experiencing stiff stellar winds that, after exposure of the N-rich layers that experienced H-burning, are seen as WN stars, and if the layers that experienced He-burning are exposed, the stars are seen as WC stars.

A milder process that strictly enjoys the right to belong to group A is astration; stars with a convective envelope transport light fragile nuclei downwards and expose them to destructive nuclear reactions. Lithium is astrated by proton capture in cool main sequence stars including the Sun.

Surface nuclear reactions, particularly for flaring M dwarfs and chemically peculiar A-type stars with magnetic fields, have a long and continuing theoretical history but without convincing observational confirmation. Here, the site of nucleosynthesis and the stellar atmosphere are collocated.

Departures from the restricted hypothesis for several classes of stars are associated with the presence of a companion star. Prime examples are the barium stars which were convincingly shown to be spectroscopic binaries by McClure

(1988) and in which the *s*-process enrichment results from transfer and acquisition of gas from the companion at the time that it was an AGB star. The companion remains in the binary system as a white dwarf. A different role for the companion is played in the case of novae where transfer of mass from the companion triggers explosive nuclear burning on the mass-receiving white dwarf. Evolution of some binary stars may conclude with the merger of the stellar remnants, a process that may proceed with either relative calm or nucleosynthetic vigor to produce a single star with surface abundance anomalies.

Stars with abundance anomalies assignable to group B obviously include the chemically peculiar B–A–F main sequence stars in which the dominant effects are the result of a competition between radiative levitation and gravitational settling, a process here termed simply "diffusion". An especially intriguing anomaly with diffusive connections is the Li-dip among F-type main sequence stars (Boesgaard & Tripicco 1986). Another sparkling appearance of diffusion occurs among He-core burning giants resident on the horizontal branch (HB) with hot (effective temperature greater than 11 500 K) stars exhibiting abundance anomalies (Behr 2003). These hot HB stars would be an example of the mixed failure mode B(A).

Another and more recent entry into group B or B(A) has been those evolved stars with an atmospheric composition depleted in refractory elements. Examples include post-AGB A-type stars such as HR 4049 and RV Tauri variables (Van Winckel 2003, 2007). Here, the diffusion is presumed to occur in a region containing dust and gas. Radiation pressure on the dust grains separates the grains from the gas. Accretion of gas but not dust by the star provides an atmosphere depleted in refractory elements. In extreme cases, the star may show an iron abundance $[Fe/H] \simeq -5$, an extremely metal-poor star, with near-solar abundances of elements of low condensation temperature such as C, N, O, S, and Zn.

In the following pages, I offer an eclectic sampling of stars with abundance anomalies to illustrate the richness of the failure modes.

## 3. Diffusion

### 3.1. The HgMn stars

The chemically peculiar (CP) stars are upper main sequence stars, spectral types B through F, with an array of abundance anomalies. A most useful introduction to CP stars is provided by Preston (1974), who subdivided the stars into the four groups CP1, CP2, CP3, and CP4. That the spectral types housing the CP stars as a class are susceptible to diffusion is not in hindsight a surprise. In more massive stars, diffusion is suppressed by the presence of stellar winds fed by the atmosphere with a velocity exceeding the very low velocities achieved under diffusion. Stars less massive than the CP stars have deeper convective envelopes such that convective mixing maintains a homogeneous envelope and atmosphere. The envelope's composition, as in CP stars, may be changed by selective diffusion across its base and into the radiative interior. But the diffusion velocities are slower than for the CP stars owing to the higher densities at the envelope's base. In addition, the reservoir (atmosphere plus envelope) of cooler stars is more massive than for the CP stars. Thus, the efficiency of diffusion is greatly reduced.

Here, I discuss aspects of the abundance anomalies offered by the HgMn stars which form Preston's class CP3. Strong lines of Mn II in the optical are one defining characteristic. Bidelman's (1962—quoted in AJ, 67, 640) identification of the strong Hg II resonance line at 3984 Å expanded the labeling of the class as HgMn stars. However, not all CP3 stars show enhanced lines of both Mn and Hg. One can find stars of similar temperature but with contrasting abundances, for example 53 Tau has a Mn overabundance near the maximum for the class but no trace of the Hg II 3984 Å line while $\chi$ Lup has a Hg overabundance by $10^5$ and a slight Mn deficiency. This "diversity" adds to the appeal of the class.

The HgMn stars with effective temperatures of 10 000–15 000 K are among the hottest of the CP stars. Many are sharp-lined. They are not, if at all, possessed of a magnetic field. They do not exhibit extreme rotational modulation of light and spectrum. To even the experienced eye, they appear to present a simpler challenge than most of the cooler CP stars. Vauclair & Vauclair (1982) declared that 'Mercury manganese stars probably represent the closest approach to the ideal situation of pure diffusion in stellar atmosphere'.

My introduction to the remarkable abundance anomalies present among upper main sequence stars most probably came from a personal account by Wal Sargent in the early 1960s. In particular, I strongly suspect that it was the story of 3 Cen A and its highly anomalous helium isotopic ratio that stuck in my head. This star is not strictly a HgMn star; with an effective temperature close to 18 000 K, it is hotter than HgMn stars. While it has a Mn abundance only slightly less than maximum abundance of HgMn stars, its Hg abundance is about 2 dex less than the maximum value (Castelli, Parthasarathy, & Hack 1997). But remember diversity!

Sargent & Jugaku (1961) measured the wavelengths of ten He I lines and reported a correlation with laboratory measurements of isotopic wavelength shifts between $^4$He and $^3$He indicating that the atmosphere was rich in $^3$He. Interpretation of the correlation led to the isotopic ratio $^3\mathrm{He}/(^3\mathrm{He} + {}^4\mathrm{He}) = 0.84 \pm 0.10$. Then, the sole benchmark for the He isotopic ratio of a normal star was the terrestrial ratio of $10^{-4} - 10^{-5}$. Today, we would assume that 3 Cen A began life with a ratio $^3\mathrm{He}/^4\mathrm{He} \simeq 2 \times 10^{-4}$, the value obtained from direct measurements of interstellar helium (Gloecker & Geiss 1996) and radio measurements of the 3.46 cm spin-flip line of $^3\mathrm{He}^+$ (Bania, Rood, & Balser 2007). Sargent & Jugaku (1961) considered 'some possible ways in which He$^3$ could be produced in large quantities in the atmosphere of 3 Centauri', i.e., a group A failure mode was assumed. Today, a group B mode is the leading suspect: diffusion augmented by light-induced drift (LID) (Atutov 1986; Michaud et al. 1979; LeBlanc & Michaud 1993; Michaud & Proffitt 1993).

Particularly challenging observations for the diffusion hypothesis are provided by the measurements of the isotopic abundance ratios for very heavy elements such as Hg. Surveys of the Hg isotopic abundances have been reported by White et al. (1976), Woolf & Lambert (1999), and Dolk, Wahlgren, & Hubrig (2003). Figure 1, from Woolf & Lambert, suffices to show the diversity of isotopic ratios: e.g., a quasi-solar mix for 74 Aqr, pure $^{204}$Hg for 46 Aql, roughly equal concentrations of $^{202}$Hg and $^{204}$Hg and an absence of lighter isotopes for HR 7775,

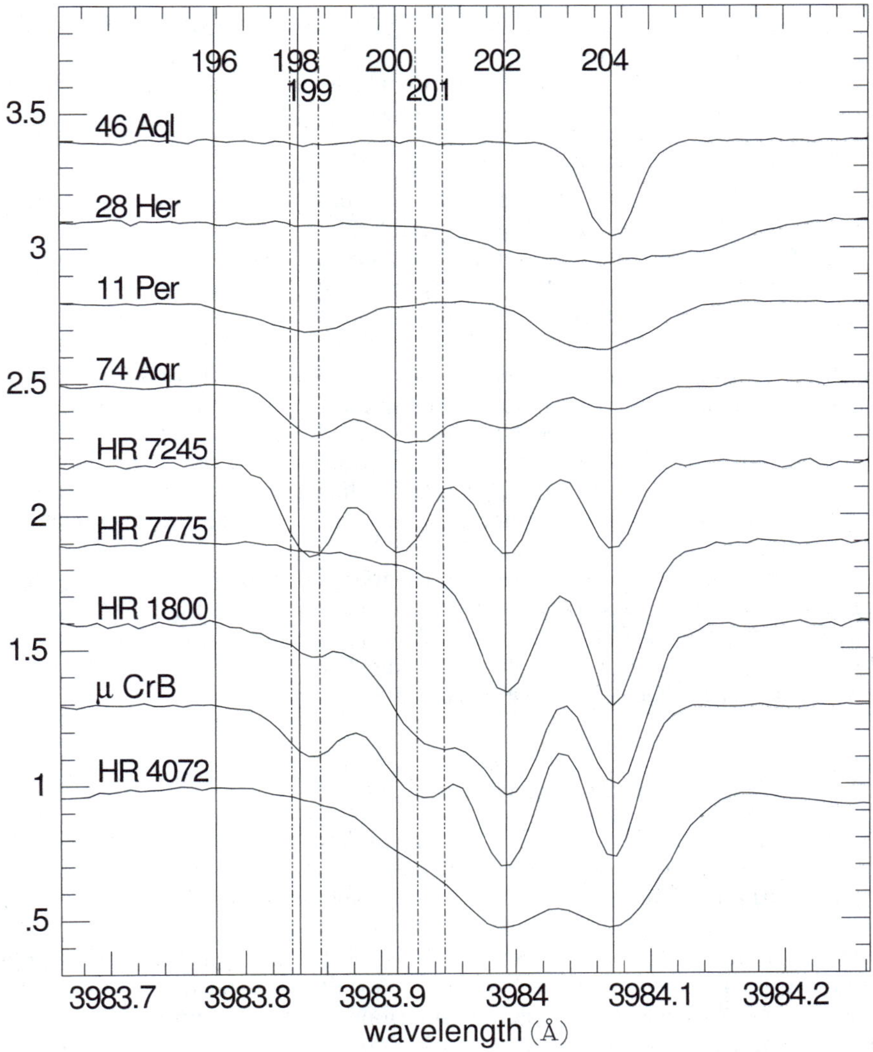


Figure 1. Observed spectra of sharp-lined HgMn stars, ordered by strength of the 3984 Å feature. Solid vertical lines indicate wavelengths for even Hg isotopic components. Dotted vertical lines indicate wavelength ranges for the hyperfine components of the two odd Hg isotopes. (*From Woolf & Lambert 1999.*)

and similar abundances of $^{198}$Hg+$^{199}$Hg and $^{204}$Hg for 11 Per and an absence of other isotopes. In all but one of the stars with isotopic ratios determined by Woolf & Lambert, $^{204}$Hg is the isotope with the greatest degree of enrichment relative to its solar abundance where only 6.9% of Hg is in the form of this, the heaviest stable Hg isotope. The maximum enrichment factor for this isotope is about $10^7$. This bias towards selective enrichment of the heaviest isotopes is shown also by Pt and Tl, although the data available are sparse for these elements (Proffitt et al. 1999).

Quantitative and even qualitative explanations for the Hg isotopic ratios are proving elusive. 'The mechanisms responsible for producing the chemical peculiarities in HgMn stars are still unclear' (Dolk et al. 2003). Diffusion has been advocated as a key factor (Michaud, Reeves, & Charland 1974—see, also, Lambert 1995). Proffitt et al. (1999) show that diffusion is unable to support the large Hg enrichment seen in $\chi$ Lup and HR 7775 (Fig. 1). It is far from obvious that diffusion can provide the isotopic mixes seen in 11 Per and HR 7245. LID has been promoted as a contributor (Aret & Sapar 2002) but observers have identified problems with this suggestion (Woolf & Lambert 1999; Dolk et al. 2003).

Observers have searched unsuccessfully for correlations between the Hg elemental and isotopic abundances and other abundances (Mn, for example), stellar mass and age (i.e., effective temperature and surface gravity. Dolk et al. (2003) note that diffusion calculations exploring the age dependence of diffusive separation (e.g., Alecian 1993; Seaton 1996) have shown that the effects should vary considerably with age and, therefore, the search for correlations with mass and age becomes complicated.

Yet another layer of complication and fascination is provided by Doppler imaging of the classical HgMn star and spectroscopic binary $\alpha$ And. Long term monitoring the Hg II 3984 Å line shows not only a non-uniform surface distribution of Hg on the HgMn star but an evolution of this distribution which the discoverers (Kochukhov et al. 2007) dub 'a weather pattern'.

### 3.2. Diffusion and the Big Picture

The admonition 'Look at the Big Picture' is heard with increasing frequency, especially from Time Allocation Committees overburdened with meritorious proposals. Although I doubt that this is always sound advice, I will concede that one might wonder if intense interest in the CP stars has a benefit beyond the narrow confines of the upper main sequence. Although a firm quantitative explanation for the CP anomalies remains elusive, these (and hot HB) stars have proven to be test beds for diffusion calculations. Hence, the prospect exists that validation of those calculations on CP stars *may* install greater confidence in diffusion calculations when they are applied to stars where the abundance effects are much smaller but the significance of the anomaly under scrutiny is crucial in a Big Picture sense.

Such an application exists in the determination of the lithium yield from the Big Bang. This determination is made using metal-poor dwarfs near the mean sequence turn-off. Spite & Spite (1982) in a seminal contribution discovered that such dwarfs have an identical Li abundance for stars with [Fe/H] less than about −1. This abundance, which defines what is commonly called the Spite plateau, has been widely identified as the Li abundance of the Big Bang. Metal-poor stars on the main sequence away from the turn-off have Li abundances less than that of the turn-off stars; this is attributable to astration, i.e., convective mixing and destruction of Li by warm protons. Unevolved warm main sequence stars with [Fe/H] $> -1$ show a Li in excess of the plateau, an increase attributed to

Li synthesis in stars and the interstellar medium (ISM) that increased the Li abundance of the gas from which [Fe/H] $> -1$ stars formed.

With the appearance of the *WMAP* results, the Li abundance expected of a standard Big Bang model is determined with good precision; the previously single unknown parameter, the ratio of baryon to photon number density, was measured from the cosmic microwave background (CMB) fluctuations. The predicted Li (all $^7$Li) abundance is $\log\epsilon(\mathrm{Li}) = 2.64\pm0.03$, a value substantially larger than is generally ascribed to the Spite plateau, e.g., Ryan et al. (2000) give $\log\epsilon(\mathrm{Li}) = 2.09^{+0.19}_{-0.13}$. This gap between the predicted and observed Li abundance has occasioned much debate—see, for example, Lambert (2004).

Loss of surface Li among Spite plateau residents through diffusion to deeper layers is one proposal to account for the gap. A test of this proposal would seem possible using stars of a globular cluster where the stars are of the same age and generally of the same initial composition. Korn et al. (2006a,b) have analyzed stars in the cluster NGC 6397 from the main sequence turn-off to the first red giant branch and compared the abundances of Li, Ca, and Fe with predictions from stellar evolutionary calculations that include diffusion. Elements (e.g., Ca and Fe) that diffuse out of the atmosphere of a main sequence star and avoid astration are restored to very nearly their full abundance by the deep convective envelope of a red giant. Lithium in main sequence stars is predicted to diffuse inwards and be subject to astration. Then, the convective envelope as it develops in a subgiant raises the surface Li abundance by a small amount before the deeper convective envelope of the giant lowers the surface Li abundance through dilution with layers of complete Li astration. Korn et al.'s observations show the predicted rise by 0.1 to 0.2 dex in the Ca and Fe abundances between the turn-off and the giant branch. The Li abundance of turn-off stars, the very slightly higher abundance in the subgiants and the sharp decrease in abundance for the giants are well reproduced by the calculations including diffusion for an initial Li abundance of of $2.54 \pm 0.1$, a value consistent with the WMAP predicted Li abundance of $2.64 \pm 0.03$. (The measured Li abundance in the turn-off stars is about 2.25, indicating a reduction of 0.3 dex attributable to diffusion.) The title of their *Nature* letter reflects Korn et al.'s (2006b) quiet confidence at having resolved the Li-gap—'A probable stellar solution to the cosmological lithium discrepancy'.

This probable solution of the cosmological lithium discrepancy is discussed further by Molaro (this volume, p. 472). Much remains to be done, as recognized by Korn et al. Key tasks required to assure confidence in abundance changes of 0.2 dex or less include development, testing, and application of atmospheric models that include hydrodynamics (so-called 3D models, Asplund 2005), critical examination of the stellar parameters across the sample, consideration of the effects of non-LTE, and understanding the physics behind 'turbulent diffusion', a free parameter in the diffusion calculations.

## 4. Hydrogen-Deficient Stars

Luminous H-deficient stars comprise an intriguing sequence from cool H-deficient carbon (HdC) stars, through the R CrB stars to the extreme He (EHe) stars at the hot end. These are extremely rare stars: for example, just five HdC stars are known. This rarity implies that the stars are either the product of a very lightly traveled branch in stellar evolution and/or are in a rapid phase of evolution.

After a sifting of ideas, two origins for these H-poor stars remain in play. In one, a post-AGB star experiences a final He-shell flash that consumes remaining hydrogen and returns the star to the vicinity of the AGB tip from which it evolves rapidly back to the white dwarf cooling track. This is the final flash or FF scenario. In the second, the luminous H-poor star results from the merger of a He white dwarf onto a C–O white dwarf, a scheme dubbed the double-degenerate or DD scenario.

An important step in assigning the H-poor stars to one scenario or the other is the provision of data on their chemical compositions. Evidence from elemental abundances is accumulating that the majority of the R CrB and EHe stars arise from the DD scenario (Asplund et al. 2000; Saio & Jeffery 2002; Pandey et al. 2006). Certainly, there are other stars for which the FF scenario provides a natural explanation, e.g., FG Sge and Sakurai's object (V4334 Sgr).

Thanks to their daunting spectra dominated by a dense thicket of molecular lines, HdC stars have largely escaped modern scrutiny; the abundance analysis of the five known HdC stars by Warner (1967) remains unsurpassed. Yet, the molecules offer a potential that atoms cannot. Thanks to the dependence of rotational and vibrational energies on the nuclear masses, substitution of one isotope for another may lead to a measurable shift in the wavelengths of molecular lines and bandheads. This opportunity was exploited recently and dramatically by Clayton et al. (2005, 2007), who obtained low resolution spectra of the CO first overtone vibration–rotation bands in spectra of HdC stars and some cool RCB stars.

Clayton et al. (2005) presented a spectrum of the HdC star HD 137613 that showed bandheads of $^{12}C^{18}O$ and $^{12}C^{16}O$ of similar strength, implying a $^{16}O/^{18}O$ ratio of about unity. This is a remarkable result given that the ISM has a ratio of about 500 and $^{18}O$ is expected to be destroyed in the course of normal stellar evolution. Additional observations and analysis were provided by Clayton et al. (2007), showing that three of the HdC stars have $^{16}O/^{18}O \simeq 0.3$ and the remaining two HdC stars are too warm to provide detectable CO bands. Clayton et al. (2007) also show that several cool RCB stars have significant amounts of $^{18}O$: $^{16}O/^{18}O \simeq 1$ to 6. Our high resolution ($\lambda/\Delta\lambda \simeq 50,000$) PHOENIX spectra of HdC and RCB stars confirm these discoveries of $^{18}O$ enrichments (A. García-Hernández et al., in preparation).

The beauty of this discovery of the $^{18}O$-rich nature of HdC stars by Clayton et al. (2005, 2007) is that it does not require supporting analysis or high resolution spectra in order to conclude that significant synthesis of $^{18}O$ has occurred. Furthermore, the nuclear physics of this synthesis is well recognized, as first noted by Warner (1967). Layers previously exposed to H-burning will be rich in $^{14}N$. Then, the $^{14}N$ may be subject to $\alpha$-capture with $^{18}O$ as the principal product provided that the gas is cooled and $^{18}O$ avoids destruction by $\alpha$-capture. Clayton et al. (2007) argue that $^{18}O$ synthesis is possible in the DD but not in the FF scenario except under very contrived constructions. In the DD scenario, a short-lived phase of nucleosynthesis may occur as the He white dwarf material is accreted onto the C–O white dwarf. A major outcome of this phase is considered to be $^{18}O$ production (Clayton et al. 2007).

Open questions remain concerning the relationship between the HdC, RCB, and EHe stars. Is there a tight evolutionary sequence? If so, is the presumption that the stars evolve from HdC → RCB → EHe correct? Is the DD scenario the sole mechanism for production of these H-deficient stars?

## 5. Dust–Gas Winnowing

Formation of dust grains in cool gas removes elements selectively from the gas. A clear example of this selectivity is shown by the composition of the local interstellar gas, which is deficient in elements expected to reside in the grains (Savage & Sembach 1996). If dust were separated from gas, a process here referred to as "winnowing", and the gas accreted onto a star and were to comprise a major fraction of the stellar atmosphere, a star with a very striking set of abundance anomalies would result.

Anomalies attributable to dust–gas winnowing have now been reported for three kinds of stars: $\lambda$ Boo stars, post-AGB A-type spectroscopic binaries, and RV Tauri variables. In these cases, the effects are large but it seems likely that dust–gas winnowing may operate less obviously to affect stars as yet unsuspected as beneficiaries of the process. Not surprisingly, theoretical ideas have been proposed for other environments and other stars. An obvious candidate is selective accretion of interstellar material (gas preferred to dust?) by stars; very metal-poor stars might be changed to just metal-poor stars.

Lambda Boo stars are main sequence A-type stars long recognized to be surprisingly metal-poor for their youth. Venn & Lambert (1990) undertook an abundance analysis of three stars, including the eponym. They showed that the compositions of the stars could be attributed to accretion of circumstellar gas. The Fe deficiency in the case of $\lambda$ Boo was [Fe/H] $\simeq -2$ and slightly less severe for 29 Cyg with normal abundances of C, N, O, and S for both stars, as expected of dust-free gas. Even Vega, a "standard" A0 star, was shown to be a mild $\lambda$ Boo star. The shallow convective envelope of early A stars is presumably a key factor in creation and maintenance of the abundance anomalies.

A metal deficiency of greater severity is found among a group of stars known as post-AGB spectroscopic binaries (those in Van Winckel 2003, 2007). The original quartet (Van Winckel, Waelkens, & Waters 1995) comprised HR 4049, HD 44179, HD 52961, and BD +39 4926 with [Fe/H] from $-3.0$ to $-4.8$ but an abundance pattern again reminiscent of interstellar gas, i.e., quasi-solar abundances of C, N, O, S, and Zn but underabundances of, for example, Si, Ca, and Fe. The pattern shows that it is the former set of elements that define the initial composition of these stars and not the latter set. Gas is imagined to be accreted onto the star from a circumbinary disk. Radiation pressure exerted by the star on the dust grains inhibits their accretion by the star and may also promote a separation of dust and gas in the disk. The dusty disk is betrayed by an infrared excess. HD 44179, also known as the Red Rectangle, is a rather special protoplanetary nebula with a striking IR excess. However, BD +39 4926 lacks an IR excess. Shallow convective envelopes are presumably a key factor to the appearance of the huge abundance anomalies.

Discovery of the third category of stars displaying the marks of dust–gas winnowing began with the analysis of the RV Tauri variable IW Car (Giridhar et

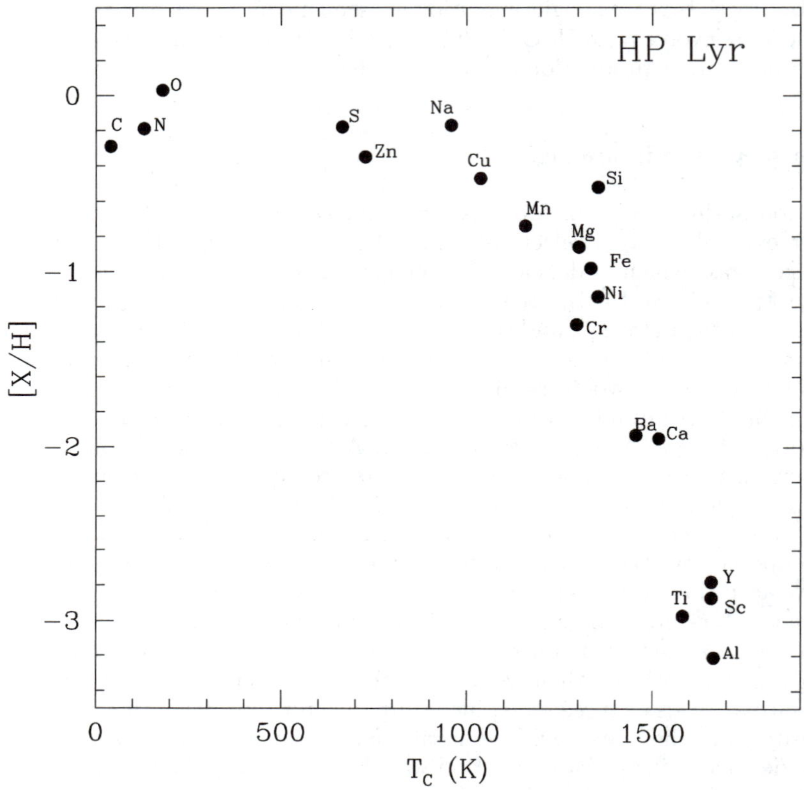


Figure 2. Abundance $[X/\mathrm{H}]$ versus condensation temperature $T_{\mathrm{cond}}$ for HP Lyr. Elements are identified by their chemical symbol. (*From Giridhar et al. 2005.*)

al. 1994). Our subsequent analyses (see Giridhar et al. 2005 for complete references to the papers) showed that the effects of dust–gas separation among RV Tau stars appear limited to the warmer stars ($T_{\mathrm{eff}} > 5000$ K) and to the stars with intrinsic metallicities $[\mathrm{Fe/H}] > -1$ or so. The reasons for these boundaries are not entirely clear but the cooler stars possess more extensive convective envelopes that dilute the accreted gas and dust–gas winnowing may be impaired in low metallicity gas where the dust-to-gas ratio is necessarily lower. The winnowing site is again presumed to be a circumbinary disk. There is increasing evidence that affected stars are spectroscopic binaries; detection of orbital velocity variations in the presence of an atmospheric pulsation demands intensive observing campaigns. How the gas is captured onto the star from the circumbinary disk in the presence of a stellar wind remains an unsolved problem. This problem has much in common with that posed by the post-AGB binaries.

A signature of a star affected seriously by dust–gas winnowing is a correlation between an element's abundance and the predicted condensation temperature

$T_{\rm cond}$ (or the abundance in interstellar gas). Estimates of $T_{\rm cond}$ depend on the initial composition and pressure of the gas and the assumption that cooling of the gas and condensation of grains occurs under equilibrium conditions. Lodders (2003) provides a comprehensive discussion of $T_{\rm cond}$ estimates for all elements for a mixture of solar composition. In some RV Tau stars, the abundance $[X/{\rm H}]$ is smoothly correlated with $T_{\rm cond}$: for example, HP Lyr (Fig. 2) from Giridhar et al. (2005) has $[X/{\rm H}] \simeq -3$ for Al, Sc, Ti, and Y with $T_{\rm cond}$ near 1600 K, $[X/{\rm H}] \simeq -1$ for Fe and Ni with $T_{\rm cond}$ near 1300 K, $[{\rm Na/H}] \simeq -0.2$ near 900 K, and $[X/{\rm H}] \simeq -0.2$ for S and Zn with $T_{\rm cond}$ near 700 K. In other cases, the $[X/{\rm H}]$ may show a spread but with no obvious trend with $T_{\rm cond}$: for example, EQ Cas also from Giridhar et al. (2005) has $[{\rm Sc/H}] \simeq -3$ but $[{\rm Ti/H}] \simeq -1$, $[{\rm Na/H}] \simeq -1.6$. Rao & Reddy (2005) brought order to chaos by noting that the $[X/{\rm H}]$ for EQ Cas were correlated with the ionization potential of the neutral atom (the "first ionization potential" or FIP). This raises the intriguing possibility that an alternative or additional process is operating. The FIP effect is a well known phenomenon in the solar corona—unexplained but thought to reflect the greater ease with which ions (low FIP elements) rather than neutral atoms (high FIP elements) are fed from the cool chromosphere into the corona. Perhaps, EQ Cas is a star where there is a selective feeding of the stellar wind. In single stars, the wind may control the abundance anomalies but where the variable is in a binary the circumbinary disk may exert control. Further study of the relative occurrence of dust–gas winnowing and the FIP effect is desirable.

## 6. Concluding Remarks

In this essay, I do not pretend to have offered anything more than a personal selection of topics related to the determination and interpretation of stellar atmospheric compositions. I offer it as a token of my affection for John and in the hope that, just perhaps, it may catch the eye of a young person looking for a truly fascinating field in which to venture.

Among the great minds at Oxford when John and I spent happy years there was Sir Cyril Hinshelwood, Nobel Laureate in chemistry. Sir Cyril once wrote, 'A common fallacy in much of the adverse criticism to which science is subjected today is that it claims certainty, infallibility and complete emotional objectivity. It would be more nearly true to say that it is based upon wonder, adventure, and hope'.

May 'wonder, adventure, and hope' beguile everyone but especially John.

**Acknowledgments.** I am pleased to acknowledge that my adventures into stellar compositions have long been supported by the Robert A. Welch Foundation of Houston, Texas.

### References

Alecian, G. 1993, ASP Conf. Ser., 44, IAU Coll. 138, Peculiar versus Normal Phenomena in A-Type and Related Stars, ed. M. M. Dworetsky, F. Castelli, & R. Faraggiana (San Francisco: ASP), p. 450

Aret, A., & Sapar, A. 2002, AN, 323, 21

Asplund, M. 2005, ARA&A, 43, 481

Asplund, M., Gustafsson, B., Lambert, D. L., & Rao, N. K. 2000, A&A, 353, 287
Atutov, S. N. 1986, Phys. Lett. A, 119, 121
Bania, T. M., Rood, R. T., & Balser, D. M. 2007, Space Sci. Rev., 130, 53
Behr, B. B. 2003, ApJS, 149, 67
Boesgaard, A. M., & Tripicco, M. J. 1986, ApJ, 302, L49
Castelli, F., Parthasarathy, M., & Hack, M. 1982, A&A, 321, 254
Clayton, G. C., Geballe, T. R., Herwig, F., Fryer, C., & Asplund, M. 2007, ApJ, 662, 1220
Clayton, G. C., Herwig, F., Geballe, T. R., Asplund, M., Tenenbaum, E. D., Engelbracht, C. W., & Gordon, K. D. 2005, ApJ, 623, L141
Dolk, L., Wahlgren, G. M., & Hubrig, S. 2003, A&A, 402, 299
Giridhar, S., Rao, N. K., & Lambert, D. L. 1994, ApJ, 437, 476
Giridhar, S., Lambert, D. L., Reddy, B. E., Gonzalez, G., & Yong, D. 2005, ApJ, 627, 432
Gloecker, G., & Geiss, J. 1996, Nature, 381, 210
Kochukhov, O., Adelman, S. J., Gulliver, A. F., & Piskunov, N. 2007, Nature Phys., 3, 526
Korn, A., Grundahl, F., Richard, O., Barklem, P. S., Mashonkina, L., Collet, R., Piskunov, N., & Gustafsson, G. 2006a, ESO Messenger, 125, 6
Korn, A., Grundahl, F., Richard, O., Barklem, P. S., Mashonkina, L., Collet, R., Piskunov, N., & Gustafsson, G. 2006b, Nature, 442, 657
Lambert, D. L. 2004, AIP Conf. Proc., 743, 206
Lambert, D. L. 1995, in From Astronomy to Astrophysics, ed. B. Sinha & M. Bhattacharya (Calcutta: Saha Inst. of Nuclear Physics), p. 48
LeBlanc, F., & Michaud, G. 1993, ApJ, 408, 251
Lodders, K. 2003, ApJ, 591, 220
McClure, R. D. 1988, in IAU Coll. 106, Evolution of Peculiar Red Giant Stars (Cambridge: CUP), p. 196
Merrill, P. W. 1952, ApJ, 116, 21
Michaud, G., Montmerle, T., Cox, A. N., Magee, N. H., Hodson, S. W., & Martel, A. 1979, ApJ, 234, 206
Michaud, G., & Proffitt, C. R. 1993, ASP Conf. Ser., 44, IAU Coll. 138, Peculiar versus Normal Phenomena in A-Type and Related Stars, ed. M. M. Dworetsky, F. Castelli, & R. Faraggiana (San Francisco: ASP), p. 439
Michaud, G., Reeves, H., & Charland, Y. 1974, A&A, 37, 313
Pandey, G., Lambert, D. L., Jeffery, C. S., & Rao, N. K. 2006, ApJ, 638, 454
Preston, G. W. 1974, ARA&A, 12, 257
Proffitt, C. R., Brage, T., Leckrone, D. S., Wahlgren, G. M., Brandt, J. C., Sansonetti, C. J., Reader, J., Johansson, S. G. 1999, ApJ, 512, 942
Rao, N. K., & Reddy, B. E. 2005, MNRAS, 357, 235
Ryan, S. G., Beers, T. C., Olive, R. A., Fields, B. D., & Norris, J. E. 2000, ApJ, 530 L57
Saio, H., & Jeffery, C. S. 2002, MNRAS, 333, 1210
Sargent, W. L. W., & Jugaku, J. 1961, ApJ, 134, 777
Savage, B. D., & Sembach, K. R. 1996, ARA&A, 34, 279
Seaton, M. J. 1996, Ap&SS, 237, 107
Spite, F., & Spite, M. 1982, A&A, 115, 357
Van Winckel, H. 2003, ARA&A, 41, 391
Van Winckel, H. 2007, Baltic Astron., 16, 112
Van Winckel, H., Waelkens, C., & Waters, L. B. F. M. 1995, A&A, 293, L25
Vauclair, S., & Vauclair, G. 1982, ARA&A, 20, 37
Venn, K. A., & Lambert, D. L. 1990, ApJ, 363, 234
Warner, B. 1967, MNRAS, 137, 119
White, R. E., Vaughan, A. H., Jr., Preston, G. W., & Swings, J. P. 1976, ApJ, 204, 131
Woolf, V. M., & Lambert, D. L. 1999, ApJ, 521, 414

*Pathways through an Eclectic Universe*
*ASP Conference Series, Vol. 390, © 2008*
*J. H. Knapen, T. J. Mahoney, and A. Vazdekis, eds.*

# Bolivia y América Latina: Tendencias de su desarrollo económico, social, y político [Bolivia and Latin America: Trends in Economic, Social, and Political Development]

C. Quiroga Blanco

An English translation follows the Spanish text.

## 1. Introducción

A fines de la década de 1980 el panorama político, social y económico de muchos países de América Latina era sombrío. Salían de un orden constitucional quebrado por las dictaduras militares. Las economías nacionales parecían fuera de control.

Sin embargo la sociedad boliviana, como otras, encontró fuerzas para ponerse de pie:

- Restableció la institucionalidad democrática,
- Reasumió el control de la economía,
- Luego de innumerables revueltas sociales y de golpes de Estado, comenzó a borrar la imagen de un país ingobernable.

Es en ese momento (1985) cuando en Washington, en alguna apacible oficina de los Organismos Internacionales, hombres como Hayeck, Friedman, personeros del Fondo Monetario Internacional, del Banco Mundial y otros expertos concluyeron que el modelo de desarrollo, el comportamiento de las economías de América Latina durante medio siglo ya no daba más. Las conclusiones de ese día se habían de conocer como el Consenso de Washington. En esas minutas se trazó el marco conceptual y operativo de las Reformas que estremecerían al Continente en los próximos 20 años y que todavía lo convulsionan. Allí está el libro de la doctrina llamada Neo-liberal. Esta doctrina adquirió categoría de verdad revelada, modificada y controlada por los mismos que la formularon. Y no hubiera alcanzado la significación que tuvo si no hubiera recibido la bendición de la primera potencia mundial y de poderosas Instituciones del Sistema Internacional.

Intentaré, simplemente, enumerar los pasos que habrían de seguirse, en los países en desarrollo, para acercarse a la tierra prometida. La primera generación se concentra en la necesidad de lograr el equilibrio macro-económico, la disciplina fiscal, el control del gasto público. Medidas que permitirían, entre otros, acorralar y liquidar a la inflación galopante en esa época. La segunda, para los ideólogos del Neo-liberalismo el Estado era el problema, la fuente de todos los males: de la inflación, burocracia, ineficiencia, del proteccionismo, y de la corrupción. En consecuencia había que destruirlo o por lo menos debilitarlo, desmantelarlo. Para ello era necesario que el rol protagónico de la economía se transfiriera al sector privado. Nadie debía preocuparse porque la mano invisible del mercado se encargaría de poner todas las cosas en su lugar. Se propuso, entonces, el libre mercado, el juego de las leyes de la oferta y la demanda para ocupar el sitio del Estado. Por ello, entre otras medidas, se concentra la mirada en la privatización de las Empresas Públicas y de los sistemas de pensiones, en la liberalización del sistema financiero, en el libre comercio de importación y exportación. Finalmente la tercera, tiene relación con la organización política de la sociedad, antes pisoteada por los regímenes dictatoriales. Se trataba de restablecer la primacía de la ley; que la democracia y los derechos humanos retornen al escenario.

Habiendo transcurrido el tiempo suficiente de la puesta en práctica de estas medidas y para tratar de entender mejor lo que acontece hoy en día en muchos países de América Latina, procuraré hacer una evaluación sintética de los aspectos positivos y negativos de estos cambios. No sin antes señalar algo obvio, las circunstancias impusieron tantas modificaciones a la doctrina que, a estas alturas, cualquier semejanza entre la realidad y las previsiones del Consenso de Washington es pura coincidencia.

## 2. La Reforma Económica

1. No cabe duda que el modelo logró controlar la inflación, que hoy parece un gatito apacible junto a una estufa. La tasa de inflación, en América Latina, se redujo de 390% en 1994 a 26% en 1995 a 22% en 1996 y hoy fluctúa entre el 2 y el 5%.

2. La presión tributaria contribuyó en gran medida a la disminución del déficit fiscal y al equilibrio macro-económico.

3. Se amplió el acceso al mercado mundial del dinero y se acentuó de manera notable el ritmo de expansión de las exportaciones de América Latina, de 146 mil millones en 1992 a 245 mil millones en 1995.

   Los defensores del modelo sustentan su optimismo en estas cifras. Sin embargo, el comportamiento de otras variables macro-económicas, como el empleo, el ahorro, la inversión y la distribución del ingreso no justifica explosiones de euforia. Me referiré solamente a algunas de estas variables.

   (a) El deterioro del mercado de trabajo. La tasa de desempleo urbano abierto pasó de 7% en 1992 a 9.4% en 1995, a 10% en 1996 y hoy, en Bolivia, fluctúa entre el 15 y 20%. Al mismo tiempo se agravó el problema del subempleo, como se demuestra en los casos de Argentina y Colombia, en los que la masa laboral subutilizada se situó en rangos próximos al 13%.

   (b) No se puede ocultar que los resultados positivos del modelo beneficiaron a pocos y los negativos afectaron particularmente a los sectores menos favorecidos de la población. A comienzos de los años 90 cerca de 200 millones de latinoamericanos (46% de la población total) no estaban en condiciones de satisfacer sus necesidades fundamentales, mientras cerca de 100 millones se encontraban en situación de extrema pobreza. El efecto de la concentración de la economía abierta ya no se discute. Lo que en verdad espanta ya no es el hecho que una parte de la humanidad viva mejor que la otra—lo que, por otra parte, no es ninguna novedad—sino la dimensión de ultraje y ofensiva de la brecha. En Bolivia el ingreso promedio de un habitante de la ciudad de Santa Cruz es dos veces y medio superior al de la ciudad de Potosí y el 10% más favorecido de la población urbana percibe el 35.4% del total del ingreso nacional, mientras el 40% más pobre recibe el 15%. En escala global la situación es todavía más chocante. Todos sabemos que los activos de 360 personas equivaldrían al ingreso anual de más de 2 mil millones de personas. Es cierto que la pobreza tiene antiguas raíces y que no es criatura de las reformas neo-liberales. Pero ya está más allá de toda comprobación empírica que la economía abierta de mercado, en América Latina, mejoró las condiciones de vida de los sectores de mayores ingresos, pero no logró resolver los problemas de los grupos menos favorecidos. En algunos casos agravó su situación, redujo su salario y acentuó la intensidad de su pobreza.

La gente ya no admite que la estabilidad macroeconómica haya traído seguridad económica. Sabe, por experiencia directa, que los primeros resultados de la estabilidad macroeconómica (desocupación, pérdida del poder adquisitivo del salario, informalidad) han causado mayor inseguridad. No la convencen de la estabilidad porque no la encuentran en sus trabajos, en la educación de sus hijos, en el cuidado de su salud. Podría decirse, sin exagerar, que la inestabilidad social, en su forma actual, ha sido uno de los primeros frutos de la estabilidad macroeconómica y eso explica el sentimiento generalizado de incertidumbre y de inseguridad a lo largo y ancho del Continente.

(c) Hay algo más. Me referiré a la vulnerabilidad externa de los países de América Latina vía la liberalización financiera impuesta por el modelo. El Fondo Monetario Internacional y el Banco Mundial condicionaron su apoyo a los países en desarrollo a la apertura acelerada de sus mercados financieros. Para esta flamante amenaza no existen todavía antídotos. Los órdenes de magnitud del mercado de capitales desafían la imaginación del hombre y la capacidad de control de los gobiernos. Las transacciones globales suman más de 1.5 trillones (millones de millones) de dólares por día. Las colocaciones de los fondos de pensión, de los fondos mutuos y de las empresas de seguros en el mercado mundial del dinero se sitúan en el orden de los 25 trillones de dólares. Ese mercado de capitales ha sustituido largamente al antiguo sistema de financiamiento para los países en desarrollo basado en aportes de los gobiernos de las naciones industrializadas. El 99% de créditos e inversiones a los mercados emergentes más importantes (México, Brasil, Argentina) provenía de recursos del mercado privado de dinero. En consecuencia, tasas de interés, tipos de cambio, competitividad internacional, niveles de crecimiento del producto, para no hablar de la felicidad de la gente, dependen ahora de la continuidad o retracción de este flujo. Si se mantiene, bien. Cuando se interrumpe súbitamente, las consecuencias son desastrosas. Los grandes especuladores deciden dónde, cuánto y en qué momento se invierte ... o se huye de un mercado. De su voluntad y temple depende el destino de mucha gente. Este hecho ha provocado una alarmante volatibilidad en los mercados de capital y una peligrosa inestabilidad en los sistemas financieros de los países en desarrollo. En 1998 los bancos retiraron 150 mil millones de dólares de los mercados emergentes y provocaron la gigantesca crisis brasilera, asiática y rusa, y una terrible recesión en el ciclo productivo de América Latina. Es notorio el carácter global y la recurrencia de las crisis financieras. Afecta a todos y se repiten con consecuencias cada vez más graves.

## 3. Las Reformas Institucionales y Sociales

Hay aspectos que están a la vista. No cabe duda, por ejemplo, que el sistema democrático llegó para quedarse. El grado de respaldo popular a la democracia varía en los distintos países de la región, pero siempre se sitúa por encima del 75% de los encuestados. Se respetan los derechos de expresión y asociación de la ciudadanía y las violaciones masivas y sistemáticas de los derechos humanos son cosa del pasado.

Sin embargo las encuestas en América Latina advertían que algo no andaba bien. En las encuestas la gente transmitía una sensación de inquietud y de inseguridad. Ni la dirección política latinoamericana, ni los expertos de los Organismos Internacionales dieron mucha importancia a estas señales. Los sistemas de alerta temprana estaban desconectados. Ya pasará, decían. Los fundamentos de la economía están bien y ahora es cuestión de tiempo. Unas escuelas aquí y unas postas sanitarias allá, harán el resto.

No fue así, lamentablemente. La sociedad decidió salir a las calles para dejar constancia de su descontento; para recordar a los ricos, a los poderosos y a los expertos, que podían manchar la alfombra de la sala, insultar a la dueña de casa y armar un gran escándalo. Eso ocurrió en San José de Costa Rica, en Quito, en Bolivia en la ciudad de Cochabamba y La Paz, en Brasil.

El tablado democrático, recién terminado de montar comenzó a crujir. Surgieron de ninguna parte líderes carismáticos y autoritarios. Revolucionarios sin revolución y tratan de cambiar el equilibrio de los astros.

En rigor de verdad, sin embargo, las marchas populares de América Latina, llenas de mestizos e indígenas, tienen algún parentesco con las de los rubios de Davos, de Seattle o Washington. Son voces del mismo grito. Son la denuncia de que no bastan el equilibrio macroeconómico, la disciplina fiscal y el crecimiento del producto interno bruto. Que la gente no está dispuesta a tolerar la acumulación de la prosperidad de un lado de la balanza y la persistencia de la miseria en el otro. Pero tienen, además, otra connotación. La democracia, en la forma en que está institucionalizada en América Latina, tampoco es suficiente. Tal vez sea porque la ciudadanía intuye que las Instituciones democráticas son patrimonio de una nueva casta, la de la denominada clase política. Que los partidos que retienen el monopolio de la delegación popular, cada vez la representan menos. La gente quiere participar directamente. ¿Cómo es eso de que yo hablo cada cuatro o cinco años por la voz de un diputado al que no he visto nunca, si puedo efectivamente hablar aquí, en la calle, en este momento?

Es que, en realidad, el viejo sistema de intermediación de la sociedad con el Estado por un solo canal, el de los Partidos Políticos, está agotado. Son percibidos por la gente ya no como puentes entre la Sociedad Civil y el Estado sino como muros infranqueables. Crece la sensación, si no la certeza, que los Partidos Políticos se interesan mucho más en el Estado que en la Sociedad, como si toda su atención estuviera centrada en COMO llegar al Poder, antes que preguntarse PARA QUE. En consecuencia la Sociedad se aleja de los Partidos y busca sus propios canales de relacionamiento con el Estado. A su vez los Partidos se cierran sobre sí mismos y se proclaman como una clase diferente. La clase política que no tendría otro objetivo que su propio poder e incluso el enriquecimiento personal de sus miembros. La corrupción, en nuestro medio, ya no sólo es una ruptura de un orden ético de la política, sino que es un factor de atraso y conflictividad social. La Sociedad organizada por fuera de los partidos quiere ser escuchada y sentarse en la mesa en la que tomen decisiones. ONGs, jóvenes, mujeres, ecologistas, indígenas, los movimientos sociales en general, tienen su verdad y quieren decirla. El diálogo entre Estado y mercado, entre Gobierno y empresarios ya no alcanza. Se plantea hoy, cómo pasar de una democracia formal y representativa (delegativa) a una democracia participativa y solidaria en la que se democratice el poder político, económico y social.

La descentralización administrativa y política es la primera respuesta a esta problemática. Pero en esta dinámica el planeta lejos de convertirse en la Aldea Global, de la célebre frase de Mac Luhan, tiende a asemejarse cada vez más a un archipiélago de miles de aldeas, cada una tratando de afirmar una identidad cultural, administrativa, política y hasta histórica para su propia parroquia.

## 4. La Reforma del Estado

La Reforma del Estado, como ya lo señalamos, fue algo así como la columna vertebral de las reformas neo-liberales. Es verdad que se había convertido efectivamente en un monstruo. La mayoría de las cosas que se dijeron de él eran ciertas: la burocracia, la ineficiencia, el clientelismo, la corrupción, etc. En consecuencia se lo desmanteló puesto que se había convertido efectivamente en un monstruo. Se transfirió al sector privado el rol protagónico de la economía. Sin embargo, el proceso de privatización, multiplicó varias veces la fuerza política de los empresarios y grupos económicos, acumulando poder en sus manos, particularmente en el de las grandes corporaciones extranjeras que compraron las empresas públicas de energía, transportes y telecomunicaciones.

Pero resulta que los problemas que hoy tenemos que resolver no son los que se atribuían a la acción del Estado, sino que resultan, debido a las reformas, de su ausencia. En consecuencia algunos países latinoamericanos se han dado a la tarea de redefinir el rol del Estado. Tarea nada fácil puesto que habrá que hacerlo al interior de una economía globalizada, en la sociedad del conocimiento, en la integración regional. Es, no cabe duda, un Estado diferente, pensado en una lógica distinta a la del pasado. De ahí que se entienda que no se trata de plantear su reestructuración, sino la de proponer su reconstrucción. Este nuevo Estado tendrá que jugar un rol central en la erradicación de la desigualdad y la pobreza (habrá, en consecuencia, que fortalecer su función redistribuidora del ingreso); tendrá que fortalecer su función regulatoria de la economía, así como su función negociadora de los intereses nacionales tanto en el sistema internacional como frente a las empresas transnacionales presentes en el país.

## 5. Algunas Tendencias de la Coyuntura Actual

El análisis antes descrito considero es el telón de fondo que permite, probablemente, entender lo que hoy acontece en Bolivia y en muchos países de América Latina donde surgen vientos de cambio sin saber a ciencia cierta cuál será su desemboque final. Enumeremos algunas tendencias:

1. El péndulo de la historia pasa de la aplicación de un modelo neo-liberal a otro de carácter socialista con fuertes rasgos nacionalistas e inclusive indigenistas. Ahí tenemos por el lado de la izquierda a países que van desde Venezuela, Ecuador, Bolivia y a una tendencia más moderada como Argentina, Brasil y Uruguay.

2. Se desencadenan procesos de Nacionalización de la explotación, comercialización y transformación de los recursos naturales que vuelven, en lo fundamental, a manos del Estado. De igual manera el rubro de las Comunicaciones.

3. Para atacar la pobreza, en particular del sector agropecuario campesino, se inician procesos de redistribución de la Tierra.

4. Se amplia la cobertura de la educación básica y técnica y de la atención a la salud. Se pone particular énfasis en la lucha contra el analfabetismo.

5. Se realizan esfuerzos significativos para redefinir y fortalecer el proceso de integración sudamericana articulando a los países que formaban parte de la Comunidad Andina de Naciones con el MERCOSUR. En ese marco:

    - Se crea la Corporación Energética en base a países productores de Hidrocarburos, como Venezuela, Bolivia, Argentina y Brasil, de manera que el petróleo y el gas sirvan para el desarrollo e integración de América del Sur y de los países del Caribe.
    - Con recursos financieros de los países anteriormente citados, se crea el Banco Sudamericano que permita liberarse, a los países del área, de las ataduras y condicionamientos del Banco Mundial y en particular del Fondo Monetario Internacional.

6. En el ámbito político se asiste al desmoronamiento de los Partidos tradicionales que sustentaron el modelo neo-liberal a la par del surgimiento de movimientos sociales y desde ellos nuevos lideratos como el de Evo Morales en Bolivia, Chávez en Venezuela, Correa en Ecuador, Lula en Brasil.

7. En países con alta densidad demográfica indígena, se tiende a revalorizar el carácter multicultural y pluriétnico de las sociedades.

8. Se convoca a referendums y elecciones (Bolivia, Venezuela, Ecuador) para la formación de Asambleas Constituyentes que redacten una nueva Carta Magna en la que se asienten los principios de la nueva organización política, económica y social de nuestras sociedades.

Todo ello, sin embargo, no sin contradicciones, enfrentamientos y tendencias todavía inciertas. Una de las cuales, probablemente la más controvertible y peligrosa, es la del totalitarismo que aparece como la reacción de un pueblo contra el liberalismo democrático, contra las Instituciones democráticas, teóricamente causantes del marasmo económico, de la pobreza, de la exclusión, de la dependencia etc., es decir, de todos los males. En consecuencia se trataría de salvar a la Nación en peligro derrotando a los enemigos Externos e Internos a través del reforzamiento del Estado a precio del copamiento político de sus Instituciones en beneficio del Poder Ejecutivo y coartando la libertad económica, política y social. Tarea que inicialmente es encomendada a un Jefe apoyado, en sus inicios, por un grupo dinámico para finalmente desembocar en un partido Único. De consolidarse esta tendencia, podemos prever días obscuros para Bolivia y los países que están en esta dinámica.

Carlos Quiroga Blanco (*left*) and John Beckman, translating.

## 1. Introduction

At the end of the 1980s the social, political, and economic outlook for many South American countries was gloomy. They were emerging from constitutions torn apart by military dictatorships. Their national economies seemed to be out of control.

In spite of this, Bolivia (along with some of the other countries in the area) found the strength to struggle to its feet:

- It re-established democratic institutions,
- It took control of its own economy again,
- After numerous social revolts and coups d'état, it started to erase its previous image of an ungovernable country.

It was at that point (1985) when in Washington, in quiet offices in certain international organizations, people such as Hayeck, Friedman, officials of the International Monetary Fund, the World Bank, and other experts concluded that the model of development and performance of the economies of Latin America for the previous half century was now exhausted. The conclusions reached at that time came to be referred to as the Washington Consensus. The published report traced out the conceptual and practical framework of the reforms which would make the continent tremble for the following 20 years and which is still producing convulsions. This report set the line which is referred to as neo-liberal, a doctrine which achieved the status of a revealed truth, modified and controlled by those who formulated it. It would never have attained the significance it did, had it not been blessed by the world's leading power and by the powerful institutions of the international system.

I will try, as simply as possible, to set out the steps which, according to this doctrine, would need to be followed, in a developing country, in order to reach the promised land. The first generation has to concentrate on the need to attain macroeconomic equilibrium, applying fiscal discipline and control of public spending. These measures would permit the countries concerned, among other things, to rein in and get rid of the galloping inflation of those times. The second step: for the ideologues of neo-liberalism the State was the problem, the source of all evil—of inflation, bureaucracy, inefficiency, protectionism, and corruption. As a result it was necessary to destroy it, or at least to weaken and dismantle it. To accomplish this the prime role in the economy had to be shifted to the private sector. Nobody need worry about this, as the hidden hand of the market would take care to put everything in its rightful place. The free market, then, the interplay of the laws of supply and demand, was to take the place of the State. To this end, among the measures proposed, were the privatization of public companies and of pension systems, the liberalization of the financial system, and free trade in importing and exporting. Finally came the third step, which was related to the political organization of societies previously trampled underfoot by dictatorial regimes. This was the reestablishment of the rule of law; democracy and human rights were to return to the political agenda.

As sufficient time has passed for these measures to have been put into practice, in an attempt to explain better what is happening today in many Latin

American countries, I will make a synthetic evaluation of the positive and negative aspects of the changes which have occurred. Before I do I must bring out an obvious point: circumstances have imposed so many modifications on the original doctrine that by now any similarity between reality and the Washington Consensus is pure coincidence.

## 2. Economic Reforms

1. There can be no doubt that the economic model did bring inflation under control. Nowadays inflation in South America is like a cat purring by the fire, as opposed to its previous status as an uncontrolled tiger. The inflation rate over the whole continent was reduced from 390% in 1994 to 26% in 1995, to 22% in 1996, and today it varies between 2% and 5%.

2. The pressure of taxation contributed in large measure to a reduction in the national debt, and to macroeconomic equilibrium.

3. Access to the world's money markets was made easier, and the rate of exports from Latin America enjoyed a notable increase, from 146 000 million dollars in 1992 to 245 000 million dollars in 1995.

   Supporters of the model base their optimism on these figures. However, the behavior of other macroeconomic variables such as employment, savings, and the distribution of income in no way justify their euphoria. I will mention only a small number of these parameters.

   (a) The deterioration in the labour market. The rate of declared urban unemployment in Bolivia went from 7% in 1992 to 9.5% in 1995 and reached 10% in 1996; today it varies between 15% and 20%. At the same time the problem of underemployment worsened, as can be seen for Argentina and Colombia, where the rate of underemployment is around 13%.

   (b) It is impossible not to see that the positive results of applying the model have benefitted only the few, and that the negative results affect most of all the least favored sectors of the population. At the beginning of the nineties around 200 million Latin Americans (some 46%) were not in a position to satisfy their basic needs, and of these 100 million were in extreme poverty. Nobody really disputes the effects of the "free" economy in fact. What is really frightening is not the fact that one section of humanity lives better than another—which in any case is nothing new—but the outrageous and offensive magnitude of the gap. In Bolivia the average income of an inhabitant of the city of Santa Cruz de la Sierra is two and a half times that of an inhabitant of Potosí, and the most favored 10% of the urban population earns 35.4% of the total national income, while the bottom 40% earns only 15% of this income. On a world scale the situation is even more shocking. We all know that the wealth owned by (the richest) 360 people is equal to the annual income of more than two thousand million of the world's poor. Of course poverty has very deep and ancient roots and has not been created by neo-liberal reforms. But there is far more than sufficient evidence that the free market economy in Latin America has improved the living conditions of those with higher incomes but has not managed to resolve the problems of the least favored groups. In some cases it made the situation worse, reduced their wages, and deepened their degree of poverty.

People no longer believe that macroeconomic stability ensures economic security. They know from direct experience that the initial results of macroeconomic stability—unemployment, loss of purchasing power, more short term contracts—have given rise to greater insecurity. The "stability" is not convincing to them because they don't find it in their work, or their health care, or in their children's education. One can say without any exaggeration that the social instability that is at present rife has been one of the first fruits of the macroeconomic stability, and this explains the general air of uncertainty and insecurity in the whole of Latin America.

(c) There is something else. I will discuss the external vulnerability of the Latin American countries due to the financial liberalization imposed by the model. The International Monetary Fund and the World Bank predicate their support to developing countries on their opening up their financial markets very rapidly. Against this recent threat there are as yet no antidotes. The scale of the capital market defies human imagination and is beyond the ability of governments to control. Transactions across the world total more than 1.5 trillion (million million) dollars per day. The placement by pension funds, mutual funds, and those of insurance companies in the world money market is of order 25 trillion dollars. This capital market has largely substituted the older systems of finance for the developing countries, which were based on direct contributions by the governments of the industrialized countries. 99% of the credits and investments in the most important developing markets (Mexico, Brazil, Argentina) come from the private credit market. As a result interest rates, exchange rates, international competitiveness, the growth rates of PIB, to say nothing of the personal happiness of the people in the countries concerned, depend nowadays on the freedom or the restriction of this flow. If it remains continuous, all's well. If it is suddenly interrupted the consequences are disastrous. The biggest speculators decide where, how much, and exactly at what moment to invest, or to flee the market. On their goodwill and nerve depends the destiny of very large numbers of people. This has caused an alarming degree of volatility in capital markets, and dangerous instability in the financial systems of developing countries. In 1998 the banks withdrew 150 thousand million dollars from developing markets, which caused the massive crisis in Brazil, in Russia, and in southeast Asia, and gave rise to a terrible recession in Latin America. The global and recurrent nature of financial crises is well known. They affect everyone, and each time they occur they produce increasingly serious consequences.

### 3. Social and Institutional Reforms

There are a number of evident features to discuss here. There is no doubt, for example, that democracy has arrived and is here to stay. The level of popular support for democracy varies among the countries in the region, but in opinion polls it always receives more than 75% approval. The rights to free speech and association of citizens are respected, and large scale, systematic violations of human rights are a thing of the past.

However, the polls in Latin America gave warnings that all was not well. In their responses the people gave an impression of worry and lack of security. Neither the rulers in Latin America nor the experts in the international organizations attributed much importance to these signs. The early warning systems were not connected. These feelings will go away, they said. The basic economic parameters are in order, and it is just a question of time. Building a few schools here and a hospital or two there would be advisable, but nothing more.

Unfortunately things were not like that. The societies opted to take to the streets to show their discontent: to remind the rich and powerful, and the experts, that they could dirty the lounge carpet, insult the mistress of the house, and raise a shindig. This occurred in San José, Costa Rica, in Quito, Ecuador, in Bolivia in the cities of Cochabamba and La Paz, and in Brazil. The democratic framework so recently set up began to creak. Charismatic authoritarian leaders began to emerge from nowhere. Revolutionaries without revolutions tried to change the equilibrium of the Universe.

If the truth be told, the popular marches in Latin America, mostly by groups of native or mixed races, bear considerable relation to those of the blond groups in Davos, Seattle, or Washington. They are voices shouting the same message. They are denouncing the fact that macroeconomic equilibrium, fiscal discipline, and the growth of gross national products are not enough. That people are not willing to tolerate the enormous accumulation of wealth on one side of the scales, and the persistence of misery on the other. But they have an additional significance. Democracy, at least in the institutional form which it takes in Latin America, is not sufficient either. Perhaps because the citizens know that the democratic institutions are dominated by a new breed, the so called "political class". That the political parties which have a monopoly of delegated popular power represent the population less and less. People wish to participate more directly. What is the value of being able to express my view once every four or five years via the voice of a parliamentarian whom I have never actually met, if I can speak effectively here and now, in the street?

The fact is that the old system in which the State mediates with society via a single channel, that of the political parties, is out of date. The parties are now seen by people not as bridges between civil society and the State, but as impenetrable walls. The sensation is growing, and becoming virtually a certainty, that the political parties are much more interested in the State than in society, as if all their attention were centered on *how* to achieve power rather than asking themselves *why*. As a result, society is moving away from the parties and is looking for its own means of relating to the State. At the same time the parties are shutting themselves off and are setting themselves up as a class apart. This political class tends to have as its main objectives maintaining itself in power and even the personal enrichment of its members. Corruption in this context is not only a breach of the ethical principles of politics, but the direct cause of backwardness and social conflict. Society, organized outside the ambit of the parties, wants to be heard, and to sit at the decision making table. NGOs, young people, women, ecologists, the indigenous population, and social movements in general see the truth in their different ways and want to express it. The dialogue between the State and the market, between Government and business is not enough. What is at stake today is the transition from a formal, representative, and delegated democracy to a participative democracy, in which the political social and economic power is democratized.

Administrative and political decentralization is the first answer to this set of problems. But even adopting these kinds of measures, the planet, far from becoming a Global Village (to use Marshall McLuhan's well known expression) is tending more and more to be an archipelago with thousands of villages, each trying to reaffirm a cultural, administrative, political, and even historical identity for its own parish.

## 4. The Reform of the State

The reform of the State was essentially the backbone of the neo-liberal reforms. It is true that the State had in many places become something of a monster. The majority of the negative descriptions given were true: the bureaucracy, the inefficiency, the power of the lobbies, the corruption, etc. As a result, it was dismantled, since indeed it had been behaving like a monster. The active role in the economy was transferred to the private sector. However, the process of privatization, not untainted by the very corruption it sought to exorcise, enhanced the political power of the businessmen and associated economic groups by a large factor. They accumulated power, particularly in the hands of the large foreign corporations, which bought the public companies in the sectors of energy, transport, and telecommunications.

But it turns out that now the problems that we must solve are not those that were attributed to the action of the State, but that due to the reforms themselves, to the absence of a State. As a result some Latin American countries have set about redefining the role of the State. This is no easy task, given the need to do it within a globalized economy, in a knowledge-based society, and in a region undergoing integration. The State must, without doubt, be different in kind, based on different rules from those previously in force. We can see that it is not a question of restructuring it, but of reconstructing it. This new State will need to play a major role in the elimination of inequality and poverty (it will, necessarily, need to strengthen its regulatory role in the redistribution of income). It will need to strengthen all its regulatory functions within the economy, as well as assume a negotiating role in favor of national interests both within the international system and face to face with the major transnational companies that operate within a given country.

## 5. Some Tendencies at the Present Juncture

The analysis above I consider a background which allows us to understand, hopefully, what is going on today in Bolivia and in many Latin American countries, where winds of change are blowing without our clear understanding of where they will take us. Let us give the details of some of these trends:

1. The historical pendulum has swung from the application of a neo-liberal model to a model with socialist characteristics but with strong nationalist, even indigenist, features. On the left of the spectrum we have countries such as Venezuela, Ecuador, and Bolivia, and more moderate versions are being applied in Argentina, Brazil, and Uruguay.

2. Processes of nationalization of the exploitation, commercialization and transformation of the natural resources of countries have been set in motion; these are being returned, as far as basic principles go, into State hands. Similar plans are afoot for communications.

3. In order to tackle poverty, particularly among the poor rural farming communities, processes of redistributing land have been initiated.

4. The systems of basic and technical education are being opened to a wider cross-section of the community, and more resources are being devoted to health care. Stress is being placed on the struggle to combat illiteracy.

5. Significant efforts are being made to redefine and strengthen the process of South American integration, linking the countries which formed the Andean Community of Nations with MERCOSUR. In this framework:

    - The *Corporación Energética* was created, by those countries which are hydrocarbon producers, such as Venezuela, Bolivia, Argentina, and Brazil, so that the oil and gas can be used for the development of the South American and the Caribbean countries.
    - With funds from the above countries the *Banco Sudamericano* was created, with the aim of liberating the countries in the zone from the ties and conditions of the World Bank and especially the International Monetary Fund.

6. In the political sphere we see the collapse of the traditional parties which sustained the neo-liberal model and at the same time the rise of social movements, and from these new leadership such as that of Evo Morales in Bolivia, Chávez in Venezuela, Correa in Ecuador, and Lula in Brazil.

7. In countries with high concentrations of indigenous populations, the multicultural and multi-ethnic nature of their societies has become valued more positively.

8. Referendums and elections are called (Bolivia, Venezuela, Ecuador) in order to form Constituent Assemblies whose purpose is to produce a new Magna Carta, on which to base the principles of a new organization of the politics, the economies and the societies of the countries concerned.

All of this, however, has not occurred without contradictions, confrontations, and many trends with uncertain outcome. One of these, probably the most negative and dangerous, is that of totalitarianism, which emerges as the reaction of a people against democratic liberalism, and against democratic institutions, which are theoretically the cause of the economic swamp, of poverty, of social exclusion and dependence, etc., in other words the cause of all evil. As a result the attempt is made to save the nation perceived to be in danger by defeating its internal and external enemies by reinforcing the State at the price of limiting the political power of its institutions and by implication strengthening the power of the executive. At the same time economic, political, and social liberties are curtailed. This task is initially in the hands of a Leader, supported at first by a very dynamic group, but eventually coalescing into single party rule. If this tendency is consolidated we can foresee dark days ahead for Bolivia and for those countries which are going along this path.

From left to right: Julia Trenado, Rosi Rodríguez, Valeria Buenrostro, and Carlos Quiroga Blanco.

*Pathways through an Eclectic Universe*
*ASP Conference Series, Vol. 390, © 2008*
*J. H. Knapen, T. J. Mahoney, and A. Vazdekis, eds.*

# Biogas at 4000 m above Sea Level: The pioneering work of *Tecnologías en Desarrollo*

John E. Beckman

*Instituto de Astrofísica de Canarias, E-38200 La Laguna, Spain*

Oliver Campero

*Tecnologías en Desarrollo, Bolivia*

**Abstract.** In this article we describe a series of projects designed to supply families in rural Bolivia with heat for their cooking needs using simple technology to derive biogas from a mixture of animal manure and water. The novelty is that the adaptation of the anaerobic fermentation process to the climatic conditions at 4000 m above sea level has, for the first time, yielded reliable systems that can be maintained by their users. In addition, the non-governmental organization *Tecnologías en Desarrollo* has implemented latrines which act as auxiliary supply for the biogenerators, and solar heated showers, all of which go to augmenting the quality of life in the arduous surroundings of the Bolivian *Altiplano*.

Bolivia is a country of contrasts. The most obvious and striking contrast is between the *Altiplano*, the Andean high altitude section of the country to the west, and the jungle lowlands of the Amazon basin to the east. No less important is the contrast between the richest 5% of the population and the poorest 50%. Bolivia is the South American country with the lowest GNP per head, and the lowest investment per head in infrastructure, while a small group enjoys a standard of living which would be the envy of any European country.

*Tecnologías en Desarrollo* (TED) is a small non-governmental organization (NGO), founded in the central Bolivian city of Cochabamba in 2001, whose aim is to make a contribution to offsetting this imbalance by bringing technical solutions to the infrastructure deficit where they are most desperately needed: in rural Bolivia, where the poverty is greatest. As a result of a series of studies carried out by students of the Universidad Mayor de San Simón (the publicly funded university in Cochabamba) during their Master's degree course (while on visits to Tenerife funded by the Canary Regional Government), the participants in TED decided that the most helpful and cost-efficient way to make a real difference to domestic energy resources in rural Bolivia would be via the introduction of biogas generators. As a result of this decision, over 1000 people are now benefiting from their efforts.

A domestic biogas generator can be a reliable and constant source of cooking and lighting energy for any family that owns at least three cows or ten pigs. The system consists of an inflatable tubular polythene chamber, of around 1 m in diameter by 5 or 6 m long. An initial input of some 2000 liters of water and about one sixth of this volume of cow or pig manure is allowed to ferment in this chamber. Within two or three weeks (this time is temperature-dependent) a steady flow of biogas is emitted, and this flow can be maintained indefinitely

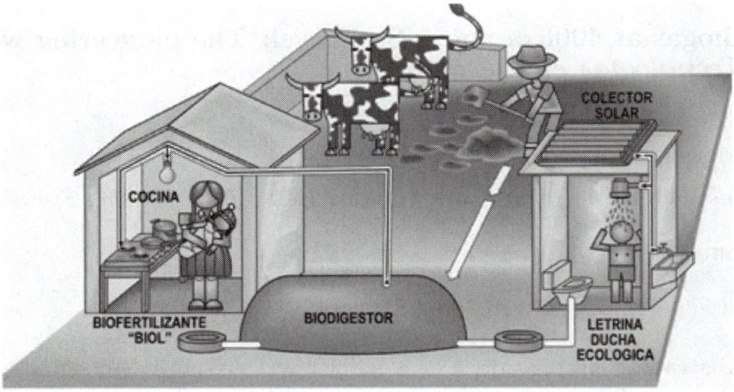


Figure 1. Schematic representation of a house equipped with the renewable energy installations supplied by TED: a biogas generator plus latrine, feeding a gas stove and light in the kitchen, and a thermal solar panel for showers.

Figure 2. An Aymara couple with a biogas stove installed in their kitchen in a dairy cooperative village, Achacachi, Bolivia.

by topping up the ferment with water and manure every couple of days, at a 5% level. Apart from the polythene chamber the rest of the supply circuit is a simple set of plastic tubes, bags, and valves, all made from materials easily acquired locally. The gas, whose active constituent is methane, is supplied to a pair of burner rings which act as a cooking stove, or to a gas mantle for lighting. A simple schematic of such a system is shown in Fig. 1, and a stove with its gas flame alight in Figure 2.

The first phase of the biogas project was carried out in a village in the province of Mizque, at 150 km from Cochabamba, and at 2000 m above sea level. Some 30 biogas generators were installed. As a pilot project, this had a number of problems, notably that the local technician we trained left the area, and that it was not practicable to keep maintenance staff available from Cochabamba. Five years later some 50% of these generators are still operating. Then came a series of diverse projects closer to Cochabamba, where TED was able to supply basic maintenance operations and resupply small parts. As a result, some 150 family generators are now operational in 15 small rural settlements. A novelty during this phase was the introduction of latrines, whose waste products feed into the existing generator, and it is now more common to supply a complete generator–latrine system than just a generator for a given family.

Figure 3. Covered adobe trench for a biogas generator at 4000 m above sea level, in Achacachi, showing the gas supply tube to the house. In the background the snowcapped Andes.

The most successful phase of TED operations to date, however, has been the installation of over 100 generators in the municipality of Achacachi, at 4000 m above sea level between La Paz and Lake Titicaca. The success is mainly due to the fact that the installations have been in small dairy cooperatives, where not only have the raw materials been available, but also a level of local technical support that has guaranteed the functioning of the generators. We at TED can claim a world first in building operational biogas plants at the 4000 m level. The positive results have been achieved by a very careful, though simple, thermal design. Black polythene and several plastic layers are used to maximize the greenhouse effect, and the generator is protected within an adobe walled trench (see Fig. 3). The generators produce gas even though the ambient temperature falls to ten degrees below zero. The combination of sun and the exothermic nature of the anaerobic digestion keeps the reaction going, and gas production is continuous and stable.

The advantages of single-family biogas generators are: (a) They avoid the need to search for wood, a task which can employ women and children up to several hours per day; (b) They save forests in the lowlands, or money in the *Altiplano*, where the alternative is bottled gas; (c) They avoid carcinogenic burning of wood or dung inside the house; (d) Especially combined with latrines, they improve family hygiene; (d) They yield, as a by-product, the liquid effluent, which is a clean, organic and flexibly efficient fertilizer ("biol"). They do not represent a high tech intrusion into the way of life of the Quechuas or Aymaras, and they are particularly welcomed by the women of the villages as a considerable aid in their daily lives.

In addition to the biogas and latrine installations, TED has developed low cost solar thermal panels to provide hot water for showers. This is particularly appreciated by the dairy farmers in Achacachi, for whom a hot shower was a luxury. We intend to develop these lines of renewable energies for small-scale rural use in Bolivia, extending our activities to cover the lowlands of eastern Bolivia around Santa Cruz de la Sierra. We feel that we have shown that the approach of introducing family-scale biogas has proved its worth and convinced some of those made sceptical by the failure of much larger scale projects in previous decades. The work will continue and is limited only by our ability to raise funding. The net cost of an installed biogas generator+latrine system is some $250 per family, and the generators have a minimum expected lifetime of 5 years. TED is working towards a scheme where self-financing replacement will be possible so that the impact on the local communities can be permanent.

*Pathways through an Eclectic Universe*
*ASP Conference Series, Vol. 390, © 2008*
*J. H. Knapen, T. J. Mahoney, and A. Vazdekis, eds.*

# Cosmology between Physics and Philosophy—Galileo to Einstein

Jochen Büttner and Jürgen Renn

*Max-Planck-Institut für Wissenschaftsgeschichte, Boltzmannstraße 22, D-14195 Berlin, Germany*

**Abstract.** The intention of the present paper is to illuminate, using the example of cosmology, some of the general processes and mechanisms involved in the development of knowledge. Rather than getting immersed in the details of the advancement of cosmological theories, we will concentrate instead on the wider context in which scientific knowledge develops. From its inception, cosmological knowledge in particular has been integrated into comprehensive worldviews. Such worldviews in turn imposed challenges on science and especially on cosmology, and, in a reciprocal development, scientific challenges were incorporated into the formation of comprehensive worldviews as presented in particular by religion. Besides this negotiation process with existing worldviews, technological challenges will be identified as another mainspring that, mediated by cultural contexts, contributed substantially to the development of cosmological knowledge.

This paper is based on research on the rise and fall of the mechanical worldview undertaken in recent years at the Max Planck Institute for the History of Science in Berlin. This research does not focus on the great heroes of the history of science but rather attempts to elaborate the general processes involved in the development of knowledge. It deals for instance with the relations between theoretical and practical knowledge or between individual insights and holistic worldviews. Even though we are focusing in this paper on two great scientists, Galileo and Einstein, we consider them as representing two important and incisive revolutions in cosmology and analyze their contributions against the background of our general approach.

The title of this paper 'Cosmology between Physics and Philosophy—Galileo to Einstein' requires further elucidation. "Philosophy" stands not for a specific academic discipline but for the wider contexts in which scientific knowledge and thus also cosmological knowledge in particular develop. We will discuss how within these contexts areas of conflict and new challenges arose that substantially shaped the development of cosmology. Central to our discussion will be the question of how comprehensive worldviews impose challenges for science and how in turn scientific challenges are incorporated in comprehensive worldviews as offered in particular by religion.

The very first cosmologies that emerged historically, mythical cosmologies, were faced with the challenge of unifying knowledge about nature and man in a convincing overall picture. Thus, the myths of creation connect religious, social and cosmological concepts of order. This connection is based not least on the ancient role of astronomical observations for agricultural planning. In the empires of antiquity such myths supported the cohesion of society. The

Babylonian myth of creation, for example, describes the emergence of spatial and temporal order from chaos. After his victory over chaos Marduk, the local god of Babylon, divided the year into twelve months and established the constellation of the stars. For the Egyptians, the world was a flat disk surrounded with water over which was arched the star-spangled body of the goddess Nut. And according to the Chinese philosophy of elements human concerns obey the same rules as the processes of nature.

Scientific cosmologies develop, as we will show in the following, against the background of the connections between different areas of knowledge and experience. Until the present day cosmology has thrived on the dialectics of knowledge gained on the one hand from terrestrial experience and on the other hand from observations of the heavens. The development of cosmology is characterized not only by the accumulation of knowledge in the respective areas in the course of history, but also by the development of structures and mechanisms serving either to separate different areas of knowledge or to tie them more closely together. The interaction of areas of knowledge within cosmology is shaped by a double dynamics. New insights into terrestrial phenomena, which can have a multiplicity of different sources, lead to changes in the perception of celestial phenomena. In turn, altered cosmological conceptions can have repercussions on theories about terrestrial phenomena that originally substantiated them. In our opinion, the history of cosmology can be written largely as the history of such interactions and their mechanisms. Without such mechanisms it would in fact ultimately remain incomprehensible—or could merely be explained by the intervention of an ominous zeitgeist—that changing external conditions of science, such as the technological upheaval of Galileo's time or later the industrial revolution in Einstein's era, could have consequences for the progress of cosmology.

In Galileo's as well as in Einstein's time, certain barriers, in each case of a historically different nature, impeded a direct transfer between different areas of knowledge. However, in both cases it was basic concepts of physical knowledge, which could not be limited to a particular domain such as the concepts of space and time, motion and inertia, that entailed a bridging of the areas. Furthermore, the comprehensive worldviews, against the background of which physical knowledge developed, almost inevitably enforced a thorough exploration of the cosmological implications of individual insights. In the historical situations under consideration these mechanisms led ultimately to a replacement of the established worldviews along with their associated cosmologies. Attempts to confine and control the radical consequences of new theories are part of the complex mechanisms of such revolutionary processes. However, avoidance strategies as a rule turned out to be rather unproductive.

The cosmological revolution of early modern times is often referred to as the "Copernican turn," implying that Copernicus' new astronomy of 1543, which placed the Sun rather than the Earth at the center of the Universe, established a new worldview that abruptly superseded the then dominant Aristotelian worldview. In reality though only a few scholars were initially involved with this alleged revolution and used Copernicus' new astronomy mostly as a simplified method to determine planetary configurations. The new knowledge of nature that accumulated in early modern times in fact only gradually entered into the inevitable conflict with the predominant worldview of the time. The one long revolution of early modern times began with Copernicus, came to a climax of conflict between science and religious dogma with Galileo, and reached a tentative conclusion with Newtonian physics and is here interpreted as the result of processes of restructuring knowledge.

That terrestrial and celestial physics belong to different realms and obey different laws is one of the basic tenets of the Aristotelian worldview. Thus the transfer of experience with terrestrial phenomena to the sphere of celestial bodies seems to be ruled out from the outset. The Aristotelian worldview describes the motions of the heavenly bodies by a model of homocentric spheres on which the heavenly bodies are fixed and at whose center the Earth is located. Each of these spheres has its own motion and simultaneously takes on the motion of the sphere that encloses it. This mechanism was supposed to explain the observed motion of the celestial bodies, in particular of the planets. Starting already in antiquity this model was considerably differentiated over time, for example, by introducing the assumption of epicycles in order to produce as exact an explanation and prediction of astronomical phenomena as possible. Mechanical models produced to instantiate the Aristotelian worldview manifest that this worldview was founded on fundamental concepts of motion based on elementary experiences. Hence, contrary to the theoretically perceived division, terrestrial and celestial phenomena obey similar principles, revealing the ultimately terrestrial origins of cosmology.

Until early modern times the Aristotelian worldview had developed into much more than a comprehensive philosophical system. Since Thomas Aquinas at the latest it had been transformed into the essential framework for any interpretation of the world on the basis of the Christian doctrine and as such allowed, for example, the explanation of multifaceted everyday experiences in line with an intuitively plausible physics as well as the explanation of subtle dogmas such as Eucharistic transubstantiation. Had Christianity merely retained its earlier naive conceptions of God and the world, instead of promoting such an expansion, diversification, and refinement of the basic principles of knowledge, it could hardly have been in a position not only to assert its message of salvation, but also to establish its worldly preeminence and intellectual dominance so effectively for such a long time. The success of the synthesis of biblical transmission and the Aristotelian worldview is perhaps nowhere as apparent as in the identification of the Christian God of creation with the Aristotelian Unmoved Mover who is responsible for the motion of the celestial spheres.

Against the background of this successful synthesis, which was of major relevance for legitimizing the Christian dogma, it seems that attempts at a modification, such as that of Copernicus, necessarily had to challenge the underlying worldview as a whole and at the same time those who drew their power from it. However, this kind of assessment underestimates the manifold mechanisms of adaptation as well as possible strategies for avoiding conflict, by means of which such commotions can be suppressed. Copernicus' theory was thus indeed not at first perceived to be heretical, and his work *De Revolutionibus*, in which he had presented this theory, was first put on the index of forbidden books by the Pope over 60 years after its publication. In fact Copernicus, himself deeply religious, had reorganized the mathematical description of the celestial motions—but in doing so had left other aspects of the prevailing worldview, for example the crystalline spheres as agents of motion, practically untouched.

Galileo is commonly seen as the figure whose heroic courage and scientific genius conferred a physical foundation to the Copernican system and who, with his astronomical discoveries with the aid of the telescope, lent this system a power of persuasion unimaginable until then, thus inevitably triggering the conflict with the Church. The Copernican system indeed confronted physics with new challenges. Why don't we actually notice the Earth moving? Galileo answered by pointing out the difference between absolute and relative motion. Relative to the Sun we move rapidly through space, but relative to the Earth we are motionless. The laws of terrestrial physics remain practically unaffected by the motion of the Earth with everything on it around the Sun. From this insight the relativity principle of classical physics eventually emerged. Can this principle be expanded to accelerated motion? We will return to this question with Einstein. Yet, the derivation of a new physics was not possible solely from the insight into the relativity of motion, let alone the establishment of a physical foundation of the Copernican system. The new cosmology that Newton eventually established at the end of the long revolution of the early modern period rather drew on additional knowledge sources of a different, terrestrial nature connected with the technological challenges of the time.

In his *Discorsi* Galileo himself identified such technological challenges of this kind as the prime source of both the new sciences and presented these in this masterpiece. In the opening passage of his work he thanked the naval engineers of the Venetian Arsenal, which may be referred to as the largest military–industrial complex of the time, for having furthered the emergence of his new theories enormously by contributing their practical knowledge. Galileo's assertion has been long dismissed as a reflexive flourish in the style of the time. Scientists at that time felt the need to communicate immediately the practical consequences of their insights in order to enjoy as it were the corresponding advancement. For this reason as well Galileo's statement has been dismissed as scientific propaganda. Yet we have pursued this matter and have been able to ascertain that Galileo was not only in close contact with engineers from the Arsenal and called in by them to give his expert advice, but that he also in turn profited from their expertise when creating his new theory of resistance to fracture of materials.

With the advent of artillery in the late Middle Ages and its rapid development since then a further challenge arose for the engineers of early modern times. The first cannons, so-called siege engines, were machines of gigantic proportions that had to be brought as near as possible to the enemy's fortification and could be fired only once with either the fortification or the cannon itself being blown asunder. However, the range of artillery soon increased and thus questions of ballistics became more and more relevant.

The problems of ballistics were also taken up by Galileo, who was the first to provide an explanation for the parabolic form of projectile trajectories. As an example of his engagement with the problems of ballistics, we would like to outline an experiment that Galileo carried out on questions of projectile motion, and which has been handed down in the form of an experimental record preserved on a folio of his *Notes on Motion*. These notes are part of the manuscript bundle Ms. Gal. 72 preserved at the Biblioteca Nazionale Centrale in Florence. As

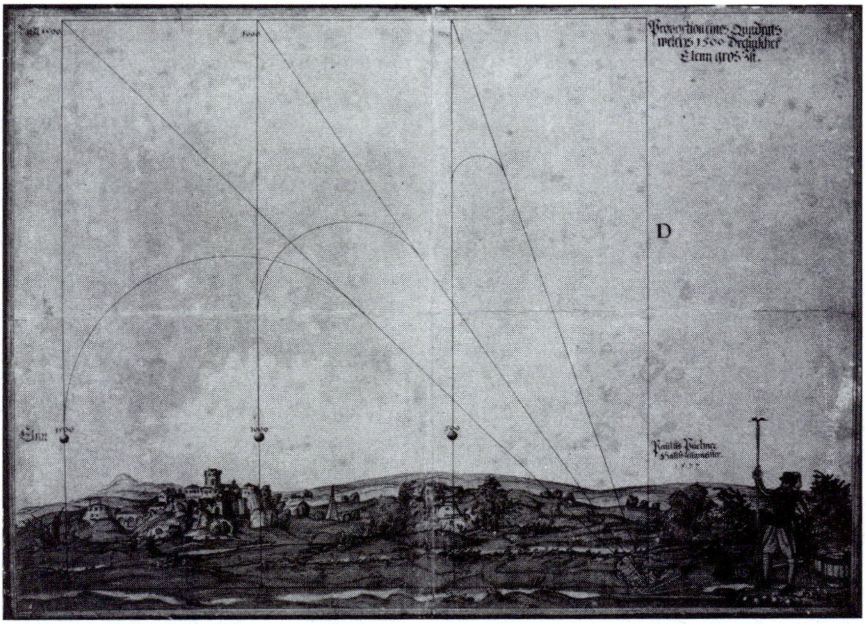


Figure 1. Geometrical construction of artillery trajectories. (*Mathematisch-Physikalischer Salon, Staatliche Kunstsammlungen Dresden.*)

emerges from the diagram and the calculations in this folio, Galileo's experiment consists in determining the range of horizontal shots from the same height and with varying initial speeds. In order to achieve this, Galileo first of all let a ball drop from different heights along an inclined plane and subsequently redirected this accelerated downward motion into a horizontal motion of uniform speed, for example, along a tabletop. At the edge of the table the ball was projected horizontally and fell along a parabola to the ground. The greater the initial speed of the projection, i.e., indirectly the higher the starting point of the ball on the inclined plane, the further away the point of impact was from the edge of the table.

A basic conceptual and practical asset of the experiment is the deflection of a motion of fall into a horizontal motion. This deflection model comprises several aspects: first it is assumed that the increase in speed during the motion of fall corresponds to Galileo's special principle of fall; second it is assumed that the speed of motion thus attained remains constant after being redirected; and third it is postulated, assuming a restricted principle of inertia, that the horizontal motion—free of external obstacles—maintains its initial speed, that is, continues uniformly. What does Galileo's theory of projectiles and in particular the experiment just outlined have to do with cosmology? As we shall see, Galileo developed his own cosmology on the basis of exactly the deflection model on which his theory of projectile motion was founded.

Galileo's attempt at a cosmology, or more precisely at a cosmogony—a theory of the creation of the Universe, at that time still essentially limited to the

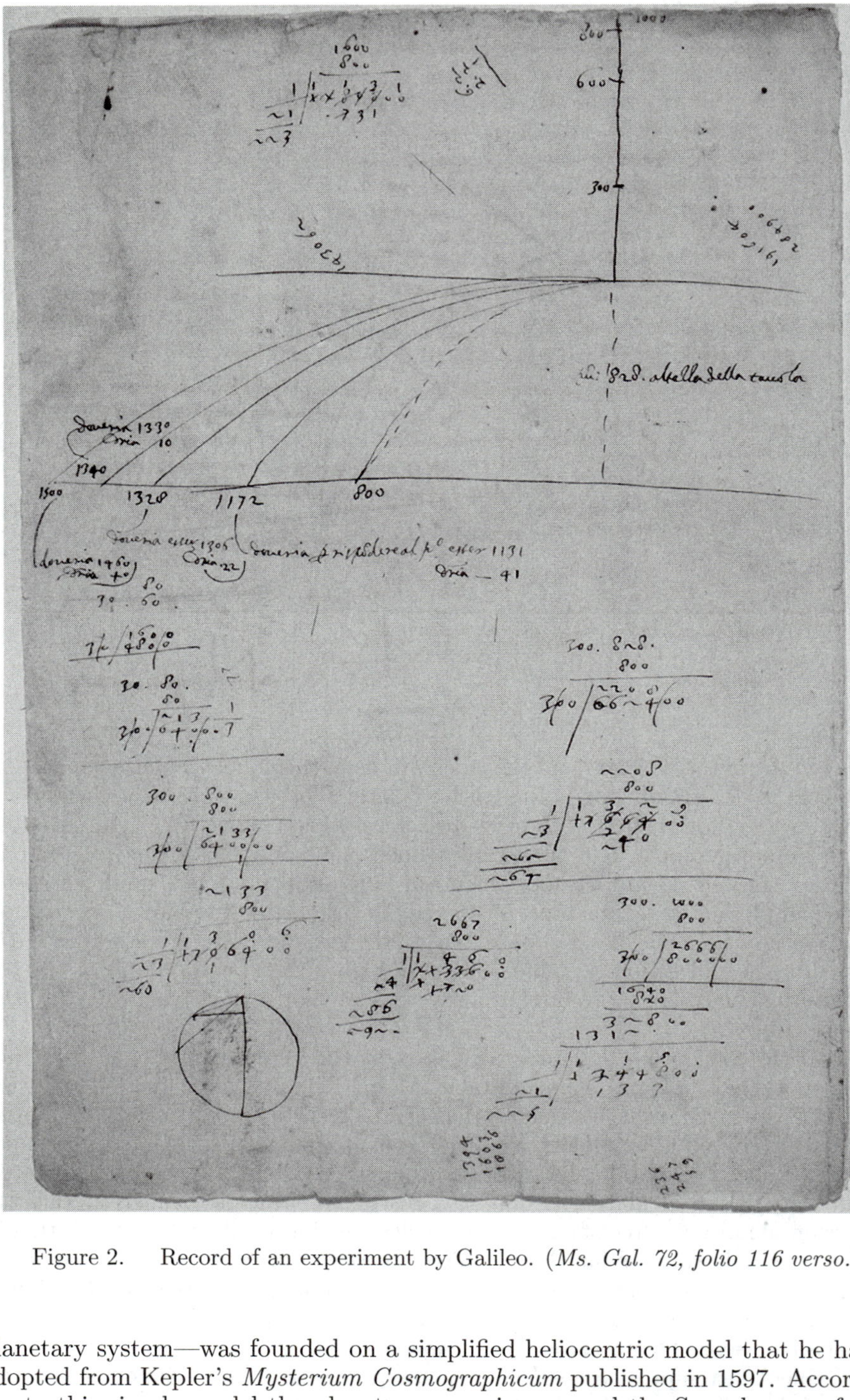


Figure 2. Record of an experiment by Galileo. (*Ms. Gal. 72, folio 116 verso.*)

planetary system—was founded on a simplified heliocentric model that he had adopted from Kepler's *Mysterium Cosmographicum* published in 1597. According to this simple model the planets are moving around the Sun along perfect circular paths. In addition Galileo took over the precise astronomical data, that is, the orbital radii and the periods of orbital motions of the planets given by Kepler. According to Galileo's cosmology, the motions of the planets resulted from the planets' falling through a particular distance to their orbits, where they were deflected and continued their motion uniformly along their circular paths thereby retaining their speeds. The explanatory value of Galileo's hypothesis lies in the assumption that a common starting point for the motion of fall of all planets can be ascertained so that his model would yield the exact orbital speeds of the planets given by Kepler.

The fundamental idea behind Galileo's cosmology thus indeed corresponds conceptually to the idea behind his experiment on projectile motion. Where in the case of the experiment the ball fell from different heights to one and the same path where it continued to move in uniform horizontal motion at different speeds, the planets according to his cosmology fell from one and the same height to different orbits where they continued to move in uniform circular motion at different speeds.

This example shows, as we have claimed above, that cosmological applications can indeed have repercussions on the underlying concepts of physics. It is in fact remarkable that Galileo's case, application of his deflection model in a cosmological context, led to a problematic generalization of the principle of inertia. Instead of generalizing the concept of inertia from uniform horizontal motion to straight motion of arbitrary direction Galileo erroneously assumed that circular motion may be perceived as an inertial motion.

Galileo presented his cosmological hypothesis in his *Dialogue* concerning the two chief world systems and once again after his condemnation by the Inquisition in the *Discorsi* and moreover alleged to have made calculations that manifest "wonderful agreement" between his hypothesis and the observations. Such calculations can indeed be identified in Galileo's *Notes on Motion.* Their interpretation shows that although Galileo using his hypothesis could not reproduce Kepler's data precisely, his calculations yielded a qualitative agreement with the astronomical observations such as greater orbital speeds for planets closer to the Sun. These qualitative agreements induced Galileo to publish his cosmological hypothesis, even though it was not fully elaborated.

In the published version Galileo's cosmology became a story about the creation of the world. Not only did he claim that the planets fell from the same place but he declared them to have been created there by the "divine architect" and to have been deflected through his "divine wisdom" into their respective paths. With this embedding of his cosmological hypothesis into religious patterns of explanation Galileo followed a pattern typical of his times, as he had already done in his early treatise *De Motu* in his description of his then still Aristotelian, geocentric-based worldview. In *De Motu* it was the divinus Opifex who from the excrement of the creation of the Universe created the Earth at its center. On the one hand, such an embedding of cosmology in religious contexts follows the established tradition according to which cosmology was an integral part of the ecclesiastic dogma. On the other hand, since the new worldviews emerging in the early modern times increasingly challenged the borders set by this dogma, the efforts to reconciliate the new insights with the prevailing religious views became ever more vital.

Just how closely associated and virtually inevitable scientific arguments and religious conflicts were at this time can be seen by the example of the unorthodox scientist Giordano Bruno who was burned at the stake as a heretic. This was also perceived by his contemporaries:

> Today he was led to the funeral pyre or stake. As the holy crucifix was held up in front of the already dying man he turned his head away with a contemptuous look and perished so wretchedly roasted, I believe in order to report in those other worlds invented by him how we Romans deal as with blasphemous and ungodly people. [...] Now I would like to hear if you must not approve such a treatment, or do you believe for instance that every man must be free to think and profess as he pleases? [...] But you may perhaps observe that the Lutherans do not teach and believe such things and therefore should be treated differently. In this I agree with you entirely and by no means do we burn a Lutheran. However we have a different opinion about your prophet Luther himself.

Given the limitations of the knowledge available in early modern times many possibilities existed to elaborate the increasing knowledge of astronomy together with theologically or philosophically based concepts to a comprehensive worldview. Kepler's insights into the laws of planetary motion are proof of the productivity of the integration of precise observations with religiously motivated philosophical speculations, even though such integrations, for instance the connection Kepler established between the structure of the planetary system and the alignment of Platonic bodies, may appear obscure from today's point of view.

At the same time, the dogmatic borders of the Church's official worldview suggested avoidance strategies that seemed apt to confine the revolutionary potential of the new knowledge. Such an avoidance strategy may by exemplified by the Tychonian world system proposed by the excellent astronomer Tycho Brahe, who recognized the strengths of the Copernican system but, on physical and theological grounds, favored a modified geocentric worldview. In contradistinction to the original model, only the Sun and the Moon circled the Earth while the other planets, as in the Copernican system, moved around the Sun. This system initially found many followers, particularly among the Jesuits as it seemed in an astute manner to unify modern astronomy with conventional beliefs. However, for the development of physics and its integration of astronomical insights this avoidance strategy remained fruitless.

Was it thus Galileo's courage that forced a paradigm shift and in the end made him a martyr for science? Widening the perspective to developments that were apparently irrelevant for cosmology, like those of practical and technical knowledge as well as the contributions of less prominent scientists, even those who served the Church, a different story emerges. The establishment of the worldview of classical physics was not the result of an erratic paradigm shift, but rather the final result of the long revolution of early modern times in which fundamental structures of knowledge as well as the relation between Church and science were changed.

From the time of the beginning of the mechanization of the worldview, which at the same time was the era of the great discoveries, we shall now turn to the final phase of its development, which was also the era in which colonialism and imperialism culminated. In the intervening time, the European nations had created networks of commerce, knowledge, beliefs and power across the globe, destroying and subduing other cultures. At the beginning, discoverers, commercial travelers and missionaries had penetrated the most distant corners of the Earth; they were soon followed by conquerors and scientists. With exploration and conquest scientific knowledge also grew, at least in part still embedded in a Christian–European worldview which, for one historical moment, may have appeared to offer (to the conquerors) a universal framework for any kind of knowledge.

Soon enough, however, the gaps between a rapidly developing science and the dogma of the Church, which had opened up in Galileo's time, precisely the gaps between ecclesiastic and political power, could no longer be bridged. At the end of the 19th century at the latest the forced marriage between European religion and science had definitively come to an end. Was the way finally free to transfer insights from terrestrial physics to cosmology, a transfer that in Galileo's time was, due to the blending of the Aristotelian natural philosophy with the Christian dogma into one comprehensive worldview, so arduous and risky, but at the same time so productive?

Through the formation of methodical and conceptual canons and a rapidly advancing specialization science had acquired a tremendously productive power. However, this progress had two sides. It was, on the one hand, the expression of an effective division of labor. Yet, on the other hand, it had become possible only at the expense of relinquishing an integration of the separate fragments of knowledge into a comprehensive scientific worldview. In the 19th century this integration rather became the domain of popular science and of individual idiosyncratic attempts to regain, on a philosophic basis, a universal worldview. Bridging the gaps between the individual areas of knowledge—in particular also between insights concerning the celestial domain and those concerning the terrestrial domain—was frequently due to mere chance or ingenious intuition, and was more often hindered than advanced by the new disciplinary order of knowledge.

Towards the end of the 19th century Karl Schwarzschild, one of the founding fathers of astrophysics, was among the few scientists who were actually successful in bridging gaps between the individual areas of knowledge. Can astronomical observations lead us to conclude that there is a possible curvature of space? This question had already been asked by Schwarzschild. He boldly speculated that non-Euclidean geometry could help to understand the structure of the cosmos. Like Copernicus, he advanced hypotheses about a new geometry of the cosmos, which, with the introduction of curved space, differed from the old geometry in a similarly radical way to how the Copernican system differed from the geocentric worldview.

Yet, from the insight that space may be curved it is not possible to derive a new physics, let alone to develop the physical foundation of a universal cosmology, which until today has been developed on the basis of Einstein's theory. The new cosmology was not only founded on the ingenious theoretical preparatory efforts of the creators of non-Euclidean geometry and of exceptional scientists such as Schwarzschild. It was also rooted in quite different, additional sources of knowledge, which—similar to the case of the long revolution of early modern times—were connected with the technical challenges of the time.

Since the mid-18th century at the latest the development of scientific knowledge on the powers of nature went hand in hand with their exploitation by rapidly developing industry. Above all it was the new technologies and the new practical knowledge associated with them that in these times confronted science with ever new challenges. Engineers concerned with research such as Sadie Carnot considered for instance the steam engine as a means for attaining knowledge. The steam engine thus was not only the motor for industrialization, as it has been referred to metaphorically, but also became a driving force for the production of scientific insights. Indeed thermodynamics, which emerged as a new sub-discipline of classical physics in the mid-19th century, can be traced back not least to the study of the steam engine. The flourishing of the chemical and the electrotechnical industries are other facets of the industrialization, which were closely connected as well with the further development of scientific knowledge.

As part of the scientific and technical infrastructure, public institutions were created such as the Physikalisch-Technische Reichsanstalt in Berlin, which endeavored to establish standards, for example, for the luminosity of light bulbs. It is here where modern considerations on the role of quantum fluctuations in the origin of the Universe ultimately have their roots, thus supporting our thesis of the terrestrial origins of cosmology. The light- and heat radiation of a light bulb posed a problem situated on the borderline between radiation theory and thermodynamics. Its solution led Max Planck to introduce, albeit involuntarily, the quantum hypothesis. The famous cosmic background radiation, a last faint echo of the Big Bang, in fact has the same properties as the heat radiation of the black body that in 1900 was measured accurately for the first time in a laboratory of the Physikalisch-Technische Reichsanstalt.

Relativity theory, the theoretical foundation of modern cosmology, owes its introduction not least to the epistemic challenges of the new technologies to which Einstein had already been exposed in his parent's electrotechnical factory. The new concepts of space and time introduced by the special theory of relativity of 1905 in fact ultimately emerged from Einstein's attempts to solve the problems of the electrodynamics of moving bodies. Such problems are posed where electromagnetic and mechanical phenomena such as motion come together, for example, in the case of the dynamo or in the case of the speculative question asked by the young Einstein of whether one can overtake a ray of light.

According to the special theory of relativity all physical processes, even those of electrodynamics, will take place in a laboratory at rest and in one moving uniformly in exactly the same way. But can this relativity principle also be applied to a laboratory that is accelerated? The falling elevator shown in Fig. 3 in fact represents such an accelerated laboratory. Evidently the physicist in the falling elevator on the right experiences weightlessness and his experiments proceed in exactly the same way as though he were in a laboratory at rest in space. He may speculate that he is really in such a laboratory at rest in space and he may be right. His colleague in the elevator on the left, on the other hand, who believes he is in an elevator at rest in which he senses gravity is possibly mistaken. In reality the force that causes his yoyo to fall may not be

Figure 3. Einstein's elevator thought experiment. (*Illustration by Laurent Taudin.*)

due to the action of gravitation, but to the fact that someone is pulling on the elevator's hoisting rope accelerating the elevator upwards. Gravity and inertial forces occurring due to accelerated motion are evidently related to each other.

Does the same hold true for the inertial forces that occur in rotational motions? Where does the force come from that swings the passenger outwards on a chairoplane? Isaac Newton posed the same question albeit with reference to the behavior of water in a rotating bucket. For Newton the deformation of the surface of the water in rotational motion was due to an inertial force. This suggested to him the existence of a mysterious absolute space with respect to which the bucket was turning. The concept of absolute space had a theological connotation for Newton representing the sensorium dei. But what could this theologically laden concept mean in the sober times of industrialization when not only theology but also mechanics had become questionable as the foundations of a worldview?

Can the action of a force be anything but the expression of an interaction between bodies that, unlike absolute space, can be sensed and empirically tested? This question was asked by the physicist and philosopher Ernst Mach, who did not want to accept Newton's metaphysical assumption of an absolute space. Newton's mysterious inertial forces with no perceptible cause should, according to Mach, be explicable as an interaction between bodies too. Could it be, therefore, that the inertial forces of a chairoplane actually derive from the action of the distant cosmic masses of stars with respect to which one is rotating? If so, a terrestrial phenomenon like the chairoplane inevitably takes on a cosmic meaning. We are thus reminded of Galileo's case in which a simple mechanical device, the deflection of the motion of fall into a horizontal projection, could take the cosmological meaning of a divine creation of a planetary system. In both cases a few basic concepts charged a terrestrial phenomenon with cosmic meaning. Only because Einstein took Mach's reflections on the underlying physical concepts seriously was he able to turn his insight into the essential connection between inertial forces and gravity—gained in the thought experiment with the elevator—into the foundation of a new cosmology.

On this basis Einstein worked out his new theory of gravitation, the general theory of relativity. After the insight into the essential connection between inertial forces and gravity of 1907 it took another eight years until in 1915 he was able to exploit this fundamental idea to develop a convincing theory of gravity. The terrestrial experiences embodied in the thought experiments with the elevator and bucket guided Einstein's arduous and sometimes dramatic search for equations to prove this theory, as is documented in his famous "Zurich Notebook", a research notebook documenting Einstein's groping for the new theory between 1912 and 1913.

He tested his hypotheses time and again using the two paradigmatic cases of rotation and linear acceleration. Yet, the elaboration of his physical considerations brought him to the limits of his mathematical capabilities, and progress was only possible through the help of his friend, the mathematician Marcel Grossmann. Some of the conceptual consequences of the mathematical elaboration of the new theory were so surprising and revolutionary that Einstein himself initially discarded the correct ansatz only to return to it years later.

Even after the completion of the theory in 1915, its cosmological embedding, which from today's viewpoint seems almost self-evident, and which, as we have seen, was already inherent in its fundamental ideas, remained at first only a philosophical—though consequential—footnote. Einstein would not abandon the Machian idea according to which inertial forces are the result of the interaction with cosmic masses. In 1917 this idea misled him into changing his original theory by introducing a cosmological constant. He did so in order to ensure that the world models derivable from the theory complied with his cosmological expectation of a static and closed Universe in which all inertia could be traced back to the interaction of matter. As in the case of Tycho Brahe, preconceived beliefs had thus led to the formation of a worldview skillfully adapted to these preconceptions.

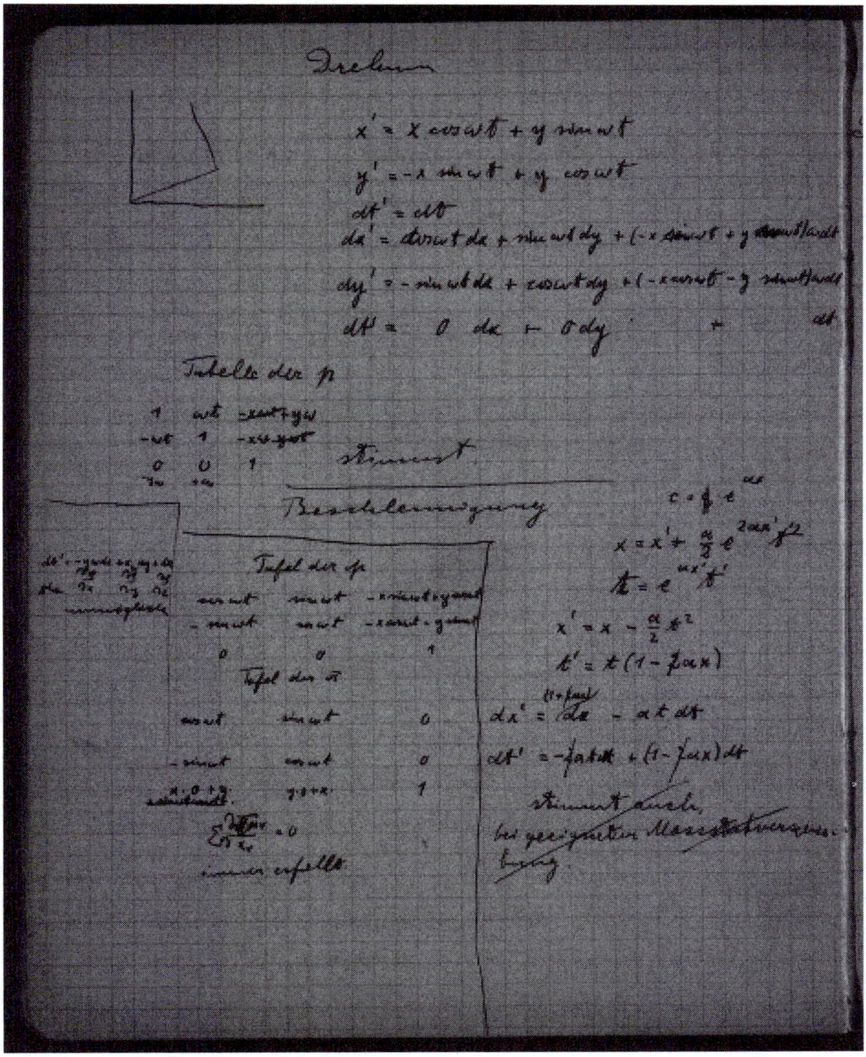

Figure 4. Rotation and linear acceleration in Einstein's "Zurich Notebook". (*Einstein Archives, Hebrew University Jerusalem; Item No. 3-006.*)

Einstein's belief in a static Universe was so firm that he erroneously considered the calculations presented in 1922 by the Russian mathematician Alexander Friedmann, which implied the possibility of a dynamic and expanding Universe, to be inaccurate. Even after Einstein had discovered his own mistake he did not want to ascribe a physical meaning to this possibility. Yet, the emergence of modern cosmology is in no way due solely to Einstein's ingenious theory and its elaboration. Without the advent of new astronomical insights the cosmological dimension of the theory would ultimately have remained an object of philosophical and mathematical speculation.

A first significant step towards an empirically based cosmology was made when the elaboration of dynamic world models started to correspond with new astronomical results, in particular the redshift of distant galaxies, which Hubble and his colleagues achieved at the end of the 1920s at the Mount Wilson Observatory in California. The cosmological upheaval of modern times must be considered to be the result of a long revolution that reached from the crisis of modern physics around 1900 to the establishment of the standard cosmological model in the 1960s. In a certain sense it remains incomplete even today. Although many of even the most basic aspects of modern cosmology remain contentious, such as the concept of dark energy and dark matter, the assumption of the Big Bang is one of the firm points of reference. It is therefore not surprising that precisely this aspect with its almost imposing parallels to the creation myths of the past occasionally became an object of ideological conflict.

Steven Hawking in 2006 told of an encounter with Pope John Paul II:

> At a conference on cosmology in the Vatican, the Pope told the delegates that it was OK to study the Universe after it began, but they should not inquire into the beginning itself, because that was the moment of creation, and the work of God. I was glad he didn't realize I had presented a paper at the conference, suggesting how the Universe began. I didn't fancy the thought of being handed over to the Inquisition, like Galileo.[1]

Is the era of Bruno and Galileo perhaps not really over, when scientists feared for their lives when crossing ideological borders? As far as cosmology is concerned Hawking's story seems to deal with an exception of anecdotal character. Today more open and tolerant relations prevail between cosmology and the Church. After all, one of the founders of modern cosmology, the physicist and priest Georges Lemaître, already warned the ecclesiastical authorities of his time not to draw impetuous dogmatic conclusions from scientific insights since such conclusions would be falsifiable. Yet the temptation to draw such conclusions of course still exists.

Indeed the peaceful coexistence of religion and science apparently does not just depend on the caution with which both sides evaluate scientific insights and their ideological or ethical implications. The conflict seems in fact to go much deeper. Influential religions do not seem to get by without interpretations of the world. They not only have to incorporate bodies of knowledge, but also have to embed them in relatively stable interpretative networks. These interpretative networks in turn are part of the processes of societal cohesion and consensus formation in which religion plays its stabilizing role, thus becoming fixed frames of reference. On the other hand science thrives on the dynamic creation of new knowledge, which inevitably challenges such reference frames time and again. At the same time the role science itself plays in processes of societal cohesion and consensus formation often competes with the role of religion in these processes.

---

[1]Cited from: Origins of the Universe: Stephen Hawking's J. Robert Oppenheimer Lecture. See http://www.berkeley.edu/news/media/releases/2007/03/16_hawking_text.shtml (10.10.2007).

The repression of conflicts between science and religion can often be more dangerous for both sides than attempts to resolve these conflicts constructively. This becomes particularly apparent in the problems arising from the increasing disparity between societies where science and technology shape the societal discourse, and societies where religion still or even increasingly takes on a leading role for life orientation. Perhaps science could learn from the great traditions of religions to not only accumulate useful knowledge but to take on the challenge to offer orientation in the world. Religion for its part could conceivably find ways of incorporating in its offer for orientation the dynamically increasing knowledge of the sciences, to give up the last fallback position of its outdated dogmas and replace inflexible and authoritarian interpretations of the world with an open attitude to knowledge and science.

*This paper was first delivered at the meeting for the 50th anniversary of the Institut der Görres Gesellschaft für Interdisziplinäre Forschung in Munich. A German version will be published in* "Kosmologie und Evolution. Woher die Welt? Woher der Mensch?" *Eds. Stephan Borrmann and Guenter Rager, Verlag Karl Alber, 2008.*

John Beckman (*left*) and Jürgen Renn.

*Pathways through an Eclectic Universe*
*ASP Conference Series, Vol. 390, © 2008*
*J. H. Knapen, T. J. Mahoney, and A. Vazdekis, eds.*

# Pathways through Interstellar Matter: From the Closest Stars to the Most Distant Quasars

Juan Vladilo

*Osservatorio Astronomico di Trieste, INAF, Italy*

**Abstract.** Observations of quasar absorption systems relevant for studies of star formation at redshift $2 \lesssim z \lesssim 4$ are briefly reviewed. Emphasis is given to the role played by dust in our understanding of the star formation history of galaxies detected as absorption systems. Local interstellar studies are used as a reference for understanding the properties of high redshift interstellar media.

## 1. Introduction

The interstellar medium (ISM) is relevant for two galactic processes discussed in the present conference: (1) it provides the gas that feeds *star formation* and (2) it collects the *feedback* from stellar radiation and ejecta. The present contribution is focused on interstellar observations obtained from absorption line spectroscopy of nearby stars and distant quasars. Thanks to their brightness, nearby stars offer the possibility of studying the local ISM in detail since they can be observed at high spectral resolution from space. High resolution is required because interstellar lines are generally narrow; space instrumentation allows us to detect the most important interstellar absorption lines, which lie in the ultraviolet spectral range. The local ISM serves as a paradigm for interstellar studies of distant galaxies. The diffuse gas of high redshift galaxies can be probed by means of high resolution spectroscopy of quasars. Thanks to the expansion of the Universe, many ultraviolet interstellar absorption lines of the intervening galaxies are redshifted to the optical spectral range, where they can be studied in detail in high resolution quasar spectra collected with 10 m-class telescopes.

The fact that this conference serves as a celebration of the astrophysical career of John Beckman offers me the opportunity to recall, in the first part of my talk, some of the local interstellar studies that I had the pleasure of doing in collaboration with John during his first years at the Instituto de Astrofísica de Canarias. In the second part I will review some recent work on the interstellar media of high redshift galaxies observed as quasar absorption line systems.

## 2. Pathways to Nearby Stars

Interstellar absorption lines were discovered accidentally, as the result of studies of high resolution stellar spectroscopy: Ca II interstellar lines were first identified in the spectrum of $\delta$ Ori because they did not share the photospheric radial velocity curve of this binary star (Hartmann 1904). It is less known that the Mg II interstellar lines of the very local gas ($\lesssim$ 10 pc from the Sun) were discovered accidentally in the Mg II chromospheric emissions of the cool star $\beta$ Hyi, in the framework of a collaboration between John Beckman, newly arrived at the IAC, and the Trieste group of stellar spectroscopy (Vladilo et al. 1985). The claim for detection of interstellar Mg II was based on the velocity separation between the photospheric/chromospheric lines and the projected velocity vector of the ISM predicted from previous studies of the gas within $\simeq 10^2$ pc of the Sun (Crutcher 1982). Observations taken with the *International Ultraviolet Explorer* (*IUE*) confirmed the interstellar nature of the Mg II absorptions detected at the edges of the stellar chromospheric emissions (Molaro, Vladilo, & Beckman 1986). Having developed this method of identification of local Mg II lines in late-type stars, [1] we analyzed the full *IUE* spectroscopic database and produced a map of the interstellar gas at $\lesssim 30$ pc from the Sun (Genova et al. 1990). We found that the nearby gas is very inhomogeneous, with regions of low density, in the direction of the Galactic anticenter and the Galactic poles, and regions of high density, in the direction of the Galactic center.

As a result of the ISM kinematics and of the inhomogeneous distribution of the gas, the local Mg II interstellar absorption shows remarkable spatial variations in radial velocity and intensity that confuse the analysis of the stellar Mg II chromospheric lines. As a by-product of our analysis of the local gas, we were able to perform a study of the Mg II chromospheric emission profiles uncontaminated by interstellar absorption (Vladilo et al. 1987). A lesson learned from this investigation is that one should beware of the subtle effects introduced by the intervening medium on the observed properties of any class of astronomical objects. If this is true for nearby stars, one should be concerned about studies of distant quasars, as we discuss in the next part of this talk.

The velocity vector of the local ISM is material that has been shocked and accelerated by stellar winds and supernovae associated with the Sco–Oph OB association (Crutcher 1982). This is an example of how the ISM bears the signature of stellar feedback and can therefore be used to trace the history of star formation in galaxies. In the next section we will show other examples of the connection between physical properties of the ISM and star formation.

## 3. Pathways to Distant Quasars

As originally predicted by Bahcall & Salpeter (1965), quasar spectra show absorption line systems originating in the diffuse gas that lies along the path to the quasar. The study of these spectra allows us to probe the physical and chemical properties of the intergalactic/interstellar medium over a large fraction of the Hubble time, back to the epoch of quasar formation, with an accuracy comparable to that achievable in UV studies of the local ISM mentioned in the previous section.

Quasar absorbers are classified according to their H I column density, $N(\mathrm{H\,I})$, and to the presence of metal lines at the same redshift of the Ly $\alpha$, if any. The weak Ly $\alpha$ lines, with column densities $10^{12} \lesssim N(\mathrm{H\,I})\,[\mathrm{atoms\,cm^{-2}}] \lesssim 10^{16}$

---

[1] Early-type stars provide a better background for the detection of interstellar absorptions than late-type stars; however, they are too rare within $\sim$30 pc.

are very numerous and produce a "forest" of narrow absorptions in the spectral region bluewards of the Ly $\alpha$ emission; no metal lines are directly associated with these absorbers. At the other extreme of the column density distribution, we have the strong Ly $\alpha$ lines with $N(\mathrm{H\,I}) \geq 2\times10^{20}$ atoms cm$^{-2}$, which are instead very rare and always associated with metal lines of high and low ionization species. These absorbers are called "damped Ly $\alpha$" (DLA) systems (Wolfe et al. 1986), because the Ly $\alpha$ line profile shows "damping wings", due to natural broadening, that extend beyond the Doppler core. There is consensus that DLA systems originate in the ISM of intervening galaxies (hereafter "DLA galaxies"). A recent review on DLA systems can be found in Wolfe, Gawiser, & Prochaska (2005). Here, we focus on the effects of dust on our understanding of the star formation history of DLA galaxies.

By analogy with studies of nearby galaxies, we expect interstellar dust to be a pervasive component of DLA galaxies. The observational evidence for the existence of dust in DLA systems has been slowly growing in the course of time. At low redshift, definitive evidence has been found in the DLA system at $z = 0.52$ towards AO 0235+164, where the dust extinction bump at 2175 Å and the silicate absorption at 9.7 $\mu$m have been detected (Junkkarinen et al. 2004; Kulkarni et al. 2007). The evidence for dust in the bulk of the DLA population is mainly based on studies of elemental depletions: by analogy with local interstellar studies we expect refractory elements, such as Fe or Cr, to be depleted from the gas phase, where they can be detected via absorption spectroscopy, to the dust component, where they escape detection with this technique; we instead expect volatile elements, such as Zn or S, to be undepleted. Evidence for differential depletion in DLA systems has been confirmed by many studies (Pettini et al. 1997, 2000; Vladilo 1998, 2004; Hou, Boissier, & Prantzos 2001; Prochaska & Wolfe 2002; Dessauges-Zavadsky et al. 2006). The existence of general trends between depletion, metallicity and $H_2$ molecular fraction (Ledoux, Petitjean, & Srianand 2003; Petitjean et al. 2006) indicates that the observed depletions are indeed due to dust, rather than to anomalous DLA abundance patterns.

To probe the properties of the high redshift dust in DLA galaxies it is important to complement the studies of differential depletion with studies of quasar extinction. From our knowledge of local dust, we expect the extinction to become more efficient with decreasing wavelength and to produce a reddening of the quasar continuum. The measurement of quasar reddening due to *individual* DLA systems is quite challenging (Ellison, Hall, & Lira 2005); reddening detections have been obtained so far for a few metal-rich systems at redshift $z \lesssim 2$ (Vladilo et al. 2006). The detection of the *mean* quasar reddening due to Ca II systems (Wild, Hewett, & Pettini 2006) and Mg II systems (York et al. 2006; Menard et al. 2007) at $z \lesssim 1$ suggests that for DLA systems it should also be possible to obtain a statistical detection. A claim of detection based on the analysis of a sample of 13 DLA/QSOs (Pei, Fall, & Bechtold 1991) was not confirmed by the study of a large set of Sloan Digital Sky Survey (SDSS) spectra of the 2nd data release (Murphy & Liske 2004). Analysis of the photometric and spectroscopic database of the 5th SDSS data release indicates that the mean quasar reddening due to DLA systems at $2.2 \lesssim z \lesssim 3.5$ has been detected; the low value of the dust-to-gas ratio obtained, $\langle A_V/N(\mathrm{H\,I}) \rangle \sim 3\times10^{-23}$ mag cm$^{-2}$, is in line with the low level of metallicity of these systems (Vladilo, Prochaska, & Wolfe 2007).

### 3.1. Dust and Star Formation in DLA Systems

The techniques used to determine the star formation rate (SFR) based on the measurement of emission lines are ineffective in DLA systems. This is due to the difficulty of disentangling the DLA galaxy against the quasar point spread function. In only a few DLA systems has it been possible to derive limits, of the order of a few $M_\odot\,\mathrm{yr}^{-1}$, from studies of the Ly $\alpha$ (Møller, Fynbo, & Fall, 2004) or H$\alpha$ (Kulkarni et al. 2001) emission. A value larger by one order of magnitude was derived in one DLA system at $z = 1.9$ from the analysis of the FUV continuum (Møller et al. 2002). The paucity of these measurements indicates the need for specially designed techniques aimed at determining the SFR in DLA systems.

A method developed by Wolfe, Prochaska, & Gawiser (2003) makes use of the column density of the C II* 1335.7 Å line to determine the [C II] 158 $\mu$m cooling rate. Assuming thermal balance, one can estimate the heating rate and so constrain the efficiency of the heating mechanisms. The main heating mechanism considered is photoelectric emission from dust grains. The efficiency of this mechanism depends on the intensity of the interstellar radiation field, which in turn is related to the SFR. With this method, two SFR solutions can be found corresponding to the thermally stable states of a two-phase medium with a cold neutral medium (CNM) in pressure equilibrium with a warm neutral medium (WNM). The average SFR per unit area found for 23 DLA systems with detected C II* line is $\langle \dot{\psi}_* \rangle \simeq 11 \times 10^{-3}\, M_\odot\,\mathrm{yr}^{-1}\,\mathrm{kpc}^{-2}$ for the CNM solution; the larger value found for the WNM solution, $\langle \dot{\psi}_* \rangle \simeq 0.21\, M_\odot\,\mathrm{yr}^{-1}\,\mathrm{kpc}^{-2}$, implies a bolometric background in excess of the observational limits (Wolfe et al. 2004). Since the efficiency of the heating mechanism considered in these computations depends on the amount of dust present in the medium, an accurate estimate of the dust-to-gas ratio is necessary to infer an accurate value of the SFR from the C II* 1335.7 Å column densities.

If the Kennicutt–Schmidt law is valid at high redshift, one can estimate the SFR in DLA systems from the measured H I column densities. This can be done by taking into account inclination effects to convert the $N$(H I) into the column density perpendicular to the galactic disk, $N_\perp$, which is used in the law $(\dot{\psi}_*)_\perp = K\,(N_\perp/N_c)^\beta$ (Kennicutt 1998). From this type of calculation, Wolfe & Chen (2006) find that 3% of the sky should be covered with extended objects brighter than $\mu_V \sim 28.4\,\mathrm{mag\,arcsec}^{-2}$, if DLAs at redshift $z \in [2.5, 3.5]$ undergo in situ star formation. To test this hypothesis these authors searched for low surface brightness features in the *Hubble* Ultra Deep Field (UDF). They found upper limits on the comoving SFR densities that are between factors of 30 and 100 lower than predictions, suggesting a reduction by more than a factor of 10 in the star formation efficiency predicted by the Kennicutt–Schmidt law at $z \sim 3$. This type of conclusion may be affected to some extent by the presence of dust: the extinction would yield a lower observed intensity and hence a lower value of the constant $K$ inferred with this method.

Constraints on the SFR history of DLA galaxies can also be obtained by comparing the elemental abundances measured from the spectra with those predicted by models of galactic chemical evolution (e.g., Matteucci, Molaro, & Vladilo 1997). Abundance ratios can be used as diagnostics of evolution if they involve two elements synthesized on different timescales, such as $\alpha$-capture elements, mainly produced by type II supernovae (SNe), and iron-peak elements, mainly produced by type Ia SNe. The $\alpha$/Fe ratio, when compared with a metallicity tracer, such as Fe/H, allow us to clarify the particular history of star formation involved. In a regime of high SFR we expect to observe enhanced $\alpha$/Fe ratios for a large interval of Fe/H, while the opposite is expected for a regime of low SFR (Matteucci 1991). In applying this type of study to DLA systems we face the problem that the measured abundance ratios may be affected by differential depletion. In fact, the $\alpha$/Fe ratios measured in DLA systems are often enhanced due to differential depletion, as is the case of the Si/Fe ratios. This apparent enhancement mimics the behavior of galaxies with relatively high SFR. However, when depletion effects are taken into account (Vladilo 2002) or volatile elements are used in the analysis (Centurión et al. 2000), or dust-free DLA systems are investigated (e.g., Molaro et al. 2000) the resulting $\alpha$/Fe ratio is generally not enhanced. The comparison with chemical evolution models suggests an origin of DLA systems in galactic regions with low or episodic SFR (Calura, Matteucci, & Vladilo 2003; Dessauges-Zavadsky et al. 2006).

The presence of dust may also affect our understanding of the star formation history of DLA systems due to the extinction of the quasar continuum. In the most extreme cases, the extinction due to dust-rich DLA systems could lead to the obscuration of the quasar (Ostriker & Heisler 1984; Fall & Pei 1989, 1993). In general, the extinction may induce a selection bias acting against the detection of dust-rich galactic regions in magnitude-limited surveys (Boissé et al. 1998; Prantzos & Boissier 2000; Vladilo & Pèroux 2005). Studies of radio-selected quasar surveys, not affected by extinction, suggest that the impact of this effect on the statistical properties of DLA systems is modest (Ellison et al. 2001; Jorgenson et al. 2006), but the size of these surveys is not large enough to reach firm conclusions. Due to the extinction bias, the mean DLA metallicity of magnitude-limited surveys may be underestimated since the dust-rich regions that are missed are likely to originate in metal-rich systems. In particular, the H I column density weighted metallicity $\langle \mathrm{M/H} \rangle_{\mathrm{w}} = [\sum_i (\mathrm{M/H})_i \times N_i(\mathrm{H\,I})]/[\sum_i N_i(\mathrm{H\,I})]$, an indicator of the mean cosmic metallicity at high redshift (Lanzetta, Wolfe, & Turnshek 1995) may be underestimated. The weighted metallicity $\langle \mathrm{M/H} \rangle_{\mathrm{w}}$ can be used to infer the global rate of star formation of the Universe at high redshift (Pei & Fall 1995; Pei, Fall, & Houser 1999). If $\langle \mathrm{M/H} \rangle_{\mathrm{w}}$ is also underestimated the global SFR is underestimated. Present estimates of this bias are still open to debate (Akermann et al. 2005; Vladilo & Péroux 2005).

### 3.2. Are We Missing the Regions of High SFR?

The study of the metallicities versus H I column densities indicates that DLA systems are concentrated *below* the value of metal column density[2] [Zn/H] + log $N(\mathrm{H\,I}) \sim 20.5$ (Boissé et al. 1998). Interstellar regions with metal column densities *above* this empirical threshold are instead predicted to exist at high

---

[2]Zinc is used as a tracer of the metals since it is believed to be undepleted. We adopt the usual definition of metallicity relative to the solar value $[X/\mathrm{H}] \equiv \log N(X)/N(\mathrm{H}) - \log(X/\mathrm{H})_{\odot}$.

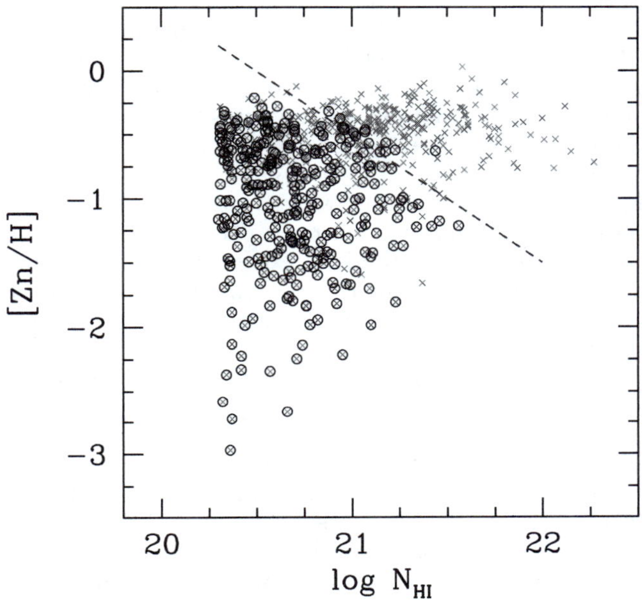


Figure 1. Metallicity versus H I column density in DLA galaxies predicted by a cosmological simulation (Nagamine et al. 2004; model Q5). Crosses: systems extracted at random from the simulation. Circles: systems observable in a DLA-QSO survey with limiting magnitude $r = 19$. An SMC-type extinction curve has been adopted to obtain the extinction in the observer's frame. Dashed line: empirical threshold [Zn/H] + log $N$(H,I) = 20.5.

redshift by models of galactic chemical evolution (Prantzos & Boissier 2000; Churches, Nelson, & Edmunds 2004) and by SPH cosmological simulations of DLA galaxies (Cen et al. 2003; Nagamine, Springer, & Hernquist 2004). The lack of observed DLA systems at [Zn/H] + log $N$(H I) $> 20.5$ has been attributed by some authors to the extinction bias that would affect DLA systems with high metallicity and $N$(H I) (Boissé et al. 1998; Vladilo & Péroux 2005). If this hypothesis is correct, we would be missing the regions of high SFR. The relation between interstellar extinction, $A_\lambda$ [mag], and metal column density is fundamental to test this hypothesis. Adopting iron as an indicator of metallicity, one can derive the following expression for the $V$-band extinction in the rest frame of the absorber

$$A_V = \langle s_V^{\rm Fe} \rangle \times f_{\rm Fe} \times ({\rm Fe/H}) \times N({\rm H\,I}), \qquad (1)$$

where (Fe/H) is the total abundance by number of iron (gas plus dust), $f_{\rm Fe}$ is the fraction of iron in dust form, and $\langle s_V^{\rm Fe} \rangle$ is a line-of-sight average of dust grain parameters (Vladilo et al. 2006). Empirical estimates of $\langle s_V^{\rm Fe} \rangle$ yield a typical value $\langle s_V^{\rm Fe} \rangle \approx 2.4 \times 10^{-17}\,{\rm mag\,cm^{-2}}$ in H I interstellar clouds, metal absorption systems and a few DLA systems at $z_{\rm a} \lesssim 2$ (Vladilo et al. 2006). By adopting this value of $\langle s_V^{\mathrm{Fe}} \rangle$, together with an empirical estimate of $f_{\mathrm{Fe}}$ (Vladilo 2004), we can use relation (1) to compute the extinction expected for DLA systems of given metal column density. This type of calculation can be applied, for instance, to test the predictions of SPH cosmological simulations. For each DLA system found at random in the cosmological box one enters the $N(\mathrm{H\,I})$ and (Fe/H) predicted by the simulation and estimates the rest-frame extinction $A_V$. The extinction in the observer's frame is then obtained adopting a normalized curve of interstellar extinction, $\xi(\lambda) = A_\lambda / A_V$. This extinction is added to the magnitude of a simulated quasar, which is assigned at random to each line of sight according to the known frequency distribution of quasar magnitudes. Lines of sight with dimmed magnitude in excess of a given magnitude limit are then discarded from the mock sample. An example of the result of this type of computation is shown in Fig. 1 for a mock survey of DLA systems at $z \simeq 3$ with limiting magnitude $r = 19$. This value is representative of the current magnitude limit of high spectral resolution metallicity surveys. One can see in Fig. 1 that the empirical threshold [Zn/H] + log $N(\mathrm{H\,I}) \lesssim 20.5$ is naturally reproduced by the extinction effect. This result suggests that DLA systems with high SFR are indeed missed. However, even after taking into account the extinction effect, the mean weighted metallicity predicted by the simulation, $[\langle \mathrm{M/H} \rangle_{\mathrm{w}}] \simeq -0.85$ dex, is too high compared to the one measured in metallicity surveys at $z \simeq 3$, $[\langle \mathrm{M/H} \rangle_w] \simeq -1.5$ dex (Prochaska, Gawiser, & Wolfe 2001). This discrepancy indicates that the SFR predicted by the cosmological simulation is too high. More refined simulations, with improved physics of star formation, are required before this method can be used to quantify the fraction of regions of high SFR missed due to the extinction bias.

**Acknowledgments.** It is a pleasure to thank the organizers, and in particular Johan Knapen, for the invitation to be part of a really eclectic experience. I am grateful to John Beckman for the trust, encouragement, and example he has given me since the very beginning of my career. This work has been supported by PRIN INAF 2006 (PI Molaro).

## References

Akerman, C. J., Ellison, S. L., Pettini, M., Steidel, C. C. 2005, A&A, 440, 499
Bahcall, J. N., & Salpeter, E. E. 1965, ApJ, 142, 1677
Boissé, P., Le Brun, V., Bergeron, J., & Deharveng, J. M. 1998, A&A, 333, 841
Calura, F., Matteucci, F., & Vladilo, G. 2003, MNRAS, 340, 59
Cen, R., Ostriker, J. P., Prochaska, J. X., & Wolfe, A. M. 2003, ApJ, 598, 741
Centurión, M., Bonifacio, P., Molaro, P., & Vladilo, G. 2000, ApJ, 536, 540
Churches, D. K., Nelson, A. H., & Edmunds, M. G. 2004, MNRAS, 347, 1234
Crutcher, R. M. 1982, ApJ, 254, 82
Dessauges-Zavadsky, M., Prochaska, J. X., D'Odorico, S., Calura, F., Matteucci, F. 2006, A&A, 445, 93
Ellison, S. L., Hall, P. B., Lira, P. 2005, AJ, 130, 1345
Ellison, S. L., Yan, L., Hook, I. M., Pettini, M., Wall, J. V., & Shaver, P. 2001, A&A, 379, 393
Fall, S. M., & Pei, Y. 1989, ApJ, 337, 7
Fall, S. M., & Pei, Y. 1993, ApJ, 402, 479
Genova, R., Molaro, P., Vladilo, G. & Beckman, J. E. 1990, ApJ, 355, 150
Hartmann, J. 1904, ApJ, 19, 268

Hou, J. L., Boissier, S., & Prantzos, N. 2001, A&A, 370, 23
Jorgenson, R. A., Wolfe, A. M., Prochaska, J. X., Lu, L., Howk, J. C., Cooke, J., Gawiser, E., & Gelino, D. M. 2006, ApJ, 646, 730
Junkkarinen, V. T., Cohen, R. D., Beaver, E. A., Burbidge, E. M., Lyons, R. W., & Madejski, G. 2004, ApJ, 614, 658
Kennicutt, R. C., Jr. 1998, ARA&A, 36, 189
Kulkarni, V. P., et al. 2001, ApJ, 551, 37
Kulkarni, V. P., York, D. G., Vladilo, G., & Welty, D. E. 2007, ApJ, 663, L81
Lanzetta, K. M., Wolfe, A. M., & Turnshek, D. A. 1995, ApJ, 440, 435
Ledoux, C., Petitjean, P., & Srianand, R. 2003, MNRAS, 346, 209
Matteucci F., 1991, in SN 1987A and Other Supernovae, eds I. J. Danziger & K. Kjar (Garching: ESO), p. 703
Matteucci, F., Molaro, P. & Vladilo, G. 1997, A&A, 321, 45
Menard, B., et al. 2007, submitted (astro-ph/0706.0898)
Molaro, P., Bonifacio, P., Centurión, M., D'Odorico, S., Vladilo, G., Santin, P., Di Marcantonio, P. 2000, ApJ, 541, 54
Molaro, P., Vladilo, G., & Beckman, J. E. 1986, A&A, 161, 339
Møller, P., Fynbo, J. U., & Fall, S. M. 2004, A&A, 422, L23
Møller, P., Warren, S. J., Fall, S. M., Fynbo, J. U., & Jakobsen, P. 2002, ApJ, 574, 51
Murphy, M. T., & Liske, J. 2004, MNRAS, 354, L31
Nagamine, K., Springer, V., & Hernquist, L. 2004, MNRAS, 348, 435
Ostriker, J. P., & Heisler, J. 1984, ApJ, 278, 1
Pei, Y. C., & Fall, S. M. 1995, ApJ, 454, 69
Pei, Y. C., Fall, S. M., & Bechtold, J. 1991, ApJ, 378, 6
Pei, Y. C., Fall, S. M., & Houser, M. G. 1999, ApJ, 522, 604
Petitjean, P., Ledoux, C., Noterdaeme, P., & Srianand, R. 2006, A&A, 456, L9
Pettini, M., Ellison, S. L., Steidel, C. C., Shapley, A. E., & Bowen, D. V. 2000, ApJ, 532, 65
Pettini, M., Smith, L. J., King, D. L., & Hunstead, R. W. 1997, ApJ, 486, 665
Prantzos, N., & Boissier, S. 2000, MNRAS, 315, 82
Prochaska, J. X., Gawiser, E., & Wolfe, A. M. 2001, ApJ, 552, 99
Prochaska, J. X., & Wolfe, A. M. 2002, ApJ, 566, 68
Vladilo, G. 1998, ApJ, 493, 583
Vladilo, G. 2002, A&A, 391, 407
Vladilo, G. 2004, A&A, 421, 479
Vladilo, G., Beckman, J. E., Crivellari, L., Franco, M. L., & Molaro, P. 1985, A&A, 144, 181
Vladilo, G., Centurión, M., Levshakov, S. A., Péroux, C., Khare, P., Kulkarni, V. P. & York, D. G. 2006, A&A, 454, 151
Vladilo, G., Molaro, P., Crivellari, L., Foing, B. H., Beckman, J. E., & Genova, R. 1987, A&A, 185, 233
Vladilo, G., & Péroux, C. 2005, A&A, 444, 461
Vladilo, G., Prochaska, J. X., & Wolfe, A. M. 2007, A&A, submitted
Wild, V., Hewett, P. C., & Pettini, M. 2006, MNRAS, 367, 211
Wolfe, A. M., & Chen, H.-W. 2006, ApJ, 652, 981
Wolfe, A. M., Gawiser, E., & Prochaska, J. X. 2005, ARA&A, 43, 861
Wolfe, A. M., Howk, J. C., Gawiser, E., Prochaska, J. X., & Lopez, S. 2004, ApJ, 615, 625
Wolfe, A. M., Prochaska, J. X., & Gawiser, E. 2003, ApJ, 593, 215
Wolfe, A. M., Turnshek, D. A., Smith, H. E., & Cohen, R. D. 1986, ApJS, 61, 249
York, D. G., et al. 2006, MNRAS, 367, 945

*Pathways through an Eclectic Universe*
*ASP Conference Series, Vol. 390, © 2008*
*J. H. Knapen, T. J. Mahoney, and A. Vazdekis, eds.*

# An Eclectic Summary of an Eclectic Conference

M. G. Edmunds

*School of Physics and Astronomy, Cardiff University, Queens Buildings, The Parade, Cardiff, CF24 3AA, UK*

**Abstract.** We have enjoyed talks and posters on a very wide range of subjects, mirroring John Beckman's eclectic research interests. What have we learned? Where do we go from here? I attempt a personal and selective overview.

## 1. Introduction

We have been celebrating John Beckman's 42 years in research. Forty-two years ago was, of course, long before the significance of that number 42 was revealed in the *Hitch-Hikers Guide to the Galaxy.* It would be fascinating to catalog all the changes since then in astronomy—to see how much discussed in this meeting was unsuspected, and which big questions would have been familiar but are still not satisfactorily answered. One great difference is in what is regarded as high redshift. At that time I think it would have been regarded as *impossible* ever to observe the redshift of a galaxy above one, indeed $z = 0.1$ *was* a high-redshift object!

## 2. The Star Formation Process

In 1965 ideas on star formation were pretty elementary too—stars had to form because they could be seen, but the physical path through some 22 orders of magnitude in density was virtually unknown. It was only with the opening up of infrared and sub-mm wavelength observations, and a working out of the physics of accretion disks, that a reasonable outline picture emerges. At this meeting Kroupa and Krumholtz have agreed that the overall sequence is

$$\text{CLOUDS} \Longrightarrow \text{CORES} \Longrightarrow \text{STARS}$$

where the initial mass function (IMF) for star formation is basically set by the mass function for the cores of the molecular clouds. That immediately begs the question of what determines the core mass function, and I see that John in a poster with Casuso revives the idea (championed historically by Larson and by Zinnecker) of statistical "fragmentation", but in a "pressurized medium". This implies some sort of random process, but with a degree of external control. Whether the subsequent mapping of mass distribution into stars is straightforward was challenged by Bonnell, who pointed out that variable amounts of accretion may occur. Kroupa showed rather convincingly that a reasonably simple two-slope IMF emerges after correction of the stellar luminosity for multiple systems and for the effect of a knee due to the $H_2$ to $H^-$ opacity transition:

$$\begin{aligned} N(M)dM &\propto M^{-1.3} && 0.08 < M/M_\odot < 0.50 \\ &\propto M^{-2.3} && 0.50 < M/M_\odot \end{aligned}$$

In accord with all my prejudices, this IMF was described as having a "non-behavior"—i.e., on average not varying significantly with conditions. One remark he made worried me though, the suggestion that there is a *discontinuity* in the IMF at 0.08 solar masses. I don't believe nature likes discontinuities, so if the discontinuity is real then there must be a very good (unknown?) reason. The fun really started in considering the IMF in star clusters. The majority of clusters are already formed gravitationally *unbound*, and although I did not hear a convincing estimate of the fraction of field stars originally formed in clusters, I gained the impression that it is large, perhaps near unity. We heard that the upper limit on the mass of stars formed might well correlate with the total mass of the cluster, Bonnell suggesting something like

$$M_{\rm max} \propto M_{\rm system}^{2/3}.$$

If this is indeed the case, then whatever regulates the size of formation of star clusters in a system also regulates (to some extent) the IMF. An obvious case would be of small galactic systems simply not being big enough to house the formation of large clusters, and hence having a lack of high mass stars. I'll comment further on possible implications of this below. Evans reminded us, though, that low and high mass star formation may be rather separate processes—the "tail" and "trunk" of the general elephant in between. Smith showed us that the influences of external pressure on star formation are accessible to study in systems like M82. She also showed the promising cluster NGC 346, which looks as though it has not yet had a supernova explosion—and is therefore a rare example of star formation for study where little feedback in energy and gas has yet occurred (except, presumably, from stellar winds). I shall stress several times the need for clarity in nomenclature, and we should perhaps be clearer in deciding whether feedback in star formation just alters the rate of star formation, or also has a more involved role in actually altering the mass function. Feedback will inevitably alter what we observe in terms of velocities and chemistry. Another target for clarity would be the meaning of "efficiency" of star formation. Is efficiency a measure of timescale or of the fraction of gas consumed (on some particular timescale)? Cores were said to be "efficient" while clouds were not—by which, I suppose, is meant that cores don't hang around once they form, they get on and form stars. Clouds do seem to hang around, with the details of the clumping process to form cores remaining rather uncertain.

*Is* there a universal IMF? Lyell's 1833 doctrine of uniformity—'natural processes were the same in the past as they are today'—allowed real progress in the study of geology and history of the Earth. Except for the possible effect of star cluster size, the observational position seems to be that the Universe is surprisingly obliging in not varying the IMF either now or in the past. Be grateful! It could be argued that physical conditions for star formation may have been very different in the past, but I suspect the variation of conditions may have not been much greater than we observe in different galaxies and parts of galaxies today. I'll accept things could have been different for the very earliest metal-free star formation, but we sometimes seem to have an almost Freudian desire to uncover deviations in the IMF when it really may be behaving itself with admirable consistency.

To return to the *prima facie* (but as yet unproven) case for the influence of cluster size on the most massive star mass, with the size of star clusters perhaps influenced by the overall star formation rate as well as physical size of a galaxy. I would like to hear more about this, for the obvious reason (Köppen, Weidner & Kroupa 2007) that the chemical yield will be affected, and it could provide a solution to the puzzling mass–metallicity relation in galaxies. The "conventional" explanation—that of outflow of gas regulated by the gravitational potential (e.g., Garnett 2002; Tremonti et al 2004) suffers from a possible weakness in that quite small variations in star formation rates between galaxies could vitiate the correlation (see, for example, Edmunds 2005)—and how does a galaxy know at what rate it should be forming stars? The "cluster size" argument might be more robust—where the system is large enough to form big clusters, the chemical yields (and hence abundances) are enhanced by the extra massive stars. It is a promising approach that needs more work.

## 3. Star Formation Rates

We were wisely warned by Calzetti about the perils of observationally estimating star formation rates without sufficiently broad wavelength coverage, but it does seem as though a combination of luminosities at H$\alpha$ and 24 $\mu$m makes a fairly robust indicator, provided that there is no significant unrecognized contribution from an AGN. James suggested that the characteristic timescale for star formation is 8 Gyr in all galaxies, irrespective of mass. González Delgado emphasized that at $z \simeq 1$ the star formation in the Universe is dominated by luminous infrared galaxies (LIRGs), highlighting the important (annoying?) role of dust in re-processing radiation at high redshifts. A major treat at the meeting was the review of the "Schmidt–Kennicutt" law for the star formation rate by Kennicutt and others. In its simplest form the star formation rate in a galaxy is proportional to a power of the gas mass surface density:

$$\mathrm{SFR} \propto \Sigma_{\mathrm{gas}}^{n}.$$

The power $n$ is about 1.5–1.6, and this relationship—remarkably—appears to hold over nearly five orders of magnitude in gas density. Kennicutt suggested that in M51 the law holds down to a resolution of 500 pc—and indeed the fine maps now available at UV, H$\alpha$, 8 $\mu$m, 24 $\mu$m, and H I suggest that even more detailed pixel-by-pixel comparisons might be worthwhile. Lundgren showed us heroic mapping of M83 in CO. The Schmidt–Kennicutt law certainly breaks down, though, at sufficiently low densities ($30\,M_\odot\,\mathrm{pc}^{-2}$ ?). This is not a complete "threshold" cut-off, and some star formation still does go on in low density (and low density parts of) disks. Although observationally difficult, it would be very good to know the nature of this low density star formation and its dependence on physical conditions. The density of the molecular component of the gas in all these studies still remains problematic because of uncertainties in conversion of CO observations to $H_2$ masses—we know it certainly is affected by metallicity, and perhaps radiation field too (e.g., Bell, Viti, & Williams 2007). This is a lingering and admittedly difficult problem, but it really ought to be solved. Future mapping of cold dust in the sub-mm (e.g., by the *Herschel* satellite, SCUBA2 and ALMA) may help.

Is the Schmidt–Kennicutt law either universal or fundamental? Kennicutt himself wisely pointed out that it is not clear whether the relation is causal with a simple physical explanation, or a result of other underlying correlations. There might be some evidence of a second parameter (perhaps metallicity) in the scatter around the basic correlation. For constructing theoretical models of galaxy formation and evolution, using a Schmidt–Kennicutt law is apparently not enough on its own (e.g., NGC 4676—The Mice; Barnes 2004)—other ingredients are needed.

Visual and H I images of M83 reminded us that the gas in galaxies can extend a long way beyond their apparent visible extent—again prompting speculation as to the possible modes of low-level star formation out there, and whether what we are seeing is gas reservoirs which (as Silk emphasized) may be a necessary component of galaxy evolution.

## 4. Star Formation Histories

We re-ran the classic debate over monolithic collapse or merger hierarchy as the best model for the formation of galaxies. Again, nomenclature sometimes had us arguing at cross purposes. Balcells suggested the definition of bulges was "in a mess", a view rather re-enforced by Peletier's response to the question of whether bulges were young or old—"Yes, No". Indeed we perhaps need a clearer definition of what is meant by "hierarchical merger" which quotes timescales, since the violent event that is monolithic collapse can look (in many computer simulations) like a fast merger hierarchy! We did not resolve the question of whether low surface brightness (LSB) galaxies formed in the same way as "normal" galaxies, but more slowly and less violently. Kissler-Patig pointed out that LSB galaxies have "normal" globular cluster systems, implying a similar history. The bimodal metallicity distributions of most globular cluster systems must be trying to tell us something about their formation history, but what? He also made the very effective point that when we look at a globular cluster (including those in our own Galaxy) we may well be looking at a $z \sim 10$ object. High redshift close to home! Several speakers rightly emphasized that a high observed alpha element/iron ratio in a stellar population (e.g., [Mg/Fe]) is not an indicator of *age* but rather an indicator of a short (i.e., rapid) star formation timescale.

I was struck that Silk needed to introduce feedback "fixes" to explain many details of galaxy formation within the ΛCDM picture—but he robustly defended his viewpoint against my antediluvian questioning of the correctness of the CDM picture. He pointed out (quite rightly) that it might be perverse to abandon a picture that is so successful at explaining large-scale structure in the Universe, even though—as has become clear—a merger tree picture of galaxy formation is not the whole story. Just how much evolution there is in the mass and structure of galaxies as a function of redshift/time has not been agreed. We heard here from Conselice and Iglesias-Páramo that massive galaxies were in place by $z \sim 2$, and that there was not much change for $z < 1$. That's consistent with Peletier's claim about bulges ('stars in more massive galaxies form quicker')—and both Sheth and Yamada (poster) telling us that the bulge/disk ratio did not change much over the last 7 Gyr in early-type spirals, while the disk scale-length remained unchanged in $L^*$ galaxies (Sheth). Not everyone agrees. Trujillo suggested that radii may change by a factor of four—implying major density changes. According to Lehnert the Lyman break systems at redshift 5 could be small systems with spheroid densities, but it is of course possible that larger early spheroid systems are hidden by dust (cf. Dunne, Eales, & Edmunds 2003).

## 5. Feedback and Infall

Yet again, there may be nomenclature confusion over what is meant by *merger* and what is meant by *infall*. I think we would be happy with assuming that a merger happens when two galaxies, both including stars, come together often with evidence of tidal disruption, while infall simply dumps gas (possibly with angular momentum) into a galaxy. Several speakers showed the power and beauty of spatially resolved spectroscopy—from FLAMES/GIRAFFE and SAURON—for studying the kinematics of stars and gas in galaxies, and it will be interesting to see whether it will be possible to calibrate tunable filters well enough to push such studies to significantly fainter magnitudes. An intriguing anthropic thought was suggested by Hammer, that the Milky Way has had an "exceptionally quiet" interaction history. But is accretion of gas necessary to explain observed star formation histories in other galaxies? There are beautiful color–magnitude diagrams such as those shown by Gallart to work on. These reach down to the main sequence turn-off, a prerequisite for reliable interpretation. We had robust debate (Combes/Silk/Shlosman), which I don't think was entirely resolved, about how much gas must come in, and particularly about whether it comes in cold, or shocks and comes in hot. Pagel pointed out that the *FUSE* results imply that there is a very significant reservoir of gas persisting to today in the remains of the Lyman-$\alpha$ forest.

I will make my usual plea for determination of *quantitative* inflow and outflow rates. Do all galaxies go through a major outflow phase like Lyman break systems? Do low-metallicity dwarf galaxies require a dark matter halo to stop them breaking apart during a strong outflow phase? Do giant galaxies with dark halos require an AGN to quench star formation and to halt accretion of gas? The influence of AGN on the early history of a galaxy *may* be critical—but I'm not convinced we yet know whether its dominant effect (through jets and radiation field) is to stimulate or suppress (e.g., Croton et al. 2006) star formation, perhaps it does both at different times! Since the beasts may lurk at the centers of most galaxies, a clear picture of their (periodic?) interaction with the rest of the system is surely a good area for both theoretical and observational advance. But real physics is needed, this should not be just another *deus ex machina* we can appeal to (like a variable IMF...) when we don't know what is going on.

## 6. H II Regions

In the "strong-line" abundance analysis of H II regions, Vílchez and Díaz reminded us that sulfur lines provide a viable $S_{23}$ alternative to the oxygen emission lines used in the $R_{23}$ index. They suggested that the maximum abundance observed in emission lines is around $12 + \log(O/H) = 8.95$, i.e., twice solar. I will not rehearse the arguments over the calibration of these abundance indicators, which were debated at another recent Canary Islands conference (*The Metal-Rich Universe*, La Palma - Israelian & Meynet 2008), other than to point out again that we need to understand what is wrong with the detailed computational models of H II regions that apparently mislead us into thinking that the abundances could be higher. Luridiana and Ercolano (poster) rightly emphasized that we still need better atomic data, particularly for primordial helium determination from very-metal poor systems. I have the impression that nitrogen-to-oxygen abundance ratios from optical and infrared studies still do not agree very well. But perhaps at least one uncertainty can be laid to rest—we heard that IFU spectroscopy on a kpc-scale in galaxies implied chemical homogeneity of the ISM to at least $\pm 0.15$ dex. No problem! That's just what theoretical mixing estimates (and most observations) have been saying for years. (Could this be yet another result of that quasi-Freudian desire to find deviations?!).

Díaz highlighted the behavior of excitation conditions in circumnuclear H II regions, giving NGC 3351 as a good example. What was particularly interesting here is the potential of using the kinematic structure of lines (e.g., line width) to separate out different conditions. It seems that some circumnuclear regions are either illuminated by selective leakage of rather hard radiation from a mainly highly obscured nucleus, or perhaps there is some unrecognized source of additional ionization.

## 7. Bars and Galaxy Evolution

The role of bars in the evolution of galaxies is a subject which has been close to John's heart. It seems that all galaxies have one at some time or other, but it remains rather unclear how influential they actually are in the overall long-term evolution. Athanassoula pointed out that we may sometimes be confused (nomenclature again!) between bulges and (perhaps destroyed) bars—indeed she suggested that there might well be *multiple* sources of bulges. Shlosman outlined the importance of understanding the influence and interaction between the visible baryonic components of galaxies, and their dark matter halos—for example the effect of a bar on the dark matter profile. Such studies are not only important for understanding galaxy morphology and evolution, but also this might be one of our few ways of observationally investigating the properties of dark matter.

Lambert's lucid exposition of processes in stars may have made us feel a little envious of the (perhaps only apparent) more "physical" feel of stellar, compared to galaxy, studies. But one cannot, I think, escape the suspicion that in twenty or thirty years' time there may well be more left of current ideas on stellar evolution than there will of our current ideas on galactic evolution. This should be a stimulating, and not a depressing, thought.

## 8. Last Things

We were all moved by Clarke's fine tribute to the ability, calm dependability and integrity of the late Eduardo Delgado. It was particularly poignant and appropriate to hear of their work on the "extreme mass ratio" problem in binaries during the week in which the world's press were announcing the discovery of a very extreme mass ratio system with an exoplanet comparable to Earth.

We wish John many more years of eclectic activity. May the inflow be of good wine, and not cause the usual dynamical bulge of middle-age, while the outflow is of wisdom.

### References


Barnes, J. E. 2004, MNRAS, 350, 798

Bell, T. A., Viti, T., & Williams, D. A. 2007, MNRAS, 378, 983

Croton, D. J., et al. 2006, MNRAS, 365, 11

Dunne, L., Eales, S. A., & Edmunds, M. G. 2003, MNRAS, 341, 589

Edmunds, M. G. 2005, A&G, 4, 12

Garnett, D. T. 2002, ApJ, 581, 1019

Israelian, G., & Meynet, G. (eds) 2008, The Metal Rich Universe (Cambridge: Cambridge University Press)

Köppen, J., Weidner, C., & Kroupa, P. 2007, MNRAS, 375, 673

Tremonti, C., et al. 2004, ApJ, 613, 898

Bernard Pagel and Mike Edmunds.

# Author Index

# General Index

# Object Index

**ASTRONOMICAL SOCIETY OF THE PACIFIC**
**(ASP)**

An international, nonprofit, scientific and educational organization
founded in 1889 established the

## ASP CONFERENCE SERIES (ASPCS)

in 1988 to publish books on recent developments in
astronomy and astrophysics.

A list of recently published volumes follows. For a complete list of all
volumes published please see our web site at
http://www.astrosociety.org/pubs.html

For electronic versions of volumes please see our e-book site at
http://www.aspbooks.org

All book orders or inquiries concerning

ASTRONOMICAL SOCIETY OF THE PACIFIC
CONFERENCE SERIES
(ASPCS)

and

INTERNATIONAL ASTRONOMICAL UNION VOLUMES
(IAU)

should be directed to

**Astronomical Society of the Pacific**
**390 Ashton Avenue**
**San Francisco CA 94112-1722 USA**

**Phone: 800-335-2624 (within USA)**
**Phone: 415-337-2126**
**Fax: 415-337-5205**

**E-mail: service@astrosociety.org**
**Web Site: http://www.astrosociety.org**
**E-book site: http://www.aspbooks.org**

# ASP CONFERENCE SERIES VOLUMES

Published by the Astronomical Society of the Pacific

**PUBLISHED: 2003**

Vol. CS 300 RADIO ASTRONOMY AT THE FRINGE, A Conference held in honor of Kenneth I. Kellermann, on the occasion of his 65th Birthday
eds. J. Anton Zensus, Marshall H. Cohen, and Eduardo Ros
ISBN: 1-58381-147-8 e-Book ISBN: 978-1-58381-243-3

Vol. CS 301 MATTER AND ENERGY IN CLUSTERS OF GALAXIES
eds. Stuart Bowyer and Chorng-Yuan Hwang
ISBN: 1-58381-149-4

Vol. CS 302 RADIO PULSARS, In celebration of the contributions of Andrew Lyne, Dick Manchester and Joe Taylor – A Festschrift honoring their 60th Birthdays
eds. Matthew Bailes, David J. Nice, and Stephen E. Thorsett
ISBN: 1-58381-151-6

Vol. CS 303 SYMBIOTIC STARS PROBING STELLAR EVOLUTION
eds. R. L. M. Corradi, J. Mikołajewska, and T. J. Mahoney
ISBN: 1-58381-152-4

Vol. CS 304 CNO IN THE UNIVERSE
eds. Corinne Charbonnel, Daniel Schaerer, and Georges Meynet
ISBN: 1-58381-153-2

Vol. CS 305 International Conference on MAGNETIC FIELDS IN O, B AND A STARS: ORIGIN AND CONNECTION TO PULSATION, ROTATION AND MASS LOSS
eds. Luis A. Balona, Huib F. Henrichs, and Rodney Medupe
ISBN: 1-58381-154-0

Vol. CS 306 NEW TECHNOLOGIES IN VLBI
ed. Y. C. Minh
ISBN: 1-58381-155-9

Vol. CS 307 SOLAR POLARIZATION 3
eds. Javier Trujillo Bueno and Jorge Sanchez Almeida
ISBN: 1-58381-156-7

Vol. CS 308 FROM X-RAY BINARIES TO GAMMA-RAY BURSTS
eds. Edward P. J. van den Heuvel, Lex Kaper, Evert Rol, and Ralph A. M. J. Wijers
ISBN: 1-58381-158-3

**PUBLISHED: 2004**

Vol. CS 309 ASTROPHYSICS OF DUST
eds. Adolf N. Witt, Geoffrey C. Clayton, and Bruce T. Draine
ISBN: 1-58381-159-1 e-Book ISBN: 978-1-58381-244-0

Vol. CS 310 VARIABLE STARS IN THE LOCAL GROUP, IAU Colloquium 193
eds. Donald W. Kurtz and Karen R. Pollard
ISBN: 1-58381-162-1 e-Book ISBN: 978-1-58381-245-7

Vol. CS 311 AGN PHYSICS WITH THE SLOAN DIGITAL SKY SURVEY
eds. Gordon T. Richards and Patrick B. Hall
ISBN: 1-58381-164-8 e-Book ISBN: 978-1-58381-246-4

Vol. CS 312 Third Rome Workshop on GAMMA-RAY BURSTS IN THE AFTERGLOW ERA
eds. Marco Feroci, Filippo Frontera, Nicola Masetti, and Luigi Piro
ISBN: 1-58381-165-6 e-Book ISBN: 978-1-58381-247-1

## ASP CONFERENCE SERIES VOLUMES

Published by the Astronomical Society of the Pacific

**PUBLISHED: 2004**

| | |
|---|---|
| Vol. CS 313 | ASYMMETRICAL PLANETARY NEBULAE III: WINDS, STRUCTURE AND THE THUNDERBIRD<br>eds. Margaret Meixner, Joel H. Kastner, Bruce Balick, and Noam Soker<br>ISBN: 1-58381-168-0 e-Book ISBN: 978-1-58381-248-8 |
| Vol. CS 314 | ASTRONOMICAL DATA ANALYSIS SOFTWARE AND SYSTEMS XIII<br>eds. Francois Ochsenbein, Mark G. Allen, and Daniel Egret<br>ISBN: 1-58381-169-9 e-Book ISBN: 978-1-58381-249-5<br>ISSN: 1080-7926 |
| Vol. CS 315 | MAGNETIC CATACLYSMIC VARIABLES, IAU Colloquium 190<br>eds. Sonja Vrielmann and Mark Cropper<br>ISBN: 1-58381-170-2 e-Book ISBN: 978-1-58381-250-1 |
| Vol. CS 316 | ORDER AND CHAOS IN STELLAR AND PLANETARY SYSTEMS<br>eds. Gene G. Byrd, Konstantin V. Kholshevnikov, Aleksandr A. Mylläri, Igor' I. Nikiforov, and Victor V. Orlov<br>ISBN: 1-58381-172-9 e-Book ISBN: 978-1-58381-251-8 |
| Vol. CS 317 | MILKY WAY SURVEYS: THE STRUCTURE AND EVOLUTION OF OUR GALAXY, The 5th Boston University Astrophysics Conference<br>eds: Dan Clemens, Ronak Shah, and Tereasa Brainerd<br>ISBN: 1-58381-177-X e-Book ISBN: 978-1-58381-252-5 |
| Vol. CS 318 | SPECTROSCOPICALLY AND SPATIALLY RESOLVING THE COMPONENTS OF CLOSE BINARY STARS<br>eds. Ronald W. Hilditch, Herman Hensberge, and Krešimir Pavlovski<br>ISBN: 1-58381-179-6 e-Book ISBN: 978-1-58381-253-2 |
| Vol. CS 319 | NASA OFFICE OF SPACE SCIENCE EDUCATION AND PUBLIC OUTREACH CONFERENCE<br>eds. Carolyn Narasimhan, Bernhard Beck-Winchatz, Isabel Hawkins, and Cassandra Runyon<br>ISBN: 1-58381-181-8 e-Book ISBN: 978-1-58381-254-9 |
| Vol. CS 320 | THE NEUTRAL ISM OF STARBURST GALAXIES<br>eds. Susanne Aalto, Susanne Hüttemeister, and Alan Pedlar<br>ISBN: 1-58381-182-6 e-Book ISBN: 978-1-58381-255-6 |
| Vol. CS 321 | EXTRASOLAR PLANETS: TODAY AND TOMORROW<br>eds. J. P. Beaulieu, A. Lecavelier des Etangs, and C. Terquem<br>ISBN: 1-58381-183-4 e-Book ISBN: 978-1-58381-256-3 |
| Vol. CS 322 | THE FORMATION AND EVOLUTION OF MASSIVE YOUNG STAR CLUSTERS<br>eds. Henny J. G. L. M. Lamers, Linda J. Smith, and Antonella Nota<br>ISBN: 1-58381-184-2 e-Book ISBN: 978-1-58381-257-0 |
| Vol. CS 323 | STAR FORMATION IN THE INTERSTELLAR MEDIUM: In Honor of David Hollenbach, Chris McKee and Frank Shu<br>eds. D. Johnstone, F. C. Adams, D. N. C. Lin, D. A. Neufeld, and E. C. Ostriker<br>ISBN: 1-58381-185-0 e-Book ISBN: 978-1-58381-258-7 |
| Vol. CS 324 | DEBRIS DISKS AND THE FORMATION OF PLANETS: A Symposium in Memory of Fred Gillett<br>eds. Larry Caroff, L. Juleen Moon, Dana Backman, and Elizabeth Praton<br>ISBN: 1-58381-186-9 e-Book ISBN: 978-1-58381-259-4 |

# ASP CONFERENCE SERIES VOLUMES

Published by the Astronomical Society of the Pacific

**PUBLISHED: 2004**

Vol. CS 325 — THE SOLAR-B MISSION AND THE FOREFRONT OF SOLAR PHYSICS, The Fifth Solar-B Science Meeting
eds. Takashi Sakurai and Takashi Sekii
ISBN: 1-58381-187-7 e-Book ISBN: 978-1-58381-260-0

Vol. CS 326 — Never Submitted by Editor
ISBN: 1-58381-188-5 LOC#: 2004118371

Vol. CS 327 — SATELLITES AND TIDAL STREAMS
eds. F. Prada, D. Martínez Delgado, and T. J. Mahoney
ISBN: 1-58381-190-7 e-Book ISBN: 978-1-58381-261-7

**PUBLISHED: 2005**

Vol. CS 328 — BINARY RADIO PULSARS
eds. F. A. Rasio and I. H. Stairs
ISBN: 1-58381-191-5 e-Book ISBN: 978-1-58381-262-4

Vol. CS 329 — NEARBY LARGE-SCALE STRUCTURES AND THE ZONE OF AVOIDANCE
eds. A. P. Fairall and P. A. Woudt
ISBN: 1-58381-192-3 e-Book ISBN: 978-1-58381-263-1

Vol. CS 330 — THE ASTROPHYSICS OF CATACLYSMIC VARIABLES AND RELATED OBJECTS
eds. J.-M. Hameury and J.-P. Lasota
ISBN: 1-58381-193-1 e-Book ISBN: 978-1-58381-264-8

Vol. CS 331 — EXTRA-PLANAR GAS
ed. Robert Braun
ISBN: 1-58381-194-X e-Book ISBN: 978-1-58381-265-5

Vol. CS 332 — THE FATE OF THE MOST MASSIVE STARS
eds. Roberta M. Humphreys and Krzysztof Z. Stanek
ISBN: 1-58381-195-8 e-Book ISBN: 978-1-58381-266-2

Vol. CS 333 — TIDAL EVOLUTION AND OSCILLATIONS IN BINARY STARS: THIRD GRANADA WORKSHOP ON STELLAR STRUCTURE
eds. Antonio Claret, Alvaro Giménez, and Jean-Paul Zahn
ISBN: 1-58381-196-6 e-Book ISBN: 978-1-58381-267-9

Vol. CS 334 — 14$^{TH}$ EUROPEAN WORKSHOP ON WHITE DWARFS
eds. D. Koester and S. Moehler
ISBN: 1-58381-197-4 e-Book ISBN: 978-1-58381-268-6

Vol. CS 335 — THE LIGHT-TIME EFFECT IN ASTROPHYSICS: Causes and Cures of the $O-C$ Diagram
ed. Christiaan Sterken
ISBN: 1-58381-200-8 e-Book ISBN: 978-1-58381-269-3

Vol. CS 336 — COSMIC ABUNDANCES as Records of Stellar Evolution and Nucleosynthesis, in honor of Dr. David Lambert
eds. Thomas G. Barnes, III and Frank N. Bash
ISBN: 1-58381-201-6 e-Book ISBN: 978-1-58381-270-9

Vol. CS 337 — THE NATURE AND EVOLUTION OF DISKS AROUND HOT STARS
eds. Richard Ignace and Kenneth G. Gayley
ISBN: 1-58381-203-2 e-Book ISBN: 978-1-58381-271-6

# ASP CONFERENCE SERIES VOLUMES

Published by the Astronomical Society of the Pacific

**PUBLISHED: 2005**

Vol. CS 338 ASTROMETRY IN THE AGE OF THE NEXT GENERATION OF LARGE TELESCOPES
eds. P. Kenneth Seidelmann and Alice K. B. Monet
ISBN: 1-58381-205-9 e-Book ISBN: 978-1-58381-272-3

Vol. CS 339 OBSERVING DARK ENERGY
eds. Sidney C. Wolff and Tod R. Lauer
ISBN: 1-58381-206-7 e-Book ISBN: 978-1-58381-273-0

Vol. CS 340 FUTURE DIRECTIONS IN HIGH RESOLUTION ASTRONOMY: A Celebration of the 10th Anniversary of the VLBA
eds. Jonathan D. Romney and Mark J. Reid
ISBN: 1-58381-207-5 e-Book ISBN: 978-1-58381-274-7

Vol. CS 341 CHONDRITES AND THE PROTOPLANETARY DISK
eds. Alexander N. Krot, Edward R. D. Scott, and Bo Reipurth
ISBN: 1-58381-208-3 e-Book ISBN: 978-1-58381-275-4

Vol. CS 342 1604–2004: SUPERNOVAE AS COSMOLOGICAL LIGHTHOUSES
eds. M. Turatto, S. Benetti, L. Zampieri, and W. Shea
ISBN: 1-58381-209-1 e-Book ISBN: 978-1-58381-276-1

Vol. CS 343 ASTRONOMICAL POLARIMETRY: CURRENT STATUS AND FUTURE DIRECTIONS
eds. Andy Adamson, Colin Aspin, Chris J. Davis, and Takuya Fujiyoshi
ISBN: 1-58381-210-5 e-Book ISBN: 978-1-58381-277-8

Vol. CS 344 THE COOL UNIVERSE: OBSERVING COSMIC DAWN
eds. C. Lidman and D. Alloin
ISBN: 1-58381-211-3 e-Book ISBN: 978-1-58381-278-5

Vol. CS 345 FROM CLARK LAKE TO THE LONG WAVELENGTH ARRAY: Bill Erickson's Radio Science
eds. Namir E. Kassim, Mario R. Pérez, William Junor, and Patricia A. Henning
ISBN: 1-58381-213-X e-Book ISBN: 978-1-58381-279-2

Vol. CS 346 LARGE SCALE STRUCTURES AND THEIR ROLE IN SOLAR ACTIVITY
eds: K. Sankarasubramanian, M. Penn, and A. Pevtsov
ISBN: 1-58381-214-8 e-Book ISBN: 978-1-58381-280-8

Vol. CS 347 ASTRONOMICAL DATA ANALYSIS SOFTWARE AND SYSTEMS XIV
ed: Patrick L. Shopbell, Matthew C. Britton, and Rick Ebert
ISBN: 1-58381-215-6 e-Book ISBN: 978-1-58381-281-5
ISSN: 1080-7926

**PUBLISHED: 2006**

Vol. CS 348 ASTROPHYSICS IN THE FAR ULTRAVIOLET: FIVE YEARS OF DISCOVERY WITH FUSE
eds. George Sonneborn, H. W. Moos, and B-G Andersson
ISBN: 1-58381-216-4 e-Book ISBN: 978-1-58381-282-2

Vol. CS 349 ASTROPHYSICS OF VARIABLE STARS
eds. Christiaan Sterken and Conny Aerts
ISBN: 1-58381-217-2 e-Book ISBN: 978-1-58381-283-9

Vol. CS 350 BLAZAR VARIABILITY WORKSHOP II: ENTERING THE GLAST ERA
eds. H. R. Miller, K. Marshall, J. R. Webb, and M. F. Aller
ISBN: 1-58381-218-0 e-Book ISBN: 978-1-58381-284-6

# ASP CONFERENCE SERIES VOLUMES

Published by the Astronomical Society of the Pacific

**PUBLISHED: 2006**

Vol. CS 351 ASTRONOMICAL DATA ANALYSIS AND SOFTWARE SYSTEM XV
eds. Carlos Gabriel, Christophe Arviset, Daniel Ponz, and Enrique Solano
ISBN: 1-58381-219-9 e-Book ISBN: 978-1-58381-285-3
ISSN: 1080-7926

Vol. CS 352 NEW HORIZONS IN ASTRONOMY: FRANK N. BASH SYMPOSIUM 2005
eds. Sheila J. Kannappan, Seth Redfield, Jacqueline E. Kessler-Silacci, Martin Landriau, and Niv Drory
ISBN: 1-58381-220-2 e-Book ISBN: 978-1-58381-286-0

Vol. CS 353 STELLAR EVOLUTION AT LOW METALLICITY: MASS LOSS, EXPLOSIONS, COSMOLOGY
eds. H. Lamers, N. Langer, T. Nugis, and K. Annuk
ISBN: 978-1-58381-221-1 e-Book ISBN: 978-1-58381-287-7

Vol. CS 354 SOLAR MHD THEORY AND OBSERVATIONS: A High Spatial Resolution Perspective, in Honor of Robert F. Stein
eds. Han Uitenbroek, John Leibacher, and Robert F. Stein
ISBN: 978-1-58381-222-8 e-Book ISBN: 978-1-58381-288-4

Vol. CS 355 STARS WITH THE B[e] PHENOMENON
eds. Michaela Kraus and Anatoly S. Miroshnichenko
ISBN: 978-1-58381-223-5 e-Book ISBN: 978-1-58381-289-1

Vol. CS 356 REVEALING THE MOLECULAR UNIVERSE:ONE ANTENNA IS NEVER ENOUGH, in Honor of the academic retirement of Jack Welch
eds. D. C. Backer, J. L. Turner, and J. M. Moran
ISBN: 978-1-58381-224-2 e-Book ISBN: 978-1-58381-290-7

Vol. CS 357 THE SPITZER SPACE TELESCOPE: NEW VIEWS OF THE COSMOS
ed. L. Armus and W. T. Reach
ISBN: 978-1-58381-225-9 e-Book ISBN: 978-1-58381-291-4

Vol. CS 358 SOLAR POLARIZATION 4
eds. R. Casini and B. W. Lites
ISBN: 978-1-58381-226-6 e-Book ISBN: 978-1-58381-292-1

Vol. CS 359 NUMERICAL MODELING OF SPACE PLASMA FLOWS: ASTRONUM-2006, 1st IGPP – CalSpace International Conference
eds. Nikolai V. Pogorelov and Gary P. Zank
ISBN: 978-1-58381-227-3 e-Book ISBN: 978-1-58381-293-8

Vol. CS 360 AGN VARIABILITY FROM X-RAYS TO RADIO WAVES
eds. C. Martin Gaskell, Ian M. McHardy, Bradley M. Peterson, and Sergey G. Sergeev
ISBN: 978-1-58381-228-0 e-Book ISBN: 978-1-58381-294-5

**PUBLISHED: 2007**

Vol. CS 361 ACTIVE OB-STARS: LABORATORIES FOR STELLAR AND CIRCUMSTELLAR PHYSICS
eds. Stanislav Štefl, Stanley P. Owocki, and Atsuo T. Okazaki
ISBN: 978-1-58381-229-7 e-Book ISBN: 978-1-58381-295-2

Vol. CS 362 THE 7TH PACIFIC RIM CONFERENCE ON STELLAR ASTROPHYSICS
eds. Young-Woon Kang, Hee-Won Lee, Kwong Sang Cheng, and Kam-Ching Leung
ISBN: 978-1-58381-230-3 e-Book ISBN: 978-1-58381-296-9

# ASP CONFERENCE SERIES VOLUMES

Published by the Astronomical Society of the Pacific

**PUBLISHED: 2007**

Vol. CS 363 THE NATURE OF V838 MON AND ITS LIGHT ECHO
eds. Romano L. M. Corradi and Ulisse Munari
ISBN: 978-1-58381-231-0 e-Book ISBN: 978-1-58381-297-6

Vol. CS 364 THE FUTURE OF PHOTOMETRIC, SPECTROPHOTOMETRIC AND POLARIMETRIC STANDARDIZATION
ed. Christiaan Sterken
ISBN: 978-1-58381-232-7 e-Book ISBN: 978-1-58381-298-3

Vol CS 365 SINS – SMALL IONIZED AND NEUTRAL STRUCTURES IN THE DIFFUSE INTERSTELLAR MEDIUM
eds. M. Haverkorn and W. M. Goss
ISBN: 978-1-58381-233-4 e-Book ISBN: 978-1-58381-299-0

Vol. CS 366 TRANSITING EXTRASOLAR PLANETS WORKSHOP
eds. C. Afonso, D. Weldrake, and Th. Henning
ISBN: 978-1-58381-234-1 e-Book ISBN: 978-1-58381-300-3

Vol. CS 367 MASSIVE STARS IN INTERACTING BINARIES
eds. Nicole St-Louis and Anthony F. J. Moffat
ISBN: 978-1-58381-235-8 e-Book ISBN: 978-1-58381-301-0

Vol. CS 368 THE PHYSICS OF CHROMOSPHERIC PLASMAS
eds. Petr Heinzel, Ivan Dorotovič, and Robert J. Rutten
ISBN: 978-1-58381-236-5 e-Book ISBN: 978-1-58381-302-7

Vol. CS 369 NEW SOLAR PHYSICS WITH SOLAR-B MISSION, The Sixth Solar-B Science Meeting
eds. K. Shibata, S. Nagata, and T. Sakurai
ISBN: 978-1-58381-237-2 e-Book ISBN: 978-1-58381-303-4

Vol. CS 370 SOLAR AND STELLAR PHYSICS THROUGH ECLIPSES
eds. Osman Demircan, Selim O. Selam, and Berahitdin Albayrak
ISBN: 978-1-58381-238-9 e-Book ISBN: 978-1-58381-304-1

Vol. CS 371 STATISTICAL CHALLENGES IN MODERN ASTRONOMY IV
eds. G. Jogesh Babu and Eric D. Feigelson
ISBN: 978-1-58381-240-2 e-Book ISBN: 978-1-58381-305-8

Vol. CS 372 15th EUROPEAN WORKSHOP ON WHITE DWARFS
eds. Ralf Napiwotzki and Matthew R. Burleigh
ISBN: 978-1-58381-239-6 e-Book ISBN: 978-1-58381-306-5

Vol. CS 373 THE CENTRAL ENGINE OF ACTIVE GALACTIC NUCLEI
eds. Luis C. Ho and Jian-Min Wang
ISBN: 978-1-58381-307-2 e-Book ISBN: 978-1-58381-308-9

Vol. CS 374 FROM STARS TO GALAXIES: BUILDING THE PIECES TO BUILD UP THE UNIVERSE
eds. Antonella Vallenari, Rosaria Tantalo, Laura Portinari, and Alessia Moretti
ISBN: 978-1-58381-309-6 e-Book ISBN: 978-1-58381-310-2

Vol. CS 375 FROM Z-MACHINES TO ALMA: (SUB)MILLIMETER SPECTROSCOPY OF GALAXIES
eds. Andrew J. Baker, Jason Glenn, Andrew I. Harris, Jeffrey G. Mangum, and Min S. Yun
ISBN: 978-1-58381-311-9 e-Book ISBN: 978-1-58381-312-6

# ASP CONFERENCE SERIES VOLUMES

Published by the Astronomical Society of the Pacific

**PUBLISHED: 2007**

Vol. CS 376 — ASTRONOMICAL DATA ANALYSIS AND SOFTWARE SYSTEM XVI
eds. Richard A. Shaw, Frank Hill, and David J. Bell
ISBN: 978-1-58381-313-3 e-Book ISBN: 978-1-58381-314-0
ISSN: 1080-7926

Vol. CS 377 — LIBRARY AND INFORMATION SERVICES IN ASTRONOMY V: COMMON CHALLENGES, UNCOMMON SOLUTIONS
eds. Sandra Ricketts, Christina Birdie, and Eva Isaksson
ISBN: 978-1-58381-316-4 e-Book ISBN: 978-1-58381-317-1

Vol. CS 378 — WHY GALAXIES CARE ABOUT AGB STARS: THEIR IMPORTANCE AS ACTORS AND PROBES
eds. F. Kerschbaum, C. Charbonnel, and R. F. Wing
ISBN: 978-1-58381-318-8 e-Book ISBN: 978-1-58381-319-5

Vol. CS 379 — COSMIC FRONTIERS
eds. Nigel Metcalfe and Tom Shanks
ISBN: 978-1-58381-320-1 e-Book ISBN: 978-1-58381-321-8

Vol. CS 380 — AT THE EDGE OF THE UNIVERSE: LATEST RESULTS FROM THE DEEPEST ASTRONOMICAL SURVEYS
eds. José Afonso, Henry C. Ferguson, Bahram Mobasher, and Ray Norris
ISBN: 978-1-58381-323-2 e-Book ISBN: 978-1-58381-324-9

**PUBLISHED: 2008**

Vol. CS 381 — THE SECOND ANNUAL SPITZER SCIENCE CENTER CONFERENCE: INFRARED DIAGNOSTICS OF GALAXY EVOLUTION
eds. Ranga-Ram Chary, Harry I. Teplitz, and Kartik Sheth
ISBN: 978-1-58381-325-6 e-Book ISBN: 978-1-58381-326-3

Vol. CS 382 — THE NATIONAL VIRTUAL OBSERVATORY: TOOLS AND TECHNIQUES FOR ASTRONOMICAL RESEARCH
eds. Matthew J. Graham, Michael J. Fitzpatrick, and Thomas A. McGlynn
ISBN: 978-1-58381-327-0 e-Book ISBN: 978-1-58381-328-7

Vol. CS 383 — SUBSURFACE AND ATMOSPHERIC INFLUENCES ON SOLAR ACTIVITY
eds. R. Howe, R. W. Komm, K. S. Balasubramaniam, and G. J. D. Petrie
ISBN: 978-1-58381-329-4 e-Book ISBN: 978-1-58381-330-0

Vol. CS 384 — 14TH CAMBRIDGE WORKSHOP ON COOL STARS, STELLAR SYSTEMS, AND THE SUN
ed. Gerard van Belle
ISBN: 978-1-58381-331-7 e-Book ISBN: 978-1-58381-332-4

Vol. CS 385 — NUMERICAL MODELING OF SPACE PLASMA FLOWS: ASTRONUM-2007
eds. Nikolai V. Pogorelov, Edouard Audit, and Gary P. Zank
ISBN: 978-1-58381-333-1 e-Book ISBN: 978-1-58381-334-8

Vol. CS 386 — EXTRAGALACTIC JETS: THEORY AND OBSERVATION FROM RADIO TO GAMMA RAY
eds. Travis A. Rector and David S. De Young
ISBN: 978-1-58381-335-5 e-Book ISBN: 978-1-58381-336-2

Vol. CS 387 — MASSIVE STAR FORMATION: OBSERVATIONS CONFRONT THEORY
eds. Henrik Beuther, Hendrik Linz, and Thomas Henning
ISBN: 978-1-58381-642-4 e-Book ISBN: 978-1-58381-643-1

## ASP CONFERENCE SERIES VOLUMES

Published by the Astronomical Society of the Pacific

**PUBLISHED: 2008**

Vol. CS 388 MASS LOSS FROM STARS AND THE EVOLUTION OF STELLAR CLUSTERS
eds. Alex de Koter, Linda Smith, and Rens Waters
ISBN: 978-1-58381-644-8 e-Book ISBN: 978-1-58381-645-5

Vol. CS 389 EPO AND A CHANGING WORLD: CREATING LINKAGES AND EXPANDING PARTNERSHIPS
eds. Catharine Garmany, Michael G. Gibbs, and J. Ward Moody
ISBN: 978-1-58381-648-6 e-Book ISBN: 978-1-58381-649-3

Vol. CS 390 PATHWAYS THROUGH AN ECLECTIC UNIVERSE
eds. J. H. Knapen, T. J. Mahoney, and A. Vazdekis
ISBN: 978-1-58381-650-9 e-Book ISBN: 978-1-58381-651-6